THE PHYSICS
OF ELECTRONIC
AND ATOMIC
COLLISIONS

Previous Proceedings
in the Series of International Conferences of
The Physics of Electronic and Atomic Collisions (ICPEAC)

	Year	Held in	Publisher	ISBN
XX	1997	Vienna, Austria	World Scientific	981-02-3425-2
XIX	1995	Whistler, Canada	AIP Conference Proceedings 360	1-56396-440-6
XVIII	1993	Aarhus, Denmark	AIP Conference Proceedings 295	1-56396-290-X
XVII	1991	Brisbane, Australia	Adam Hilger	0-7503-0167-8
XVI	1989	New York, USA	AIP Conference Proceedings 205	0-88318-390-0
XV	1987	Brighton, England	North Holland/Elsevier	0-444-87083-0

Other Related Titles from AIP Conference Proceedings

477 Atomic Physics 16: Sixteenth International Conference on Atomic Physics
Edited by William E. Baylis and Gordon W. F. Drake, May 1999, 1-56396-752-9

463 Photoacoustic and Photothermal Phenomena: Tenth International Conference
Edited by F. Scudieri and M. Bertolotti, March 1999, 1-56396-805-3

454 Resonance Ionization Spectroscopy: Ninth International Symposium
Edited by J. C. Vickerman, I. Lyon, N. P. Lockyer, and J. E. Parks, December 1998,
1-56396-810-X

438 Lectures on the Physics of Highly Correlated Electron Systems
Edited by Ferdinando Mancini, July 1998, 1-56396-789-8

434 Atomic and Molecular Data and Their Applications: ICAMDATA - First
International Conference
Edited by Peter J. Mohr and Wolfgang L. Wiese, June 1998, 1-56396-751-0

416 Similarities and Differences Between Atomic Nuclei and Clusters: Toward a
Unified Development of Cluster Science
Edited by Y. Abe, I. Arai, S. M. Lee, and K. Yabana, December 1997, 1-56396-714-6

To learn more about these titles, or the AIP Conference Proceedings Series, please visit the webpage
http://www.aip.org/catalog/aboutconf.html

THE PHYSICS
OF ELECTRONIC
AND ATOMIC
COLLISIONS

XXI International Conference

Sendai, Japan   July 1999

EDITORS

Yukikazu Itikawa
Inst. of Space and Astronautical Science, Sagamihara

Kazuhiko Okuno
Tokyo Metropolitan University, Tokyo

Hiroshi Tanaka
Sophia University, Tokyo

Akira Yagishita
Inst. of Materials Structure Science, KEK, Tsukuba

Michio Matsuzawa
The University of Electro-Communications, Tokyo

AMERICAN
INSTITUTE
OFPHYSICS

Melville, New York
AIP CONFERENCE PROCEEDINGS ■ 500

Editors:

Yukikazu Itikawa
Institute of Space and Astronautical Science
3-1-1 Yoshinodai, Sagamihara 229-8510
JAPAN

E-mail: itikawa@pub.isas.ac.jp

Kazuhiko Okuno
Department of Physics
Tokyo Metropolitan University
1-1 Minami-Ohsawa, Hachioji, Tokyo 192-0397
JAPAN

E-mail: okuno@phys.metro-u.ac.jp

Hiroshi Tanaka
Department of Physics
Sophia University
7-1 Kioicho, Chiyodaku, Tokyo 102-8554
JAPAN

E-mail: h_tanaka@hoffman.cc.sophia.ac.jp

Akira Yagishita
Institute of Materials Structure Science
KEK
1-1 Oho, Tsukuba, Ibaraki 305-0801
JAPAN

E-mail: yagisita@kekvax.kek.jp

Michio Matsuzawa
Department of Applied Physics and Chemistry
University of Electro-Communications
1-5-1 Chofu-ga-oka, Chofu-shi, Tokyo 182-8585
JAPAN

E-mail: michio@pc.uec.ac.jp

L.C. Catalog Card No. 99-069861
ISBN 1-56396-777-4
ISSN 0094-243X
Printed in the United States of America

CONTENTS

PLENARY

COLLISIONS INVOLVING PHOTONS

*Italicized name indicates author who presented the paper.

COLLISIONS INVOLVING ELECTRONS AND POSITRONS

*Italicized name indicates author who presented the paper.

*Italicized name indicates author who presented the paper.

COLLISIONS INVOLVING HEAVY PROJECTILES

*Italicized name indicates author who presented the paper.

*Italicized name indicates author who presented the paper.

ix

PREFACE

The 21st International Conference on the Physics of Electronic and Atomic Collisions (21st ICPEAC) was held in Sendai, Japan from 22 July to 27 July 1999. Twenty years ago, the 11th ICPEAC was successfully held in Kyoto, Japan. This was the first ICPEAC held in the Asian region and outside North America and Europe. This strongly stimulated the research activity in Japan in the field of atomic collision physics. As a result, we could organize the 21st ICPEAC in Japan, the second in Asia and the third outside of North America or Europe.

The Conference was attended by 565 delegates from 37 countries/region including 103 student participants. In addition, 89 registered accompanying persons participated in the social programs in which they enjoyed excellent cross-cultural experiences.

This book contains the written version of invited talks, including 4 plenary talks, 9 review talks, and 41 progress reports. It also contains 14 special reports selected from the contributed papers and presented at oral sessions. The abstracts of 797 contributed papers were received; 792 of them including 28 post-deadline papers are included in the Book of Abstracts of Contributed Papers, which, together with the present volume, forms the official scientific records of the 21st ICPEAC.

It should be noted that one third of the contributed papers come from Asian countries. Furthermore, a satellite meeting was held in Beijing, China, and a related meeting was held in Hanoi, Vietnam after the main conference. We hope that the 21st ICPEAC has given a strong stimulus to the research activities in the field of atomic collision physics in many Asian countries.

September   1999

Michio Matsuzawa

Chair, The Local Organizing Committee of the 21st International Conference on the Physics of Electronic and Atomic Collisions

SPONSORS

The 21st International Conference on the Physics of Electronic and Atomic Collisions
was held under the auspices of
  The Society for Atomic Collision Research, Japan
  The Physical Society of Japan
  The Chemical Society of Japan.

This Conference
 received sponsorship and financial support from
   The International Union of Pure and Applied Physics (IUPAP)
 was supported by
   Grant-in-Aid for Scientific Research
   from The Ministry of Education, Science, Sports and Culture, Japan.

This project was executed partly with a grant from
   The Commemorative Association for the Japan World Exposition (1970).

The Local Committee gratefully acknowledges financial assistance from
   Sendai City
   The Atomic Energy Society of Japan
   Japan Atomic Energy Research Institute

   Matsuo Foundation
   Inoue Foundation for Science
   Casio Science Promotion Foundation
   The Iwatani Naoji Foundation
   The Kao Foundation for Arts and Sciences
   The Daiwa Anglo-Japanese Foundation
   Science Measurements Research Foundation
   The Asahi Glass Foundation.
   Ishikawa Foundation for Carbon Science and Technology

International Conference on the Physics of Electronic and Atomic Collisions
Executive Committee (1997-1999)

International Conference on the Physics of Electronic and Atomic Collisions
General Committee (1997-1999)

XXI International Conference on the Physics of Electronic and Atomic Collisions
Local Organizing Committee

Co-Chairperson :

M.Matsuzawa

Y.Hatano

Y.Itikawa

Y.Sato

Members:

Y.Awaya	K.Ohno
T.Fujimoto	S.Ohtani
Y.Fujimura	M.Okunishi
K.Hino	K.Okuno
Y.Itoh	I.Shimamura
K.Kameta	H.Sugai
Y.Kanai	I.H.Suzuki
Y.Kato	H.Takagi
M.Kimura	H.Tanaka
N.Kobayashi	H.Tanuma
T.Koizumi	S.Tsurubuchi
N.Kouchi	K.Ueda
N.Mikami	M.Ukai
K.Mima	S.Watanabe
K.Motohashi	A.Yagishita
H.Nakamura	Y.Yamazaki
K.Ohmori	

PLENARY

Manipulating Molecules with Intense Lasers

Keith Codling, Jan H Posthumus and Leszek J Frasinski

JJ Thomson Physical Laboratory
Whiteknights, Reading, RG6 6AF, UK

Abstract. Molecules exposed to intense, sub-picosecond laser fields exhibit phenomena such as above threshold ionisation, high harmonic generation, bond softening, bond hardening (vibrational trapping) and alignment. They can be deflected, accelerated, focused, trapped and spun. More generally, one can manipulate the outcome of a uni- or bimolecular reaction by the coherent superposition of two or more lasers (coherent control) or by using a sequence of ultrashort laser pulses. This article will take a brief and necessarily limited look at some of these methods of molecular manipulation.

INTRODUCTION

The development of the chirped pulse amplification technique some years ago (1) led to the possibility of producing focused laser intensities well in excess of 10^{15} W cm^{-2}. This meant that the electric field of the laser could now exert a force on the electron that is of the order of, or indeed can greatly exceed, the forces that bind valence electrons in atoms and molecules. Not surprisingly, ground state atoms and molecules can easily be ionised using such laser intensities.

Above about 10^{13} W cm^{-2} atoms exhibit the phenomenon of above threshold ionisation (ATI), where more photons are absorbed than the minimum required to ionise the atom (2). The electron kinetic energy spectrum, the photoelectron spectrum, consists of a number of peaks separated by the laser photon energy. Moreover, harmonic generation can be observed to very high harmonics and with efficiencies much larger than can be accounted for by lowest order perturbation theory (3).

Molecules, of course, also exhibit photoelectron peaks associated with ATI (4) and high harmonics are observed with similar efficiencies (5). But in addition, due to the extra degrees of freedom, molecules exhibit a phenomenon that cannot occur in atoms, namely above threshold dissociation (ATD). This phenomenon was initially observed in H_2 (6), where it manifested itself in the proton spectrum as a series of peaks separated by half the photon energy.

As the laser intensity is increased, atoms such as Xe or molecules such as N_2 or CO can be multiply ionised. In the multiphoton picture, this would inevitably require many (≥ 100) low energy photons (A Ti: sapphire laser has a wavelength of about 800 nm or an energy

CP500, *The Physics of Electronic and Atomic Collisions*, edited by Y. Itikawa, et al.
© 2000 American Institute of Physics 1-56396-777-4/00/$17.00

of 1.6 eV.). One might, therefore, expect to be able to describe the multiple ionisation process in terms of a classical field ionisation model and indeed, in the case of atoms, calculated appearance intensities for specific charge states agree well with experiment (7). In their first experiments on the multielectron dissociative ionisation (MEDI) of N_2, Frasinski et al (8) suggested that their fragment ion spectra might be explained using a similar field ionisation approach.

Even in the simplest diatomic molecules, the multiphoton ionisation process is considerably more complex than in atoms because ionisation and dissociation occur on roughly similar time scales, depending on the molecule concerned and the rise time of the laser. It was pointed out in the above publication on N_2 that one could learn about the dissociation dynamics on a femtosecond timescale, even if a picosecond laser was used. It was also noted that the very act of dissociation enhances the ionisation rate.

Moreover, a diatomic molecule has an axis of symmetry and it was argued that, because of the molecule's elongated shape, the potential difference created by the laser E-field must be larger along its axis than at right angles to it. Consequently the molecule should be more easily ionised when its axis happens to be along the E-field direction. Since the molecule has no time to rotate, the ions should also be ejected preferentially along the laser E-field. The experiments on N_2 showed a strong peaking of fragment ions along the E-field, perfectly consistent with this simple 'geometrical' argument. But it was soon realised that a molecule in a strong laser field is likely to experience a torque that would tend to align the molecule. Indeed, the process of laser-induced alignment is one of the simpler examples of manipulating molecules and this aspect will be discussed in a later section.

It was also found, quite surprisingly, that the energies of the fragment ions were independent of laser pulse rise time (9,10). One would have expected that, if the rise time were to be increased, then the dissociating ions would reach a larger inter-ion distance, R, before further ionisation by the increasing laser field and therefore the kinetic energies would reduce. Equally surprising was the fact that the kinetic energy release of all $(Q_1, Q_2,)$ channels were, to a good approximation, a *constant fraction*, C_m, of the Coulomb energy at $R = R_e$, the neutral molecule internuclear separation. The simplest explanation of this behaviour was that the transient molecular ions were stabilised by the laser field at a critical distance R_c (= R_e/C_m). That is

$$E_c(eV) = C_m \frac{14.4\, Q_1 Q_2}{R_e} = \frac{14.4\, Q_1 Q_2}{R_c} \tag{1}$$

This idea of laser-induced stabilisation, or vibrational trapping, will be expanded upon in a later section.

There are, of course, other ways that an intense, sub-picosecond laser field can be used to manipulate the dynamics of dissociation. As well as the above bond-hardening, there is the phenomenon of bond-softening (6); this will also be discussed in a later section. There is the possibility of trapping, where the molecule is highly localised in space (11). The ability to coherently control the outcome of a dissociation or dissociative ionisation process is presently of great interest and that will also be discussed. Finally, Coulomb explosions are an inherent part of the MEDI process and there are suggestions to use Coulomb explosion imaging to follow the development of a dissociation process or chemical reaction (12).

Before discussing these methods of manipulating molecules, it seems appropriate to consider first two particular aspects, one experimental, the other theoretical. The fragment ion energies are usually determined by ion time-of-flight (TOF) mass spectroscopy and this technique will be discussed briefly in the next section. In the following section the field ionisation, Coulomb explosion model will be introduced, prior to applying it in an attempt to explain certain aspects of the interaction of intense lasers with simple molecules.

EXPERIMENTAL TECHNIQUES

Before we show results of the MEDI process, we will describe briefly the approach used to determine the kinetic energies of the various fragment ions and how one can unambiguously correlate these ions with their transient molecular ion precursors. A conventional ion TOF spectrometer was used in a number of the experiments outlined in this article. The laser light is focused by an on-axis parabolic mirror on to an effusive beam of H_2 molecules, for example. Assuming the E-vector of the linearly polarised laser light lies along the axis of the drift tube and, furthermore, that the ions are ejected preferentially along the E-vector, then two groups of ions are produced; these are seen in Figure 1. The 'forward' ions, H_f^+, with initial velocities towards the drift tube and microchannel plate detector, arrive first. The 'backward' ions, H_b^+, are reversed by the extraction field, U, and arrive at the detector at a slightly later time.

The TOF, t, of an ion of mass, m, and charge, q, is given by the simple formula:

$$t = A\left(\frac{m}{qU}\right)^{\frac{1}{2}} + B\left(\frac{mv_\parallel}{qU}\right),\tag{2}$$

where A and B are constants and $v_\parallel$ is velocity of the ion along the drift tube axis. The difference in TOF between forward and backward ions gives the kinetic energy of the ion.

A second shorter drift tube, with a higher collection efficiency, is used to correlate the various fragment ions. Along the x and y axes of Figure 1 is shown the conventional, time-averaged ion TOF spectrum of H_2. The parent H_2^+ ions are evident at about 1 µs, as are the forward and backward H^+ ions. The question as to which of these H^+ ions are to be associated with the dissociative ionisation channel, $H_2 \rightarrow H + H^+$, labelled (0, 1), and which

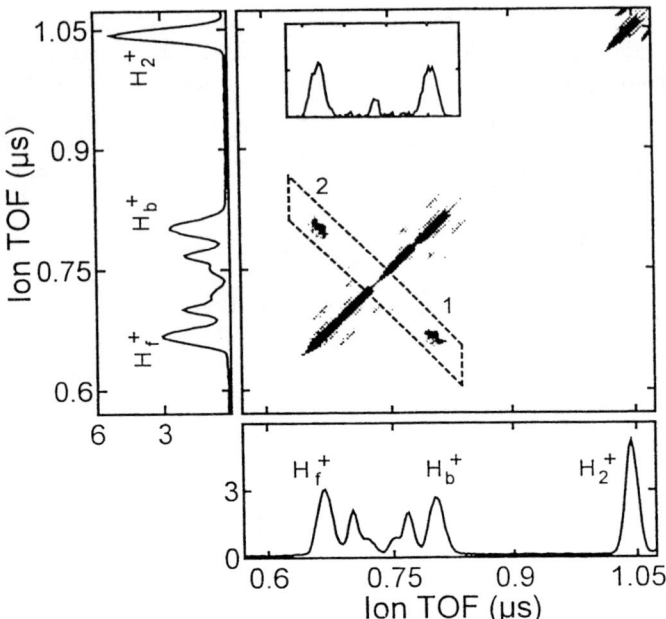

FIGURE 1. Covariance map of H_2 subjected to laser pulses of wavelength 750 nm, pulse duration 55fs and peak intensity of $\simeq 10^{15}$ W cm^{-2}.

with the Coulomb explosion channel $H_2 \rightarrow H^+ + H^+$, labelled (1, 1), can be determined by using the covariance mapping technique (13). The islands on the covariance map are clearly associated with correlated forward and backward H^+ ions and hence the peak centred at about 0.8 μs must involve the $H^+ + H^+$ channel. The remaining peaks are associated with correlations between H^+ ions and undetected H atoms.

THE FIELD IONISATION, COULOMB EXPLOSION MODEL

The field ionisation Coulomb explosion model has been discussed in a relatively recent paper (14) and therefore the ideas will be described only briefly. Figure 2 illustrates the process of field ionisation for the dissociating molecular ion I_2^+, that is two I^+ ions and a single outer electron. We assume that the axis of the molecular ion is aligned with the laser E-field. At an inter-ion separation of 5 a.u., a laser intensity of 5.3×10^{13} W cm^{-2} is required for over-the-barrier ionisation (bottom left). For a separation of R = 7.5 a.u. (not shown), the inner barrier rises to localise the electron in one of the two wells (15). Between 7.5 and 9.8 a.u. the electron is Stark-shifted and a much lower laser intensity of 1.2×10^{13} W cm^{-2} is required for field ionisation (bottom centre). Above 10 a.u., however, the central barrier rises to inhibit ionisation and at R = 14 a.u. the 'appearance' intensity for the (1, 1) channel rises to 3×10^{13} W cm^{-2} (bottom right). Indeed, beyond 20 a.u. the dissociating ions behave almost as free ions, virtually unaffected by the neighbouring ion.

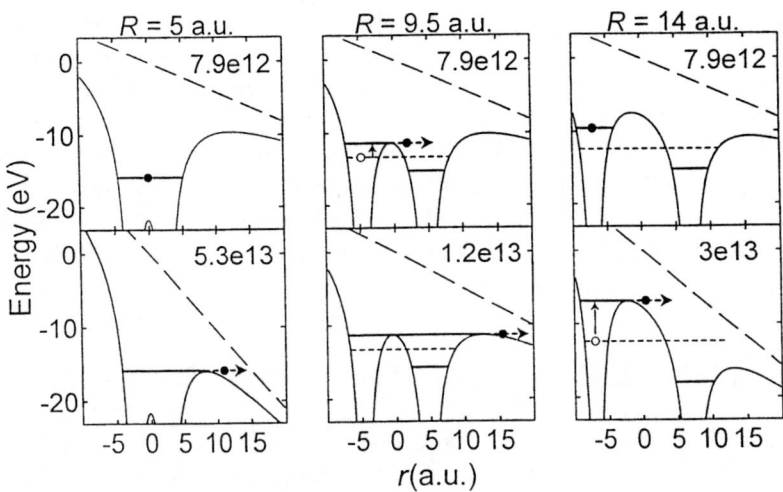

FIGURE 2. Double well potentials for I_2^+ at three internuclear distances and a number of laser intensities.

In summary, one sees a well-defined minimum in the appearance intensity for this channel. More interesting, is the fact that the ion separation that maximises the field ionisation process is *virtually the same* for all (Q_1, Q_2) channels; this is shown in Figure 3. This value of R can be associated with the critical distance, R_c, introduced earlier. One can now calculate the classical trajectories for molecules situated in various regions of the laser focus. The dash-dot curves in Figure 3 trace out the temporal evolution of the laser intensity versus inter-ion separation in four regions of the focus for two pulse lengths, 150 and 400fs.

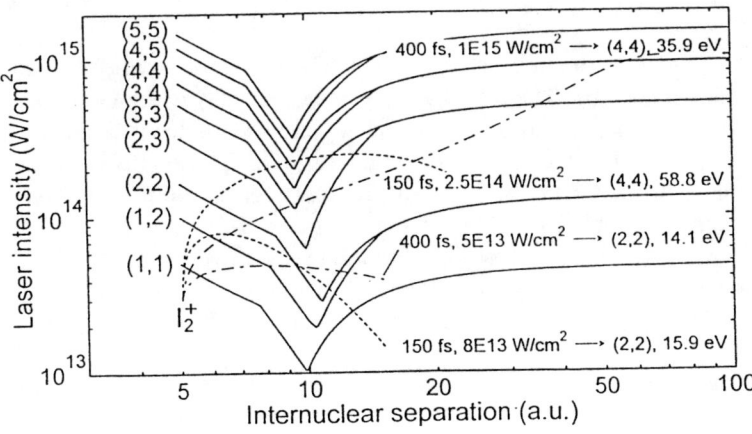

FIGURE 3. Threshold intensities of the (Q_1, Q_2) channels of I_2 (solid curves) and four trajectories (dashed curves).

The process of over-the-barrier ionisation depicted in Figure 2 occurs each time a specific trajectory crosses an appearance intensity curve. One sees, for example, that with a 150 fs laser pulse of intensity 8×10^{13} W cm^{-2}, the (2, 2) channel is created at R $\approx$ 9.5 a.u., giving a dissociation energy of 15.9 eV. With a 400 fs pulse of intensity 5×10^{13} W cm^{-2}, the same channel has an energy release of 14.1 eV. That is, the energy release is virtually independent of pulse rise time, as observed experimentally. Moreover, the fact that the kinetic energy release in all channels of I_2 are found experimentally (16) to be a constant percentage of the Coulomb energy at R_e (about 70% in the case of I_2) finds explanation within this simple model.

MANIPULATING THE DISSOCIATIVE IONISATION PROCESS

As discussed briefly in the introduction, there are a number of ways in which an intense, sub-picosecond laser can modify the behaviour of a molecule. Some of these will be discussed in further detail in the following sections.

Laser-Induced Dissociation

When a molecule is subjected to an intense laser field, the process of above threshold dissociation (ATD) occurs (17). As mentioned earlier, the signature of ATD would be expected to be a series of ion peaks separated, in the case of a homonuclear diatomic molecule, by half the photon energy. However, in order to explain the ion TOF spectra in detail, one must consider the various bound and unbound states of the molecular ion to be 'dressed' by the laser field. The result is that these multiphoton couplings soften the molecular bond, resulting in ion fragments with lower kinetic energies than one might otherwise expect.

Figure 4 shows the $1s\sigma_g$ (bound) and $2p\sigma_u$ (unbound) potential energy curves for H_2^+ dressed by the photon field (6). The curves are repeated vertically with one-photon energy spacings, shown by the broken lines. These diabatic curves are distorted by the photon-molecular ion interactions into the solid adiabatic curves. One photon dissociation appears as tunnelling through the barrier at 2.3Å to give protons of about 0.5eV energy. In addition, the bound state potential curve is crossed by the u (N-3) state. Dissociation along this curve leads to protons with an energy of about 2.8eV. However, if the adiabatic route is chosen, fragments have an energy about 1.65eV, as if only two photons were absorbed. In fact three photons are absorbed, as required by the selection rule, and one emitted.

These three routes are shown by the bold arrows and explain the three peaks observed experimentally at a wavelength of 532 nm (2.33eV), and shown alongside. The phenomenon has been described as bond softening, since the potential curves flatten or 'soften' in the vicinity of a multiphoton resonance.

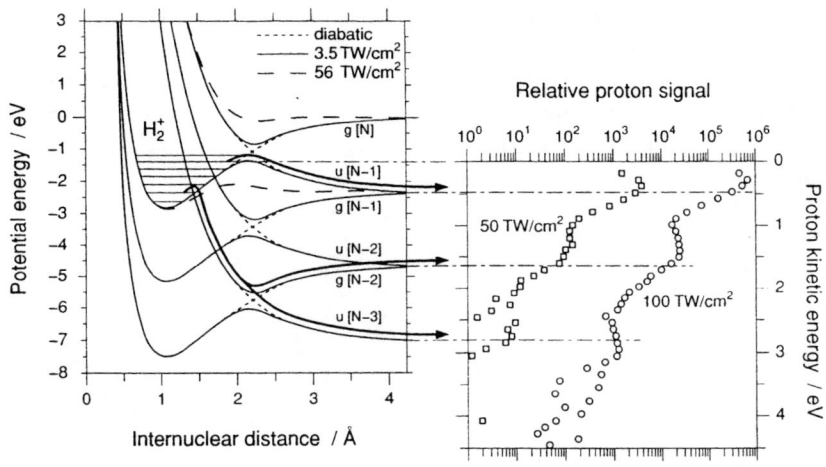

FIGURE 4. The $1s\sigma_g$ and $2p\sigma_u$ potential energies curves of H_2^+ dressed by an intense laser of wavelength 532 nm. The bold lines trace the adiabatic and diabatic routes to dissociation, resulting in the three groups of protons in the spectrum alongside.

Laser-Induced Stabilisation

As pointed out in the introduction, experiments in the early '90's gave results for the fragment ion energies that could be interpreted in terms of laser-induced stabilisation at a critical distance, R_c. In the lighter molecules the experimentally determined energies were about 45% of the Coulomb energy at R_c, whereas in the heavier molecules the value was about 70%. Schmidt et al (18) noted that for all molecules studied, the Coulomb explosion occurred at an internuclear distance given by $R_c = 2.2\ (R_e)^{\frac{1}{2}}$. They pointed out that the idea of light-induced bound states had been introduced by Federov et al (19) and suggested that the trapping might be created by charge-resonance or charge-transfer couplings.

Population trapping in dressed states of H_2^+ was claimed to have been seen by Zavriyev et al (20) using a laser of 769 nm wavelength and 160 fs pulse length. The dressed state picture of Figure 4 shows a temporary bound state (a well) created at the avoided crossing of g (N) and u (N-3) and ionisation from the vibrational states of H_2^+ supported by this well to the $H^+ + H^+$ repulsive state was claimed to be the reason for observing vibrational structure on the 'Coulomb explosion' feature in their TOF spectra.

In fact, subsequent measurements have failed to reproduce this structure and calculations suggest that at the peak laser intensities of 10^{15} W cm^{-2}, creation of a temporary bound state of H_2^+ is unlikely to have occurred (21). As the H and H^+ particles reach the critical distance, R_c, enhanced ionisation on the rising edge of the laser pulse should lead to a structureless (1, 1) feature, see Figure 1 and also Figure 5. It is the inherent stability of H_2 that is responsible for observing this (1, 1) channel so strongly, rather than vibrational trapping.

The fact that the Coulomb explosion (1, 1) peak is created primarily on the rising edge of the laser pulse and the bond-softening peaks (1ω and 2ω) on the falling edge has been verified by a recent double-pulse experiment (22). Figure 5 shows the laser beam fed through a Mach-Zehnder-like optical arrangement so that two linearly polarised, but orthogonal beams, A and B, are focused in the vacuum chamber, one with an intensity about twice the other. The delay line was adjusted to a precision of a few femtoseconds.

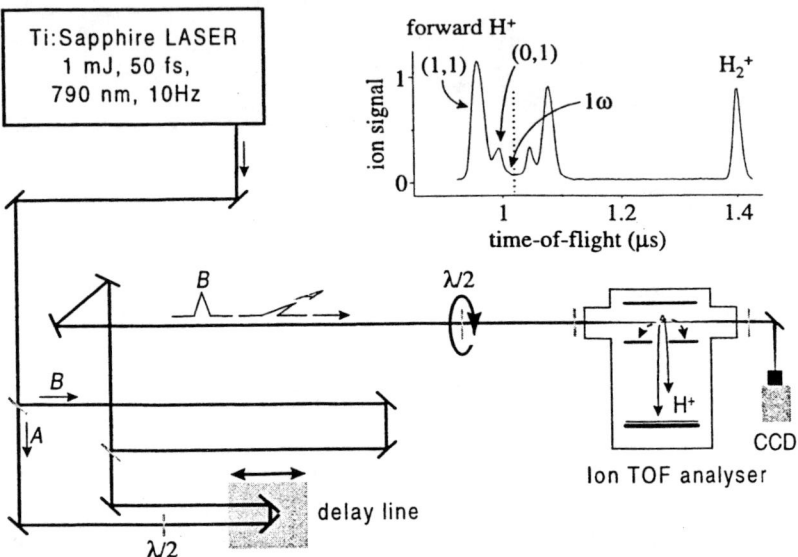

FIGURE 5. Optical arrangement to produce two pulses, A and B, of orthogonal polarisation and variable time delay, τ. The inset shows an ion TOF spectrum for H_2 produced by B.

The angular distribution of the fragments were measured by simultaneously rotating the polarisation of both beams with a half-wave plate located just in front of the vacuum chamber and taking ion TOF spectra at intervals of 2°. Momentum maps of the energy versus angle for the forward ions (23) obtained by rotating the polarisation through 360° are shown in Figure 6 for six delays. At the side of each distribution are shown the two laser beams in perspective, with their time delays. Analysis of these distributions, which provide information with *sub-pulse-length* time resolution, confirms that the Coulomb explosion occurs on the rising edge and bond softening on the falling edge of the pulse.

Vibrational trapping does in fact occur in H_2^+ if the appropriate experimental conditions are realised. We have recently observed bond hardening at an intensity of about 10^{14} W cm^{-2} using 792 nm laser pulses of duration ranging from 45 to 500 fs. The effect can be understood in terms of a light-induced potential well created at twice the normal equilibrium internuclear distance by an adiabatic mixing of 1- and 3- photon resonances (24). In this case, however, the trapped population dissociates into H + H$^+$ when the potential well

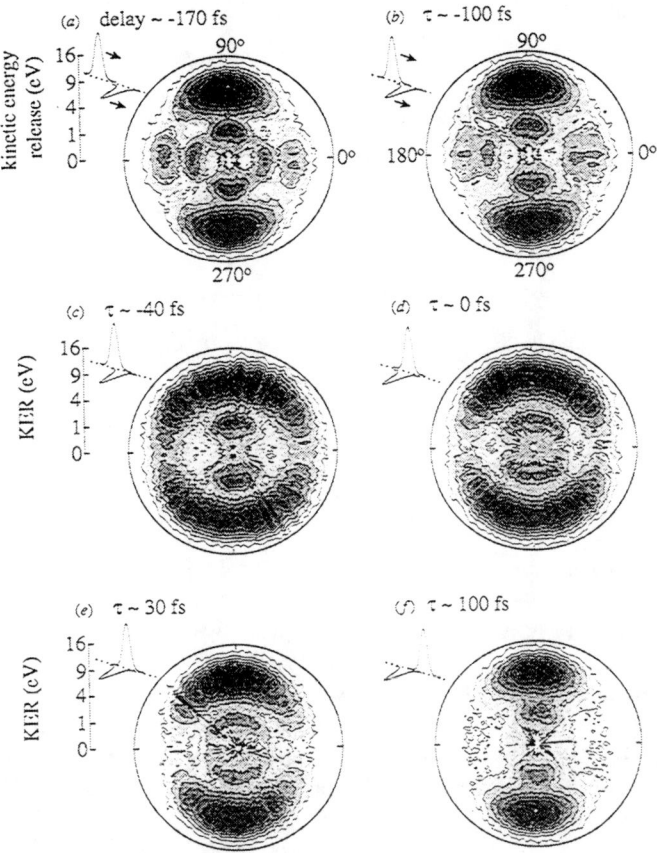

FIGURE 6. Momentum maps for six different time delays, τ, showing both the angular and energy distributions for forward H^+ ions.

becomes convex on the trailing edge of the pulse. The release of the nuclear wavepacket from the potential well, as exemplified by a shift in the 1ω peak, can be explained in terms of absorption of photons from the high energy end of the laser spectrum and re-emission at the low energy end, a dynamic Raman effect within the laser bandwidth. The interpretation of this shift is supported by previous theoretical work (17), but see also (25).

A second example of trapping involves the use of the third harmonic of the Ti:sapphire laser at 266 nm to observe the process of 'zero-photon dissociation' in H_2^+ (26). Its observation relied upon a combination of relatively short pulses (~250 fs), short wavelength and high intensity. Figure 7 shows an ion TOF spectrum with two groups of protons, ones with energies between 0 and 0.5eV and others with a peak around 3eV; both are associated with the (0, 1) channel. The 3eV peak is produced by bond-softening and is accompanied by the net absorption of one photon.

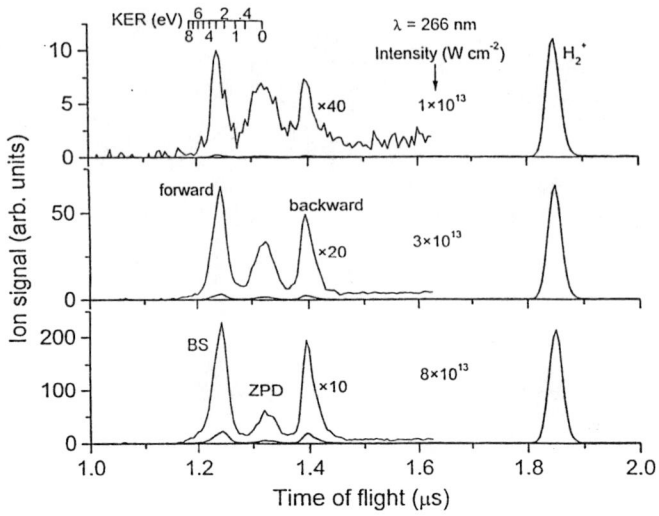

FIGURE 7. H_2 ion TOF spectra showing bond softening (BS) and zero-photon dissociation (ZPD).

However, the low energy peak is attributed to a vibrational wavepacket of H_2^+ which is temporarily trapped in the potential well created by the upper adiabatic curve. The wavepacket partially escapes as the bottom of the well is pushed upwards in response to an increase in laser intensity on the rising edge of the pulse. If the pulse is not too short, the trapped population can escape so that it is not trapped again on the falling edge of the pulse. No net number of photons is absorbed by the H_2^+ ion, hence the term 'zero-photon dissociation'

Although there were clear experimental reasons to countenance the possibility of laser-induced stabilisation or vibrational trapping in N_2, CO, Cl_2, etc, it was hard to conceive of a mechanism that could stabilise a Cl_2^{8+} or I_2^{10+} transient ion (16, 18). The field ionisation, Coulomb explosion model, described in an earlier section, was introduced in an attempt to explain the data on I_2, and hopefully other molecules, in a more acceptable way. As we have seen, the inter-ion separation that maximises ionisation is roughly the same for *all* channels and one can therefore explain the dissociation energies in terms of extremely rapid sequential ionisation near R_c, rather than laser-induced stabilisation.

It should be pointed out that the phenomenon of enhanced ionisation at a specific internuclear separation was consistent with earlier quantum mechanical calculations (27, 28) and it was important to check directly whether this enhancement actually does occur. Constant et al (29) employed 80 fs pulses at 625 nm to create the slowly dissociating (0, 2) channel in I_2. A second pulse, which caused the (0, 2) to (1, 2) transition, was delayed with respect to the first. We used a laser of 750 nm and 55 fs duration and the results of these experiments (30) are shown in Figure 8 (a). For delays ranging from 100 to 480 fs one observes an extra peak (arrowed) due to the creation of the (1, 2) channel.

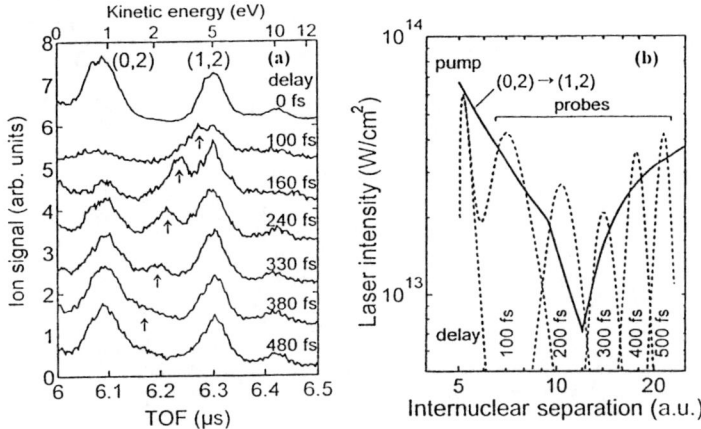

FIGURE 8. (a) I^{2+} fragments in the I_2 ion TOF spectrum, the arrow denoting the (0,2) to 1,2) transition; (b) Pump-probe curves showing why the TOF of the arrowed peak decreases with increasing delay.

The curves in Figure 8 (b) explain why the peak moves to lower energies (lower TOFs for the backward ions) as the delay is increased. Calculations indicated that the second pulse should be most effective when the delay is about 250 fs, that is when the I and I^{2+} fragments arrive at R_c; experimentally the greatest enhancement was found to occur at a delay of about 200 fs, in reasonably good agreement with theory. In fact the existence of a critical distance is now well documented in a number of molecules (31).

Laser-Induced Charge-Asymmetry

There has been considerable discussion in the past with regard to the charge-symmetry of the fragmentation process. Earlier disagreements stemmed from an inability to unambiguously correlate the ion fragments. In the context of the field ionisation model and a laser field that is reversing its polarity roughly every femtosecond, it seems obvious that a transient molecular ion such as I_2^{4+} should fragment symmetrically, since a lower E-field is required to produce this outcome. Using covariance mapping and a laser with 150fs pulses, Hatherly et al (16) saw very little evidence of the (1,3) channel.

The first convincing evidence for charge-asymmetric fragmentation was the observation of the previously mentioned (0, 2) channel in I_2. One might have expected the (1, 1) channel to dominate, but I_2^+ ions become strongly polarised in a laser E-field that is almost strong enough to induce field ionisation and, for a small range of internuclear separations and laser intensities, one can expect to observe the (0, 2) channel. That is, transfer of electron charge from the up-field to the down-field potential well occurs. At this point two electrons must be oscillating back and forth between the two wells.

13

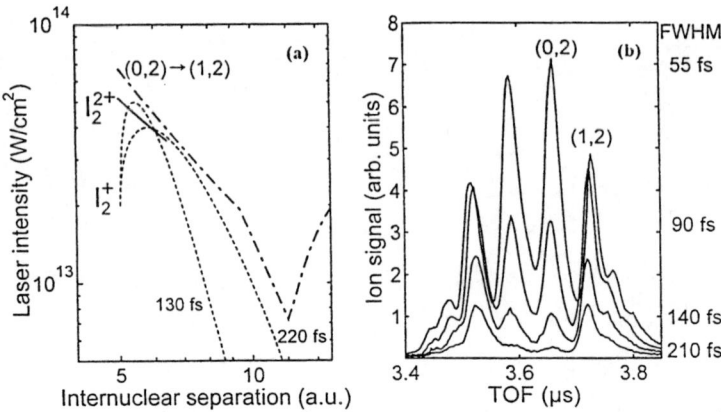

FIGURE 9. (a) Threshold intensity curve for the (0,2) to (1,2) transition and trajectories for 130 and 220fs pulses; (b) The (0,2) channel decreases steadily with increasing pulse width.

To make this point, Figure 9 (a) shows examples of trajectories for two pulse lengths, 130 and 220 fs. Both show dissociation along the (0, 1) potential curve until they cross the I_2^{2+} appearance intensity curve (the solid line). At this pioint transfer to the (0, 2) channel can occur (14). In order to continue to observe the (0, 2) channel, one must ensure that the trajectory does not cross the (0, 2) → (1,2) appearance intensity curve (the dash-dot curve) at larger values of R. If the pulse width exceed 220 fs, the (1, 2) channel is created and the (0, 2) channel disappears. Figure 9 (b) shows the experimental data and the (0, 2) channel has virtually disappeared at 210 fs, in agreement with the field ionisation model.

Laser-Induced Alignment

In their first publication on the MEDI of N_2, Frasinski et al (8) suggested that the strong peaking of fragment ions along the laser E-field might be explained in purely geometrical terms, see introduction. It was then pointed out that the peaked angular distribution in I_2 could not be explained in this way (32) but required dynamic, laser-induced alignment. Subsequently double-pulse experiments on CO using 30 ps pulses (10) and on I_2 using sub-100fs pulses (33) appeared to confirm that laser-induced alignment occurs in both instances. In fact the interpretation of these experiments is questionable since the phenomenon of enhanced ionisation at R_c was not appreciated at the time and focal volume effects created certain ambiguities.

We have recently performed experiments of I_2, N_2 and H_2 and, in I_2 in particular, have attempted to determine the extent to which angle-dependent enhanced ionisation, a purely geometrical effect, could explain the highly anisotropic angular distributions. In this case a laser pulse length of about 50 fs was used. In fact the angular distributions obtained as a function of laser intensity could be explained assuming random orientation of the I_2 molecules and essentially no laser-induced rotation (34). This is in good agreement with experiments using 80 fs pulses, where the I_2 ion TOF spectra obtained with linearly polarised light were compared with those using circularly polarised light (35). In contrast, the Cl_2 TOF spectra indicated dynamic alignment by the laser field.

We have also measured angular distributions for H_2 and found them considerably sharper than those of I_2 (34). The considerable difference in behaviour of the two molecules, H_2 and I_2, has been confirmed by recent double-pulse (pump-probe) experiments (36). The experimental arrangement was quite similar to that shown in Figure 5, except that the half-wave plate was removed. In this case laser pulse A preceded pulse B by about 1 ps.

As before, A was linearly polarised with its E-field orthogonal to the axis of the TOF analyser and therefore H^+ fragments produced by these pulses were ejected perpendicular to the drift tube and missed the detector, see Figure 5. Beam B was polarised with its E-field along the axis and therefore forward and backward H^+ ions were detected. In the present experiment beam B was used to probe *spatially* the effect of beam A. In order to improve the spatial resolution, the probe beam had a sharper focus but A and B had roughly

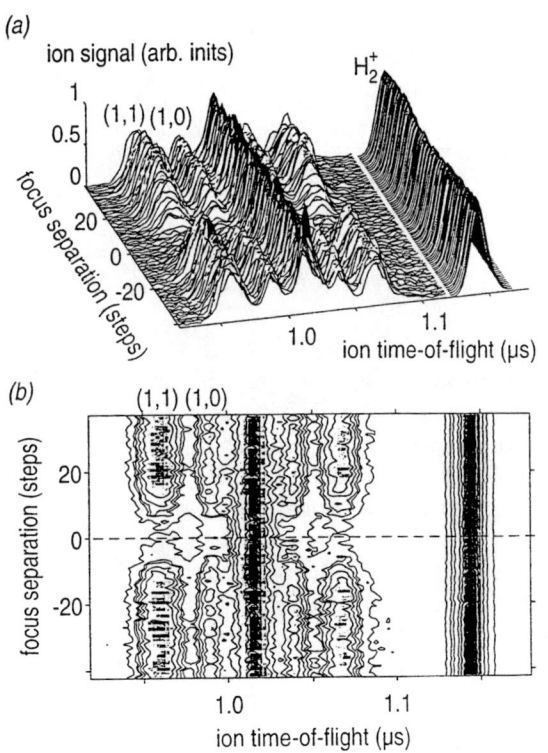

FIGURE 10. (a) H_2 ion TOF spectrum as a function of spatial separation of pump and probe. When the foci overlap, the energetic ions disappear; (b) Contour plots of data in (a).

the same peak intensity. The pump focus was scanned systematically through the probe focus by step-wise rotation of a mirror in the arm of beam A.

At each step an ion TOF spectrum was recorded. Typical results are shown in Figure 10 (a) and (b). The TOF spectra show five H^+ peaks, two forward and two backward peaks produced by beam B and one central peak of low energy fragments, produced by beam A; the forward peaks are labelled (1, 0) and (1, 1). Note that the fragments produced by beam B disappear almost completely when the beams overlap significantly. This shows that the first pulse orients and fragments *all* H_2 molecules and sends them in the orthogonal direction. These molecules must have been rotated parallel to the E-field on a very short timescale, since they were initially randomly oriented.

When the same double pulse experiment was performed with I_2, quite different results were obtained. The forward and backward ions did not disappear when the foci overlapped; virtually no dynamic alignment occurs in the case of I_2 using 50 fs pulses.

The high degree of alignment achieved in H_2 underlines the potential of a strong laser field as a general tool for aligning molecules. The degree of alignment is high compared to standard optical methods and because the process is non-linear, the focusing of the laser beam localises the alignment on a micrometer scale in all three dimensions. Of course here we have detected the alignment by ionising the molecule but it is reasonable to assume that a molecule will align if the intensity stays below the ionisation and dissociation thresholds and if the pulse duration is sufficiently long (37).

Laser-Induced Trapping

In a recent article Corkum et al (38, 39) discussed how an intense laser might be used to deflect and trap molecules by means of non-resonant Stark shifts. The dependence of the Stark shift with respect to the intensity profile of the laser focus determines the spatial force on the molecule whilst its dependence on the orientation of the molecule with respect to the laser polarisation determines the torque. Through these forces, one can in principle control position, orientation and linear and/or angular velocity.

Experiments are described which used a focused CO_2 laser (10.6 μm) to exert an induced dipole force in order to create a deflection of a beam of neutral CS_2 molecules. The resulting velocity change was monitored by ionising the molecules with a tightly focused 80 fs laser beam (625 nm) and observing the distribution of flight times of the CS_2^+ ions. The molecules that passed near the centre of the focus of the CO_2 laser experienced a displacement that caused them to meet at a common position. The focal length of the 'molecular lens' was about 500 μm.

The suggestion was that the powerful trapping and cooling techniques developed for atoms would not be applicable to molecules due to their complex energy level structure and weak transition moments and that the technique described above had potential for two-and three-dimensional trapping. However, recent developments seem to suggest that molecules can be trapped in conventional magneto-optical traps. One starts with Cs atoms in a conventional trap and uses a laser to cause photo association from the $^3\Sigma_u^+$ (6s + 6s) state to the O_g^- (6s + 6p) state (40). Evolution on the excited state is followed by spontaneous decay to bound vibrational levels of the ground triplet state or back to a pair of atoms.

It has recently been proposed that the efficiency of cold molecule formation can be enhanced by using the Wilson scheme for population inversion by a chirped pulsed laser field (41). This would achieve almost complete transfer of colliding particles from the electronic ground state onto the excited state (42). Unfortunately, the molecules would end up in vibrationally excited states of the ground state and so another femtosecond laser would be required, coupled with appropriate pulse shaping or pulse sequencing techniques, in order to manipulate the molecular motion so that the molecules would finish up in the v=0 state (43). Feedback control may also be required.

Coherent Control of Dissociation Dynamics

A few years ago the fundamental (1053 nm) and the second harmonic (527 nm) of a Nd:YLF laser were used in an experiment on phase control in the two-colour photo dissociation of HD (44). The aim was to control the spatial asymmetry of the dissociation channels $H + D^+$ and $H^+ + D$. We performed similar experiments on H_2 (21) with laser pulses of less than 100 fs, that is almost three orders of magnitude shorter than used in the earlier experiment and an order of magnitude more intense. Despite these large differences in experimental conditions, and the fact that shorter wavelengths were used, namely 750 and 375 nm, the results were quite similar in that the ions appeared to be emitted counter-intuitively. That is the ions were emitted in a direction opposite to that in which the maximum of the combined two-colour laser E-field pointed, see Figure 11.

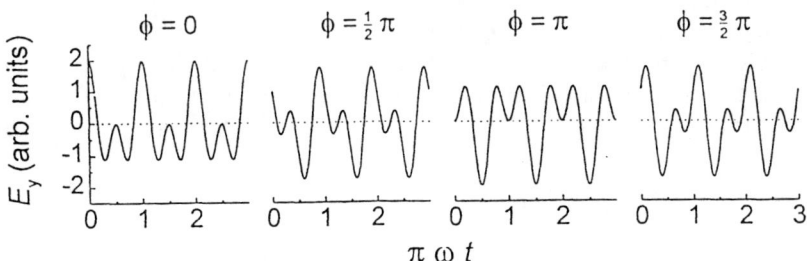

FIGURE 11. The superposed, two-colour laser E-field $\cos(\omega t) + \cos(2\omega t + \phi)$ for equal intensities and four phase differences.

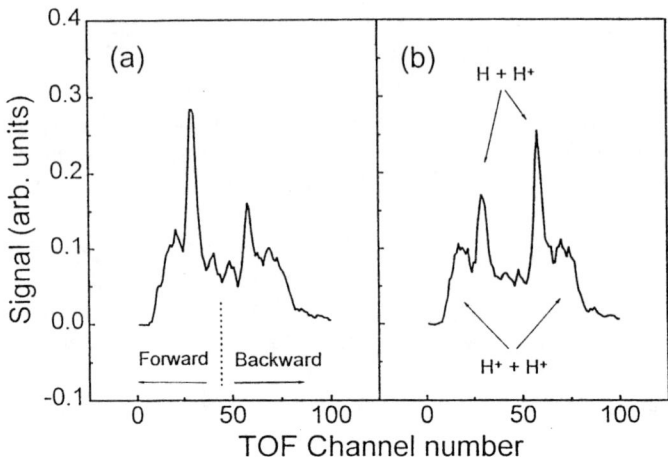

FIGURE 12. The proton emission changes from (a) predominantly foward to (b) predominantly backward when φ is changed by π.

The experiments utilised an interferometer of the Michelson type to create a phase difference between fundamental and second harmonic and filters were used to ensure that the two had similar intensities. In order to restrict the production of the (1, 1) channel to the centre of the laser focus, both beams were attenuated to give peak intensities of about 10^{14} W cm^{-2}. Figure 12 (a) shows enhancement of the forward protons. A change of π in the relative phases caused the backward proton emission to dominate, see Figure 12 (b).

In such a short pulse experiment it was impossible to define the direction of the peak in the superposed fields. To achieve this end, an electron drift tube was located opposite the ion drift tube and the extraction field directed the electrons and ions to their respective detectors. Assuming that the electrons behaved intuitively, and there is apparently now some doubt about that (45), the ions were found to be ejected counter-intuitively.

It had been suggested earlier that coherent superposition in the high intensity regime could be utilised to orient rather than align a diatomic molecule. It was argued that, by using an appropriately long wavelength, it ought to be possible to induce different asymmetries in the H$^+$ and D$^+$ photofragmentation of HD and thereby control the spatial separation of these isotopes (46). As far as we are aware, this has yet to be achieved experimentally.

A fundamental goal in chemistry is to achieve active control of chemical reactions. One approach is to use quantum mechanical interference, as first proposed by Brumer and Shapiro (47), to modify reaction pathways. Experimental realisation of this approach has been demonstrated in atoms and small molecules (48, 49) and a simple example of the use of this technique is described above. A somewhat different approach makes use of the rapid developments in ultra-short laser pulse technology. For example, the 'pump-dump' technique (50) has been realised experimentally by a number of groups (51, 52).

Both schemes are based on a limited number of optimisable parameters and in complex systems these are insufficient. In principle the electric field of the laser pulse should be designed for the particular molecule, such that the amplitudes of the different interfering vibrational modes add up in a given bond some time after photoabsorption, causing a break (53). Optimal control theory is used (54) to calculate the appropriate electric fields. However, for large molecules this approach fails because the potential energy surfaces are not known sufficiently accurately and it has proved difficult to generate the complex laser fields.

An alternative is to include the experimental output in the optimization procedure. A suitable algorithm forms an ultra-short laser pulse and uses the output as feedback in a learning algorithm. In this case one is 'teaching lasers to control molecules' (55). In a relatively recent example (56), femtosecond laser pulses were modified in a computer-controlled pulse shaper. Ionic fragments from molecular dissociation of Fe $(CO)_5$ were recorded by a TOF spectrum which then modified the pulse shape in order to optimise a particular ion yield or branching ratio, as indicated in Figure 13 (a).

The pulse shaper was set up as a zero dispersion compressor (57), with a liquid crystal spatial light modulator in its Fourier plane. The modulator contained 128 rectangular pixels whose refractive index could be changed independently. In this way different optical paths could be introduced to the spatially separated spectral components, resulting in a shift of the relative phases. The emerging pulses were then focused, as usual, into the TOF spectrometer. These tailored femtosecond pulses were able to optimise the branching ratios of different orgometallic photodissociation reaction channels. For example, the ratio of Fe $(CO)_5^+$/Fe$^+$ could be maximised (solid blocks) or minimised (open blocks) as shown in Figure 13 (b).

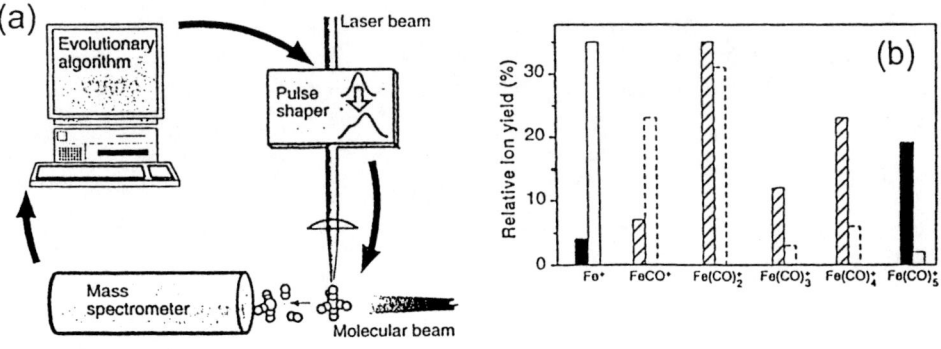

FIGURE 13 (a) The pulsed laser ionises the molecule, the signal from the ion TOF mass spectrometer is fed to the controlling evolutionary computer algorithm to optimise the branching ratio; (b) The ratio of Fe $(CO_5)^+$/Fe$^+$ can be maximised (solid blocks) or minimised (open blocks).

Coulomb Explosion Imaging

Coulomb explosions are, in a real sense, central to the MEDI process and there is now the possibility of utilising Coulomb explosion imaging (CEI) to map out the time evolution of a photodissociation process or, indeed, a chemical reaction. The requirement is for a laser of short pulse length (10 fs or less, depending on the molecule, but 5fs lasers are now possible (58) and with high intensity and repetition rate). In the photodissociation experiment, a pulse-probe experiment is envisaged. The first laser could cause a neutral molecule (ABC) to dissociate, with a certain branching ratio, into AB + C and A + BC. The second pulse would cause rapid multiple ionisation. Coulomb explosion would occur and the vector correlation of the various ions using ion imaging could reconstruct their initial positions. The experiment could then be performed as a function of pulse-probe time difference to determine the time evolution of the dissociation process.

Timed CEI offers advantages over x-ray (59) and electron diffraction (60) methods for determining time dependent structural changes during a photochemical process in that the x-ray pulses are about 300 fs in length, the electron pulses about 20 ps. There are problems inherent in the CEI technique, in that the atoms can move during the ionisation process, there may be deviation from perfect Coulomb behaviour, there will be a dependence on laser polarisation and on internuclear distance (as discussed earlier) and there will be an influence of the laser field on the ions themselves (35). However, a number of these problems are minimised with heavy atoms and high charge states. A recent publication suggests that CEI retains the ability to faithfully reflect the fastest nuclear motion, even in the case of protons (12).

ACKNOWLEDGEMENTS

We are pleased to acknowledge the Engineering and Physical Sciences Research Council (UK) for their financial support. Special thanks go to Drs P F Taday and A J Langley of Rutherford Appleton Laboratory for their expert assistance with the relevant experiments described here.

REFERENCES

1. Strickland, D., and Morou, G., *Opt.Commun.* **56**, 219 (1985).
2. Agostini, P., Fabré, F., Mainfray, G., et al, *Phys.Rev.Lett.,* **42**, 1127 (1979).
3. McPherson, A., Gibson, G., Jara, H., et al, *J.Opt.Soc.Am.,* **B4**, 595 (1987).
4. Gibson, G.N., Freeman, R.R., and McIlrath, T.J., *Phys.Rev.Lett.* **67**, 1230 (1990).
5. Liang, Y., Augst, S., Chin, S.L., et al, *J.Phys. B*, **27**, 5119 (1994)
6. Bucksbaum, P.H., Zavriyev, A., Muller, H.G., and Schumacher, D.W., *Phys. Rev.Lett.* **64**, 1883 (1990).
7. Augst, S., Strickland, D., Meyerhofer, D.D., et al, *Phys.Rev.Lett.,* **63**, 2212 (1989).
8. Frasinski, L.J., Codling, K., Hatherly, P.A., et al, *Phys.Rev.Lett.,* **58**, 2424 (1987).
9. Cornaggia, C., Lavancier, J., Normand D., et al, *Phys.Rev. A,* **44**, 4499 (1991).
10. Normand, D., Lompré. L.A., and Cornaggia, C., *J.Phys.B*, **25**, L497 (1992).
11. Friedrich, B., and Herschbach, D., *Phys.Rev.Lett.,* **74**, 4623 (1995).
12. Chelkowski, S., Corkum, P.B., and Bandrauk, A.D., *Phys.Rev.Lett.,* **82**, 3416 (1999).
13. Frasinski, L.J., Codling, K., and Hatherly, P.A., *Science,* **246**, 973 (1989).
14. Posthumus, J.H., Giles, A.J., Thompson, M.R., and Codling, K., *J.Phys.B*, **29**, 5811 (1996).
15. Posthumus, J.H., Frasinski, L.J., Giles, A.J., and Codling, K., *J.Phys.B*, **28**, L349 (1995).
16. Hatherly, P.A., Stankiewicz, M., Codling, K., et al., *J.Phys.B*, **27**, 2993 (1994).
17. Giusti-Suzor, A., Mies, F.H., DiMauro, L.F., et al, *J.Phys.B*, **28**, 309 (1995).
18. Schmidt, M., Normand, D., and Cornaggia, C., *Phys.Rev.A*, **50**, 5037, (1994).
19. Federov, M.V., Kudrevatova, O.V., Makarov, V.P., and Samokhin, A.A., *Opt.Commun.,* **13**, 299 (1975).
20. Zavriyev, A., Bucksbaum, P.H., Squier, J., and Saline, F., *Phys.Rev.Lett.,* **70**, 1077 (1993).
21. Thompson, M.R., Thomas, M.K., Taday, P.F., et al, *J.Phys.B*, **30**, 5755, (1997).
22. Posthumus, J.H., Plumridge, J., Taday, P.F., et al, *J.Phys.B*, **32**, L93 (1999).
23. Hishikawa, H., Iwamae, A., Hoshina, K., et al, *Chem.Phys.Lett.,* **282**, 283 (1998).
24. Frasinski, L.J., Posthumus, J.H., Plumridge, J., et al, *Phys.Rev.Lett.,* submitted.
25. Numico, R., Keller, A., and Atabek, O., *Phys.Rev.A*, **56**, 772 (1997).
26. Posthumus, J.H., Plumridge, J., Frasinski, L.J., et, *Phys.Rev.Lett.,* submitted.
27. Seideman, T., Ivanorv, M.Yu., and Corkum, P.B., *Phys.Rev.Lett.,* **75**, 2819 (1995).
28. Zuo, T., and Bandrauk, A.D., *Phys.Rev.A*, **52**, R2511 (1995).
29. Constant, E., Stapelfeldt, H., and Corkum, P.B., *Phys.Rev.Lett.,* **76**, 4140 (1996).
30. Posthumus, J.H., Codling, K., Frasinski, L.J., and Thompson, M.R., *Laser Phys.,* **7**, 813 (1997).
31. Sanderson, J., Thomas, R.V., Bryan, W.A., et al, *J.Phys.B*, **30**, 4499 (1997).
32. Strickland, D.T., Beaudoin, Y., Dietrich, P., and Corkum, P.B., *Phys.Rev.Lett.,* **68**, 2755 (1992).
33. Dietrich, P., Strickland, D.T., Laberge, M., and Corkum, P.B., *Phys.Rev.A*, **47**, 2305 (1993).

34. Posthumus, J.H., Plumridge, J., Thomas, M.K., et al, *J.Phys.B*, **31**, L553 (1998).
35. Ellert, C., Stapelfeldt, H., Constant, F., et al, *Phil.Trans.R.Soc.Lond.A*, **356**, 329 (1998).
36. Posthumus, J.H., Plumridge, J., Frasinski, L.J., et al, *J.Phys.B.*, **31**, L985 (1998).
37. Sakai, S., Safven, C.P., Larsen, J.J., et al, *J.Chem.Phys.*, **110**, 10235 (1999).
38. Corkum, P.B., Ellert, C., Mehendale, M., et al, *Faraday Disc.*, **113**, (1999) in press.
39. Stapelfeldt, H., Sakai, H., Constant, E., and Corkum, P.B., *Phys.Rev.Lett.*, **79**, 2787 (1997).
40. Fioretti, A., Comparat, D., Crubellier, A., et al, *Phys.Rev.Lett.*, **80**, 4402 (1998).
41. Cao, J., Bardeen, C.J., and Wilson, K.R., *Phys.Rev.Lett*, **80**, 1406 (1998).
42. Vala, J., Dulieu, O., Masnou-Seeuws, F., et al, *XXI ICPEAC*, **SA162**, Sendai (1999).
43. Tannor, D.J., Kosloff, R., and Bartana, A., *Faraday Disc*, **113**, (1999) in press.
44. Sheehy, B., Walker, B., and Di Manro, L.F., *Phys.Rev.Lett.*, **74**, 4799 (1995).
45. Bandrauk, A.D., private communication.
46. Charron, E., Giusti-Suzor, A., and Mies, F.H., *Phys.Rev.Lett.*, **75**, 2815 (1995).
47. Brumer, P., and Shapiro, M., *Chem.Phys.Lett.*, **126**, 541 (1986).
48. Chen, C., Yin, Y.Y., and Elliot, D.S., *Phys.Rev.Lett.*, **64**, 507 (1990).
49. Park, S.M., Lu, S.P., and Gordon, R.J., *J.Chem.Phys.*, **94**, 8622 (1991).
50. Tannor, D.J., and Rice, S.A., *J.Chem.Phys.*, **83**, 5013 (1985).
51. Baumert, T., Grosser, M., Thalweiser, R., and Gerber, G., *Phys.Rev.Lett.*, **67**, 3753 (1991).
52. Potter, E.D., Herek, J.L., Pederson, S., et al, *Nature*, **355**, 66 (1992).
53. Shi, S., Woody, A., and Rabitz, H., *J.Chem.Phys.*, **88**, 6870 (1988).
54. Kosloff, R., Rice, S.A., Gaspard, P., et al, *Chem.Phys.*, **139**, 201 (1989).
55. Judson, R.S., and Rabitz, H., *Phys.Rev.Lett.*, **68**, 1500 (1992).
56. Assion, A., Baumert, T., Bergt, M., et al, *Science*, **282**, 919 (1998).
57. Martinez, O.E., *IEEEJ.Quantum Electron*, **24**, 2530 (1988).
58. Nisoli, M., Silvestri, S.D., Svelto, O, et al, *Opt.Lett.*, **22**, 522 (1997).
59. Rischel, C., Rousse, A., Uschmann, I., et al, *Nature*, **390**, 490 (1997).
60. Williamson, J.C., Cao, J., Ihee, H., et al, *Nature*, **386**, 159 (1997).

Collisions at nanokelvin temperatures in Bose-Einstein condensates

Wolfgang Ketterle and Chandra Raman

Department of Physics and Research Laboratory of Electronics, Massachusetts Institute of Technology, Cambridge, Massachusetts 02139, USA

Abstract. Bose-Einstein condensed atomic gases are a new class of quantum fluids. They are produced by cooling a dilute atomic gas to nanokelvin temperatures using laser and evaporative cooling techniques. In this paper we review developments in Bose-Einstein condensation, emphasizing how this new quantum fluid has become a laboratory for the study of collisions at ultralow energy and of collective effects in light-atom and atom-atom interactions. Magnetic fields have been used to modify the scattering length for atomic collisions. Spinor condensates were created, with a spin structure determined by spin relaxation collisions and external magnetic fields. We have used light scattering to study collective excitations and observed superradiant light emission. Dissipation was studied by dragging a repulsive, blue-detuned laser beam through the fluid, as well as by inducing collisions between condensates.

I INTRODUCTION

When a gas of bosonic atoms is cooled below a critical temperature T_c, a large fraction of the atoms condenses in the lowest quantum state. This phenomenon was first predicted by Albert Einstein in 1925 [1] and is a consequence of quantum statistics. Atoms at temperature T and mass m can be regarded as quantum-mechanical wavepackets which have an extent on the order of a thermal de Broglie wavelength $\lambda_{dB} = (2\pi\hbar^2/mk_BT)^{1/2}$. When atoms are cooled to the point where λ_{dB} is comparable to the interatomic separation, the atomic wavepackets "overlap" and the indistinguishability of particles becomes important At this temperature, bosons undergo a quantum-mechanical phase transition and form a Bose-Einstein condensate, a coherent cloud of atoms all occupying the same quantum mechanical state. The transition temperature and the peak atomic density n are related as $n\lambda_{dB}^3 \simeq 2.612$. The quest for Bose-Einstein condensation has a long history and is nicely summarized in various contributions to the 1998 Varenna summer school [2].

The realization of Bose-Einstein condensation (BEC) in dilute atomic gases [3–7] achieved several long-standing goals. First, neutral atoms were cooled into the lowest energy state, thus exerting ultimate control over the motion and position of atoms, limited only by Heisenberg's uncertainty relation. Second, a coherent

CP500, *The Physics of Electronic and Atomic Collisions*, edited by Y. Itikawa, et al.

macroscopic sample of atoms all occupying the same quantum state was generated, leading to the realization of atom lasers, devices which generate coherent matter waves. Third, degenerate quantum gases were produced with properties quite different from the quantum liquids ^{3}He and ^{4}He. This provides a testing ground for many-body theories of the dilute Bose gas which were developed many decades ago but never tested experimentally [8]. BEC of dilute atomic gases is a macroscopic quantum phenomenon with similarities to superfluidity, superconductivity and the laser [9]. More generally, atomic Bose-Einstein condensates are a new "nanokelvin" laboratory where interactions and collisions at ultralow energy can be studied.

II BASIC TECHNIQUES

The realization of Bose-Einstein condensation requires techniques to cool gases to sub-microkelvin temperatures and atom traps to confine them at high densities and to keep them away from the hot walls of the vacuum chamber. Over the last 15 years, such techniques were developed in the atomic physics and low-temperature communities [2]. The MIT experiment uses a multistage cooling process to cool hot sodium vapor down to temperatures where the atoms form a condensate. A beam of sodium atoms is created in an atomic beam oven at a density of about 10^{14} atoms per cm^{-3}, similar to the eventual density of the condensate. The gas is cooled by nine orders of magnitude from 600 K to 1 μK first by slowing the atomic beam, followed by optical trapping and laser cooling, then by magnetic trapping and evaporative cooling.

Table 1 shows how these cooling techniques together reduce the temperature of the atoms by a factor of a billion. The phase space density enhancement is almost equally distributed between laser cooling and evaporative cooling, providing about six orders of magnitude each. Bose-Einstein condensation can be regarded as "free cooling," as it increases the quantum occupancy by another factor of about a million without any extra effort. This reflects one important aspect of BEC: the fractional population of the ground state is no longer inversely proportional to the number of states with energies smaller than $k_B T$, but quickly approaches unity when the sample is cooled below the transition temperature.

Atom clouds are observed either by absorptive or dispersive techniques. In the first case, the shadow cast by the atom cloud is imaged onto a CCD camera. In the latter case, dispersively scattered photons are collected creating an image of the spatially varying index of refraction. The BEC phase transition can be directly observed in the spatial domain [10]. Fig. 1 shows a series of such spatial images above and below the phase transition. They show the sudden appearance of a high-density core of atoms in the center of the distribution—the Bose-Einstein condensate. Lowering the temperature further, the condensate number grows and the thermal wings of the distribution become shorter. Finally, the temperature drops to the point where only the central peak remains.

Similarly, the BEC phase transition can be observed by imaging the shadow

TABLE 1. Multi-stage cooling to BEC in the MIT experiment. Through a combination of optical and evaporative cooling, the temperature of a gas is reduced by a factor of 10^9, while the density at the BEC transition is similar to the initial density in the atomic oven (all numbers are approximate). In each step shown, the ground state population increases by about 10^6.

	Temperature	Density (cm^{-3})	Phase-space density
Oven	500 K	10^{14}	10^{-13}
Laser cooling	50 μK	10^{11}	10^{-6}
Evaporative cooling	500 nK	10^{14}	1
BEC			10^7

cast by an atom cloud which expands ballistically after suddenly switching off the magnetic trap. The signature of BEC is the sudden appearance of a slow component with anisotropic expansion [3,4]. This can be regarded as observing BEC in momentum space.

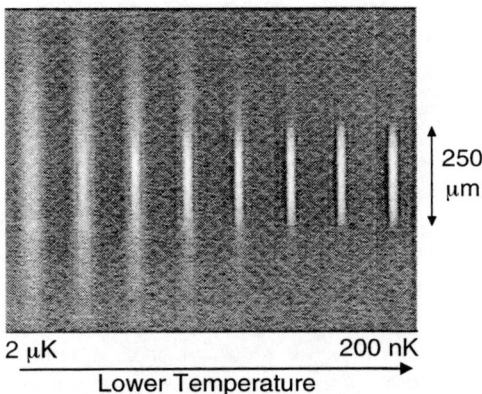

250 μm

2 μK   200 nK

Lower Temperature

FIGURE 1. Phase contrast images of a trapped sodium gas across the BEC phase transition. As the final radio frequency used in evaporative cooling is lowered, the temperature is reduced (left to right). Images show the onset of Bose condensation, the growth of the condensate fraction and contraction of the thermal wings, and finally a pure condensate with no discernible thermal fraction. The axial and radial frequencies are about 17 and 230 Hz, respectively.

III ATOM LASERS AND QUANTUM FLUIDS

Research on gaseous BEC can be divided into two areas: In the first (which could be labeled "The atomic condensate as a coherent gas", or "Atom lasers"), one would like to have as little interaction as possible—almost like photons in an optical laser. Thus the experiments are preferentially done at low densities. The Bose-Einstein condensate serves as an intense source of ultracold coherent atoms for experiments in atom optics, in precision studies or for explorations of basic aspects of quantum mechanics. The second area could be labeled as "BEC as a new quantum fluid" or "BEC as a many-body system". The focus here is on the interactions between the atoms which are most pronounced at high densities.

In the spirit of the talk at ICPEAC, we will first summarize some highlights of BEC and then focus on the role of low-energy atomic collisions. The selected examples illustrate both aspects of BEC. Multi-component condensates and the evidence for a critical velocity are part of our study of BEC as a new quantum fluid, whereas output couplers and coherence are aspects of atom lasers.

A   Spinor Bose-Einstein condensates

In a magnetic trap, the atomic spin adiabatically follows the direction of the magnetic field. Thus, although alkali atoms have internal spin, their Bose-Einstein condensates are described by a scalar order parameter similar to the spinless super-fluid ^{4}He. One exception is the two-component condensate which was discovered by the Boulder group, when they trapped atoms in both the upper and lower hyperfine states of ^{87}Rb [11]. This observation was surprising because a large rate of inelastic collisions had been expected for this system. The suppression of these spinflip collisions turned out to come from a fortuitous equality in the scattering lengths in the two hyperfine states.

A general method for creating multi-component condensates is to employ an optical trap that can confine condensates with arbitrary orientations of the spin thus liberating the spin as a new degree of freedom [12]. Our group used an optical trap to study condensates with arbitrary population in the three orientations $m = 1, 0, -1$ of the ground state of sodium which has a total spin $F = 1$ [13]. These condensates have a three-component vectorial order parameter. A variety of new phenomena have been predicted for such spinor condensates such as spin textures, spin waves, and the coupling between atomic spin and superfluid flow [14–16]. Such phenomena cannot occur in condensates with a single component order parameter such as in ^{4}He and more closely resemble the complex features of the superfluid phases of ^{3}He.

If the components are not coupled (i.e. transformed into each other), they can be regarded as multi-species condensates ("condensate alloys"). Both the group in Boulder and our group have studied the dynamics of the phase-separation of these components [17,18]. We observed long-lived metastable structures which could tunnel through each other and reach the equilibrium configuration [19]. By selecting two of the three states of the $F = 1$ spinor condensates, we could realize two-component condensates which were either miscible or immiscible [13]. Multi-component condensates are promising systems for the study of interpenetrating superfluids, a long-standing goal since the early attempts in 1953 using ^{4}He-^{6}He mixtures [20]. New phenomena arise when the three components are coupled by spinflip collisions, as displayed in Fig. 2.

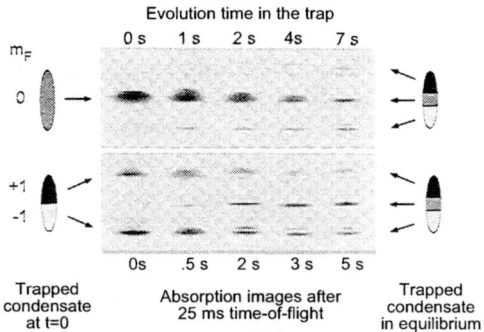

FIGURE 2. Observation of coupled spinor condensates. The components of the condensate are states with different orientations ($m = -1, 0, 1$) of the total spin $F = 1$ confined in an optical trap. These components are coupled by an anti-ferromagnetic spinflip interaction which drives them into an equilibrium domain structure. After a variable holding time, the condensate was analyzed by time-of-flight absorption imaging. During the ballistic expansion, a magnetic-field gradient acted as a Stern-Gerlach filter and separated the components with different spin orientation as indicated by the arrows. The upper image shows how a condensate, initially in a pure $m = 0$ state, developed a spin domain structure. The same equilibrium state was reached when the condensate started in an equal mixture of $m = +1$ and -1 states [13,18] .

B   Evidence for a critical velocity in a BEC

The existence of a macroscopic order parameter implies superfluidity of a gaseous condensate. Observing frictionless flow is a challenge given the small size of the system and its metastability. We have taken a step towards this goal by studying dissipation when an object was moved through the fluid [21]. This is in direct analogy with the well-known argument by Landau [22] and the vibrating wire experiments in superfluid helium [23]. Instead of a massive macroscopic object we used a blue detuned laser beam which repelled atoms from its focus to create a moving boundary condition.

The beam created a "hole" with a diameter of 13 μm which was scanned back and forth along the long axis of the cigar-shaped condensate (Thomas-Fermi diameters of 45 and 150 μm in the radial and axial directions, respectively). After exposing the condensate to the scanning laser beam for about one second, the final temperature was determined. As a function of the velocity of the scanning beam, we could distinguish two regimes of heating separated by a critical velocity. For low velocities, little dissipation was observed, and the condensate appeared immune to the presence of the scanning laser beam. For higher velocities, the heating increased, until at a velocity of about 6 mm/s the condensate was almost completely depleted after the stirring. The cross-over between these two regimes was quite pronounced and occurred at a velocity of about 1.6 mm/s which was a factor of four smaller than the speed of sound at the peak density of the condensate (Fig. 3).

These observations are in qualitative agreement with numerical calculations based on the non-linear Schroedinger equation which predict the onset of vortex nucleation at such subsonic velocities [24–26]. Because of surface effects and the non-zero temperature, we expect dissipation even at low velocities and a smooth crossover between low and high dissipation. More precise measurements of the heating should allow us to study these finite-size and finite-temperature effects.

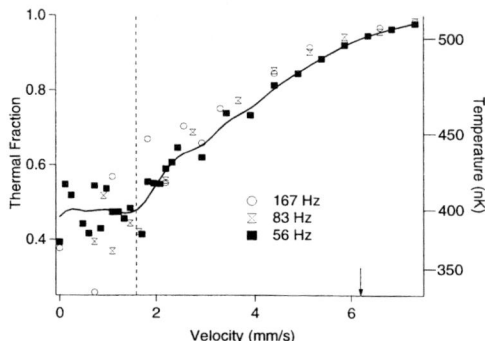

FIGURE 3. Evidence for a critical velocity. Shown is the final temperature after a laser beam was scanned through the condensate at variable velocity for 900 ms using different scan frequencies. The dashed line separates the regimes of low and high dissipation. The peak sound velocity is marked by an arrow. The data series for 83 and 167 Hz showed large shot-to-shot fluctuations at velocities below 2 mm/sec. The solid line is a smoothing spline fit to the 56 Hz data set to guide the eye. Figure taken from ref. [21] .

C   Atom lasers

In an ideal gas, Bose condensed atoms would all occupy the same single-particle ground-state wavefunction. This picture is largely valid even when weak interactions are included. They lead to admixtures of other configurations of typically 1 % or less for the alkali condensates. This is in contrast to liquid helium where this correction (called quantum depletion) is about 90 %. This means that even for the interacting gases, the atoms can all be regarded as having the same single-particle wavefunction with 99 % accuracy. Consequently, gaseous Bose-Einstein condensates can serve as sources of coherent atomic beams called atom lasers. Essential aspects of atom lasers are coherence, the output coupling and the gain process.

The coherence of the condensate was demonstrated by our group in 1997 when two condensates in a double-well potential were released from the trap and allowed to expand. They displayed a high-contrast interference pattern in their overlap region [27]. Interference between many condensates was observed by Mark Kasevich's group at Yale, when they generated a condensate in a multiple-well optical potential and saw interference between them. The observed temporal oscillations were related to Josephson oscillations [28]. Coherence in multi-component condensates was demonstrated at Boulder [29]. A recent spectroscopic measurement of the coherence length of a condensate is illustrated in Fig. 8 [30].

An output coupler for a Bose-Einstein condensate was realized by our group in 1997 by using pulsed radio-frequency radiation to flip the spin of a fraction of the condensed atoms into an untrapped state which fell downward by gravity [31]. The atoms were shown to be coherent, and the system constituted a pulsed atom laser. Recently, there has been a lot of excitement about more advanced output couplers. T.W. Hänsch's group in Munich was able to expose a magnetically shielded condensate to continuous radiofrequency radiation and realized a cw output coupler [32]. The Gaithersburg group replaced the radio-frequency transition by an optical Raman transition. The photon recoil pushed the atoms out of the trap horizontally, realizing a directional output coupler [33]. The Yale group saw self pulsing atom emission from an array of condensates held by an optical lattice which can be regarded as a "mode-locked" atom laser [28].

The gain mechanism of an atom laser is analogous to that in the optical laser: bosonic stimulation by the coherent matter wave. Bosonic stimulation was observed in the formation of the condensate at MIT [34], during the four-wave mixing experiments at Gaithersburg [35] and most dramatically in the build-up of "superradiant" pulses of matter-waves [36].

What will atom lasers be used for? The Gaithersburg group has used the condensate as a superior atom source with its high brightness, small rms momentum and excellent initial spatial localization. Their experiments include several studies of diffraction of atoms by light [37], an important element in atom interferometers. In the optical domain, the laser is crucial for nonlinear optics. Similarly, atom lasers are crucial for non-linear atom optics. In contrast to photons, however, atoms have no need for a non-linear medium—their interactions provide the non-linearity. A beautiful example is the recent experiment in Gaithersburg, where three condensates collided and formed a fourth condensate by four-wave mixing [35].

Condensates can be a highly nonlinear media not only for matter waves, but also for light. This was dramatically demonstrated recently by Lene Hau and collaborators at the Rowland institute in Cambridge, when they slowed the speed of light to 17 m/s using the condensate as a dense cold medium [38]. Ultimately, atom lasers may replace conventional atomic beams in applications like precision measurements of fundamental constants, tests of fundamental symmetries, atom optics (in particular, atom interferometry and atom holography) and precise deposition of atoms.

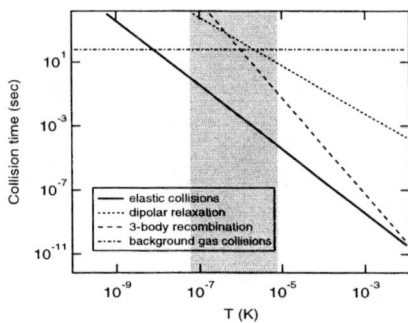

FIGURE 4. Mean collision time for several elastic and inelastic processes in a sodium gas as a function of temperature at the critical density for Bose-Einstein condensation. The "BEC window," where the lifetime of the sample exceeds 0.1 seconds and the rate of elastic collisions is faster than 1 Hz is shaded. This figure uses a scattering length of $50\,a_0$ and rate coefficients for two- and three-body inelastic collisions of $10^{-16} \mathrm{cm}^3 \mathrm{s}^{-1}$ and $6 \cdot 10^{-30} \mathrm{cm}^6 \mathrm{s}^{-1}$ respectively.

IV   ATOM-ATOM COLLISIONS

Laser cooling is based on the properties of single atoms, whereas evaporative cooling relies on elastic collisions which establish thermal equilibrium. A requirement for evaporative cooling is a favorable ratio between the elastic collision rate and the inelastic collision rate (which leads to trap loss and heating). Collisional properties of the relevant atoms with respect to evaporative cooling are discussed in refs. [39,40]. Several review papers [41–43] summarize the field of cold collisions.

Figure 4 shows the situation for sodium. The elastic and several inelastic rates are plotted vs. temperature (at the density of the BEC transition). The region where the elastic rate is much larger than the inelastic rate spans most of the temperature range displayed. Within this region, we have shaded the "BEC window" where the lifetime of the sample exceeds 0.1 sec and where the rate of elastic collisions is faster than 1 Hz. The fact that it covers several orders of magnitude in temperature and density allows studies of BEC over a large range of parameters. In our experiments, we have realized condensates with densities between 2×10^{13} and 3×10^{15} cm^{-3} and crossed the BEC transition at temperatures from 100 nK to 5 μK. BEC can be studied at lower and higher densities by expanding or compressing a condensate after it has been formed.

Although this window is in agreement with early estimates of collisional properties [44], it is fortunate that this window exists at all. We know now that collisional properties at ultralow temperatures are more complicated than we first expected. Experiments suggest that there may be no window at all for magnetically trapped cesium. The group of Dalibard in Paris discovered an s-wave resonance in cesium which increases the elastic cross section at low temperatures and favors evaporative cooling [45]. However, they also found that the gas undergoes severe dipolar relaxation when placed in the upper hyperfine state $(F = 4)$ [46]. This rate is 3 orders of magnitude higher than for the other alkalis and prevents evaporative cooling to phase space densities below 10^{-5}. Therefore, prospects for Bose-Einstein condensation in cesium are severely restricted.

^{85}Rb has a very small elastic cross section at intermediate temperatures which makes it hard to reach runaway evaporative cooling [48]. Differences found in trap loss and heating for ^{87}Rb atoms in the upper and lower hyperfine state are not understood [49]. In hindsight, it was very fortunate that sodium and ^{87}Rb happened to have an ideal combination of properties for BEC.

The lifetime of condensates is usually limited by the rate of three-body collisions. This decay was studied for rubidium [50,51] and sodium [12]. For sodium, the measured rate coefficient of $1.1 \cdot 10^{-30} \text{cm}^6/\text{s}$ implies a lifetime of about 10 sec for a condensate in a magnetic trap at a typical peak density of $3 \cdot 10^{14}$ cm^{-3}. The highest densities observed in optical traps ($3 \cdot 10^{15}$ cm^{-3}) were already limited by the three-body decay.

V   MODIFICATION OF ATOMIC COLLISIONS BY FESHBACH RESONANCES

All the essential properties of Bose condensed systems—the cross-section for elastic collisions, the formation and shape of the condensate, the nature of its collective excitations and statistical fluctuations, the dynamics of solitons and vortices—are determined by the strength of the atomic interactions. At low temperatures, these interactions are controlled solely by the s-wave scattering length for elastic collisions between atoms. But unlike many chemical properties which govern the behavior of condensed-matter systems, the scattering lengths are not immutable, but can vary dramatically near a zero-energy collisional resonance, either a shape resonance [52] or, of current experimental relevance, a Feshbach resonance [53].

A Feshbach resonance occurs when a quasi-bound molecular state has an energy equal to that of two colliding atoms. Such a resonance strongly affects both elastic and inelastic collisions such as dipolar relaxation [54,55] and three-body recombination. Feshbach resonances have been studied in the past by varying the collisional energy to correspond to the fixed energy of a quasi-bound state [56]. For ultracold atoms (in a Bose condensate), the near-zero collisional energy is fixed, thus one must bring the quasi-bound state energy down to zero. Proposals have been made to do so with external magnetic [54,55], optical [57,58], rf [59], and electric fields [52].

Recently, Feshbach resonances for ultracold atoms have been found in the presence of external magnetic fields. Our group discovered several Feshbach resonances in sodium and directly confirmed their strong effect on the interaction energy of a condensate [60], demonstrating how minute changes in a magnetic field can bring about dramatic changes in a macroscopic system. Observations in ^{85}Rb were made with non-condensed atoms via photoassociation spectroscopy, which probes the relative wavefunction between colliding atoms [61,62], and via measurements of the elastic collision cross section [63].

Theoretical calculations predicted Feshbach resonances for sodium at high fields and in hyperfine states that cannot be magnetically trapped [64,65]. Thus, the optical trap was indispensable for our study of Feshbach resonances in sodium condensates.

To locate the Feshbach resonances, we first placed the optically trapped condensate into the desired hyperfine state, and then ramped up the magnetic field to as high as 1200 G while repeatedly probing the atoms with phase-contrast imaging. At specific values of the magnetic field, a rapid loss of atoms was observed. This rapid loss was expected at a Feshbach resonance, due to either the collapse of a condensate as the scattering length turns negative, or to increased inelastic collision rates. Furthermore, the losses were observed only for atoms in a specific hyperfine state, again indicating a collisional resonance. This was used to locate three resonances near 853 G, 907 G, and 1195 G [60,66].

To definitively identify the Feshbach resonances, we looked for the true "smoking gun" phenomenon: a drastic change in the interaction energy of a Bose condensate. The scattering length was determined by measuring the condensate number and the width of the expanding condensate in time-of-flight images. Since the kinetic energy of the expansion is identical to the mean-field energy, the scattering length can be obtained from such measurements. Results are shown in fig. 5 as a function of the magnetic field for the resonance at 907 G. Near a Feshbach resonance at a field B_0, the scattering length a was predicted to vary dispersively as a function of magnetic field B [64]:

$$a = \tilde{a} \left(1 + \frac{\Delta B}{B_0 - B} \right) \tag{1}$$

where $\tilde{a}$ is the scattering length away from the resonance, and ΔB is the width of the resonance which is determined by the strength of coupling between the quasi-bound state and the free-particle state of the incoming atoms. Our measurements verified this dispersive shape, and demonstrated a factor of 10 change in the scattering length by this collisional resonance.

Our observation confirms the theoretical predictions about "tunability" of the scattering length with the prospect of "designing" atomic quantum gases with novel properties. For example, by setting $a \approx 0$, one can create a condensate with essentially non-interacting atoms. By setting $a < 0$ one can make a once stable ($a > 0$) condensate suddenly unstable, and observe the collapse of a macroscopic wavefunction in a controlled manner [67–69], akin to the less controlled collapse of a ^{7}Li condensate [5]. Rapid variations of a may also lead to novel forms of collective oscillations [68]. Sweeps across the resonance may pass condensed atoms adiabatically into molecular condensates [70]. Further, Feshbach resonances are also predicted to exist for collisions between atoms of unlike atomic species or hyperfine state. Such resonances can be used to tune interspecies interactions, making multi-component Bose condensates overlap or phase-separate. Finally, Feshbach

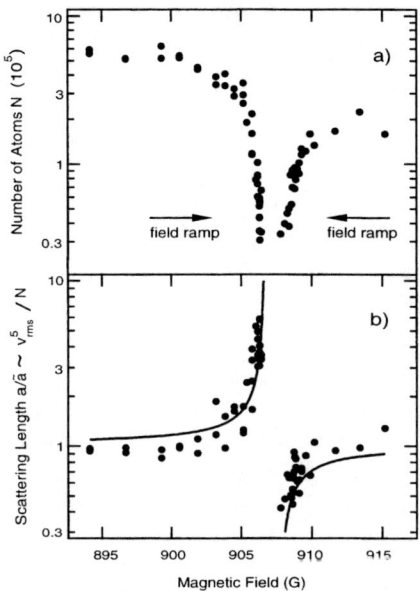

FIGURE 5. Observation of a ^{23}Na Feshbach resonance at 907 G. (a) Rapid number losses in sweeps close to the resonance from the low- and high-field sides located the resonance. (b) Measuring the scattering length based on time-of-flight images of expanding condensates confirmed its dispersive variation. The solid line is a fit based on eq. (1). Reprinted by permission from *Nature*[60], copyright 1998, Macmillan Magazines Ltd.

resonances may also be important in atom optics for modifying the atomic interactions in an atom laser, or more generally for controlling non-linear atom-optical coefficients.

However, our experiments suggest there may be severe limitations to the use of Feshbach resonances due to the concomitant increase in the rates of inelastic collisions and trap loss. The rapid losses by which the Feshbach resonances were observed were exclusively the result of these increased rates, since the region of negative scattering length which would cause a collapse of the condensate due to *elastic* collisions was never accessed. We ascribe these losses to three-body collisions, since dipolar relaxation, the mechanism for two-body decay, is prohibited for atoms in the lowest energy hyperfine state where two of these resonances occur. The rate constant for three-body decay was found to increase on both sides of the resonance, rising by more than three orders of magnitude [66]. The surprisingly rapid loss when the magnetic field was swept through the Feshbach resonance has now been convincingly explained by several theories as being due to a transfer from an atomic to a molecular condensate [71,72]. However, the direction of the sweep was chosen in such a way that it moved the molecules above the dissociation

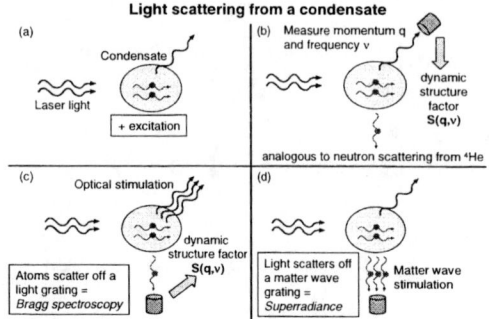

FIGURE 6. Light scattering from a Bose-Einstein condensate. When a photon is scattered, it transfers momentum to the condensate and creates an excitation (upper left). Therefore, an analysis of the scattered light allows the determination of the dynamic structure factor, in close analogy to neutron scattering experiments with superfluid helium (upper right). The signal is greatly increased by stimulating the light scattering by a second laser beam and detecting the scattered atoms (lower left)—this is the scheme for Bragg spectroscopy. Light scattering can also be stimulated by adding a coherent atomic field (lower right). This led to superradiant scattering of light and atoms.

limit. The opposite direction of the sweep should stabilize the molecules in highly vibrational states.

VI COLLISIONS BETWEEN PHOTONS AND CONDENSATES: LIGHT SCATTERING FROM BOSE-EINSTEIN CONDENSATES

In the early '90s, before Bose-Einstein condensation was realized in atomic gases, there were lively debates about what a condensate would look like. Some researchers thought it would absorb all the light and would therefore be "pitch black", some predicted it would be "transparent" (due to superradiant line-broadening [73]), others predicted that it would reflect the light due to polaritons [74] and be "shiny" like a mirror.

All the observations of Bose condensates have employed scattering or absorption of laser light. Until recently, the observations have been consistent with the assumption that a Bose condensate is a cold dilute cloud of atoms that scatters light as ordinary atoms do. On resonance, the condensate strongly absorbs the light, giving rise to the well-known "shadow pictures" of expanding condensates where the condensate appears black. For off-resonant light, the absorption can be made negligibly small, and the condensate acts as a dispersive medium bending the light like a glass sphere. This regime has been used for non-destructive in-situ imaging of Bose-Einstein condensates (see Fig. 1).

Our group has recently looked more closely at how coherent atoms interact with coherent light. Light scattering imparts momentum to the condensate and creates an excitation (Fig. 6). Consequently, the coherence and collective nature of excitations in the condensate can strongly affect the optical properties. As we discuss here, the use of light scattering to characterize atomic Bose condensates is analogous to the use of neutron scattering in the case of superfluid helium [75,76].

Since the light scattered from a sample containing only 10^7 atoms is hard to detect when it is distributed over the full solid angle, we used a second laser beam to stimulate the scattering of light with a frequency and direction, which was pre-determined by the laser beam rather than post-determined by analyzing scattered light (Fig. 6). This scheme, which we call Bragg spectroscopy, establishes a high-resolution spectroscopic tool for Bose-Einstein condensates which is sensitive to the momentum distribution of the trapped condensate as well as the effects of interactions [30]. We studied Bragg scattering in two regimes differing by the amount of momentum transfer.

Bogoliubov theory predicts that for a momentum transfer which is smaller than the speed of sound (times the atomic mass) phonons are excited, whereas for larger momentum transfer, the excitations are free-particle like (Fig. 7). In the regime of large momentum transfer, the impulse approximation is valid, and the resonance shows a Doppler broadening due to the zero-point motion of the condensate, i.e. it can be used to measure the momentum distribution of the condensates as pursued for superfluid ^{4}He [76,77]. More generally, Bragg spectroscopy can be used to determine the dynamic structure factor $S(\mathbf{q}, \nu)$ over a wide range of frequencies ν and momentum transfers $\mathbf{q}$ [78].

Bragg spectroscopy was realized by exposing the condensate to two off-resonant laser beams with a frequency difference ν. The intersecting beams formed a moving interference pattern from which atoms could scatter when the Bragg condition was fulfilled (i.e. energy and momentum were conserved). The momentum transfer q is given by $q = 2\hbar k \sin(\vartheta/2)$, where ϑ is the angle between the two laser beams with wavevector k. Figures 8 and 9 summarize the results for large and small scattering angles, probing both the phonon and free-particle regime. In the regime of low momentum transfer, light scattering was observed to be dramatically reduced (Fig. 9). In this regime, where atoms cannot absorb momentum "individually" but only collectively, the suppression arises from destructive interference of two excitation paths. The suppression provides dramatic evidence for the presence of correlated momentum excitations in the many-body condensate wavefunction. A similar suppression would occur when a sufficiently dense condensate scatters light spontaneously—turning a "pitch-black" condensate transparent!

A condensate which reflected the incident light was encountered when it was illuminated with a single intense laser beam [36]. When a condensate has scattered a photon, an imprint is left in the form of long-lived excitations. These excitations form a periodic density modulation which diffracts light into the same direction as the first scattered photon. The more photons have been scattered into a certain

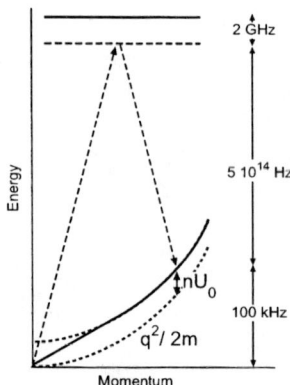

FIGURE 7. Probing the dispersion relation of a Bose-Einstein condensate by off-resonant light scattering. The Bogoliubov dispersion relation is phonon-like (linear) for small momenta. For large momenta, it is particle-like (quadratic) with a mean-field shift $nU_0 \approx 2$ kHz indicated in the figure. The momentum is transferred by scattering of visible light which is about 2 GHz detuned from the atomic resonance.

direction, the larger is the density modulation left behind and the larger the increase of the scattering rate into this direction. This self-acceleration of scattering can be described as bosonic stimulation of the scattering by the population of the final (quasi-particle) state (see Fig. 6).

The gain for this process is highest when the light is scattered along the long axis of the cylindrically shaped condensate and leads to the generation of directed beams of atoms (Fig. 10). This is accompanied by directed emission of light—a new form of superradiance where a density modulation spontaneously develops which makes the condensate "reflect" light like a mirror.

VII   CONDENSATE-CONDENSATE COLLISIONS

Bragg and Raman scattering has been used to realize output couplers for atom lasers. The realization of atom lasers with a large flux of atoms may require the use of much larger condensates. This raises the question of how the outcoupled atoms penetrate the condensate. Most theories on output couplers only include the coherent interactions between two modes, the condensate and the output mode. The coherent coupling between discrete modes lead for example to four-wave mixing [35]. However, when all other final states for elastic scattering are included, matter waves are attenuated by elastic collisions with a cross section of $8\pi a^2$ and $4\pi a^2$ for atoms in the same or in different internal states, respectively, where a denotes the scattering length.

In preliminary measurements, we used counter-propagating laser beams and

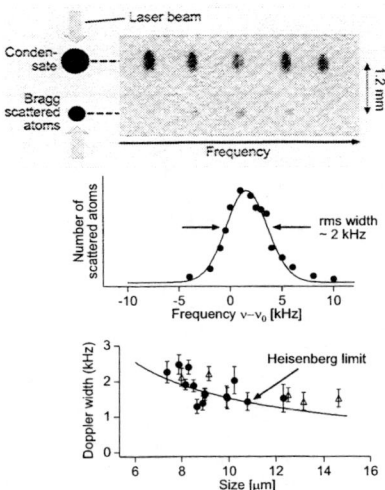

FIGURE 8. Measuring the momentum distribution of a condensate [30]. A condensate was exposed to two counterpropagating laser beams. Atoms absorbed a photon from one beam and were stimulated to re-emit it by the other beam, resulting in the transfer of recoil momentum to the atoms, as observed in ballistic expansion using absorption imaging after 20 ms time-of-flight (upper part). The number of Bragg scattered atoms showed a narrow resonance when the difference frequency between the two laser beams was varied (upper and middle part). The width of the resonance is caused by Doppler broadening and is therefore proportional to the condensate's momentum uncertainty Δp. It was determined for various sizes Δx of the condensate. The agreement with the Heisenberg limit $\Delta p \approx \hbar/\Delta x$ proves that the Doppler width of the resonance is only due to the zero-point motion of the condensate, or equivalently, that the coherence length of the condensate is equal to its physical size. This demonstrates that a condensate is one "coherent matter wave"! An analogous measurement in the time domain has been done by W.D. Phillips' group in Gaithersburg [79] .

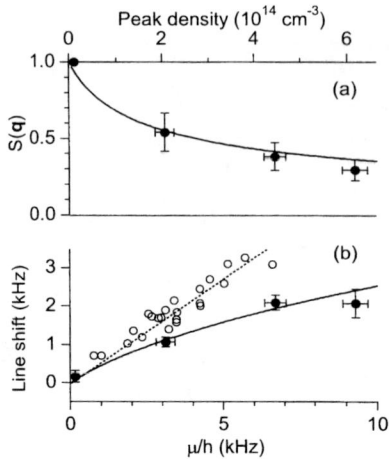

FIGURE 9. (a) Static structure factor $S(\mathbf{q})$ and (b) shift of the line center from the free-particle resonance. $S(\mathbf{q})$ is the ratio of the line strength at a given chemical potential μ to that observed for free particles. As the density and μ increase, the structure factor is reduced, and the Bragg resonance is shifted upward in frequency. Solid lines are predictions of a local-density approximation for light scattering by 14 degrees. The dotted line indicates a mean-field shift of $4\mu/7h$ as measured in the free-particle regime using a scattering angle of 180 degrees. Figure taken from Ref. [80]

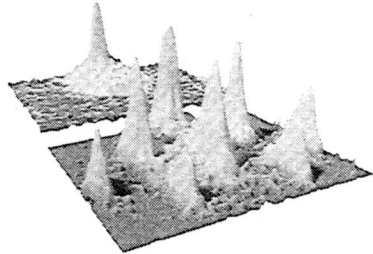

FIGURE 10. Superradiance and matter-wave amplification in a Bose-Einstein condensate [36]. The figure shows the velocity distribution of the atoms after a condensate (highest peak at the back of each image) was illuminated by laser light. Normal Rayleigh scattering (rear image) is random, and the velocity distribution was smeared out in the direction of the incident light. In contrast, collective "superradiant" scattering created several highly directional "atom laser" beams (front image). The images are absorption images of a cloud which expanded ballistically after the light scattering. The two horizontal axes represent two velocity components and the vertical axis is the column density of the observed atoms. The closest spacing between the peaks is $\sqrt{2}$ times the single-photon recoil velocity.

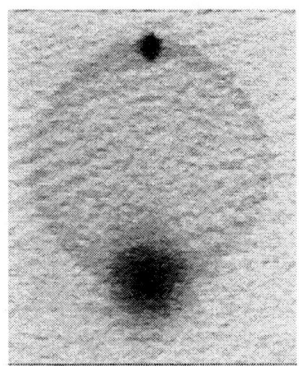

FIGURE 11. Observation of elastic collisions between the condensate (lower black dot) and Bragg diffracted atoms (upper spot). The image is taken after a time-of-flight of 30 ms, and shows the velocity distribution after the collision. The products of the collision are distributed over a sphere in momentum space leading to the observed halo. The height of the image is 3.2 mm.

drove either a two-photon Bragg or Raman transition, thus coupling out either $m_F = -1$ or $m_F = 0$ atoms from the $m_F = -1$ condensate (see also [33,37]). A clear signature of collisions between the out-coupled atoms and the condensate is a halo in time-of-flight pictures since momentum conservation requires the final momenta to be distributed on the surface of a sphere (Fig. 11). Typically, the collisional density along the radial and axial directions for our condensates is 1 and 10, respectively. A quantitative study should result in a determination of the scattering length which can be deduced directly from the number of out-coupled atoms which survive the passage through the condensate. Preliminary measurements already indicate that the surviving fraction is higher for collisions between atoms in different hyperfine states, which signals the absence of the exchange term in the scattering between "unlike" atoms.

VIII CONCLUSIONS

Bose-Einstein condensation gives us access to a variety of inelastic and elastic low-energy collision processes. Due to the low kinetic energy of the condensed atoms, they only involve s-waves in the input channel. Pure s-wave scattering sets in at microkelvin temperatures—therefore, in principle, these collision processes could have been studied already with laser-cooled atoms. However, they only became observable at the much higher density of BECs—condensates have densities which are typically a thousand times higher than in laser-cooled atom clouds.

Another fascinating aspect of condensates is that the same interaction can be observed as a coherent coupling and as an incoherent collisional process (usually described by a rate equation). The mean-field energy of condensates is an energy shift of the many-particle ground state. The same matrix element is responsible for elastic collisions in the thermal cloud. In spinor condensates, the anti-ferromagnetic coupling between atoms with different spin orientation led to the formation of spin domains. The same interaction is responsible for spin relaxation which is a loss process in magnetic traps [39].

The rapid pace of developments in atomic BEC during the last few years has taken the community by surprise. After decades of an elusive search nobody expected that condensates would be so robust and relatively easy to manipulate. Also, nobody imagined that such a simple system would pose so many challenges, not only to experimentalists, but also to our fundamental understanding of physics. The list of future challenges is long and includes the complete characterization of elastic and inelastic collisions at ultralow temperatures, the exploration of super-fluidity, vortices, and second sound in Bose gases, the study of quantum-degenerate molecules and Fermi gases, the development of practical "high-power" atom lasers, and their application in atom optics and precision measurements.

We gratefully acknowledge the contributions of our collaborators M.R. Andrews, A.P. Chikkatur, D.S. Durfee, A. Görlitz, S. Gupta, Z. Hadzibabic, S. Inouye, M. Köhl, C.E. Kuklewicz, M.-O. Mewes, H.-J. Miesner, R. Onofrio, T. Pfau, D.E. Pritchard, D.M. Stamper-Kurn, J. Stenger, C.G. Townsend, N.J. van Druten, and J. Vogels. This work was supported by the ONR, NSF, ARO, NASA, and the David and Lucile Packard Foundation.

REFERENCES

1. A. Einstein, Sitzungsber. Preuss. Akad. Wiss. Bericht 3, 18 (1925).
2. Proceedings of the "Enrico Fermi" Summer School on Bose-Einstein Condensation, Varenna, Italy, 1999, in press. See cond-mat/9904034 for the contribution by W. Ketterle, D.S. Durfee,D.M. Stamper-Kurn, cond-mat/9903109 for the contribution of E. Cornell, J.R. Ensher, and C.E. Wieman, and physics/9812038 for the contribution of D. Kleppner, T.J. Greytak, T.C. Killian, D.G. Fried, L. Willmann, D. Landhuis, and S.C. Moss.
3. M. Anderson, J. Ensher, M. Matthews, C. Wieman, and E. Cornell, Science **269**, 198 (1995).
4. K. Davis, M.-O. Mewes, M. Andrews, N. van Druten, D. Durfee, D. Kurn, and W. Ketterle, Phys. Rev. Lett. **75**, 3969 (1995).
5. C. Bradley, C. Sackett, and R. Hulet, Phys. Rev. Lett. **78**, 985 (1997).
6. D. Fried, T. Killian, L. Willmann, D. Landhuis, S. Moss, D. Kleppner, and T. Greytak, Phys. Rev. Lett. **81**, 3811 (1998).
7. BEC home page of the Georgia Southern University, http://amo.phy.gasou.edu/bec.html.
8. K. Huang, in *Studies in Statistical Mechanics*, edited by J. de Boer and G. Uhlenbeck (North-Holland, Amsterdam, 1964), Vol. II, pp. 3–106.

9. A. Griffin, D. Snoke, and S. Stringari, *Bose-Einstein Condensation* (Cambridge University Press, Cambridge, 1995).

10. M. Andrews, M.-O. Mewes, N. van Druten, D. Durfee, D. Kurn, and W. Ketterle, Science **273**, 84 (1996).

11. C. Myatt, E. Burt, R. Ghrist, E. Cornell, and C. Wieman, Phys. Rev. Lett. **78**, 586 (1997).

12. D. Stamper-Kurn, M. Andrews, A. Chikkatur, S. Inouye, H.-J. Miesner, J. Stenger, and W. Ketterle, Phys. Rev. Lett. **80**, 2072 (1998).

13. J. Stenger, S. Inouye, D. Stamper-Kurn, H.-J. Miesner, A. Chikkatur, and W. Ketterle, Nature **396**, 345 (1998).

14. T.-L. Ho, Phys. Rev. Lett. **81**, 742 (1998).

15. T. Ohmi and K. Machida, J. Phys. Soc. Jap. **67**, 1822 (1998).

16. C. Law, H. Pu, and N. Bigelow, Phys. Rev. Lett. **81**, 5257 (1998).

17. D. Hall, M. Matthews, J. Ensher, C. Wieman, and E. Cornell, Phys. Rev. Lett. **81**, 4531 (1998).

18. H.-J. Miesner, D. Stamper-Kurn, J. Stenger, S. Inouye, A. Chikkatur, and W. Ketterle, Phys. Rev. Lett. **82**, 2228 (1999).

19. D. Stamper-Kurn, H.-J. Miesner, A. Chikkatur, S. Inouye, J. Stenger, and W. Ketterle, Phys. Rev. Lett. **83**, 661 (1999).

20. L. Guttman and J. Arnold, Phys. Rev. **92**, 547 (1953).

21. C. Raman, M. Köhl, R. Onofrio, D. Durfee, C. Kuklewicz, Z. Hadzibabic, and W. Ketterle, Phys. Rev. Lett. **83**, 2502 (1999).

22. K. Huang, *Statistical Mechanics* (Wiley, New York, 1987).

23. C. Castelijns, K. Coates, A. Guénault, S. Mussett, and G. Pickett, Phys. Rev. Lett. **56**, 69 (1985).

24. T. Frisch, Y. Pomeau, and S. Rica, Phys. Rev. Lett. **69**, 1644 (1992).

25. C. Huepe and M.-E. Brachet, C.R. Acad. Sci. Paris Série II **325**, 195 (1997).

26. T. Winiecki, J. McCann, and C. Adams, Phys. Rev. Lett. **82**, 5186 (1999).

27. M. Andrews, C. Townsend, H.-J. Miesner, D. Durfee, D. Kurn, and W. Ketterle, Science **275**, 637 (1997).

28. B. Anderson and M. Kasevich, Science **282**, 1686 (1998).

29. D. Hall, M. Matthews, C. Wieman, and E. Cornell, Phys. Rev. Lett. **81**, 1543 (1998).

30. J. Stenger, S. Inouye, A. Chikkatur, D. Stamper-Kurn, D. Pritchard, and W. Ketterle, Phys. Rev. Lett. **82**, 4569 (1999).

31. M.-O. Mewes, M. Andrews, D. Kurn, D. Durfee, C. Townsend, and W. Ketterle, Phys. Rev. Lett. **78**, 582 (1997).

32. I. Bloch, T. Hansch, and T. Esslinger, Phys. Rev. Lett. **82**, 3008 (1998).

33. E. Hagley, L. Deng, M. Kozuma, J. Wen, K. Helmerson, S. Rolston, and W. Phillips, Science **283**, 1706 (1999).

34. H.-J. Miesner, D. Stamper-Kurn, M. Andrews, D. Durfee, S. Inouye, and W. Ketterle, Science **279**, 1005 (1998).

35. L. Deng, E. Hagley, J. Wen, M. Trippenbach, Y. Band, P. Julienne, J. Simsarian, K. Helmerson, S. Rolston, and W. Phillips, Nature **398**, 218 (1999).

36. S. Inouye, A. Chikkatur, D. Stamper-Kurn, J. Stenger, D. Pritchard, and W. Ketterle, Science **285**, 571 (1999).

37. M. Kozuma, L. Deng, E. Hagley, J. Wen, R. Lutwak, K. Helmerson, S. Rolston, and W. Phillips, Phys. Rev. Lett. **82**, 871 (1999).
38. L. Hau, S. Harris, Z. Dutton, and C. Behroozi, Nature **397**, 594 (1999).
39. W. Ketterle and N. van Druten, in *Advances in Atomic, Molecular, and Optical Physics*, edited by B. Bederson and H. Walther (Academic Press, San Diego, 1996), Vol. 37, pp. 181–236.
40. D. Heinzen, to appear in Proceedings of the "Enrico Fermi" Summer School on Bose-Einstein Condensation, Varenna, Italy (1999).
41. B. Verhaar, in *Atomic Physics*, edited by D. Wineland, C. Wieman, and S. Smith (AIP, New York, 1995), Vol. 14, p. 351.
42. J. Weiner, in *Advances in Atomic, Molecular, and Optical Physics*, edited by B. Bederson and H. Walther (Academic Press, San Diego, 1995), Vol. 35, pp. 45 – 78.
43. T. Walker and P. Feng, in *Advances in Atomic, Molecular, and Optical Physics*, edited by B. Bederson and H. Walther (Academic Press, San Diego, 1994), Vol. 34, pp. 125–170.
44. D. Pritchard, in Electronic and atomic collisions : invited papers of the XIV International Conference on the Physics of Electronic and Atomic Collisions, Palo Alto, California, 24-30 July, 1985, edited by D. Lorents, W. Meyerhof, and J. Peterson (Elsevier, New York, 1986), pp. 593–604.
45. M. Arndt, M. B. Dahan, D. Guéry-Odelin, M. Reynolds, and J. Dalibard, Phys. Rev. Lett. **79**, 625 (1997).
46. J. Söding, D. Guery-Odelin, P. Desbiolles, G. Ferrari, and J. Dalibard, Phys. Rev. Lett. **80**, 1869 (1998).
47. D. Guéry-Odelin, J. Söding, P. Desbiolles, and J. Dalibard, Europhys. Lett. **44**, 25 (1998).
48. E. Cornell, private communication.
49. C. Myatt, Ph.D. thesis, University of Colorado, 1997.
50. E. Burt, R. Ghrist, C. Myatt, M. Holland, E. Cornell, and C. Wieman, Phys. Rev. Lett. **79**, 337 (1997).
51. J. Söding, D. Guéry-Odelin, P. Desbiolles, F. Chevy, H. Inamori, and J. Dalibard, preprint, cond-mat/9811339.
52. M. Marinescu and L. You, Phys. Rev. Lett. **81**, 4596 (1998).
53. H. Feshbach, Annals of Physics **19**, 287 (1962).
54. E. Tiesinga, A. Moerdijk, B. Verhaar, and H. Stoof, Phys. Rev. A **46**, R1167 (1992).
55. E. Tiesinga, B. Verhaar, and H. Stoof, Phys. Rev. A **47**, 4114 (1993).
56. H. Bryant, B. Deiterle, J. Donahue, H. Sharifian, H. Tootoonchi, D. Wolfe, and P. Gram, Phys. Rev. Lett. **38**, 228 (1977).
57. P. Fedichev, Y. Kagan, G. Shlyapnikov, and J. Walraven, Phys. Rev. Lett. **77**, 2913 (1996).
58. J. Bohn and P. Julienne, Phys. Rev. A **56**, 1486 (1997).
59. A. Moerdijk, B. Verhaar, and T. Nagtegaal, Phys. Rev. A **53**, 4343 (1996).
60. S. Inouye, M. Andrews, J. Stenger, H.-J. Miesner, D. Stamper-Kurn, and W. Ketterle, Nature **392**, 151 (1998).
61. P. Courteille, R. Freeland, D. Heinzen, F. van Abeelen, and B. Verhaar, Phys. Rev. Lett. **81**, 69 (1998).

62. F. van Abeelen, D. Heinzen, and B. Verhaar, Phys. Rev. A **57**, R4102 (1998).

63. J. Roberts, N. Claussen, J. Jr., C. Greene, E. Cornell, and C. Wieman, Phys. Rev. Lett. **81**, 5109 (1998).

64. A. Moerdijk, B. Verhaar, and A. Axelsson, Phys. Rev. A **51**, 4852 (1995).

65. F. van Abeelen and B. Verhaar, private communication.

66. J. Stenger, S. Inouye, M. Andrews, H.-J. Miesner, D. Stamper-Kurn, and W. Ketterle, Phys. Rev. Lett. **82**, 2422 (1999).

67. M. Ueda and A. Legget, Phys. Rev. Lett. **80**, 1576 (1998).

68. Y. Kagan, A. Muryshev, and G. Shlyapnikov, Phys. Rev. Lett. **81**, 933 (1998).

69. C. Sackett, H. Stoof, and R. Hulet, Phys. Rev. Lett. **80**, 2031 (1998).

70. P. Tommasini, E. Timmermans, M. Hussein, and A. Kerman, preprint, cond-mat/9804015.

71. F. van Abelen and B. Verhaar, Phys. Rev. Lett. **83**, 1550 (1999).

72. V. Yurovsky, A. Ben-Reuven, P. Julienne, and C. Williams, Phys. Rev. A **60**, R765 (1999).

73. J. Javanainen, Phys. Rev. Lett. **72**, 2375 (1994).

74. H. Politzer, Phys. Rev. A **43**, 6444 (1991).

75. P. Nozières and D. Pines, *The Theory of Quantum Liquids* (Addison-Wesley, Redwood City, CA, 1990).

76. P. Sokol, in *Bose-Einstein Condensation*, edited by A. Griffin, D. Snoke, and S. Stringari (Cambridge University Press, Cambridge, 1995), pp. 51–85.

77. P. Hohenberg and P. Platzman, Phys. Rev. **152**, 198 (1966).

78. T. Greytak, in *Quantum Liquids*, edited by J. Ruvalds and T. Regge (North-Holland, New York, 1978), pp. 121–165.

79. W. Phillips, personal communication.

80. D. Stamper-Kurn, A. Chikkatur, A. Grlitz, S. Inouye, S. Gupta, D. Pritchard, and W. Ketterle, Phys. Rev. Lett. **83**, 2876 (1999).

Dissociative Recombination and Excitation in Ion Storage Rings

Mats Larsson

Department of Physics
Stockholm University
P.O. Box 6730
S-113 85 Stockholm
Sweden

Abstract. The application of ion storage rings to the study of electron-molecular ion interaction has led to an experimental breakthrough. The development since the first experiments with molecular ions in storage rings about seven years ago, which in themselves represented a big leap forward, has been striking, and was impossible to envision at the outset. The development has been driven by advances in accelerator physics, detector technology, challenging applications in astrophysics and atmospheric physics, and by a close interplay with theory. Despite the remarkable progress, many important questions remain unanswered. For example, even for someone with a good knowledge of molecular physics it may come as a surprise that it is far from understood how the simplest polyatomic molecule H_3^+ recombines with electrons, and it remains an experimental controversy at what rate it recombines.

INTRODUCTION

Ion storage rings were first applied to problems in molecular physics in 1992, and the first results emerging from storage ring experiments with molecules appeared in a single issue of the *Physical Review Letters* (1–3). They all concerned dissociative recombination of molecular ions with electrons, with HeH^+ being studied in TARN II in Japan (1), HD^+ in the Test Storage Ring (TSR) in Germany (2), and H_3^+ in CRYRING, Sweden (3). These experiments proved ion storage rings to be excellent tools for the study of dissociative recombination. The results showed clear evidence that vibrationally cold molecular ions could be obtained during storage by means of emission of infrared radiation (1–3), that it was possible to electron cool a stored beam of molecular ions (2, 3), that new features in the cross section appeared when the collision energy was increased above a few eV (1–3), and that absolute cross sections could be obtained (3). The rapid development since the first experiments is described in (4, 5). An attempt to review the whole field of atomic and molecular physics in storage rings was made in 1995 (6), but since then the field has grown so

CP500, *The Physics of Electronic and Atomic Collisions*, edited by Y. Itikawa, et al.
© 2000 American Institute of Physics 1-56396-777-4/00/$17.00

much that a single review of the field has become impractical. Instead several reviews of selected subfields have appeared; in addition to (4, 5), reviews on highly charged ions (7), negative ions (8), and laser spectroscopy of stored ions (9) have appeared.

Interlude: The 50th anniversary of dissociative recombination

A plenary talk on dissociative recombination at the XXI ICPEAC seems like an appropriate way of celebrating the 50th anniversary of the first published experimental evidence that electrons recombine rapidly with molecular ions (10). The manuscript was received by *Physical Review* on August 5, 1949, that is, almost exactly fifty years ago with respect to the XXI ICPEAC in Sendai. The evidence for rapid electron-ion recombination came from an experiment aimed at the study of ambipolar diffusion of ions in an afterglow plasma following a microwave discharge, and was carried out by Biondi and Brown at the Research Laboratory of Electronics at MIT. It is not clear from the paper whether Biondi and Brown were aware of the work by Bates and Massey (11) where they discussed possible recombination processes in the Earth's ionosphere. Massey and Bates had earlier dismissed dissociative recombination (12, 13) on the grounds that the process should be far too slow to account for atmospheric observations, but faced with the lack of any other conceivable process, they concluded that dissociative recombination might, after all, be rapid (11). Thus, when the results of the microwave discharge experiment were published (10), dissociative recombination had been given a great deal of thought for more than a decade by the leading atomic collision physicists of the time without the emergence of any conclusive view.

Bates was inspired by afterglow experiments by Biondi and Brown, but it was an earlier paper of theirs (14) that led him to propose that the recombination coefficient observed in the helium afterglow (1.7×10^{-8} cm^{3}s^{-1} at 300 K (14)) was due to He$_2^+$ rather than He$^+$ (15). The dimer ion would, Bates argued (15), recombine much faster than the atomic ion, in agreement with experiment (14). It was later found out that the helium results were flawed by the effect of diffusion loss to the walls (16). Later experiments showed that He$_2^+$ recombines slowly (17), and the final twist in the tail came as late as 1999, when Carata, Orel and Suzor-Weiner (18) used the multichannel quantum defect theory (MQDT) to calculate the recombination coefficient of He$_2^+$ in its zeroth vibrational level to be 6.1×10^{-11} cm^{3}s^{-1} at 300 K.

The helium afterglow results (14) and the afterglow results for hydrogen, nitrogen and oxygen (10) convinced Bates that dissociative recombination is a rapid process and led to his seminal paper (19) in which he explains the mechanism for dissociative recombination.

A final word in this historical interlude: the papers of Biondi and Brown (10, 14) were far from the earliest on recombination involving positively and negatively charged particles. The general problem of the disappearance of electrons from a gas discharge plasma had been the subject of much work even before the Second World War (20).

DISSOCIATIVE RECOMBINATION AND EXCITATION

Dissociative recombination (DR) is the process wherein a free electron is captured by a molecular ion in a radiationless transition that makes the electron enter a bound molecular orbital, followed by stabilization of the capture process by molecular dissociation. Since dissociation is an extremely fast process, of the order of tens of femtoseconds, stabilization of electron capture by dissociation is far more effective than radiative stabilization, which is operative in dielectronic recombination of atomic ions. Dissociative recombination can be described for the generic molecular ion AB^+ as

$$AB^+ + e^-(\varepsilon) \rightarrow AB^{**} \rightarrow A + B \tag{1}$$

$$\rightarrow A^- + B^+ \tag{2a}$$

where A and B are atoms or molecules. Process 2a is resonant ion pair formation (RIP), a process which proceeds through the same compound state, AB^{**}, as DR. Figure 1 shows how dissociative recombination and RIP can be described for a diatomic molecule by the use of potential energy curves.

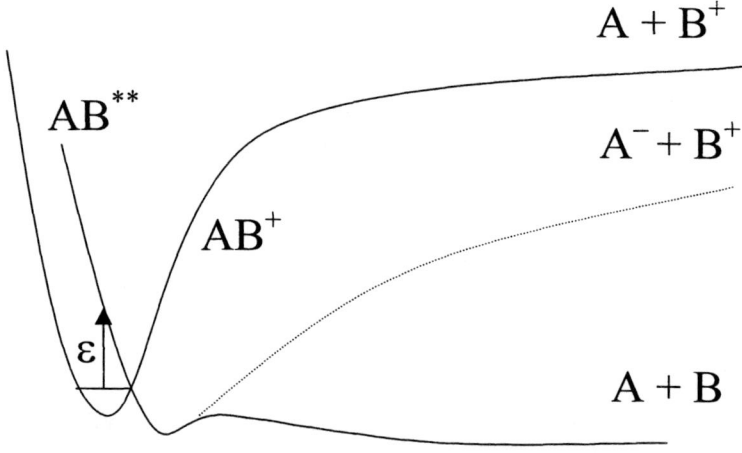

FIGURE 1. The molecular ion AB^+ captures an electron with kinetic energy ε to form the compound state AB^{**}, sometimes also known as the resonant state. The potential energy curve for the neutralized and electronically doubly excited molecule AB^{**} is strongly repulsive, which renders the two atoms A and B to rapidly depart from each other. When the potential energy is lower than the ionization potential of AB, autoionization is prohibited and the electron capture process is stabilized. The cross section for dissociative recombination is essentially proportional to $1/\varepsilon$. Resonant ion pair formation occurs when the outgoing flux continues on the dotted curve to $A^- + B^+$.

Fig. 1 is the direct mechanism of dissociative recombination proposed by Bates (19), except that he did not include the resonant ion pair channel. One can show that the direct mechanism, under some circumstances, leads to a thermal rate coefficient α_{DR} which is proportional to $T_e^{-1/2}$, where T_e is the electron temperature (21). Fig. 1 is of course an oversimplification. In order for the dissociating molecule to have a choice between a neutral channel and an ion pair channel, we must include the Rydberg states converging to the ground electronic state of AB^+. Bardsley (22) introduced the indirect mechanism in order to account for experimental rate coefficients that departed from the temperature dependence suggested by a quantal treatment of the direct mechanism (21). The indirect mechanism involves a sequence of two radiationless transitions, where the electron is first captured into a vibrationally excited Rydberg state, which is then subsequently predissociated by the same state that is reached in the direct process (i.e., AB^{**} in Fig. 1). A unified mathematical treatment of DR using MQDT including both the direct and indirect processes was developed by, in particular, Giusti (23). Her unified treatment of DR was well adapted for comparisons with results from the merged electron-molecular ion beams experiments that were started in McGowan's laboratory around 1976 (24). The parallel development of experiment and theory over the period until 1993, just at the start of the storage ring experiments, is well described in one review (25) and two books (26, 27).

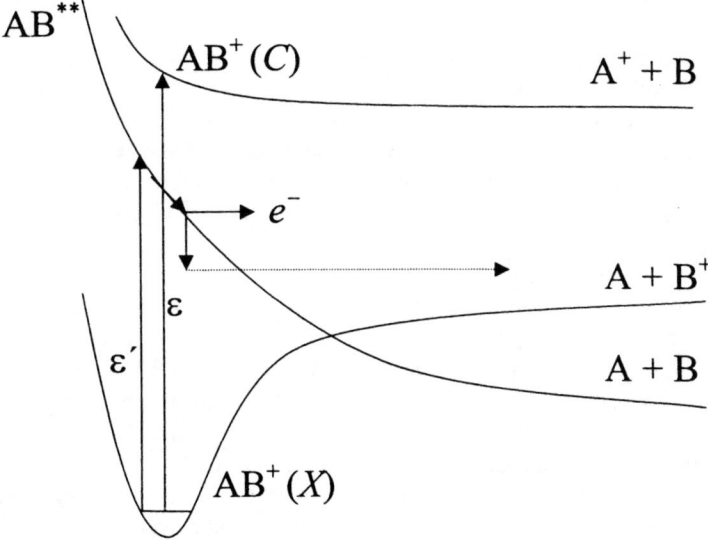

FIGURE 2. Schematic illustration of reactions (2b) and (2c). The first process, (2b), involves a direct excitation from the electronic ground state of AB^+ to the repulsive state D, which leads to dissociation into $A^+ + B$. The second process is more complex. The first step is identical with DR. Before the molecule has a chance to dissociate, autoionization into the vibrational continuum of AB^+ occurs, which leads to dissociation into $A + B^+$. Thus, there is a competition between resonant dissociative excitation and DR. The arrow leading to e^- is used to indicate autoionization.

Dissociative excitation can be described as

$$AB^+ (X, v=0) + e^- (\varepsilon) \rightarrow AB^+ (D) + e^- \rightarrow A^+ + B + e^- \qquad (2b)$$

$$AB^+ (X, v=0) + e^-(\varepsilon') \rightarrow AB^{**} \rightarrow AB^+ (X, v=cont.) + e^- \rightarrow A + B^+ + e^- \qquad (2c)$$

where X and C are different electronic states in AB^+ and v is the vibrational quantum number. Strictly, RIP also belongs to the class of dissociative excitation processes since the number of positive and negative charges is preserved in the reaction. This has been emphasized by the labeling (2a, 2b, 2c). Figure 2 exemplifies processes (2b) and (2c) with potential energy curves. Process (2c) is called resonant dissociative excitation and was first observed in CRYRING for HeH^+ (28) and D_3^+ (29).

Dissociative excitation has been much less studied than dissociative recombination. In this article I will give one example; for a comprehensive review, the reader is well served by the recent book chapter by Dunn and Djuric (30).

ION STORAGE RINGS

Dissociative recombination cross sections are basically inversely proportional to the incident electron energy. Thus, in order to study the process in detail one must use a technique which allows cross section measurement at very low centre-of-mass energy at high resolution. The merged-beams technique is then the obvious method of choice. Apart from the access to very low centre-of-mass collisions, the geometry of the merged beams allows the reaction products to be collected with high efficiency. Provided that the overlap between the two beams can be measured in addition to the particle flux in each beam, absolute cross sections can be measured. A comprehensive review of merged beams in atomic and molecular collision physics has recently been published (31)

Ion storage rings represent the latest development in merged-beams techniques. A storage ring combines a source for ion production, an acceleration device, and a ring system in which a beam of particles is confined by magnetic fields. Some rings also operate as synchrotrons by means of an accelerator cavity in the ring and synchronous ramping of the bending magnets. The stored beam is kept in a vacuum system which is kept at a very low pressure ($\sim 10^{-11}$ Torr), and it is controlled by a periodic structure of bending and focusing magnets. The four rings that have been used for studies of DR and DE are ASTRID at the University of Aarhus, Denmark (32), CRYRING at the Manne Siegbahn Laboratory of Stockholm University, Sweden (33), TARN II at the High Energy Accelerator Research Organization, Tanashi, Japan (34), and TSR at the Max-Planck-Institut für Kernphysik, Heidelberg, Germany (35). Figure 3 shows part of the CRYRING facility at the Manne Siegbahn Laboratory.

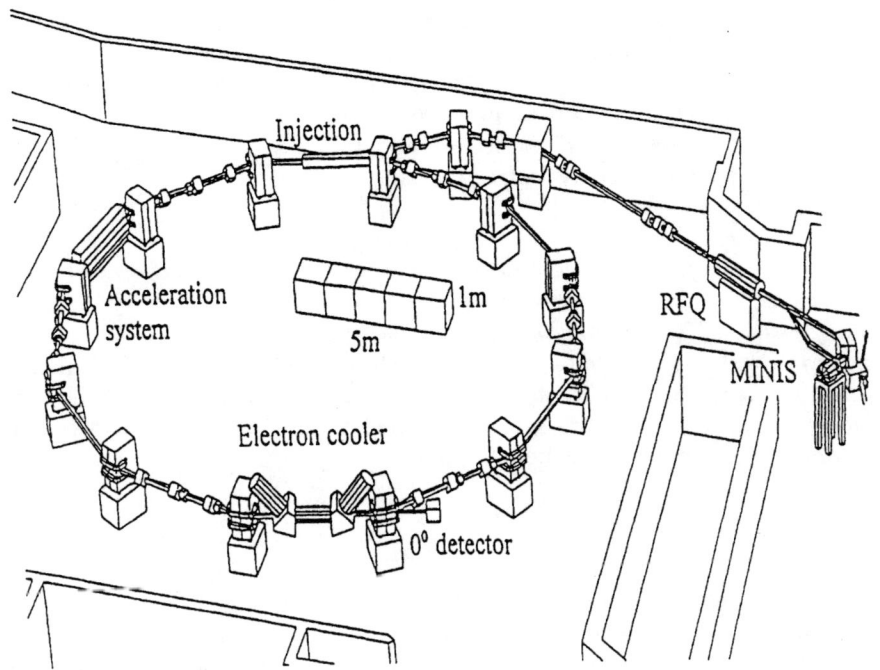

FIGURE 3. CRYRING at the Manne Siegbahn Laboratory, Stockholm University. Not shown in this figure is a cryogenic electron beam ion source for production of highly charged ions and experimental beam lines associated with this ion source. CRYRING is 51.6 m in circumference and has twelve bending magnets and six-fold symmetry. Six of the straight sections are left without focusing and correction magnets. The remaining six sections contain quadrupole magnets (four poles) as focusing elements and sextupoles magnets (six poles) for second-order corrections to the beam orbit. In addition to its function as beam cooling device, the electron cooler also serves as target for recombination experiments. Neutral reaction products are detected by means of various detectors in the area labeled "0^0 detector", whereas charged reaction products are detected by detectors in other positions.

The most important advantages offered by the ion storage rings for studies of dissociative recombination and excitation, in addition to those given above for the merged-beams technique, are:

- high-quality beams by means of electron cooling, which gives stored ion beams of narrow velocity spread, low divergence and small cross-sectional area (cooled beams are not necessary for all experiments, but are crucial for some, e.g. ref. (36)).
- recirculation of ions in the ring so that high effective currents are obtained
- the possibility of storing molecular ions for tens of seconds so that vibrational excitations are removed by infrared emission
- high beam energy, in the MeV region, which significantly reduces electron capture from rest gas molecules; for light ions the DR signal is detected against a zero background

- high beam energy, which allows separation of particles with different energies by means of energy sensitive detectors
- an ion beam which is completely overlapped by an approximately homogeneous electron target; hence, not need for measurement of overlap factor
- for TSR: the possibility to extract the stored ion beam simultaneously with a DR measurement and use the extracted part of the beam to monitor the internal vibrational state distribution

The ion beam is cooled by the electrons when the velocity difference between the two beams, the detuning velocity (v_d), is zero and the laboratory electron-beam energy is E_{cool}. The detuning energy is related to the detuning velocity through the relation $E_d = m_e v_d^2/2$ (nonrelativistically). When DR data are taken at a desired detuning energy, the electron cooler cathode voltage is changed so that the laboratory electron-beam energy becomes E_{meas}. The detuning energy and the two laboratory electron beam energies are related through

$$ E_d = \left(\sqrt{E_{meas}} - \sqrt{E_{cool}} \right)^2 . \tag{3} $$

It should be noted that the electron-beam energy is not equal to the cathode voltage times the electron charge since the electron space charge potential must be taken into account (37). The interaction of the electron beam and the molecular ion beam via a velocity dependent cross section $\sigma(v)$ results in a rate of interaction per unit time equal to

$$ R = n_e \langle v\sigma \rangle l N_i C^{-1} , \tag{4} $$

where n_e is the electron density, N_i is the number of stored ions, l is the interaction length, C the circumference of the storage ring, and $\langle v\sigma \rangle = \alpha(v_d)$ (in units of cm^{3}s^{-1}) the rate coefficient in the electron cooler. The rate coefficient $\alpha(v_d)$ can be expressed in the usual way in terms of an integral over the relative velocity distribution. Owing to the large mass difference between an ion and an electron, the velocity spread in the ion beam can usually be neglected. The resolution in the experiment is thus dependent on the electron-velocity distribution, which is a two-component Maxwell distribution. The longitudinal electron-velocity spread is compressed by the acceleration of the electrons, whereas the transverse electron-velocity spread essentially is determined by the cathode temperature (~1000 K; $kT_{e\perp} \approx 100$ meV).

A significant improvement was made when Danared et al. (38) developed the adiabatic expansion technique, which reduced the electron temperature by a factor corresponding to the expansion ratio. This technique has been implemented at all four storage rings, and CRYRING and TARN II now have electron coolers with a transverse electron temperature of just a few meV.

DISSOCIATIVE RECOMINATION AND EXCITATION OF HD⁺

Dissociative recombination of the simplest molecular ion, H_2^+ and its isotopomers, has played an important role for two decades as the benchmark system for comparison of experiment and theory (we note in passing that afterglow techniques are not applicable to H_2^+ because of its rapid conversion to H_3^+). The comparison of single-pass merged beams data and MQDT calculations for H_2^+ in the early 1990s represented the culmination of a decade long effort (39, 40).

Contributions from ion storage rings to DR of HD⁺

Already the very first storage ring experiment on HD^+ (the preferred isotopomer because of its dipole moment, which ensures complete vibrational relaxation) discovered new modes of DR at high electron energy (~9 eV), which occur through Rydberg states converging to HD^+ $(2p\sigma_u)$ (2) Further combined theoretical and experimental work have elucidated how the recombination proceeds (41, 42).

The low-energy region, below 1 eV, has also been devoted much attention, in particular since the development of the adiabatic expansion technique (38), which has allowed the resonances which occur as a result of the interference between the direct and indirect mechanisms to be revealed. Takagi (43) had already shown that rotational coupling must be included in MQDT calculations, so an averaging over several rotational levels were needed in order to facilitate a comparison of theory and experiment (42, 44). The agreement between the different resonances observed at TARN II (42) and CRYRING (44) is very good, and recently also comparison on the absolute level has shown very good agreement (45). A comparison of results from the different storage rings is in preparation. The theoretical calculations (42, 44, 46) give larger cross sections than obtained in ring experiments (42, 44–46).

The ion storage ring technique in combination with imaging detectors has allowed the question of the state of excitation of the molecular fragments to be addressed. Pioneering experiments had already been performed by Dunn and co-workers when they used optical methods to determine the state of excitation of a deuterium atom resulting from DR of D_2^+ (47, 48). A different approach is used in storage rings. In the collision process $HD^+ + e^- \rightarrow H(n) + D(n') + E_{KER}$, where KER stands for kinetic energy release, information about the quantum states of the molecular fragments can be obtained by measuring E_{KER}. Figure 4 illustrates the principles.

The imaging technique was pioneered at storage rings by Zajfman et al. (49) in their study of product state distributions in DR of HD^+ using an imaging detector consisting of multichannel plates linked to a phosphor screen and a charged-coupled-device (CCD) camera. They found a complete dominance of the H(1s) + D(n=2) limit (or vice versa) at $E_d = 0$ eV, a conclusion also reached in imaging experiments at other rings (46, 50). At low energy, the HD $(2p\sigma_u)^2 \, ^1\Sigma_g^+$ resonant state completely dominates recombination of HD^+. The results show that the $(2p\sigma_u)^2 \, ^1\Sigma_g^+$ state adiabatically dissociates to $H(1s) + D(n=2)$ and that no formation of $H(1s) + D(1s)$ occurs. Further experiments at TSR (51), now with an imaging detector extended to supply also information about the difference in arrival time between correlated particles, have shown that when the electron energy is sufficient for $H(1s) + D(n \geq 3)$ channels to open, a redistribution of the outgoing flux into these new channels occur (51). At $E_d = 2.35$ eV all channels up to $n=6$ are open; at this energy no single channel dominates and no channel has diminished. The experimental results were reproduced by a Landau-Zener treatment of how dissociation along the $(2p\sigma_u)^2 \, ^1\Sigma_g^+$ resonant state occurs (51).

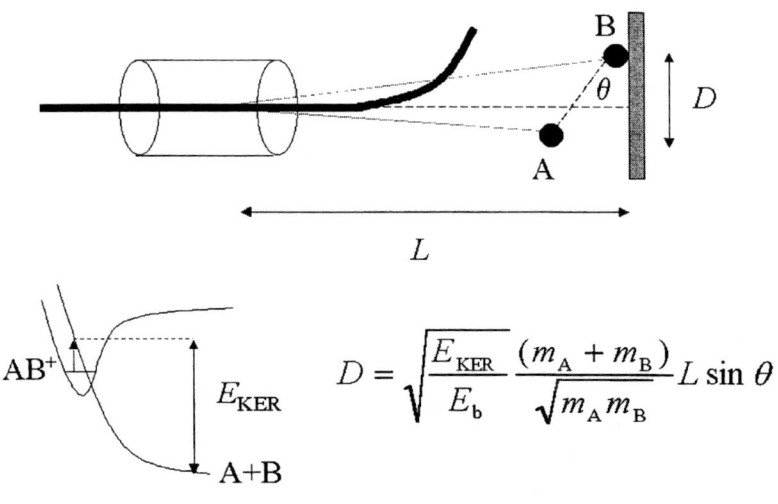

$$D = \sqrt{\frac{E_{KER}}{E_b}} \frac{(m_A + m_B)}{\sqrt{m_A m_B}} L \sin \theta$$

FIGURE 4. A schematic illustration of the principles behind the two-dimensional imaging technique. Dissociative recombination of AB^+ takes place in the electron-ion interaction region. An imaging detector at a distance L from the point where the recombination occurred detects the molecular fragments A and B. The projected distance between A and B is given by the formula in the figure, where E_b is the total ion beam energy. When $\theta = 90^0$, the maximum value of D is obtained. Extension to three-dimensional imaging can be made by measuring the difference in time-of-arrival between particles A and B (52, 53).

The presence of molecules occupying vibrationally excited levels has often been a problem in DR experiments. It can be inferred from theoretical calculations that vibrationally excited levels can have a recombination coefficient very different from the zeroth vibrational level. Until recently, however, there were no experiments that could measure the DR cross section or rate coefficient of a *specific* vibrationally excited level (naturally there are many results in the literature of DR and DE measurements of molecular ions populating an *unknown* distribution of vibrational levels). A powerful technique that allows simultaneous measurement of the vibrational population and the DR cross section was recently developed at the TSR and applied to HD^+ (54). A beam of 2 MeV HD^+ was stored in the TSR and allowed to interact with the electron target at $E_d = 0$ eV, and the recombination products were detected with a two-dimensional imaging system as a function of storage time. Simultaneously, a fraction of the stored beam was extracted and sent into a Coloumb explosion imaging (CEI) system (55), which allowed a measurement of the vibrational population. Combining the two sets of data gave a determination of the DR cross sections for vibrationally excited levels with respect to $v = 0$. Good agreement with theory was found for all levels up to $v = 7$, with the exception of $v = 3$ and 5. The rather poor agreement for these levels could indicate that the theoretical results are very sensitively depending on the location of the $(2p\sigma_u)^2\ ^1\Sigma_g^+$ state with respect to the electronic ground state of HD^+ (55, 56).

Resonant ion pair formation in DE of HD⁺

Of the three processes classified as dissociative excitation, (2a), (2b), and (2c), the direct DE process (2b) is the most well studied (30). There appears to be only three studies of process (2a), resonant ion pair formation, presented in the literature (57–59), and only one that could make claim to concern a vibrational ground state ion (59). A broad, structureless peak with an onset of 5 eV was observed in a single-pass experiment with H_3^+ (59).

CRYRING was used to study resonant ion pair formation in electron collisions with HD^+ ($v = 0$) (60). Negative deuterium atoms, D^-, formed in the reaction $HD^+ + e^- \rightarrow H^+ + D^-$ were separated from the HD^+ beam in the bending magnet after the electron cooler section and detected with a surface barrier detector (for geometric reasons, the H^- ions from the $H^- + D^+$ channel could not be detected). The results shown in Figure 5 came as surprise and part of the spectrum resisted explanation for some time. First, it is noteworthy that the RIP cross section is only a few percent of the DR cross section, and recorded against a zero background. Second, the dissociation energy of HD^+ is 2.6677 eV, which means that peaks 5–14 occur at electron energies larger than this energy. Third, there are peaks in region 3–5 eV, where neither bound nor repulsive Rydberg states are present in the Franck-Condon region of HD^+ ($v = 0$).

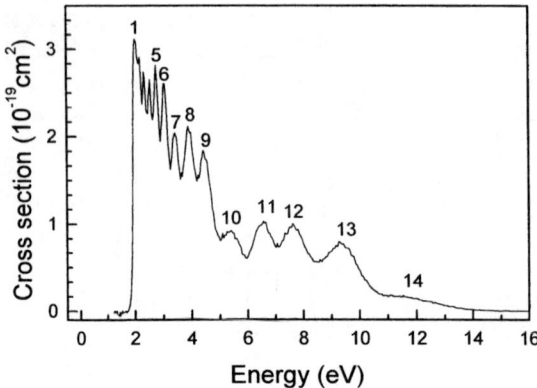

FIGURE 5. Resonant ion pair formation cross section for the $HD^+ + e^- \rightarrow H^+ + D^-$ channel obtained by detection of D^- in CRYRING (reproduced from ref. (60)).

A preliminary explanation for the resonances emerged from simple Landau-Zener-Stückelberg calculations (60, 61). Figure 6 is an attempt to schematically illustrate the mechanism that gives rise to the RIP resonances by the simplest possible diagram of potential curves that still preserve some reality. Much more realistic curves are given in (60, 61).

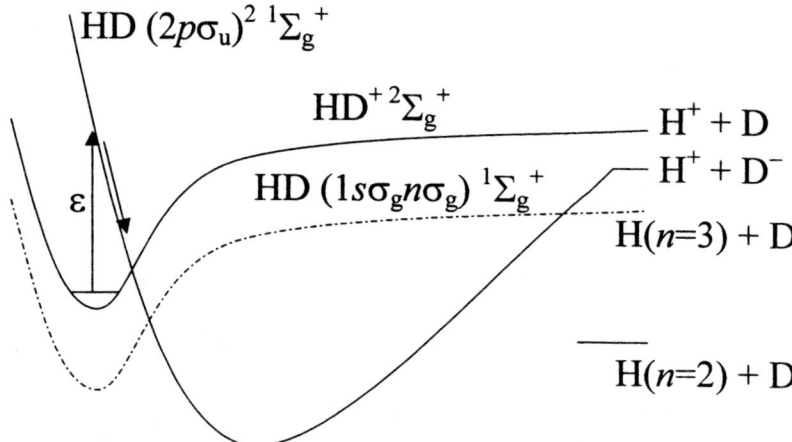

FIGURE 6. Naive illustration of the origin of the RIP resonances in HD^+. The curves are drawn as diabatic potentials, and in this representation the resonant state, $(2p\sigma_u)^2\ ^1\Sigma_g^+$, dissociates to $H^+ + D^-$. On its way to the ion pair limit, the resonant state crosses twice a Rydberg state of the same molecular symmetry. At each crossing point, the dissociating molecule can continue to dissociate along the curve it is following, or switch to the crossing curve. The resonances derive from interference effects of different pathways leading to the ion pair limit. In reality, of course, there is not one but an infinite number of Rydberg states that is crossed by the resonant state at short internuclear distance.

In the initial step, the free electron is captured by HD^+ and enters the $2p\sigma_u$ orbital while simultaneously causing the excitation $1s\sigma_g \to 2p\sigma_u$, which leads to the formation of the resonant state HD $(2p\sigma_u)^2$ $^1\Sigma_g^+$, exactly as in dissociative recombination. We know from the imaging work (46, 49, 50, 51) that when the electron energy is lower than $H(n = 3) + D$, the dissociating molecule follows adiabatically the $(2p\sigma_u)^2$ $^1\Sigma_g^+$ potential curve to the $H(n = 2) + D$ (or $H + D(n = 2)$) limit. We also know from (51) that when the electron energy is increased so that higher dissociation limits are opened, part of the outgoing flux is redirected to these limits. The ion pair limit is located 1.91 eV above HD^+ ($v = 0$), just above $H(n = 4) + D$. In a diabatic framework, the resonant state correlates with the ion pair limit. As illustrated in Fig. 6 in a very simplistic way, the outgoing flux ending on the ion pair limit can either follow the $(2p\sigma_u)^2$ $^1\Sigma_g^+$ curve diabatically to $H^+ + D^-$, or switch to the Rydberg state at the first crossing and then switch back to $(2p\sigma_u)^2$ $^1\Sigma_g^+$ at the second crossing at large internuclear distance. In reality, of course, there is an infinite number of Rydberg states converging to HD^+ $^2\Sigma_g^+$, and also other states in HD that can be operative in the initial capture process when the electron energy is a few eV. By including these states, preliminary Landau-Zener-Stückelberg calculations (60, 61) have been able to reproduce both the positions and absolute cross sections for the observed RIP peaks quite well. More theoretical work is needed, but the proof-of-principle seems established. An experimental observation of also the $H^- + D^+$ channel would be interesting. A photoionization study of HD observed the H^- signal to be twice the D^- signal, but this was in an energy range of only 0.6 eV above threshold (62).

DR OF NONHYDRIDE DIATOMIC IONS

Much of the ion storage ring work concerning nonhydride diatomic molecules have concerned the atmospheric molecules O_2^+, N_2^+ (63, 64), and NO^+ (65), but also CO^+ and CN^+ (66) have been studied.

Dissociative recombination of CO^+ ($v = 0$) with zero-energy electrons can lead to four exothermic channels:

$$CO^+ + e^- \to C(^3P) + O(^3P) + 2.92 \text{ eV} \qquad (5a)$$

$$\to C(^1D) + O(^3P) + 1.66 \text{ eV} \qquad (5b)$$

$$\to C(^3P) + O(^1D) + 0.96 \text{ eV} \qquad (5c)$$

$$\to C(^1S) + O(^3P) + 0.24 \text{ eV} \qquad (5d)$$

The final-state branching ratios into (5a) – (5d) can be measured with the imaging technique described earlier. This was done at CRYRING using a stored beam of $^{13}CO^+$ in order to avoid contamination from N_2^+ (67). Figure 7 shows a measurement of the distribution of distances in mm between carbon and oxygen atoms following DR in the electron target in CRYRING. The projected distance distributions, $P(D)$, where D is given in Fig. 4, have been cut off by means of the timing information, so that essentially only fragments flying apart perpendicular to the electron beam axis are shown. The right flank of the peaks has a shape arising from the uncertainty as to where the recombination event occurs, which equals the length of the electron target.

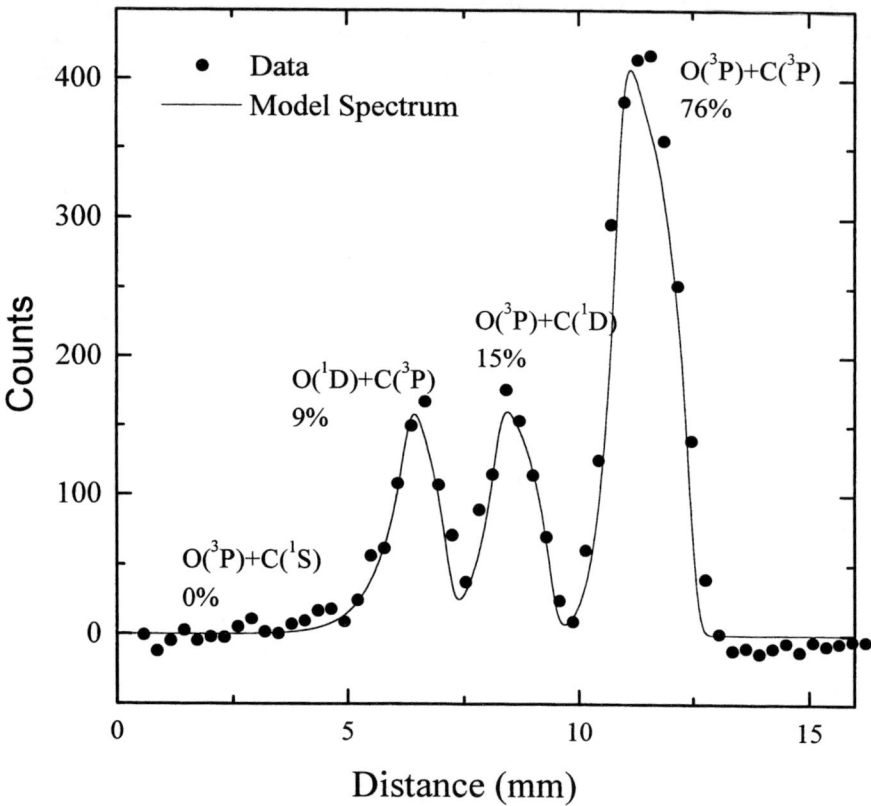

FIGURE 7. Distance spectrum for determination of final state branching ratios in DR of CO^+. The full drawn line shows a fit with correction for timing cut off and the fact that some particles missed the detector (reproduced from ref. (67)).

The CO^+ experiment shows that 76% of the recombination events leads to the formation of ground state products. When the electron energy was increased to 0.4, 1.0 and 1.5 eV the ground state fraction decreased and other channels increased. In addition, the $C(^1D) + O(^1D)$ channel, endothermic by 0.3 eV for zero-energy electrons, received some flux. The example given here illustrates how one can the combination of the merged-beams technique in a storage ring with imaging technique to obtain information about final state branching ratios.

A similar technique was used at ASTRID (68) to attack the long-standing problem with the green airglow at 557.7 nm in the Earth's atmosphere between 200 and 300 km, which derives from emission of $O(^1S)$ products from dissociative recombination of O_2^+. There had been problems to reconcile the quantum yields of 0.0016 and 0.0012 for $O(^1S)$ calculated by Guberman (69) and Guberman and Giusti-Suzor (70) with rocket and ground based observations (see e.g. ref. (71) for a brief review). The ASTRID experiment gave an $O(^1S)$ quantum yield of 0.05 for DR of O_2^+ ($v = 0$) (68), and, almost simultaneously, Guberman (72) was able to propose a new mechanism that explained the higher yield. The new mechanism involves spin-orbit coupling between the $^3\Sigma_u^-$ and $^1\Sigma_u^+$ states of O_2. The $^3\Sigma_u^-$ state has a favourable vibrational overlap with the zeroth vibrational level of the electronic ground state of O_2^+, but produces no $O(^1S)$ upon dissociation, whereas the $^1\Sigma_u^+$ state is the sole supplier of $O(^1S)$ but with an unfavourable overlap with O_2^+ ($v = 0$). Guberman (72) showed that initial electron capture into the $^3\Sigma_u^-$ state followed by transfer via a Rydberg state of mixed $^3\Sigma_u^-/^1\Sigma_u^+$ character to the $^1\Sigma_u^+$ state would give a quantum yield more than an order of magnitude larger than his earlier calculated values, and in quite good agreement with the storage ring experiment.

The ASTRID experiment was complicated by fact that the O_2^+ ions used in the experiment populated a broad range of vibrational levels; not even the use of $^{18}O^{16}O^+$ gave sufficient dipole moment to allow radiative cooling. New experiments have been performed at CRYRING (73) with implementation of the following improvements: (*i*) the use of a hollow cathode ion source that delivers vibrationally cold O_2^+; (*ii*) ultracold electron beam, which allows the $O(^1S)$ quantum yield to be measured in very small electron energy steps; (*iii*) an improved imaging detector with increased signal to noise. These experiments have allowed a detailed comparison with theory (73). A remarkable feature with the $O(^1S)$ quantum yield is that it varies radically over a narrow range of electron energies, from 0.04 at 1 meV to 0.0 at 10 meV. The possibility of using the green airglow as a probe of the electron temperature is presently being explored.

NEW MODE OF DISSOCIATIVE RECOMBINATION

The case study: HeH$^+$

The examples I have given so far all rely on a favourable crossing between the ion state and the resonant state. The question is what happens when the crossing is unfavourable, so that the vibrational overlap is very small, or when the ion potential curve is not crossed by any neutral state. It was for a long time assumed that the absence of a favourable crossing was synonymous with a negligible DR rate coefficient, or at least one that was determined by radiative stabilization, i.e., of the order of 10^{-11} cm^{3}s^{-1}. The absence of a curve crossing near $v = 0$ in the electronic ground state of H$_3^+$ was used, quite understandably, as an argument in favour of the flowing afterglow result that gave a DR rate coefficient of 10^{-11} cm^{3}s^{-1} (74).

The non-crossing case was discussed in connection with CH$^+$ by Giusti-Suzor (75), and later Bates introduced a multi-step mechanism that would not require a curve crossing (76). The best studied example of a molecule lacking a curve crossing in its electronic ground state is HeH$^+$. Experiments using the flowing afterglow/Langmuir probe (FALP) technique (74) and the single-pass merged beams technique (77) gave results differing by two orders of magnitude, the merged beams data being the larger. A series of experiments in TARN II (1, 78, 79), CRYRING (80–84), and TSR (83) has firmly established that HeH$^+$ recombines quite effectively; for ^{4}HeH$^+$, a thermal rate coefficient of 3×10^{-8} cm^{3}s^{-1} was obtained at 300 K (82). The mechanism has been identified as coupling by the nuclear kinetic-energy operator of the HeH$^+$ plus electron system with Rydberg states close to the left turning point of the zeroth vibrational level of HeH$^+$ (85, 86). Characteristic of the mechanism is that indirect DR through the Rydberg states has a larger cross section than direct DR along the repulsive ground state of HeH (85). Imaging experiments show at DR of HeH$^+$ with thermal energy electrons leads to the formation of He(^1S) + H(n=2), with no trace of the formation of H($1s$) (78, 82, 83). An interesting observation is that when the He(^1S) + H(n=3) channel opens, the flow of flux is completely redirected into this channel (83). A wave packet approach has been used to theoretically study recombination of HeH$^+$ also at higher electron energies (> 5 eV) in terms of DR (87), DE (88) and final state distributions (89). Cross sections given in (87–89) must be corrected by a factor of 2 (90).

The enigma: H$_3^+$

No other molecule has generated so much controversy concerning its DR rate coefficient as H$_3^+$. A comprehensive review of the history is not possible here (see e.g. ref. (91)); it suffices to note that measurements with the FALP technique at one point gave a value as low as 10^{-11} cm^{3}s^{-1} (74, 92), whereas a stationary afterglow experiment gave 1.5×10^{-7} cm^{3}s^{-1} (93). The spread among experimental results has been narrowed, as shown in Table 1, but there is not yet consensus. It is noteworthy that the DR cross section measured in CRYRING (3,95) recently has been reproduced

TABLE 1. Measured H$_3^+$ Dissociative Recombination Coefficient at 300 K

α(H$_3^+$, $v = 0$) (10^{-7} cm^{3}s^{-1})	Method	Reference
1.6	Infrared absorption spectroscopy	94
1.15	Ion storage ring; CRYRING	95
0.78	FALP	96
0.1-0.2	FALP	97
~0.1	FALP	98
1.2	Single-pass merged beams	99

by measurements in TARN II (45).

The two FALP measurements giving a low value for the DR rate coefficient agree on the magnitude of α, but propose different explanations for their observations (97, 98). Both experiments found a non-linear decay of the plasma, which was explained by the presence of vibrationally excited H_3^+ in the early part of the plasma in (97), and by three-body recombination in (98). The non-linear decay was not observed in the third FALP experiment (96). Attempts to enhance the DR rate for D_3^+ by adding an electric field of 30 Vcm^{-1} in the interaction region failed (29, 100).

For H_3^+ ($v = 0$) recombining with 0 eV electrons, the following channels are open:

$$H_3^+ (v = 0) + e^-(\varepsilon) \rightarrow H(1s) + H(1s) + H(1s) \qquad (6a)$$

$$\rightarrow H_2(^1\Sigma_g^+) + H(1s) \qquad (6b)$$

$$\rightarrow H_3^* \quad . \qquad (6c)$$

The branching ratios into these channels were measured in CRYRING and found to be 75% for (6a), 25% for (6b), and 0% for (6c) (101). The electronic ground state of H_3^+ is crossed by the resonant state $^2A_1$ close to the outer turning point of $v = 3$. Thus, for 0 eV electrons, this state plays no role in DR of H_3^+ ($v = 0$). However, if the electron energy is increased, the $^2A_1$ state becomes accessible for DR, and this leads to an increase of the branching into channel (6b) (101). In the same way, DR of H_3^+ ($v \geq 3$) is dominated by the $^2A_1$ state, and this leads to an increase of the branching to $H_2 + H$ when vibrationally hot ions are used (102, 103). This is in particular apparent in the TARN II experiment (103), where the breakup as a function of storage time was studied.

The H_3^+ ($v = 0$) DR rate coefficient has in particular come into focus during the last couple of years because of the indications that chemical models of interstellar molecular clouds may have bistable solutions (104), and because of the discoveries of H_3^+ in interstellar molecular clouds (105, 106).

The most important step at this point is to establish a mechanism that could account for a DR rate coefficient around 10^{-7} cm^3s^{-1} for H_3^+ ($v = 0$). Recombination through the $^2A_1$ state is impossible for thermal energy electrons. The only other alternative is a mechanism similar to that in HeH^+. Theoretical work along this line is presently in progress (108). The direct mechanism is too slow to account for any of the data in Table 1, even the lowest values. But just as in HeH^+, it could be that the indirect mechanism enhances the cross section substantially. This brings a new role to the indirect mechanism. In the past it was regarded as a mechanism that gave rise to a number of window resonances in a cross section that was dominated by the direct mechanism (75). Now it emerges as something that possibly could explain a cross section much larger than anticipated on grounds of the direct mechanism alone.

CONCLUSIONS

The intention of this article has been to give the reader some flavour of the excitement that ion storage rings have brought to the field of electron-molecular interaction. A full review was neither possible nor desired; when ref. (5) is published it will give a comprehensive coverage of dissociative recombination in storage rings, and a partial coverage of dissociative excitation (for a comprehensive review of DE, see (30)). Interaction of electrons and negative molecular ions has become an interesting area of research, but was completely left out of this article (see, however, e.g. ref. (8)).

Important aspects of dissociative recombination have only been mentioned in passing in this article, such as branching ratios in recombination of polyatomic molecular ions. This is a very important problem for the chemistry of interstellar molecular clouds, and much progress has been made in recent years at the storage rings, but also in flowing afterglow machines. The experimental results will pose serious challenges to theory.

It should finally be pointed out that one very important aspect of the ion storage rings is that there are several of them. This means not only a healthy competition, but also the possibility to compare results from different rings.

ACKNOWLEDGMENTS

I am thankful to all my collaborators and colleagues, to numerous to be mentioned here (but their names can be found in the references). This work has been supported by the Swedish Natural Science Research and the Göran Gustafsson Foundation. The article was written during a visit to T. Oka at The University of Chicago, and I would like to thank him for his generous hospitality.

REFERENCES

1. Tanabe, T., Katayama, N., Inoue, N., Chida, K., Arakaki, Y., Watanabe, T., Yoshizawa, M., Ohtani, S., and Noda, K., *Phys. Rev. Lett.* **70**, 422 (1993).
2. Forck, P., Grieser, M., Habs, D., Lampert, A., Repnow, P., Schwalm, D., Wolf, A., and Zajfman, D., *Phys. Rev. Lett.* **70**, 426 (1993).
3. Larsson, M., Danared, H., Mowat, J.R., Sigray, P., Sundström, G., Broström, L., Filevich, A. Källberg, A., Mannervik, S., Rensfelt, K.-G., and Datz, S., *Phys. Rev. Lett.* **70**, 430 (1993).
4. Larsson, M., *Annu. Rev. Phys. Chem.* **48**, 151 (1997).
5. Larsson, M., "Dissociative Electron-Ion Recombination Studies Using Ion Synchrotrons", in *Adv. Ser. Chem. Phys.: Photoionization and Photodetechment*, ed. Ng, C.-Y., Singapore: World Scientific, 1999 (in press).
6. Larsson, M., *Rep. Prog. Phys.* **58**, 1267 (1995).
7. Mokler, P.H. and Stöhlker, Th., *Adv. At. Mol. Opt. Phys.* **37**, 297 (1996).
8. Andersen, L.H., Andersen, T., and Hvelplund, P., *Adv. At. Mol. Opt. Phys.* **38**, 155 (1998).

9.   Kühl, T., *Adv. At. Mol. Opt. Phys.* **40**, 114 (1999).
10. Biondi, M.A. and Brown, S.C., *Phys. Rev.* **76**, 1697 (1949).
11. Bates, D.R. and Massey, H.S.W., *Proc. R. Soc. London Ser. A* **192**, 1 (1947).
12. Massey, H.S.W., *Proc. R. Soc. London Ser. A* **163**, 542 (1937).
13. Bates, D.R., Buckingham, R.A., Massey, H.S.W., and Unwin, J.J., *Proc. R. Soc. London Ser. A* **170**, 322 (1939).
14. Biondi, M.A. and Brown, S.C., *Phys. Rev.* **75**, 1700 (1949).
15. Bates, D.R., *Phys. Rev.* **77**, 718 (1950).
16. Biondi, M.A. and Bardsley, J.N., *Adv. At. Mol. Phys.* **6**, 1 (1970).
17. Ferguson, E.E., Fehsenfeld, F.C., and Schmeltekopf, A., *Phys. Rev.* **138**, 381 (1965).
18. Carata, L., Orel, A.E., and Suzor-Weiner, A., *Phys. Rev. A* **59**, 2804 (1999).
19. Bates, D.R., *Phys. Rev.* **78**, 492 (1950).
20. Loeb, L.B., *Fundamental Processes of Electrical Discharge in Gases*, New York: John Wiley and Sons, Inc., 1939.
21. Bates, D.R. and Dalgarno, A., "Electronic recombination", in *Atomic and Molecular Processes*, ed. Bates, D.R., New York: Academic Press, 1962, p. 245.
22. Bardsley, J.N., *J. Phys. B* **1**, 365 (1968).
23. Giusti, A., *J. Phys. B* **13**, 3867 (1980).
24. McGowan, J.Wm., Caudano, R., and Keyser, J., *Phys. Rev. Lett.* **36**, 1447 (1976).
25. Mitchell, J.B.A., *Phys. Rep.* **186**, 215 (1990).
26. Mitchell, J.B.A. and Guberman, S.L., eds., *Dissociative Recombination: Theory, Experiment and Applications*, Singapore: World Scientific, 1989.
27. Rowe, B.R., Mitchell, J.B.A., and Canosa, A., eds., *Dissociative Recombination: Theory, Experiment, and Applications*, NATO ASI Series B: Physics Vol. 313, New York: Plenum Press, 1993.
28. Strömholm, C., Semaniak, J., Rosén, S., Danared, H., Datz, S., van der Zande, W.J., and Larsson, M., *Phys. Rev. A* **54**, 3086 (1996).
29. Le Padellec, A., Larsson, M., Danared, H., Larson, Å., Peterson, J.R., Rosén, S., Semaniak, J., and Strömholm, C., *Phys. Scripta* **57**, 215 (1998).
30. Dunn, G.H. and Djuric, N., "Electron Impact Dissociative Excitation and Ionization of Molecular Ions", in *Novel Aspects of Electron-Molecule Scattering*, ed. Becker, K., Singapore: World Scientific, 1998, p. 241.
31. Phaneuf, R.A., Havener, C.C., Dunn, G.H., and Müller, A., *Rep. Prog. Phys.* **62**, 1143 (1999).
32. Møller, S.P., in *Conference Record of the IEEE Particle Accelerator Conference*, ed. Berkner, K., New York: IEEE press, 1991, p. 2811.
33. Abrahamsson, K., Andler, G., Bagge, L., Beebe, E., Carlé, P., Danared, H., Egnel, S., Ehrnstén, K., Engström, M., Herrlander, C.J., Hilke, J., Jeansson, J., Källberg, A., Leointein, S., Liljeby, L., Nilsson, A., Paál, A., Rensfelt, K.-G., Rosengård, U., Simonsson, A., Soltan, A., Starker, J., af Ugglas, M., and Filevich, A., *Nucl. Instr. Meth. Phys. Res.* **B79**, 269 (1993).
34. Tanabe, T., Noda, K., Honma, T., Kodaira, M., Chida, K., Watanabe, T., Noda, A., Watanabe, S., Mizobuchi, M., Yoshizawa, M., Katayama, T., Muto, H., and Ando, A., *Nucl. Instr. Meth. Phys. Res.* **A307**, 7 (1991).
35. Krämer, D. Bisoffi, G., Blum, M., Friedrich, A., Geyer, Ch., Holzer, B., Heyng, H.W., Habs, D., Jaeschke, E., Jung, M., Ott, W., Pollock, R.E., Repnow, R., and Schmitt, F., *Nucl. Instr. Meth. Phys. Res.* **A287**, 268 (1990).
36. Datz, S., Larsson., M., Strömholm, C., Sundström, G., Zengin, V., Danared, H., Källberg, A., and af Ugglas, M., *Phys. Rev. A* **52**, 2901 (1995).
37. Kilgus, G., Habs, D., Schwalm, D., Wolf, A., Badnell, N.R., and Müller, A., *Phys. Rev. A* **46**, 5730 (1992).
38. Danared, H., Andler, G., Bagge, L., Herrlander, C.J., Hilke, J., Jeansson, J., Källberg, A., Nilsson, A., Paál, A., Rensfelt, K.-G., Rosengård, U., Starker, J., and af Ugglas, M., *Phys. Rev. Lett.* **72**, 3775 (1994).
39. Van der Donk, P., Yousif, F.B., Mitchell, J.B.A., and Hickman, A.P., *Phys. Rev. Lett.* **67**, 42 (1991).

61

40. Schneider, I.F., Dulieu, O., and Giusti-Suzor, A., *J. Phys. B* **24**, L289 (1991).
41. Strömholm, C., Schneider, I.F., Sundström, G., Carata, L., Danared, H., Datz, S., Dulieu, O., Källberg, A., af Ugglas, M., Urbain, X., Zengin, V., Suzor-Weiner, A., and Larsson, M., *Phys. Rev. A* **52**, R4320 (1995).
42. Tanabe, T., Katayama, I., Kamegaya, H., Chida, K., Arakaki, Y., Watanabe, T., Yoshizawa, M., Saito, M., Haruyama, Y., Hosono, K., Hatanaka, K., Honma, T., Noda, K., Ohtani, S., and Takagi, H., *Phys. Rev. Lett.* **75**, 1066 (1995).
43. Takagi, H., *J. Phys. B* **26**, 4815 (1993).
44. Schneider, I.F., Strömholm, C., Carata, L., Urbain, X., Larsson, M., and Suzor-Weiner, A., *J. Phys. B* **30**, 2687 (1997).
45. Tanabe, T., Chida, K., Watanabe, T., Arakaki, Y., Takagi, H., Katayama, I., Haruyama, Y., Saito, M., Nomura, I., Honma, T., Noda, K., and Hoson, K., "Dissociative Recombination at the TARN II Storage Ring", in *Dissociative Recombination: Theory, Experiment and Applications IV*, eds. Larsson, M., Mitchell, J.B.A., and Schneider, I.F., Singapore: World Scientific (in preparation).
46. Andersen, L.H., Johnson, P.J., Kella, D., Pedersen, H.B., and Vejby-Christensen, L., *Phys. Rev. A* **55**, 2799 (1997).
47. Phaneuf, R.A., Crandall, D.H., and Dunn, G.H., *Phys. Rev. A* **11**, 528 (1975).
48. Vogler, M. and Dunn, G.H., *Phys. Rev. A* **11**, 1983 (1975).
49. Zajfman, D., Amitay, Z., Broude, C., Forck, P., Seidel, B., Grieser, M., Habs, D., Schwalm, D., and Wolf, A., *Phys. Rev. Lett.* **75**, 814 (1995).
50. van der Zande, W.J., Semaniak, J., Zengin, V., Sundström, G., Rosén, S., Strömholm, C., Datz, S., Danared, H., and Larsson, M., *Phys. Rev. A* **54**, 5010 (1996).
51. Zajfman, D., Amitay, Z., Lange, M., Hechtfischer, U., Knoll, L., Schwalm, D., Wester, R., Wolf, A., and Urbain, X., *Phys. Rev. Lett.* **79**, 1829 (1997).
52. Amitay, Z. and Zajfman, D., *Rev. Sci. Instr.* **68**, 1 (1997).
53. Rosén, S., Peverall, R., ter Horst, J., Sundström, G., Semaniak, J., Sundqvist, O., Larsson, M., de Wilde, M., and Zande, W.J., *Hyperfine Interact.* **115**, 201 (1998).
54. Amitay, Z., Baer, M., Dahan, M., Knoll, L., Lange, M., Levin, J., Schneider, I.F., Schwalm, D., Suzor-Weiner, A., Vager, Z., Wester, R., Wolf, A., and Zajfman, D., *Science* **281**, 75 (1998).
55. Zajfman, D., Amitay, Z., Krohn, S., Wolf, A., Schwalm, D., Wester, R., Knoll, K., Lange, M. Levin, J., and Urbain, X., "Dissociative Recombination Cross Section of Simple Molecular Ions: From Initial States to Final States", in *Dissociative Recombination: Theory, Experiment and Applications IV*, eds. Larsson, M., Mitchell, J.B.A., and Schneider, I.F., Singapore: World Scientific (in preparation).
56. Schneider, I.F. and Suzor-Weiner, A., (cited in ref. 55 as personal communication).
57. Peart, B. and Dolder, K.T., *J. Phys. B* **8**, 1570 (1975).
58. Peart, B., Forest, R.A., and Dolder, K.T., *J. Phys. B* **12**, 3441 (1979).
59. Yousif, F.B., Van der Donk, P., and Mitchell, J.B.A., *J. Phys. B* **26**, 4249 (1993).
60. Zong, W., Dunn, G.H., Djuric, N., Larsson., M., Greene, C.H., Al-Khalili, A., Neau, A., Derkatch, A.M., Vikor, L., Shi, W., Le Padellec, A., Danared, H., and af Ugglas, M., *Phys. Rev. Lett.* **83**, nr. 5, in press (1999).
61. Larson, Å., Greene, C.H., Orel, A.E., Schneider, I.F., and Suzor-Weiner, A., "Theoretical Study of Ion-Pair Formation in Collisions of Electrons with HD^{+}", in *Dissociative Recombination: Theory, Experiment and Applications IV*, eds. Larsson, M., Mitchell, J.B.A., and Schneider, I.F., Singapore: World Scientific (in preparation).
62. Chupka, W.A., Dehmer, P.M., and Jivery, W.T., *J. Chem. Phys.* **63**, 3929 (1975).
63. Kella, D., Johnson, P.J., Pedersen, H.B., Vejby-Christensen, L., and Andersen, L.H., *Phys. Rev. Lett.* **77**, 2432 (1996)
64. Peterson, J.R., Le Padellec, A., Danared, H., Dunn, G.H., Larsson, M., Larson, Å., Peverall, R., Strömholm, C., Rosén, S., af Ugglas, M., van der Zande, W.J., *J. Chem. Phys.* **108**, 1978 (1998).
65. Vejby-Christensen, L., Kella, D., Pedersen, H.B., and Andersen, L.H., *Phys. Rev. A* **57**, 3627 (1998).

66. Le Padellec, A., Mitchell, J.B.A., Al-Khalili, A., Danared, H., Källberg, Λ., Larson, Å., Rosén, S., af Ugglas, M., Vikor, L., and Larsson, M., *J. Chem. Phys.* **110**, 890 (1999).

67. Rosén, S., Peverall, R., Larsson, M., Le Padellec, A., Semaniak, J., Larson, Å., Strömholm, C., van der Zande, W.J., Danared H., and Dunn, G.H., *Phys. Rev. A* **57**, 4462 (1998).

68. Kella, D., Vejby-Christensen, L., Johnson, P.J., Pedersen, H.B., and Andersen, L.H., *Science* **276**, 1530 (1997); *ibid.* **277**, 167 (1997).

69. Guberman, S.L., *Nature* **327**, 408 (1987).

70. Guberman, S.L. and Giusti-Suzor, A., *J. Chem. Phys.* **95**, 2602 (1991).

71. Bates, D.R., *Adv. At. Mol. Opt. Phys.* **34**, 427 (1994).

72. Guberman, S.L., *Science* **278**, 1276 (1997).

73. Peverall, R., Rosén, S., Peterson, J.R., Larsson, M., Vikor, L., Semaniak, J., Le Padellec, A., Danared, H., af Ugglas, M., Al-Khalili, A., Bobbenkamp, R., Maurellis, A., Guberman, S.L. and van der Zande, W.J., "The Dissociative Recombination of O_2^+ and the Oxygen Green Airglow: A Surprising Dependence on Electron Temperature", presented as poster at the Fourth International Conference on Dissociative Recombination, Nässlingen, Stockholm, Sweden, June 1999.

74. Adams, N.G. and Smith, D., in *Rate Coefficients in Astrochemistry*, eds. Millar, T.J. and Williams, D.A., Dordrecht: Kluwer Academic, 1988, p. 173.

75. Giusti-Suzor, A., "Recent Developments in the Theory of Dissociative Recombination and Related Phenomena", in *Atomic Processes in Electron-Ion and Ion-Ion Collisions*, ed. Brouillard, F., NATO ASI Series B: Physics Vol. 145, New York: Plenum Press, 1986, p. 223.

76. Bates, D.R., *J. Phys. B* **25**, 5479 (1992).

77 Yousif, F.D. and Mitchell, J.B.A., *Phys. Rev. A* **40**, 4318 (1989).

78. Tanabe, T., Katayama, I., Inoue, N., Chida, K., Arakaki, Y., Watanabe, T., Yoshizawa, M., Saito, M., Haruyama, Y., Hosono, K., Honma, T., Noda, K., Ohtani, S., and Takagi, H., *Phys. Rev. A* **49**, R1531 (1994).

79. Tanabe, T., Katayama, I., Ono, S. Chida, K., Watanabe, T., Arakaki, Y., Haruyama, Y., Saito, M., Odagiri, T., Hosono, K., Noda, K., Honma, T., and Takagi, H., *J. Phys. B* **31**, L297 (1998).

80. Sundström, G., Datz, S., Mowat, J.R., Mannervik, S., Broström, L., Carlson, M., Danared, H., and Larsson, M., *Phys. Rev. A* **50**, R2806 (1994).

81. Mowat, J.R., Danared, H., Sundström, G., Carlson, M., Andersen, L.H., Vejby-Christensen, L., af Ugglas, M., and Larsson, M., *Phys. Rev. Lett.* **74**, 50 (1995).

82. Strömholm, C., Semaniak, J., Rosén, S., Danared, H., Datz, S., van der Zande, W., and Larsson, M., *Phys. Rev. A* **54**, 3086 (1996).

83. Semaniak, J., Rosén, S., Sundström, G., Datz, S., Danared, H., af Ugglas, M., Larsson, M., van der Zande, W.J., Amitay, Z., Hechtfischer, U., Grieser, M., Repnow, R., Schmidt, M., Schwalm, D., Wester, R., Wolf, A., and Zajfman, D., *Phys. Rev. A* **54**, R4617 (1996).

84. Al-Khalili, A., Danared, H., Larsson, M., Le Padellec, A., Peverall, R., Rosén, S., Semaniak, J., af Ugglas, M., Vikor, L., and van der Zande, W.J., *Hyperfine Interact.* **114**, 281 (1998).

85. Guberman, S.L., *Phys. Rev. A* **49**, 4318 (1994)

86. Sarpal, B.K., Tennyson, J., and Morgan , L., *J. Phys. B* **27**, 5943 (1994).

87. Orel, A.E., Kulander, K.C., and Rescigno, T.N., *Phys. Rev. Lett.* **74**, 4807 (1995).

88. Orel, A.E. and Kulander, *Phys. Rev. A* **54**, 4992 (1996).

89. Larson, Å. and Orel, A.E., *Phys. Rev. A* **59**, 3601 (1999).

90. Orel, A.E., "Wave Packet Studies of Dissociative Recombination and Dissociative Excitation of Molecular Ions", in *Dissociative Recombination: Theory, Experiment and Applications IV*, eds. Larsson, M., Mitchell, J.B.A., and Schneider, I.F., Singapore: World Scientific (in preparation).

91. Dalgarno, A., *Adv. At. Mol. Opt. Phys.* **32**, 57 (1994).

92. Adams, N.G. and Smith, D., "Recent Advances in the Studies of Reaction Rates Relevant to Interstellar Chemistry", in *Astrochemistry: proceedings of the 120th Symposium of the International Astronomical Union*, eds. Vardaya, M.S. and Tarafdar, S.P., Dordrecht: D. Reidel Publishing Company, 1987, p. 1.

93. Macdonald, J.A., Biondi, M.A., and Johnsen, R., *Planet. Space Sci.* **32**, 651 (1984).

94. Amano, T., *J. Chem. Phys.* **92**, 6492 (1990).

95.   Sundström, G., Mowat, J.R., Danared, H., Datz, S., Broström, L., Filevich, A., Källberg, A., Mannnervik, S., Rensfelt, K.-G., Sigray, P., af Ugglas, M., and Larsson., M., *Science* **263**, 785 (1994).

96.   Laubé, S., Le Padellec, A., Sidko, O., Rebrion-Rowe, C., Mitchell, J.B.A., and Rowe, B.R., *J. Phys. B* **31**, 2111 (1998).

97.   Smith, D. and Spanel, P., *Int. J. Mass Spectrom. Ion Proc.* **129**, 163 (1993).

98.   Gougousi, T., Johnsen, R., and Golde, M.F., *Int. J. Mass Spectrom. Ion Proc.* **149/150**, 131 (1995).

99.   Yousif, F.B., Rogelstad, M., and Mitchell, J.B.A., in *Atomic and Molecular Physics: 4th US/Mexico Symposium*, eds. Alvarez, I., Cisneros, S., and Morgan, T.J., Singapore: World Scientific, 1995, p. 343.

100.   Larsson, M., Danared, H., Larson, Å., Le Padellec, A., Peterson, J.R., Rosén, S., Semaniak, J., and Strömholm, C., *Phys. Rev. Lett.* **79**, 395 (1997).

101.   Datz, S., Sundström, G., Biedermann, Ch., Broström, L., Danared, H., Mannervik, S., Mowat, J.R., and Larsson, M., *Phys. Rev. Lett.* **74**, 896 (1995).

102.   Mitchell, J.B.A., Forand, J.L., Ng, C.T., Levac, D.P., Mitchell, R.E., Mul, P.M., Clayes, W., Sen, A., and McGowan, J.Wm., *Phys. Rev. Lett.* **51**, 885 (1983).

103.   Tanabe, T., Katayama, I., Kamegaya, H., Chida, K., Arakaki, Y., Watanabe, T., Yoshizawa, M., Saito, M., Haruyama, Y., Hosono, K., Hatanaka, K., Honma, T., Noda, K., Ohtani, S., and Takagi, H., in *Dissociative Recombination: Theory, Experiment and Applications III*, eds. Zajfman, D., Mitchell, J.B.A., Schwalm, D., and Rowe, B.R., Singapore: World Scientific, 1996, p. 84.

104.   Lee, H.-H., Roueff, E., Pineau des Forêts, G., Shalabiea, O.M., Terzieva, R., and Herbst, E., *Astron. Astrophys.* **334**, 1047 (1998).

105.   Geballe, T.R. and Oka, T., *Nature* **384**, 334 (1996).

106.   McCall, B.J., Geballe, T.R., Hinkle, K.H., and Oka, T., *Science* **279**, 1910 (1998).

107.   Schneider, I.F., Larsson, M., Suzor-Weiner, A., and Orel, A.E., "Dissociative Recombination of H_3^+ and Predissociation of H_3", in *Dissociative Recombination: Theory, Experiment and Applications IV*, eds. Larsson, M., Mitchell, J.B.A., and Schneider, I.F., Singapore: World Scientific (in preparation).

Collisions of slow multicharged ions with atoms, molecules, clusters and surfaces

Reinhard Morgenstern, Thomas Schlathölter, Ronnie Hoekstra

KVI Atomic Physics, Rijksuniversiteit Groningen
Zernikelaan 25, 9747 AA Groningen, Netherlands

Abstract. Dynamic processes induced by the high potential energy of multiply charged ions have been investigated in a large number of collision systems during the last years. We give a review of these activities with special emphasis on the developments since the last ICPEAC in Vienna 1997.

INTRODUCTION

When highly charged ions (HCI) approach neutral matter, a situation far from stability arises. A large amount of potential energy is available – for bare Ar ions e.g. more than 10 keV – and a variety of electronic processes is induced. The target electrons "see" in fact a deep potential well and one might be reminded of the situation of a waterfall, where the water cascades down. So the question is, what happens with this potential energy, and what are the mechanisms. In the early eighties Bárány et al[1] and Niehaus[2] developed a classical over-the-barrier model to give a quantitative description of charge transfer processes in ion-atom collisions. In this model the electrons initially bound at the target spread out over the whole quasi molecule as soon as the Coulomb barrier between target and projectile has dropped below their binding energy. With a certain chance the electrons will stay with the projectile when the collision partners recede from each other. Part of the electrons are thus transferred to the projectile in a quasi-resonant way, i.e. with the binding energy conserved at the moment of transfer. Essentially this implies that highly excited orbitals are populated in the projectile. The classical over-the-barrier model has been extended by Burgdörfer et al[3] to cover highly charged ion-surface collisions. Also here the target electrons are assumed to be resonantly transferred to excited projectile orbitals, so that a multiply excited, so-called "hollow" atom or ion is formed.

The transfer of electrons to the projectile can be regarded as the first step in a series of subsequent processes. Many investigations have been dedicated to these processes using atoms, molecules, fullerenes, clusters and surfaces as targets. A broad range of experimental techniques have been used, including the analysis of emitted electrons or photons, scattered and recoiling collision partners or fragments formed during the collisions. A general review can be found in the book edited by Lin[4]. A recent review of

CP500, *The Physics of Electronic and Atomic Collisions*, edited by Y. Itikawa, et al.

these activities has been given by Cederquist et al[5]. Regarding ion-surface collisions a rather detailed overview on this has been given by Arnau et al[6].

In the following we will discuss some selected results from various groups which can be regarded as characteristic examples, documenting recent progress in the field. We begin with a short chapter on atoms and then proceed to surfaces, which were subject of a large variety of investigations during the last years. Finally we will discuss C_{60} targets as examples for molecules and clusters.

ATOMS

In the past, collisions of multiply charged ions with various atoms have been investigated by many different methods, including translational spectroscopy as well as the spectroscopy of emitted photons or electrons. In recent years another method has been developed by Schmidt-Böcking and coworkers[7], namely 'cold target recoil ion momentum spectroscopy' (COLTRIMS). In this method the final charge state of the collision partners, their kinetic energy gain or loss and their scattering angle are determined. These quantities yield information on the number of transferred electrons, the final states of the collision partners and on the specific path, which was followed through the system of potential curves. Although only a very small fraction of the total kinetic energy gain goes to the target, it is advantageous to analyze the recoiling target particle: the small amount of energy gained is large with respect to the particle's initial energy, especially when the target is cooled (supersonic jet or – in future experiments – laser cooling). A measurement of the target particle's momenta parallel and perpendicular to the primary projectile beam – measured with time-of-flight analysis and modern two-dimensional position sensitive detectors – gives direct information on energy gain and scattering angle.

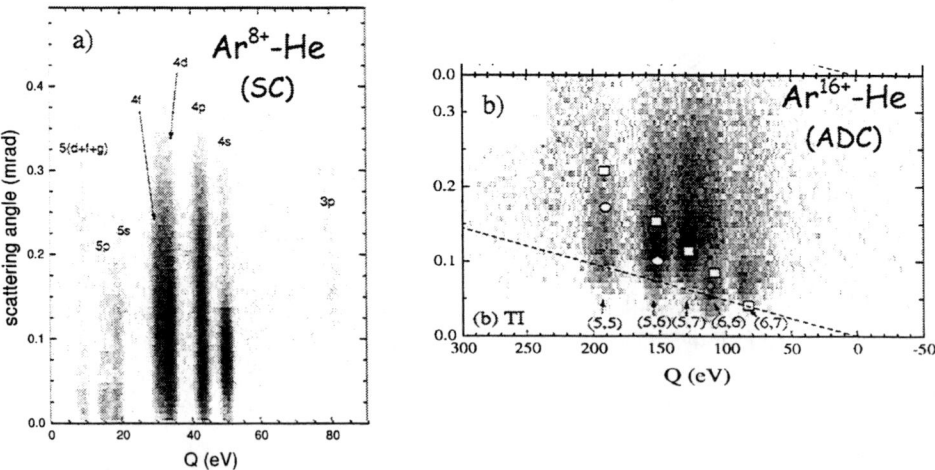

FIGURE 1. Scattering events as a function of scattering angle and energy gain for a) single (SC) and b) autoionizing double capture (ADC) in Ar^{q+} + He collisions (from Abdallah et al, refs. 8 and 9).

We will discuss an example of ion-atom collisions exploiting this relatively young method. Experimental results for single capture in Ar^{8+}-He (Abdallah et al[8]) and double capture in Ar^{16+}-He collisions (Abdallah et al[9]) are shown in FIGURE 1. Obviously the energy resolution is sufficient to distinguish e.g. capture into the various $4l$ orbitals in case of single capture (fig.1a).

Regarding double electron capture a remarkable result can immediately be deduced from figure 1b, namely that the two electrons in this case are preferentially not captured simultaneously, but rather sequentially: the straight line in fig.1b indicates the scattering angle which would be expected for simultaneous capture. The fact that most events are observed for larger scattering angles can be ascribed to the fact that the first capture takes already place at large distances, whereafter the system follows a repulsive Coulomb curve instead of the more or less flat potential curve for the initial ion-atom system. Of course there are other examples where simultaneous capture prevails and electron correlation has to be invoked to interpret the results (see e.g. Chesnel et al[10]), however in most cases investigated so far, sequential electron capture is the most important mechanism.

SURFACES

The other extreme of a target, namely one with a nearly infinite supply of electrons, are surfaces. Collisions of highly charged ions on surfaces have attracted a lot of attention during the last years mainly for two reasons. First of all because of the formation of so-called "hollow atoms" which are intriguing objects from an atomic physics point of view, and secondly because of the envisaged possibilities of surface modifications. Regarding the latter point one hopes to exploit the high potential energy of HCI to disturb the electronic structure in such a way that also the geometrical structure is modified or that particles are desorbed from the surface. A most appealing feature hereby is the fact that for sufficiently low kinetic projectile energies the modifications can be limited to the topmost layer, without damage in deeper layers. In order to reach that goal one needs a profound understanding of the electronic processes, which take place during the interaction of various surfaces.

Metal surfaces

Especially for metal surfaces a rather good understanding has evolved on the basis of the classical over-the-barrier model[3]. An artists view of the most relevant processes is shown in FIGURE 2. As the HCI approaches the surface it is attracted by its image charge. At a critical distance $R_c=(2q)^{1/2}/W$ (with q the projectile charge and W the work function of the surface) the potential barrier has sufficiently dropped to allow an over-the-barrier transfer of electrons from the conduction band into resonant projectile levels. As a rule of thumb one can assume that transfer proceeds into orbitals with principal quantum number $n \approx q$. Up to that point the projectile has already gained a kinetic energy of $\Delta E(R_c)=0.25Wq^{3/2}/2^{1/2}$. In the following time-interval until 'touch-down' on the surface – which due to the image acceleration is generally less than about 100 fs – a large variety of electronic processes takes place. Additional electrons are captured, Auger and Coster Kronig transitions occur, resulting in the emission of fast and slow electrons, electronic states are promoted to energies above the Fermi level due to the image shift, leading to resonant ionization, i.e. re-feeding of electrons into unoccupied states of the conduction band, and – dependent on the atomic number Z of the projectile - also radiative transitions may occur. The net effect of all these processes is a more or less continuous shrinking of the electron cloud under simultaneous emission of electrons and photons. Eventually the projectiles dive into the electron sea of the metal, where the conduction band electrons are rearranged to mimic an atomic shell with the corresponding binding energy [11]. Mostly the inner shell vacancies of the projectiles are still present at this time, since typical Auger decay rates are too low to allow a complete relaxation during the approach of the HCI to the surface. Sometimes the continuous rearrangement of the electron cloud is somewhat artificially divided into the formation of hollow atoms of the first, second and third generation 'above', 'at' and 'below' the surface respectively. However there are no 'sudden' processes separating these regions from each other.

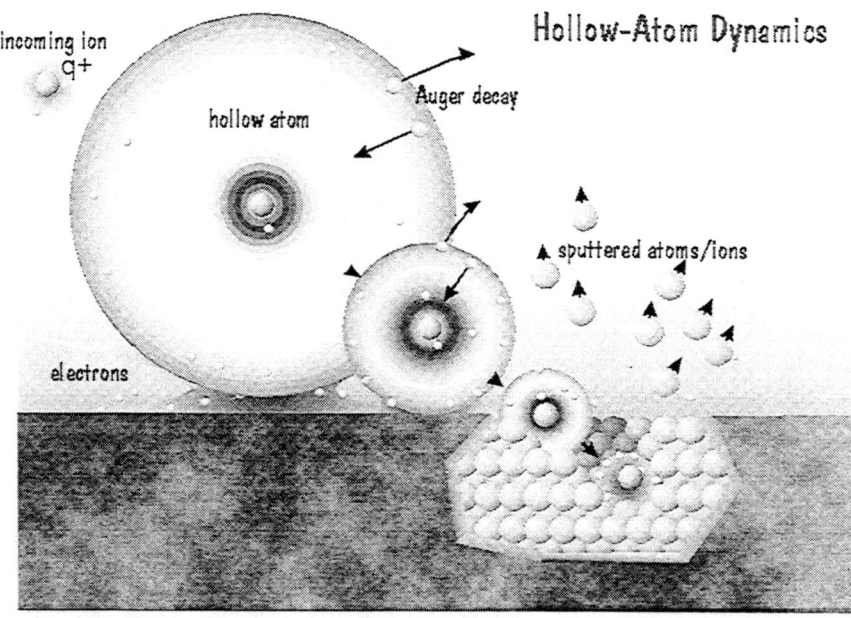

FIGURE 2. An artist's view of processes induced by the approach of a highly charged ion to a surface.

The most quantitative proof for the scenario sketched above comes from energy gain measurements. Winter et al[12] and Meyer et al[13] have e.g. measured the scattering angles of projectiles, grazingly incident on very flat metal surfaces. They found a characteristic deviation from specular reflection, which is due to the fact that the projectiles are bent towards the surface on the incoming part of the trajectory only. On the outgoing part there is no deflection because the projectile – being neutralized – does not induce an image charge anymore. The energy gains derived from this deviation from specular reflection are in excellent agreement with the value of $\Delta E=(1/3)Wq^{3/2}/2^{1/2}$, predicted by the over-the-barrier model.

Several methods have been exploited to obtain more detailed information about the electronic processes taking place between capture of the first electron and the formation of the subsurface hollow atom. For low-Z projectiles (Z<18) radiative transitions play a minor role and most detailed information is contained in the multiplicity- and energy distributions of electrons emitted during Auger cascades. The first steps in these cascades, corresponding to transitions between outer orbitals, supposedly give rise to low-energy electrons, whereas the last steps – KLL transitions in case of bare or hydrogenlike projectile ions – give rise to electrons of several hundred eV.

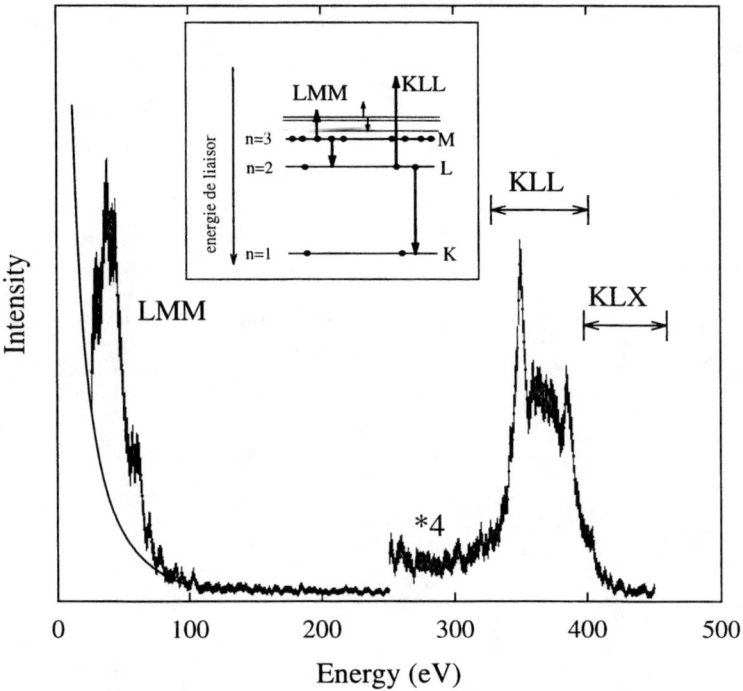

FIGURE 3. Energy spectum of electrons resulting from N^{6+} + Ni(110) collisions.

Regarding the 'early' steps, information can be gained from the low energy electrons[14], which are at least partly emitted during a rearrangement in the outer shells. Niemann et al[15] have measured energy distributions of such electrons. At low collision energies (e.g. for N^{6+} on Au at 90 eV) the electron energy distributions and absolute intensities were found to be in fair agreement with the predictions of the classical over-the-barrier model, whereby only 'above-surface' emission processes were taken into account. With increasing collision energies however the measured intensities were significantly higher than the predicted ones, implying that a considerable fraction of low-energy electrons is emitted below the surface. Lemell et al[16], investigating 450 eV Ar^{8+} on Au collisions by measuring electron emission statistics in coincidence with projectiles at well-defined scattering angles, found indeed that high multiplicities of slow electrons are only observed for subsurface trajectories. These electrons may partly be due to Coster Kronig transitions[17], whereby a rearrangement of L-shell electrons is accompanied by the emission of electrons from outer shells or from the conduction band. As a remarkable result however Lemell et al found that even the yield of electrons emitted above the surface can not completely be accounted for when theoretically calculated Auger rates are used in the model calculations. Therefore the question arises whether there are mechanisms which could speed up the emission of low energy electrons. We will come back to this point in connection with HCI collisions on C_{60}.

The last steps in the Auger cascades can again be investigated be analyzing the emitted electrons. A typical energy spectrum of such electrons arising from N^{6+} collisions on a Ni(110) surface is shown in FIGURE 3 (Limburg et al[18]). One can clearly distinguish the LMM and KLL Auger electrons around 50 eV and 380 eV respectively. A comparison with Hartree Fock atomic structure calculations[19] shows that the sharp peaks at the low- and high energy side of the KLL structure are due to transitions from a completely neutralized projectile with 2 and 5 electrons in the L-shell respectively. A Doppler shift analysis moreover shows that the 350 eV peak can be ascribed to emission processes above the surface[20]. At higher collision energies, i.e.

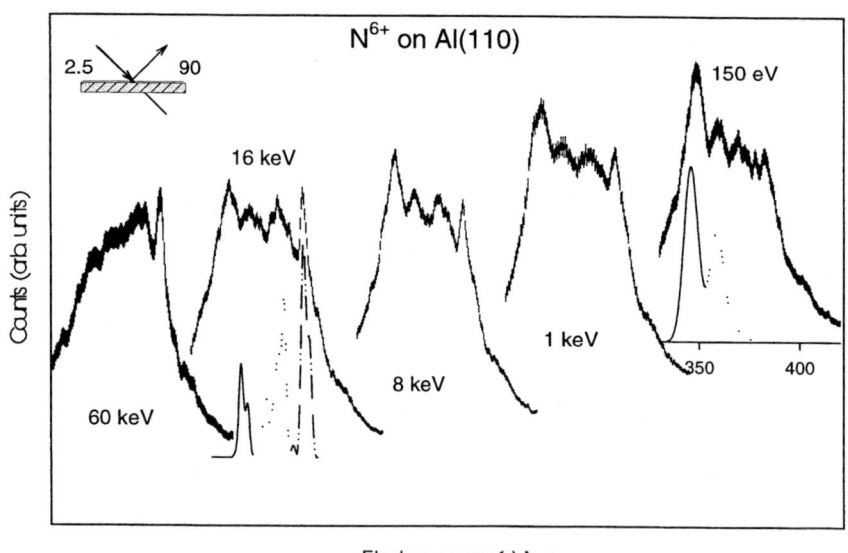

FIGURE 4. KLL part of electron spectra at various collision energies.

when less time is available between first capture and 'touch down', this peak disappears. This is illustrated in FIGURE 4 where the KLL part of energy spectra resulting from N^{6+} collisions on Al(110) at different energies is shown.

In the past there was a long lasting discussion about the question, via which mechanisms the electrons arrive in the L-shell. Andrä[21] pointed out very early that normal Auger rates are too slow to explain L-shell population. Folkerts and Morgenstern[22] concluded from a comparison of LMM and KLL Auger intensities, that in addition to Auger cascades a 'side-feeding' mechanism must be active to fill the L-shell. Limburg et al[23] and Köhrbrück et al[24] developed cascade models in which not only Auger cascades but also also direct L-shell filling processes in binary projectile-target atom collisions were taken into account. By using Auger rates calculated by Diez Muino et al[25] (for more recent values see Diez Muino et al[26]) they could satisfactorily reproduce the measured electron energy spectra.

The fact that the various electronic processes at metal surfaces are reasonably well understood opens the possibility for applications, in which HCI are used as probes, e.g. to study surface magnetism. In the KLL Auger electron spectra one can clearly identify peaks due to the decay of singlet and triplet states respectively. If at a ferromagnetically ordered surface preferentially electrons of one spin direction are present, a preferential population of triplet states should take place. Since – dependent on the angle of impact and the collision energy – electrons are captured from a very limited area of a few Angstroms only, a short range ordering of the surface electrons could in this way be detected. In that respect this method could have a significant advantage as compared to other methods.

Insulating surfaces

The various experiments with metal surfaces have led to a reasonable understanding of the relevant processes. However, from a point of view of surface modifications metal surfaces are the least suitable targets: electronic relaxation is much too fast to allow a reconstruction of the surface due to the intermediate disturbance of the electronic structure. For many groups this was one of the reasons to investigate semiconducting or insulating surfaces. Whereas semiconductor surfaces turned out to behave very similarly as metal surfaces[18], insulators show characteristic differences. Remarkably enough though, experimentally determined energy gains for HCI approaching insulating surfaces[27] were nearly the same as those for metals. In an extension of the classical over-the-barrier model, which takes the various additional effects at insulator surfaces into account, Hägg et al[28] could solve this puzzle. They showed that the similarity of energy gains is due to two counteracting effects. The smaller electronic response of LiF reduces the effective image charge. On the other hand the "switch off" of the image charge is delayed, i.e. happens closer to the surface, due to the higher binding energy of the electrons (12 eV instead of ca 5 eV for metals). Therefore a reduced image charge acts over an increased pathlength, leading to nearly the same energy gains.

A similar treatment of processes in front of insulating surfaces, called 'dynamical classical over-the-barrier model', was given by Ducrée et al[29]. FIGURE 5 shows a comparison of measured and calculated energy gains for HCI in front of LiF and KI surfaces respectively. Two theoretical curves are given in both cases, based on the two respective assumptions that capture induced surface charges are compensated instantly (q=0) or only after the projectile has receded. The assumption of q=0 seems to be more appropriate for LiF, and q>0 for KI respectively. Future investigations will have to clarify the reason for these differences.

More pronounced differences between metal- and insulator surfaces are seen in electron energy spectra. Limburg et al[30] found that the sharp peaks, observed for metal surfaces and ascribed to above-surface electron emission, are absent for the LiF target (see FIGURE 6a). This triggered a lively discussion[31] whether at all hollow atoms are formed above insulator surfaces, and which of the different properties of LiF as compred to metals are responsible for the spectral change.

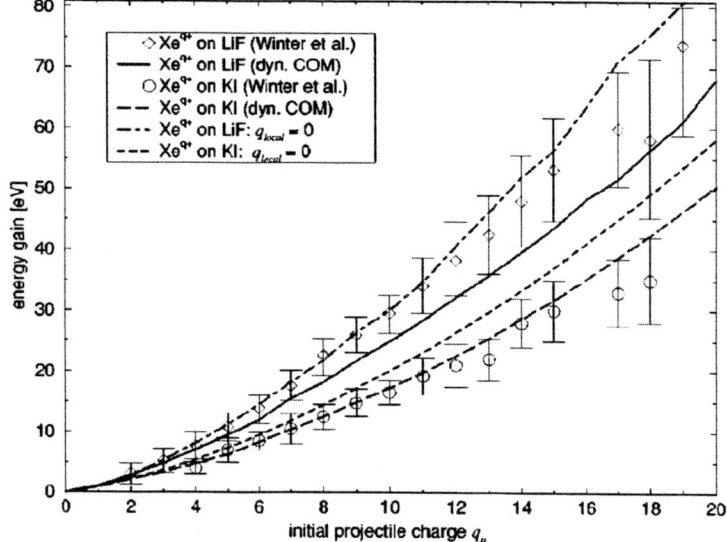

FIGURE 5. Energy gain of ions in front of insulator surfaces (from ref 29).

When replacing Au by LiF, one does not only change the electric conductivity of the target, but at the same time also the binding energy of the electrons, the electronic band structure and the surface density of states. All these properties may influence the neutralization dynamics. In order to identify the most relevant parameters Khemliche et al[32] investigated Au targets with a varying coverage of LiF up to one monolayer. Corresponding spectra are shown in FIGURE 6b. The spectra from monolayer covered surfaces resemble closely those of bulk LiF targets and Khemliche et al concluded from this that the bandstructure – which is not yet developed at a monolayer coverage

– can not be responsible for the modified spectral features. Rather the higher electron binding energy in LiF of ca 12 eV seems to be the most influential parameter.

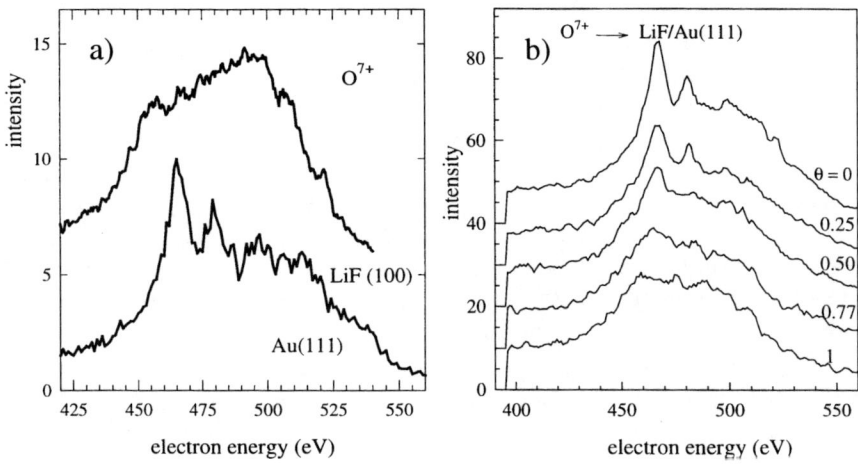

FIGURE 6. a) Comparison of spectra from Au and LiF , b) spectra from LiF covered Au surfaces with coverage up to 1 monolayer

Surface modifications

An enhanced sputtering or desorption of surface particles driven by the potential energy of HCI has long been discussed, especially in view of possible applications, aiming at a modifications of surface layers without bringing damage to deeper layers. In the past atomic displacements due to the potential energy of very highly charged ions were observed[33] at insulating mica surfaces. However, clear evidence for 'potential sputtering' has only been seen recently. One of the major problems to quantitatively observe such a sputtering is the fact that most of the sputtered particles are neutral[34]. Mainly two methods have successfully been applied to tackle this problem. The group. Neidhart et al[35] and Sporn et al[36] used a quartz crystal microbalance to measure the decrease of LiF of SiO_2 upon bombardment with various Ar and Xe ions (up to $q=14$ and 27 respectively). Schenkel et al[37] on the other hand used catcher foils, which were placed in front of a GaAs target. Ablation rates were subsequently determined from catcher surface coverages of Ga and As. Results of these experiments are shown in FIGURE 7. Clearly the sputtering yield increases with the charge, i.e. the potential energy of the projectiles, whereas the kinetic energy has only a minor effect. Schenkel et al have also measured the yields of ionized particles, and as can be seen these yields are indeed two orders of magnitude lower than those for sputtered neutrals.

Of course now the question arises, which mechanisms are responsible for this potential sputtering. The most intuitive picture is that of Coulomb explosion, caused by an electron depletion of an extended volume which is expected to increase with the projectile charge[38]. Cheng and Gillaspy[39] have performed molecular dynamics simulations for a region of about 300 Si⁺ ions at a Si[111] surface, assuming that reneutralization of the ions is neglibible during the explosion time. For the examples shown in FIGURE 7 however, Coulomb explosion can be excluded. First of all the reneutralization rates are such that the ions will be neutralized long before the explosion has taken place. Also Coulomb explosion implies that most of the sputtered particles are ejected as ions. Especially the measurements of Schenkel et al shown above clearly demonstrate that this is not the case.

For the sputtering of alkali halides and silicon oxide Aumayr et al[40] have shown that the model of defect mediated desorption, originally developed for electron stimulated sputtering, is most suitable to explain the observations. In this model it is assumed that defects like electron-hole pairs produced by the primary process become localized, so-called self-trapped excitons. These decay into colour centers, which diffuse to the surface and lead to evaporation of neutrals.

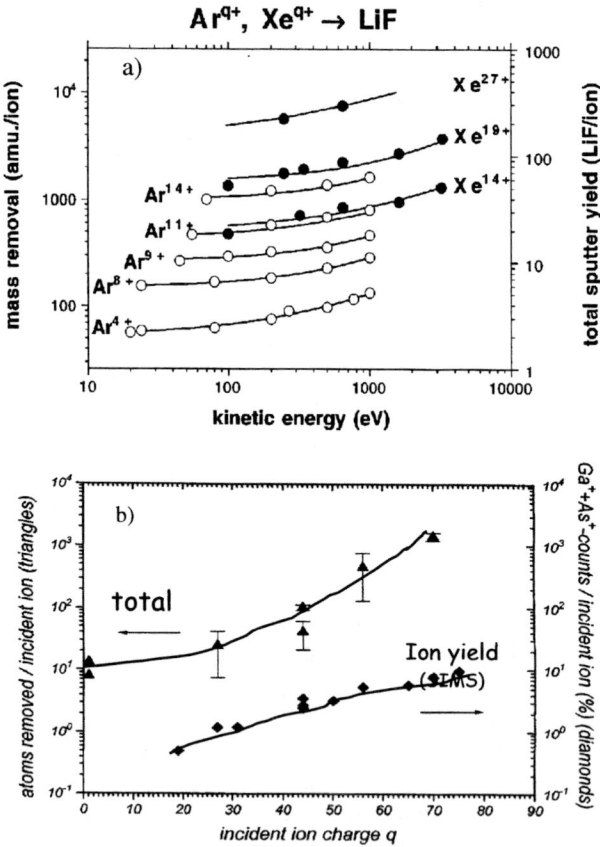

FIGURE 7. Sputtering yields of a) Sporn et al (ref.36) and b) Schenkel et al.(ref.37. modified figure)

For GaAs this model is not applicable and Schenkel et al discuss that high density electronic excitations – similarly as in laser ablation – reduces the structural stability sufficiently to result in a sputtering of neutrals.

In an interesting application Hamza et al[41] have recently exploited the fact that a large number of atoms – up to 1400 in the example shown in FIGURE 7b – is sputtered from a very small area (with a diameter of only about 40 nm) upon impact of a single HCI. They investigated a SiO_2 microchip containing a tungsten pattern on its surface. By means of coincident detection of several sputtered particles they could show that SiF contaminations were only sputtered in conjunction with tungsten particles. In this way the contaminations could clearly be localized near the W deposits.

FULLERENES AND CLUSTERS

Generally, one of the major problems regarding the observation of hollow atoms is the fact that in a typical surface scattering experiment they are destroyed upon touch down on the surface, which – due to the image charge acceleration – happens less than typically 100 fs after their creation. Yamazaki et al[42] have tried to avoid this problem by transmitting HCI through small capillaries. A certain fraction of the projectiles is then expected to approach the walls sufficiently to form hollow atoms, but - especially near the end of the capillary – leave it before they are destroyed. X-ray spectra observed in these experiments were in fact dominated by emission from transmitted metastable ions formed in the capillaries. Another possibility to avoid the destruction of hollow atoms is to form them during distant collisions with clusters or fullerenes.

During the last years C_{60} fullerene molecules have become fashionable as targets for HCI collisions because they bridge the gap between atoms and surfaces. They resemble atomic targets because they can supply only a limited number of electrons, and the ionization potential increases monotonically with each additional electron captured. Capture processes will therefore preferentially take place sequentially. On the other hand they exhibit properties similar to a solid surface because a practically unlimited supply of electrons is available, and the ionization potentials are significantly lower than for atoms. Therefore on the average more electrons will be transferred, and they will end up in higher orbitals. 'Hollow atom' formation - mainly discussed for solid targets – can thus also take place in HCI collisions on C_{60}. Moreover, due to their polarisability C_{60} molecules build up an image charge[43], which in the incoming channel causes an acceleration of the collision partners towards each other. This is reflected in the scattering angle and can be exploited to identify the impact parameter at which a certain collision took place. Last but not least a theoretical description of the electron dynamics is easier for C_{60} than for a solid, and consequently a large number of theoretical papers has been devoted to HCI collisions on C_{60}.

The most frequently used method to investigate such collisions experimentally is a time-of-flight mass spectrometric measurement of the ionized target particles and their fragments. FIGURE 8 shows such a spectrum[44] obtained for 12.5 keV collisions of O^{5+} on C_{60}. We can distinguish three different groups of particles: (i) the intact C_{60} cages in different charge states, i.e. the most prominent peaks in FIGURE 8. (ii) the so-called evaporation peaks $(C_{60-2m})^{q+}$, which result from an evaporation of C_2 units from $(C_{60})^{q+}$ or C_2^+ units from $(C_{60})^{(q+1)+}$, and finally (iii) the small fragments C^+, C_2^+, C_3^+ etc (at the left side of the spectrum), which mainly result from a more or less complete destruction of the C_{60} cage. By means of coincidence measurements it was recently shown that these groups of fragments can be ascribed to specific collision events at well defined impact parameters. The intact $(C_{60})^{q+}$ ions e.g. are mainly due to 'soft' collisions at large impact parameters where q electrons are captured into highly excited projectile orbitals without any energy transfer to the molecule. Evaporation- and fragmentation peaks on the other hand are mainly due to close collisions, in particular those during which the projectile penetrates through the C_{60} cage. We will now in the first place discuss the 'soft' collisions and the corresponding electron dynamics, and thereafter the 'penetrating' collisions.

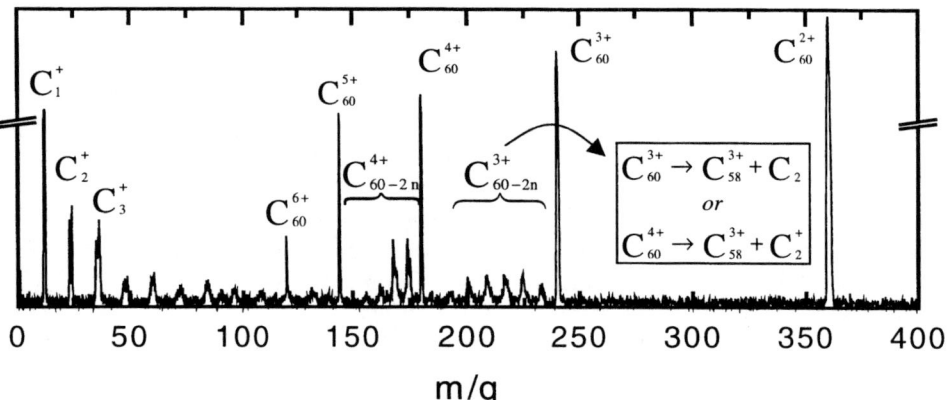

FIGURE 8. Time-of-flight mass spectrum of C_{60} fragments, obtained from collisions of O^{5+} ions on C_{60}.

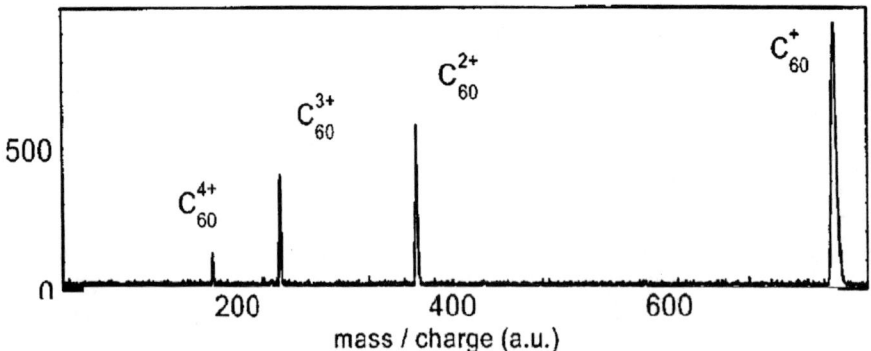

FIGURE 9. Mass/charge spectrum of fullerene ions, produced in 16 keV Ar^{8+}/C_{60} collisions[45], measured in coincidence with Ar^{7+} scattered projectiles (fig 2 in Opitz et al, ref 45).

'Soft' collisions can experimentally be selected by means of coincidence measurements. FIGURE 9 shows a fragment spectrum[45] resulting from 16 keV Ar^{8+} collisions on C_{60}, taken in coincidence with charge exchanged Ar^{7+} projectiles. Such a coincidence measurement does by no means imply that only one electron has been captured: from the measured target charge state it is obvious that up to 4 electrons have initially been captured. However, for large impact parameters electrons are preferably captured into highly excited orbitals, from where all but one are subsequently re-emitted during Auger processes[46]. Such collision events are just selected by looking for coincidences with Ar^{7+}. In this way it is guaranteed that only 'soft' collisions at large impact parameters contribute to the spectra. During the gentle removal of electrons from the C_{60} apparently little or no energy is transferred to the fullerene cage and only $(C_{60})^{q+}$ peaks occur in the corresponding spectra. Collisions with HCI are obviously an excellent tool for 'gentle' electron removal. So it is not amazing that the most highly charged C_{60} ions observed so far were produced by HCI impact[47,44]. Plagne and Guet[48] have recently shown theoretically that with sufficiently highly charged ions it should be possible to ionize metal clusters close to their Raleigh charge limit without significant heating.

During such 'soft' collisions the C_{60} behaves more or less like an atomic target. More detailed information about such collisions can therefore be obtained similarly as for atoms by measuring the energy gain and the angular distribution of the corresponding projectiles. This has recently been done by Selberg et al[49] and by Walch et al[50] for Ar^{8+} on C_{60} collisions. Especially the energy gain measurements provide insight into the electron dynamics during the collision. FIGURE 10 shows an energy gain spectrum measured by Selberg et al[49] for collisions of Ar^{8+} on C_{60}. It exhibits a remarkable peak structure which so far has only partly be interpreted in a quantitative way[51,52].

In the following we will suggest a very straightforward mechanism to explain the various peak positions. An essential assumption for this is based on the experience of HCI-surface collisions, namely that Auger processes are very fast and that a considerable fraction of Auger electrons is already emitted during the collision. If we make the assumption – indeed a crude one – that at the critical radius where a certain number of r electrons has been captured, instantaneously a certain number of n electrons are re-emitted so that only $s=r-n$ remain with the projectile, we can easily calculate the corresponding energy gain resulting from the Coulomb repulsion. The first prominent peak is clearly due to single electron capture into the $7s$ orbital (with the shoulder on the lower energy-gain side resulting from capture into $7l$ with higher l values. The peaks correponding to higher energy gains can all be assigned to collision events characterized by charge pairs $(r,q-s)$ indicating the effective charges r of the C_{60} target and $q-s$ of the projectile. The only fit-parameters to arrive at the peak positions indicated in the figure, are critical distances of R= 20, 17.9, 15.2 and 10.7 a.u. for transferring 2, 3, 4, and 5 electrons respectively. One might argue that these values are not in agreement with those obtained by other methods. However, since in the past various approaches have yielded quite different sets of critical radii (see e.g. table 1 given by Opitz et al [45]) one should at present not take these values too seriously.

This also implies that it is still too early to answer the question whether the remaining charged $(C_{60})^{q+}$ behaves as a conducting sphere or whether the charges are localized – either at the side where the projectile is passing or on the opposite side[43].

It should be noted that a variation of the Auger decay times involved has no major influence on the various peak positions. For the extreme cases of very low or very high decay rates one finds peaks corresponding to the energy gains of either q_1q_2/R or $q_1(q_2+1)/R$ respectively. For intermediate decay rates one finds peaks at both of these positions! A variation of the decay rate mainly influences the relative height of these peaks.

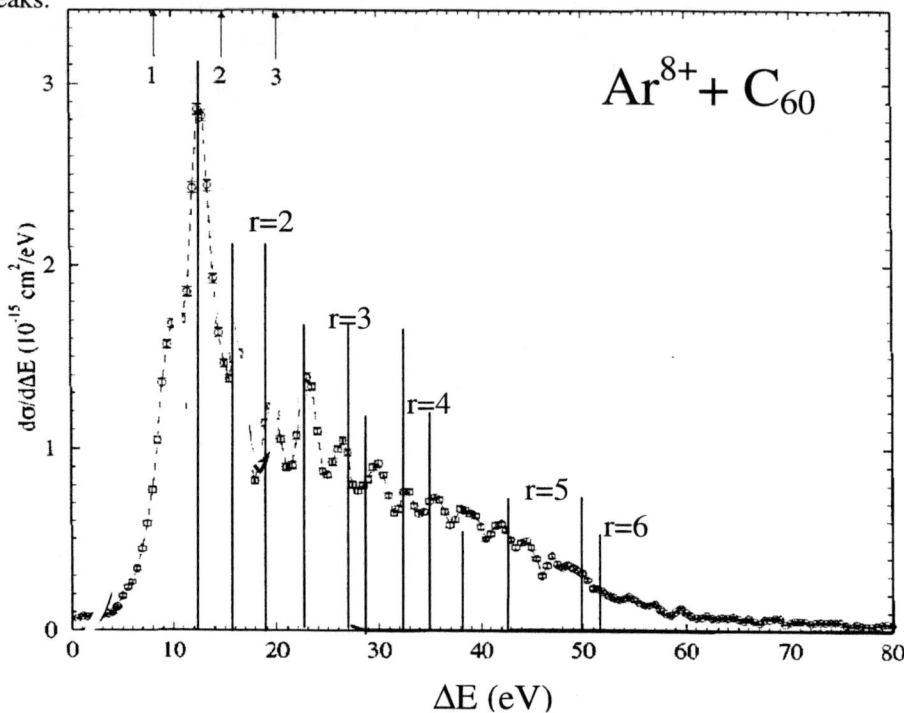

FIGURE 10. Energy gain spectrum of $Ar^{8+} + C_{60}$ measured by Selberg et al[49]. The bars indicate peak positions as expected from Coulomb repulsion (see text).

The occurrence of the large number of peaks in FIGURE 10 can therefore be taken as an indication that Auger processes take place at relatively small internuclear distances, i.e. during the collision. This is in fact in good agreement with the finding of Martin et al[53] that the number of electrons 'active' during the collision is much higher than the projectile charge q: for $Xe^{25+} + C_{60}$ collisions at $v = 0.38$ *a.u.* they found e.g. a number of 60 active electrons. It is also reminiscent of the extremely high electron yields observed for ion surface collisions, and it clearly implies that Auger processes have to take place at such small internuclear distances that subsequent additional capture processes are still possible. According to traditional estimates Auger rates are significantly lower and most of the electrons are expected to be emitted only after the collision partners have receded to very large internuclear distances (see e.g fig. 3 of ref.52). Such estimates, e.g. by Thumm[54] are generally based on Hartree Fock calculations, assuming electrons in stationary states, in which electron correlation plays only a minor role. Although these calculations predict correctly the formation of highly unstable hollow ions, which lose nearly all of their electrons via Auger processes, the predicted decay times are much too long! At this moment we can only speculate that electrons captured during the collision follow a highly correlated motion with a much stronger mutual interaction and therefore significantly increased Auger decay rates.

One could of course try to further check the suggestions above by calculating the angular distributions and comparing the results with corresponding measurements of Walch et al[50]. However, such a comparison is by no means straightforward, because the influence of the image charge acceleration on the incoming part of the trajectories has to be taken into account[50]. The exact magnitude of the image charge acceleration is not known because the electronic response of a C_{60}^{q+} on a sub-fs scale is not well known.

'Penetrating' collisions will not only result in ionization of the fullerene, but also cause vibrational excitation due to elastic collisions between the nuclei and to electronic excitation during inelastic interactions with the electron cloud. Often this will lead to a destruction of the C_{60} cage and one of the interesting questions is, how this destruction takes place and whether it can be influenced by the way in which energy is deposited into the molecule. To investigate these questions Schlathölter et al[55] have investigated the collision energy dependence of the fragmentation patterns for He^+ - C_{60} collisions. They found that with increasing collision velocity the 'small fragments' (see FIGURE 8) were increasing, whereas the 'evaporation' peaks were decreasing.

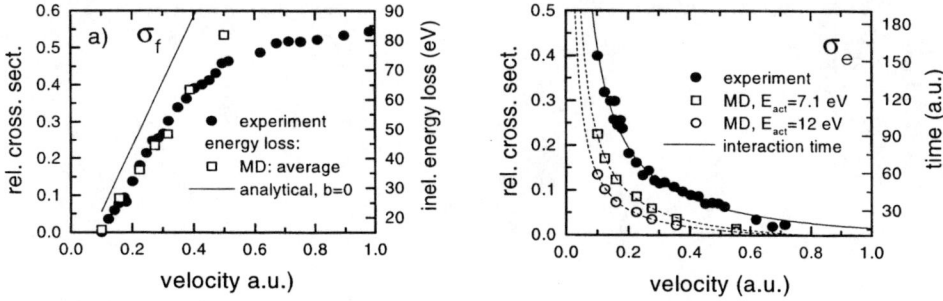

FIGURE 11. Comparison of calculated cross sections with measured signals. for a) 'evaporative' peaks; b) multi-fragmentation into small fragments. Full points always give experimental data, open points reflect the molecular dynamics calculations (from ref. 58)

In order to understand this behavior they performed molecular dynamics calculations in which the energy transfer to the C_{60} was modeled. Transfer of vibrational energy was calculated by a Monte Carlo method in which trajectories of the projectile and all 60 carbon atoms were followed by a numerical solution of the classical Newton equations, using realistic three-body potentials[56] for the fullerene and a screened Coulomb potential for the He-C interaction[57]. For the transfer of electronic energy a friction model was used which was originally developed to describe electronic stopping of nuclei in solids. From these calculations it was found that the transfer of vibrational energy via elastic collisions exhibits the same trend as the integral over the 'evaporation' peaks, and that the electronic excitation is more or less in pace with the 'small fragment' signals. This is shown in FIGURE 11, where experimental data are compared with molecular dynamics calculations. The similar collision energy dependence of cross sections on the one hand and energy deposition into the fullerene according to the various mechanisms on the other hand suggests that the molecule 'remembers' the excitation path, implying that fragmentation can be steered!

A remarkable feature of the friction model is the fact that the inelastic energy loss depends in an oscillatory way on the atomic number of Z of the projectile: around Z-values corresponding to rare gas atoms (Z=10, 18, ...) the inelastic loss shows a minimum.

If energy deposition into the electronic degree of freedom is in fact responsible for the destruction of the C_{60} cage structure, one should be able to observe also a projectile-Z dependent oscillation in the yield of small fragments. Corresponding measurements have been performed by Hadjar et al[58], and the result is shown in FIGURE 12. We see in fact a characteristic oscillatory variation of the small-fragment yield, with minima around Z=10 and Z=18.

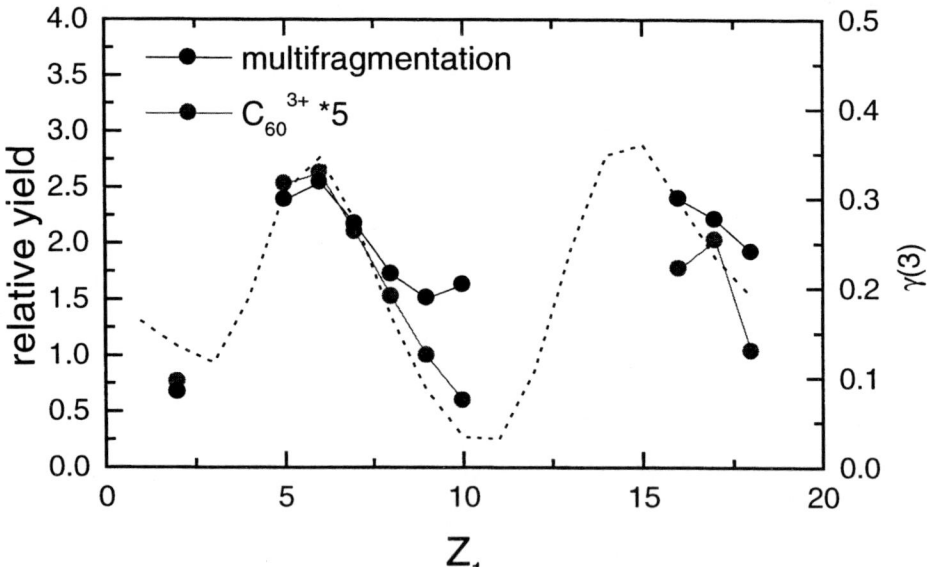

FIGURE 12. Z-dependent oscillations in the yield of small fragments

Regarding the so-called 'evaporation peaks' Martin et al [59,60] performed coincidence studies which clarified the relevant fragmentation pathways. In particular they could answer the question, whether the $(C_{60-2n})^{q+}$ fragments are due to evaporation of neutral C_2 units from a $(C_{60})^{q+}$ mother ion, or due to an asymmetric fission of $(C_{60})^{(q+1)+}$ into charged fragments $(C_{60-2n})^{q+}$ and C_2^+. Some fragment spectra, taken in coincidence with a well defined number of ejected electrons are shown in FIGURE 13. One can see that $(C_{60-2n})^{3+}$ predominantly decays via neutral evaporation and only to a small fraction via asymmetric fission, while $(C_{60-2n})^{4+}$ predominantly decays via asymmetric fission. With increasing charge of the parent C_{60}^{q+} ion however asymmetric fission was found to become more important and to be even dominant for $q > 5$.

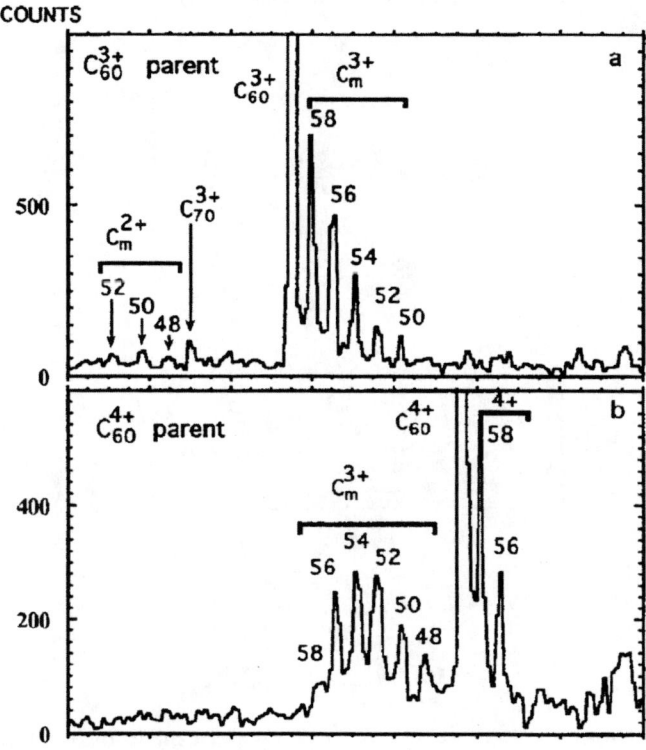

FIGURE 13. Coincidence spectra of evaporation peaks (from fig 3 in Martin et al, ref.59)

Moreover these coincidence measurements allow to observe a delayed decay of the vibrationally excited fullerenes. A scatter plot of coincidences between $(C_{60-2n})^{q+}$ and small carbon fragments is shown in FIGURE 14. One can clearly observe the 'tails' which indicate a longer time of flight for the small fragments and a shorter one for the $(C_{60-2n})^{q+}$ ions - a fact which can be ascribed to the fact that these fragments formed a complex in the first phase of their acceleration. From these data Martin et al[59] estimated a decay time in the order of 100 ns. This value exceeds the typical vibration times of a fullerene (e.g. period of the breathing mode ~ 0.1 ps) by several orders of magnitude. On the other hand, delayed fragmentation is known to occur up to several μs after an excitation process. To obtain a deeper understanding of the underlying mechanisms a coincidence measurement of the above mentioned decay times with the kinetic energy or the scattering angle of the projectile would be helpful. Such an experiment would unambiguously link the excitation energy to fragmentation processes and respective decay times, thus giving a deep insight into the interaction dynamics.

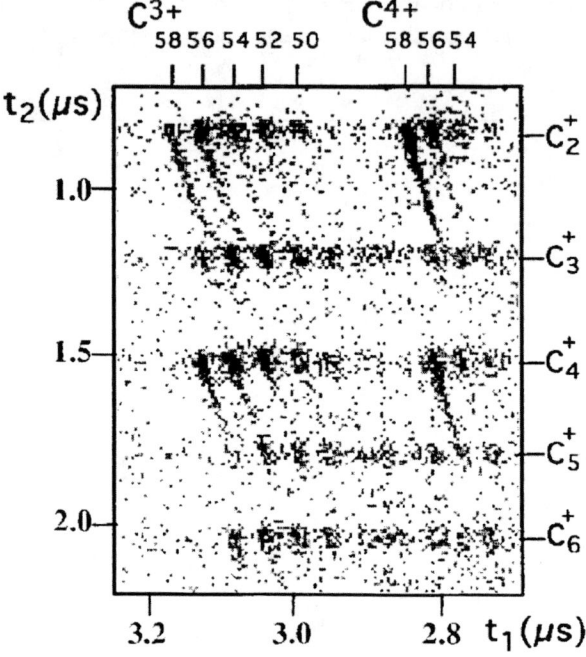

FIGURE 14. Scatterplot of C_{60} fragments (from ref.59)

CONCLUSIONS

Since the previous ICPEAC large progress has been made in the understanding of electron dynamics during collisions of highly charged ions on various targets. Especially the use of C_{60} targets has shed new light on this field. It can be foreseen that in the coming years collisions of highly charged ions are not only a topic of scientific research, but that they will be exploited for applications such as investigating surface magnetism or modifying surfaces in a controlled way.

ACKNOWLEDGEMENT

The authors acknowledge the support of their collegues, especially the PhD students who performed most of the experimental work in Groningen. The work performed in

Groningen is part of the research program of the "Stichting voor Fundamenteel Onderzoek der Materie" (FOM). Th. Schlathölter is grateful for a Marie-Curie fellowship of the EU (ERBFMB1CT961704)

REFERENCES

[1] A. Bárány, G. Astner, H. Cederquist, H.Danared, S. Huldt, P. Hvelplund, A. Johnson, H. Knudsen, L. Liljeby, K.-G. Rensfeldt, Nucl. Instr. Meth. B **9**, 397 (1985)

[2] A. Niehaus, J.Phys.B: At. Mol.Opt. Phys, **19**, 2925-2937 (1986)

[3] J. Burgdörfer, P. Lerner, F.W. Meyer, Phys.. Rev. A **44**, 5647 (1991)

[4] C.D. Lin: Review of Fundamental Processes and Applications of Ions and Atoms (World Scientific, Singapore, 1993)

[5] H. Cederquist, A. Fardi, K. Haghihat, A. Langereis, H.T. Schmidt, S.H. Schwartz, Phys. Scripta **T80** 46-51(1999)

[6] Arnau et al, Surface Science Reports **27**, 113-240 (1997)

[7] J. Ullrich, R. Moshammer, R. Dörner, O. Jagutzki, V. Mergel, H. Schmidt-Böcking, L. Spielberger, Topical Review, J.Phys.B: At.Mol.Opt. Phys, **30**, 2917 (1997)

[8] M.A. Abdallah, W. Wolff, H.E. Wolf, E. Sidkey, E.Y. Kamber, M. Stöckli, S.D. Lin, C.L. Cocke, Phys.Rev. A **57** (1998) 4373-4378), Abdallah et al , ibid (1998)

[9] M.A. Abdallah, W. Wolff, H.E. Wolf, E.Y. Kamber, M. Stöckli, C.L. Cocke, Phys.Rev. A **58**, 2911-2919 (1998)

[10] J.-Y. Chesnel, B. Sulik, H. Merabet, C. Bedouet, F. Fremont, X. Husson, M. Grether, A. Spieler, N. Stolterfoht, Phys. Rev. A **57**, 3546-3553 (1998)

[11] A. Arnau, P. Zeijlmans van Emmichoven, J.I. Juaristi, E. Zaremba, Nucl. Instr. Meth. B **100**, 279 (1995)

[12] H. Winter, C. Auth, R. Schuch, E. Beebe, Phys. Rev. Lett. **71**, 1939-1942 (1993)

[13] F.W. Meyer, L. Folkerts, H.O. Folkerts, S. Schippers, Nucl. Instr. and Meth. B **98** 441-444 (1995)

[14] HP. Winter, F. Aumayr, Topical Review, J.Phys.B: At.Mol.Opt.Phys. **32**, R39-R65 (1999)

[15] D. Niemann, M. Grether, A. Spieler, N. Stolterfoht, C. Lemell, F. Aumayr, HP. Winter, Phys. Rev. A**56**, 4774-4780 (1998)

[16] C. Lemell, J. Stöckli, J. Burgdörfer, G. Betz, HP. Winter, F. Aumayr, Phys.Rev. Lett.**81**, 1965-1968 (1998)

[17] J. Limburg, J.Das, S. Schippers, R. Hoekstra, R. Morgenstern, Phys. Rev. Lett.**73**, 786-789 (1994)

[18] J. Limburg, J.Das, S. Schippers, R. Hoekstra, R. Morgenstern, Surf. Sci. **313**, 355-364 (1994)

[19] S. Schippers, J. Limburg, J.Das, R. Hoekstra, R. Morgenstern, Phys. Rev. A **50**, 540-552 (1994)

[20] J. Das, R. Morgenstern, Phys.Rev.A **47**, R755-R758 (1993)

[21] H.J. Andrä, Nucl.Instr. Meth. B **43**, 306 (1989)

[22] L. Folkerts and R. Morgenstern, Europhys.Lett.**13**, 377 (1990)

[23] J. Limburg, S. Schippers, I. Hughes, R. Hoekstra, R. Morgenstern, S. Hustedt, N. Hatke, W. Heiland, Phys.Rev. A **51**, 3873-3882 (1995)

[24] N. Stolterfoht, A. Arnau, M. Grether, R. Köhrbrück, A. Spieler, R. Page, A. Saal, J. Thomaschewski, J. Bleck-Neuhaus, Phys. Rev. A **52**, 445-456 (1995)

[25] R. Diez Muino, N. Stolterfoht, A. Arnau, A. Salin, P. Echenique, Phys. Rev. Lett. **76**, 4636 (1996)

[26] R. Diez Muino, A. Salin, N. Stolterfoht, A. Arnau, P. Echenique, Phys. Rev. A **57**, 1126 (1998)

[27] Ch. Auth, H. Winter, Phys. Lett A **217**, 119 (1996)

[28] L. Hägg, C.O. Reinhold, J. Burgdörfer, Phys.Rev. A **55**, 2097-2108 (1997)

[29] J.J. Ducrée, F. Casali, U. Thumm, Phys.Rev. A **57**, 338-350 (1998)

[30] J. Limburg, S. Schippers, R. Hoekstra, R. Morgenstern, H. Kurz, F. Aumayr, HP. Winter, Phys.Rev.Lett. **75**, 217 -220 (1995)

[31] J.-P. Briand et al, Phys. Rev. Lett. **77**, 1452 (1996), Aumayr et al, Phys.Rev. Lett. **76**, 2590 (1997) and Briand et al, Phys.Rev. Lett. **79**, 2591 (1997)

[32] H. Khemliche, T. Schlathölter, R. Hoekstra, R. Morgenstern, S. Schippers, Phys. Rev. Lett. **81**, 1219 (1998)

[33] D. Schneider, M.A. Briere, M.W. Clark, J. McDonald, J. Biersack, W. Siekhaus, Surf. Sci. **294**, 403-408 (1993)

[34] S.T. de Zwart, T. Fried, D.O. Boerma, R. Hoekstra, A.G. Drentje, A.L. Boers, Surf. Sci. **177**, L939-L946 (1986)

[35] T. Neidhart, F. Pichler, F. Aumayr, HP. Winter, M. Schmid, P.Varga, Phys.Rev. Lett. **74**, 5280-5283 (1995)

[36] M. Sporn, G. Libiseller, T. Neidhart, M.Schmid, F. Aumayr, HP. Winter, P.Varga, Phys.Rev. Lett. **79**, 945-948 (1997)

[37] Schenkel et al, Phys.Rev. Lett. **81**, 2590 (1998)

[38] I. Bitenskii, E. Parilis, S. Della-Negra, Y. LeBeyec, Nucl. Instum. Meth. Phys. Res. B **72**, 380 (1992)

[39] H.-P. Cheng, J.D. Gillaspy, Phys.Rev. B **55**, 2628-2636 (1997)

[40] F. Aumayr, J. Burgdörfer, P. Varga, HP. Winter, Comments At. Mol. Phys. **34**, 201-219 (1999)

[41] A.V. Hamza, T. Schenkel, A.V. Barnes, D.H. Schneider, J. Vac. Sci. Techn. A **17**, 303 (1999)

[42] S. Ninomiya, Y. Yamazaki, F. Koike, H. Masuda, T. Azuma, K. Komaki, K. Kuroki, M. Sekiguchi, Phys. Rev. Lett. **78**, 4557 (1997)

[43] A. Bárány , C.J. Setterlind, Nucl. Instrum. and Methods in Phys. Res. B **98**, 184-186 (1995)

[44] T. Schlathölter, R. Hoekstra, R. Morgenstern, J. Phys. B: At.Mol.Opt. Phys. **31** 1321-1331 (1998)

[45] J. Opitz, H. Lebius, B.Saint, A. Jacquet, B.A. Huber, Phys.Rev. A **59** 3562-3568 (1999)

[46] G. de Nijs, R. Hoekstra and R. Morgenstern, J.Phys.B: At.Mol.Opt.Phys. **29,** 6143-6153 (1996)

[47] J. Jin, H. Khemliche, M.H. Prior, Phys. Rev. A **53**, 615-618 (1996)

[48] L. Plagne, C. Guet, Phys.Rev. A **59** 4461-4469 (1999)

[49] N. Selberg, A. Bárány, C. Biedermann, C.J. Setterlind, H. Cederquist, A. Langereis, M.O. Larsson, A. Wännström, P. Hvelplund, Phys. Rev. A **53** 874 (1996)

[50] B.Walch, U. Thumm, M. Stöckli, C.L. Cocke, S. Klawikowski, Phys. Rev. A **58** 1261-1266 (1998)

[51] U. Thumm, A. Bárány, H. Cederquist, L. Hägg, C.J. Setterlind, Phys. Rev. A **56** 4799-4806 (1997)

[52] U.Thumm, Comments At. Mol. Phys. **34** 119-140 (1999)

[53] S. Martin, L. Chen, A. Denis, J. Désesquelles, Phys. Rev. A **59** R1734-R1737 (1999)

[54] U. Thumm, Phys.Rev. A **55**, 479-487 (1997)

[55] O. Hadjar, T. Schlathölter, J. Manske, R. Hoekstra, R. Morgenstern, Int.J.Mass Spectr.**192**, 245-257 (1999)

[56] D.W. Brenner, Phys. Rev. B **42** 9458 (1990)

[57] G. Moliere, Z.Naturforsch. Teil A **2**, 133 (1947)

[58] T. Schlathölter, R. Hoekstra, R. Morgenstern, Phys.Rev. Lett. **82**, 73-76 (1999)

[59] S. Martin, L. Chen, A. Denis, J. Désesquelles, Phys. Rev. A **57** 4518-4521 (1998)

[60] L. Chen, J. Bernard, A. Denis, S. Martin, J. Désesquelles, Phys. Rev. A **59** 2827-2835 (1999)

COLLISIONS INVOLVING PHOTONS

Theory of Photodetachment from Negative Ions

Vadim K. Ivanov

Department of Experimental Physics,
St.Petersburg State Technical University, St.Petersburg 195251, Russia

Abstract. This short survey is devoted to some achievements in theoretical study of negative ion photodetachment over recent years. The results of photodetachment cross section calculations performed within various theoretical approaches are discussed. Main attention is centered on the resonances in the single-electron photodetachment where the manifestation of many-electron effects is essential.

INTRODUCTION

Photodetachment from negative ions has been an area of intense experimental and theoretical investigation in recent years (see (1-7) for example). The current interest to negative ions is related to the fact that they possess a variety of properties, not found in neutral atoms and positive ions. Negative ions are formed due to the short range potential (polarization potential $\sim r^{-4}$) and exhibit weakly bound systems with binding energies up to 3.6 eV. Because of that there is no Rydberg series of excited states in negative ions, they have typically a few discrete excited states or don't have them at all. Besides, many-electron collective effects in negative ions are often more substantial than in neutral atoms. These effects sometimes play a dramatic role in photodetachment processes (6) because the many-electron correlational interactions are considered in the neutral core field and not against the background of an unshielded attractive Coulomb potential generated by a nucleus as occurs for a neutral atom. Therefore, all the problems related to negative ions turn out to be essentially many-body problems.

Though the photodetachment processes in negative ions have been studied much less than those in neutral atoms, in recent years, new experimental data have been obtained and have already supplied theorists with very interesting information about many-electron effects in negative ions. On the other hand, the theoretical study of photodetachment is a good test for various approaches relevant to atomic system investigations. The many-electron correlations may be taken into account in different ways. At present, there are several many-body methods, which are known to have been applied successfully to the photoabsorption and photodetachment of negative ions. Main attention in theoretical study is concentrated on many-body effects in resonance

CP500, *The Physics of Electronic and Atomic Collisions*, edited by Y. Itikawa, et al.
© 2000 American Institute of Physics 1-56396-777-4/00/$17.00

photodetachment from negative ions. This short review is devoted to some recent achievements in this study.

THEORETICAL BACKGROUND

General presentation

We study the following single-photon detachment process

$$\gamma + A^- \rightarrow \begin{cases} A + e^- \\ A^* + e^- \end{cases} \tag{1}$$

which results in the final state by a neutral atom (in the ground or excited state) and a photoelectron. Photodetachment cross section is defined by the standard formula (atomic units are used):

$$\sigma_{i \rightarrow f} = 4\pi^2 \alpha_o \omega \left| \left\langle f |\hat{d}| i \right\rangle \right|^2 \tag{2}$$

where $\alpha_o = \frac{1}{137}$, ω is photon energy, $\left\langle f |\hat{d}| i \right\rangle$ is the dipole amplitude of the transition from the initial state i to the final one f.

Up to the end of 1980s the majority of photodetachment cross section calculations were completed within the restricted independent-electron approximations. The simplest methods used the fact that an additional electron spends the most time at a large distance from nucleus, where its motion may be considered in the polarization potential field $V_{pol}(r) = -\frac{\alpha}{2r^4}$, where α is the atomic polarizability. The importance of this potential for negative ions is related to the fact that this potential is responsible for the binding of an extra electron. The single-electron wavefunctions determined in this potential are used in calculations of the dipole matrix element. Within this rather simple approach nevertheless, one obtains the series of very general results for negative ion photodetachment. Thus, the threshold behavior of photodetachment cross section obeys to the Wigner law (8):

$$\sigma_{nl \rightarrow \varepsilon l'} \sim \varepsilon^{(l'+1/2)} = (\omega + E_{nl})^{l'+1/2} \tag{3}$$

Here n, l are the principle quantum number and electron orbital momentum of the initial state with energy E_{nl}; ε, l' are the photoelectron energy and orbital momentum in the final state. Further, the independent-particle methods obtain the typical behavior of partial photodetachment cross sections with a shape resonance. Thus, for s-electron photodetachment using the asymptotic behaviour of ground-state wavefunction and plane waves for outgoing electron one obtains the simple analytic formula for partial cross section

$$\sigma_{ns} = \frac{4}{3} 2^{3/2} \pi N_{ns} \alpha_0 B^2 \sqrt{-2E_{ns}} \frac{(\omega - E_{ns})^{3/2}}{\omega^3} \tag{4}$$

where N_{ns} is a number of s-electrons, B is asymptotic wavefunction parameter. This analytical formula describes qualitatively the energy dependence of outer-electron photodetachment cross section: the threshold behavior agrees with the Wigner law for outgoing p-wave (usually the Wigner law is valid for a very narrow energy range lasting from the tenths meVs to several μeVs (9)), then with the growing of photon energy the cross section passes the maximum value at $\omega = -2\,E_{ns}$ and then decreases as $\omega^{-1.5}$. Note that in practice all independent-particle models predict the cross section energy dependence for photodetachment from any subshell similar to one described by Equation (4).

The series of calculations within more sophisticated model potential methods, including the Hartree-Fock (HF) method or its simplified modifications, (see for example (1) and references therein) has been performed for photodetachment from negative ions with np^q outer subshells. Using the model potential to account for the polarization interaction between an extra electron and neutral core, Robinson and Geltman (10) have obtained the typical behavior of np-subshell photodetachment cross section with a shape resonance.

However, it is obvious that the single-electron methods are restricted by the account of only static effects on the electron initial and final states. Within its frame the dynamic effects of many-electron interactions, i.e. the dependence of this interaction on the photon energy ω, are neglected. Therefore, it is difficult to expect that any independent-particle method could describe the photodetachment processes in whole energy range, especially for many-electron negative ions, where the interelectron effects are significant. This is why all modern calculations of photodetachment phenomena are based on many-electron methods.

Many-electron processes

All interelectron interactions, beyond independent-particle approaches, in particular the HF approximation, present the many-body effects. Considering an independent-electron picture as a zero approximation and the residual part of interelectron interaction perturbatively it is possible to perform the qualitative classification of the many electron effects (6).

1) Dynamic polarization effects. In the absence of the Coulomb force between the electron and neutral atom, the polarization of the core by the external electron is very important. Most of the negative ions are formed due to the polarization interaction between additional electron and neutral atom. After photon absorption an outgoing electron is scattered by the neutral atom field and all the peculiarities of elastic electron scattering are reflected in the photodetachment process too. So, the dynamic polarization interaction in negative ions plays extremely important role both in the initial (ground) and final (continuum spectrum) states.

2) Intrachannel (intrashell) and interchannel (intershell) many-electron interaction. The intrashell correlations are responsible for the value of the cross section especially for many-electron subshells. The dynamic intershell correlations lead to the interference between the direct photoionization process and photoionization via the excitations of other subshells or shells.

3) Core-relaxation processes during photoabsorption and electron escape. The interaction between a photoelectron and the core with the vacancy is dynamically screened by the other core electrons. Physically, it means that the outgoing electron

moves in the field of the atomic core, which is altered because of relaxation of all other electrons to the final states. It is obvious that the rearrangement effects are more significant for small energy photoelectrons escaping from inner shells.

Because it is sometimes difficult to distinguish the contributions of all many-body effects separately, they may manifest themselves simultaneously and influence each other too, therefore this classification is rather conventional despite of natural physical meaning. Besides, these effects are taken into consideration sometimes within the many-body approaches where similar classification is not easy to introduce.

Multi-Configuration Hartree-Fock and Close-Coupling Methods

The main idea of the Multi-Configuration Hartree-Fock (MCHF) is to represent the total wavefunction of ground and excited states of an N electron system as a linear combination of wavefunctions corresponding to different configurations of atomic system:

$$\Psi(\vec{r}_1,...,\vec{r}_N) = \sum_q C_q \psi_q(\vec{r}_1,...,\vec{r}_N) \qquad (5)$$

Here ψ_q is the Slater determinant which consists of the single-electron HF wavefunctions. The coefficients C_q of this expansion are obtained from the minimization of the total atom energy with the additional condition of the orthonormality for ψ_q. The C_q determines the amplitude of an q-configuration admixture to the total wavefunction. The final state wavefunctions may be also represented by the expansion in different excitations. Usually one can choose a few outer electrons involved in the photoionisation process, and consider the different states of these electrons above the frozen core with closed subshells (11,12).

The close-coupling approximation is rather similar to the MCHF method and usually considers a wave function of the N+1 electron system as the following expansion:

$$\Psi(\vec{r}_1,...,\vec{r}_N,\vec{r}_{N+1}) = \hat{A} \sum_j \psi_j(\vec{r}_1,...,\vec{r}_N)\varphi_j(\vec{r}_{N+1}) \qquad (6)$$

where $\hat{A}$ is an antisymmetrization operator, the index j denotes the set of eigenstates ψ_j of N electron target and appropriate scattering orbitals φ_j. This approach in early years was developed for electron scattering calculations and was applied rather often to the negative ion problems (13,14).

At present the close-coupling approximation is used very often in combination with the MCHF, when the initial state is considered within the MCHF and the interchannel interaction in the final state are taken into account within the close-coupling method.

R-matrix and MQDT methods

The R-matrix and Multichannel Quantum Defect Theory (MQDT) are very effective and popular methods in applications to negative ion calculations (15-22). Within these methods all atomic space is divided into two regions, internal (r < a) and external (r > a) ones, by the sphere of radius a. The wave function of an atom within the r < a region is expanded in the limited set of eigenfunctions of an modified Shrodinger equation. At a large distance from the nucleus the photoelectron moves in the polarization potential, so the solution for the r >a region can be represented analytically. The solutions for the different parts of the space are matched at the sphere r = a. The R-matrix approximation is presently applied to calculations of photodetachment from negative ion He⁻, Li⁻, B⁻, C⁻, Ca⁻ (15-18).

The semi-empirical method, which combines the eigenchannel R-matrix method and generalized MQDT, was developed for the calculations of negative ion photodetachment (19-22). The small-scale eigenchannel R-matrix approach is used to construct a two-electron (or more) variational basis. Inside the R-matrix sphere the valence electrons are assumed to move in the model potential describing the interaction with a core, parameters of which are fitted to reproduce the experimentally measured energy levels of the neutral atom. Then a two-electron variational basis is constructed using the one-electron radial eigenfunctions calculated within the finite R-matrix reaction volume with the selected potential. Outside the R-matrix sphere there is assumed to be only a single electron, and thus only single detachment processes are considered. The semi-empirical eigenchannel R-matrix method is very effective in describing the double excited resonances associated with higher thresholds.

The Many Body Perturbation Theory approaches

Methods, based on the Many Body Perturbation Theory (MBPT), have been successfully applied to atomic photoionisation studies for many years (23) and more recently to negative ion photodetachment (5,6). Thus, the ions with closed ns^2 and np^6 subshells were considered within non relativistic (24,25) and relativistic (26,27) versions of the Random Phase Approximation with Exchange (RPAE and RRPA, respectively). These approximations, being self-consistent methods, include an important infinite subsequence of the perturbation theory series and describe the collective dynamical response of an atomic system in a weak external field (23). However, these methods neglect the dynamical polarization interaction between additional electron and the atomic core and screening effects.

There is the other method within the many-body perturbation theory to include *ab initio* the dynamical polarizability of an atomic core. This method (28-30) is based on the Dyson equation, which determines the self-energy part of the single-electron Green function. The wavefunction $\varphi_E(\bar{r})$, describing the electron state with energy E, satisfies the equation, which can be obtained from the Dyson equation for the single-electron Green function:

$$\hat{H}^{HF}\varphi_E(\bar{r}) + \int \Sigma_E(\bar{r},\bar{r}')\varphi_E(\bar{r}')d\bar{r}' = E\varphi_E(\bar{r}) \qquad (7)$$

Here $\hat{H}^{HF}$ is the static Hartree-Fock Hamiltonian of the atom and $\Sigma_E(\bar{r},\bar{r}')$ plays the role of the energy-dependent non-local polarization potential.

To take into account the dynamic interchannel and the polarization interactions and relaxation processes simultaneously the method, which combines the DEM and RPAE, has been developed (31,32). Note that the dynamic polarization potential is included for the both final and intermediate states in each interacting channel. This method has been applied successfully to photodetachment from B⁻, C⁻ and Cr⁻ negative ions (31-33).

PHOTODETACHMENT FROM OUTER SUBSHELLS

Negative ions with closed subshells

The negative ions of the first group of the periodic system attract much attention both experimentalists and theorists. Thus, almost all existing theoretical methods were applied to the description of photodetachment from the simplest system H⁻ (see (34,35) and references therein). Main attention there is paid to photodetachment processes with excitation of the other electron when neutral hydrogen is left in an excited state (35).

The outer-shell photodetachment of alkali-metal negative ions (...$(n-1)p^6ns^2$ ^1S) have been examined by different methods (14,15,19,21). All calculations yield the near-threshold shape resonance in the ns^2 cross section, the strong many-electron correlations influencing the near-threshold behavior of the cross section and resonance peak value. However, the most interesting many-body effects in alkali-metal negative ions manifest themselves in the appearance of Wigner cusps. These resonance features are due to new opening channels and quasi-stationary states in elastic photoelectron scattering in neutral atom field. As an example, Figure 1 shows the $2s^2$ photodetachment cross section for Li⁻ in the vicinity of the 2^2P threshold. The results of two of the most recent *ab initio* calculations of total cross section (15,36) are in a good agreement with experiment (37). Similar resonance features are obtained for photodetachment from all other alkali-metal negative ions (14,19,21), although for heavier ions it is necessary to take into account the spin-orbit interaction which produces the cusps at the $nP_{1/2}$ and $nP_{3/2}$ thresholds.

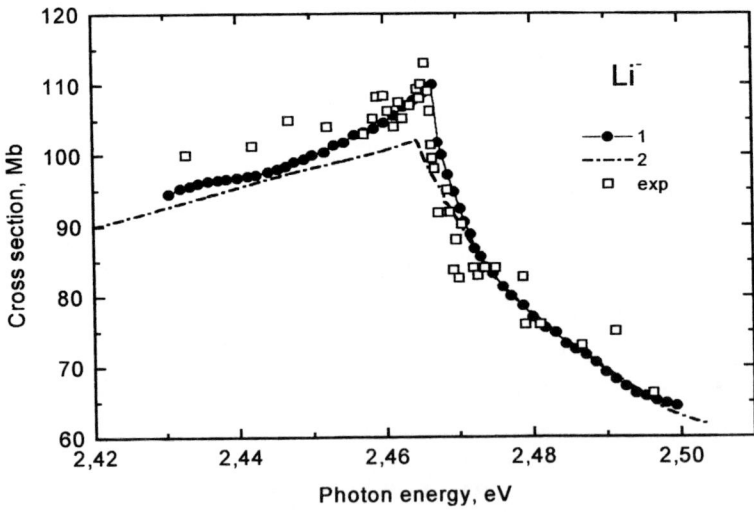

FIGURE 1. Total photodetachment cross section of Li⁻ near the 2p ^2P threshold. Open squares - experimental data (37); 1 - the K-matrix calculations (36); 2 - the multichannel R-matrix method (15).

The variety of resonance features is revealed not only in total, but partial photodetachment cross sections. Recently, the outer s-electron photodetachment process with excitation of the other s-electron has been studied both experimentally (38) and theoretically (21) in Na⁻ negative ion. The results of calculations of the Na(4s) partial cross section within the eigenchannel R-matrix theory is presented in Figure 2 and one can see many resonances related to opening channels. Note that this theoretical cross section reproduces rather well relative experimental data in this region (38). The other Na(nl) partial photodetachment cross sections have similar behavior with photon energy (21).

The photodetachment from negative Cu⁻, Ag⁻, Au⁻ ions (...(n-1)d^{10}ns^2 ^{1}S) was studied not so intensively as that from alkali metal negative ions. Nevertheless a few experimental data (39) and theoretical results, obtained within the R-matrix approach (40) and within the model RPAE version (41), for Cu⁻ have found the near threshold shape resonance, the position and shape of which depend strongly on the polarisation

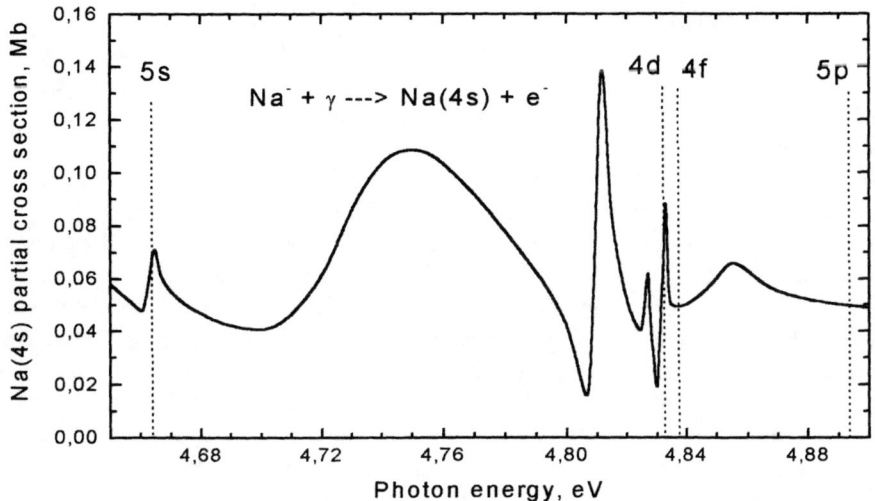

FIGURE 2. The partial Na (4s) photodetachment cross section of Na⁻ obtained within the eigenchannel R-matrix approximation (21). The vertical dashed lines indicate the locations of the thresholds.

interaction between the outgoing electron and core electrons. The RPAE calculations give also the typical shape resonance for photodetachment from 5s^2-subshell of Ag⁻ (24). These results do not show the characteristic features, which were found for alkali metal ions, although the recent experiment for Au⁻ (42) reveals two very interesting window resonances near photodetachment threshold.

Photodetachment from negative ions of halogens with closed outer ns^{2}np^6 subshells have been studied theoretically in a wide range of photon energies within the RPAE and RRPA methods (24,26,27). Since the begining of the 90s there are no new data on photodetachment obtained both experimentally and theoretically for these ions. The former calculations (see references in (6)) have found the similarity of total and partial photodetachment cross sections to those in neutral atoms of noble gases. Though it should be noted that the intrashell and intershell correlations, as well as rearrangement effects for inner-shell photodetachment, are more essential for negative ions. Thus, a study of photodetachment from negative halogens is of interest, for example for ns^2 partial and total cross section in the vicinity of inner $(n-1)d^{10}$ subshell in heavier ions.

Negative ions with one open shell

The main features of negative ions with open shells are related to the producing of different terms in the final state and to the formation of quasi-bound states in the vicinity of inner-shell thresholds. All these lead to the necessity of a multi-channel description of a photodetachment process with more complicated interchannel interaction taken into account. The formation of quasi-bound states produces the resonant structure in a photodetachment cross section. Originally the nature of the latter resonances is due to the inner-shell electron transitions to a vacant state of an outer open subshell of negative ion.

There are only a few negative ions formed by attaching one electron to atom with closed shells. This new class of negative ions: Ca⁻, Sr⁻, Ba⁻ with ...ns^2np electron structure (Pd⁻ with ...$(n-1)d^{10}ns$ electron structure may be also refereed to that class) being discovered about ten years ago are studied very intensively (see (3,6,7) and references therein). There are a few calculations of photodetachment cross sections for these ions (5,11,18,29,43). The polarization interaction is very important to bind an extra electron to atom with closed shells and to describe the motion of photoelectron in continuum spectra. The neutral heavy alkaline-earth atoms possess large polarizabilities, therefore the core polarization influences the outgoing electron, and that leads to the well known low-energy d-wave resonances in elastic electron scattering. Besides that, it was shown that the interchannel interaction, namely the influence of inner ns electrons, is very important in outer p-electron photodetachment. Figure 3 presents the results of three different calculations and comparison with experimental data (44-46) for Ca⁻ photodetachment. It is shown (5,11) that the maximum in cross section near the threshold is due to the d-wave of the outgoing electron and the position of minimum is very sensitive to the polarization interaction between photoelectron and the core and the intershell correlations. The recent R-matrix calculations (18) (the authors have corrected the cross section published in (18) and reproduced in (6) multiplying that by factor 1/3) have been performed for the larger photon energy range. Comparing with experimental data (44,45) the R-matrix results show the correct position of the minimum and confirm the existence of the p-wave shape resonance just above Ca 4s4p ¹P threshold. There are two absolute experimental values of the photodetachment cross section (46). Therefore, the relative experimental data of (45) obtained in a wide energy range (dotted line in Fig.3) may be normalized on those absolute measurements. Then, these data reveal the huge resonance with the peak value about 450 Mb near the ¹P threshold. The R-matrix calculations produce smaller peak value and larger width. The nature of this resonance is originally due to the strong transition of 4s electron to

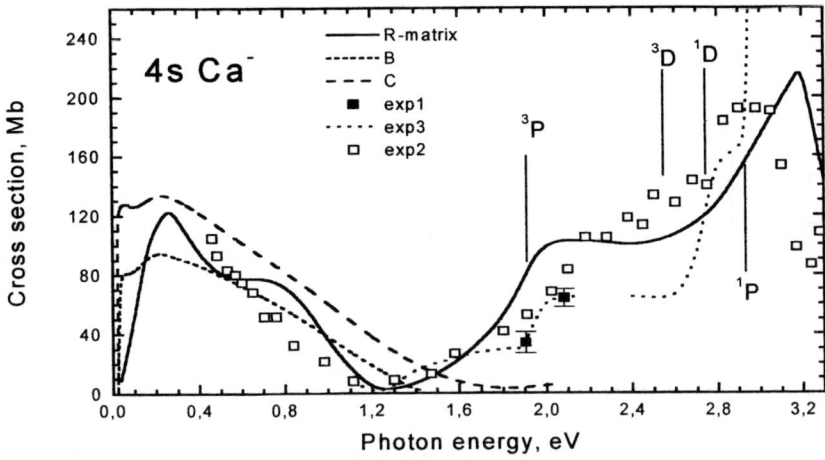

FIGURE 3. Photodetachment of Ca⁻. Theory: dashed line - the DEM (5); short-dashed line - the MCHF (11); solid line - R-matrix (18). Experiment: open squares - normalized data of (44); full squares - absolute measurements of (46); dotted line - averaged data of (45), normalized on absolute cross section of (46). The vertical lines indicate the locations of the thresholds.

vacant states in 4p subshell although the "4s4p" excited state of Ca⁻ is quasi-bound one and lies in the continuum. The big cross section value agrees very well with the fact that the discrete transition 4s → 4p in neutral Ca has a large oscillator strength.

More dramatic changes due to the polarization effects could be expected for Sr⁻ and Ba⁻ photodetachment (5,6,29,43), where the d-wave resonances are more pronounced, although the preliminary results of R-matrix calculations in Sr⁻ (43) show approximately the same value of photodetachment cross section at the threshold.

The resonance behavior of photodetachment in the vicinities of inner subshell thresholds are studied in negative ions with open outer p-subshells. The first prediction of resonance features for ions with half-filled np^3 subshells was made using the spin-polarized version of the RPAE (24,25). Figure 4 presents the photodetachment cross section of Si⁻ (...$3s^2 3p^3$ ^4S). Calculations were performed with taking into account the interaction between outer $3p^3$ electrons only and the intershell interaction between 3p and 3s electrons. The latter leads to the "window"-type resonance just before the 3s (...$3s3p^3$ ^4P) threshold. The experimental measurements (47) have confirmed the existence of strong intershell effects, leading to "window" resonance, however the comparison between the data and different calculations shows the difference in the position and the width of this resonance. One can see that the experimental resonance lies between two RPAE calculations, performed with the frozen-core wavefunctions and taking into account the static rearrangement effects. It means that besides intershell correlations and rearrangement effects the other many-electron processes should be taken into consideration as performed recently for the B⁻ (16,31) and C⁻ photodetachment (17,33).

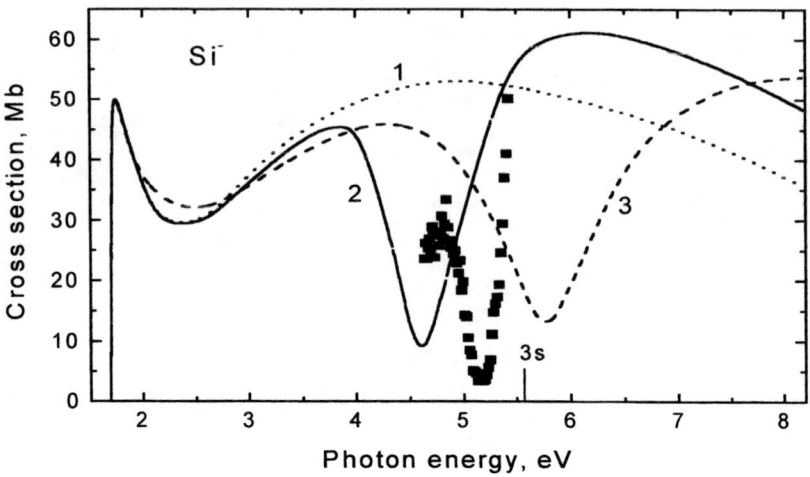

FIGURE 4. Photodetachment cross section of Si⁻. 1 - the $3p^3$ cross section with the intrashell correlations taken into account only; 2 - the SP RPAE results with inclusion of the "$3s3p^4$" resonance (25); 3 - the same with account the static rearrangement effects (5). Experimental data from (47).

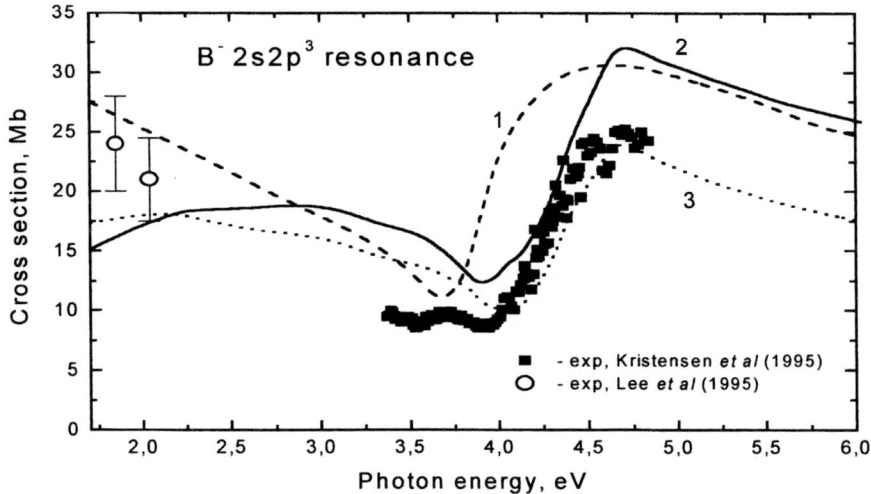

FIGURE 5. Photodetachment cross section of B⁻ in the vicinity of the 2s threshold. Experiment: open circles - (50); squares - (48). Theory: 1 - the R-matrix method (16); 2 and 3 - the combined many-body theory method (31), the length and velocity forms, respectively.

For both negative ions the rather sharp resonances near the 2s threshold were expected (see references in (6)). However, then the experimental study has not found any resonance structure in the C⁻ photodetachment (49) within the 5.23-6.04 eV range of photon energies, as was predicted, and has revealed the salient feature in the B⁻ photodetachment in the 3.37-4.83 eV range (48). As for C⁻ the calculations within the R-matrix method (16) and the many-body theory approach (33) have shown later that the resonances structure should appear at larger energies just above the experimental data. However, this resonance is not so sharp, as was expected, due to the dynamic polarization and relaxation effects. The theory predicts the resonance behavior of C⁻ photodetachment cross section near the 2s threshold which is rather similar to the resonance feature found for the B⁻ photodetachment (Fig. 5). The same methods, the R-matrix and the combined many-body approach (the DEM + the RPAE), were applied to B⁻ photodetachment calculations, the results of which are presented in Figure 4 in comparison with absolute cross section obtained experimentally (48,50).

Negative ions with more than one open subshells

Among negative ions with two and more open subshells the He⁻ (1s2s2p ⁴P) metastable ion takes a very important place and therefore attracts much attention from both experimentalists and theorists. In recent years the interest to He⁻ calculations (12,22,51-54) is renewed due to new experimental data. The main features of the total He⁻ photodetachment cross section may be seen in Figure 6. Agreement between the

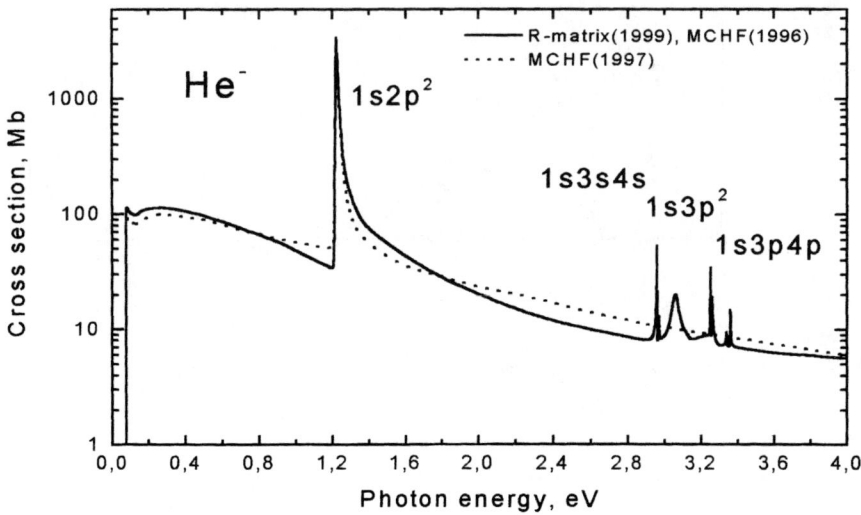

FIGURE 6. The total photodetachment cross section of He⁻ 1s2s2p ⁴P. The full curve - the MCHF (12) and the R-matrix (51), the dotted line - the MCHF (52).

MCHF (12) and the R-matrix (51) for the total cross section in whole region permits to suggest that the total photodetachment cross section for He⁻ is known to an accuracy of a few percent.

The 2p photodetachment cross section reveals the shape resonance just after the 2p threshold. This shape resonance is formed by the centrifugal barrier for εd photoelectron and the polarization potential. The calculations within the various approaches reproduce the 2p cross section with similar shape and slightly different magnitudes in the maximum, but all results are in fairy agreement with experimental data (see (6,12,51) and references therein). Note that the polarization interaction between electron and core in initial and final states plays the most important role in the description of the He⁻ photoabsorption cross section.

In the vicinity of the 2^3P He⁻ threshold both experiment (55) and theory (56) have revealed the enormous peak in the photodetachment cross section, associated with the shape resonance in the 2s → εp (^4P) transition to the continuum spectrum. The latter is due to the $1s2p^2$ (^4P) quasi-bound state of He⁻, which appears to lie just above its parent He (2^3P) state.

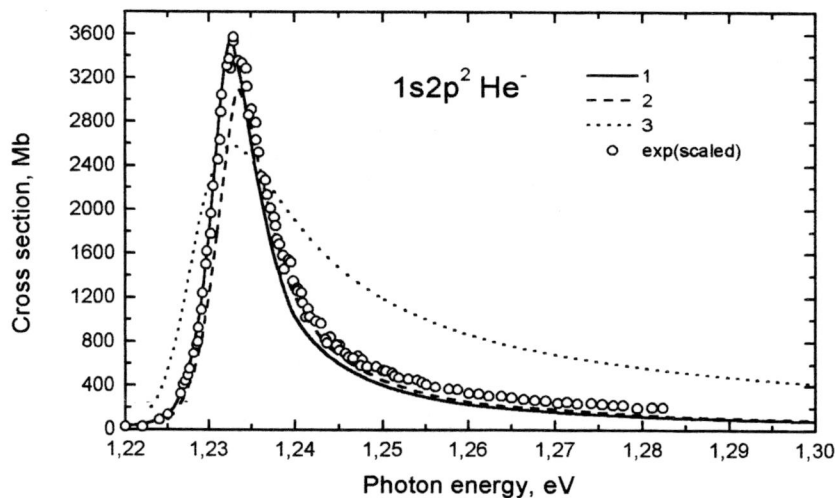

FIGURE 7. The 2s photodetachment cross section for He⁻. Theory: 1 - the R-matrix (51); 2 - the MCHF (12); 3 - the DEM (30). Experiment: open-circles (57).

The comparison of the results of recent calculations with latest experimental data (57) on 2s photodetachment cross section is shown in Figure 7.

In Figure 6 one can see the various and rich resonance structure near the range 2.9-3.4 eV which associated with photodetachment from He⁻ accompanied by the excitation of the other electron into discrete states. The latter process is completely determined by the many-electron correlations. The series of the experimental measurements (58-60) and the calculations (12,22,51,54) within different methods has been recently performed to define the partial and total photodetachment cross section in this region. Figure 8 presents the results for partial photodetachment cross sections $\sigma(2s)$ and $\sigma(2p)$ in the vicinity of the 1s3s4s ^{4}S resonance. Both the MCHF and eigenchannel R-matrix calculations are in a good agreement between each other and with relative experimental data (58,59). The R-matrix results (51) for the $\sigma(2p)$ are also give similar resonance behavior with small shift to higher energies.

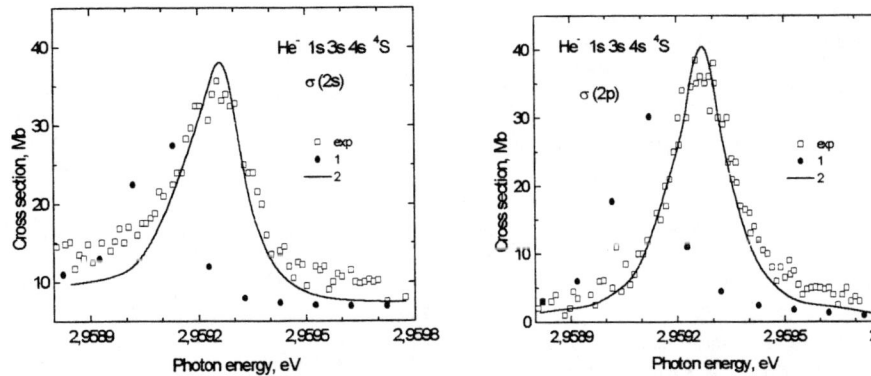

FIGURE 8. Partial 1s2s He and 1s2p He cross sections in the vicinity of 1s3s4s ^{4}S He⁻ resonance. 1 - the MCHF (12); 2 - the eigenchannel R-matrix (22); open squares - the relative experimental data (58,59), normalized at calculated cross section.

CONCLUDING REMARKS

The study of photodetachment from negative ions reveals the large variety of many-body effects which manifest themselves in photoprocesses with negative ions. Because of important role of many-body effects in negative ion photodetachment the physics of these processes is very interesting and attractive. Though good agreement between experiment and theory has been reached in many cases the negative ion photodetachment is investigated essentially less than the photoionization of neutral atoms.

In this paper we have concerned only outer-shell photodetachment, although the photodetachment from inner subshells is of a great interest also (see (6) and references therein). The many-body effects in these processes are mainly related to interchannel interactions and relaxation effects, the influence of which is sometimes more significant than for neutral atoms. However, there are no experimental data to testify the correctness of results obtained by theory because the study using synchrotron radiation is presently rather difficult due to the small sample densities.

To better understand the processes of the interaction of electromagnetic radiation with negative ions and to test present theoretical methods it is necessary to compare the absolute experimental data and results of calculations over a wide photon range.

REFERENCES

1. Esaulov, V. A., *Ann. Phys.(Paris)* **11**, 493-592 (1986).
2. Andersen, T., *Phys. Scripta* **T24**, 23-35 (1991); **T59**, 230 (1995).
3. Andersen, T., Andersen, H. H., Balling, P., Kristensen, P., and Petrunin, V. V., *J. Phys. B: At. Mol. Opt. Phys.* **30**, 3317-3332 (1997).
4. Buckman, S. J., and Clark, C. W., *Rev. Mod. Phys.* **66**, 539-655 (1994).
5. Ivanov, V. K., "Collective phenomena in negative ion photodetachment", in *Correlations in clusters and related systems. New perspectives on the many-body problem*, Singapore: Wold Scientific Publishing, 1996, pp. 73-91.
6. Ivanov, V. K., *J. Phys. B: At. Mol. Opt. Phys.* **32**, R67-R101 (1999).
7. Andersen, T., Haugen, H. K., and Hotop, H., *J. Phys. Chem. Ref. Data*, to be published (1999).
8. Wigner, E. P., *Phys.Rev.* **73**, 1002 (1948).
9. Hotop, H., "Threshold behaviour of photodetachment and electron attachment processes", in *Proceedings of the International Workshop on Photoionization 1992. Berlin, Germany*, New York: AMS Press, 1993, pp. 85-88.
10. Robinson, E. J., and Geltman, S., *Phys.Rev.* **153**, 4-8 (1967).
11. Froese Fischer, C., and Hansen, J. E., *Phys.Rev.A* **44**, 1559-64 (1991).
12. Xi, J., and Froese Fischer, C., *Phys.Rev. A* **53**, 3169 (1996); *A* **59**, 307 (1999).
13. Schulz, G. J., *Rev.Mod.Phys.* **45**, 378 (1973).
14. Moores, D. L., and Norcross, D. W., *Phys.Rev.* **A 10**, 1646-1657 (1974).
15. Ramsbottom, C. A., Bell, K. L., and Berrington, K. A., *J.Phys.B:At.Mol.Opt.Phys.* **26**, 4399-4408 (1993); **27**, 2905-2918 (1994).
16. Ramsbottom, C. A., and Bell, K. L., *J.Phys.B:At.Mol.Opt.Phys.* **28**, 4501-4508 (1995); *ibid* **29** 3009-3015 (1996).
17. Miura, N., Noro, T., and Sasaki, F., *J.Phys.B:At.Mol.Opt.Phys* **30**, 5419 (1997).
18. Yuan, J., and Fritsche, L., *Phys.Rev.A* **55**, 1020-1027 (1997).
19. Greene, C. H., *Phys.Rev. A* **42**, 1405-1415 (1990).
20. Pan, C., Starace, A. F., and Greene, C. H., *J.Phys.B:At.Mol.Opt.Phys* **27**, L137 (1994); *Phys.Rev. A* **53**, 840 (1996).
21. Liu, C.-N., and Starace, A. F., *Phys.Rev. A* **59**, 3643-3654 (1999).
22. Liu, C.-N., and Starace, A. F., "Identification of two-electron resonances in He⁻ photodetachment", in *Proceedings of the 21st International Conference on the Physics of Electronic and Atomic Collisions Abstract of Contributed Papers, Sendai, Japan*, 1999, p. 122; to be published.
23. Amusia, M. Ya., *Atomic Photoeffect*, Plenum Press, 1990, ch. 4-5.
24. Amusia, M. Ya., Gribakin, G. F., Ivanov, V. K., and Chernysheva, L. V., *Izv. AN SSSR Ser.Fiz.* **50**, 1274-1278 (1986) (in Russian); *J.Phys.B:At.Mol.Opt.Phys.* **23**, 385- 391 (1990).
25. Gribakin, G. F., Gribakina, A. A., Gul'tsev, B. V., and Ivanov, V. K., *J.Phys.B:At.Mol.Opt.Phys.* **25**, 1757-1772 (1992).
26. Radojevic, V., Kelly, H. P., and Johnson, W. R.., *Phys.Rev. A* **35**, 2117-2121 (1987).
27. Radojevic, V., and Kelly, H. P., *Phys.Rev.A* **46**, 662-665 (1992).
28. Chernysheva, L. V., Gribakin, G. F., Ivanov, V. K., and Kuchiev, M. Yu., *J.Phys.B:At.Mol.Opt.Phys.* **21**, L419-L425 (1988).
29. Gribakin, G. F., Gul'tsev, B. V., Ivanov, V. K., and Kuchiev, M. Yu., *J.Phys.B:At.Mol.Opt.Phys.* **23**, 4505-4519 (1990).

30. Ivanov, V. K., Kashenock, G. Yu., Gribakin, G. F., and Gribakina, A. A., *J.Phys.B:At.Mol.Opt. Phys.* **29**, 2669-2687 (1996).

31. Kashenock, G. Yu., and Ivanov, V. K., *J.Phys.B: At.Mol.Opt.Phys.* **30**, 4235-4253 (1997).

32. Ivanov, V. K., Krukovskaya, L. P., and Kashenock, G. Yu., *J.Phys.B:At.Mol.Opt.Phys.* **31**, 239-247 (1998).

33. Kashenock, G. Yu., and Ivanov, V. K., *Phys.Lett. A* **245**, 110-116 (1998).

34. Abrashkevich, A. G., and Shapiro, M., *Phys.Rev. A* **50**, 1205-1217 (1994).

35. Sadeghpour, H. R., Greene, C. H., and Cavagnero, M., *Phys.Rev. A* **45**, 1587-1595 (1992).

36. Moccia, R., and Spizzo, P., *J.Phys.B:At.Mol.Opt.Phys.* **23** 3557 (1990).

37. Bae, Y. K., and Peterson, J. A., *Phys.Rev. A* **32**, 1917-1920 (1985).

38. Haeffler, G., Kiyan, I. Yu., Hanstorp, D., Davies, B. J., and Pegg, D. J., *Phys.Rev. A* **59**, 3655 (1999).

39. Balling, P., Brink, C., Andersen, T., and Haugen, H. K., *J.Phys.B: At.Mol.Opt.Phys.* **25**, L565 (1992).

40. Scheibner, K. F., and Hazi, A. U., *Phys.Rev.* **A 38**, 539 (1988).

41. Ivanov, V. K., Ipatov, A. N., and Krukovskaya, L. P., *Opt.Spectr.* **83**, 726-732 (1997).

42. Champeau, R. J., Cribellier, A., Marescaux, D., Pavolini, D., and Pinard, J., *J.Phys.B:At.Mol.Opt. Phys.* **31**, 741-749 (1998).

43. Yuan, J., "R-matrix photodetachment cross section of Sr⁻ ions", in *Proceedings of the 21st International Conference on the Physics of Electronic and Atomic Collisions Abstract of Contributed Papers, Sendai, Japan*, 1999, p. 121.

44. Heinicke, E., Kaiser, H. J., Rackwitz and Feldmann, D., *Phys.Lett.* **50A**, 265 (1974).

45. Walter, C. W., and Peterson, J. R., *Phys Rev.Lett.* **68** 2281 (1992).

46. Lee, D. H., Poston, M.B., Hanstorp, D., Berzinsh, U., and Pegg, D. J., "Photodetachment of the Ca⁻ ion", in *Proceedings of the 21st International Conference on the Physics of Electronic and Atomic Collisions Abstract of Contributed Papers, Sendai, Japan*, 1999, p. 120.

47. Balling, P., Kristensen, P., Stapelfeldt, H., Andersen, T., and Haugen, H. K., *J.Phys.B:At.Mol.Opt. Phys.* **26**, 3531-3539 (1993).

48. Kristensen, P., Andersen, H. H., Balling, P., Steele, L. D., and Andersen, T., *Phys.Rev. A* **52**, 2847-2851 (1995).

49. Haeffler, G., Hanstorp, D., Kiyan, I. Yu., Ljungblad, U., Andersen, H. H., and Andersen, T., *J.Phys.B: At.Mol.Opt. Phys.* **29**, 3017-3022 (1996).

50. Lee, D. H., Tang, C. Y., Thompson, J. S., Brandon, W. D., Ljungblad, U., Hanstorp, D., Pegg, D. J., Dellwo, J., and Alton, G. D., *Phys.Rev A* **51**, 4284 (1995).

51. Ramsbottom, C. A., and Bell, K. L., *J.Phys.B:At.Mol.Opt. Phys.* **32**, 1315 (1999).

52. Kim, D.-S., Zhou, H.-L., and Manson, S. T., *Phys.Rev. A* **55**, 414-25 (1997).

53. Zhou, H.-L., Manson, S. T., Vo Ky, L., Hibbert, A., and Berrington, K.A., "R-matrix calculation of the photodetachment of the He⁻ 1s2s2p 4P in the region of the 1s threshold", in *Proceedings of the 21st International Conference on the Physics of Electronic and Atomic Collisions Abstract of Contributed Papers, Sendai, Japan*, 1999, p. 123.

54. Brandefelt, N., and Lindroth, E., *Phys.Rev. A* **59**, 2691-2696 (1999).

55. Hodges, R. V., Coggiola, M. J., and Peterson, J. R., *Phys.Rev. A* **23**, 59 (1981).

56. Hazi, A. U., and Reed, K., *Phys.Rev. A* **24**, 2269 (1981).

57. Walter, C. W., Seifert, J. A., and Peterson, J. R., *Phys.Rev. A* **50**, 2257-2262 (1994).

58. Klinkmuller, A. E., Haeffler, G., Hanstorp, D., Kiyan, I. Yu., Berzinsh, U., Ingram, C. W., Pegg, D. J., and Peterson, J.R., *Phys.Rev. A* **56**, 2788 (1997).

59. Klinkmuller, A. E., Haeffler, G., Hanstorp, D., Kiyan, I. Yu., Berzinsh, U., and Pegg, D. J., *J.Phys.B:At. Mol.Opt. Phys.* **31**, 2549-2557 (1998).

60. Kiyan, I. Yu., Berzinsh, U., Hanstorp, D., and Pegg, D. J., *Phys.Rev.Lett.* **81**, 2874 (1998).

101

Molecules in Intense Laser Fields – From Charge Resonance Enhanced Ionization to Laser Coulomb Explosion

André D. Bandrauk[*]

Laboratoire de Chimie Théorique, Faculté des Sciences
Université de Sherbrooke, Que., J1K 2R1, Canada

Abstract. Exact numerical solutions of the time dependent Schroedinger equation, TDSE, of one and two electron molecules and ions allow us to investigate the highly nonlinear, nonperturbative behavior of such systems in the presence of ultrashort (t < 50 femtoseconds), intense (I $\geq 10^{14}$ W/cm^2) laser pulses. Quasistatic models of ionization lead to a qualitative understanding of a universal phenomenon occurring in these extreme conditions: Charge Resonance Enhanced Ionization, CREI. Exact non-Born Oppenheimer TDSE simulations of the dissociative-ionization of H_2^+ lead to the conclusion that CREI is responsible for anomalous kinetic energy distributions observed in Coulomb Explosion experiments, i.e., Coulomb explosions are non-Franck-Condon and occur at critical internuclear distances R_c and geometries. Analytic formulas are obtained for R_c which are shown to arise from charge resonance or transfer induced at the maximum field strength of the laser fields.

INTRODUCTION

The investigation of the interaction of atoms with ultrashort intense laser pulses in the multiphoton, thus in the nonlinear and nonperturbative regime has lead to the discovery of new nonperturbative phenomena such as above-threshold ionization, ATI, tunneling ionization, stabilization, high order harmonic generation, etc. and has recently been documented in a comprehensive review by Gavrila (1). These new phenomena are the results of highly nonlinear optical physics which cannot be described by perturbative models, but are amenable to qualitative and sometimes quantitative understanding in terms of classical concepts of plasma physics, such as quasistatic models (2-3).

In the case of molecules, similar nonlinear nonperturbative processes have been recently discovered. The presence of the extra degree of freedom arising from nuclear motion has necessitated the introduction of new concepts such as laser-induced avoided potential crossings, above threshold dissociation, ATD (the analogue of atomic ATI) (4). Thus in the molecular case, different time scales are of major importance: electronic-attosecond (atts - 10^{-18}s) and nuclear-femtosecond (fs – 10^{-15} s). In the present review we will discuss two highly nonlinear multiphoton processes which occur in molecules. Charge Resonance Enhanced Ionization, CREI, and Coulomb Explosions, CE. It will be shown that quasistatic models of field ionization can also be

CP500, *The Physics of Electronic and Atomic Collisions*, edited by Y. Itikawa, et al.
© 2000 American Institute of Physics 1-56396-777-4/00/$17.00

applied to understand qualitatively these highly nonperturbative phenomena of laser-molecule interactions.

Surprisingly, although high laser intensities produce highly nonperturbative phenomena, simple classical models can be used to explain the resulting intense field atomic physics (1-3) or molecular spectroscopy (4). In the case of atomic ionization, a quasistatic laser field induced Coulomb barrier suppression model allows for predicting atomic tunneling ionization probabilities (2-3) and even harmonic generation, HG, plateaus (5). A similar quasistatic model which includes the multilevel Coulomb potentials of electrons in molecules can be used to explain enhanced ionization at critical internuclear distances R_c (6-8) and angles θ_c (9), as well as high order HG in molecules (10-11).

A useful parameter, γ, called the Keldysh parameter (1), (12), helps separate nonlinear multiphoton processes, either atomic or molecular into two regimes, the multiphoton and tunneling regimes.

γ is defined as

$$\gamma = \left(I_p / 2U_p\right)^{1/2}, \quad U_p = e^2 \mathcal{E}_0^2 / 4m\omega^2 \quad , \tag{1}$$

where I_p is the ionization potential, U_p is a plasma physics concept corresponding to the ponderomotive energy of an electron of charge e and mass m in the presence of a field of maximum amplitude $\mathcal{E}_0$, frequency ω and phase ϕ, (13),

$$\mathcal{E}(t) = \mathcal{E}_0(t) \cos(\omega t + \phi) \quad . \tag{2}$$

For $\gamma > 1$, typical of high frequencies (e.g. UV, X-Ray) one has the regime of *multiphoton* ionization where the ionization rate is usually proportion to I^n, where n is the multiphoton order, $I = c\mathcal{E}_0^2 / 8\pi$ is the maximum field intensity. The regime ($\gamma > 1$) corresponding to small ponderomotive energies U_p therefore suggests a perturbative approach for describing the ionization. For $\gamma < 1$, where field effects as measured by U_p are greater than atomic or molecular electronic energies I_p, it is found generally that ionization rates become less and less dependent on wavelength. This is usually described as the *tunneling* ionization regime where now ionization can be simply modeled as tunneling of the ionizing electron through field-induced (laser + Coulomb) static barriers (2-3). When applied to explain saturation intensities of ion yields, it is found that such a quasistatic model works well for both atoms and molecules (14).

The earliest numerical evidence of unusual ionization rates of molecules in the nonperturbative regime (15) was interpreted in terms of laser-induced localization of the electron in the molecule, as a result of large charge resonance effects (6,16) and was therefore called *Charge Resonance Enhanced Ionization*, CREI. Other interpretations of enhanced molecular ionization extended the atomic quasistatic ionization barrier-suppression model to molecules (6-9). Recent exact numerical simulations via the TDSE, Born-Oppenheimer and non-Born Oppenheimer, the latter with moving nuclei (17), have enable us to relate nuclear kinetic energy anomalies in laser induced Coulomb explosions, LICE, (18-22) to the CREI phenomenon. Finally there is now complete experimental confirmation of CREI in the diatomics I_2 (23) and H_2 (22), (24)

103

and D_2 (25) by pulse-probe laser experiments. We describe in this paper the theoretical models and calculations which explain the highly nonlinear phenomenon CREI, and possible applications of molecular Coulomb explosions, CE.

NUMERICAL CALCULATIONS

Intense static field dissociative ionization of simple molecules such as H_2 was considered as early as 1961 by Hiskes (26) and 1975 by Hanson (27). Current experiments are being done using intense frequency dependent laser fields, which enable one to achieve even higher field intensities, exceeding the atomic unit (a.u.) of electric-field $\mathcal{E}_o$ and corresponding intensity I_o (4), (13),

$$\mathcal{E}_o = 5.14 \times 10^9 \text{V/cm} \quad ; \quad I_o = c\mathcal{E}_o^2 / 8\pi = 3.51 \times 10^{16} \text{ W/cm}^2 \quad . \quad (3)$$

Nonperturbative calculations are therefore required to model laser-molecule interactions at such high intensities and these can be obtained from numerical solutions of the appropriate TDSE. We have performed the first Born-Oppenheimer, i.e., with static nuclei in 3-D (15) and non-Born Oppenheimer with moving nuclei in 3-D and 1-D calculations of dissociative-ionization of H_2^+ (17). Exact static nuclei calculations have also recently been obtained for the one-electron symmetric and nonsymmetric linear H_3^{2+}, in 3-D (28) and 1-D (29), and the 1-D two electron systems H_2, H_3^{2+} (9) in order to study geometrical effects on CREI.

As an example, we write the complete 1-D one-electron TDSE for H_2^+ for linearly polarized laser fields, equation (2), parallel to the internuclear (z,R), axis: (17), (22),

$$\frac{i\hbar \partial \varphi(z,R,t)}{\partial t} = \hat{H}(z,R,t)\varphi(z,R,t) \quad , \quad (4)$$

$$\hat{H}(z,R,t) = \hat{H}_R(R) + V_c(z,R) + \hat{H}_z(z) \quad ,$$

$$\hat{H}_R = \frac{\hbar^2}{m_p} \frac{\partial^2}{\partial R^2} + 1/R \quad ; \quad \hat{H}_z = -\beta \frac{\partial^2}{\partial z^2} - \kappa e z \mathcal{E}(t) \quad ,$$

$$\beta = \frac{(2m_p + m_e)}{4m_p m_e} \hbar^2 \quad ; \quad \kappa = 1 + \frac{m_e}{2m_p + m_e} \quad ,$$

where m_e, m_p are respectively the electron and proton masses, z and R the corresponding electron and nuclear coordinates, $\mathcal{E}(t)$ the laser pulse (equation (2)).

$$V_c(z,R) = -\frac{e^2}{|1+z\pm R/2|^{1/2}} \tag{5}$$

is a regularized Coulomb potential in 1-D between the electron and the two nuclei situated at $\pm R/2$ along the z-axis. The Hamiltonian (4-5) represents the *exact* three-body Hamiltonian obtained after separation of the center-of-mass motion (4). The TDSE (4) is solved numerically using high-order split-operator methods (31). Thus complete numerical solution of the electron-nuclear wavefunction $\varphi(z,R,t)$ allows for calculations of nonperturbative ionization rates in the static (fixed nuclei) Born-Oppenheimer limit (15) and for a study of complete electron-nuclear dynamics during the photodissociation-ionization process with moving nuclei, therefore the exact non-Born Oppenheimer regime (17). In the latter full dynamical case, projection onto electronic Volkov states (13), i.e., the exact wavefunction of a free electron in an electromagnetic field $\mathcal{E}(t)$, allows for calculation of electronic kinetic energy spectra, the so called molecular ATI spectra and nuclear kinetic energy spectra which are comprised of ATD and CE nuclear energies.

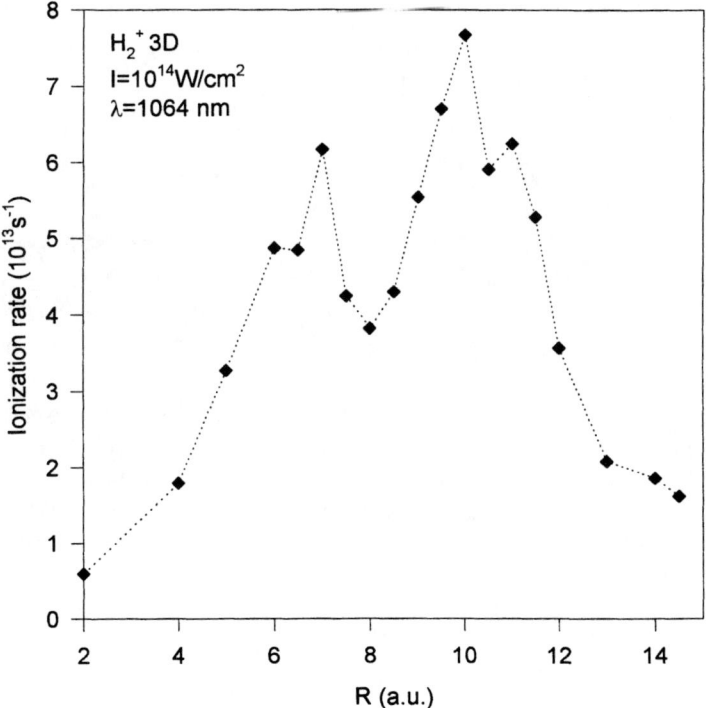

FIGURE 1. Ionization rates (10^{13} s^{-1}) for H_2^+ (3-D) as a function of internuclear distance R at $I = 10^{14}$ W/cm^2, $\lambda = 1064$ nm.

We illustrate typical Born-Oppenheimer, static nuclei ionization rates as a function of inter-proton distance R for the one-electron system, H_2^+, in Figure 1 above. The results were obtained from exact 3-D calculations for the laser parameters $I = 10^{14}$ W/cm^2, $\lambda = 1064$ nm, five cycle rise of the field from zero to the peak strength $\mathcal{E}_o = (8\pi I/c)^{1/2}$ followed by 20 cycles of constant field strength $\mathcal{E}_o$. The ionization rates $\Gamma(s^{-1})$ are obtained from the exponential decrease of the total probability or norm $N(t)$ after integrating over the electronic volume dv in a 3-D numerical grid with absorbing boundaries (15),

$$\ln N(t) = -\Gamma t \quad ; \quad N(t) = \int |\varphi(z,R,t)|^2 \, dv \quad . \tag{6}$$

The asymptotic, large R ionization rates converge to the atomic H (1s) rate. In all cases calculated so far one observes clear maxima in the ionization rates at critical distances $R_c = 7$ and 10 a.u. for H_2^+. Such maxima have been observed in all our numerical simulations to date at critical distances R_c for the linear systems $H_2^+, H_2, H_3^{2+}, H_3^+, H_4^{3+}, H_4^{2+}, H_5^{4+}, H_3^{3+}$ (31) and nonlinear H_3^{2+} (9). These numerical results confirm that enhanced ionization at critical distances R_c and goemetries θ_c is a *universal* phenomenon which we explain in the next section in terms of a quasistatic above barrier ionization model.

CREI CRITICAL DISTANCES

Enhanced ionization of molecules in ultrashort (t < 50 fs) intense ($I \geq 10^{14}$ W/cm^2) laser pulses at large internuclear distances, now called critical distances R_c, was first discovered as early as 1993 in our exact numerical simulations of the ionization rates of H_2^+ in search of optimum conditions for high order HG (6). The two ionization maxima at $R_c = 7$ and 10 a.u., Fig. 1, which have now been confirmed experimentally (24) were originally attributed to *laser induced localization*, a concept which also has been used for laser-driven quantum wells (33). Figure 2 shows that the second ionization maximum at $R_c = 10$ a.u. can also be explained by a quasistatic model (6-8), originally applied to atoms (2-3) to explain tunneling ionization. In the molecular case, the static energy levels, illustrated in Figure 2 are the lowest occupied molecular orbital, HOMO, and the highest unoccupied molecular orbital, LUMO and these were obtained by numerical solutions of the TDSE in a static field of strength $\mathcal{E}_o$ corresponding to the intensity $I = 10^{14}$ W/cm^2 (6). In the field free case these are the $1\sigma_g$ and $1\sigma_u$ bonding and antibonding MO's of H_2^+ which asymptotically dissociate to $(1s_1 \pm 1s_2)/\sqrt{2}$ combinations of 1s atomic orbitals respectively, on protons 1 and 2, producing a charge resonance, CR, transition moment $(1\sigma_g/z/1\sigma_u) = R/2$, as predicted in 1939 by Mulliken (34). In the presence of the field these levels are coupled radiatively by the electronic Rabi frequency (in a.u.) $\Omega_R = \mathcal{E}_o R/2$. At large internuclear distances where the two MO's are quasidegenerate, these levels are separated by the energies $\pm \Omega_R$. It is to be noted that this corresponds to the difference in potential energy $\mathcal{E}_o R$ experienced by the electron at opposite ends of the molecule when the protons are separated by the internuclear distance R. This static (Stark) energy separation of the HOMO and LUMO is illustrated in Fig. 2, where one readily observes that at $R \simeq 9$ a.u. the field modified LUMO is degenerate with the maximum of the right barrier created by the net electrostatic field (we set $e = 1$ and q is the nuclear charge),

$$V(z,R) \equiv -q/|\frac{R}{2} \pm z| + \mathcal{E}_o z \quad . \tag{7}$$

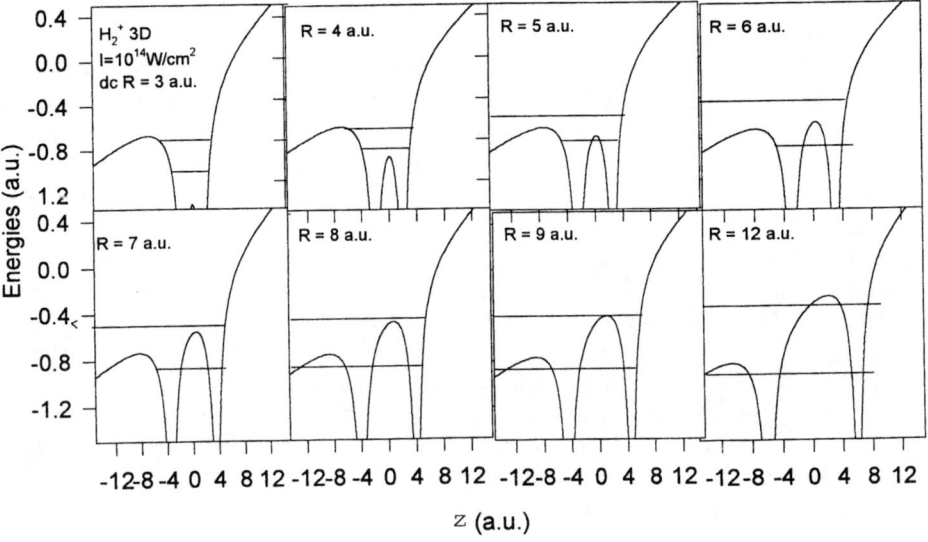

FIGURE 2. 3-D H_2^+ LUMO (highest) and HOMO (lowest) molecular orbital energies for different R's in a static field $\mathcal{E}_o = (8\pi I/c)^{1/2}$ at $I = 10^{14}$ W/cm^2. z is the electron coordinate.

Thus for distances $4 < R < 10$ a.u., the LUMO is situated above all barriers, whereas for $R = 10$ a.u., it is trapped by the total barrier created by the net static potential V, equation (7). This readily explains the sudden drop of the ionization rate towards the atomic H(1s) value for $R > 10$ a.u. as observed in Fig. 1. The sharp rise of the ionization rate from the equilibrium distance $R_e \simeq 2$ a.u. to the first peak at $R = 7$ a.u. is due to the decrease of the ionization potential and increasing radiative excitation of the LUMO from the HOMO as a function of R (35). Thus at $R \simeq 8$ a.u., one is in a diabatic excitation regime, where the populations of the LUMO and HOMO are equal on average and are thus equally depopulated upon ionization after half-cycles of the field

(6), (36). The first maximum at R $\simeq$ 7 a.u. can be explained in terms of electron localization in either well by the laser field. This is a dynamical effect which suppresses the electron tunneling when the field is zero (6). In summary, three time scales are operative at the R_c distance: the ionization time $\tau_c = 1/\Gamma$, the field oscillation period $\tau_L = 2\ \pi/\omega$, and finally the tunneling time τ_t. The latter is modified by the field and becomes zero at the laser induced-electron localization condition (6,16), (33)

$$J_o(2\Omega_R/\omega) = 0 \quad, \tag{8}$$

where J_o is a zeroth order Bessel function, Ω_R is the electronic Rabi frequency equal to $\frac{1}{2}$ the electronic Stark energy difference $\mathcal{E}_oR$ of the electron at the positions of the protons and ω is the laser frequency. The sharp ionization peaks at 7 and 10 a.u. can be shown to correspond to such laser electron localization (see Fig. 3 in ref. (6)) as well as the barrier trapping seen for R $\geq$ 9 a.u. in Fig. 2.

An approximate estimate of the distance R_c where the LUMO (antibonding MO) becomes trapped by the net static barrier V, equation (7) was proposed earlier by Codling et al. (37) as the internuclear distance where the two field free MO's become quasidegenerate at large distances and are at the top of the mid-point of the symmetric field free Coulomb barrier V_c, equation (5). Using the approximation that the ionization potential of an atomic ion with charge (q-1) is q I_p, where I_p is the first atomic ionic potential, then one can define the critical point as the above degeneracy point between the two MO's and the maximum of V_c, which gives $R_c = 3/I_p$, independent of charge q (8,19). The motivating assumption for this approximation was that for R > R_c, the MO's fall in energy below the maximum of V_c at R/2 and thus electron localization should occur (8,37). However as pointed out at the beginning of this section, there are three important time scales, t_i for ionization, t_L for the laser oscillation and t_t for tunneling through the barrier V_c at zero field. For very large charges q, t_t will become negligible and one expects tunneling to become unimportant.

The above approximation neglects the important field charge resonance energy displacements of the MO's through the electronic Rabi frequency $\mathcal{E}_oR/2$ as pointed out first in ref. (16). A more accurate value of R_c is obtained by introducing the laser field interaction between the two doorway states, HOMO (bonding) and LUMO (antibonding) originating from the same asymptotic pair of orbitals, e.g., H(1s). This results in HOMO-LUMO energy separation equal to $\mathcal{E}_oR$, i.e., the classical potential energy difference between the two potential wells, Fig. 1. Therefore for R > R_c, the upper level E_+ is below the maximum of the total V, equation (7) and is trapped, i.e., ionization is suppressed, whereas for R < R_c, E_+ is free and can rapidly ionize. This result can be shown to be independent of nuclear charge q and depends little on field strength $\mathcal{E}_o$ (7). Thus including the charge resonance field displacement $\mathcal{E}_oR/2$ in the energy E_+, we get

$$E_+(R) = -\ q\ I_p - q/R + \mathcal{E}_oR/2 \quad, \tag{9}$$

and finding R_c by equating $E_+(r)$ to V(z,R), equation (7) gives the result

$$R_c = 4/I_p \quad. \tag{10}$$

108

Since $I_p = 0.5$ a.u. for H, then $R_c = 8$ a.u., the average of the two maxima in Fig. 1. The insensitivity of R_c to q and $\mathcal{E}_o$ comes from the fact that the largest maximum of $V(z,R)$ approaches $\mathcal{E}_o R/2$ for large q, thus canceling the charge resonance or Stark energy shift $\mathcal{E}_o R/2$ of the LUMO, i.e., E_+. We reiterate that the two peaks in Figure 1 are also coincident with the zeroes of J_o in equation (8) for H_2^+. Thus for small nuclear charges such as in H_2^+, both above barrier ionization and laser induced localization are operative.

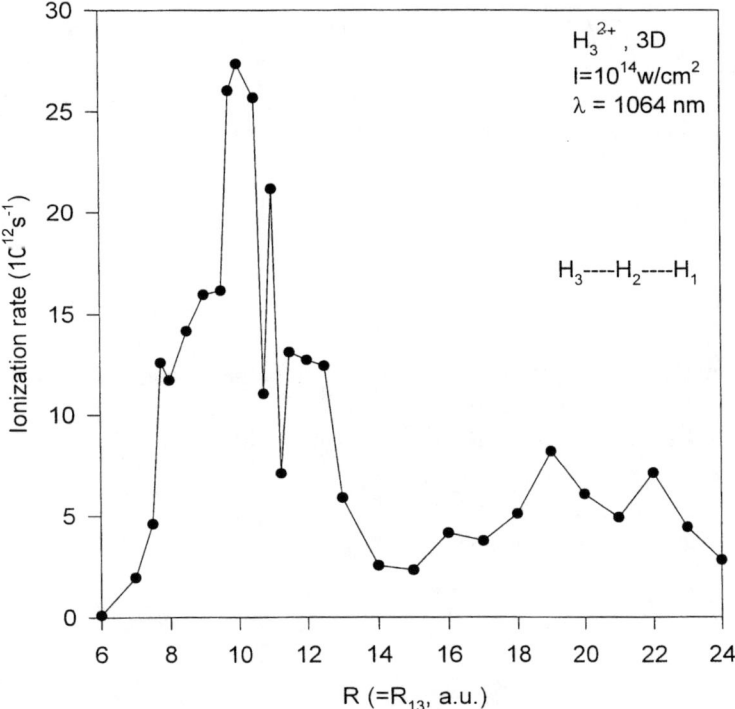

FIGURE 3. Ionization rates (10^{12} s^{-1}) for H_3^{2+} (3-D) as a function of $R = R_{13}$ at $I = 10^{14}$ W/cm^2, $\lambda = 1064$ nm.

Extending this static ionization model to triatomics such as H_3^{2+} involves consideration now of three doorway states, (9), (28). Both for the linear geometry, Fig. 3, (28-29) and the nonlinear molecule (9) sharp ionization maxima occur at $R_c \simeq 10\text{-}11$ a.u. where R is the total bond length of the linear molecule. For the nonlinear-geometry an ionization maximum occurs also at the bond angle $\theta_c = 87°$ (9). Thus the CREI mechanism is again operative through the field induced displacement of the three doorway MO's. We consider here the linear case, Fig. 4, where the three MO's are given by the LCAO method (38),

$$1\sigma_g = 2^{-1/2}\left[1s_2 + 2^{-1/2}(1s_1 + 1s_3)\right] \ ,$$
$$1\sigma_u = 2^{-1/2}\left[1s_1 - 1s_3\right] \ , \tag{11}$$
$$2\sigma_g = 2^{-1/2}\left[1s_2 - 2^{-1/2}(1s_1 + 1s_3)\right] \ ,$$

where $1s_2$ is the central atomic orbital. The transition moments $\mu = <\sigma_g|z|\sigma_u>$ can readily be shown to be $R/(2)^{3/2}$ in contrast to H_2^+ where $\mu = R/2$. In each case such divergent moments are due to charge resonance effects (34). In the presence of a static field of amplitude $\mathcal{E}_o$, one obtains three new field induced states (see Fig. 3) with energies (28),

$$E_{1(3)} = -I_p - \frac{4}{R} - (+)\mathcal{E}_o R/2 \ ,$$
$$E_2 = -I_p - 4/R \ , \tag{12}$$

where the middle level E_2 remains undisplaced with energy corresponding to the atomic ionic potential – I_p perturbed by two adjacent Coulomb potentials – $2/R$. Efficient ionization of E_2 will occur at values of R_c whenever $E_2(R_c) = V(-R/2,-z)$, i.e.

$$I_p + \frac{4}{R} = \frac{1}{z} + \frac{1}{z+R/2} + \frac{1}{z+R} + \mathcal{E}_o(z+R/2) \ , \tag{13}$$

which gives the result

$$R_c = 5/I_p \ , \tag{14}$$

in agreement with the numerical results (see Fig. 3) at $I = 10^{14}$ W/cm^2, $\lambda = 1064$ nm laser excitation. This result can also be obtained by considering the value of R at which the unperturbed middle level E_2 of energy – I_p – R/4, is degenerate with the field free barriers at $z = \pm R/4$,

$$I_p + \frac{4}{R} = 2/(R/4) + 1/(R/2 + R/4) = 28/3 R \ , \tag{15}$$

which gives $R_c = 16/3 \ I_p$. Inserting $I_p = 0.5$ a.u. give $R_c \simeq 5.5$ a.u. in excellent agreement with the major ionization peak of H_3^{2+} illustrated in Fig. 3. The field induced $R_c = 5$ a.u., equation (14) is thus equal to the free field estimate R_c, obtained above. The coincidence of these two results comes from the equality of the classical energy shifts of the potential maxima V_1, V_3 (Fig. 4) with the charge resonance energy shifts of the levels E_1 and E_3. Thus both classical and quantum energy shifts cancel each other rendering the values of the CREI critical distance R_c nearly independent of charge q and field strength $\mathcal{E}_o$ as in H_2^+.

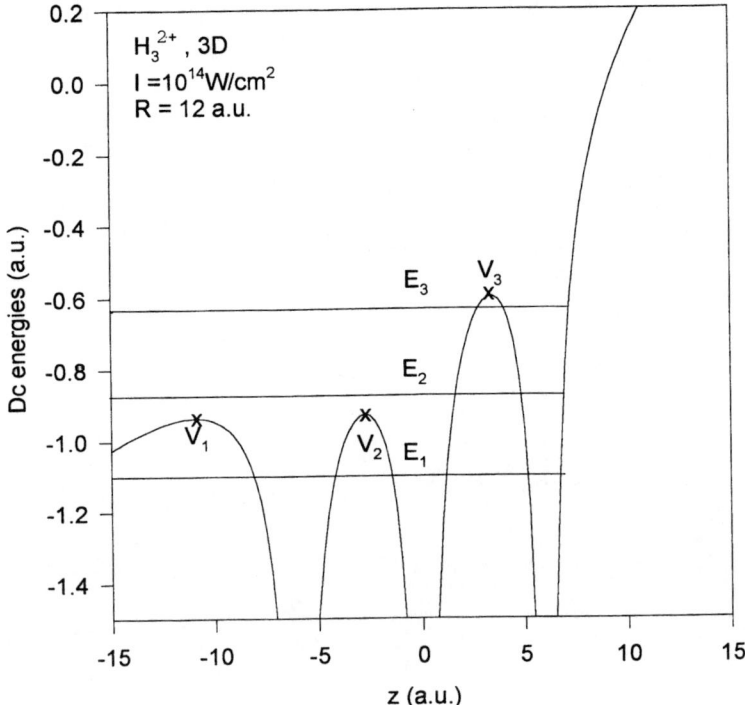

FIGURE 4. Static (Dc) molecular orbital energies for H_3^{2+} (3-D) in a field $\mathcal{E}_o = (8\pi I/c)^{1/2}$ at $I = 10^{14}$ W/cm², $\lambda = 1064$ nm.

The above one-electron symmetric models always leads to symmetrically charged products. Recent experiments using very short (30 fs) pulses indicate preponderance of asymmetric charged products (39-40). We generalize next the quasistatic model to diatomics with one electron in the presence of two nuclei of charge q_1 and q_2 giving rise to different ionization potentials $I_p(1)$ and $I_p(2)$ (28). The electron energy in atom (1) is then given for large R approximately by,

$$\varepsilon = - (I_p(1) + q_2/R) \ , \tag{16}$$

where the atomic electronic energy $-I_p(1)$, is perturbed by the neighboring Coulomb potentials $-q_2/R$. An electron at a position z with respect to the center of the system, $z = 0$, will evolve in the net Coulomb potential similar to equation (7)

$$V(z) = - q_1/|z - R/2| - q_2/|z + R/2| + \mathcal{E}_o z \quad . \tag{17}$$

For large q_1, q_2 as in the previous discussion one can neglect the static laser field contribution $\mathcal{E}_o z$ as a first approximation. The field free barrier maximum occurs at

$$z_m = \frac{R}{2}\left[\frac{q_1^{1/2} - q_2^{1/2}}{q_1^{1/2} + q_2^{1/2}}\right] \quad , \tag{18}$$

and corresponding

$$V(z_m, R) = -\frac{1}{R}\left(q_1^{1/2} + q_2^{1/2}\right)^2 \quad . \tag{19}$$

Equating the molecular electron energy ϵ , equation (16) to $V(z_m R)$, (19), gives the analytic field independent result

$$R_c = \left[q_1 + 2(q_1 q_2)^{1/2}\right]/I_p(1) \quad . \tag{20}$$

For $q_1 = q_2 = q$, i.e., the symmetric case (e.g., H_2^+, $q_1 = q_2 = 1$), the above equation leads to the result

$$R_c = 3q/I_p(q) \quad . \tag{21}$$

Setting $I_p(q) = qI_p$, where I_p is the first ionization potential and the q electrons are in the same electronic shell or principle quantum number n, gives a charge independent $R_c = 3/I_p(8)$. This neglects the charge resonance displacements as discussed above. Including these effects gives $R_c = 4/I_p$, the CREI critical distance (6,7).

The nonsymmetric model, $q_1 \neq q_2$, can be applied to two electron systems such as for e.g., the H_2^+ - H_2^+ system illustrated in Fig. 5. Recent high intensity LICE experiments on D_2 clusters result in complete ionization of these clusters and neutrons were detected from collision of high energy deuterons (41). We show that since in clusters the average interatomic or molecular distance is $\sim$ 6 a.u. (42), these systems are indeed already at the CREI critical distance thus inducing highly efficient ionization. Thus if we consider one H_2^+ with charge $q_1 = 1$ perturbing the neighboring electron on an H_2^{2+} ($q_2 = 2$), and since $I_p(1) = I_p(H_2) \simeq 0.7$ a.u. then from equation (20) we obtain $R_c \simeq 6$ a.u. for the H_2^+ - H_2^+ system, in good agreement with the lower intensity

$(10^{14}\ \text{W/cm}^2)$, Fig. 5a. R_c is displaced to shorter values at higher intensities, Fig. 5b but remains fixed at $\simeq 6$ a.u. for higher I_p's as in He_2^{2+}, Fig. 5c.

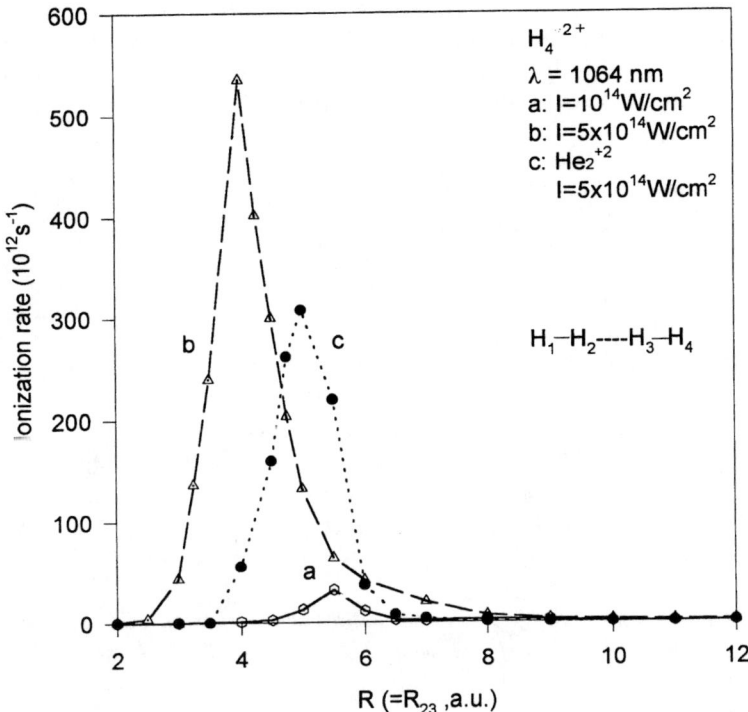

FIGURE 5. Ionization rates of H_4^{2+} (1-D) as a function of $R = R_{23}$, the inner interproton distance at $\lambda = 1064$ nm: a) $I = 10^{14}$ W/cm^2; b) $I = 5 \times 10^{14}$ W/cm^2; c) He_2^{2+}, $I = 5 \times 10^{14}$ W/cm^2.

The above considerations have neglected charge resonance effects which we have shown to be important for the one-electron systems H_2^+, H_3^{2+}. Two electron systems, due to electron correlation involve three essential or doorway states (30), (43), (44). The electronic wavefunctions φ of these doorway states dissociate into atomic configurations with atomic orbitals a or b on one or the other atom (43),

$$\varphi_1(1,2) = \sigma_g(1)\sigma_g(2) \xrightarrow{R\rightarrow\infty} [a(1)a(2) + b(1)b(2) + a(1)b(2) + b(1)a(2)]/2 \ ,$$

$$\varphi_2(1,2) = [\sigma_g(1)\sigma_u(2) + \sigma_g(2)\sigma_u(1)]/2^{1/2} \xrightarrow{R\rightarrow\infty} [a(1)a(2) - b(1)b(2)]/2^{1/2} \ , \qquad (22)$$

$$\varphi_3(1,2) = \sigma_u(1)\sigma_u(2) \xrightarrow{R\rightarrow\infty} [a(1)a(2) + b(1)b(2) - a(1)b(2) - b(1)a(2)]/2 \ .$$

Both ground φ_1 and second excited states φ_3 are linear combinations of *ionic* (aa,bb) atomic configurations and *covalent* (ab) configurations whereas the first excited state φ_2 is purely ionic upon dissociation. The corresponding transition moments can be shown to be (28), (34), $<\varphi_1|z|\varphi_2> = <\varphi_2|z|\varphi_3> = R/2^{1/2}$ as these are all $<\sigma_g|z|\sigma_u>$ single electron moments and thus are identical to the H_2^+ case, the typical charge resonance system (34). Since all three states are coupled in pairs by the same radiative interaction $\mathcal{E}_oR/2^{1/2}$, then three new field induced states are obtained at large field strengths $\mathcal{E}_o$ by replacing the 1s atomic orbitals by the MO's φ in equations (11-12):

$$\epsilon_+ = \mathcal{E}_oR \quad : \; ^{\prime} \, ^{-1/2}[\varphi_2 - 2^{1/2}(\varphi_1 + \varphi_3)] \xrightarrow{R\to\infty} b(1)b(2) \quad,$$

$$\epsilon_0 = 0 \qquad : \varphi_0 = [\varphi_1 - \varphi_3]/2^{1/2} \xrightarrow{R\to\infty} [a(1)b(2) + b(1)a(2)]/2^{1/2} \quad, \qquad (23)$$

$$\epsilon_- = -\mathcal{E}_oR \quad : \varphi_- = 2^{-1/2}[\varphi_2 + 2^{-1/2}(\varphi_1 + \varphi_3)] \xrightarrow{R\to\infty} a(1)a(2) \quad.$$

We note that the levels $\epsilon_\pm$ have energies corresponding to the classical Stark shifts $\pm\,\mathcal{E}_oR$. Thus charge resonance effects, which take into account molecular structure and hence electron delocalization via tunneling through internal Coulomb barriers (e.g. Fig. 2,4), predict from equation (23) that in the presence of an intense static field $\mathcal{E}_o$ the ground φ_- state become ionic, corresponding to the creation of the asymmetric charge transfer state, $A^{+q+1} - A^{+q-1}$, whereas the first excited state φ_0 is the symmetric state $A^{+q} - A^{+q}$. The ionic state is stabilized by the classic Stark energy - $\mathcal{E}_oR$.

We next include the two-electron field-charge resonance effects obtained in equation (23) into the charge transfer process $A^{+q} - A^{+q} \to A^{+q+1} - A^{+q-1}$. The initial total energy is

$$E_{qq}(R) = -2I_p(q) + \frac{q^2}{R} , \qquad (24)$$

where each electron of energy $- I_p(q)$ is under the influence of the Coulomb attraction from the neighbouring ion $-q/R$ and q^2/R is the ion-ion repulsion. The final energy of the charge transfer state in the field is

$$E^{(R)}_{q+1,q-1} = -I_p(q-1) - I_p(q) + \frac{(q+1)(q-1)}{R} - \mathcal{E}_oR , \qquad (25)$$

where we have added the last term $-\mathcal{E}_oR$ corresponding to the new field electronic energy ϵ_-, equation (23). The energy difference between these two states is then

$$E^{(R)}_{q+1,q-1} - E^{(R)}_{qq} = \Delta I_p - \frac{1}{R} - \mathcal{E}_oR , \qquad (26)$$

and

$$\Delta I_p = I_p(q) - I_p(q-1) \quad . \tag{27}$$

Most efficient charge transfer followed by rapid ionization of the less stable A^{+q-1} fragment will occur at the crossing distance R_c of the symmetric $A^{+q} - A^{+q}$ and charge transfer configuration $A^{+q+1} - A^{+q-1}$ in the field, i.e., when $E_{q+1,q-1} = E_{qq}$. The solution of this equation is

$$R_c = \left[\Delta I_p + \left(\Delta I_p^2 - 4\mathcal{E}_o \right)^{1/2} \right] / 2\mathcal{E}_o \quad , \tag{28}$$

with a minimum

$$R_c = 2/\Delta I_p \quad . \tag{29}$$

Equations (28-29) predict a decrease of R_c with increasing field strength $\mathcal{E}_o$ as observed in Fig. 5. Applying this result to the two electron systems, H_2 and H_3^+ (29) for which $\Delta I_p \simeq I_p (H) = 0.67$ a.u. (this is the I_p for 1D-H atom) gives $R_c \simeq 6$ a.u. This corresponds to the CREI critical distances R_c found from numerical static field calculations. For the $H_2^+ - H_2^+$ and He_2^{2+} systems, in Fig. 5, $\Delta I_p \simeq 1$ a.u., which gives an $R_c \simeq 4$ a.u. in good agreement with Fig. 5. Finally applying equation (28) to the highly charged system I_2^{10+} (39), for which $\Delta I_p = I_p(5) - I_p(4) = 0.77$ a.u., one obtains $R_c = 6$ a.u. at $I = 4.5 \times 10^{14}$ W/cm^2. This corresponds to the experimentally observed distance at which highly ionized fragments are produced with ultrashort pulses (39) and is shorter than the CREI R_c distance for the I_2^{2+} system. For the latter, since $\Delta I_p \simeq 0.4$ a.u., $R_c \simeq 10$ a.u. in agreement with the experimental $10 < R_c < 12$ a.u., (24).

CONCLUSION

We have shown in the previous section, based on accurate numerical calculations of the TDSE for one and two electron molecular systems that charge resonance effects, corresponding to charge transfer between different ends of a molecule give rise to critical distances where ionization is enhanced an this is the distance at which Coulomb explosion, CE, occurs. We call this Charge Resonance Enhanced Ionization, CREI, in order to emphasize this important effect which occurs in electron delocalized systems such as molecules but not in atoms. Odd electron systems such as $H_n^{(2n-1)+}$ are adequately described by two doorway states, the HOMO and LUMO which are coupled radiatively by the field-charge resonance energy $\mathcal{E}_o R/2$ thus localizing the electron on one atom. Even electron systems, such as H_n^{2n+} involve at least three doorway states. In this latter case, for intense fields, a new ground state is produced which is the ionic structure corresponding to complete charge transfer of one electron to one atom, for e.g., $H_2 \rightarrow H^+ H^-$. This state is stabilized by the field-molecule interaction $-\mathcal{E}_o R$. In the one electron case, or odd electron systems, over barrier escape or trapping of the LUMO determines the critical distance R_c for CREI (Figs 2,4). For two electron systems or even charged molecular ions, it is the crossing of the asymmetric ionic charge transfer state with the symmetric dissociative state which determines the critical distance R_c where efficient charge transfer can occur followed by rapid ionization of the transferred electron at large internuclear distances. For H_2^+ and odd charged diatomic systems, $R_c \simeq 4/I_p$ where I_p is the first ionic potential of the atomic dissociation fragments. For more delocalized one electron systems such as H_3^{2+}, $R_c \simeq 5/I_p$ where R is the total length of the molecule. For two electron systems, charge resonance effects involving three doorway states produce also a minimum $R_c \simeq 4/\Delta I_p$, where ΔI_p is the difference in ionization potential between a fragment A^{+q} and its charge transfer state A^{+q-1}.

Over barrier charge transfer models have a long history in Rydberg and ion-ion collisions, (45) explaining the relative ease of creating inverted electron populations with possible applications to X-Ray lasers (46). Recently similar models have been used to explain charge-ion-surface electron transfer collisions (47-48). The influence of short intense laser pulses in such charge exchange processes have not yet been considered. The present CREI models suggest that electron transfer processes will be considerably modified if not enhanced in the presence of short laser pulses and may be even be controlled by combinations of different laser frequencies and phases (49).

ACKNOWLEDGMENTS

We thank the following collaborators, S. Chelkowski, H. Yu, T. Zuo and colleagues, S.L. Chin, P.B. Corkum, G.N. Gilson for their input and suggestions which have helped shape this article.

REFERENCES

1. Gavrila, M., *Atoms in Intense Fields*, New York, Academic Press, 1992.
2. Corkum, P.B., Burnett, N.H., and Brunel, F., in ref. (1), p. 109.
3. Corkum, P.B., Burnett, N.H., and Brunel, F., *Phys. Rev. Lett.* **62**, 1259 (1989).
4. Bandrauk, A.D., *Molecules in Laser Fields*, New York, M. Dekker Publisher, 1994.
5. Corkum, P.B., *Phys. Rev. Lett.* **71**, 1994 (1993).
6. Zuo, T., and Bandrauk, A.D., *Phys. Rev.* **A52**, 2511 (1995).
7. Chelkowski, S., and Bandrauk, A.D., *J. Phys.* **B28**, L723 (1995).
8. Seiderman, T., Ivanov, M.Y., and Corkum, P.B., *Phys. Rev. Lett.* **75**, 2819 (1995).
9. Bandrauk, A.D., and Ruel, J., *Phys. Rev.* **A59**, 2153 (1999).
10. Bandrauk, A.D., Chelkowski, S., and Constant, E., *Phys. Rev.* **A56**, 2537 (1997).
11. Bandrauk, A.D., and Yu, H., *Phys. Rev.* **A59**, 2511 (1995).
12. Keldysh, L., *Sov. Phys. JETP*, **20**, 1307 (1965).
13. Krainov, V.P., Reiss, H., and Smirov, B.M., *Radiative Processes in Atomic Physics*, New York, J. Wiley & Sons, 1997.

14. Walsh, T., Ilkov, F.A., Decker, J., and Chin, S.L., *J. Phys.* **B27**, 3767 (1994).
15. Chelkowski, S., Zuo, T., and Bandrauk, A.D., *Phys. Rev.* **A46**, 5342 (1992).
16. Zuo, T., Chelkowski, S., and Bandrauk, A.D., *Phys. Rev.* **A48**, 3837 (1993).
17. Chelkowski, S., Foisy, C., and Bandrauk, A.D., *Phys. Rev.* **A54**, 1176 (1998); **A52**, 2977 (1995).
18. Schmidt, M., Normand, D., and Cornaggia, C., *Phys. Rev.* **A50**, 5037 (1994).
19. Posthumus, J.H., Frasinski, L.J., Giles, A.J., and Codling, K., *J. Phys.* **B28**, L349 (1995).
20. Cornaggia, C., *Phys. Rev.* **A54**, 2535 (1996).
21. Normand, D., and Schmidt, M., *Phys. Rev.* **A53**, 1958 (1996).
22. Walsh, T., Chin, S.L., Chelkowski, S., and Bandrauk, A.D., *Phys. Rev.* **A58**, 3922 (1998).
23. Constant, E., Stapelfeldt, H., and Corkum, P.B., *Phys. Rev. Lett.* **78**, 4140 (1996).
24. Gibson, G.N., Li, M., Guo, C., and Neva, J., *Phys. Rev. Lett.* **79**, 202 (1997).
25. Trump, C., Rothke, H., and Sandner, W., *Phys. Rev.* **A59**, 2858 (1999).
26. Hiskes, J.R., *Phys. Rev.* **122**, 1207 (1961).
27. Hanson, G.R., *J. Chem. Phys.* **62**, 1161 (1975).
28. Yu, H., and Bandrauk, A.D., *Phys. Rev.* **A56**, 685 (1997); *J. Phys.* **B31**, 1533 (1998).
29. Hu, S.X., Qu, W.X., and Xu, Z.Z., *J. Phys.* **B31**, 1523 (1998).
30. Yu, H., and Bandrauk, A.D., *Phys. Rev.* **A53**, 3290 (1996).
31. Bandrauk, A.D., and Shen, H., *J. Chem. Phys.* **99**, 1185 (1993).
32. Bandrauk, A.D., *Comm. Atom. Molec. Phys.*, to appear, 1999.
33. Gomez, J.M., and Plata, J., *Phys. Rev.* **A45**, 6954 (1992).
34. Mulliken, R.S., *J. Chem. Phys.* **7**, 20 (1939).
35. Kawata, I., Kono, H., and Fujimura, Y., *Chem. Phys. Lett.* **289**, 546 (1998); *J. Chem. Phys.* **110**, 11152 (1999).
36. Zuo, T., Chelkowski, S., and Bandrauk, A.D., *Phys. Rev.* **A49**, 3943 (1994).
37. Codling, K., Frasinski, L.J., and Hatherly, P.A., *J. Phys.* **B22**, L321 (1989).
38. Slater, J.C., *Quantum Theory of Molecules and Solids*, vol. I, New York, McGraw-Hill (1963).
39. Gibson, G.N., Li, M., Guo, C., and Nibarger, J.P., *Phys. Rev.* **A58**, 4723 (1998).
40. Guo, C., Li, M. and Gibson, G.N., *Phys. Rev. Lett.* **82**, 2492 (1999).
41. Ditmire, T., *et al.*, *Nature* **398**, 489 (1999).
42. Purnell, J., Snyder, E.M., Wei, S., and Castleman, A.W., *Chem. Phys. Lett.* **229**, 333 (1994).
43. Ref. 38, table 4-1, p. 63.
44. Rojo, A.G., and Mahan, G.O., *Phys. Rev.* **B47**, 1794 (1993).
45. Ostrovsky, V.N., *J. Phys.* **B28**, 3901 (1995).
46. Kato, Y., Takuma, H., and Daido, L., *X-Ray Lasers*, Bristol, UK: IOP Pub. Ltd., 1998.
47. Burgdörfer, J., Lerner, P., and Meyer, F.W., *Phys. Rev.* **A44**, 5674 (1991)..
48. Ducrée, J.J., Casali, F., and Thumm W., *Phys. Rev.* **A57**, 338 (1998).
49. Bandrauk, A.D., and Yu, H., *Internatl. J. Mass. Spectrometry*, to appear 1999.

Photodetachment of negative ions through doubly excited states

E. Lindroth, N. Brandefelt and A. Bürgers

Department of Atomic Physics, Stockholm University, Frescativ. 24, S-104 05 Sweden

Abstract. The possibility to form doubly excited states give rise to resonances in photodetachment spectra. The states are fragile, dominated by electron correlation and require a careful theoretical treatment. Here recent calculations on He$^-$ and Li$^-$ are discussed and the resonances are compared to those in the well studied H$^-$- ion.

INTRODUCTION

When an electron attach to an atom to form a negative ion it is the dipole field induced in the atom by the electron itself which makes binding possible. In atoms the Coulomb field leads to the presence of an infinite number of bound states. The dipole field in negative ions can in contrast only support a finite number of bound states, the existence of only one single bound state is very common and e.g. He$^-$ has no true bound state at all. Generally negative ions have, however, several resonance states. Resonance states are doubly (or multiply) excited autodetaching states which show up as resonances e.g. in photodetachment experiments. These states are strongly affected by electron correlation and independent particle models cannot even give a qualitative description. The fact that doubly excited states in negative ions are fragile and behave as true many-particle states makes them interesting as prototype correlated systems.

For H$^-$ the dipole field is unusually strong. This is due to the almost perfect degeneracy of the excited energy levels in the hydrogen atom which allows complete mixing of states of different parity in the presence of the electric field provided by the extra electron. The dipole potential will in this case decay as $1/r^2$ and it is possible to show [1] that with such a potential one would expect an infinite series of states converging exponentially to the degenerate thresholds, i.e. the m:th state below a threshold is bound relative this threshold with $\Delta E_m = exp(-km)$. In other negative ions, which have non-degenerate levels in the parent atom, the dipole potential decays as $1/r^4$ and the series of resonances is truncated and converge even faster to the threshold. In reality the series of states will be truncated also in H$^-$ due to the small splitting in the thresholds caused by relativistic and radiative

CP500, *The Physics of Electronic and Atomic Collisions*, edited by Y. Itikawa, et al.
© 2000 American Institute of Physics 1-56396-777-4/00/$17.00

corrections. This situation has been studies quantitatively recently and calculations show that below the H($n = 2$) threshold the $^1P^o$ series is truncated already after the third member [2–4]. Experimentally the truncation of the series is hard to study since the higher members of the series get harder and harder to photo-excite from the ground state. The second member of the $^1P^o$ series below the H($n = 2$) thresholds resonances was even so recently measured [5].

Doubly excited states can be classified from their total angular momentum and parity. These quantum numbers are, however, not enough to explain why some states are rather long lived while others have several order of magnitude shorter lifetime or why certain states give rise to prominent photodetachment resonances while other states of the same symmetry have a very small excitation cross section. Since independent particle models fail to describe doubly excited states in general and negative ions in particular, a lot of effort has been invested into the search for sets of other quantum numbers which could give a good description of the state (i.e. they should be approximately good quantum numbers) and also give a physical explanation of the properties of the states.

H$^-$ is the prototype negative ion and being a pure two-electron system it is also possible to treat "exactly" and has become a favourite test case for theoretical approaches. The efforts to develop new sets of quantum numbers have consequently focused on two-electron systems. In this progress report we will discuss other few electron negative ions; He$^-$ and Li$^-$. We will focus on the comparison with H$^-$ and the question to which extent the understanding of H$^-$ can be translated to these systems. First the alternative sets of quantum numbers are discussed and then the properties of the three systems are discussed in more detail.

CLASSIFICATION

One classification scheme developed to understand doubly excited states in two-electron systems is the $(K, T)^A$ quantum numbers. K and T were first introduced by Herrick and Sinanoglu [6] in a group theoretical study of double excitation in He and describe angular correlation. The approximate quantum number T is approximately the projection of the total angular momentum of the electron pair onto the inter-electronic axis, $\mathbf{r}_{12}$. K is related to the angle between the electrons; $\cos \theta_{12} \approx -K/N$ [7], where N is the threshold to which a series of resonances is converging. Radial correlation is commonly described by the approximate quantum number A, which was suggested by Lin [8]. The value of A=+1(-1) is often referred to as an in-phase(out-of-phase) radial oscillation of the two electrons about the nucleus. More precisely the wave functions describing states with $A = +1(-1)$ have an anti node (node) on $r_1 = r_2$.

In photodetachment studies of H$^-$ above the H($n = 2$) threshold only certain doubly excited states give rise to prominent resonances, see e.g. the experimental study of photodetachment below the H($n = 5 - 8$) thresholds in Ref. [9] and the related theoretical study in Ref. [10]. The states which are seen in the experiment are of $A = +1$ character and have among those states the largest possible K-value (corresponding to maximum angle between the electrons). The complete dominance of $A = +1$ can be explained from the fact that the ground state has $A = +1$ character and the probability to photo-excite to a state with different A is very small. Also the dominance of excitation to the maximum K-states can be explained with the help of new classification schemes. This was first noted by Feagin and Briggs [11] in their molecular orbital description of two electron systems. In this approach e.g. H^- is described in accordance with the H_2^+ molecular ion; the roles of the electrons and the nucleus are interchanged, the inter-electronic distance r_{12} is treated as the adiabatic coordinate and molecular quantum numbers are deduced. Although the derivation is very different compared to that of the (K, T)-scheme [6] it is possible to show that there is a one to one correspondence between the two classification schemes [11]. One of the molecular quantum numbers n_λ (sometimes also called ν), which counts the number of elliptical nodal surfaces, can be written as $n_\lambda = (1/2)(N - K - T - 1)$ [11] and was shown to be approximately conserved in radiative transitions [11,12]. The ground state has $n_\lambda = 0$ and since the maximum K is given by $K_{max} = N - 1 - T$ [6] a state with $K = K_{max}$ has also $n_\lambda = 0$, which explains why excitation to such states dominates.

How well do the new classification schemes describe real doubly excited states? Eigenstates to the K and T quantum numbers, $| \{n_1 n_2 KT\}LS \rangle$ can be written as a sum of coupled hydrogenlike basis states $| \{n_1 \ell_1 n_2 \ell_2\}LS \rangle$ [6]. An expansion of the wave function for a doubly excited state in H^- in hydrogenlike basis functions can thus be rewritten as an expansion in the $| \{n_1 n_2 KT\}LS \rangle$ basis. While the expansion in the hydrogenlike $| \{n_1 \ell_1 n_2 \ell 21\}LS \rangle$ basis is a sum over a large number of configurations (of which none can be said to be dominating) more or less all contributions in the $| \{n_1 n_2 KT\}LS \rangle$ expansion come generally from basis states with one specific K and one specific T [13]. Note that it is the $\ell_1 \ell_2$-label which is replaced by the more adequate KT-label, in both pictures the hydrogenic n-value of especially the "outer" electron has completely lost its meaning in the doubly excited state, leading to admixtures from a large number of $n_1 n_2$-configurations.

In conclusion; for H^- a classification scheme exist which gives a description close to the true states and in addition provides a framework within which the behaviour of the states can be understood. There is, however, no reason why this scheme should hold when there is a structured core as in the case of He^- or Li^-, on the contrary the pure Coulombic potential between the core and the electrons is a key feature in the group theoretical derivation of the K and T quantum numbers.

In Ref. [14] the result from a calculation of the expectation values of $\cos \theta_{12}$ and the use of $K \approx -N \langle \cos \theta_{12} \rangle$ was compared with the result of a direct projection of the accurate two-electron wave functions onto pure (K, T) states to investigate both the validity of $K \approx -N \langle \cos \theta_{12} \rangle$ and to which extent a departure from an integer value indicates mixing of (K, T) states. It was shown that the expectation value of the inter-electronic angle provides a simple way to identify a state and to determine the purity of it. Here this expectation value will be used to investigate the validity of the classification scheme for He^- and Li^-.

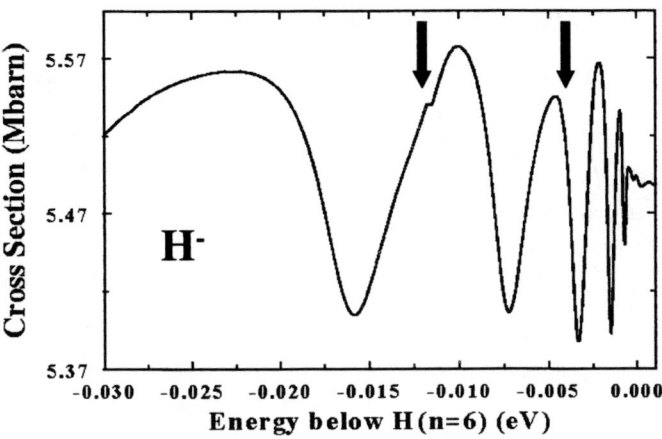

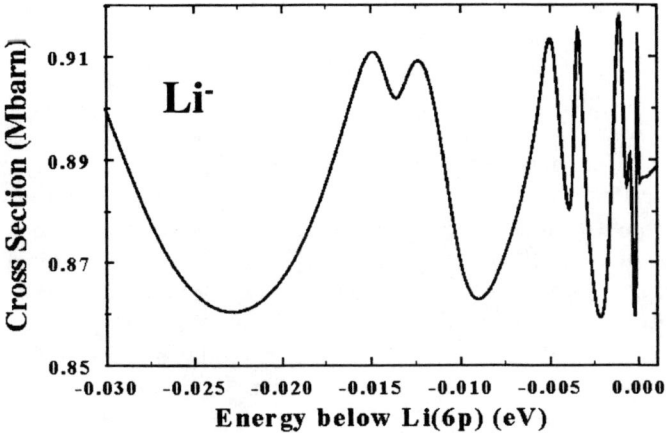

FIGURE 1. Total photodetachment cross section for H$^-$ and Li$^-$ below the $6p$-thresholds. In H$^-$ the prominent resonances are all of $A = +1$ character and have the largest possible K-value, in this case they are $(K, T)^A = (4, 1)^+$- states. Calculations give, however, also resonances belonging to other series. The arrows mark two states of $(K, T)^A = (3, 0)^-$-character. The first of these is visible in the theoretical spectrum shown here, but cannot be distinguished in experiments. The second cannot be seen even in the calculated photoabsorption spectrum with the scale used here. The Li$^-$ spectrum is rather similar to the H$^-$ spectrum, but shows a double peak structure. One possible explanation for this structure is that it is the $(K, T)^A = (3, 0)^-$-states, present but "invisible" in H$^-$, which in Li$^-$ are less pure and thus get larger photo excitation cross section and consequently become clearly visible.

THE NEGATIVE LITHIUM ION

The Li$^-$-ion can be viewed as a quasi two-electron system. As in H$^-$ the two outer electrons move in the field from a core which have a net charge of one unit. The extended core in the case of Li$^-$ lifts, however, the degeneracy of the different $n\ell$ - states in the parent atom and this change the resonance structure. Below the lower thresholds, where the splitting is substantial, the change is drastic [15–17], but below higher thresholds the situation becomes more and more similar to that in H$^-$ since the splitting of the thresholds decreases. This has earlier been discussed up to the $n = 6$ threshold by Starace and coworkers in Ref:s [18,19].

Photodetachment in the energy region between the Li(6s) and Li(6p) thresholds was very recently investigated experimentally [20] and it is interesting to compare H$^-$ and Li$^-$ in this range. Is this a region where Li$^-$ starts to show H$^-$ -like properties? Fig. 1 shows the calculated total photodetachment cross section for both ions. The calculation is described in [20,21]. The horizontal scales show the energy relative the 6p-thresholds and the threshold are further aligned in order to simplify the comparison. In H$^-$ the prominent resonances, which are also seen in experiments [9] are, as mentioned above, all of $A = +1$ character and have among those the largest possible K-value, in this case they are $(K,T)^A = (4,1)^+$- states. The calculations yield, however, also resonances belonging to other series. The thick arrows mark two states of $(K,T)^A = (3,0)^-$-character. The first of these is visible in the theoretical spectrum shown here, but cannot be distinguished in the experiment. The second is not seen even in the calculated photoabsorption spectrum with the scale used in Fig. 1. The Li$^-$ spectrum is rather similar to the H$^-$ spectrum, but shows a double peak structure. One possible explanation for this structure is that it is the $(K,T)^A = (3,0)^-$-states, present but "invisible" in H$^-$, which in Li$^-$ are less pure and thus get larger photo excitation cross section and consequently become clearly visible. To investigate this possibility we have calculated the expectation value of $\cos\theta_{12}$ and used the relation $K \approx -N\langle\cos\theta_{12}\rangle$ to guess the approximate quantum number of the states. The K, T and A quantum numbers are not independent of each other and for a state of $^1P^o$ symmetry below the $N = 6$ threshold the states with even K have $T = 1$ and $A = +1$ and those with odd K have $T = 0$ and $A = -1$, thus it is enough to know K in order to label a state with $(K,T)^A$. The result is shown in Fig. 2. The circles show the regular behaviour for H$^-$ [22]. Most resonances are well described by the $(K,T)^A$ scheme; the calculation of $-N\langle\cos\theta_{12}\rangle$ yield close to integer numbers and this is even more true when the resonances closest to threshold are considered. The diamonds show some of the resonances in Li$^-$. Here the result is less regular; $-N\langle\cos\theta_{12}\rangle$ yield non-integer numbers indicating a mixing of $(K,T)^A$ states. However, the calculated K-values indicate clearly that even in Li$^-$ there are at least two series and these will be analogous to the $(K,T)^A = (4,1)^+$ and $(K,T)^A = (3,0)^-$ series in H$^-$. In conclusion it makes sense to use the schemes developed for pure two-electron systems also on these highly excited states of Li$^-$, but the states are less pure which result in resonances which are unexpected from the experience gained on H$^-$.

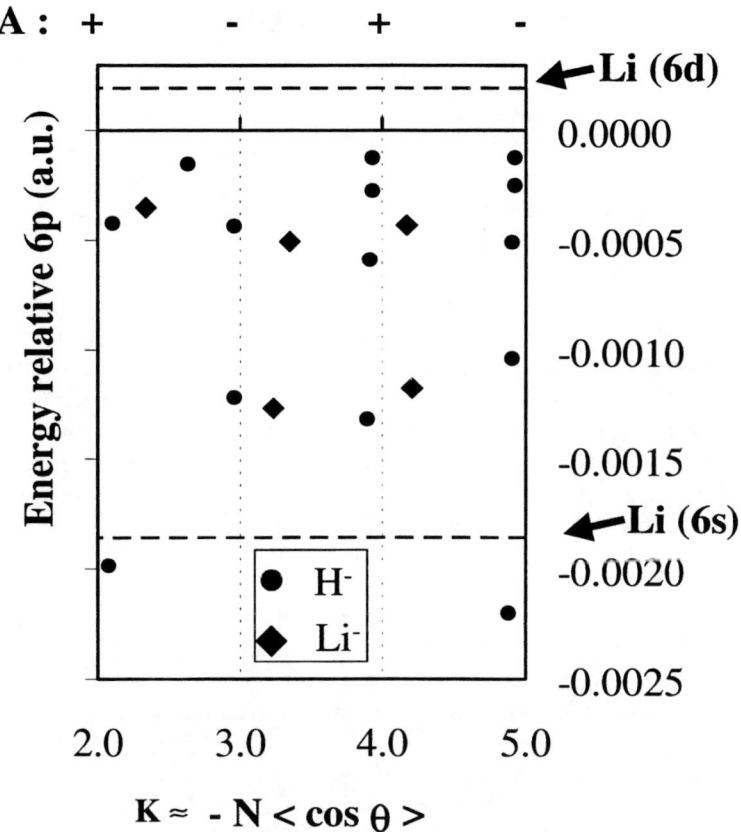

FIGURE 2. Calculated values of $-N\langle\cos\theta_{12}\rangle$ for resonant states in H$^-$(circles) and Li$^-$(diamonds) which shows to which extent the approximate quantum number K describes the state, $K \approx -N\cos\theta_{12}$. The K, T and A quantum numbers are not independent of each other and for a state of $^1P^o$ symmetry below the $N = 6$ threshold the states with even K have $T = 1$ and $A = +1$ and those with odd K have $T = 0$ and $A = -1$. It is thus enough to know K in order to label a state with $(K,T)^A$. Most resonances in H$^-$ are well described by the $(K,T)^A$ scheme; the calculation of $-N\langle\cos\theta_{12}\rangle$ yield close to integer numbers and this is even more true when the resonances closest to threshold are considered. In Li$^-$ the result is less regular; $-N\langle\cos\theta_{12}\rangle$ yield non-integer numbers indicating a mixing of $(K,T)^A$ states. However, the calculated K-values indicate clearly that in Li$^-$ there are at least two series and these are analogous to two of the series in H$^-$; the $(K,T)^A = (4,1)^+$ and $(K,T)^A = (3,0)^-$ series.

THE NEGATIVE HELIUM ION

He$^-$ has no bound state, but the lowest energy $^4P^o$ state is meta-stable since it is bound below the lowest triplet state in Helium, $(1s2s)\,^3S$, and can only autodetach by a spin flip. This long lived state can be used for photodetachment studies and its small binding energy make possible the use of laser techniques and consequently very high accuracy can be achieved. Recently several such studies of autodetaching states in He$^-$, above the single ionization threshold but below the double ionization threshold, have been published [23–26]. Theoretical investigations of He$^-$ in this region had been carried out even before the experimental results were available; Themelis and Nicolaides [27] had used an approach combining the multiconfigurational Hartree-Fock method and complex rotation, Xi and Froese-Fisher [28] had used multiconfigurational Hartree-Fock to describe the target states in the helium atom and a B-spline basis to construct and couple the open channels, and Bylicki [29] had employed complex rotation and a large basis set of so called r_{ij}-correlated configurations. Nevertheless several resonances found in the experimental investigation had not been predicted by the calculations and this stimulated additional efforts; in Ref. [21] the present authors used complex rotation combined with B-spline basis functions to construct and diagonalize the three-particle matrix, Ramsbottom and Bell used the multichannel R-matrix method [30] and very recently Liu and Starace. [31] have calculated a large number of resonances in He$^-$ using the eigenchannel R-matrix method within the "two-active-electron" model. In addition to measured, but previously unidentified, resonances both Ref. [21] and Ref. [31] also find new still not experimentally detected resonances. The knowledge about resonances in He$^-$ is thus still in a developing stage.

Here we want to investigate He$^-$ in the same way as Li$^-$ was investigated to understand if the region currently studied is sufficiently highly excited that He$^-$ starts to behave similarly to H$^-$. The investigated states in He$^-$ are two states of $^4S^e$ symmetry and they are compared to those of $^3S^e$ - symmetry in H$^-$. A H$^-$-like situation is expected in He$^-$ when the thresholds in the parent He atom is close enough that they start to look degenerate from the view point of the resonance. However, the lowest energy resonance below He($1s4s\,^3S$) is bound relative the threshold with much less energy than the splitting of the thresholds and it is unlikely that it can be well described within the $(K,T)^A$ scheme. The calculation of $K \approx -N\langle\cos\theta_{12}\rangle$ gives also a very irregular results. An electron-pair in $L = 0$ can only form states with $T = 0$ (since T is a projection of $\mathbf{L}$). Below the $n = 4$ threshold only odd K;s are then possible [6]. This is also what is found for H$^-$. The result for the lowest energy resonance below He($1s4s\,^3S$) is instead $-N\langle\cos\theta_{12}\rangle \approx 2$. In Fig. 3. the circles show the regular results for H$^-$ and the diamonds the results for He$^-$. The lowest energy resonance below He($1s4s\,^3P$) is on the other hand bound relative the threshold with more or less the same amount of energy as is given by the splitting of the He($1s4s\,^3P$) and He($1s4s\,^3D$) thresholds. The result $-N\langle\cos\theta_{12}\rangle \approx 1.3$ might be seen as a non-pure $K = 1$ state, but the conclusion is still that this region is not highly excited enough to be truly H$^-$-like.

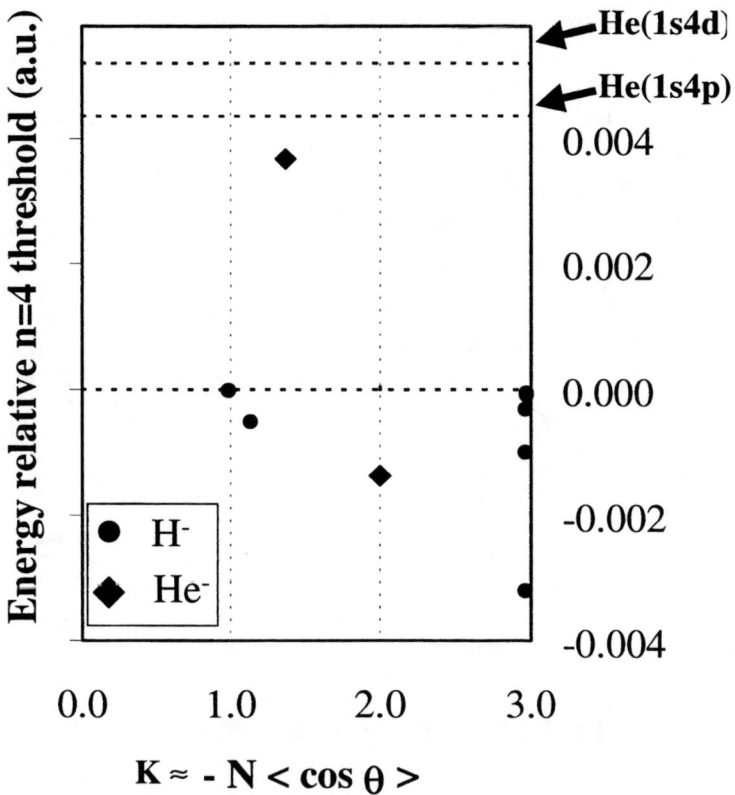

FIGURE 3. The circles show the result for H$^-$ relative the H($n = 4$) threshold while the diamonds show the result for He$^-$ relative the He($1s4s^3S$) threshold. The relative positions of nearby helium thresholds are also shown. The He$^-$ resonance below He($1s4s^3S$) is bound relative the threshold with much less energy than the splitting of the thresholds and it is unlikely that it can be well described within the $(K,T)^A$ scheme. The calculation of $K \approx -N\langle\cos\theta_{12}\rangle$ gives also a very irregular results: $-N\langle\cos\theta_{12}\rangle \approx 2$, although $(K,T)^A$ states with $L = 0$ and with parallel spins below the $n = 4$-threshold should have odd K-quantum number only.

In Ref. [31] Liu and Starace investigate the $(K, T)^A$ -character of the same states in a different manner. They isolate the bound state part of the resonance and plot the electron density e.g. in spheriodal coordinates, λ and μ (related to the molecular description discussed above) for a fixed value of the hyperspherical radius $R = \sqrt{r_1^2 + r_2^2}$. From the number of nodes in λ they then deduce the molecular quantum number $n_\lambda = (1/2)(N - K - T - 1)$. With this method Liu and Starace do find states of regular $(K, T)^A$ -character. This result seams contradictory to the findings presented here. However, Liu and Starace look at the electron density for a given value of the hyperspherical radius, that for which the wave function has maximum amplitude. This part of the wave function might show the most regular behaviour. In the present investigation the whole wave function is used. Also this that only the bound part of the wave function is considered in Ref. [31] might lead to a more regular behaviour.

CONCLUSIONS

In Li$^-$ experiments have reached a level of excitation where the system starts to look H$^-$-like, but the deviation from a true point-charge core is still leading to appearance of resonances not seen in H$^-$. In He$^-$ the investigated energy range is not excited enough for a H$^-$-like behaviour.

ACKNOWLEDGEMENTS

Financial support for this research was received from the Swedish Natural Science Research Council (NFR) and the European Union's TMR Programme contract ERBFMBICT961473.

REFERENCES

1. M. Gailitis and R. Damburg, Proc. Phys. Soc. **82**, 192 (1963).
2. E. Lindroth, A. Bürgers, and N. Brandefelt, Phys. Rev. A **57**, R685 (1998).
3. T. Purr and H. Friedrich, Phys. Rev. A. **57**, 4279 (1998).
4. M. K. Chen, J. Phys. B **32**, L487 (1999).
5. H. H. Andersen *et al.*, Phys. Rev. Lett **79**, 4770 (1997).
6. D. R. Herrick and O. Sinanoglu, Phys. Rev. A **11**, 97 (1975).
7. D. R. Herrick, Adv. Chem. Phys. **52**, 1 (1983).
8. C. D. Lin, Phys. Rev. A **29**, 1019 (1984).
9. P. G. Harris *et al.*, Phys. Rev. Lett. **65**, 309 (1990).
10. H. R. Sadeghpour and C. H. Greene, Phys. Rev. Lett. **65**, 313 (1990).
11. J. M. Feagin and J. S. Briggs, Phys. Rev. Lett **57**, 984 (1986).
12. J. M. Rost and J. S. Briggs, J. Phys. B **23**, L339 (1990).
13. N. Brandefelt, MSc Thesis, Stockholm University(unpublished), 1996.

14. A. Bürgers, N. Brandefelt, and E. Lindroth, J. Phys. B **31**, 3181 (1998).
15. U. Berzinsh *et al.*, Phys. Rev. Lett. **74**, 4795 (1995).
16. E. Lindroth, Phys. Rev. A **52**, 2737 (1995).
17. C. Pan, A. Starace, and C. Greene, Phys. Rev. A **53**, 840 (1996).
18. C. Pan, A. F. Starace, and C. H. Greene, J. Phys. B **27**, L137 (1994).
19. C. Liu and A. F. Starace, Phys. Rev. A **53**, 4997 (1998).
20. G. Haeffler *et al.*, to be published.
21. N. Brandefelt and E. Lindroth, Physical Review A **59**, (1999).
22. A. Bürgers and E. Lindroth, submitted to Eur. Phys. J. D.
23. C. W. Walter, J. A. Seifert, and J. R. Peterson, Phys. Rev. A **50**, 2257 (1994).
24. A. E. Klinkmüller *et al.*, Phys. Rev. A **56**, 2788 (1997).
25. A. E. Klinkmüller *et al.*, J. Phys. B. **31**, 2549 (1998).
26. I. Y. Kiyan, U. Berzinsh, D. Hanstorp, and D. J. Pegg, Phys. Rev. Lett. **81**, 2874 (1998).
27. S. I. Themelis and C. A. Nicolaides, J. Phys. B **28**, L379 (1995).
28. J. Xi and C. Froese-Fischer, Phys. Rev. A **53**, 3169 (1996).
29. M. Bylicki, J. Phys. B **30**, 189 (1997).
30. C. A. Ramsbottom and K. L. Bell, J. Phys. B **32**, 1315 (1999).
31. C. N. Liu and A. F. Starace, submitted to Phys. Rev. A.

Spin- and angle-resolved Auger spectroscopy of Xenon

G. Snell[#‡], B. Langer[+], E. Kukk[#‡], and N. Berrah[#]

[#] *Western Michigan University, Department of Physics, Kalamazoo, MI 49008*
[‡] *Lawrence Berkeley National Laboratory, University of California, Berkeley, CA 94720*
[+] *Max-Born-Institut, 12489 Berlin, Germany*

Abstract. The angular distribution and spin polarization of the Xe $N_{4,5}O_{2,3}O_{2,3}$ Auger electrons was measured with linearly and circularly polarized synchrotron radiation in a wide photon energy range, covering the Cooper minimum of the $4d$ photoionization cross section. Using the framework of the two-step model of Auger decay the photoion alignment and orientation were determined.

INTRODUCTION

Study of the 4d photoionization of xenon has been a show case subject of inner-shell ionization for more than three decades and a benchmark experiment for theoretical models. Despite the large number of partial cross section and angular distribution studies [1] only very few experiments were able to derive information beyond these quantities in order to determine the corresponding dipole matrix elements and their relative phases [2-4]. The most complete set of information was derived from a coincidence experiment between $4d_{5/2}$ photo- and NOO Auger electrons [3, 4] at 94.5 eV and 132 eV photon energies and from spin-resolved measurements at 94 eV [2].

The Xe 4d photoionization cross section shows some characteristic features over the photon energy, such as the broad maximum of the shape resonance which peaks at approximately 100 eV (22 Mb) and the Cooper minimum around 185 eV (0.3 Mb). To obtain detailed information about which processes play an important role at different photon energies, a complete characterization of the photoionization process should be performed over a broad energy range. For this purpose several photoionization dynamical parameters have to be measured, in addition to the already known cross section σ, branching ratio ρ and angular distribution parameter β. These quantities have been measured from threshold (67.5 eV) up to 250-280 eV (see Refs. in [1, 2]).

In the above mentioned 'complete' experiments [2-4] the Auger decay process following the 4d photoionization was used to obtain additional information. The photoionization process usually leaves the singly charged photoion in a polarized state, i.e. with an uneven population of the magnetic sublevels. The decay of a polarized photoion may lead to an anisotropic angular distribution and also to spin polarization of Auger elec-

CP500, *The Physics of Electronic and Atomic Collisions*, edited by Y. Itikawa, et al.
© 2000 American Institute of Physics 1-56396-777-4/00/$17.00

trons. In the two-step model of Auger decay, where the photoionization and Auger decay processes are assumed to proceed subsequently and independently of each other, the angular anisotropy parameter and the spin polarization parameters of the Auger electron can be factorized into a parameter describing the anisotropy of the primary hole state and into an 'intrinsic' parameter describing the Auger decay itself. With known values of the intrinsic parameters the alignment A_{20} and the orientation A_{10} of the primary hole state can be determined from the angular distribution and the spin polarization of Auger electrons, respectively. Alignment and orientation are proportional to the electric quadrupole moment and magnetic dipole moment of the photoion and are directly connected to the amplitudes of the dipole matrix elements. While alignment can be created by any kind of particle or photon impact, orientation requires excitation by polarized particle impact or by circularly polarized photons.

A thorough investigation of the angular distribution of the Auger decay process, which follows the $4d$ photoionization, was done only up to 142 eV photon energy [5, 6]. Between 150 and 190 eV four data points with large uncertainties exist [7]. A direct determination of the photoion orientation was done only at 94eV photon energy from the spin polarization of the $N_{4,5}O_{2,3}O_{2,3}$ Auger electrons [8]. The major difficulty in performing angular distribution and spin polarization measurements of the NOO Auger electrons above 150 eV photon energy is the relatively low $4d$ photoionization cross section (<1 Mb).

In the present paper we report about spin polarization measurements of the Xe $N_{4,5}O_{2,3}O_{2,3}$ Auger electrons with circularly polarized light from threshold up to 540 eV photon energy. We also measured the angular distribution of the Xe $N_{4,5}O_{2,3}O_{2,3}$ Auger electrons in the photon energy range 80-250 eV. From these measurements we could derive the orientation and the alignment of the Xe $4d^{-1}$ hole states over a broad photon energy range including the important region of the Cooper minimum.

EXPERIMENT

The experiments were carried out at the Advanced Light Source (ALS) storage ring at Lawrence Berkeley National Laboratory.

The angular distribution measurements were performed at the AMO beamline 10.0.1. The light is generated in a 4.5 m long 10 cm period undulator and monochromatized by a spherical grating monochromator. Gratings with 925 lines/mm and 2100 lines/mm were used for operating at photon energies below and above 160 eV, respectively. The electron spectra were measured using an end station designed for gas-phase angle-resolved studies and based on a Scienta SES-200 hemispherical electron analyzer [9, 10]. The analyzer is rotatable in a plane perpendicular to the propagation direction of the beam of linearly polarized photons (the degree of linear polarization is estimated to be higher than 99%), allowing electron angular distribution studies. We performed measurements at the angles of 0°, 54.7° and 90°. The analyzer was operated at the constant pass energy of 40 eV with the electron energy resolution of 25-30 meV.

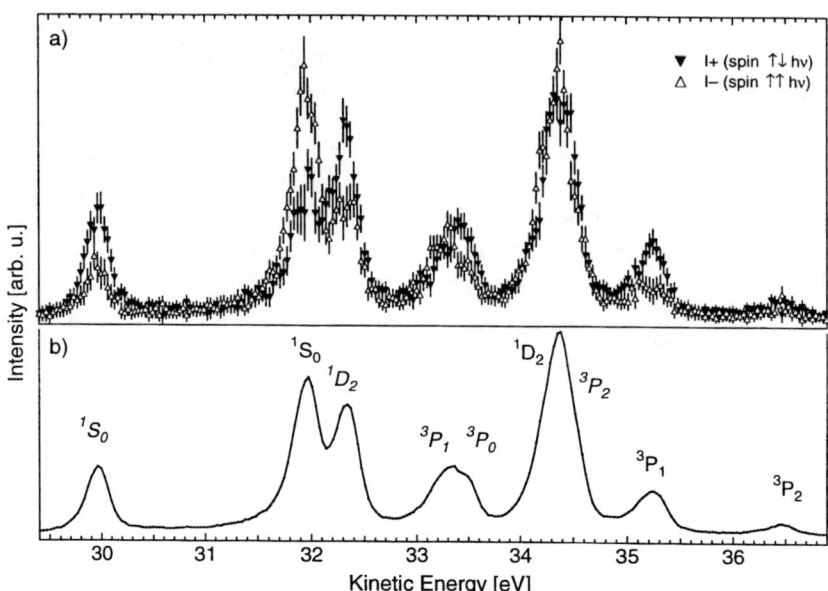

FIGURE 1. Xe $N_{4,5}O_{2,3}O_{2,3}$ Auger spectra recorded at 141 eV photon energy with circularly polarized radiation. The photon bandpass was approx. 4 eV. a) spin-resolved intensity; b) total intensity. The annotations in italic denote the N_5 initial hole states.

The spin-resolved measurements were carried out at the new Elliptical Polarization Undulator (EPU) beamline (BL 4.0.2) of the ALS [11]. The EPU is a pure permanent-magnet device modeled after the design of Sasaki and Carr. It has a period length of 5 cm and 37 full periods. It produces high-flux, high-brightness beams of circularly, elliptically or linearly polarized radiation in the VUV and soft x-ray range. Since for Auger electron spectroscopy no high photon resolution is needed, we could utilize the unmonochromatized beam of the fundamental of the undulator, which resulted in a bandwith of $\approx 3\%$. The photon flux behind a focusing mirror was estimated to be 10^{16} photons/ $(s \cdot 100 \text{ mA})$ in the full undulator fundamental. An additional benefit was that in the circular polarization mode, the EPU produces no higher harmonics, thereby avoiding the need for filters. The spin-resolved electron spectra were recorded by a new spectrometer system consisting of a time-of-flight (TOF) electron energy analyzer combined with a retarding field Mott polarimeter [12, 13]. This instrument allows very effective data acquisition, because all electron lines in the TOF spectrum are spin-analyzed simultaneously with a high signal to noise ratio. The electron spin polarization component parallel to the photon beam at an emission angle of 90° with respect to the light beam was measured. A spin-resolved Xe $N_{4,5}O_{2,3}O_{2,3}$ Auger spectra is shown in Fig. 1.

RESULTS AND DISCUSSION

The spin polarization component in the given geometry for completely circularly polarized light is given by [14, 15]

$$P = \frac{(\gamma_1/2 - \beta_1)A_{10}}{1 + (\alpha_2 A_{20})/4} \tag{1}$$

and the angular distribution parameter is given by [15]

$$\beta = \alpha_2 A_{20}. \tag{2}$$

A_{10} and A_{20} are the orientation and alignment parameters of the primary hole state; β_1, γ_1 and α_2 are the Auger intrinsic parameters, which depend on the Coulomb matrix elements. A_{20} is given here with respect to the direction of the electric field vector $\mathbf{E}$. Equations (1) and (2) already incorporate the factorization mentioned above, which is possible if the two-step model of Auger decay is valid. For transitions to a $^1S_0$ final ionic state the intrinsic parameters α_2, γ_1 and β_1 are fixed geometrical factors as only one partial wave can be emitted: $(\gamma_1/2 - \beta_1) = 2/\sqrt{5}$, $\alpha_2 = -1$ for N_4 holes and $(\gamma_1/2 - \beta_1) = -9/\sqrt{105}$, $\alpha_2 = -1.07$ for N_5 holes [15, 16]. Thus from the spin polarization and angular distribution of the $N_4O_{2,3}O_{2,3}$ and $N_5O_{2,3}O_{2,3}$ $^1S_0$-Auger lines the orientation and alignment of the primary hole states can be determined.

The description of the photoionization process in the framework of the dipole approximation limits the possible values of the orbital angular momenta of the outgoing photoelectrons according to the selection rules. In the case of the $4d$ ionization, the electrons can leave the atom as εp or εf continuum waves. In the nonrelativistic approximation there is no further spin-orbit splitting of these waves in the continuum. The alignment and orientation tensors of the $4d$ hole state depend only on the ratio of the dipole amplitudes [17]

$$A_{20}^{5/2} = -\frac{1}{5}\sqrt{\frac{2}{7}}\frac{2 + 7\lambda^2}{1 + \lambda^2}, \quad A_{20}^{3/2} = -\frac{1}{10}\frac{2 + 7\lambda^2}{1 + \lambda^2}, \tag{3}$$

$$A_{10}^{5/2} = \frac{1}{2}\sqrt{\frac{7}{15}}\frac{3\lambda^2 - 2}{1 + \lambda^2}, \quad A_{10}^{3/2} = \frac{3}{4\sqrt{5}}\frac{3\lambda^2 - 2}{1 + \lambda^2} \quad \text{with } \lambda = \sqrt{\frac{2}{3}}\frac{R_{4d,\varepsilon p}}{R_{4d,\varepsilon f}}. \tag{4}$$

The orientation is given here for completely circularly polarized light. In the nonrelativistic approximation the ratio A_{10} and A_{20} for the two spin-orbit components is a fixed value [15, 17]:

$$A_{20}^{5/2}/A_{20}^{3/2} = \sqrt{8/7}, \quad A_{10}^{5/2}/A_{10}^{3/2} = \sqrt{28/27}. \tag{5}$$

The extrema of the alignment and orientation are reached, when either the εp or the εf wave completely vanishes. In the case of $A_{20}^{5/2}$ these extrema are -0.214 and -0.748 and in the case of $A_{10}^{5/2}$ they are -0.683 and 1.025 for $R_{4d,\varepsilon p}=0$ and $R_{4d,\varepsilon f}=0$, respectively.

Alignment

Figure 2 shows the alignment A_{20} of the Xe $4d^{-1}\,^2D_{5/2}$ hole state together with other experimental data and a theoretical curve. During data analysis the validity of the nonrelativistic approximation was assumed (cf. Eq. (5)). At 173.5 eV and 176.1 eV a partial overlap of the $N_{4,5}O_{2,3}O_{2,3}$ Auger group with the 4p photolines increased the uncertainties of the A_{20} values. The agreement between all the experimental data is very good up to 142 eV photon energy. At higher energies a considerable discrepancy between our results and those of Southworth *et al.* [7] can be seen. The latter have large uncertainties and do not show a minimum, while our data show a strong dip.

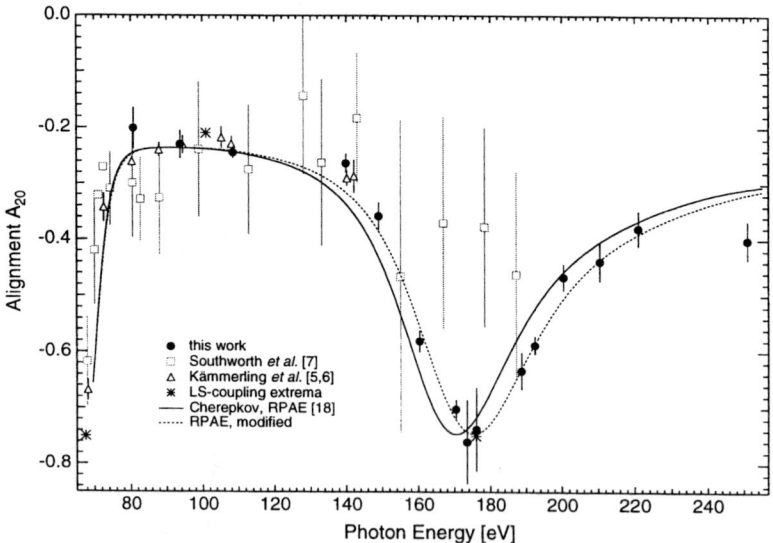

FIGURE 2. Alignment A_{20} of the Xe $4d^{-1}\,^2D_{5/2}$ hole state. The stars show the extreme values of A_{20} in LS-coupling for a vanishing f-wave (at threshold and in the Cooper minimum) and vanishing p-wave in the cross section maximum (see also text). The dashed curve was obtained by stretching the calculation of Cherepkov to match the A_{20} minimum.

The overall shape of the theoretical curve of Ref. [18] (which uses the nonrelativistic approximation) is similar to the experimental data, giving the position of the minimum at 6 eV lower energy than the experiment. To get a better idea about the course of the A_{20} curve, we modified the RPAE curve by stretching it to higher energies in order to match the minimum in the experimental data. In this way we obtain a good overall agreement with a slight deviation around 145 eV. This deviation might be due to the fact that the $4p$ threshold is approximately at 146 eV photon energy and this was not taken into account during the calculations.

As described above, A_{20} can reach two extrema for vanishing p- or f-waves. Due to the potential barrier for f-electrons near threshold ($E_{bind}(4d_{5/2})=67.54$ eV), the outgoing p-wave dominates the photoionization. At increasing energies, the f-electrons overcome the barrier and carry almost all the cross section in the maximum of the shape resonance (~100 eV). In the Cooper minimum $R_{4d,\varepsilon f}$ goes through zero, i.e. only the p-wave is present at ~176 eV. Whereas the first two extrema were confirmed by the previous experiments, the present data proves the existence of the alignment minimum in the Cooper minimum of the cross section.

Orientation

Figure 3 shows the orientation A_{10} of the Xe $4d^{-1}$ hole state as a result of a preliminary analysis. Since the orientation for the two spin-orbit components differs only slightly (cf. Eq. (5)), their mean value is presented. The horizontal bars of the data points correspond to the approximately 3% bandwith of the undulator fundamental. The error bars contain both systematical and statistical uncertainties, the latter being significant in the Cooper minimum around 180eV. The open circles do not represent directly measured A_{10} values, they were calculated from the published matrix elements of Refs. [3, 4]. The data point at 132eV agrees very well with our results.

The overall shape of the theoretical curve of Ref. [18] is similar to the experimental data, giving the position of the maximum at somewhat lower energy than the experiment. The dashed curve in Fig. 3 was obtained by stretching the RPAE curve the same amount as in Fig. 2 for the alignment. This curve gives a good agreement with our results, with the largest deviation for the data point at 174eV photon energy. Besides the large statistical uncertainty, an averaging over the a 5eV bandwidth could also lower the measured value at this point.

Similarly to Fig. 2, the possible extrema of A_{10} in LS-coupling are also shown in Fig. 3. Close to threshold no measurements of A_{10} exist. In the shape resonance maximum around 100eV the measured value is very close to the minimum connected to a vanishing p-wave. In the Cooper minimum, where A_{10} reaches its maximum due to the vanishing f-wave, the measured data is compatible with the predicted LS-coupling value.

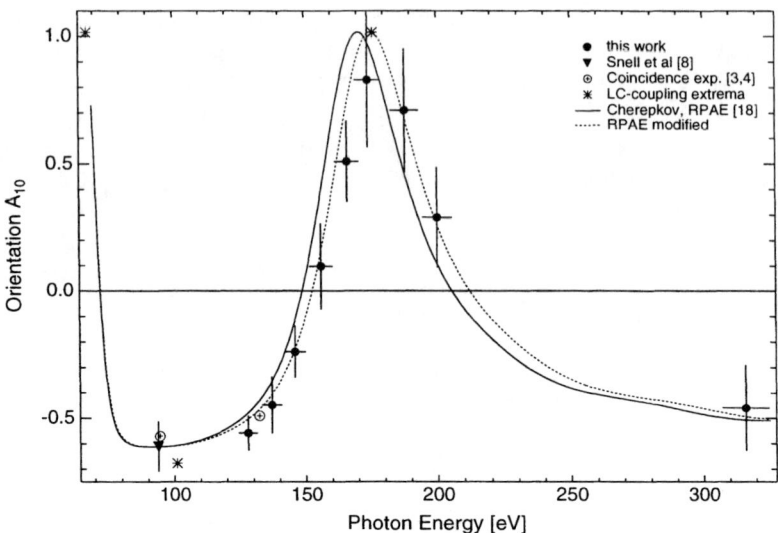

FIGURE 3. Orientation A_{10} of the Xe $4d^{-1}$ hole state. The stars show the extreme values of A_{10} in LS-coupling for a vanishing f-wave (at threshold and in the Cooper minimum) and vanishing p-wave in the cross section maximum (see also text). The dashed curve was obtained by stretching the theoretical RPAE curve the same amount as in Fig. 2.

SUMMARY

We measured the angular distribution and spin polarization of the Xe $N_{4,5}O_{2,3}O_{2,3}$ Auger electrons in a wide photon energy range, covering the Cooper minimum of the $4d$ ionization. Using the framework of the two-step model of Auger decay:

i) The alignment A_{20} and orientation A_{10} of the $4d^{-1}$ ionic state was determined at different excitation energies. The pronounced minimum A_{20} of and maximum of A_{10} in our data experimentally proves for the first time the vanishing of the εf component of the outgoing electron wave. This is in good agreement with the theoretical prediction of an RPAE calculation [18], taking into account a small deviation in the position of the predicted minimum/maximum.

Our results show that both the alignment and orientation can be well described by a nonrelativistic model and they prove that the amplitude $R_{4d,\varepsilon f}$ crosses zero in the Cooper minimum of the $4d$ photoionization cross section. An important implication of the present result is, that with $R_{4d,\varepsilon f}=0$ at 176 eV all dynamical parameters of photoionization have fixed values (e.g. $\beta=0.2$ [2]) at this energy. This might serve as a benchmark or calibration during measurement.

ACKNOWLEDGEMENTS

We would like to thank the staff of the ALS, especially Dr. J. D. Bozek and Dr. A. Young, for providing excellent working conditions. We are grateful to W.-T. Cheng for his help during the measurements. This work was supported by the Department of Energy, Office of Science, Basic Energy Sciences, Chemical Sciences Division. The Advanced Light Source is supported by DOE, Materials Sciences Division. B.L. acknowledges the support of the Alexander von Humbold-Foundation.

REFERENCES

1. U. Becker and D. A. Shirley, *VUV- and Soft X-ray Photoionization*, eds. U. Becker and D. A. Shirley, New York: Plenum Press, 1996, pp. 148.

2. G. Snell, B. Langer, M. Drescher, N. Müller, B. Zimmermann, U. Hergenhahn, J. Viefhaus, U. Heinzmann, and U. Becker, *Phys. Rev. Lett.* **82**, 2480 (1999).

3. B. Kämmerling and V. Schmidt, *J. Phys. B* **26**, 1141 (1993).

4. S. J. Schaphorst, Q. Qian, B. Krässig, P. v. Kämpen, N. Scherer, and V. Schmidt, *J. Phys. B* **30**, 4003 (1997).

5. B. Kämmerling, B. Krässig, and V. Schmidt, *J. Phys. B* **23**, 4487 (1990).

6. B. Kämmerling, Inaugural thesis, Universität Freiburg, 1991 (unpublished).

7. S. Southworth, U. Becker, C. M. Truesdale, P. H. Kobrin, D. W. Lindle, S. Owaki, and D. A. Shirley, *Phys. Rev. A* **28**, 261 (1983).

8. G. Snell, M. Drescher, N. Müller, U. Heinzmann, U. Hergenhahn, and U. Becker, *J. Phys. B* **32**, 2361 (1999).

9. N. Berrah, B. Langer, A. Wills, E. Kukk, J. D. Bozek, A. Farhat, and T. W. Gorczyca, *J. Electron Spectrosc. & Relat. Phenom.* **101-103**, 1 (1999).

10. E. Kukk, J. D. Bozek, T. D. Thomas, T. X. Carroll, L. J. Saethre, J. A. Sheehy, P. W. Langhoff, and N. Berrah, Proceedings of the XXI. ICPEAC, Sendai, Japan (1999).

11. T. Young, A. Robinson, G. Snell, and N. Berrah, *Synchrotron Rad. News* **12**, 31 (1999).

12. G. C. Burnett, T. J. Monroe and F. B. Dunning, *Rev. Sci. Instrum.* **65**, 1893 (1994).

13. N. Müller, R. David, G. Snell, R. Kuntze, M. Drescher, N. Böwering, P. Stoppmanns, S.-W. Yu, U. Heinzmann, J. Viefhaus, U. Hergenhahn, and U. Becker, *J. Electron Spectrosc. Relat. Phenom.* **72**, 187 (1995).

14. K.-N. Huang, *Phys. Rev. A* **22**, 223 (1980).

15. N. M. Kabachnik and O. V. Lee, *J. Phys. B* **22**, 2705 (1989).

16. B. Lohmann, U. Hergenhahn, and N. M. Kabachnik, *J. Phys. B* **26**, 3327 (1993).

17. E. G. Berezhko, N. M. Kabachnik, and V. S. Rostovsky, *J. Phys. B* **11**, 1749 (1978).

18. N. A. Cherepkov, *J. Phys. B* **12**, 1279 (1979); N. A. Cherepkov and B. Zimmermann (private communication).

Circular Dichroism of Angular Correlation Patterns in Two-step and Direct Double Photoionization

Kouichi Soejima

Graduate School of Science and Technology, Niigata University
Igarashi-ninocho 8050, Niigata-shi, Niigata 950-2181, JAPAN

Abstract. Angular correlation patterns of N_5-$O_{23}O_{23}$ $^1S_0$ Auger electrons in coincidence with $4d_{5/2}$ photoelectrons in two-step double photoionization of xenon and those of two emitted photoelectrons in coincidence with each other in direct double photoionization of helium have been studied with elliptically polarized photons. Circular dichroism in the angular distribution patterns has been reported in both cases.

INTRODUCTION

Circular dichroism in double photoionization of atoms and molecules has attracted both experimental and theoretical attention in recent years because intense circularly polarized photon beams have become available at synchrotron radiation facilities. The circular dichroism is characterized by difference between processes induced for right and left circularly polarized photons. The double photoionization can be classified into two cases, i.e., a two-step process with the sequential emission of a photoelectron and an Auger electron, and a direct process with simultaneous emission of two electrons. The Circular dichroism can be observed for both cases. For the two-step process, Schmidt (1) predicted that the Circular Dichroism in the Angular Distribution (CDAD) of Auger electrons can exist if they are measured in coincidence with the photoelectrons. And Kabachnik and Schmidt (2) demonstrated CDAD for calcium using two-step model calculation. The first observation of the CDAD for the two-step double photoionization of xenon was performed by our group in 1996 (3). Whereas, in the direct process, a strong circular dichroism has been predicted by Berakdar and Klar (4) for helium. And Berakdar *et al.* (5) suggested that the CDAD becomes a new manifestitation of studying electron correlation. The first experimental evidence for the circular dichroism was reported by Viefhaus *et al.* (6) in 1996. They showed a good example of the circular dichroism on energy sharing between the two photoelectrons at three relative emission angles. In 1998, Mergel *at al.* (7) reported preliminary results on the CDAD for helium. Recently, the studies in the direct double photoionization of helium have been reported using circularly polarized radiation from helical undulator by our group (8). The recent experimental results of the helium double photoionization were good agreement with convergent close-coupling calculation (9). And it is proved

CP500, *The Physics of Electronic and Atomic Collisions*, edited by Y. Itikawa, et al.

that observation of the circular dichroism is a new and useful approach to understand the dynamical effects, i.e., electron correlation, in double photoionization. Especially, circular dichroism gives us direct information about phase difference between symmetric and anti-symmetric dipole amplitudes of the direct double photoionization process.

In this paper, the experimental results of the angular correlation patterns between emitted electrons are presented for;

(I) Two-step double photoionization of xenon;

$$Xe + h\nu \rightarrow Xe^+(4d_{5/2}^{-1}) + e_{ph}$$
$$\rightarrow Xe^{2+} + e_{Auger}(N_5 - O_{23}O_{23}:{}^1S_0)$$

and

(II) Direct double photoionization of helium;

$$He + h\nu \rightarrow He^{2+} + e_1 + e_2.$$

EXPERIMENTAL SETUP

The experiments were performed at the helical undulator (10) beamline BL-28A equipped with the constant deviation monochromator (11) of the Photon Factory storage ring in Tsukuba. The fundamental of helical undulator radiation provides soft X-ray with high photon flux and a high degree of circular polarization. Kimura *et al.*

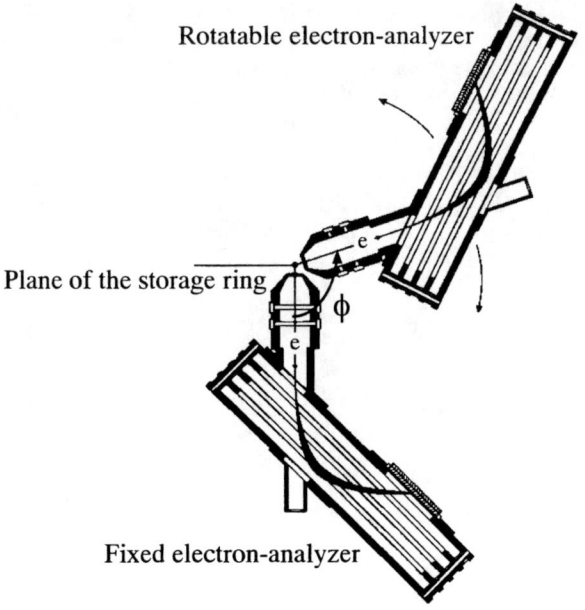

FIGURE 1. The schematic diagram of experimental apparatus.

137

measured the Stokes parameters S_1 and S_3 which describe the degree of linear and circular polarization, respectively at photon energy of 97 eV (12). They obtained the results; for the standard mode, S_1=-0.2, S_2=0, and S_3=-0.95 in the frame tilted by +135°±2° relative to the plane of the storage ring, and for the left-handed mode, S_1=+0.2, S_2=0, and S_3=-0.95 in the frame tilted by +44°±2°. The polarization parameters were not measured in the course of our experiments. But we reasonable assumed that the Stokes parameters are the same as Kimura et al. (12) determined. Because the degree of circular polarization at the first harmonic is independent on the undulator gap (10).

The experimental apparatus has been previously described in detail (13) and the setup is shown in Fig. 1. It consists of two parallel plate energy analyzers placed in a plane perpendicular, polar angle θ=90°, to the photon beam z-axis. One of the analyzers placed with a fixed position, azimuthal angle ϕ=-90°, and the other is rotatable around the z-axis from ϕ=-10° to 180°. The azimuthal angle ϕ is measured counterclockwise from the storage ring plane as seen by as observer facing the photon beam. The acceptance angle and energy resolution were estimated as $\Delta\phi$=±5.3° and ΔE=0.7 eV, respectively. The standard electronic circuit of detection system for coincidence measurement has been used.

RESULTS AND DISCUSSION

For the double photoionization, the complete information is provided by energy- and angle-resolved triply differential cross section (TDCS), which can be obtained from coincidence measurement for two emitted electrons. The experimental TDCSs are presented for two-step double photoionization of xenon at 110 eV photon energy and for the direct double photoionization of helium at 88 eV photon energy. The shape of TDCS is dramatically dependent on experimental conditions, i.e., angle of e_1, light polarization, excess energy, and energy sharing.

For partial light polarization, it is convenient to introduce a coordinate frame in which the Z- and X-axis points into the light direction and the major axis of the polarization ellipse, respectively. In this frame, introducing two contributions $TDCS_X$ and $TDCS_Y$, which refer to linear polarization along the X- and Y-axis, and the other two contributions $TDCS_R$ and $TDCS_L$, which refer to right and left circular polarization, the measured TDCS can be written as the incoherent sum of these four contributions (14) as follows;

$$TDCS = \frac{1}{2}(TDCS_X + TDCS_Y) + \frac{S_1}{2}(TDCS_X - TDCS_Y) + \frac{S_3}{2}(TDCS_R - TDCS_L). \quad (1)$$

The parametrization for $TDCS_{X,Y}$ and $TDCS_{R,L}$ has been derived by several authors (2,15,16,17). The number of out going waves for the two-step process is different from that for the direct process. In the case of two-step process, the angular orbital momenta l of emitted electrons are restricted to a few values. Whereas, such restriction do not exist for direct process.

Two-step Process

Photoionization in xenon leading to a $4d^{-1}\,^2D_{5/2}$ final ionic state and the subsequent N_5-$O_{2,3}O_{2,3}$ $^1S_0$ Auger decay was selected because this process was deeply investigated in the complete experiment using linearly polarized light . The photon energy was tuned to 110 eV, because of obtaining angular distribution of Auger electrons free from the post-collision interaction effect (18).

The experimental results are shown for the two-step double photoionization of xenon in Fig.2 (a) and (b) (3). The intensities of true coincidences are plotted in a polar diagram with statistical error bars. We confirmed that the coincidence rates for right and left elliptically polarized photons were same within error bars at opposite to the photoelectron direction. In comparing Fig. 2 (a) and (b), it is clearly demonstrated that the angular distribution of N_5-$O_{2,3}O_{2,3}$ Auger electrons in coincidence with the $4d_{5/2}$ photoelectron is different for right (a) and left (b) elliptically polarized lights. The electron correlation patterns for right (a) and left (b) elliptically polarized lights have reflection symmetry relative to photoelectron direction as predicted by Schmidt (1) and Kabachnik and Schmidt (2). This is an example of circular dichroism for two-step double photoionization.

Parametrization for two-step process

In the two-step double photoionization, Kämmeling and Schmidt (19) has already done the parametrization of TDCS, and it is expressed in our experimental set-up as

$$TDCS = A_0 + A_2 \cos 2\phi + A_4 \cos 4\phi + B_2 \sin 2\phi + B_4 \sin 4\phi . \qquad (2)$$

The coefficients A_i and B_i depend on the dipole matrix elements of the $4d_{5/2}$ photoionization, on the Auger decay anisotropy coefficient, on the angle of the photoelectron emission, and also on the Stokes parameters of the incident light.

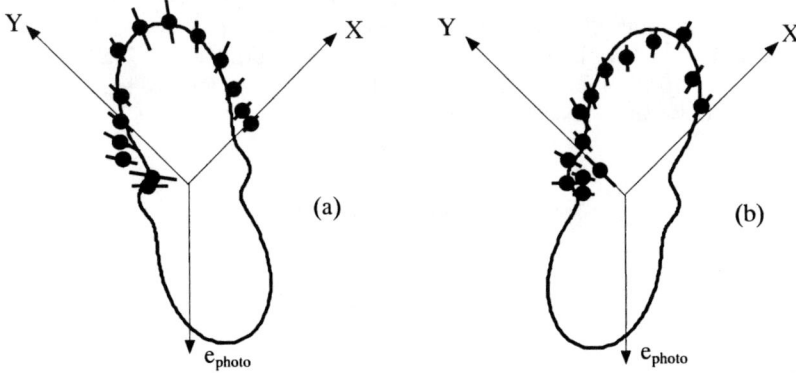

FIGURE 2. The experimental TDCS for two-step double photoionization of xenon with S_1=-0.20, S_3=+0.95 (a) and S_1=+0.20, $S3$=-0.95 (b).

Photoionization of a $4d_{5/2}$ electron is described by three dipole matrix elements of D_+, D_0 and D_- which belong to the three photoionization channels of $\varepsilon f_{7/2}$, $\varepsilon f_{5/2}$ and $\varepsilon p_{3/2}$. And the Auger decay is described by only one matrix element which belongs to an $\varepsilon d_{5/2}$ partial wave of the Auger electron. Since the photoelectron was detected at a fixed direction, the included trigonometric functions in Eq.(2) are determined by the orbital angular momentum ($l=2$) of Auger electron. It is also noted that they are restricted to even values.

The theoretical values of the dipole matrix elements of D_+, D_0, D_- and phase difference between them can be obtained from the literature of Johnson and Cheng (20). Therefore, one can obtain the angular correlation patterns from first principle based on Eq. (2) using values of coefficients A_i and B_i which can be calculated from the dipole matrix elements, Auger decay anisotropy coefficient and light polarization parameters. The results of theoretical calculations (21) are expressed by solid curves in Fig. 2. The angular correlation patterns are well reproduced by the theoretical curves. Especially, the large lobe in the angular correlation patterns agree well with the theoretical prediction.

Direct Process

In the double photoionization of helium, the photon energy was tuned to 88 eV, which corresponds to the excess energy of 9 eV. The available energy of 9 eV can be shared by the emitted electrons. The angular distributions were measured under the condition of equal energy sharing, $E_1=E_2=4.5$ eV, and unequal energy sharing, $E_1=3$ eV, $E_2=6$ eV (1:2) and $E_1=1$ eV, $E_2=8$ eV (1:8). The experimental results of TDCS under the condition of the equal energy sharing are shown in Fig. 3 (a) and (b). And the experimental results of TDCS under the unequal energy sharing condition are shown in Fig. 4 (a), (b) for 1:2 energy sharing and (c), (d) for 1:8 energy sharing.

Parametrization for direct process

In the direct double photoionization, the parametrization of the TDCS derived by Malegat *et al.* (17) is applied to our experimental condition. Applying the experimental condition of X-axis along the +45°, Y-axis along the +135°, and mutual angle $\phi=\phi_1-\phi_2$ between ejected electrons, ϕ_1 of e1 emission angle is -135°, the TDCS can be expressed as

$$TDCS = \left\{ (1 + \cos\phi)|M_g|^2 + (1 - \cos\phi)|M_u|^2 \right\}$$
$$+ S_1 \sin\phi \left\{ (1 - \cos\phi)|M_u|^2 - (1 + \cos\phi)|M_g|^2 - 2\cos\phi \operatorname{Re}(M_g M_u^*) \right\}$$
$$- S_3 \sin\phi \left\{ 2\operatorname{Im}(M_g^* M_u) \right\}. \tag{3}$$

Here, the two complex amplitudes of M_g and M_u, which describe the dynamical effect in the double photoionization, are unknown function of kinetic energy E_1, E_2 and

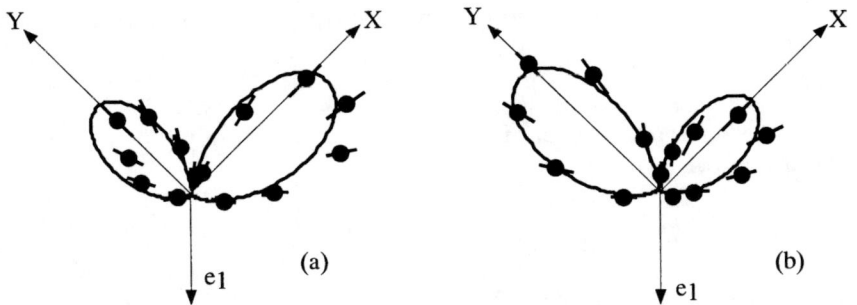

FIGURE 3. The experimental TDCS of helium double photoionization under the equal energy sharing of $E_1=E_2=4.5$ eV; (a) $S_1=-0.20$ and $S_3=+0.95$, (b) $S_1=+0.20$ and $S_3=-0.95$. The full curves present the best-fitted curves.

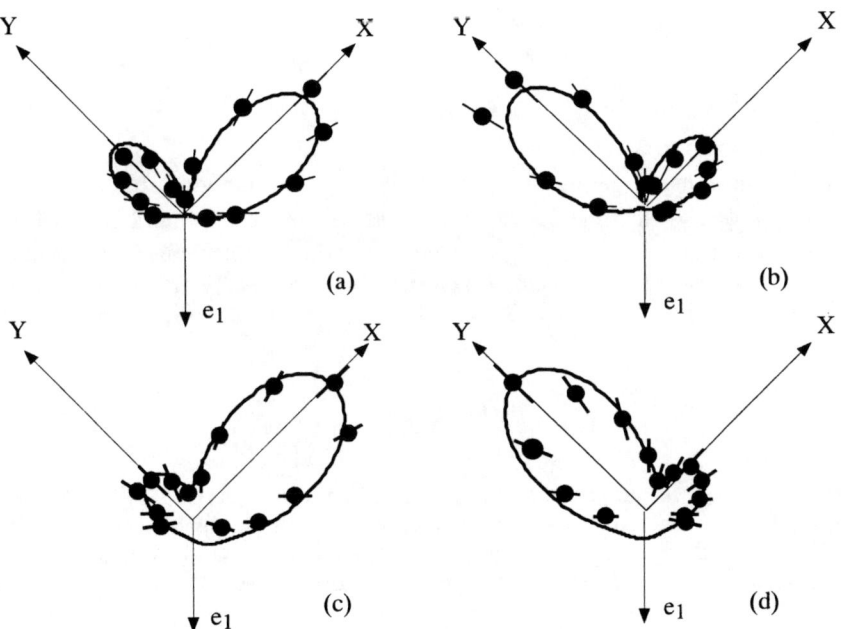

FIGURE 4. The experimental TDCS of helium double photoionization under the unequal energy sharing of $E_1=3$ eV and $E_2=6$ eV; (a) $S_1=-0.20$ and $S_3=+0.95$, (b) $S_1=+0.20$ and $S_3=-0.95$, and of $E_1=1$ eV and $E_2=8$ eV; (c) $S_1=-0.20$ and $S_3=+0.95$, (d) $S_1=+0.20$ and $S_3=-0.95$. The full curves present the best-fitted curves.

mutual angle ϕ. Subscripts g and u means symmetric (g) and unti-symmetric (u) in the exchange of E_1 and E_2, i.e., $M_g(E_1,E_2,\phi)=+M_g(E_2,E_1,\phi)$ and $M_u(E_1,E_2,\phi)=-M_u(E_2,E_1,\phi)$.

In the equal energy sharing case, the unti-symmetric amplitude M_u goes to zero and the TDCS becomes independent of S_3. Therefore, the TDCS can be expressed as the simple equation of

$$TDCS = (1 + \cos\phi)(1 - S_1 \sin\phi)|M_g|^2, \tag{4}$$

where the correlation function of $|M_g|^2$ can be approximated by a Gaussian function peaked at $\phi=180°$ as $|M_g|^2 = \exp\{-4\ln2(180°-\phi)^2/\Gamma^2\}$. By data analysis using Eq. (4), we have obtained the results of $\Gamma=85°$ in the Gaussian function and best-fitted curves in Fig. 3 (a) and (b). The difference between Fig. 3 (a) and (b) can be attributed to the only sign of S_1 as can be seen from Eq. (4). The difference on the sign of S_1, i.e., the difference for two orthogonal linear polarizations will be called "Linear Dichroism (LD)" in the TDCS.

If we make the sum of the TDCS for right ($S_1=-0.20$, $S_3=+0.95$) and left ($S_1=+0.20$, $S_3=-0.95$) elliptically polarized lights, we obtain the simple equation of TDCS(sum) which is expressed only by the first term of Eq. (3);

$$TDCS(sum) = 2 \times \left\{ (1 + \cos\phi)|M_g|^2 + (1 - \cos\phi)|M_u|^2 \right\}. \tag{5}$$

For the unequal energy sharing case, the $|M_u|^2$ plays an important role, because the TDCS(sum) at $\phi=180°$ has the appreciable intensity due to the contribution of $|M_u|^2$, $4|M_u|^2$, in contrast to the equal energy sharing case. Based on the theoretical results of Maulbetsch and Briggs (22) and Kazansky and Ostrovsky (23), the $|M_g|^2$ is approximated by the Gaussian function mentioned above and a constant term A_i for the variable ϕ;

$$|M_g|^2 = A_0 + A_1 \exp\left[-4\ln2(180° - \phi)^2/\Gamma^2\right]. \tag{6}$$

For the $|M_u|^2$ function, the partial wave l expansion formula derived by Malegat *et al.* (17) is applicable. The continuum pair wavefunction can be obtained by an unlimited coupling of individual orbital momenta ($\varepsilon s\varepsilon p$, $\varepsilon p\varepsilon d$, $\varepsilon d\varepsilon f$,...). In our analysis, the l_{max} is truncated by 3, the $|M_u|^2$ function is described by

$$|M_u|^2 = B_0 + B_1 \cos\phi + B_2 \cos^2\phi + B_3 \cos^3\phi + B_4 \cos^4\phi, \tag{7}$$

where the coefficients of B_i are the function of the electron energies E_1 and E_2. Then, the least-squares fitting are performed using Eqs. (5), (6) and (7), coefficients A_i and B_i can be determined. Fig.5 (a) and (b) show the $|M_g|^2$ and $|M_u|^2$, which are determined by the fitting procedure, for the energy sharing of 1:2 and 1:8, respectively. The $|M_u|^2$ is much smaller than the $|M_g|^2$ for both cases. And $|M_u|^2$ of 1:8 energy

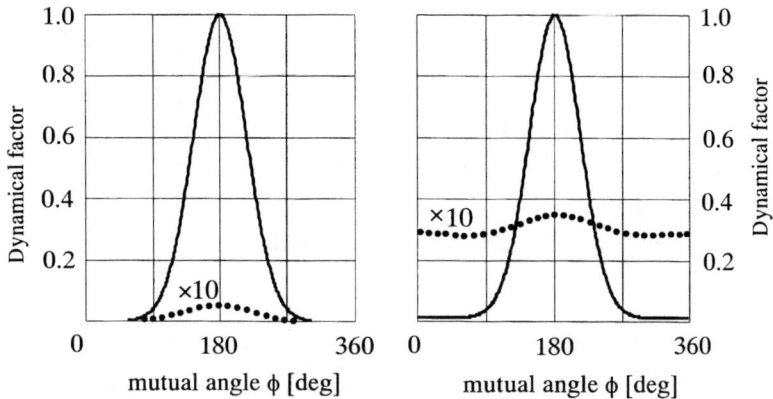

FIGURE 5. Full and dotted curves present the dipole amplitudes $|M_g|^2$ and $10\times |M_u|^2$, respectively. (a): 1:2 energy sharing, (b): 1:8 energy sharing.

sharing is much larger than that of 1:2 energy sharing.

By introducing the phases δ_g and δ_u for the complex amplitudes of M_g and M_u, $Re(M_gM_u{}^*)$ in Eq.(3) is written as $|M_g||M_u|\cos[-(\delta_u-\delta_g)]$, and $Im(M_uM_g{}^*)$ is $|M_g||M_u|\sin(\delta_u-\delta_g)$. Then we can parameterize the TDCS by the phase difference $(\delta_u-\delta_g)$, since the $|M_g|^2$ and $|M_u|^2$ have been determined as shown in Fig.5. A least-squares fit of the TDCS expressed by Eqs.(3), (6) and (7) to the experimental data sets has been performed using $(\delta_u-\delta_g)$ as only one free parameter, here the values of Γ, A_i, and B_i have been fixed. The best fitted curves are shown in Fig 4, and the phase difference of $(\delta_u-\delta_g)$ are obtained $(\delta_u-\delta_g)=200°\pm12°$ for the 1:2 energy sharing, and $(\delta_u-\delta_g)=199°\pm13°$ for the 1:8 energy sharing.

To demonstrate how the polarization properties of the incident light influence the measured TDCS, the contributions from the first term (Σ), the second term (LD), and the third term (CD) of Eq. (3) are indicated separately in Fig. 6. Surprisingly, the contribution from LD is larger than that from CD even though S_1 is much smaller than S_3 for 1:2 energy sharing. Whereas, for 1:8 energy sharing, the contribution from CD increases and that becomes larger than the contribution from LD. It is also noted that Σ, which give the direct information about $|M_u|^2$, at $\phi=180°$ rapidly increase for more asymmetric energy sharing. As can be understood from Fig.6, we can construct the TDCS including solely the dependence on the S_3 by subtracting the S_1 part from the experimental data. Fig. 7 shows the *pure* "Circular Dichroism" in TDCS. It is evidently caused by the interference term of $Im(M_uM_g{}^*)$ of Eq. (3). It is well understand that M_u plays an essential role for occurring the circular dichroism. Magnitude of the circular dichroism becomes larger for more asymmetric energy sharing. The phase difference between M_g and M_u and $|M_g|^2$ are almost independent on the energy sharing condition by our analysis. Therefore, we can say that the circular dichroism is proportion to the $|M_g|$ and $|M_u|$ as can be understood from Eq. (3). Then, it can be concluded that the increasing of circular dichroism is responsible for the increasing $|M_u|^2$ as seen in Fig. 5.

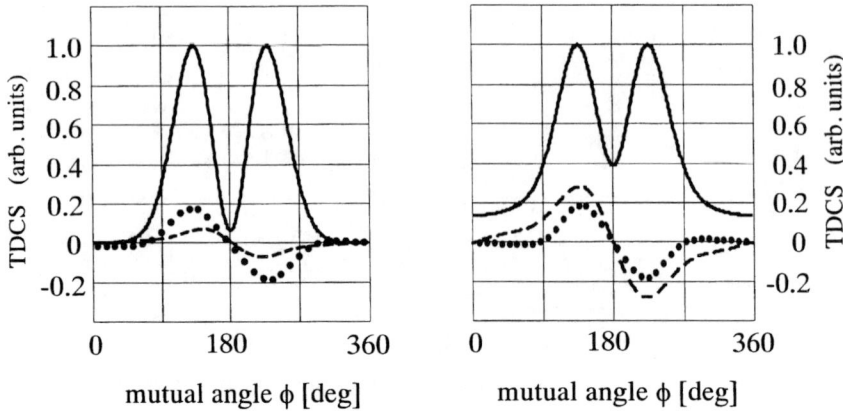

FIGURE 6. The contributions from the first term (full curves), the second term (dotted curves), and the third term (dashed curves) of Eq. (6) for $S_1=-0.20$ and $S_3=+0.95$. (a): 1:2 energy sharing, (b): 1:8 energy sharing. For the case of $S_1=+0.20$ and $S_3=-0.95$, the sign of the contributions from the second and third terms are inverted.

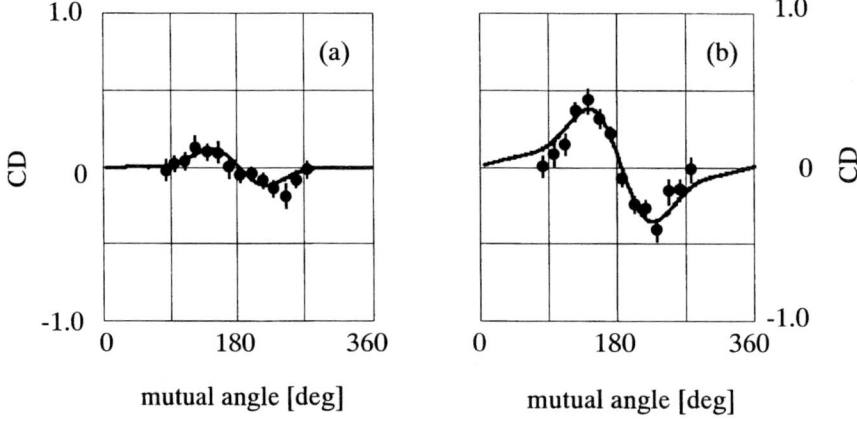

FIGURE 7. "Pure" circular dichroism of TDCS. (a):1:2 energy sharing, (b):1:8 energy sharing.

SUMMARY

In the two-step double photoionization, we can determine the ratios of the coefficients A_i/A_0 and B_i/A_0 of Eq. (2) from experimental angular correlation patterns. If these results are combined with the experimental data for the photoionization cross section, for the asymmetry parameter and for the alignment parameter, the dipole matrix elements and their relative phases can be evaluated for all photoionization channels, which is called complete experiments. And we will be able to discuss the validity of the theoretical dipole matrix elements calculated RRPA by Johnson and Cheng (20).

In the direct double photoionization, by taking the polarization properties of the incident light into account and by applying the parametrization of Malegat *et al.* (17), we have succeeded in, for the first time, deriving the symmetric amplitude of M_g, anti-symmetric amplitude of M_u and phase difference $(\delta_u\text{-}\delta_g)$ between them from our experimental data. Because these parameters give the full description for the dynamical effects in the double photoionization continuum, the present approach opens a door toward the further step to elucidate the general Coulombic three-body problem.

ACKNOWLEDGMENTS

The author is grateful to Profs. A. Danjo, K. Okuno, and A. Yagishita for discussions and Mr. Y. Oguma, Mr. S. Motoki and Mr. Nishimura for their assistance in the experiment. I also would like to acknowledge T. Koide and Y. Miyauchi for their excellent support to use the helical undulator beamline BL-28A at the Photon Factory. This study was supported by a Grant-in Aid for Scientific Research from the Ministry of Education, Science and Culture of Japan. This work has been performed under approval of the PF Advisory Committee (Proposal No. 95G413 and 97G352).

REFERENCES

1. V. Schmidt, Nucl. Instrum. and Methods **B87**, 241 (1994).
2. N. M. Kabachnik and V. Schmidt, J. Phys. B: At. Mol. Opt. Phys. **28**, 233 (1995).
3. K. Soejima *et al.*, J. Phys. B: At. Mol. Opt. Phys. **29**, L367 (1996).
4. J. Berakdar and H. Klar, Phys. Rev. Lett. **69**, 1175 (1992).
5. J. Berakdar *et al.*, J. Phys. B: At. Mol. Opt. Phys. **26**, 1463 (1993).
6. J. Viefhaus *et al.*, Phys. Rev. Lett. **77**, 3975 (1996).
7. V. Mergel *et al.*, Phys. Rev. Lett. **80**, 5301 (1998).
8. K. Soejima *et al.*, Phys. Rev. Lett. **83**, 1546 (1999).
9. A. Kheifets *et al.*, J. Phys. B: At. Mol. Opt. Phys. **32**, L501 (1999).
10. H. Kitamura, *Insertion Device Handbook 1990 Photon Factory*, KEK Report 89-24, Tsukuba.
11. Y. Kagoshima *et al.*, Rev. Sci. Instrum. **63**, 1289 (1992).
12. H. Kimura *et al.*, Rev Sci. Instrum. **66**, 1920 (1995).
13. K. Soejima *et al.*, J. Korean Phys. Soc. **32**, 368 (1998).
14. S. J. Schaphorst *et al.*, J. Electron Spectrosc. **76**, 2229 (1995).
15. H. Klar and M. Fehr, Z. Phys. D **23**, 295 (1992).
16. N. L. Manakov *et al.*, J. Phys. B: At. Mol. Opt. Phys. **29**, 2711 (1996).
17. L. Malegat *et al.*, J. Phys. B: At. Mol. Opt. Phys. **30**, 251 (1997).
18. B. Kämmerling and V. Schmidt, J. Phys. B: At. Mol. Opt. Phys. **26**, 1141 (1993).
19. B. Kämmerling and V. Schmidt, Phys. Rev. Lett. **67**, 1848 (1991).
20. W. R. Johnson and K. T. Cheng, Phys. Rev. A **46**, 2952 (1992).
21. S. J. Schaphorst and V. Schmidt, (private communication).
22. F. Maulbetsch and J. S. Briggs, J. Phys. B: At. Mol. Opt. Phys. **27**, 4095 (1994).
23. A. K. Kazansky and V. N. Ostrovsky, Phys. Rev. A **51**, 3698 (1995).

Recent Electron Angular Distribution Measurements in (γ,2e) of D₂ and Helium

T J Reddish, J.P Wightman and S Cvejanovic

Physics Department, Newcastle University, Newcastle upon Tyne, NE1 7RU, U.K.

Abstract. This report highlights the progress on photodouble ionisation of H_2/D_2 using coincidence techniques and synchrotron radiation. The results will be compared to recent theoretical studies, whose methods are also outlined, and with corresponding measurements in helium.

INTRODUCTION

Photodouble ionisation (PDI) is of fundamental interest because it results in unbound charged particles that are subject to the long-range Coulomb force. In the case of the H_2 molecule, the two-electron escape process occurs in the presence of *two* ionic centres, unlike the classic 3-particle continuum problem that is typified by helium. Investigations of the angular distributions of the two escaping electrons in atomic systems have clearly established the importance of electron correlation in calculating the triple differential cross section (TDCS) (e.g.: 1-9). In the case of helium, a bare nucleus is left behind and so no process other than direct double ionisation, in which the two electrons escape simultaneously, can occur. In contrast, all heavier atoms are also subject to other indirect processes involving intermediate configurations (i.e. ionic and neutral states converging to higher double ionisation thresholds), which also lead to the production of doubly charged ions. These processes are mirrored in diatomic systems, but there is also the inevitable dissociation into two ionic fragments during the Coulomb explosion. This occurs relatively quickly for H_2, but is temporarily impeded in larger dication systems due to local potential minima that can even support vibrational levels (e.g.: 10).

This report will outline both the experimental (11-14) and theoretical (15-17) progress on PDI of molecular hydrogen. During the last 3 years, the angular correlations of the two electrons have been investigated experimentally using coincidence techniques and synchrotron radiation. Further experiments are already underway, demonstrating that this exciting field is expanding quickly as increasingly more sophisticated experimental methods are introduced to cope with the low cross

CP500, *The Physics of Electronic and Atomic Collisions*, edited by Y. Itikawa, et al.
© 2000 American Institute of Physics 1-56396-777-4/00/$17.00

sections. Results of experimental studies in D_2 will be compared to similar work on helium, these being the simplest systems for double ionisation. (D_2 is used in preference to H_2 for experimental reasons; there is a higher number density in the interaction region for D_2 due to differences in the gas flow properties (13).)

EXPERIMENTAL PROGRESS IN PDI OF H_2/D_2

The following expression is the well-known form of the helium TDCS in the case of *equal* energy ($E_1 = E_2$) sharing conditions:

$$\text{TDCS}_{\text{He}} \sim |g|^2 ((\hat{k}_1.\hat{\varepsilon}) + (\hat{k}_2.\hat{\varepsilon}))^2 \qquad (1)$$

where the $\hat{k}_1, \hat{k}_2, \hat{\varepsilon}$, are the electron and polarisation directions (1). The angular factor is multiplied by a correlation term, which is usually approximated as Gaussian function with an excess energy-dependent half-width, $\theta_{1/2}(E)$:

$$|g(E, \theta_{12})|^2 = e^{-4\ln(2)(180-\theta_{12})^2/\theta_{1/2}^2} \qquad (2)$$

The shape of the TDCS given by this parametrisation - and indeed all the theoretical

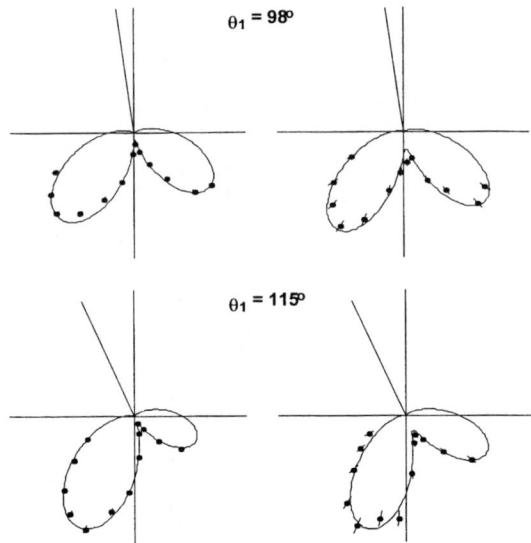

FIGURE 1. (γ,2e) angular distributions in the plane orthogonal to the photon beam for He (left) and D_2 (right) with $E_1 = E_2 = 10\text{eV}$. The fixed electron is centred on the stated θ_1 values, with respect to the major axis of the polarisation ellipse ($S_1 = 0.67 \pm 0.03$), and the angular distribution of the other electron is shown in polar form. The full curves for helium show fits using equations 1 and 2, with a $\theta_{1/2}$ value of $91 \pm 3°$, while the curves for D_2 use equations 2 and 3, with $g_\Pi/g_\Sigma \sim -2.1$ and $\theta_{1/2} = 77 \pm 3°$. All the curves include integration over the detection solid angles. (Figure from (15), reprinted with permission from IOP Publishing Limited).

approaches (1-9) - is in good agreement to the available equal energy data at all excess energies.

(γ,2e) studies of D_2 with $E_1 = E_2 = 10eV$ (11-13) have revealed the electron angular distribution to be remarkably similar to that of helium for the same kinematic conditions, in that the TDCS[1] is dominated by two lobes separated by a deep minimum (see Figure 1). The most noticeable difference is that the lobes are significantly closer together in D_2 than for helium. This difference has been quantified by two independent groups, using the above helium equations, giving $\theta_{1/2}$ values of $77 \pm 3°$ for D_2, compared with $91 \pm 3°$ for He. The direct application of those formulae to a diatomic molecule is, of course, without foundation. However, at the time of the first experiments there were no theoretical predictions, and this was the most logical starting point. A further difference can also been seen in Figure 1, which is highlighted in the D_2/He TDCS ratios shown in Figure 2. They reveal a pronounced peak at a mutual angle of 180°, the region where a node is expected in both TDCS's due to symmetry properties associated with these PDI transitions (1,2,15-17). The merit of this data representation is that the ratios are independent of variations in the toroidal analyser's angular efficiency because the measurements of

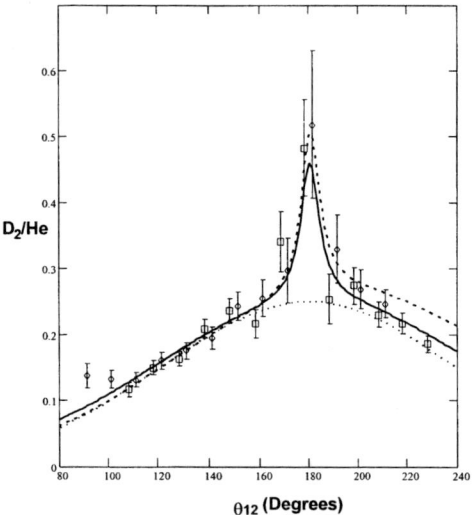

FIGURE 2. The D_2/He ratio of the coincidence yields as a function of the mutual angle (θ_{12}) between the two 10eV electrons for $\theta_1 = 98°$ (), 115° ((). The curves show the corresponding ratios computed from the fits also shown in Figure 1: $\theta_1 = 98°$ (full curve), 115° (broken curve). The dotted curve under the peak shows the ratio of the two Gaussian correlation factors used for D_2 and He, with $\theta_{1/2}$ values of $77 \pm 3°$ and $91 \pm 3°$, respectively. See (15,16) for a full discussion. (Figure from (15), reprinted with permission from IOP Publishing Limited).

[1] For diatomic molecules, the term "TDCS" is adopted here for convenience. Strictly, it should be the Quadruple Differential Cross Section due to the 'Franck-Condon' energy spread of the ions (12).

the two gases were obtained under nearly identical spectrometer tuning conditions (see (12) for full details). This apparent 'filling-in' of the node in the measured D_2 TDCS has also been observed using conventional methods (13) and is a consequence of the finite detector solid angles.

In addition to the published $(\gamma,2e)$ experiments, ion-electron coincidence studies of D_2 have been recently performed using the cold-target recoil-ion momentum spectroscopy (COLTRIMS) technique (14). Energy distributions of the electrons and the D^+ ions following double ionisation with a 58.8eV photon are shown in Figure 3. This photon energy is ~7.7eV above the nominal double ionisation potential (DIP) for D_2, corresponding to a vertical transition from the equilibrium internuclear separation to the purely repulsive Coulomb potential curve. As the $D^+ + D^+$ dissociation limit is ~19.4eV below the nominal DIP, this results in a total energy of 27.1eV for the four charged particles. Within the axial-recoil approximation, both ions will have the equal, but opposite, final momenta whose values depend on the internuclear separation at the moment of double ionisation. Together with the ground vibrational state distribution, this results in a broad energy spread centred around ~9.7eV as shown in Figure 3 (centre). Dörner et al (14) noticed the discrepancy of the measured ion distribution with the simple reflection approximation. This is due to the conservation of momentum and energy, which restricts the ion energies associated with one fast electron and so skews the net ion energy distribution. Figure 3 (right) shows the associated electron energy distribution. The corresponding curve for helium would be nearly uniform between $0 \rightarrow 7.7eV$ (perhaps slightly distorted to a symmetric U-shaped distribution). The distinct upper limit is not present in the molecular case, as the double ionisation energy is effectively a function of internuclear separation. The analysis of the ion and energy distributions is done within the Born-Oppenheimer approximation. The possibility of ion-electron interactions, or genuine 4-particle dynamics, during the double ionisation process is a tantalising idea needing further theoretical and experimental consideration.

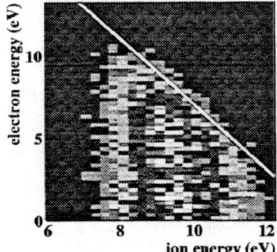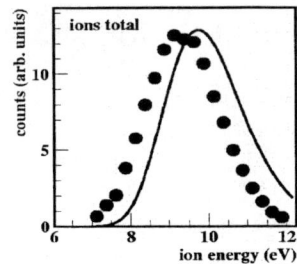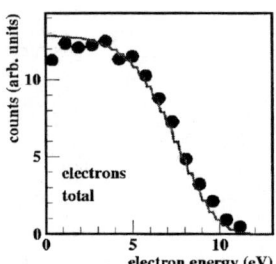

FIGURE 3. (Left): Surface plot showing the energy distribution of one of the two electrons verses the D^+ ion energy. The diagonal line indicates the total available energy, as discussed in the text. (Centre): The ion energy distribution integrated over all electron energies obtained from the surface plot. The solid curve shows the simple reflection approximation from the D_2 ground state. (Right): The electron energy distribution is also obtained from the surface plot by integration over all ion energies. Both distributions have full integration over the particles emission angles (Figure from (14)).

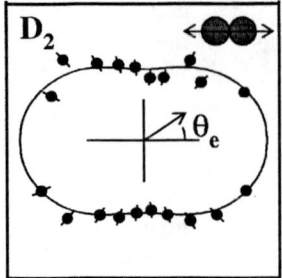 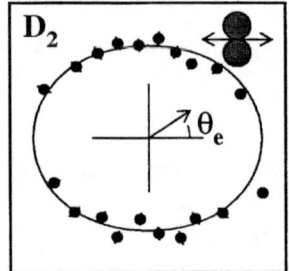 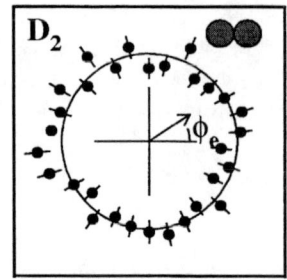

FIGURE 4. Polar plots of the electron angular distributions for one of the two electrons from photodouble ionisation of D_2 with $E_\gamma = 58.8eV$ for 'fixed-in-space' molecules. The data are integrated over all the electron energies shown in Figure 3. In the left-hand figure, the molecular axis is parallel to the horizontal polarisation vector (ε) and the full curve - to guide the eye - has $\beta = 0.4$. The central and right-hand figures *both* have the molecular axis perpendicular to ε. The polar distribution (centre) has $\beta = 0.14$, while the azimuthal distribution around ε (right) is isotropic ($\beta = 0$). The corresponding helium distribution, integrated over all the electron energies has $\beta = 0$. (Figure from (14)).

Angular distributions for one of the two electrons have also been determined by the COLTRIMS technique. Moreover, the momentum mapping gives the *direction* of the molecular axis ($\hat{R}$) at the moment of photodouble ionisation. The results of these 'fixed-in-space' molecular studies are shown in Figure 4. There is a measurable difference between the electron angular distributions (integrated over all possible electron energies) for the two, orthogonal molecular orientations shown. If such differences are present in these spectra, which also average over all the directions of the undetected electron, one should expect even more significant differences for future fixed-in-space, energy-resolved (γ,2e) studies. The surprising result is the apparent azimuthal symmetry shown in the $\hat{R} \perp \varepsilon$ case (right-hand figure).

THEORETICAL PROGRESS IN PDI OF H_2/D_2

Feagin (15,16) recently examined the form of the TDCS for diatomic molecules building on earlier theoretical studies for atoms (1-7). His helium-like description begins with a molecular dipole excitation amplitude is given by $f(\mathbf{R_N}) = \varepsilon.\mu$, where μ is the molecular transition dipole moment and $\mathbf{R_N}$ is the orientation of the internuclear axis at the instant of photoionisation (hence the proton recoil direction). The components of the photon polarisation ε' along the symmetry axis of the molecule are determined and the molecular response is expressed by a *helium-like* transition moment μ' in a *molecular* frame. Excitations that are parallel (g_Σ) and perpendicular (g_Π) to the internuclear axis are distinguished and finally the amplitude is transposed back to the laboratory frame.

For comparison with the available $E_1 = E_2$ (γ,2e) data, Feagin's general expression for the TDCS was integrated over all internuclear orientations resulting in:

$$\text{TDCS}_{D_2} \sim \frac{4\pi}{15}\left[2|g_\Sigma|^2 + 7|g_\Pi|^2 + 6\,\text{Re}\left(g_\Pi^* g_\Sigma\right)\right]((\hat{k}_1.\hat{\varepsilon}) + (\hat{k}_2.\hat{\varepsilon}))^2$$

$$+ \frac{4\pi}{15}|g_\Sigma - g_\Pi|^2|\hat{k}_1 + \hat{k}_2|^2 \qquad (3)$$

where $\hat{k}_1, \hat{k}_2$ are the electrons' directions. The first term has the same angular dependence to that of helium and equation (3) reduces to that form (equation (1)) when $g_\Sigma = g_\Pi$. As $|\hat{k}_1 + \hat{k}_2|^2 = 4\cos^2(\frac{1}{2}\theta_{12})$, the second, or *molecular*, term depends on the mutual angle (θ_{12}), rather than the angle with respect to the polarisation vector, $\hat{\varepsilon}$. This break in symmetry about the polarisation axis is also manifest in a relaxing of the selection rules that have established for atomic photodouble ionisation (1,2d). In particular, the vector identity associated with the first (or *atomic*) term, $\vec{k}_1.\hat{\varepsilon} = -\vec{k}_2.\hat{\varepsilon}$, defines a conical surface about the polarisation direction on which the TDCS is identically zero when $E_1 = E_2$ (see Figure 5). This surface also contains the back-to-back emission condition ($\vec{k}_1 = -\vec{k}_2$ or $\theta_{12} = 180°$), which is a further selection rule - a consequence of parity and exchange symmetry in a $^1S^e \rightarrow {}^1P^o$ (M = 0) transition. This latter condition is the only one that is satisfied by both terms in equation (3) and this has measurable consequences in the vicinity of the node as the detectors have finite solid angles (see Figures 1 and 2). More importantly, the lack of the node at $\theta_2 = \pi + \theta_1$ (see Figure 5) in D_2 results in a significant increase in the measured TDCS in this angular region, in comparison to helium. This is particularly evident at $\theta_1 = 144°$ (see Figure 6) as the amplitudes of the two characteristic lobes are a sensitive function of θ_1, as well as the degree of linear polarisation, S_1.

To obtain the D_2 TDCS, one needs to know the functional form of the g_Σ and g_Π amplitudes. This has yet to be investigated theoretically, but one can initially assume they are both Gaussian functions with, for simplicity, the same $\theta_{1/2}$ value for both amplitudes - but different to that of helium. This latter approximation is justified when concentrating on the TDCS shape in the vicinity of the node (Figures 1 and 2), where the Gaussian functions are peaked, but is questionable when considering the

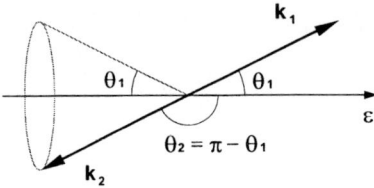

FIGURE 5. Diagram of the conical surface defined by $\vec{k}_1.\hat{\varepsilon} = -\vec{k}_2.\hat{\varepsilon}$ for an arbitrary position of $\vec{k}_1$. The surface given by $\theta_2 = \pi - \theta_1$ contains the 'back-to-back' emission condition $\theta_{12} = 180°$, which is the well-known node position observed in helium TDCS - see Figure 1. (Figure from (16), reprinted with permission from IOP Publishing Limited).

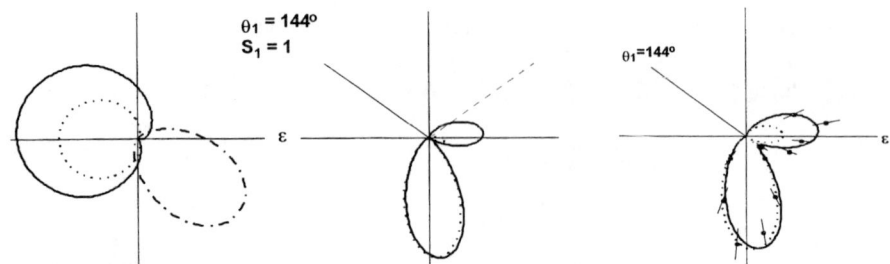

FIGURE 6. The comparison of the D_2 TDCS for $\theta_1 = 144°$ using both atomic and molecular formalism with the *same* $\theta_{1/2}$ of 77°. (Left): Polar plots showing the Gaussian correlation function (dash-dot) and the angular functions - helium (dots) and D_2 (solid) - from equations 1 and 3, respectively. With g_Π/g_Σ of ~-2.1, the D_2 function no longer has symmetry about the polarisation axis (horizontal). (Centre): The product of the angular and correlation terms for D_2 (solid) and He-*like* (dots). The dashed line shows the position of the extra node in helium, which effectively suppresses the TDCS between $\theta_2 = \pi \pm \theta_1$. (Right): $(\gamma,2e)$ data for $E_1 = E_2 = 10eV$, Stokes parameter $S_1 = 0.67 \pm 0.03$ from (11). The solid and dotted curves are molecular and atomic forms of the TDCS, respectively. The curves are normalised at their peak values and include integration over the detection solid angles. (Figure from (16), reprinted with permission from IOP Publishing Limited).

overall shape of the D_2 TDCS - despite the good level of agreement shown in Figure 6c. As Scherer *et al* (13) point out: "even in the Born-Oppenheimer approximation one can expect the Coulomb repulsion between the electrons to decrease or increase, depending on whether one of the protons is between the electrons or not".

Feagin (15,16) also showed that the amplitude ratio, $\eta = g_\Pi/g_\Sigma$ can be related to the *ion* β_N parameter via the approximation:

$$\beta_N \approx \frac{2(1-|\eta|^2)}{1+2|\eta|^2} \qquad \text{or} \qquad |\eta|^2 \approx \frac{2-\beta_N}{2(1+\beta_N)} \qquad (4)$$

An analysis of the $(\gamma,2e)$ data gave $\eta \sim -2.1$ ($\pm \sim 0.5$) corresponding to a β_N of -0.69 (± 0.13). This agreed with earlier - and intrinsically more accurate - ion-ion coincidence experiments which gave a β_N value of $-0.71(\pm 0.05)$ at the same photon energy of 71eV (18). This implies that within the assumptions of this model, one can use the measured β_N values to determine the g_Π/g_Σ ratio at different photon energies, as shown in Figure 7. The ratio appears to peak at ~ 2.3 near the nominal DIP and gradually reduces towards higher photon energies. It will be interesting to know the high-energy limit: does $\eta \to 1$, corresponding to an isotropic distribution ($\beta_N = 0$) and, incidentally, He-like equations, or does β_N change sign - and if so - at what photon energy? This question and others will no doubt be addressed in future work.

This helium-like description of PDI in H_2/D_2 has proved surprisingly successful in comparison with the data currently available. Its success rests on the dominance (~96%) of the $^1S^e$ component in a single-centre expansion of the ground state

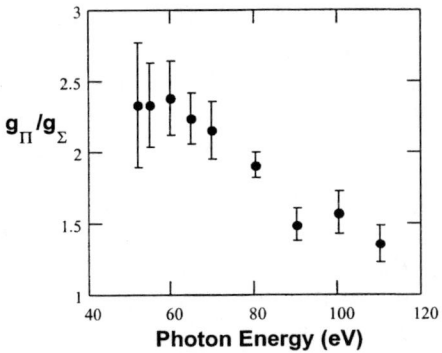

FIGURE 7. A plot of the g_Π/g_Σ ratio as a function of photon energy for H_2 obtained using the ion β_N parameter data from (18) and equation (4).

wavefunction (19,20). Indeed, this dominance can also explain the azimuthal symmetry shown in Figure 4c. Even so, this approach is incomplete, as it provides no predictions concerning the half-widths ($\theta_{1/2}$) of the associated Gaussian correlation functions. A further interesting observation has also arisen from these experimental studies. The $\theta_{1/2}$ values at an excess energy (E) of 20eV are 91° ($\pm3°$) for He and 77° ($\pm3°$) for D_2, as determined by ourselves (11,12) and confirmed by (13). Dörner *et al* (14) have inferred an average $\theta_{1/2}$ value from their β_e parameter measurements and found it to be significantly *larger* ($\sim120°$) for an E of 7eV, where the corresponding value for helium is $\sim86°$. As the $\theta_{1/2}$ value is a measure of the 'strength' of the electron correlation, that these D_2 values appear to fluctuate with excess energy about those of helium is indeed a surprising result that warrants further investigation. It could also imply, as Figures 4a,b and Dörner *et al* (14) infer, that $\theta_{1/2}$ is not the same for both g_Σ and g_Π amplitudes, or that the simple Gaussian is not an appropriate correlation function for diatomic systems. If either is the case then these results from planar (γ,2e) experiments cannot be readily compared with those from the COLTRIMS study without a full three-dimensional analysis.

Walter and Briggs (17) have recently shown that the (γ,2e) angular distributions for the symmetric $E_1 = E_2$ case are intrinsically more complex in diatomics than for helium. When the electronic final state is simply the product of (non-interacting) plane waves of momenta $\underline{k}_i$, the matrix element T^e was shown analytically to be:

$$T^e \propto \hat{\underline{\varepsilon}}.(\underline{k}_1 + \underline{k}_2)2\cos(\tfrac{1}{2}(\underline{k}_1 - \underline{k}_2).\underline{R}) \qquad (5)$$

where $\underline{R}$ is the relative vector of the two (fixed) nuclei. (A similar cosine factor occurs in *single* ionisation and the two-centre interference occurs when the de Broglie wavelength of the escaping electron is small compared to the internuclear separation.) While this expression collapses to the helium form (equation 1) when $R = 0$, it can be seen that the presence of a molecular axis produces an extra 'interference' term. This term produces a factor that is still a function of θ_{12} after integrating over all $\underline{R}$ orientations. Even when more complex final-state descriptions are used, the results contain features that can be correlated with the above molecular orientation dependence.

In their approach, the final-state interaction between the charged particles is represented by Coulomb functions. They begin with a product of two 2C functions for non-interacting electrons (i.e. a 4C function) to describe the electron-nuclei interaction. The electron-electron repulsion is accounted for by multiplying the 4C wavefunction by a two-body repulsive Coulomb factor to give a 5C wavefunction. As such, this represents the logical extension of their 3C wavefunctions to describe the atomic TDCS (2). Their study investigates the differences between the 4C and 5C wavefunctions and their effects on the electron angular distributions for both randomly orientated and 'fixed-in-space' molecules. At present, the calculations have just been compared with (γ,2e) D_2 data (see Figure 8) revealing a marked narrowing of the main lobes in comparison to helium and with the data. It is also interesting to observe the additional small lobes, thought to be due to higher-L components, which remain even after integrating over all nuclear orientations. The presence of the exact node at $\vec{k}_1 = -\vec{k}_2$ is evident, as no integration over detection solid angles has been performed. Moreover, these calculations are for R fixed at the equilibrium separation. What is not yet clear however, is precisely how the R-averaging over the ground state

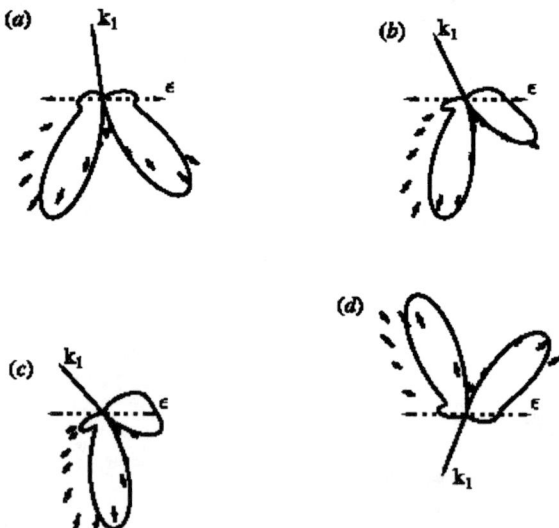

FIGURE 8. The calculated 5C angular distributions for $E_1 = E_2 = 10$eV electrons compared with the experiments: Data for (a-c) are from (12), those in (d) are from (13). (Figure from (17), reprinted with permission from IOP Publishing Limited).

distribution smoothes the features in the predicted TDCS, either for (γ,2e) or the COLTRIMS experiments. Nevertheless, this landmark work is the first to seriously consider electron correlation for photodouble ionisation in a two-centred system.

SUMMARY

This brief overview has highlighted the experimental and theoretical progress in this area since the last ICPEAC meeting. It is hoped that further theoretical studies, using a variety of methods, will be forthcoming giving detailed predictions for future experiments. For example, circular dichroism experiments will no doubt receive close attention following recent work by Reddish and Feagin (16). As the experimental techniques become even more sophisticated and sensitive, one can expect (γ,2e + D^+) studies for fixed-in-space molecules. Moreover, if the ion channel is energy-resolved, this would effectively result in studies that are a function of internuclear separation. This level of detail is foreseeable and will provide stringent tests for developing theories in this the most fundamental molecular double ionisation system.

ACKNOWLEDGMENTS

We acknowledge the financial support of EPSRC and the Leverhulme Trust and would like to thank Mike MacDonald for his valuable help at Daresbury. We have enjoyed our discussions with colleagues on this topic and would particularly like to thank Jim Feagin for his close theoretical involvement with us.

REFERENCES

1. Huetz, A. *et al, J. Phys. B.* **24**, 1917 (1991)
2. Maulbetsch, F. and Briggs, J.S., *J. Phys. B.* **26**, 1679, L647 (1993); **27**, 4095 (1994); **28** 551 (1995)
3. Kazansky, A.K. and Ostrovsky, V.N., *J. Phys. B.* **27** 447 (1994); **28** 1453, L333 (1995)
4. Pont, M. and Shakeshaft, R., *Phys. Rev. A*, **51** R2676, (1995); *J. Phys. B* **28** L571 (1995).
5. Lablanquie, P. *et al, Phys Rev Letts* **74** 2192 (1995)
6. Feagin, J.M., *J. Phys. B.* **29** L551 (1996)
7. Malegat, L. *et al, J. Phys. B.* **30** 251,263 (1997)
8. Lucey, S. *et al, J. Phys. B.* **31** 1237(1998)
9. Kheifets, A.S. and Bray, I., *J. Phys. B.* **31** L447 (1998)
10. Hall, R. *et al, Phys Rev Letts* **68** 2751 (1992)
11. Reddish, T.J. *et al, Phys Rev Letts* **79** 2438 (1997)
12. Wightman, J.P. *et al, J Phys B* **31** 1753 (1998)
13. Scherer, N. *et al, J Phys B* **31** L817 (1998)
14. Dörner, R. *et al, Phys Rev Letts* **81** 5776 (1998)
15. Feagin, J.M., *J Phys B* **31** L729 (1998)
16. Reddish, T.J. and Feagin, J.M., *J Phys B.* **32** 2473 (1999)
17. Walter, M. and Briggs, J.S., *J. Phys. B.* **32** 2487 (1999)
18. Kossmann, H. *et al, Phys. Rev. Letts.* **63** 2040 (1989)
19. Joy, H.W. and Parr, R.G., *J Chem Phys* **28** 448 (1958)
20. Bishop, D.M., *Mol Phys* **6** 305 (1963)

Beyond the Dipole Approximation: Angular Distribution Effects in the 1s Photoemission from Small Molecules

D. W. Lindle, O. A. Hemmers, H. Wang, P. Focke,[a] I. A. Sellin,[a] J. D. Mills,[b] J. A. Sheehy,[b] and P. W. Langhoff[c]

Department of Chemistry, University of Nevada, Las Vegas, Nevada, 89154-4003, USA
[a]Department of Physics, University of Tennessee, Knoxville, TN 37996
[b]Air Force Research Laboratory, AFRL/PRS, Edwards, AFB, CA 93524-7680
[c]Department of Chemistry, Indiana University, Bloomington, IN 47405

Abstract. Over the past two decades, the dipole approximation has facilitated a basic understanding of the photoionization process in atoms and molecules. Recent experiments on the 1s inner shells of small molecules at relatively low photon energies (≤ 1000 eV) show strong nondipole effects. They are significant and measurable at energies close to threshold, in conflict with a common assumption that the dipole approximation is valid for photon energies below 1 keV.

INTRODUCTION

The electric-dipole (E1) approximation [1], applied to photoionization, leads to the well-known expression for the differential cross section [2],

$$\frac{d\sigma}{d\Omega} = \frac{\sigma}{4\pi}\left[1 + \frac{\beta}{2}\left(3\cos^2\theta - 1\right)\right] \tag{1}$$

which describes the angular distribution of photoelectrons from a randomly oriented sample created by 100% linearly polarized light. Here, σ is the partial photoionization cross section, and θ is the angle between the vector of the outgoing electron and the vector of linear polarization. The parameter β completely describes the angular distribution of photoelectrons, within the dipole approximation. In this approximation, all higher-order interactions, such as electric-quadrupole (E2) and magnetic-dipole (M1), are neglected. This assumption is justified by the argument that the strengths of the E2 and M1 interactions relative to electric-dipole effects are approximately equal to the ratio of the photoelectron's velocity to the speed of light [3], a ratio which is small except at very high energies.

CP500, *The Physics of Electronic and Atomic Collisions*, edited by Y. Itikawa, et al.
© 2000 American Institute of Physics 1-56396-777-4/00/$17.00

Over the past two decades, the dipole approximation has facilitated a basic understanding of the photoionization process in atoms and molecules [2], as well as the application of photoelectron spectroscopy to a wide variety of condensed-phase systems. The first hint of deviations from the dipole approximation was provided by Krause [4] in measurements using unpolarized x-rays [5]. A small deviation from the expected dipolar angular distribution at photon energies between 1 and 2 keV was observed and attributed to the influence of E2 and M1 interactions. These lowest-order, non-electric-dipole corrections to the dipole approximation lead to so-called *nondipole* effects in the angular distributions of photoelectrons, described by [6]

$$\frac{d\sigma}{d\Omega} = \frac{\sigma}{4\pi}\left[1 + \frac{\beta}{2}\left(3\cos^2\theta - 1\right) + \left(\delta + \gamma\cos^2\theta\right)\sin\theta\cos\phi\right] \tag{2}$$

for 100% linearly polarized light. The nondipole angular-distribution parameters γ and δ are attributable to interference terms between electric-dipole and electric-quadrupole interactions. Figure 1 describes the geometry and the angles θ and ϕ.

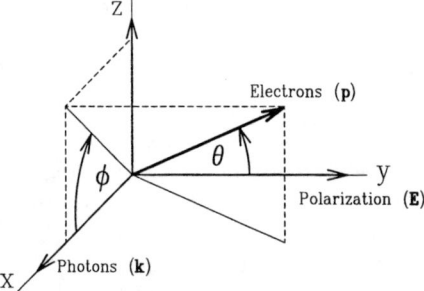

Figure 1. Geometry applicable to photoelectron angular-distribution measurements using polarized light. θ is the polar angle between the photon polarization vector ε and the momentum vector **p** of the photoelectron. ϕ is the azimuthal angle defined by the photon propagation vector **k** and the projection of **p** into the x-z plane.

More-recent measurements [7,8], focussing on noble-gas core levels (Ar K and Kr L) and photon energies above 2 keV, have begun to investigate nondipole effects in photoelectron angular distributions in more detail. In contrast, the present experiment concentrates on the N_2 $N1s$ and CO $C1s$ inner shells at relatively low photon energies (300 to 700 eV). Nondipole effects are observed to be large and highly energy dependent in this region, especially close to core-level thresholds, in conflict with a common assumption in applications of photoelectron spectroscopy; namely, that the dipole approximation is valid for photon energies below 1 keV. The potential significance of these findings is nicely illustrated by comparison of the present results for the N_2 and the CO γ_{1s} parameters with theories for atomic nitrogen and atomic carbon [9], where the influence of nondipole effects are expected to be much smaller.

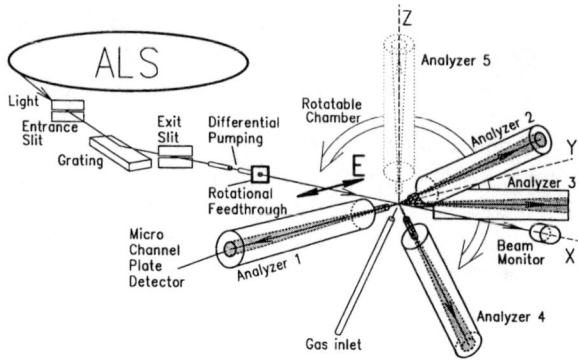

Figure 2. Experimental schematic of the electron time-of-flight system. Light from the ALS storage ring passes through beamline optics into a differential-pumping section. The chamber and analyzers can rotate around the photon beam for more accurate electron angular-distribution measurements.

EXPERIMENT

The experiments were performed on undulator beamline 8.0, [10], which covers the 100-1500 eV photon-energy range. The monochromator entrance slit was set to 70 μm and the exit slit to 100 μm yielding very high flux, because high photon resolution was not needed. During the measurements the ALS operated at 1.9 GeV in two-bunch mode with a photon pulse every 328 ns. Four time-of-flight (TOF) electron analyzers, equipped with microchannel plates for electron detection, collect spectra simultaneously at different angles. The total electron flight paths are 437.5 mm, and the analyzers have a full cone acceptance angle of 5.4°.

The interaction region is formed by an effusive gas jet intersecting the photon beam which has a diameter of less than 1 mm. Energy resolution of the TOF analyzers with a focus size of 1 mm is 1% of the electron kinetic energy. Each spectrum was collected for about 600 s. The gas samples were obtained either commercially (CO) or directly from ambient air (N_2). A mixture of the sample with xenon was used sometimes because Xe has an abundance of Auger lines below 100 eV kinetic energy, which provide excellent internal calibration for each spectrum.

RESULTS

Figure 3 shows two superimposed spectra, both taken at the magic angle θ=54.7°, but at different φ angles. The spectra were measured close to the C1s threshold (296 eV) and are scaled to the area of the Xe NOO Auger lines and the obvious intensity differences between the CO C1s peaks in the two spectra are due entirely to nondipole

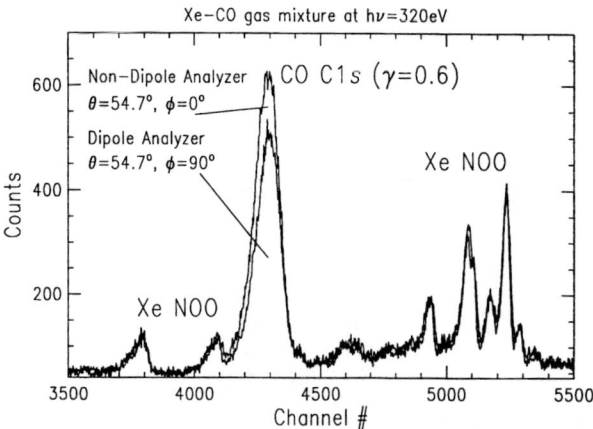

Figure 3. Photoelectron spectra of a CO-Xe mixture measured at a photon energy of 320 eV. One spectrum was taken with the dipole magic-angle analyzer and the other spectrum with the nondipole analyzer. The spectra are normalized to the Xe NOO Auger lines. The intensity differences in the CO C1s lines between the two analyzers is due entirely to nondipole effects.

effects because both spectra are at the magic angle where the β parameter has no influence.

For the dipole magic-angle analyzer the differential cross section in Eq. (2) reduces to the partial cross section; **E2** and **M1** effects vanish in the φ=90° plane even if relativistic effects are included [11]. For the nondipole analyzer,

$$\frac{d\sigma}{d\Omega} = \frac{\sigma}{4\pi}\left[1 + \sqrt{\frac{2}{27}}(3\delta + \gamma)\right] = \frac{\sigma}{4\pi}\left[1 + \sqrt{\frac{2}{27}}\zeta\right] \qquad (3)$$

which simplifies further for s subshells [6,12] in the non-relativistic approach where δ vanishes. We are using ζ=3δ+γ for measurements that don't resolve the δ and γ parameters of the angular distributions. In the case of molecular effects it is not clear if the δ parameter for s-shells vanishes near threshold.

With our experimental geometry, it is possible to measure the ζ parameter for s subshells directly, if the degree of linear polarization is known, by using the two magic angle analyzers. The data points for CO and N_2 in Figures 4 and 5 show strong nondipole contributions with maxima of ζ=1.2.

The difference between CO and N_2 lies in the position of the maxima. For N_2 the maximum is about 60 eV above the N_2 1s ionization threshold much higher than the maximum of the dipole shape resonance at about 420 eV. The maximum of the ζ parameter for the CO C1s is close to the maximum of the dipole shape resonance at 305 eV.

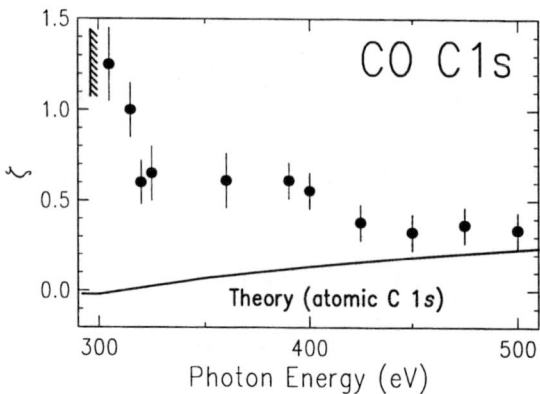

Figure 4. Electron angular anisotropy parameter ζ for the CO C1s photoline from threshold to hv= 500 eV. The theoretical curve for atomic carbon is from Lajohn and Pratt [9].

A qualitative explanation for the behavior of ζ can be obtained from the following model. Just as molecular β values can change rapidly with photon energy for large differences in polarization components for ionization along and perpendicular to a molecular axis (due to a resonance, for example), so also ζ values can behave similarly but with greater sensitivity to the difference in polarization components because of the higher power of the transition moment coordinate involved. Thus, the observed molecular ζ effects may be universal.

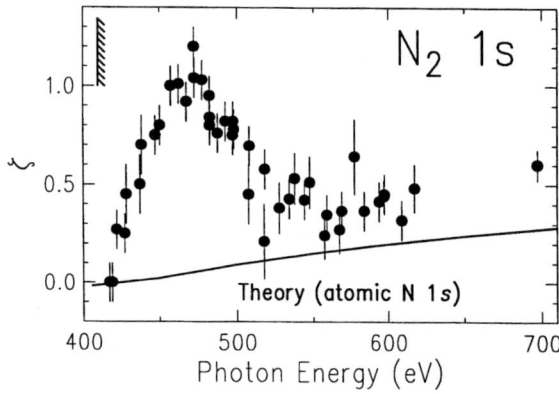

Figure 5. Electron angular anisotropy parameter ζ for the N$_2$ 1s photoline from threshold to hv= 700 eV. The theoretical curve for atomic nitrogen is from Lajohn and Pratt [9].

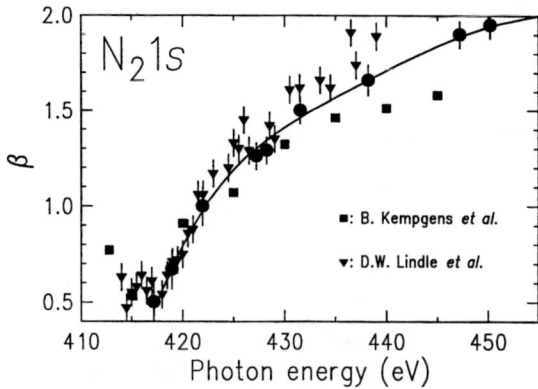

Figure 6. Electron angular anisotropy parameter β for the N_2 1s photoline from threshold to hv= 450 eV. Measurements by Kempgens [14] and Lindle [15] did not take non-dipolar effects into account and deviate from our measurements.

The variations of β (and ζ) with photon energy in atoms are due to the interference of different partial waves ($p \to s$ and d, for example) whereas in molecules this can be due to the interference of the polarization components (1s $\to p\sigma$ and $p\pi$ in N_2, for example).

One consequence of these strong molecular nondipole effects near threshold is the possibility of influences on previous measurements of β-parameters, as demonstrated in Fig. 6. If the β-parameter is not measured in the plane perpendicular to the direction of the light (and linear polarized light) the intensities used to determine β are influenced by the forward/backward intensities of ζ. These intensities reduce or increase the β values as shown in Fig. 6 for Kempgens and Lindle. Larger values of ζ lead to larger deviations in β.

The present results illustrate that any photoemission experiment, whether on gases, solids, or surfaces, can be influenced by nondipole effects at relatively low photon energies, pointing to a general need for caution in interpreting angle-resolved photoemission data.

ACKNOWLEDGMENTS

The authors thank the staff of the ALS and the IBM, LBNL, LLNL, the University of Tennessee, and Tulane University collaboration for their support. This research is funded by the NSF (PHY-9303915), the DOE Nevada EPSCoR. The ALS is supported by the U.S. DOE through the Materials Science Division, Office of Basic Energy Sciences, Office of Energy Research at the Lawrence Berkeley National Laboratory under contract No. DE-AC03-76SF00098.

REFERENCES

1. Bethe, H. A. and Salpeter, E. E., *Quantum Mechanics of One- and Two-Electron Atoms*, Berlin: Springer-Verlag, 1957.
2. Manson, S. T., and Dill, D., *Electron Spectroscopy: Theory, Techniques, and Applications*, New York: Academic, 1978, Vol. 2, edited by Brundle, C. R. and Baker, A. D. (Academic, New York, 1978).
3. Cooper, J. and Zare, R. N., *J. Chem. Phys.* **48**, 942 (1968).
4. Krause, M. O., *Phys. Rev.* **177**, 151 (1969).
5. For unpolarized incident light, $\beta/2$ is replaced by $\beta/4$ in Eq. (1), and θ is measured between the propagation vectors of the photon and the photoelectron. Otherwise, the essential physics is the same.
6. Cooper, J. W., *Phys. Rev. A* **42**, 6942 (1990); **45**, 3362 (1992); **47**, 1841 (1993).
7. Krässig, B., Jung, M., Gemmell, D. S., Kanter, E. P., LeBrun, T., Southworth, S. H., and Young, L., *Phys. Rev. Lett.* **75**, 4736 (1995).
8. Jung, M., Krässig, B., Gemmell, D. S., Kanter, E. P., LeBrun, T., Southworth, S. H., and Young, L., *Phys. Rev. A* **54**, 2127 (1996).
9. Lajohn, L. and Pratt , R. H., (private communication).
10. Perera, R. C. C., *Nucl. Instrum. Methods* **A319**, 277 (1992).
11. Scofield, J. H., *Phys. Rev. A* **40**, 3054 (1989); *Phys. Scripta* **41**, 59 (1990).
12. Amusia, M. Ya., Arifov, P. U., Baltenkov, A. S., Grinberg, A. A., and Shapiro, S. G., *Phys. Lett.* **47A**, 66 (1974); Amusia, M. Ya., Baltenkov, A. S., Grinberg, A. A., and Shapiro, S. G., *Sov. Phys.-JETP* **41**, 14 (1975); Amusia, M. Ya. and Cherepkov, N. A., *Case Studies in Atomic Physics*, Amsterdam: North-Holland, 1975, Vol. 5.
13. Hemmers, O., Whitfield, S. B., Glans, P., Wang, H., Lindle, D. W., Wehlitz, R., and Sellin, I. A., *Rev. Sci. Instrum.* **69**, 3809 (1998).
14. Kempgens, B., Kivimäki, A., Neeb, M., Köppe, H.M., Bradshaw, A.M., and Feldhaus, J., *J. Phys. B* **29** (1996).
15. Lindle, D. W., Truesdale, C.M., Kobrin, P.H., Ferrett, T.A., Heimann, P.A., Becker, U., Kerkhoff, H.G., and Shirley, D.A., *J. Chem. Phys.* **81** (1984).

New Insights into Molecular Structure and Dynamics Using Soft X-ray Electron Spectroscopy

E. Kukk[1,2], J.D. Bozek[2], T.D. Thomas[3], T.X. Carroll[4], L.J. Saethre[5], J.A. Sheehy[6], P.W. Langhoff[7], and N. Berrah[1]

[1] *Department of Physics, Western Michigan University, Kalamazoo, MI 49008- 5151*
[2] *Lawrence Berkeley National Laboratory, University of California, Berkeley, CA 94720*
[3] *Department of Chemistry, Oregon State University, Corvallis, OR 97331*
[4] *Keuka College, Keuka Park, New York 14478*
[5] *Department of Chemistry, University of Bergen, N-5007 Bergen, Norway*
[6] *Air Force Research Laboratory, AFRL/PRS, Edwards AFB, CA 93524-7680*
[7] *Department of Chemistry, Indiana University, Bloomington, IN 47405*

Abstract. The combination of high-resolution electron energy analyzers and 3rd generation synchrotron radiation sources opens up possibilities to study molecular inner-shell photo- and Auger electron emission in a new level of detail. Even weak molecular perturbations of the energy levels and angular emission patterns can be studied. In this report, some examples of such studies are given based on recent experiments at the Advanced Light Source, using a Scienta SES-200 electron spectrometer. Examples of carbon $1s$ photoemission of ethyne, sulphur $2p$ photoemission of carbonyl sulphide, and resonant Auger electron emission of carbon dioxide are presented.

INTRODUCTION

Following the introduction of 3rd generation soft x-ray synchrotron radiation sources, there has been a continuous development of modern high-resolution electron spectroscopic experiments. Soft x-ray spectroscopies of gaseous samples often suffer from the inherent problem of a very weak signal from a diffuse target and one is often forced to sacrifice high photon and electron energy resolution for acceptable count rates. 3rd generation electron storage rings (ALS in the U.S.A., BESSY II in Germany, MAX II in Sweden, ELETTRA in Italy) with their high photon brightness have been a critical improvement for atomic and molecular electron spectroscopy, allowing the investigation of subtle effects and small energy splittings that were beyond reach a few years ago.

The benefits of this experimental progress become evident in the case of molecular inner-shell studies. Whereas the electronic structure of the valence shells in

CP500, *The Physics of Electronic and Atomic Collisions*, edited by Y. Itikawa, et al.

molecules bears little resemblance to that of its atomic constituents, inner shells retain many of their atomic properties. It has become apparent over time, however, that there are subtle disturbances of the core orbitals caused by the molecular surroundings. These molecular perturbations are often manifested by relatively small energy splittings, due to minor overlap of electron density with neighboring atoms, anisotropy of the potential at the site of the atom etc. Obtaining accurate quantitative data about these effects has proven to be a difficult task. The experimental resolution requirements in this relatively high photon energy regime are demanding. Also, the molecular effects on the core electrons are usually small, often on the order of the lifetime broadening of the spectral lines. However, these small disturbances of the atomic character of core electrons can provide unique information about the molecule - chemical bonding and bond lengths, potential energy surfaces or photoemission characteristics in the molecular frame.

In the following, a brief overview of some recent results in molecular inner-shell photoelectron spectroscopy is given. These experiments were performed at the Advanced Light Source (ALS) using a new end-station, designed for gas-phase studies.

EXPERIMENT

The experiments were performed at the Advanced Light Source at Beamline 10.0.1, with the general layout as shown in Fig.1. Soft x-ray radiation is created for the beamline by a 10 cm period, 4.55 m long undulator, one of the brightest photon

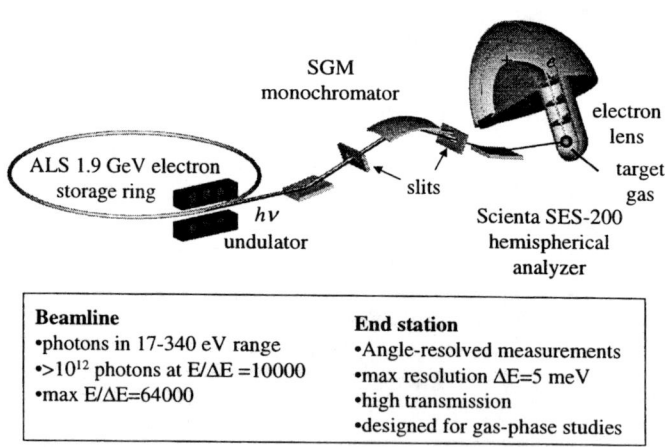

Beamline
•photons in 17-340 eV range
•>10^{12} photons at E/ΔE =10000
•max E/ΔE=64000

End station
•Angle-resolved measurements
•max resolution ΔE=5 meV
•high transmission
•designed for gas-phase studies

FIGURE 1. Schematic diagram of the experimental setup for gas-phase studies at Beamline 10.0.1 at the Advanced Light Source.

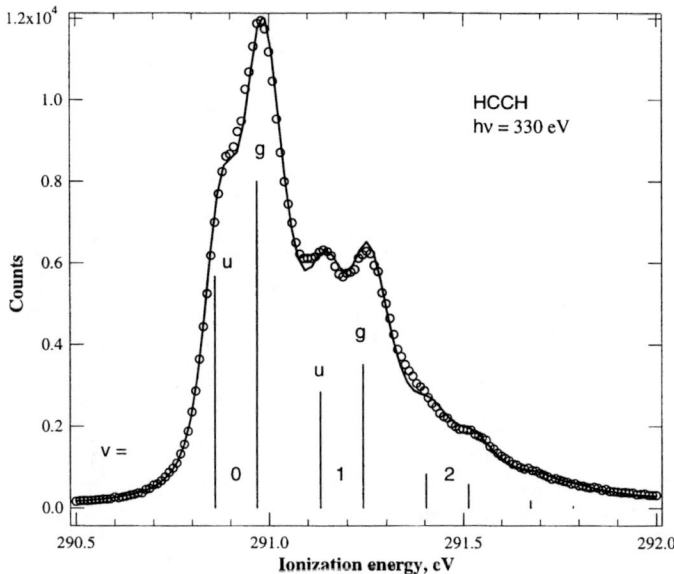

FIGURE 2. Carbon $1s$ photoelectron spectrum of ethyne, decomposed into σ_g and σ_u components by least-squares curve fitting, from Ref. [6].

sources in its energy range. The radiation is monochromatised by a spherical grating monochromator (SGM) equipped with three interchangeable gratings of 380, 925 and 2100 lines/mm groove density. Its range of operation is from 17 to 340 eV and the beamline delivers more than 10^{12} photons/sec to the end-stations over most of this range at 10,000 resolving power.

The gas-phase high-resolution end-station [1] is based on a Scienta SES-200 hemispherical energy analyser, rotatable around the axis of the photon beam in order to perform angle-resolved studies. The target gas is introduced into a gas-cell with differentially pumped openings for the photon beam. The end-station is separated from the UHV of the beamline by several differential pumping stages, permitting the use of sample gas pressures up to 10^{-2} torr within the gas-cell. In order to utilize the high resolution of the analyzer in full, the interaction area is surrounded by efficient magnetic shielding and the gas cell is equipped with electrodes to compensate for surface potential gradients [2]. In its present configuration, the spectrometer has achieved electron energy resolution of 5.7 meV under the most stringent settings.

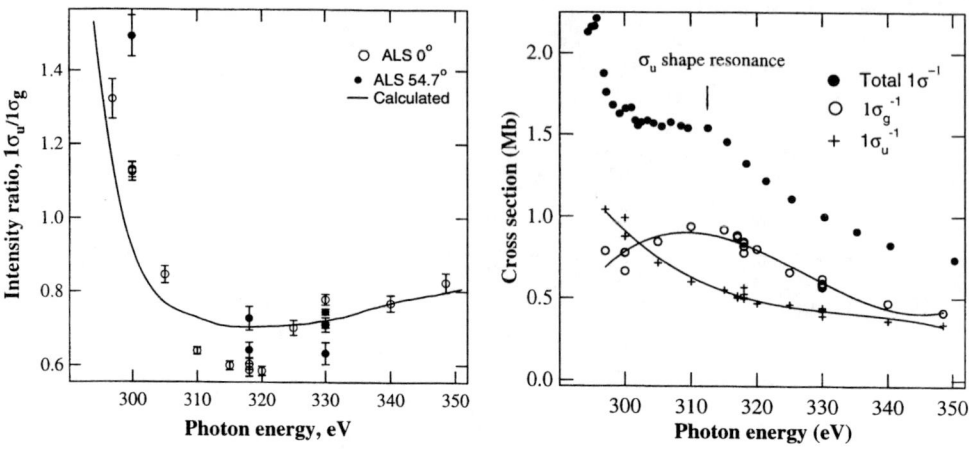

FIGURE 3. Branching ratios (left) and partial cross sections (right) of the σ_g and σ_u components of carbon $1s$ photoionization of ethyne [6]. Data for the total $1s(\sigma)$ cross-section are from ref. [4].

RECENT RESULTS AND THEIR INTERPRETATION

Ethyne

Ethyne (C_2H_2) is a linear molecule with two chemically equivalent carbon atoms. The carbon $1s$ photoemisson spectrum displays a vibrational progression characteristic of the core-ionized state, but there is no chemical shift distinguishing the two carbons. However, the structure of the spectrum shown in Fig. 2 cannot be assigned entirely to vibrational transitions. It is apparent from the high-resolution spectrum in Fig. 2, that the spectrum consists of two vibrational series, shifted with respect to each other by 105 meV, as has been noted in a previous study [3]. The splitting arises because of the very short length of the carbon-carbon triple bond and the consequent enhanced overlap of the C $1s$ orbitals. The two series in the spectrum can be assigned to the *gerade* ($1\sigma_g$) and *ungerade* ($1\sigma_u$) combinations of the two $1s$ wavefunctions, marked in Fig. 2 by g and u, correspondingly. The present experiment acieved a combined photon and electron energy resolution of 42 meV, so that the natural line width of 105 meV becomes the main factor prohibiting better separation of individual lines. The high photon flux of the beamline made it possible to record the spectra with sufficient statistics to allow reliable determination of the u/g branching ratios.

The left panel of Fig. 3 shows the u/g branching ratio as a function of photon energy, determined from a series of carbon $1s$ photoelectron spectra. The branching ratio goes through a minimum at around 315 eV. Such behavior can be understood by comparing the present results with the carbon $1s$ total photoionization cross-

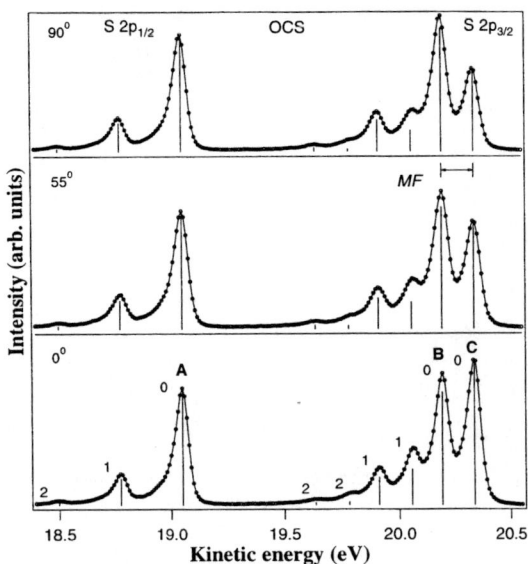

FIGURE 4. Sulphur $2p$ photoelectron spectra of carbonyl sulphide, taken at three different angles relative to the polarization plane at 191 eV photon energy.

section, which shows a maximum around the same photon energy (Fig. 3, right panel). The assignment of the enhancement in the total absorption cross-section above the C $1s$ threshold to shape resonance has been questioned by Kempgens et al. [4]. Identifying a shape resonance feature is important in the light of a suggestion that its position can be used as a ruler to measure molecular dimensions [5]. Combining the present data with the results of Ref. [4], the total cross-section in Fig. 3 [4] can be decomposed into $1\sigma_u$ and $1\sigma_g$ partial cross-sections. It is evident that the maximum in the total cross-section coincides with the maximum in $1\sigma_g$ cross-section. This allows us to identify the maximum as a σ_u-type shape-resonance and determine its position. The complete study is available in ref. [6].

Carbonyl sulphide

The sulphur $2p$ photoelectron spectrum of carbonyl sulphide (OCS), shown in Fig. 4, is another example of molecular effects upon the photoionization of atomic inner shells. The two main structures in the figure are due to the 1.2 eV spin-orbit splitting between the sulphur $2p_{1/2}$ and $2p_{3/2}$ orbitals. The S $2p_{1/2}$ peak is accompanied by a single vibrational progression assigned to the C-O stretch of the molecule. In the $2p_{3/2}$ component, additional features appear due to the reduced symmetry of the linear molecule compared to the free atom. The molecular field

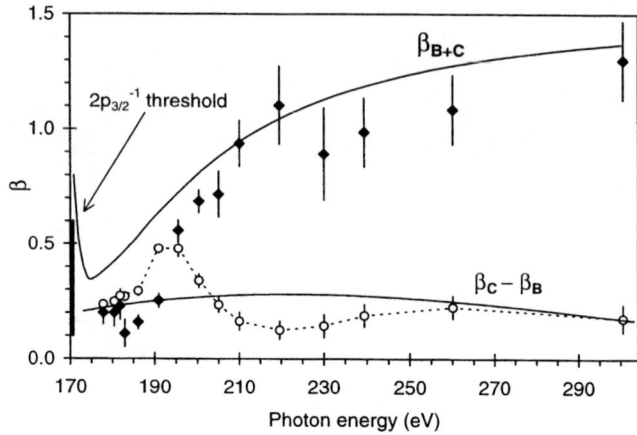

FIGURE 5. Anisotropy parameters of sulphur $2p_{3/2}$ photoemission in OCS.

in which the sulphur atom is embedded is anisotropic and therefore the degeneracy of the levels with different m_j is removed. The S $2p_{3/2}$ structure consists of two vibrational progressions, corresponding to the $2p_{3/2,\pm3/2}$ (B) and $2p_{3/2,\pm1/2}$ (C) molecular field split components, separated by 145 meV [7].

The molecular field split components B and C can also be expressed in terms of σ and π molecular orbitals. It is then apparent that the orbital corresponding to B is oriented perpendicular to the molecular axis, whereas C corresponds to the orbital with electron density mostly along the molecular axis.

The most interesting feature of Fig. 4 is the change of the relative intensities of the B and C lines in the spectra measured at different emission angles relative to the polarization plane, indicating that these two molecular field split components have different angular distributions. We studied this effect over a broad photon energy range above the $2p$ ionization threshold. The resulting anisotropy (β) parameter for the whole S $2p_{3/2}$ group and the difference of the β-parameters of the C and B components is shown in Figure 5. The β-value for peak C is clearly larger than that for peak B over the whole energy range. The principal difference between the B and C components is the different orientation of corresponding orbitals within the molecule, which affects the transitions to continuum orbitals (which, penetrating into the molecule, should be also described using molecular orbital formalism). The anisotropy parameters of the components B and C therefore provide access to molecular-frame information, which is usually lost in non-coincidence studies by averaging over randomly oriented ensemble of molecules.

In the present case, molecular frame information is conveyed to us due to the opportunity to study individual molecular field split components in the spectrum.

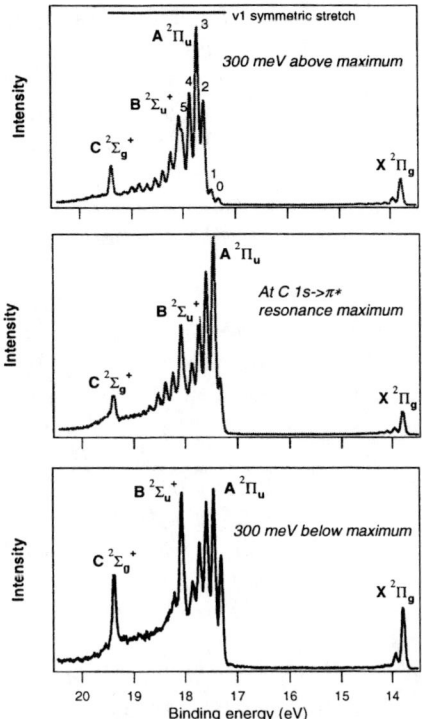

FIGURE 6. Resonant Auger decay spectra of carbon $1s \rightarrow \pi^*$ excitations in CO_2, measured at the photon energies below, at, and above the absoprtion maximum.

This method is an alternative to coincidence measurements, where the orientation of the molecular axis is determined by detecting the products of molecular dissociation together with photo- or Auger electrons [8]. It has the advantage of a much higher count rate and is also applicable to nonlinear molecules, where the determination of the molecular orientation through the dissociation fragments is not possible.

Carbon dioxide

The pre-threshold region of the carbon $1s$ photoabsorption is dominated by the excitations to the π_u unoccupied molecular orbital, which give rise to a broad and structureless peak in the spectrum [9]. The $1s^{-1}\pi_u$ core-excited configuration of CO_2 can have bent or linear equilibrium geometries. The energy splitting (Renner-Teller splitting) between the bent(A_1) and linear(B_1) states results in a complicated vibrational structure of different vibrational modes, hidden under the broad absorption peak.

The core-excited state decays via an Auger process mainly to the $A^2\Pi_u$ state of CO_2^+. The vibrational structure of the Auger electron spectrum reflects the properties of both the intermediate core-excited and final state potential energy curves through the Frank-Condon factors between the corresponding vibrational wavefunctions. We have measured the Auger electron spectra at a number of differents photon energies across the C $1s \to \pi_u$ absorption peak.

Fig. 6 displays three resonant Auger spectra, measured below, at, and above the absorption maximum. These, together with a number of spectra at other photon energies, show that:

(i) The main structure of the $A^2\Pi_u$ band can be assigned to the symmetric stretch mode of CO^+, based on valence photoelectron spectra [10].

(ii) At the photon energies below the absorption maximum, the intensity ratios of the vibrational lines within the $A^2\Pi_u$ band remain essentially unchanged. Only the total intensity changes. In this region, the C $1s$ excitations to the bent state take place and it appears that the equilibrium C-O bond length of the ground and excited A_1 state are very similar so that the lowest v_1 vibrational level is populated preferentially. To account for a rather broad photon energy range of these excitations, other vibrational modes must be involved.

(iii) In the region around and above the maximum, the vibrational envelope of the $A^2\Pi_u$ band changes rapidly, transferring intensity to higher vibrational levels of the final state as the photon energy increases. Here, the C $1s$ excitations to the linear (B_1) state dominate and the Auger transitions seem to map the population of the vibrational levels of the initial (core-excited) state to the final state levels almost one-to-one. In this case, the bond lengths of the excited and final ionic states must be close to each other, whereas they are rather different for the ground and excited states, allowing a number of symmetric stretch vibrational levels to be excited.

(iv) At the higher binding energy side of the spectra recorded below the absorption maximum, a tail of increased background level can be seen, not present in the spectra taken above the absorption maximum. This tail can be assigned to the transitions between high vibrational levels of the bending mode. Since the ground-to-excited state transitions occur between linear and bent geometries, strong excitations of the bending mode vibrations can be expected. Our calculations show that the Frank-Condon factors are the largest for the vibrational levels of the A_1 state near the saddle point of the potential energy curve at 180° O-C-O bond angle. The Auger decay from these levels has a dual character. A part of the Auger intensity goes to the lowest level of the bending mode of the final state. These transitions are accompanied by a distinct progression of symmetric stretch mode. Another part of the intensity goes to the high levels of the final state bending vibrations due to increased overlap near the classical turning points. Since a number of close-lying levels can be reached at once, the symmetric stretch progression accompanying these transitions is smeared out and they are seen as a low kinetic (high binding) energy background in the Auger electron spectrum.

As seen above, the vibrational structure of the Auger electron spectra depends strongly on the character of the initial core-excited state. In particular, the Auger spectra of the linear initial state together with high-resolution photoabsorption spectrum provide sufficient information to determine the spacing and energies of the vibrational levels of that state. Also, the magnitude of the Renner-Teller splitting between the two geometries of the core-excited state can be determined [11].

CONCLUSIONS

The above examples of molecular inner-shell studies show that, despite experimental difficulties and inherent restrictions like lifetime broadening, detailed quantitative information can be obtained on molecular effects upon the photoionization of the inner shells of the constituent atoms. This information can be of great importance in understanding more general molecular properties. In the case of ethyne, resolving the *gerade-ungerade* splitting in the carbon $1s$ photoelectron spectrum and studying the branching ratios of the two components provided proof of the existence of a shape resonance in the continuum near the C $1s$ ionization threshold. In the case of carbonyl sulphide, studying the angular distribution of the well-resolved molecular field split components in the S $2p$ photoemission provides a new way of accessing molecular body-frame information. In the case of carbon dioxide, resonant Auger electron spectra help to study the complicated vibrational structure and Renner-Teller splitting of the C $1s^{-1}\pi_u$ core-excited state.

ACKNOWLEDGEMENTS

This work was funded by the Divisions of Chemical and Material Sciences of the Office of Energy Research of the U.S. Department of Energy. The authors thank the staff of the ALS for help during the experiments.

REFERENCES

1. N. Berrah et al., J. Electron Spectrosc. Relat. Phenom. **101**, 1 (1999).
2. P. Baltzer *et al.*, Rev. Sci. Instrum. **64**, 2179 (1993).
3. B. Kempgens et al., Phys. Rev. Lett. **79**, 3617 (1997).
4. B. Kempgens et al., Phys. Rev. Lett. **79**, 35 (1997); J. Chem. Phys. **107**, 4219 (1997).
5. A. Hitchcock et al., J. Chem. Phys. **80**, 3927 (1984); J. Stöhr et al., Phys. Rev. Lett. **53**, 1684 (1984); F. Sette et al., J. Chem. Phys. **81**, 4906 (1984).
6. T.D. Thomas et al., Phys. Rev. Lett. **82**, 1120 (1999).
7. M.R.F. Siggel et al., J. Chem. Phys. **105**, 9035 (1999).
8. E. Shigemasa *et al.*, Phys. Rev. Lett. **80**, 1622 (1998).
9. G.R. Wight and C.E. Brion, J. Electron Spectrosc. Relat. Phenom. **3**, 191 (1974).
10. P. Baltzer et al., J. Chem. Phys. **104**, 8922 (1996).
11. E. Kukk et al., *to be published.*

SR-Pump and Laser-Probe Experiments for the Photofragmentation Dynamics of Atoms and Molecules

Koichiro Mitsuke

Institute for Molecular Science, Myodaiji, Okazaki 444-8585, Japan

Abstract. Synchrotron radiation-laser combination techniques developed at UVSOR in Okazaki are employed for probing ionic and neutral photofragments produced in the vacuum ultraviolet. First, $N_2^+(X\,{}^2\Sigma_g^+, v_X = 0$ and $1)$ resulting from direct ionization or autoionization of N_2 or N_2O is detected by laser induced fluorescence excitation spectroscopy in the wavelength region of the $B\,{}^2\Sigma_u^+ \leftarrow X\,{}^2\Sigma_g^+$ transition. The rotational distribution of N_2^+ is considered to be determined by the change in the rotational quantum number in photoionization and by the partitioning of the excess energy impulsively released in dissociation. Second, $S(3s^23p^4\,{}^3P_2)$ produced by predissociation of Rydberg states of OCS is detected by resonance enhanced multiphoton ionization spectroscopy.

INTRODUCTION

When polyatomic molecules ABC are electronically excited by absorbing synchrotron radiation in the vacuum ultraviolet (VUV) or soft X-ray (SX) region, various types of ionic and neutral photofragmentation may frequently occur (1-5):

$$\text{ABC} + h\nu \xrightarrow{-e^-} \text{ABC}^+ \rightarrow \text{AB}^+ + \text{C} \qquad \text{(direct ionization)} \quad (1)$$

$$\xrightarrow{-e^-} \text{ABC}^+ \xrightarrow{-e^-} \text{ABC}^{2+} \rightarrow \text{AB}^+ + \text{C}^+ \quad \text{(normal Auger)} \quad (2)$$

$$\rightarrow \text{ABC}^* \xrightarrow{-e^-} \text{ABC}^+ \rightarrow \text{AB}^+ + \text{C} \quad \text{(autoionization or participant}$$
$$\text{resonance Auger)} \quad (3)$$

$$\rightarrow \text{ABC}^* \xrightarrow{-e^-} \text{ABC}^{+*} \rightarrow \text{AB}^+ + \text{C} \quad \text{(spectator resonance Auger)} \quad (4)$$

$$\rightarrow \text{ABC}^* \rightarrow \text{AB}^* + \text{C} \left[\rightarrow \text{AB} + \text{C} + h\nu\right] \quad \text{(neutral dissociation I)} \quad (5)$$

$$\rightarrow \text{ABC}^* \rightarrow \text{AB}^* + \text{C} \xrightarrow{-e^-} \text{AB}^+ + \text{C} \quad \text{(neutral dissociation II)} \quad (6)$$

Here, ABC^{2+} in process 2 represents a doubly charged ion whose holes exist at valence orbital(s), and ABC^{+*} in process 4 a satellite state with two-hole-one-particle configuration which corresponds to the final state of photoemission shakeup. A

CP500, *The Physics of Electronic and Atomic Collisions*, edited by Y. Itikawa, et al.
© 2000 American Institute of Physics 1-56396-777-4/00/$17.00

superexcited state ABC* in processes 3 - 6 is defined as a neutral state lying above the first ionization potential of ABC and is produced by promotion of an electron from the second outermost or deeper orbital to a Rydberg or unoccupied valence orbital.   Since a superexcited state is buried in multiple ionization and dissociation continua, it decays into various pathways in a very short period. Superexcitation involving a valence electron may lead to autoionization [process 3] or neutral dissociation [processes 5 and 6], while that involving an inner-shell electron is accompanied by participant or spectator resonance Auger [processes 3 and 4] or, in some cases, by neutral dissociation [processes 5 and 6]. The neutral fragment AB* produced by direct dissociation or predissociation of ABC* often contains enough internal energy to decay radiatively by emitting UV or visible fluorescence (6) as denoted in brackets of process 5. Alternatively, autoionization of AB* can take place [process 6] if its internal energy exceeds the lowest ionization energy of AB (7).

Knowledge concerning ionic and neutral photofragments is important for clarifying dissociation mechanisms governing the above processes. Moreover, qualitative description can be achieved on potential energy surfaces of dissociative states and on the strength of nonadiabatic coupling with neighboring excited states. Combining such information with quantum yields for other probable pathways allows us to obtain comprehensive understanding of dynamical behaviors of excited states formed after absorbing a VUV or SX photon: *i.e.*, excited ions [ABC^+ or ABC^{+*} in processes 1, 3, and 4], doubly-charged ions [ABC^{2+} in process 2] and superexcited molecules [ABC* in processes 5 and 6]. For larger molecules, fragmentation processes of excited states attract considerable attention, in a close connection with site- or state-specific bond rupture induced by photoabsorption. In spite of their ubiquity, ionic and neutral fragmentations in the VUV or SX regions have been studied by only a few experimental methods; most of them had been already introduced in the late 1960s into the field of molecular physics. Detection of AB^+ produced by ionic photofragmentation has been performed mainly by photoionization mass spectrometry or kinetic energy analysis with an energy analyzer. It is not feasible so far to directly observe the vibrational and rotational distribution of the fragment ions. In the case of neutral photofragmentation, one cannot tell even the molecular formula of a fragment, to say nothing of its internal distribution, if the fragment does not emit a photon or electron.

The present paper describes synchrotron radiation-laser combination experiments that have been performed by our group for the past several years (8-11). Two methods of probing fragments have been developed, i.e. laser induced fluorescence (LIF) excitation spectroscopy and resonance enhanced multiphoton ionization (REMPI) spectroscopy, both of which have been widely accepted as most convenient to study properties of photofragments formed by laser photolysis in the visible and UV regions. In the present study, we utilized these methods to observe fragments resulting from processes 1, 3, and 5 by scanning a wavelength of a probe laser.

In general, pump-probe experiments combining synchrotron radiation and a laser quite differ from those using two lasers operating synchronously at lower repetition rate (1 - 100 Hz). Most of difficulties in performing the former type of experiment arise essentially from the temporal structure of synchrotron radiation pulses characterized by a frequency of ~10^8 Hz (9). Let us assume that a monochromatized synchrotron radiation from a bending magnet is used as a photolysis source to produce a neutral fragment, which is thereupon probed by the REMPI method with a pulsed dye laser. The REMPI count rate of the probe signal is affected by a time-averaged number density of photolysis products, which is extremely low owing to their short average residence time (10^{-6} to 10^{-7} s) in a probing region. As a consequence, the signal count rate is predicted to be 11 - 12 orders of magnitude lower than that obtainable with a laser photolysis source such as a 10 mW excimer laser. In order to overcome the low count rate, synchrotron radiation and a laser are combined in the present study at the beam line connected with an undulator which provides two or three orders of magnitude higher photon flux than a bending magnet. In addition, a quadrupole ion trap is incorporated into the LIF spectroscopy of ion fragments to increase their time-averaged number density.

As a prototypical case of LIF excitation spectroscopy, observation is made on $N_2^+(X\,^2\Sigma_g^+)$ ions produced by synchrotron radiation photoionization of N_2 or N_2O. On the other hand, REMPI spectroscopy has been invoked to observe sulfur atoms $S(3s^2 3p^4\,^3P_J, J = 0, 2)$ produced by photodissociation of Rydberg states of OCS. Two different laser systems are adopted for the different spectroscopic methods.

EXPERIMENTAL METHODS

All experiments were carried out at the beam line 3A2 of the UVSOR facility in the Institute for Molecular Science. Details of the undulator radiation and laser system are

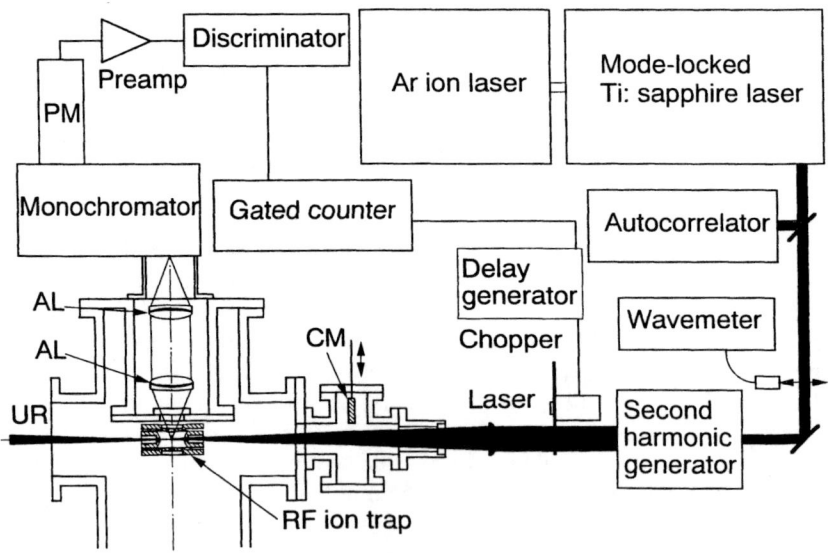

FIGURE 1. Schematic diagram of the apparatus for LIF excitation spectroscopy of fragments produced by synchrotron radiation photoionization. PM, photomultiplier; AL, spherical achromatic lenses; UR, monochromatized undulator radiation; CM, gold-mesh current monitor.

described elsewhere (8,9,11). The fundamental light of the undulator radiation was dispersed by a monochromator of constant deviation grazing-incidence type with 2.2 m focal length. The monochromatized light was focused onto the intersection region of ca. 1×1 mm^2 where a sample gas was expanded under an effusive jet condition. Commercial high purity gases were used without further purification. A typical photon intensity and spectral resolution of the synchrotron radiation were 7.8×10^{14} photons s^{-1} cm^{-2} and 30 meV (FWHM), respectively. The temporal profile of the undulator pulse is represented by a Gaussian function with FWHM of approximately 400 ps.

LIF Excitation Spectroscopy of Ionic Fragments

The molecular beam of N_2 or N_2O intersected at $90°$ with the monochromatized undulator light inside an ion trap as shown in Fig. 1. Only $N_2^+(X\ ^2\Sigma_g^+, A\ ^2\Pi_u)$ ions were produced, because the undulator photon energy E_{SR} was set below the ionization threshold for the formation of $N_2^+(B\ ^2\Sigma_u^+, v_B = 0)$ from N_2 or N_2O. The laser introduced coaxially with the undulator light probed $N_2^+(X\ ^2\Sigma_g^+, v_X = 0$ or 1) ions by means of LIF excitation spectroscopy. The laser wavelength was scanned in the region of 389 - 392 nm or 386 - 389 nm for the $(B\ ^2\Sigma_u^+, v_B = 0) \leftarrow (X\ ^2\Sigma_g^+, v_X = 0)$ or $(B\ ^2\Sigma_u^+, v_B = 1) \leftarrow (X\ ^2\Sigma_g^+, v_X - 1)$ transition, respectively. The fluorescence was monitored at 427.8 ± 4 nm or 423.6 ± 4 nm which agrees with the central wavelength in the rotational distribution of the transition $(X\ ^2\Sigma_g^+, v_X = 1) \leftarrow (B\ ^2\Sigma_u^+, v_B = 0)$ or $(X\ ^2\Sigma_g^+, v_X = 2) \leftarrow (B\ ^2\Sigma_u^+, v_B = 1)$, respectively. The fluorescence was dispersed by another monochromator and detected with a photomultiplier. The output of the photomultiplier was fed into the input of a dual-channel gated photon counter. An optical chopper modulating the laser beam with a 50% duty ratio is used to alternate the data acquisition cycle of the laser on and off at 400 Hz in combination with the photon counter. Thereby, we are able to compensate the transient changes or long term drifts in the experimental conditions, such as a monotonic decrease in the synchrotron radiation photon flux and the variation of the sample pressure.

We have chosen the second harmonic of a Ti:sapphire laser for the LIF experiments. Its wide tunability in the UV to visible region contributes towards our diverse selection of target molecules and their vibronic states. The pulse energy, duration, and energy resolution of the second harmonic of the Ti:sapphire laser were 0.13 nJ pulse^{-1} (2.6×10^8 photons pulse^{-1}), 9 ps, and 1.6 cm^{-1}, respectively, at 389 - 392 nm, when N_2^+ fragments produced from N_2 were observed by LIF spectroscopy. These values were changed to 1.1 nJ pulse^{-1} (2.2×10^9 photons pulse^{-1}), 1.2 ps, and 11 cm^{-1}, respectively, in the case of LIF spectroscopy of N_2^+ from N_2O. Only a small amount of the laser photon flux is available for excitation of a single rotational level of $N_2^+(X\ ^2\Sigma_g^+, v_X = 0)$ because of a narrow Doppler width (0.06 cm^{-1}), so that the effective photon flux is evaluated to be and $\sim 1 \times 10^7$ photons pulse^{-1} in both cases. The laser can operate synchronously with a master oscillator (90.115 MHz) for a main RF cavity on the storage ring (8); in the present study the definite synchronization was unnecessary, because the interval time of synchrotron radiation pulse is considerably shorter than the average residence time (2.1×10^{-6} s) of the primarily produced $N_2^+ (X\ ^2\Sigma_g^+)$.

REMPI Spectroscopy of Neutral Fragments

Figure 2 shows a schematic diagram of the apparatus for REMPI spectroscopy of neutral fragments produced by synchrotron radiation excitation. Carbonyl sulfide OCS molecules were subjected to irradiation of monochromatized undulator radiation (E_{SR} = 13 - 17 eV) which was crossed perpendicularly with a molecular beam discharged from a pulsed nozzle. Produced ions were repelled toward electrode L1 by a field of 11 V/cm between electrodes L1 and L2. A part of neutral sulfur atoms produced by VUV excitation were sampled through an aperture of electrode L2 and ionized by a dye laser focused on the space between electrodes L2 and L3. Ions produced by [2+1]-REMPI were extracted and mass-separated by using a quadrupole mass filter equipped with a channeltron electron multiplier. The second harmonic of a XeCl-laser-pumped dye laser (Coumarin 540A) was used for probing S ($3s^2 3p^4\ ^3P_J$, J = 0, 2). A typical pulse energy is 2 mJ pulse^{-1}. The spectral resolution of the probe laser was 0.0015 nm. The laser is made to operate at 10 Hz synchronously with the pulsed nozzle.

RESULTS AND DISCUSSION

LIF Spectroscopy of N_2^+ Produced from N_2

Figure 3(a) shows an LIF spectrum of the ($B\ ^2\Sigma_u^+$, v_B = 0) ← ($X\ ^2\Sigma_g^+$, v_X = 0) transition of N_2^+ from N_2 at E_{SR} = 15.983 eV. The rotational lines in the R branch are clearly resolved, while those in the P branch are overlapped because there exists a band head and because the spacing between the lines is narrower than the spectral resolution. Rotational temperature T of N_2^+ ($X\ ^2\Sigma_g^+$, v_X = 0) can be evaluated by simulating the experimental points in Fig. 3(a) with theoretical intensity distribution of rotation bands convoluted with the laser spectral width of 1.6 cm^{-1} (FWHM). In Fig. 3(b), a calculated spectrum at T = 300 K is indicated by a solid line. Experimental data points appear to gather around this curve. By simulation, the rotational temperature is estimated to be 300 ± 20 K. This suggests that no marked rotational excitation occurs in N_2^+ ($X\ ^2\Sigma_g^+$,

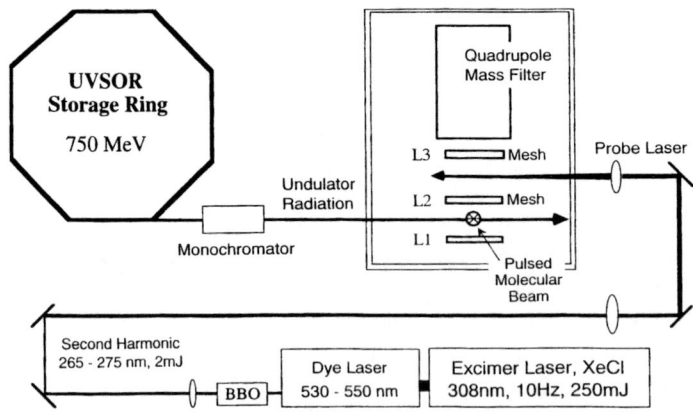

FIGURE 2. Schematic diagram of the apparatus for REMPI spectroscopy of neutral fragments produced by synchrotron radiation excitation. BBO, β-barium borate crystal; L1-L3, ion-lens system.

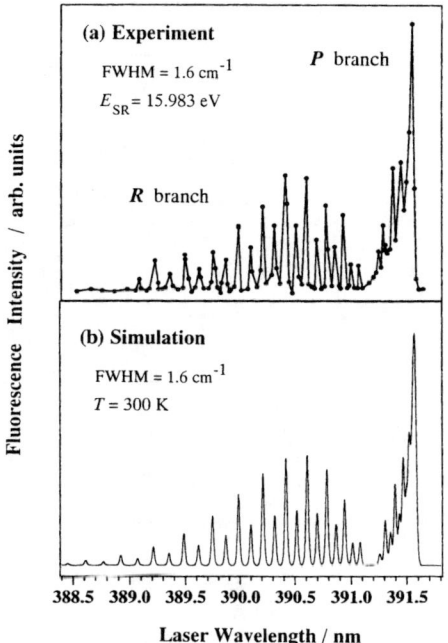

(a) Experiment

FWHM = 1.6 cm^{-1}

E_{SR} = 15.983 eV

P branch

R branch

(b) Simulation

FWHM = 1.6 cm^{-1}

T = 300 K

388.5 389.0 389.5 390.0 390.5 391.0 391.5

Laser Wavelength / nm

Fluorescence Intensity / arb. units

FIGURE 3. (a) LIF spectrum of the N_2^+ (X $^2\Sigma_g^+$, $v_X = 0$) fragments produced from N_2. (b) Simulated spectrum at $T = 300$ K.

$v_X = 0$) from N_2: a change in the rotational quantum numbers ΔN is so small during direct ionization or autoionization as not to alter the distribution curve. This statement is partly supported by the rotational line profiles in a high-resolution He I photoelectron spectra reported by Baltzer et al. (12). For direct ionization of N_2 (X $^1\Sigma_g^+$, $v = 0$) → N_2^+ (X $^2\Sigma_g^+$, $v_X = 0$), any transitions between odd-N and odd-N rotational levels are allowed for the *ortho* N_2, and those between even-N and even-N levels are allowed for the *para* N_2, since the N_2 molecule is homonuclear. Nevertheless, they found that the absolute value of ΔN is no more than 2. Their result is compatible with our LIF spectrum.

We have measured relative partial cross sections for the formation of the $v_X = 0$ and 1 levels of $N_2^+(X$ $^2\Sigma_g^+)$ as a function of E_{SR}, by fixing the laser wavelength at the maximum positions of the *P* branch—viz. 391.54 nm for (B $^2\Sigma_u^+$, $v_B = 0$) ← (X $^2\Sigma_g^+$, $v_X = 0$) and 388.53 nm for (B $^2\Sigma_u^+$, $v_B = 1$) ← (X $^2\Sigma_g^+$, $v_X = 1$). The resultant curves with resolution of 0.24 nm are illustrated in Fig. 4. The spectrum in Panel (a) for $v_X = 0$ accords well with the corresponding partial photoionization cross section curve previously reported (13,14). The appearance energy of the fluorescence yields is found to be 79.6 nm, in good agreement with the ionization threshold for the $v_X = 0$ state. The two spectra manifest peak features ascribable to transitions to autoionizing Rydberg states converging to $N_2^+(A$ $^2\Pi_u)$. But these spectra are different in relative intensities of the peaks. The most noticeable difference is that the relative intensity of the peak centering at 77.0 - 77.1 nm is higher in the $v_X = 1$ curve than in the $v_X = 0$ curve. From the assignments given by Shaw et al. (15) this peak should be identified as resulting from autoionization of the $4d\delta_g$ $^1\Pi_u$, $v = 1$ and/or $3d\delta_g$ $^1\Pi_u$, $v = 4$ state. The partial cross section for the formation of $N_2^+(X$ $^2\Sigma_g^+$, $v_X)$ through a particular vibrational level v_d of the Rydberg state is expected to be proportional to a Franck-Condon Factor $|\langle v_d|v_X\rangle|^2$, where $|v_d\rangle$ and $|v_X\rangle$ denote the vibrational wave functions for the Rydberg state and N_2^+, respectively. The ratio of $|\langle v_d=1|v_X=1\rangle|^2$ to $|\langle v_d=1|v_X=0\rangle|^2$ is estimated to be 0.138, whereas that of $|\langle v_d=4|v_X=1\rangle|^2$ to $|\langle v_d=4|v_X=0\rangle|^2$ to be 8 (16). Hence, it is likely that autoionization of the $3d\delta_g$ $^1\Pi_u$, $v = 4$ state gives rise to the strong peak at 77.0 - 77.1 nm in the $v_X = 1$ curve.

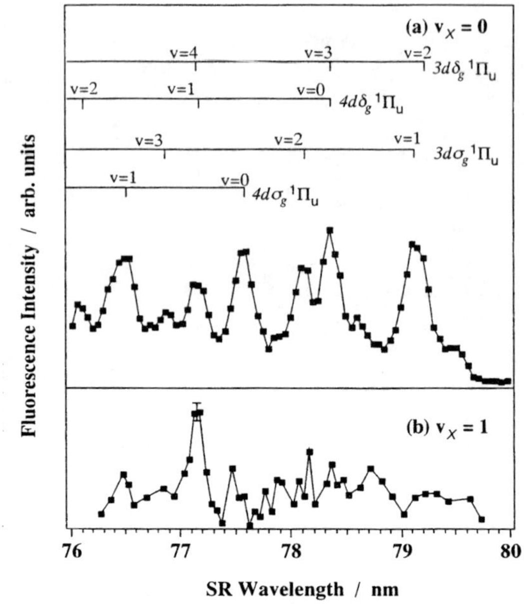

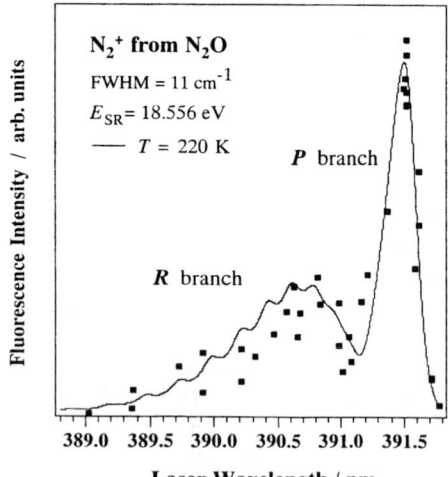

FIGURE 4. Yield curves of N_2^+ (X $^2\Sigma_g^+$) at (a) $v_X = 0$ and (b) $v_X = 1$ levels from N_2 obtained by plotting the LIF count rate as a function of E_{SR}. Tic marks indicate the excitation energies for the vibrational progressions of the $nd\sigma_g$ $^1\Pi_u$ and $nd\delta_g$ $^1\Pi_u$ Rydberg states converging to $N_2^+(A$ $^2\Pi_u)$.

LIF Spectroscopy of N_2^+ Produced from N_2O

Figure 5 shows an LIF spectrum of $N_2^+(X$ $^2\Sigma_g^+$, $v_X = 0$) from N_2O at $E_{SR} = 18.556$ eV (66.8 nm). The undulator photon energy is tuned to the excitation energy for the $3d\pi$ Rydberg state converging to $N_2O^+(C$ $^2\Sigma^+)$, so that N_2^+ is considered to be dissociated from N_2O^+ state(s) produced by autoionization of the $3d\pi$ state. The two maxima centered at 391.5 and 390.8 nm are ascribed to the P and R branches, respectively, for the (B $^2\Sigma_u^+$, $v_B = 0$) ← (X $^2\Sigma_g^+$, $v_X = 0$) transition. Because of low spectral resolution of the laser (~11 cm^{-1}), the rotational lines are heavily overlapped. The rotational temperature of $N_2^+(X$ $^2\Sigma_g^+$, $v_X = 0$) is estimated to be 200 - 230 K by fitting the observed spectrum to the theoretical intensity distribution of rotation bands convoluted with the laser spectral width. Here, a single Boltzmann distribution is assumed to be applicable to the relative intensities of the rotational levels of N_2^+ produced by dissociation of N_2O^+. The solid line in Fig. 5 represents a distribution at an optimum temperature of

FIGURE 5. LIF spectrum of the N_2^+ (X $^2\Sigma_g^+$, $v_X = 0$) fragments produced from N_2O. — Simulated spectrum at $T = 220$ K.

220 K. Evidently, the dissociation of N_2O^+ leads to reduction of the rotational energy of the N_2^+ fragments.

To discuss the intramolecular energy partitioning upon dissociation, we need information at least on the excess energy and geometrical change of N_2O^+ on the way of the dissociation coordinate. Berkowitz and Eland have proposed that, from the $nd\pi$ states converging to $N_2O^+(C\ ^2\Sigma^+)$, the formation of $N_2^+\ (X\ ^2\Sigma_g^+)$ proceeds through autoionization to $N_2O^+(B\ ^2\Pi)$ which is subsequently predissociated by a repulsive state (17). Thus, an excess energy released upon dissociation ranges from 0.4 to 1.8 eV. This can be deduced from the energy difference between the dissociation limit (17.25 eV) and the vibrational state of $N_2O^+(B\ ^2\Pi)$ (17.65 – 19.1 eV) formed by autoionization (18). It is accepted that $N_2O^+\ (X\ ^2\Pi,\ A\ ^2\Sigma^+,$ and $C\ ^2\Sigma^+)$ have linear equilibrium geometries. In contrast, no tangible evidence has been obtained in favor of a linear geometry of the vibrational manifolds of $N_2O^+(B\ ^2\Pi)$; rather a bent structure is more plausible, since a shake-up satellite with a configuration of $(2\pi)^{-2}(3\pi)^1$ may cause substantial perturbation to the main $(1\pi)^{-1}$ configuration of the $B\ ^2\Pi$ state. Such a strong vibronic coupling was first suggested by Köppel et al. (19) who analyzed complicated features appearing in the band in the photoelectron spectrum of N_2O by *ab initio* Green's function calculations. If the $(2\pi)^{-2}(3\pi)^1$ configuration has a strong effect on the $B\ ^2\Pi$ state, this state must have a bent equilibrium geometry.

A simple-minded consideration on the kinematics of the dissociation of bent $N_2O^+(B\ ^2\Pi)$ would predict an efficient excess energy transfer to the rotational degrees of freedom of N_2^+ fragments and their extensive rotational excitation. However, our result of Fig. 5 is completely opposed to this prediction. For more quantitative treatment, we estimate the rotational energy of N_2^+ as a function of the NNO bond angle θ of N_2O^+ using the procedure first introduced by Levene and Valentini (20). Here, the *modified impulsive model* is postulated and the contribution of the bending motion of the parent $N_2O^+(B\ ^2\Pi)$ is disregarded. We assume the excess energy of $0.4\ \text{eV} = 3200\ \text{cm}^{-1}$, the lower limit mentioned in the preceding paragraph, and the initial rotational energy of $N_2O^+\ (B\ ^2\Pi)$ of 200 cm^{-1}. As shown in Fig. 6, at $\theta = 180°$, only ca. 25 % of the initial rotational energy is deposited in the N_2^+ fragment as its rotational energy. Two curves in Fig. 6 designated as "High *J*" and "Low *J*" represent θ-dependences of the rotational energy of N_2^+.

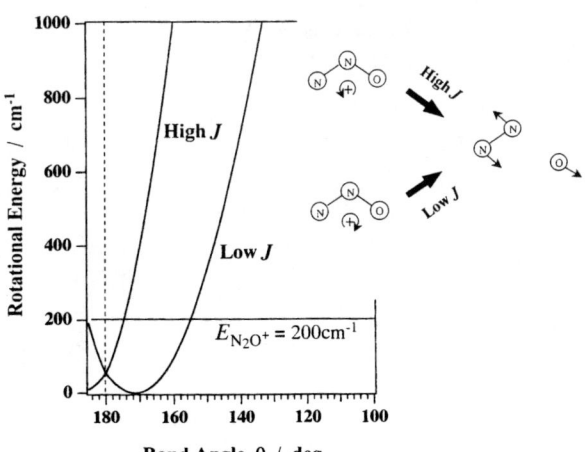

FIGURE 6. Rotational energy of $N_2^+\ (X\ ^2\Sigma_g^+)$ produced by dissociation of $N_2O^+(B\ ^2\Pi)$ as a function of the bond angle calculated on the basis of the *modified impulsive model*.

calculated under the condition that the rotational angular momentum of the parent $N_2O^+(B\,^2\Pi)$ is parallel or antiparallel, respectively, to the N_2^+ rotational angular momentum generated by the impulsive energy release. In the low J limit, the average rotational energy of the N_2^+ fragments becomes smaller at $\theta > 155°$ than that of $N_2O^+(B\,^2\Pi)$ or, that is to say, than that of the primary ground state N_2O molecules. In the high J limit, the rotational energy of N_2^+ exceeds the initial value, as long as the dissociation occurs from a bent $N_2O^+(B\,^2\Pi)$. In the preceding paragraph, we suggested that the $B\,^2\Pi$ state has a bent structure. It is therefore likely that the low J limit is the case for the dissociation of $N_2O^+(B\,^2\Pi)$. This consideration implies strong vector correlation between the rotational angular momentum of the parent $N_2O^+(B\,^2\Pi)$ and the N_2^+ rotational angular momentum arising from the dissociative energy release. In the present case, dissociation probability is higher when these two vectors are antiparallel than when they are parallel. In other words, dissociating $N_2O^+(B\,^2\Pi)$ has a tendency to maximize the orbital angular momentum. We believe that a subtle difference of the effective repulsive potentials experienced by the dissociating N_2O^+ causes a crucial shift or deformation of the conical intersection with the potential energy surface of the bound $N_2O^+(B\,^2\Pi)$ state.

Another interpretation for the relatively low rotational temperature of $N_2^+(X\,^2\Sigma_g^+, v_X = 0)$ is a short lifetime of $N_2O^+(B\,^2\Pi)$ with respect to predissociation. If the lifetime is much shorter than the period of its bending vibration, predissociation occurs from $N_2O^+(B\,^2\Pi)$ with almost linear geometry and a large part of the excess energy is transferred to the relative translational energy of N_2^+ and O. Maier and coworkers have measured the line width of vibrational levels of $N_2O^+(B\,^2\Pi)$ and obtained the lifetime of ca. 60 fs (21). Hence, it is improbable that the predissociation completes before the parent ion starts to bend.

REMPI Spectroscopy of S from OCS

Figure 7 shows a REMPI spectrum of S $(3s^23p^4\,^3P_2)$ formed at $E_{SR} = 16.5$ eV as a function of the probe laser wavelength in the region involving the two-photon transitions, $S(3s^23p^35p\,^3P_{J'}) \leftarrow S(3s^23p^4\,^3P_2)$. The width of the peak is governed by the laser spectral resolution of 0.0015 nm and Doppler broadening of 0.0012 nm. Moreover, the maximum at 269.290 nm is considered to comprise two peaks resulting from transitions $S(3s^23p^35p\,^3P_1) \leftarrow S(3s^23p^4\,^3P_2)$ and $S(3s^23p^35p\,^3P_2) \leftarrow S(3s^23p^4\,^3P_2)$. These two peaks are not separated because of a low signal-to-background ratio. We could also detect the REMPI signal of $S(3s^23p^4\,^3P_0)$ at 271.375 nm where the two-photon

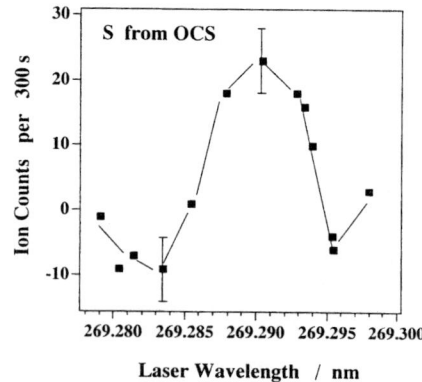

FIGURE 7. REMPI spectrum of the S $(3s^23p^4\,^3P_2)$ fragments produced from OCS.

transition from $S(3s^23p^4\ {}^3P_0)$ to $S(3s^23p^35p\ {}^3P_0)$ is allowed. Next, we fixed the laser wavelength at the maximum of Fig. 7, and measured the REMPI signal intensity as a function of the energy of E_{SR} in the range of 13 - 17 eV. The resultant spectrum is considered to represent a relative photodissociation cross section curve for the formation of $S(3s^23p^4\ {}^3P_2)$ from OCS. Band features existing at 13 - 16 eV and 16 – 17 eV are attributed to the Rydberg series converging to the $B\ {}^2\Sigma^+$ and $C\ {}^2\Sigma^+$ states, respectively, of OCS^+. In brief, we can obtain direct evidence for the formation of nonfluorescing and non-autoionizing neutral species dissociated from molecular superexcited states, by means of pump-probe spectroscopy combining synchrotron radiation and laser.

ACKNOWLEDGMENTS

The experimental work from our laboratory described in this paper involves the dedicated effort of my collaborators, Dr. Masakazu Mizutani and Mr. Hiromichi Niikura. This work has been supported by a Grant-in-Aid for Scientific Research (Grant No. 10640504) from the Ministry of Education, Science, Sports, and Culture, Japan and by a Grant for Scientific Research from Matsuo Foundation.

REFERENCES

1. Berkowitz, J., *Photoabsorption, Photoionization, and Photoelectron Spectroscopy*, New York: Academic Press, 1979, ch. 6.
2. Ng, C. Y. (Ed.), Vacuum *Ultraviolet Photoionization and Photodissociation of Molecules and Clusters*, Singapore: World Scientific, 1991.
3. Beswick, A. (Ed.), *Synchrotron Radiation and Dynamic Phenomena, AIP Conference Proceedings No. 258*, New York: Am. Inst. of Phys., 1992, ch. I.
4. Berkowitz, J., Rühl, E., and Baumgärtel, H., in *VUV and Soft X-Ray Photoionization*, Becker, U., and Shirley, D. A. (Eds.), New York: Plenum Press, 1996, ch. 7.
5. Nenner, I. and Morin, P., in *VUV and Soft X-Ray Photoionization*, Becker, U., and Shirley, D. A. (Eds.), New York: Plenum Press, 1996, ch. 9.
6. Ukai, M., Kameta, K., Machida, S., Kouchi, N., Hatano, Y., and Tanaka, K., *J. Chem. Phys.* **101**, 5473-5483 (1994).
7. Hikosaka, Y., Hattori, H., Hikida, T., and Mitsuke, K., *J. Chem. Phys.* **107**, 2950-2961 (1997).
8. Mizutani, M., Tokeshi, M., Hiraya, A., and Mitsuke, K., *J. Synchrotron Rad.* **4**, 6-13 (1997).
9. Mizutani, M., Niikura, H., Hiraya, A., and Mitsuke, K., *J. Synchrotron Rad.* **5**, 1069-1071 (1998).
10. Mitsuke, K., Mizutani, M., Niikura, H., and Iwasaki, K., *Rev. Laser Engin.* **26**, 458-462 (1998).
11. Niikura, H., Mizutani, M., and Mitsuke, K., submitted to *Chem. Phys. Letters*.
12. Baltzer, P., Karlsson, L., and Wannberg, B., *Phys. Rev. A* **46**, 315-317 (1992).
13. Haworth, A., Wilden, D. G., and Comer, J., *J. Electron Spectrosc. Relat. Phenom.* **37**, 291-299 (1985).
14. Holland, D. M. P., and West, J. B., *J. Phys. B* **20**, 1479-1485 (1987).
15. Shaw, D. A., Holland, D. M. P., MacDonald, M. A., Hopkirk, A., Hayes, M. A., and McSweeney, S. M., *Chem. Phys.* **166**, 379-391 (1992).
16. Lofthus, A., and Krupenie, P. H., *J. Phys. Chem. Ref. Data* **6**, 113-307 (1977).
17. Berkowitz, J., and Eland, J. H. D., *J. Chem. Phys.* **67**, 2740-2752 (1977).
18. Dehmer, P. M., Dehmer, J. L., and Chupka, W. A., *J. Chem. Phys.* **73**, 126-133 (1980).
19. Köppel, H., Cederbaum, L. S., and Domcke, W., *Chem. Phys.* **69**, 175-183 (1982).
20. Levene, H. B., and Valentini, J. J., *J. Chem. Phys.* **87**, 2594-2610 (1987).
21. Danis, P. O., Wyttenbach, T., and Maier, J. P., *J. Chem. Phys.* **88**, 3451-3455 (1988).

Ultrafast Structural Deformation of Polyatomic Molecules in Intense Laser Fields

Kaoru Yamanouchi,* Akiyoshi Hishikawa, Atsushi Iwamae, and Shilin Liu

Department of Chemistry, School of Science, The University of Tokyo
7-3-1 Hongo, Bunkyo-ku, Tokyo 113-0033, Japan

Abstract. The momentum vector distributions of fragment ions produced through the Coulomb explosion of small molecules such as NO, CO_2, NO_2, and H_2O in intense laser fields (~1 PW/cm^2) are measured by the mass-resolved momentum imaging (MRMI) technique. For NO, the MRMI maps for a single (p,q) pathway, $NO^{(p+q)+} \rightarrow N^{p+} + O^{q+}$ (p, q = 1~3), are extracted from the observed MRMI maps on the basis of the momentum matching of the N^{p+} and O^{q+} ion pair. In the MRMI maps for the fragment ions produced from CO_2, NO_2, and H_2O, their ultrafast structural deformation both along the stretching coordinate and along the bending coordinate is identified. The $\angle$O-C-O angle distribution of CO_2 spreads significantly (FWHM ~40°), and the $\angle$O−N−O bond angle of NO_2 increases toward a linear configuration within the ultrashort duration of the laser pulse. This type of deformation is also identified for H_2O. These structural deformations are most reasonably interpreted as a consequence of the formation of *light-dressed* potential energy surfaces.

INTRODUCTION

One of the characteristic features of ultrashort laser pulses is their capability of generating an extremely intense light field. It is now possible to increase the magnitude of the laser field as large as a Coulombic electric field in a hydrogen atom. In such an intense laser field, *light-dressed* potential energy surfaces (LDPESs) are formed by the interaction among the molecular potentials shifted by an energy of photons.[1] Therefore, nuclear dynamics of molecules is expected to be governed by LDPESs, which have been investigated intensively for one electron system of H_2^+. Recent studies of molecules which have more than one electrons demonstrated that a variety of fundamental dynamics are associated with the formation of LDPESs in intense laser fields.[2-8]

When molecules are irradiated by the intense laser light whose magnitude is comparable with the valence electric-field in atoms and molecules, the phenomenon called the Coulomb explosion occurs, in which multiply charged atomic fragments having a large released momentum are produced.[1-12] The molecular Coulomb

CP500, *The Physics of Electronic and Atomic Collisions*, edited by Y. Itikawa, et al.
© 2000 American Institute of Physics 1-56396-777-4/00/$17.00

explosion in intense laser fields is known to exhibit two characteristic features, i.e. (i) ultrafast geometrical deformation occurring during the short laser-pulse duration, and (ii) anisotropic ejection of the atomic and molecular fragment ions representing the alignment of the parent molecular ions with respect to the laser polarization direction.

Recently, in order to investigate the ultrafast nuclear dynamics of molecules in intense laser fields, we introduced a novel method called mass-resolved momentum imaging (MRMI).[2-8] In the MRMI method, atomic and molecular fragment ions ejected with a large released momentum are detected by a time-of-flight (TOF) mass spectrometer, and the momentum as well as angular distributions of the ejected ion species are obtained by rotating the direction of the laser polarization with respect to the detection axis of the TOF tube. Due to the high resolving power of our TOF mass spectrometer, atomic and molecular ions with different charge numbers are observed separately. The resultant momentum and angular distributions of the charged species are plotted either on the two-dimensional (2D) momentum plane as a contour map or in the form of three-dimensional (3D) intensity distribution on the momentum plane. By the analysis of the imaging maps, it becomes possible to extract geometrical structure of molecules just before the Coulomb explosion, from which we discuss the ultrafast geometrical deformation occurring within an intense laser pulse.

In the present study, by referring to our recent studies of ultrashort dynamics of diatomic[2,3] and triatomic[2,6-8] molecules in intense laser fields using the MRMI approach, we report how the deformation of their geometrical structure occurs and how these phenomena are interpreted in terms of the formation of LDPESs.

EXPERIMENT

The details of our experimental set-up were presented previously.[2-8] Briefly, femtosecond laser pulses at $\lambda \sim 800$ nm generated by a mode-locked Ti-sapphire laser (Spectra Physics Tsunami + Millennia) were introduced into a regenerative amplifier system (BM Industry Alpha 10B/S) to obtain high-power short-pulsed laser light at a repetition rate of 10 Hz. After a pulse compression, a laser-pulse duration of 100 fs was achieved with a total energy of up to 50 mJ/pulse.

The laser beam was focused by a quartz lens onto a skimmed pulsed molecular beam of a sample gas in the region between the extraction parallel repeller plates of a linear TOF mass spectrometer.

In our TOF mass spectrum with typical mass resolution of $m/\Delta m \sim 620$, ion species with different charge numbers were resolved well with no temporal overlap. The TOF mass spectrum with a high S/N ratio was obtained by accumulating the spectra for $\sim 1 \times 10^3$ laser shots using a digital oscilloscope (LeCroy 9370) at a 1 GHz sampling rate. When the pulsed valve was not operated, the background pressure in the main chamber was 2×10^{-8} Torr and that in the TOF tube was 1×10^{-8} Torr. During the experiment, the pressure in the main chamber was kept sufficiently low ($< 2 \times 10^{-7}$ Torr) in order to avoid the space charge effect.

For constructing the MRMI maps, the TOF mass spectra were taken at different laser polarization angles by rotating the laser polarization using a zero-order half-wave plate, which was introduced after the pulse compression stage of the regenerative amplification system. The half-wave plate was rotated manually or automatically with a small angle interval of ~6°.

MRMI AND SP-MRMI MAPS: APPLICATION TO NO

The momentum-scaled (MS) TOF spectra were measured for the atomic fragment ions, N^{p+} and O^{q+} (p,q= 1-3), produced in the intense laser field (~1.4 PW/cm^2) from the (p,q) Coulomb explosion pathways of NO, i.e. $NO^{z+} \rightarrow N^{p+} + O^{q+}$ ($z = p+q$), and they were used to construct mass-resolved momentum imaging (MRMI) maps.[5]

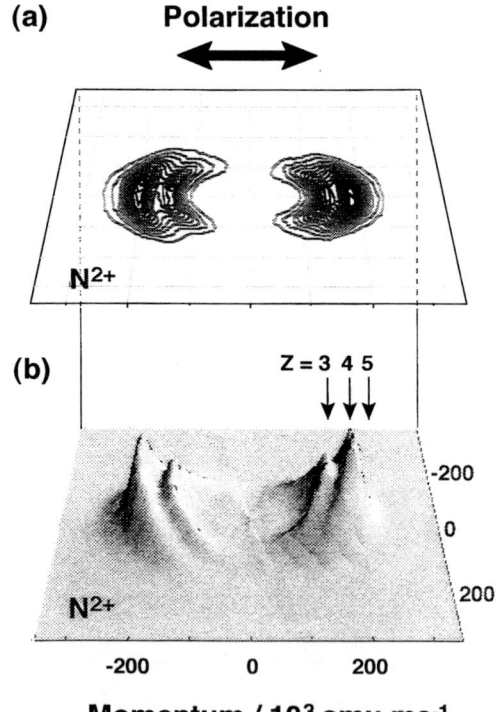

FIGURE 1. The 2D (upper) and 3D (lower) MRMI maps of N^{2+} produced via the Coulomb explosion of NO^{z+} ($z = 3 - 5$) formed in intense laser fields.

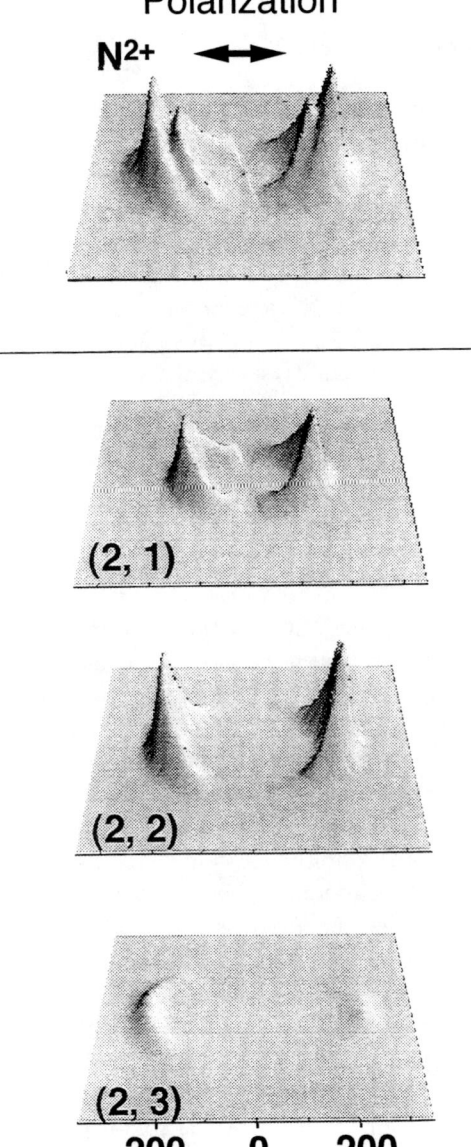

FIGURE 2. The observed MRMI map of N^{2+} produced via the Coulomb explosion of NO^{z+} in intense laser fields and its decomposition into the SP-MRMI maps for the $(p, q) = (2,1)$, $(2,2)$, and $(2,3)$ pathways.

Figure 1 shows the 2D and 3D MRMI maps of the N^{2+} channel formed through the $(p,q) = (2,1)$, $(2,2)$, and $(2,3)$ explosion pathways from the parent NO^{z+} ions with the total charges of $z = 3$, 4, and 5, respectively.

In order to extract momentum and angular distributions for the respective (p,q) Coulomb explosion pathways from the obtained MRMI maps, the least-squares fits using the Gaussian momentum distribution functions were performed for all the MS-TOF spectra obtained at the different laser-polarization angles, and the resultant Gaussian distributions for each (p,q) pathway were transformed into the corresponding single-pathway (SP)-MRMI map.

In Fig. 2, the extraction of the SP-MRMI maps from the observed MRMI map is shown in the 3D representation for the N^{2+} channel formed from NO^{z+}. It is clearly seen that the observed MRMI map is decomposed into the three MRMI maps representing the $(2,1)$, $(2,2)$, and $(2,3)$ pathways, and that the narrower angular distributions for the (p,q) explosion pathways were derived for a larger total charge z of NO^{z+}.

SPREAD OF BOND ANGLE DISTRIBUTION IN CO_2

The upper panels of Fig. 3 show the observed MRMI maps for the fragment atomic ions, C^{q+} ($q = 2$, 3), produced after the (p, q, r) Coulomb explosion of CO_2, i.e., $CO_2^{z+} \rightarrow O^{p+} + C^{q+} + O^{r+}$ ($z = p+q+r$), at the field intensity of 1.1 PW / cm^2.[6] These C^{q+} ($q = 2$, 3) channels have an elliptical pattern substantially extending perpendicular to the laser polarization with a peak at the zero momentum. This shows that the C^{2+} and C^{3+} ions gain only small released momenta even though they are formed from the highly charged parent ions, and that they are ejected more preferentially in the direction perpendicular to the laser polarization vector.

In order to derive quantitative information concerning the structure of CO_2^{z+} prior to the dissociation, we performed a trial-and-error simulation of the MRMI maps of all the atomic fragment ions by taking the following steps: (i) the released momenta of fragment ions for a given molecular geometry are calculated, and they are converted into the MRMI maps for a single (p, q, r) explosion pathway by taking account of the distributions of $R = R(C-O)$ and $\gamma = \angle O-C-O$, (ii) the MRMI map for a given fragment ion is synthesized by adding the relevant SP-MRMI maps with their weights estimated from the observed yields of O^{p+}, and (iii) the geometrical parameters R and γ are determined as a function of z on the basis of the trial-and-error comparison of the synthesized and observed MRMI maps for all the fragment ions.

The bond length $R(C-O)$ determined through the analysis of the MRMI maps of all the atomic fragment ions exhibits a gradual increase as z increases, which is consistent with the recent studies of diatomic molecules, N_2 and NO, and triatomic molecules, NO_2 and H_2O in an intense laser fields.

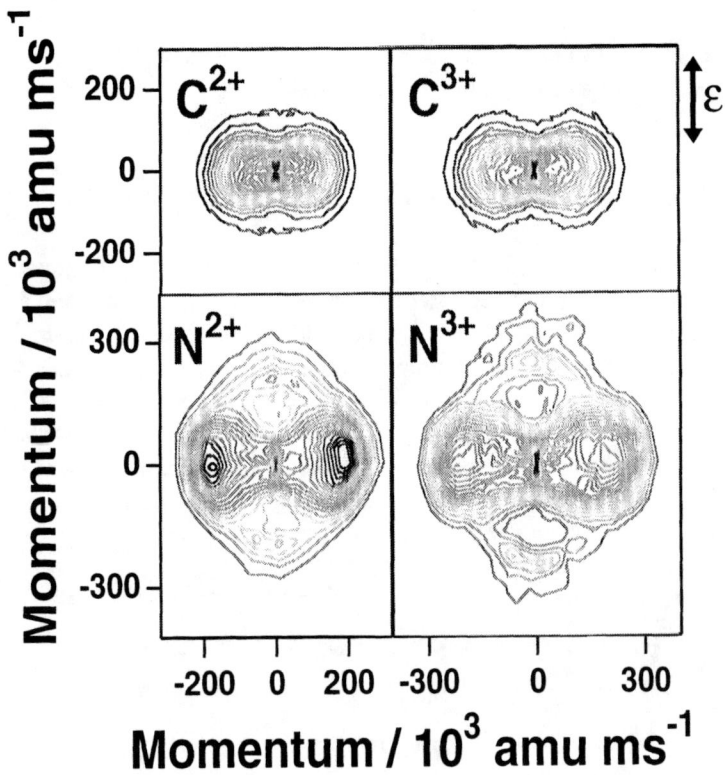

FIGURE 3. The MRMI maps of C^{2+} and C^{3+} formed from CO_2 (upper panels) and those of N^{2+} and N^{3+} formed from NO_2 (lower panels) in intense laser fields. The direction of the laser polarization vector, ε, is represented by a vertical arrow.

It was also found that the Gaussian width σ_γ of the bond angle distribution becomes $\sigma_\gamma = 50 \sim 30°$ for $z = 3 \sim 9$. Considering the mean amplitude of bending, $\sigma_\gamma = 12.5°$, in the ground vibrational level of the $\tilde{X}^1\Sigma_g^+$ state of neutral CO_2, the present results clearly show that a substantially broad γ distribution centered at the linear configuration is induced in the intense laser field. The present observation is in agreement with the report by Cornaggia,[9] who assumed a simple triangular γ distribution and derived its FWHM to be $40°$ for $z = 3 - 6$.

The observed broad γ distributions would be ascribed to the laser-induced population transfer to an excited state having a bent equilibrium; i.e., the linear ground and the excited bent state are coupled strongly by the intense laser field to form a significant avoided crossing resulting in a pair of adiabatic LDPESs (Fig.4). It is

187

expected that the potential barriers of the lower component of the resultant adiabatic LDPESs along the bond-angle coordinate are lowered at a high field intensity to cause potential softening along the bond-angle coordinate, which causes the ultrafast nuclear motion toward the bent structure within a laser-pulse duration.

BENT-TO-LINEAR DEFORMATION OF NO_2

The lower panels of Fig. 3 show the observed MRMI maps for the fragment atomic ions, N^{q+} (q = 2, 3), produced after the three-body (p, q, r) Coulomb explosion of NO_2, i.e., $NO_2{}^{z+}$ -> O^{p+} + N^{q+} + O^{r+}, at the field intensity of 1.0 PW/cm^2.[7] Both the N^{2+} and N^{3+} channels have a substantially elongated pattern with two peaks at large released momenta ($\sim$ 200×10^3 amu m/s) in the perpendicular direction to the laser polarization. The characteristic fragmentation patterns exhibit a contrast with those observed for C^{q+} (q = 2,3) ions formed from CO_2 shown in the upper panels in Fig. 3, where elliptical distributions with a peak at the zero momentum were observed.

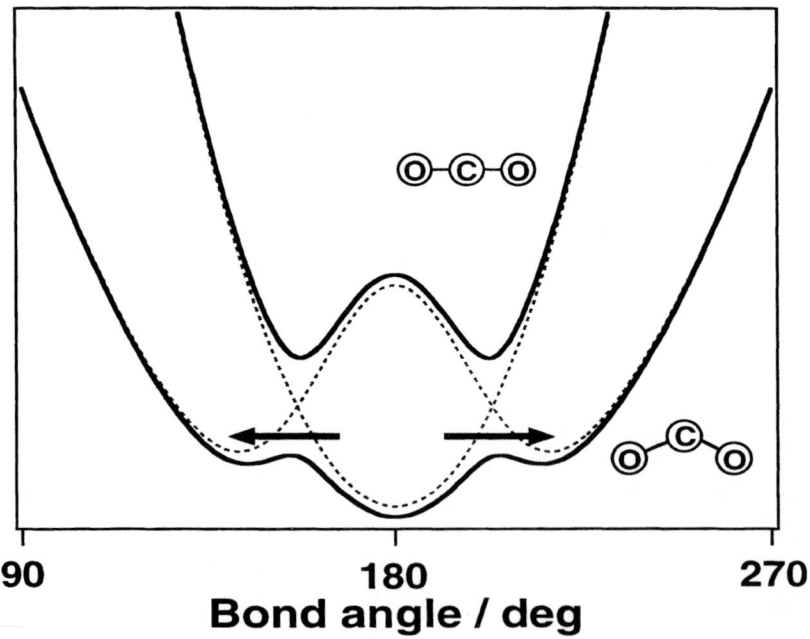

FIGURE 4. A schematic diagram of the formation of the *light-dressed* potential energy surfaces of CO_2 in intense laser fields (one dimensional cut along the bending coordinate).

The large momentum imposed on N^{q+} ($q = 2, 3$) along the perpendicular direction of the polarization vector suggests that the $NO_2{}^{z+}$ ions take a bent skeletal geometry just before the three-body explosion processes with its a-axis along the laser polarization vector.

A closer inspection of the MRMI map of N^{2+} reveals that the momentum distribution extends substantially along the coordinate perpendicular to the laser polarization (horizontal coordinate) toward the center of the map from the peaks at ~200 $\times 10^3$ amu m/s. The distribution at the zero momentum reaches about a half as large as that for the two peaks.

If symmetric charge separation pathways are assumed, the zero released-momentum imposed on the central atom results exclusively from the linear geometry of the parent molecular ion. Therefore, the present finding suggests that the probability distribution in the linear geometrical configuration substantially increases from the original equilibrium bent structure of the ground state of NO_2 ($\gamma_e = 134.1°$).

The substantial increase in the probability distribution at the linear configuration is more clearly seen in the MRMI map of N^{3+} shown in Fig. 3, where the momentum distribution is almost flat along the horizontal axis with the highest peak located at the zero momentum region. The observed momentum distribution of N^{q+} ($q = 2,3$) in Fig. 3 could be regarded as a direct evidence for a substantial deformation of NO_2 induced by the intense laser field not only along the stretching coordinate but also along the bending coordinate.

In order to derive more quantitative information, we performed a trial-and-error simulation of the MRMI maps of all the atomic fragment ions in a similar manner as in the case of CO_2. For the simplification of the analysis, we modeled the mean bond angle as a linear function of the charge z, i.e., $\gamma_0(z) = c_0 + c_1 z$, where c_0 and c_1 are constants. As the optimized parameters, $c_0 = 105°$ and $c_1 = 8.33$, were obtained from the MRMI map of N^{2+}. The best fit MRMI map for N^{2+} reproduces well the observed map in the entire momentum region. It was also found that the same $\gamma_0(z)$ function provides a good fit for N^{3+}.

Three low lying excited states, $\tilde{A}^2B_2$, $\tilde{B}^2B_1$, $\tilde{C}^2A_2$, of NO_2 are accessible from the ground $\tilde{X}^2A_1$ state through one- or two-photon absorption of near-IR light used in the present study. Since the photon-molecule interaction is most effective for a pair of electronic states coupled with a single photon transition, the one-photon light-induced coupling between the $\tilde{X}^2A_1$ and $\tilde{A}^2B_2$ states would dominate over the other higher-order photon-molecule interactions, resulting in the formation of the LDPES through the avoided-crossing between these two low-lying electronic states. Since the equilibrium bond angles of the $\tilde{X}^2A_1$ and $\tilde{A}^2B_2$ states are $\gamma_e = 134°$ and $102°$, respectively, the lower component of the pair of the LDPESs would provide force to bent NO_2 from $\gamma_e = 134°$ to a smaller bond angle. The significantly small bond-angle γ_0 ($z=0$) = $c_0 = 105°$ derived from the simulation could be regarded as an evidence of the formation of the LDPESs between the $\tilde{X}^2A_1$ and $\tilde{A}^2B_2$ states.

At higher laser-field intensities, the two-photon vibronic transition from $\tilde{X}^2A_1$ to

$\tilde{B}^2B_1$ would become important. Since the $\tilde{B}^2B_1$ state has a linear equilibrium structure, i.e., $\gamma = 180°$, the lower component of the two LDPESs formed by the mixing of the $\tilde{X}^2A_1$ and $\tilde{B}^2B_1$ would drive the originally bent NO_2 toward the linear configuration.

EXPLOSION DYNAMICS OF H_2O

The MRMI maps of H^+, O^+ and O^{2+} produced through the Coulomb explosion of H_2O in the intense laser field were measured using the ultrashort laser pulses with wavelength of $\lambda \sim 800$ and 400 nm.[8]   For both wavelengths, the two Coulomb explosion processes of H_2O, i.e., (i) $H_2O^{3+} \rightarrow H^+ + O^+ + H^+$ and (ii) $H_2O^{4+} \rightarrow H^+ + O^{2+} + H^+$ were clearly identified.

The MRMI maps show that the H^+ ions are ejected mainly in the direction parallel with the laser polarization vector. As an example, the MRMI map of H^+ observed when $\lambda \sim 400$ nm is shown in Fig. 5. From the analysis of the MRMI patterns of the fragment ions, the geometrical structure of H_2O^{3+} and H_2O^{4+} just before the Coulomb explosion as well as the extent of their alignment along the laser polarization vector were obtained.

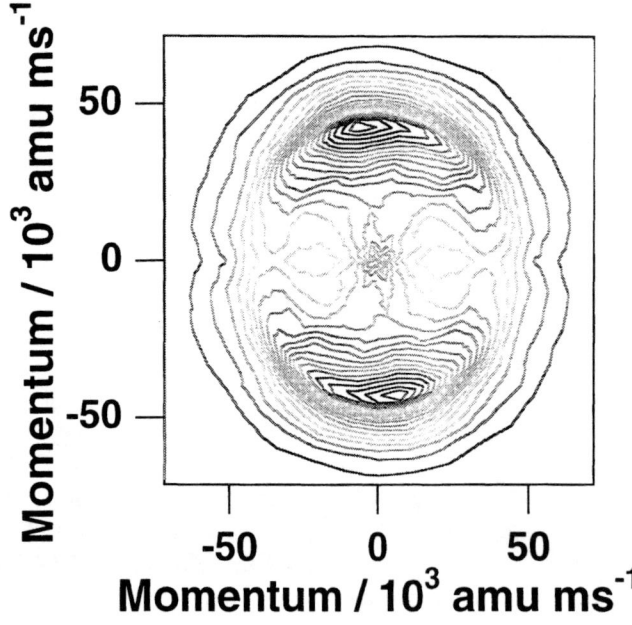

FIGURE 5.   The observed MRMI map of H^+ produced from the Coulomb explosion of H_2O in intense laser fields of $\lambda \sim 400$ nm.

In a similar manner as the analyses of the MRMI maps described above for CO_2 and NO_2, the measured MRMI maps of H^+, O^+, and O^{2+} ions were simulated to reproduce their momentum distributions simultaneously. When light pulses of $\lambda \sim 800$ nm are used, it was shown from the trial-and-error simulation that (i) the O-H bond lengths of H_2O^{3+} and H_2O^{4+} just before the Coulomb explosion are respectively 1.7 and 2.0 times longer than that of the neutral H_2O in the electronic ground state, and (ii) the bond angle $\gamma (= \angle H\text{-}O\text{-}H) \sim 180°$ and its distribution width is $\sigma_\gamma = 60°$ for both parent ions. This bond-angle widening from that of neutral H_2O ($\gamma = 104.45°$) could also be interpreted as a phenomenon caused by the ultrafast geometrical deformation of H_2O^+ on the *light-dressed* potential energy surface.[13]

ACKNOWLEDGEMENTS

The authors thank Drs. K. Hoshina and M. Kono for their valuable discussion and assistance in the experiments. The present work has been supported by the CREST (Core Research for Evolutionary Science and Technology) fund from Japan Science and Technology Corporation.

REFERENCES

1. Bandrauk, A. D., *Molecules in Intense Laser Fields*, New York: M. Dekker Pub., 1993.
2. Hishikawa, A., Iwamae, A., Hoshina, K., Kono, M., and Yamanouchi, K., *Chem. Phys. Lett.* **282**, 283 (1998).
3. Hishikawa, A., Iwamae, A., Hoshina, K., Kono, M., and Yamanouchi, K., *Chem. Phys.* **231**, 315 (1998).
4. Hishikawa, A., Iwamae, A., Hoshina, K., Kono, M., and Yamanouchi, K., *Res. Chem. Intermed.* **24**, 765 (1998).
5. Iwamae, A., Hishikawa, A., and Yamanouchi, K., *submitted to J. Phys. B* .
6. Hishikawa, A., Iwamae, A., and Yamanouchi, K., *Phys. Rev. Lett.* **83**, 1127 (1999).
7. Hishikawa, A., Iwamae, A., and Yamanouchi, K., *J. Chem. Phys.* **111** (1999), *in press.*
8. Liu, S., Hishikawa, A., Iwamae, A., and Yamanouchi, K., unpublished.
9. Cornaggia, C., Normand, D., and Morellec, J., *J. Phys. B: At . Mol. Opt.* **25** 415 (1992).
10. Posthumus, J. H., Plumridge, J., Taday, P. F., Sanderson, J. H., Langley, A. J., Codling, K., and Bryan, W. A., *J. Phys. B: At . Mol. Opt.* **32** L93 (1999).
11. Sanderson, J. H., El-Zein, A., Bryan, W. A., Newell, W. R., Langley, A. J., and Taday, P. F., *Phys. Rev. A* **59**, R2567 (1999).
12. Constant, E., Stapelfelt, H., and Corkum, P. B., *Phys. Rev. Lett.* **76**, 4140 (1996).
13. Rottke, H., Trump, C., and Sandner, W., *J. Phys. B: At . Mol. Opt.* **31**, 1083 (1998).

Progress Towards Chirped Pulse Dissociation of Molecules

Donna Strickland, Zhuhong Zhang and Adam M. Deslauriers

Department of Physics, Guelph-Waterloo Program for Graduate Work in Physics <G/W>,
University of Waterloo, Ontario, Canada N2L 3G1

Abstract.   Experimental investigations of coherently controlling molecular dynamics are currently hampered by the lack of high intensity, mid-infrared laser sources. This paper describes our approach to generating short pulse, high energy mid-infrared laser pulses. The laser will be used to investigate chirped dissociation of molecules.

CHIRPED DISSOCIATION OF MOLECULES

The ability to selectively break any bond of a polyatomic molecule has been a long sought after goal of physical chemists.[1]   With the advent of the laser, energy deposition could be localized into a specific bond by wavelength selection. However, it was soon realized that although the energy was deposited into a single bond, it was quickly redistributed to other bonds through IntraVibrational Relaxation (IVR) processes and the molecule still dissociated via the weakest bond. Since one photon can only efficiently increase a vibrational mode by one quanta, dissociation by laser interaction is necessarily a high order multiphoton process. The quantum levels of a vibrational manifold are not equally spaced and so laser radiation at one wavelength cannot efficiently couple energy between all levels. Short laser pulses have the large spectral bandwidths needed to be resonant over a large number of level spacings.   However, the vibrational quantum level can decrease as well as increase through interaction with a photon, greatly decreasing the overall efficiency of energy absorption. Linear absorption of photons leads to a lengthy absorption time. For the energy deposition time to be reduced to less than the IVR time, the laser energy will have to be absorbed by a near instantaneous, high order nonlinear process.

One possible nonlinear process that theorists have proposed is chirped dissociation.[2]   The idea is to climb a vibrational ladder by interaction with a frequency chirped laser pulse. The theoretical details of chirped dissociation have

CP500, *The Physics of Electronic and Atomic Collisions*, edited by Y. Itikawa, et al.
© 2000 American Institute of Physics 1-56396-777-4/00/$17.00

been described in a number of papers.[2,3] For the purpose of this paper, we will give a simplified, physical description of chirped dissociation. The initial spectral frequency of the laser pulse is slightly higher than the energy spacing of the two lowest vibrational levels. The laser frequency is gradually swept through the transition frequency. After a characteristic time, all the molecules will have climbed adiabatically one step up the ladder of vibrational states. At that time, the laser frequency is changed to be swept through the next transition. After each successful climb up the ladder, the laser frequency is changed in order to favour the process of climbing up the ladder rather than climbing down. By remaining upwardly resonant throughout the process, the ladder can be climbed in a short time, on the order of picoseconds, greatly reducing the chance of energy leaking into the other bonds.

Climbing a quantum ladder was first demonstrated using electronic transitions.[4] This experimental work showed that three levels 5s-5p-5d of a rubidium atom could indeed be climbed using frequency swept pulses. Although, the experiment showed that both red to blue as well as blue to red chirps could cause the ladder to be climbed. The results could be explained using a dressed state picture. Very recently, a vibrational ladder of the ground state of NO has been climbed to the fourth vibrational quantum level, using chirped mid-infrared pulses from a Free Electron Laser (FEL).[5] For the molecular case, the experiments showed that only the intuitively correct blue to red chirp led to an increase in population of the fourth level. The paper also showed that the chirp rate is important because of competing rotational effects. Neither the bandwidth nor the energy of the FEL pulses were sufficient to climb the vibrational ladder any higher. Molecular control experiments are lagging the atomic experiments mainly because of the lack of short pulse, high intensity lasers operating in this region of the spectrum.

Chirped dissociation requires high intensities on the order of 10^{13}W/cm^2 and large spectral bandwidths.[2] The required frequency chirp rates necessitate pulse durations on the order of several picoseconds. The beam diameter can be focused to wavelength dimensions, but in the mid-infrared this corresponds to 10μm diameters. The pulse energy must therefore be on the order of 100μJ in order to achieve an intensity of 10^{13}W/cm^2. We are currently developing a mid-infrared laser source in our lab to achieve these intensities. Two outputs of an amplified two colour mode-locked laser will be difference frequency mixed to generate ~100 fs, 100μJ pulses in the mid-infrared region from 5 to 10 μm.

Two Colour Amplified Laser System

A number of dual wavelength, mode-locked Ti:sapphire oscillators have recently been developed.[6,7,8,9] It has been shown that the two wavelengths can be independently tuned and that the two mode-locked trains are synchronized to better than 2 fs.[6] We are currently developing a Chirped Pulse Amplification[10] (CPA) system to simultaneously amplify pulses from both beams of a two colour oscillator in a single regenerative amplifier.

A block diagram of the entire system is shown in Figure 1. Our oscillator is identical to that described by Dykaar[6] except that we use different mirrors in order to have the two wavelengths widely separated. Two oscillator cavities are built around a single Ti:sapphire crystal, that is pumped by two 6.5 W argon ion laser beams. The two cavities share the focussing mirrors but have separate back mirrors and output couplers. The shared focussing mirrors are centred at 825 nm. The short wavelength mirror set is centred at 770 nm and the long wavelength mirrors are centred at 880nm. All the mirrors have high reflectivity over a 150 nm bandwidth. The two wavelengths of the mode-locked oscillator have been tuned to be as close as 830 nm and 860 nm and as far apart as 770 and 890 nm.

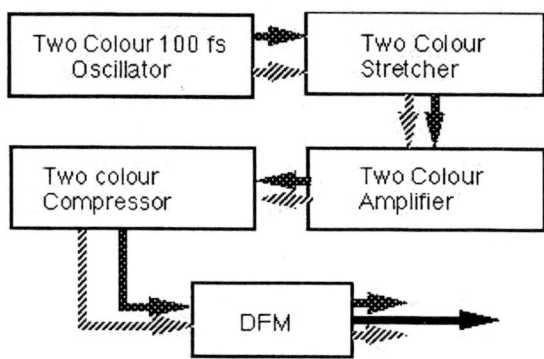

FIGURE 1. Block diagram of two colour chirped pulse amplification and difference frequency mixing system.

The two beams from the oscillator are sent to a dual wavelength pulse stretcher. The two beams are sent to a single grating at two different input angles, such that they copropagate through a standard single grating pulse stretcher. By having the beams have different incident angles, the optics of the stretcher need only to be large enough to accommodate a single colour's bandwidth. The gratings have 1200g/mm and the focussing mirror has a 1m focal length. The pulses are stretched to ~ 200ps.

The two beams from the stretcher are then combined using a 50% beam splitter and transmitted through a Faraday isolator and injected into a single regenerative amplifier. The amplifier is pumped by 35 mJ of 532 nm radiation from a 10 Hz, Q-switched, frequency doubled Nd:YAG laser that can generate 300 mJ pulses. The remaining 265 mJ of pump energy will be used in the future to pump a multipass amplifier. All the amplifier mirrors have 110 nm bandwidth, centred at 855 nm, allowing amplification from 800 to 910 nm. The gain of the Ti:sapphire gain medium peaks at 780 nm. Operating in single pulse mode, the amplifier can deliver

3 mJ at 800 nm and 0.5 mJ at 890 nm. A set of three prisms is used in the amplifier cavity to separate the two wavelengths in order to put a variable loss in the high gain beam line. Equivalent round trip gains for the two wavelengths are maintained to give equal pulse energies for the two wavelengths. An amplified spectrum is shown in Figure 2, for the oscillator generating wavelengths peaked at 800 and 860 nm. At these wavelengths, the regenerative amplifier generates 0.75 mJ pulses at each wavelength. The high gain ratio of 10^5 leads to some gain narrowing of the two spectra.

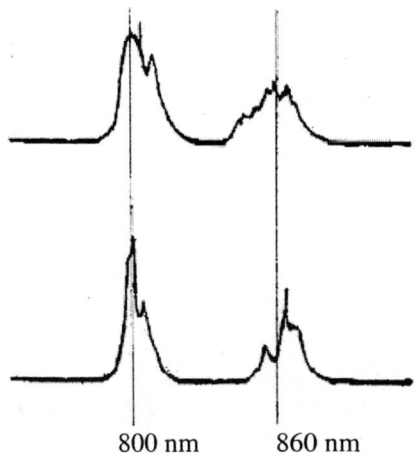

800 nm   860 nm

FIGURE 2. Upper trace is the spectrum of the two colour oscillator. Lower trace is the amplified spectrum from the regenerative amplifier.

After amplification, the pulses are recompressed in a two colour compressor. The compressor uses three gratings. The dual wavelength input beam is incident on the first grating. The two different colours then propagate to separate gratings and back mirrors. The compression of either beam can then be independently tuned. Also the timing of the two pulses can be made to overlap by translating the position of one of the back mirrors.

Assuming Gaussian profiles, the pulse width of the oscillator pulses are 110 and 85 fs, for 800 and 860 nm respectively. After amplification and recompression the pulse durations are 135 and 120 fs for the short and long wavelengths respectively. The beams experience a 35% energy loss propagating through the compressor.

Mid-Infrared Generation

We are planning on generating high intensity mid-infrared pulses by difference frequency mixing the two outputs of our two colour CPA system. A number of different techniques of Optical Parametric Amplification (OPA) or Difference Frequency Mixing (DFM) of femtosecond pulses have been tried. A second source of femtosecond mid-infrared radiation is a Free Electron Laser (FEL). By either technology, the current state-of-the-art technology generates 2μJ energy, subpicosecond pulses at wavelengths of 5μm.[11,12]

For the DFM case, the mid-infrared energy is limited for sub-picosecond pulses for two reasons. First, the photon conversion efficiency of the DFM process decreases with pulse duration due to group velocity mismatch. The pulses tend to walk off each other limiting the nonlinear interaction length. Secondly, two synchronized high intensity pulses are required to generate the mid-infrared. To achieve the 2μJ mid-infrared pulses, a 1mJ, 2.7ps pulse was mixed with a 10nJ, 300fs pulse.[11] One advantage of our proposed system over the other difference frequency mixing systems is that we will have two intense short input pulses. After the multipass amplifier, pulses in both beams will have 5-10 mJ of energy.

One disadvantage of our system is that both wavelengths are near the 750 nm absorption edge of the nonlinear crystal, $AgGaS_2$ and we could be limited by Two Photon Absorption (TPA). However, difference frequency mixing similar wavelengths in the range of 750 to 1000nm has been published.[13] The paper quoted that the intensity limit for TPA in $AgGaS_2$ was high (~100 GW/cm^2), which is more than sufficient to give good conversion efficiency.

We must also address the limitation due to pulse walk-off. To minimize the effects of group velocity mismatch, we will lengthen one pulse simply by leaving the longer wavelength pulse slightly chirped. The duration of mid-infrared pulses will be given by the shorter duration pump pulse.

Our goal is to achieve the 100μJ energy pulses, with ~150 fs pulse durations.

Chirped Dissociation Experiments

The generated mid-infrared laser pulses will be used to investigate climbing a vibrational ladder. We will initially investigate NO as it is a well studied molecule. The goal is to see if we can dissociate the molecule without first causing ionization. We first need to generate chirped mid-infrared pulses. Initially, we too will use simple linear chirps as Noordam's group did.[5] Since Noordam has already shown that indeed the blue to red chirp is the proper direction, a linear frequency sweep can be accomplished with a pair of parallel gratings.[14] A linear chirp is however not the ideal and mid-infrared pulse shaping will be required. There are now a number pulse shapers to manipulate both the amplitude and phase of visible and near infrared optical pulses. A leading technology uses acousto-optic modulators in a grating dispersion line.[6] Acousto-optic modulators are available out to the mid-infrared, although the diffraction efficiency is significantly lower than for visible radiation.

Since we do not have a narrow line, tunable laser source available for spectroscopic studies of the excited molecule, we will use the method of optically driven Coulomb Explosion Spectroscopy[15] (CES) to determine the excitation level reached, by interaction with the chirped mid-infrared photons. The mid-infrared pulses will be focussed into the interaction region of a Time-Of-Flight (TOF) mass spectrometer. CES requires a short intense probe pulse, that is synchronized with the mid-infrared pulse. There will be sufficient intensity left in the short pump pulse used for DFM to be used as the ionizing pulse. In the coherent control experiment, the chirped, mid-infrared pulse would first excite the molecule to a high vibrational level. Then after a short time delay, the high intensity pulse highly ionizes the molecule. The molecule will explode because of the Coulomb repulsion between the charged atoms. A Time-Of-Flight (TOF) mass spectrometer measures the kinetic energy, resulting from the Coulomb potential energy, which is a function of the bond length, at the time of ionization. The bond length is a measure of the vibrational level reached by the chirped pulse.

The initial goal of the proposed work is to try to dissociate simple diatomic molecules without ionization occurring. Future work will investigate manipulating atoms in larger molecules and preparing molecules in non thermodynamic configurations. The ultimate goal of this research is to achieve dissociation of any strong molecular bond, while leaving all the weaker bonds intact.

ACKNOWLEDGMENTS

The authors are grateful to Photonics Research Ontario, Natural Science Engineering Research Council, the Canada Foundation for Innovation and the Alfred P. Sloan Foundation for their support.

REFERENCES

1.   Zewail, A. and Bernstein, R., *Chem Eng. New* **66**, 24-43 (1988).
2.   Chelkowski, S., Bandrauk, A.D. and Corkum, P.B., *Phys. Rev. Lett.* **65**, 2355-2358 (1990).
3.   Yuan, J.M. and Liu, W.K., *Phys Rev. A* **57**, 1992-2001 (1998).
4.   Broers, B., van Linden van den Heuvell, H. B. and Noordam, L.D., *Phys. Rev. Lett.* **69**, 2062-2065 (1992).
5.   Maas, D.J., Vrakking, M.J.J., and Noordam, L.D., *Phys. Rev. A* **60**, 1351-1362 (1999).
6.   Dykaar, D.R., and Darack, S.B., *Opt. Lett.* **18** , 634-636 (1993).
7.   de Barros, M.R.X., and Becker, P.C., *Opt. Lett.* **18** , 631-633 (1993).
8.   Evans, J.M., Spence, D. E., Burns, D. and Sibbett, W., *Opt. Lett.* **18**, 1074-1076 (1993).
9.   Leitenstorfer, A., Fürst, C. and Laubereau, A., *Opt. Lett.* **20**, 916-918 (1995).
10.   Strickland, D. and Mourou, G., *Opt. Commun.* **56**, 219-221 (1985).
11.   Laenen, R., Simeonidis, K. and Laubereau, A., *J. Opt. Soc. Am. B* **15**, 1213-1217 (1998).
12.   Oepts, D., van der Meer, A.F.G. and van Amersfoort, P.W., *Infrared Phys. Technol.* **36**, 297-308 (1995).
13.   Hamm, P., Lauterwasser, C., and Zinth, W., *Opt. Lett.***18**, 1943-1945 (1993).
14.   Treacy, E.B., *IEEE J. Quantum Electron.* **5**, 454-457 (1969) .
15.   Ellert, CH., Stapelfeldt, H., Constant, E., Sakai, H., Wright, J., Rayner, D.M. and Corkum, P.B., *Phil. Trans. R. Soc. Lond. A*, **356**, 329-344 (1998).

Formation of cold Cs$_2$ molecules through photoassociation

A. Fioretti [†], C. Drag, D. Comparat, B. Laburthe Tolra, O. Dulieu,
A. Crubellier, C. Amiot, F. Masnou-Seeuws, and P. Pillet [1]

Laboratoire Aimé Cotton [2], CNRS II, Bât 505, Campus d'Orsay, 91405 Orsay cedex, France
[†] INFM, Dipartimento di Fisica, Università di Pisa, Via Buonarroti 2, 56126 Pisa, Italy,
and IFAM, via del Giardino 7, 56126 Pisa, Italy

Abstract. We report on the formation of translationally cold Cs$_2$ ground-state molecules through photoassociation of laser-cooled cesium atoms in a magneto-optical trap. The cold molecules are obtained after spontaneous decay of photoassociated molecules, and detected after pulsed-laser photoionization into Cs$_2^+$ ions. Photoassociation spectra of cesium attractive molecular states below the $6S_{1/2} + 6P_{3/2}$ dissociation limits are reported, obtained with both ion detection and trap-loss method. A temperature as low as 20^{+15}_{-5} has been measured for the molecular cloud.

INTRODUCTION

The experimental techniques of laser cooling of atoms in the mK-μK range, as well as the manipulation of atomic samples based on radiative forces, are now well established [1]. Their extension to molecules, which would open interesting new perspectives in chemistry, metrology and quantum optics, is however very difficult because of the lack of closed two-level transition for performing efficient cooling. Molecules subject to resonant radiation would be optically pumped into "dark" levels well before cooling is attained. Clever but rather complex laser schemes to overcome this difficulty have been proposed so far [2], and have not been experimentally demonstrated up to now.

On the contrary, an interesting specific scheme for the formation of cold molecules is to start from a cold and dense atomic sample, to "glue" together colliding pairs of free atoms into excited dimers through molecular photoassociation (PA) [3], and to wait for their spontaneous decay into cold ground-state molecules. The molecules should then maintain the same translational temperature of the starting atomic sample.

---

[1] for informations and related material e-mail to: pillet@sun.lac.u-psud.fr
[2] Laboratoire Aimé Cotton is associated with Université Paris-Sud.

CP500, *The Physics of Electronic and Atomic Collisions*, edited by Y. Itikawa, et al.
© 2000 American Institute of Physics 1-56396-777-4/00/$17.00

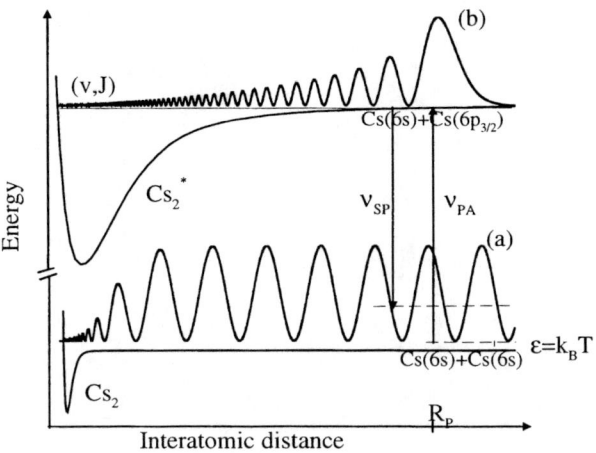

FIGURE 1. Principle of molecular PA of cold atoms with spontaneous decay back into two free atoms. The absorption and the emission occur both at large internuclear distances.

Unfortunately, the spontaneous decay step is generally very ineffective in creating ground-state cold molecules, due to the small values of the Franck-Condon factors. Excited molecules produced by photoassociation can be viewed as pairs of atoms at large internuclear distance, loosely bound by dipole-dipole interaction. It is therefore not surprising that they decay back mostly into two free atoms (Fig. 1). In this case, part of the vibrational energy of the excited molecule is converted into kinetic energy of the two free atoms, that can usually escape from the trap, yielding a trap-loss signal.

In some cases however, the decay into bound, ground-state molecules is not negligible. This happens either when PA populates vibrational levels well below the dissociation limit, in which the vibrational motion does not extend at very large internuclear distances, or alternatively when PA populates levels corresponding to "long-range molecules" [4], in which all the vibrational motion takes place between intermediate and very large internuclear distances. At the inner turning point the excited molecule has a favorable Condon point for decaying into bound molecules. The latter case, depicted in Fig. 2, is indeed much more efficient in producing cold molecules, because it relies on PA close to the dissociation limit, which involves a larger number of colliding pairs.

The first experimental observation of cold molecules has been obtained in our group, where Cs_2 molecules falling out of a magneto-optical trap (MOT) have been observed [5]. Cold molecules in their triplet ground-state were observed as the decay product of the 0_g^- long-range excited state. Later on, ground-state

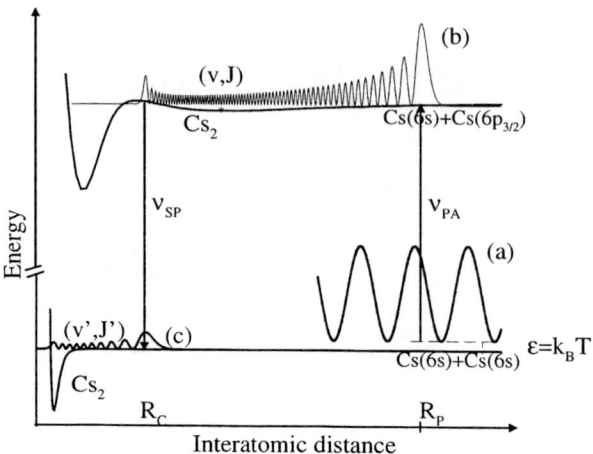

FIGURE 2. Principle of molecular PA of cold atoms with formation of cold molecules. The presence of an external long-range well allows the emission of photons at a Condon point at intermediate distances, leading to a bound-bound transition.

potassium molecules in their singlet state have also been observed [6].

Optical trapping of cold cesium molecules has subsequentely been demonstrated [7] in a far off resonance dipole trap with a CO_2 laser. A very different scheme, which is not based on laser cooling techniques, successfully produced and trapped cold molecules. CaH molecules have been cryogenically cooled through collisions with helium buffer gas, and loaded into a magnetic trap [8]. The temperature attained in this case is in the hundreds mK range, i.e. four orders of magnitude larger than with the PA scheme.

In this paper, we report a selection of our results on cold cesium molecules production, and on Cs_2 photoassociation spectroscopy.

EXPERIMENTAL SETUP

The complete experimental setup is shown in Fig. 3. The central part of the experiment is a magneto-optical trap for cesium atoms [9], realized in a high vacuum ($< 10^{-9}$ torr) stainless steel cell. In its center three pairs of counterpropagating laser beams, all split from a single laser, orthogonally cross in the zero of a quadrupolar magnetic field configuration, created by two pair of anti-Helmoltz coils. The trapping laser is a c.w., single mode diode laser of 150 mW maximum output power, injection locked to a master one. The master laser is another 50 mW single mode diode laser, linewidth narrowed in an external cavity configuration ($\Delta\nu_T < 1$ MHz),

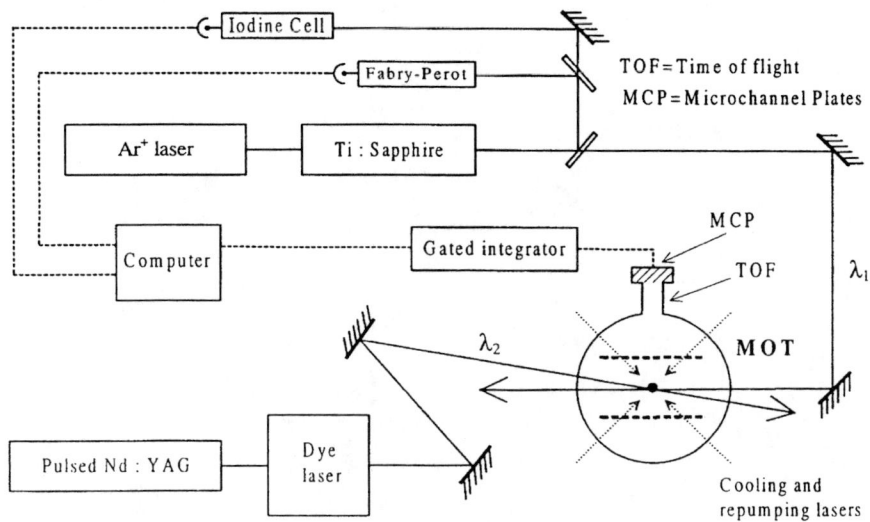

FIGURE 3. Scheme of the experimental setup.

whose wavelength λ_T is tuned 13 MHz on the red of the D_2 cesium atomic transition at 852 nm. The trapping laser, together with the quadrupolar magnetic field, provides cooling and trapping of the atoms in the zero of the magnetic field. A second laser, resonant with the $6S_{1/2}(F = 3) \rightarrow 6P_{3/2}(F' = 4)$ transition, is superposed on two of the arms. It avoids optical pumping of the atoms into their lower $S_{1/2}(F = 3)$ state, where they are not trapped.

In typical conditions, $3 \; 10^7$ cesium atoms are continuously stored into the MOT, in a volume of a fraction of mm^2, with a peak density of the order of 10^{11} cm^{-3}, and at a temperature of $T_{at} = 100 - 200 \; \mu K$, i.e. of the order of the so-called Doppler limit. By applying a sub-Doppler cooling phase, i.e. by reducing the intensity and increasing the detuning of the trapping laser, the atomic temperature can be transiently reduced to $T_{at} \sim 20 \; \mu K$.

Two additional lasers can be applied to the trap. The first one is a c.w., monomode Ti-saphire laser, pumped by a 20 W Ar$^+$ ion laser, with maximum output power of 1.5 W, and linewidth $\Delta \nu_1 \sim 1$ MHz. Its wavelength λ_1 can be tuned anywhere in the red vicinity of the D_2 line, and continuously scanned over a 30 Ghz range. This laser, fucused on a waist of nearly 500 μm, provides molecular PA of cold colliding cesium pairs. The frequency reference for this laser is provided by recording the absorption spectrum in a iodine cell, and by a Fabry-Perot interferometer. The second one is a pulsed dye laser, pumped by the second or the third harmonic of a Nd-YAG laser (7 ns pulse duration, 10 Hz repetition rate). It produces pulses of nearly 1 mJ energy which, focused on a waist of nearly 1 mm, provide partial ionization of both atomic and molecular species in the MOT.

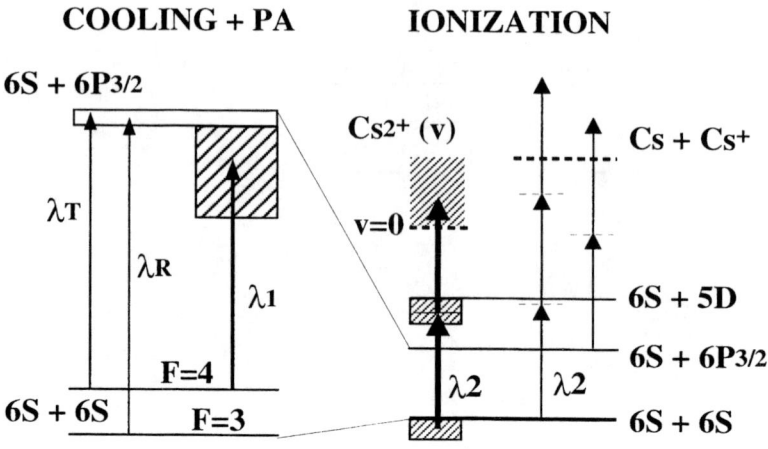

FIGURE 4. Cesium level scheme, with relevant laser transitions.

After the laser pulse is shot an electric field, applied at the MOT position by a pair of metallic grids, accelerates the ions onto a pair of microchannel plates (MCP). The mass difference between atomic and molecular ions makes them arrive onto the MCP with time-of-flights in the ratio $1 : \sqrt{2}$. As we are interested in detecting Cs_2 molecules on a background of cesium atoms which is expected to exceed them by a factor $\sim 10^5$, we choose the pulsed laser wavelength λ_2 in order to maximize the ionization process of molecules into molecular ions and minimize the direct atomic ionization. This is obtained by choosing a two-photon ionization of Cs_2 molecules, resonantly enhanced through the $a^3\Sigma_u^+ \rightarrow 2^3\Pi_g$ intermediate molecular band [10], correlated with the $6S_{1/2} + 5D_{3/2,5/2}$ atomic asymptote. Atomic ions are produced by non-resonant two-photon and three-photon ionization of respectively excited-state and ground-state atoms. The atomic signal can be reduced by switching off the trapping laser some μs prior to the ionizing pulse, leaving all the atoms into their ground-state at the shooting of the laser. The relevant atomic levels, the corresponding laser transitions, and the ionization processes are depicted in Fig. 4.

COLD MOLECULES FORMATION

When the PA laser wavelength λ_1 is tuned in the red vicinity of the D_2 line, for instance at a 10 GHz detuning, and the pulsed dye laser wavelength λ_2 is set near 720 nm, a molecular ion signal is detected. It is clearly separate from the atomic one, and can be independently recorded through a boxcar integrator.

We first analyzed the molecular ion signal by applying the PA laser only during 15 ms per cycle, and by varying the delay of the ionizing pulse with respect to the PA one. The results are shown in Fig. 5a. We observe that the signal

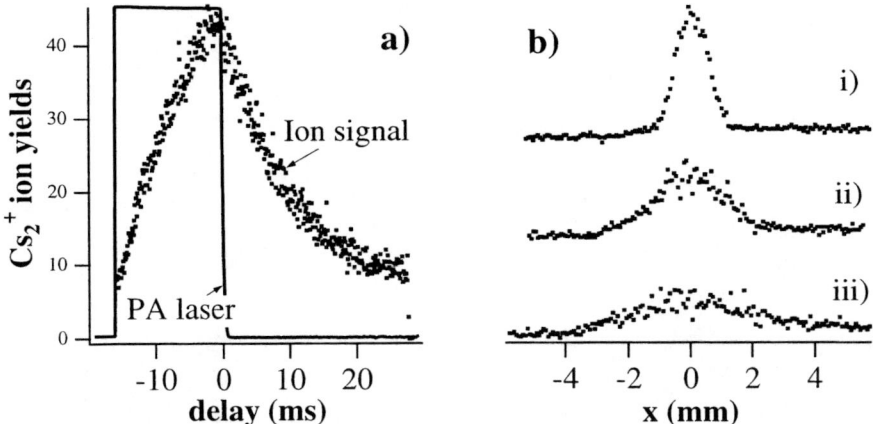

FIGURE 5. (a) Ion signal as a function of the delay between the ionizing laser pulse and the switch-off of the cw PA laser. (b) Spatial analysis of the molecular density by varying the ionizing laser horizontal position: (i) at the MOT position, (ii) 2.7 mm below, and (iii) 5.4 mm below. The corresponding temperature is $T_{\mathrm{mol}} = 85 \pm 15$ μK for an atomic estimated temperature $T_{\mathrm{at}} \sim 30$ μK.

reaches its maximum and decreses with timescales of the order of 10 ms. As this timescale is more than 5 order of magnitude larger than typical molecular excited-state lifetimes, the signal must be due to metastable species; in fact it corresponds to the time in which the molecules leave the ionization region by ballistic expansion in gravity. By scanning the wavelength λ_2 of the dye laser, we identify the $a^3\Sigma_u^+ \rightarrow 2^3\Pi_g$ band and we can unambiguously assign the signal as coming from cold ground-state molecules in their triplet $a^3\Sigma_u^+$ state.

The temperature of the molecular sample has been measured in two different ways, exploiting the fact that Cs_2 molecules, which are not held by the trap, fall ballistically as they are produced. In a first method, we continuously operated the MOT and the PA laser, and we recorded the ion yield while varying the horizontal position of the ionizing laser. This procedure has been repeated at various heigths below the MOT and, from a fit of the spatial spreading, we deduced the initial molecular temperature. A typical measurement is shown in Fig. 5b, corresponding to a temperature $T_{\mathrm{mol}} = 85 \pm 15$ μK. With this method, the measured molecular temperature resulted slightly larger than the estimated atomic one [5]. We demonstrated that this is just an artifact of the measurement procedure, and can be ascribed to the additional force exerced by the magnetic field gradient on the falling molecules. In a second type of measurements, we produced cold molecules through PA just for a short interval of 3 ms per cycle, and then analyzed temporally their fall below the trap, after having switched off the magnetic field. In the latter case, the measured molecular temperature coincides exactly with the atomic

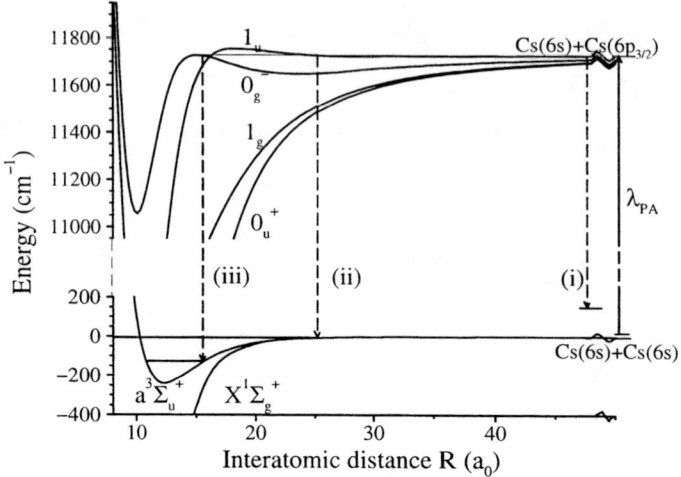

FIGURE 6. Relevant Cs_2 potential curves, deduced from refs. [11,12]. Line (i) represent spontaneous emission towards continuum states, with dissociation of the molecule; lines (ii), and (iii) represent spontaneous emission towards bound states, with formation of cold molecules.

one measured with the same method, leading to values as low as $T_{mol} = 20^{+15}_{-5}$ μK, which are, to our knowledge, the lowest molecular temperature ever reported.

The production of cold molecules in cesium through PA is explained by looking at Cs_2 excited-state potential curves below the $6S_{1/2} + 6P_{3/2}$ dissociation limit (see Fig. 6). Among the attractive molecular states, we find two states, the 0_g^- and the 1_u with a double well structure. In each state the outer well supports bound vibrational levels corresponding to long-range molecules, which are expected to efficiently decay into ground-state dimers. On the contrary, the two other electronic states, the 0_u^+ and the 1_g are standard molecular wells, whose high lying bound levels are not expected to produce cold molecules. This explanation can be tested by performing photoassociation spectroscopy.

We finally remark that a significant molecular production is observed also without any PA laser [5,7], as shown in Fig. 5a. This effect can be ascribed either to PA produced directly by the trapping and repumping lasers through multiple reexcitation processes [5], or to three-body recombination inside the MOT [7]. Specific experiments must be performed in order to discriminate between these two hypotheses.

PHOTOASSOCIATION SPECTROSCOPY

Photoassociation spectroscopy of alkali dimers has led in recent years, to a detailed knowledge of the long-range part of the excited-state molecular potential curves, and also to get detailed information about their atomic collisional processes at very low energies [13].

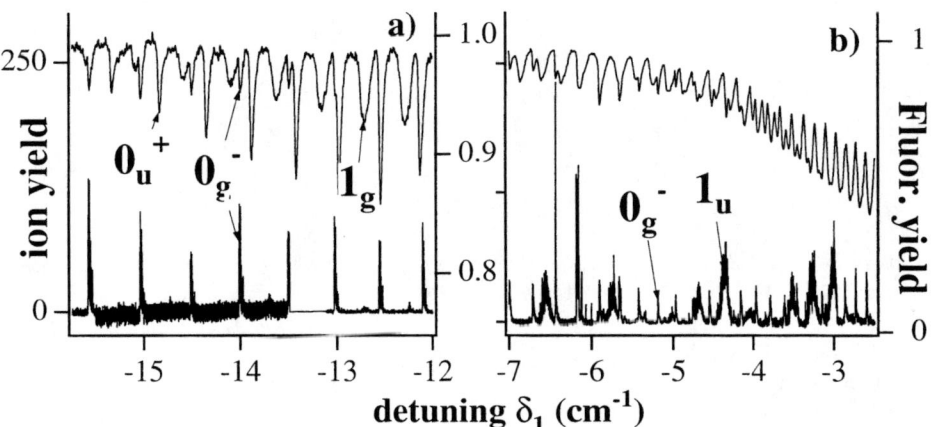

FIGURE 7. Cs_2^+ ion signal (lower signal) and trap fluorescence yield (upper signal) versus the detuning δ_1 of the PA laser. The origin is at the $6^2S_{1/2}(F = 4) \rightarrow 6^2P_{3/2}(F' = 5)$ atomic transition.

In the cesium case, the first PA spectra have been obtained in our group, by recording the molecular ion yield as a function of the PA laser wavelength λ_1 [5]. When the PA laser is resonant with a molecular excited level that can decay into cold molecules, a sharp increase of the ion yield is obtained. In further experiments, also the total number of trapped atoms, which is proportional to the trap fluorescence, was recorded, giving rise to trap-loss spectra [14].

By comparing these two types of spectra, it is possible to test the explanations on molecules formation formulated in the preceeding section. PA spectra in two specific energy ranges are shown in Fig. 7. While the ion spectrum shows only two vibrational series, assigned to the 0_g^- and to the 1_u states, the trap loss spectrum shows clearly three vibrational series, assigned to the 0_g^-, 0_u^+, and 1_g states [3]. In particular, in energy ranges where the trap-loss produced by the PA laser on the latter three states is comparable (see Fig. 7a), no evidence of molecular production is found through PA on the 0_u^+ and 1_g states, which are not long-range states [4]. In these states, the molecular vibration is highly asymmetric: excited molecules

---

[3] As in the energy range where transitions corresponding to the 1_u state are observed (Fig. 7b) all the vibrational series overlap, it cannot clearly be stated whether the 1_u state produces trap-loss or not.

spend most of their time at long-range, then come close and quickly rebound away, not allowing time for decay at intermediate distances into ground-state molecules.

The detailed high resolution PA spectroscopy of the $0_g{}^-$ and of the 1_u states has been discussed in two separate publications [15,16], while the trap-loss spectrum has been discussed in ref. [14].

In conclusion, we have reported on the observation of translationally cold cesium molecules at temperatures as low as $T_{mol} = 20~\mu K$. They are created in by spontaneous decay from photoassociated dimers inside a magneto-optical trap. These molecules are also rotationally cold but are expected to be vibrationally hot. They are produced essentially in the triplet ground-state, and in the last bound levels of the singlet state. The production of cold molecules should open new perspectives in chemistry, metrology and quantum physics.

REFERENCES

1. Chu S., *Rev. Mod. Phys.* **70**, 685 (1998); Cohen-Tannoudji C., *ibid.* **70**, 707 (1998); Phillips W.D., *ibid.* **70**, 721 (1998).
2. Bahns J.T., Stwallwy W.C., and Gould P.L., *J. Chem. Phys.* **104**, 9689 (1996).
3. Thorsheim H.R., Weiner J., and Julienne P.S., *Phys. Rev. Lett.* **58**, 2420 (1987).
4. Stwalley W.C., Uang Y.H., Pichler G., *Phys. Rev. Lett.* **41**, 1164 (1975).
5. Fioretti A., Comparat D., Crubellier A., Dulieu O., Masnou-Seeuws F., and Pillet P., *Phys. Rev. Lett.* **80**, 4402 (1998).
6. Nikolov A.N., Eyler E.E., Wang X.T., Wang H., Stwalley W.C., and Gould P.L., *Phys. Rev. Lett.* **82**, 703 (1999).
7. Takekoshi T., Patterson B.M., Knize R.J., *Phys. Rev. Lett.* **81**, 5105 (1998).
8. Weinstein J.D., de Carvalho R., Guillet T., Friedrich B., and Doyle J.M., *Nature* **395**, 148 (1998).
9. Monroe C., Swann W., Robinson H., and Wieman C., *Phys. Rev. Lett.* **65**, 1571 (1990)
10. Pichler G., Milošević S., Veža D., and Beuc R., *J. Phys. B: At. Mol. Phys.* **16**, 4619 (1983).
11. Spiess N., Ph.D. thesis, Universität Kaiserslautern (1989), unpublished.
12. Marinescu M., and Dalgarno A., *Phys. Rev. A* **52**, 311 (1995).
13. Weiner J., Bagnato V.S., Zilio S., and Julienne P.S. *Rev. Mod. Phys.* **71**, 1 (1999).
14. Comparat D., Drag C., Fioretti A., Dulieu O., and Pillet. P., *J. Mol. Spectr.* **195**, 229 (1999).
15. Fioretti A., Comparat D., Drag C., Amiot C., Dulieu O., Masnou-Seeuws F., and Pillet P., *Eur. Phys. J. D* **5**, 389 (1999).
16. Comparat D., Drag. C., Laburthe Tolra B., Fioretti A., Pillet P., Crubellier A., Dulieu O., and Masnou-Seeuws F., submitted to Eur. Phys. J. D (1999).

Coherent control of molecular processes application to cooling internal degrees of freedom

Ronnie Kosloff and Allon Bartana

Department of Chemistry and the Fritz Haber Institute for Molecular Dynamics, Hebrew University of Jerusalem, Israel

David J. Tannor

Department of Chemical Physics, Weizmann Institute of Science, 76100 Rehovot, Israel

Optimal control theory is applied to laser cooling of molecules. The objective is to cool vibrations, using shaped pulses synchronized with the spontaneous emission. The optimal mechanism is found to operate by a "vibrationally selective coherent population trapping". The trapping condition is that the instantaneous phase of the laser is locked to the phase of the transition dipole moment of $v = 0$ with the excited population. The molecules that reach $v = 0$ by spontaneous emission arc then trapped, while the others are continually repumped.

I. INTRODUCTION

The basic idea of coherent control is to employ the wave properties of matter and to create a positive interference in the desired channel and a destructive interference in all other accessible channels. Initially the means to achieve this goal were either in the time domain [1] or in the frequency domain [2]. It was realized that the different viewpoints can be integrated using optimal control theory [3] to a time-frequency phase space picture [5,4]. Since quantum coherence is the key to control a crucial issue is the competition between coherent control and decoherence. This issue is of particular importance in processes that cannot be carried out by unitary transformations. An important example of such a process is cooling.

The problem of laser cooling of molecular translations is a difficult one [7–12]. The existence of internal rotational and vibrational states of molecules complicates the situation considerably. These internal states serve as "heat traps". In the case of atoms it is possible to have a nearly closed set of states such that the spontaneous emission comes back down to states that are immediately reexcited by the pump frequency. In molecules, the spontaneous emission will in general come down to all these internal levels; amplitude in these excited rotational and vibrational states is lost from the cooling cycle, leading to significant losses in efficiency. As a result of these difficulties, there has far less experimental progress in producing ultra-cold molecules than atoms. It is worth noting that in the current experiment that have produced translationally cold molecules, via photoassociation of cold atoms, the molecules were produced vibrationally hot [12–16].

CP500, *The Physics of Electronic and Atomic Collisions*, edited by Y. Itikawa, et al.

Subrecoil laser cooling of atoms relies on the vanishing of the absorption from the target state and accumulation of population through spontaneous emission to this state i.e. the dark state. Using these principles extremely cold atoms have been obtained by **v**elocity **s**elective **c**oherent **p**opulation **t**rapping (VSCPT) [17–19] and Raman cooling [20].

The target in this study, is to cool molecules by creating a dynamical dark state. Optimal control theory is employed to optimize the use of the stimulated transitions while maintaining the target state decoupled from the field. For this to happen a general principle is invoked in which the phase of the field is locked to an internal phase of the molecule. This phase locking condition is the key to the creation of the dark state and is called "vibrationally selective coherent population trapping".

II. THE SYSTEM

The molecular system to be cooled consists of a ground and an excited electronic state, both of which depend on the internal nuclear coordinates. The radiation induces transitions between these surfaces. The state of such a system is defined in the domain of the direct product between nuclear and electronic coordinates. Using pseudo-spin notation the density operator becomes:

$$\hat{\rho} = \hat{\rho}_g \otimes \hat{\mathbf{P}}_g + \hat{\rho}_e \otimes \hat{\mathbf{P}}_e + \hat{\rho}_c \otimes \hat{\mathbf{S}}_+ + \hat{\rho}_c^\dagger \otimes \hat{\mathbf{S}}_- = \begin{pmatrix} \hat{\rho}_e & \hat{\rho}_c \\ \hat{\rho}_c^\dagger & \hat{\rho}_g \end{pmatrix} \qquad (2.1)$$

where $\hat{\rho}_g(\hat{\mathbf{r}})$, $\hat{\rho}_e(\hat{\mathbf{r}})$ are the populations on the ground and excited surfaces, with $\hat{\mathbf{r}}$ the nuclear coordinate, $\hat{\rho}_c(\hat{\mathbf{r}})$ is the nuclear coherence between the two electronic surfaces, and $\hat{\mathbf{P}}_{e/g}$ is the projection on the upper and lower surface and $\hat{\mathbf{S}}_\pm$ are the electronic transition raising and lowering operators.

The evolution of the system contains a unitary part generated by the Hamiltonian and a dissipative evolution caused by spontaneous emission. The equation of motion of the state of the system becomes:

$$\frac{\partial \hat{\rho}}{\partial t} = -\frac{i}{\hbar} \left[\hat{\mathbf{H}}, \hat{\rho} \right] + \mathcal{L}_D(\hat{\rho}) \quad , \qquad (2.2)$$

or equivalently for an operator $\hat{\mathbf{A}}$:

$$\frac{d\hat{\mathbf{A}}}{dt} = \frac{i}{\hbar} \left[\hat{\mathbf{H}}, \hat{\mathbf{A}} \right] + \mathcal{L}^\dagger{}_D(\hat{\mathbf{A}}) + \frac{\partial \hat{\mathbf{A}}}{\partial t} \quad , \qquad (2.3)$$

The Hamiltonian $\hat{\mathbf{H}}$ is partitioned into a molecular part and an induced time dependent part describing the stimulated interaction with the radiation field:

$$\hat{\mathbf{H}} = \hat{\mathbf{H}}_{\mathbf{g}} \otimes \mathbf{P}_{\mathbf{g}} + \hat{\mathbf{H}}_{\mathbf{e}} \otimes \hat{\mathbf{P}}_{\mathbf{e}} - \epsilon(t)\hat{\mu} \otimes \hat{\mathbf{S}}_+ - \epsilon(t)^*\hat{\mu} \otimes \hat{\mathbf{S}}_- = \begin{pmatrix} \hat{\mathbf{H}}_{\mathbf{e}} & -\epsilon(t)\hat{\mu} \\ -\epsilon(t)^*\hat{\mu} & \hat{\mathbf{H}}_{\mathbf{g}} \end{pmatrix} \quad (2.4)$$

where $\hat{H}_{e/g} = \hat{P}^2/2m + \hat{V}_{e/g}$ are the surface Hamiltonians, $\epsilon(t)$ is a time dependent EM field, and $\hat{\mu} = \hat{\mu}(\mathbf{r})$ is the electronic transition dipole operator.

The generator of the dissipative dynamics, $\mathcal{L}_D$, embodies the effect of spontaneous emission. The form of $\mathcal{L}_D$ in Eq. (2.2) describes Markovian evolution cast into the Lindblad semigroup form [21]:

$$\mathcal{L}_D(\hat{\rho}) = \hat{F}\hat{\rho}\hat{F}^\dagger - \frac{1}{2}\left\{\hat{F}^\dagger\hat{F}, \hat{\rho}\right\} \tag{2.5}$$

where $\hat{F}$ is an operator defined on the Hilbert space of the system. Specifically, for spontaneous emission this operator, $\hat{F} = \Gamma(\hat{r}) \otimes \hat{S}_-$, describes a transition from the excited to the ground electronic surface. The density of states of the photon bath in an isotropic space is proportional to ν^3, the cube of the emission frequency reflecting the density of states of the photon bath. Assuming a constant dipole function, $\Gamma(\hat{r})$ obeys the relation: $\Gamma(\hat{r}) = \gamma\Delta(\hat{r})^{3/2} = \gamma(\hat{V}_e - \hat{V}_g)^{3/2}$. Other dissipative processes which contain transitions to the dark state are also sufficient for closing the cooling loop.

III. THE DARK STATE

The target of cooling is any single vibrational quantum state. To simplify the notation the target state is chosen as the ground vibrational level on the ground electronic surface defined: $|0\rangle\langle 0| \otimes \hat{P}_g$. The change in time in the occupation of this state using Eq. (2.3) becomes:

$$\frac{d\left\langle |0\rangle\langle 0| \otimes \hat{P}_g \right\rangle}{dt} = \frac{i}{\hbar}\left\langle \left[\hat{H}, |0\rangle\langle 0| \otimes \hat{P}_g\right]\right\rangle + \left\langle \mathcal{L}^\dagger{}_D(|0\rangle\langle 0| \otimes \hat{P}_g)\right\rangle \quad . \tag{3.1}$$

For this state to be dark, the commutation relation with the Hamiltonian should vanish. The dark state projection operator commutes with the stationary part of the Hamiltonian, while the commutation relation with the transient part of the Hamiltonian leads to the condition:

$$2Imag\left\{\left\langle \hat{\mu}|0\rangle\langle 0| \otimes \hat{S}_+\right\rangle \epsilon(t)\right\} = 0 \quad . \tag{3.2}$$

Eq. (3.2) imposes a condition on the locking of the phase of the field to the phase of the projection of the transition dipole on the ground vibrational state:

$$2\left|\left\langle \hat{\mu}|0\rangle\langle 0| \otimes \hat{S}_+\right\rangle\right| |\epsilon(t)| \sin\left(\phi_{\mu|0\rangle\langle 0|} + \phi_\epsilon\right) = 0 \quad . \tag{3.3}$$

Eq. (3.3) vanishes if the sum of the phase angles, $\phi_{\mu|0\rangle\langle 0|} + \phi_\epsilon$, is equal to $n\pi$ where n is an integer. This condition is fulfilled by the explicit relation:

$$\epsilon(t) = \frac{\left\langle \hat{\mu}|0\rangle\langle 0| \otimes \hat{S}_-\right\rangle}{\left|\left\langle \hat{\mu}|0\rangle\langle 0| \otimes \hat{S}_-\right\rangle\right|}c(t) \tag{3.4}$$

where $c(t)$ is a real function of time which can be either positive or negative. Eq. (3.2-3.4) are equivalent statements of the dark state condition.

IV. COOLING STRATEGY: LOCAL OPTIMIZATION

Once the trapping dark state has been constructed, the cooling cycle can be completed by exciting transitions from the ground electronic surface to the excited electronic surface while maintaining the locking conditions.

To optimize the cooling rate the maximum excitation rate is sought. The rate of population excitation from the ground surface to the excited surface is obtained by inserting the projection $\hat{\mathbf{P}}_g$ into the Hamiltonian part of the Heisenberg equation of motion [7,22]:

$$\frac{d\left\langle \hat{\mathbf{P}}_g \right\rangle_H}{dt} = 2Imag\left\{ \left\langle \hat{\mu} \otimes \hat{\mathbf{S}}_+ \right\rangle \epsilon(t) \right\} \quad . \tag{4.1}$$

Inserting the locking condition Eq. (3.4) for $\epsilon(t)$ leads to:

$$\frac{d\left\langle \hat{\mathbf{P}}_g \right\rangle_H}{dt} = 2Imag\left\{ \left\langle \hat{\mu} \otimes \hat{\mathbf{S}}_+ \right\rangle \frac{\left\langle |0\rangle\langle 0|\hat{\mu} \otimes \hat{\mathbf{S}}_- \right\rangle}{\left| \left\langle |0\rangle\langle 0|\hat{\mu} \otimes \hat{\mathbf{S}}_- \right\rangle \right|} c(t) \right\} \quad . \tag{4.2}$$

A monotonic depletion of the ground surface population $\frac{d\langle \hat{\mathbf{P}}_g \rangle_H}{dt} < 0$ is achieved if:

$$c(t) = -2Imag\left\{ \left\langle \hat{\mu} \otimes \hat{\mathbf{S}}_+ \right\rangle \frac{\left\langle |0\rangle\langle 0|\hat{\mu} \otimes \hat{\mathbf{S}}_- \right\rangle}{\left| \left\langle |0\rangle\langle 0|\hat{\mu} \otimes \hat{\mathbf{S}}_- \right\rangle \right|} \right\} \quad . \tag{4.3}$$

The locking condition and the monotonic depletion of the ground surface specify completely, at a time t, the transient external field $\epsilon(t)$. The external power employed is proportional to the instantaneous molecular response, for this reason the approach is termed local in time optimization [7].

V. OPTIMAL CONTROL APPROACH: GLOBAL OPTIMIZATION

Optimal control theory (OCT) is an alternative route to find the cooling strategy without imposing locking conditions. The target of the cooling process is defined as the projection operator $\hat{\mathbf{A}} = |0\rangle\langle 0| \otimes \hat{\mathbf{P}}_g$. The expectation of the target is to be maximized under the restriction of finite time interval $\{0, t_f\}$. The field at intermediate times is therefore unrestricted.

OCT seeks the optimal field as a variation problem of an objective J which is a functional of the field $\epsilon(t)$:

$$J = tr\{\hat{\rho}(t_f) \hat{\mathbf{A}}\} \tag{5.1}$$

Two constraints are imposed:

- 1. The evolution of the system has to be governed by the Liouville von-Neumann equation.

- 2. The total field energy has to be minimized.

These two constraints lead to the modified objective

$$\bar{J} = tr\{\hat{A}\hat{\rho}(t_f)\} - \int_0^{t_f} tr\left\{\left(\frac{\partial\hat{\rho}}{\partial t} - \mathcal{L}(\hat{\rho})\right)\hat{B}\right\}dt - \lambda\int_0^{t_f}|\epsilon|^2 dt \qquad (5.2)$$

$\hat{B}$ is an operator Lagrange multiplier and λ is a scalar Lagrange multiplier. An extremum for the objective $\bar{J}$ is found by a variation of $\bar{J}$ with respect to $\delta\hat{\rho}$, $\delta Real(\epsilon)$ and $\delta Imag(\epsilon)$. The variation leads to equations of motion for the state $\hat{\rho}$ and the retarded objective operator $\hat{B}$:

$$\frac{\partial\hat{\rho}}{\partial t} = \mathcal{L}(\hat{\rho}). \qquad , \qquad \frac{\partial\hat{B}}{\partial t} = -\mathcal{L}^\dagger(\hat{B}) \qquad , \qquad (5.3)$$

subject to an initial condition for the state $\hat{\rho}(0)$ and a final condition for the Lagrange multiplier-retarded objective operator: $\hat{B}(t_f) = \hat{A}$. The structure of Eq. (5.3) represents two counter currents, the density operator $\hat{\rho}$ carries the information from the initial state $\hat{\rho}(0)$ forward in time and the retarded objective operator $\hat{B}$ propagates the information on the final objective $\hat{A}$ backward in time. In Hamiltonian dynamics the two currents carry the full information from the boundaries. In dissipative dynamics information is lost during the forward propagation of the initial state and the backward propagation of the objective.

To solve the control equation an iterative procedure was employed based on Krotov's method [23], similar to the method described previously [8] with the addition of a dissipative Liouville operator. The information from both currents is used to calculate the field for the next iteration [8].

VI. RESULTS

The cooling mechanism and performance of the local and global optimization, are compared in a particular molecular model. The model system is composed of two displaced Morse electronic surfaces. A hot thermal initial state $\hat{\rho}(0)$ with an inverse temperature of $\beta = 22.6$ was chosen. At this temperature the population on the ground vibrational level, $\nu = 0$, is $\sim 50\%$ and it decreases to $\sim 10^{-3}$ at $\nu = 9$.

In the OCT calculations, the iterative procedure was stopped when the improvement of the target projection of an additional run was smaller than 10^{-5}.

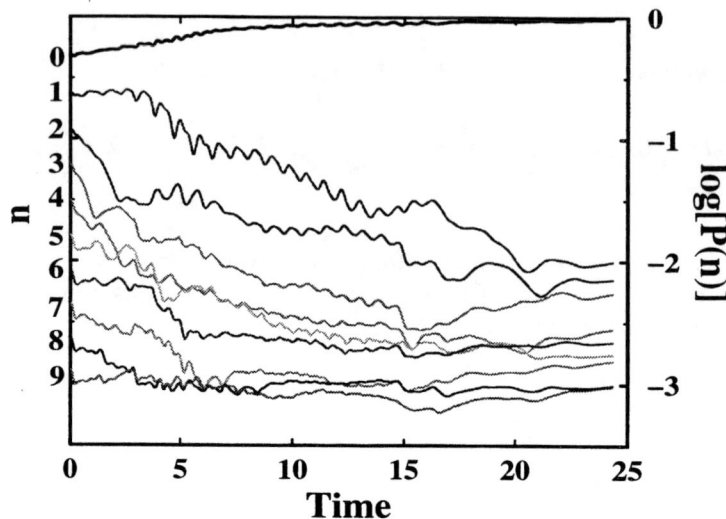

FIG. 1. The log of the projections of the vibrational levels on the ground electronic surface $\log \left\langle |\nu\rangle\langle\nu| \otimes \hat{P}_g \right\rangle$ as a function of time, in vibrational periods. The initial Boltzmann distribution is reflected by the even displacement between the log of the projections of the different states at $t = 0$.

Significant cooling was obtained within a period of 25 vibrational periods, the projection on the target state increased from 50% to 62% for the local optimization and to 97% for the OCT global approach. All other measures of cooling show similar tendencies for example the Von Neuman entropy decreases from a value of 1.4 to 0.25 for the OCT case. In all cases studied, the global approach outperformed the local method significantly.

Figure 1 shows the progress in cooling as a function of time for the OCT calculation. While the ground state population increases almost monotonically the population of all other states decreases. The small oscillations are due to cycling of population to the excited electronic surface. This cycling is a manifestation of the interplay between the creation of coherence by the stimulating radiation and destruction of coherence by spontaneous emission.

The field creating the optimal result is shown in figure 2. In the time domain the field is composed of a series of pulses. In the frequency domain the filed shows a prominent peak red shifted from the 00 line at $\nu = 1.996$. The intensity at the 00 transition is zero in analogy to the Raman cooling of atoms [20]. As time progresses the low frequencies are eliminated which means as seen in Fig. 1 that the high vibrational states are cooled first.

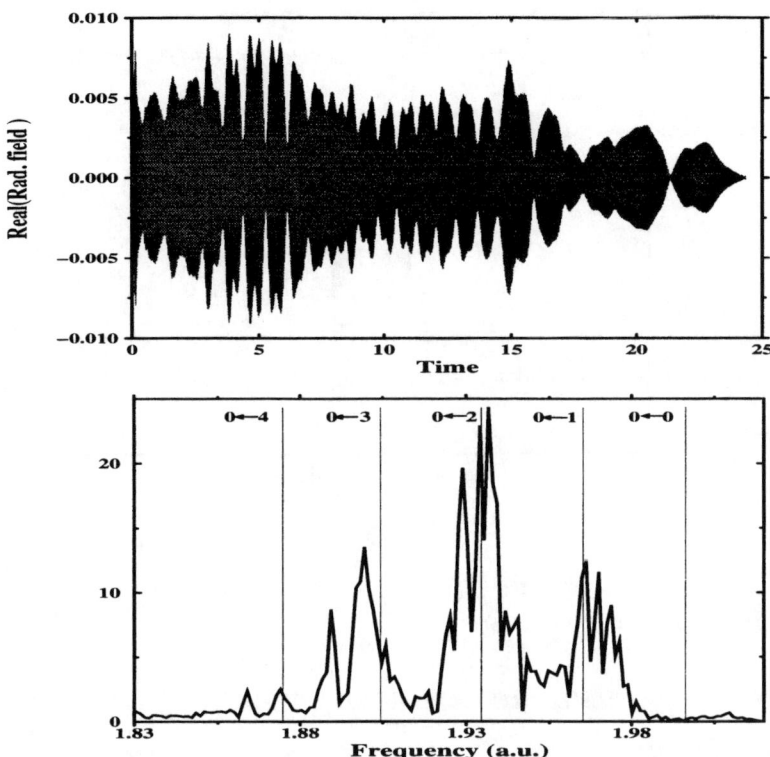

FIG. 2. The optimal cooling field. a) the pulse as a function of time b) The pulse spectrum. The vertical lines indicate the position of the transitions from $\nu = 0 - 4$ levels on the ground surface to the $v = 0$ level on the excites surface.

The most crucial comparison between the local and global optimization approaches is by examining the sum of the instantaneous phase angles of the filed and dark sate dipole projection: $\varphi = \phi_{\mu|0\rangle\langle 0|} + \phi_\epsilon$ shown in Fig. 3. After a short transient time the local optimization imposes a strict condition on this angle $\varphi = n\pi$. A priori the global optimization does not restrict the phase angle, nevertheless this angel has a very strong propensity for the locking condition $\varphi = 2n\pi$. For this angle the phase of the field and the phase of the dark sate dipole projection oppose each other. The other locking condition where n is odd is absent from the OCT mechanism. Examining the phase angle in the local optimization shows that the odd angles becomes unstable when the cooling progresses which explains their absence in the OCT global optimization.

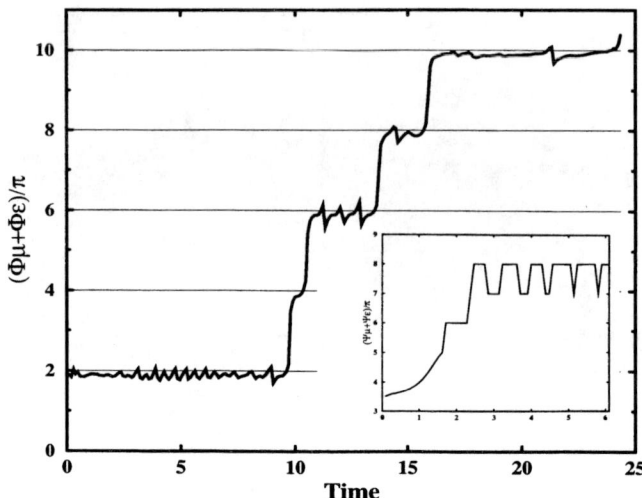

FIG. 3. Sum of the phase angles $\phi_{\mu|0\rangle\langle0|} + \phi_\epsilon$ as a function of time obtained for OCT. The inset shows the sum for local control

VII. ROTATIONAL COOLING

The rotational Hamiltonian is influenced by the fact that the light can change the angular momentum of the system. The following Hamiltonian is used:

$$\hat{H} = \hat{H}_0 + \hat{V}_t \quad ; \quad \hat{H}_0 = \begin{pmatrix} \hat{H}_e & 0 \\ 0 & \hat{H}_g \end{pmatrix} \tag{7.1}$$

where: $\hat{H}_{e/g} = 2\pi\hat{B}_{e/g}\hbar l(l+1)$, and

$$\hat{V}_t = \begin{pmatrix} 0 & \mu_x \cdot \epsilon_x(t) + \mu_y \cdot \epsilon_y(t) + \mu_z \cdot \epsilon_z(t) \\ \mu_x \cdot \epsilon_x^*(t) + \mu_y \cdot \epsilon_y^*(t) + \mu_z \cdot \epsilon_z^*(t) & 0 \end{pmatrix} \tag{7.2}$$

Fig. 4 shows the optimal control solution of the Rotational cooling. The mechanism of cooling is very different than the one found for vibrational cooling. The projection on the ground state stays low for most of the duration of the process and then it jumps abruptly to a value close to one. On the other hand the projection on the largest eigenvalue of the density operator increases monotonically as well as the Reni entropy $tr\{\hat{\rho}^2\}$

214

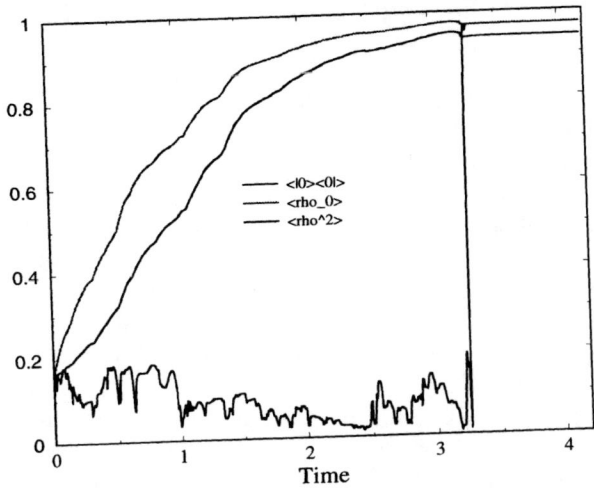

FIG. 4. The projection on the ground state $tr\{|0\rangle\langle0|\}$ the Reni entropy $tr\{\hat{\rho}^2\}$ and the largest eigenvalue of the density operator $\hat{\rho}$ as a function of time, for an optimal control solution.

VIII. DISCUSSION/CONCLUSIONS

The application of optimal control theory to cooling molecules has revealed an intriguing new physical mechanism: the instantaneous phase locking of the laser pulse to the phase of the transition dipole moment of $v = 0$ with the excited population. These conditions lead to a dark vibrational target state which eventually accumulates all the molecular population. For rotation a different mechanism is at play. The selection rules imposed by angular momentum conservation are probably the reason for this difference.

The rate limiting step of cooling then becomes the rate of spontaneous emission. It is important to state that the phase locking conditions are dynamic; as a result the field has a complicated time dependence. One can therefore ask if such fields can be calculated and created with sufficient precision to obtain significant cooling. A combination of pulse shaping techniques [24–27] and feedback control may give a positive answer to this question [28–30].

Acknowledgments

This research was supported by the US-Israel Science foundation and the US-Navy under contract number N00014-91-J-1498. The Fritz Haber Research Center is supported by the Minerva Gesellschaft für die Forschung, GmbH München, FRG.

REFERENCES

[1] D. J. Tannor and S. A. Rice, J. Chem. Phys. **83** 5013 (1985); D. J. Tannor R. Kosloff and S. A. Rice, J. Chem. Phys. **85** 5805 (1986).

[2] M. Shapiro and P. Brumer, J. Chem. Phys. **84** 4103 (1986).

[3] M. A. Dahleh A. M. Peirce and H. Rabitz, Phys. Rev **A 37** 4950 (1988).

[4] P. Brumer and M. Shapiro, Annu. Rev. Phys. Chem., **43**, 257 (1992).

[5] S. Rice, Science, **258**, 412 (1992).

[6] M. H. Anderson, J. R. Ensher, M. R. Matthews, C.E. Wienman and E. A. Cornell, Science, **269**, 198 (1995).

[7] Allon Bartana, Ronnie Kosloff and David J. Tannor, J. Chem. Phys., **99**, 196–210 (1993).

[8] Allon Bartana, Ronnie Kosloff and David J. Tannor, J. Chem. Phys., **106**, 1435–1448 (1997).

[9] Y. B. Band and P.S. Juliene, Phys. Rev. A, **51**, r4317 (1995).

[10] J. T. Bahns W. C. Stwalley and P.L. Gould, J. Chem. Phys., **104**, 9689 (1996).

[11] A. Vardi, D. Abrashkevich, E. Frishman and M. Shapiro, J. Chem. Phys., **107**, 6166 (1997).

[12] A. Fioretti, D. Comparat, A. Crubellier, O. Dulieu, F. Masnou-Seeus and P. Pillet, Phys. Rev. Lett., **80**, 4402 (1998).

[13] A. Fioretti, D. Comparat, C. Drag, C. Amiot, O. Dulieu, F. Masnou-Seeuws and P. Pillet, *Photoassociative Spectroscopy of the Cs_2 0_g^- Long-Range State*, Eur. Phys. D **5**, 389-403 (1998).

[14] T. Takekoshi, B. M. Patterson and R. J. Knize, *Observation of Optically Trapped Cold Cesium Molecules*, Phys. Rev. Lett. **81**, 5105-5108 (1998).

[15] J.P. Shaffer, W. Chalupczak, N. P. Bigelow, *Photoassociative ionization of heteronuclear molecules in a novel two-species magneto-optical trap*, Phys. Rev. Lett. **82**, 1124 (1999).

[16] A. N. Nikolov, E. E. Eyler, X. T. Wang, j. Li, H. Wang, W. C. Stwalley, P. L. Gould, *Observation of ultracold ground-state potassium molecules* Phys. Rev. Lett. **82**, 703 (1999).

[17] A. Aspect, E. Arimondo, D. Kaiser, N. Vansteenkiste and C. Cohen-Tannoudji, Phys. Rev. Lett., **61**, 826 (1988).

[18] F. Bardou, J. P. Bouchaud O. Emile A. Aspect, and C. Cohen-Tannoudji, Phys. Rev. Lett., **72**, 203 (1994).

[19] J. Reichel, F. Bardou, M. Ben Dahan, E. Peik, S. Rand, C. Salomon and C. Cohen-Tannoudji, Phys. Rev. Lett., **75**, 4575 (1995).

[20] M. Kasevich and S. Chu, Phys. Rev. Lett., **69**, 1741 (1992).

[21] G. Lindblad, Commun. Math. Phys., **33**, 305 (1973).

[22] R. Kosloff, A. Dell Hammerich, and D. Tannor, Phys. Rev. Lett., **69**, 2172–2175 (1992).

[23] D. Tannor, V. Kazakov and V. Orlov , In J. Broeckhove and L. Lathouwers, editor, *Time Dependent Quantum Molecular Dynamics*, volume NATO ASI Ser. B 299, page 347, Plenum Press, (1992).

[24] M. Dahleth W.S. Warren, H. Rabitz, Science, **259**, 1581 (1993).

[25] A. M. Weiner J. P. Heritage E. M. Kirschner, J. Opt. Soc. Am. B, **5**, 1563 (1988).

[26] K. A. Nelson A. M. Weiner D. E. Leaird G. P. Wiederrecht, J. Opt. Soc. Am. B, **8**, 1264 (1991).

[27] C. W. Hillegas, J. X. Tull, D. Goswami, D. Strickland and W. S. Warren, Opt. Lett., **19**, 737 (1994).

[28] R. S. Judson and H. Rabitz, Phys. Rev. Lett., **68**, 1500 (1992).

[29] C. J. Bardin, V. V. Yakoblev, K. R. Wilson, S. D. Karpenter P. M. Weber w. S. Warren, Chem. Phys. Lett., **280**, 151 (1997).

[30] D. Meshulach, D. Yelin and Y. Silberberg, J. Opt. Soc. Am. B, **15**, 1615 (1998).

Absolute Photoionization Cross Section for C$^+$

F. Folkmann, J.E. Hansen*, H. Kjeldsen, H. Knudsen,
M.S. Rasmussen, J.B. West† and T. Andersen

Institute of Physics and Astronomy, University of Aarhus, 8000 Aarhus C, Denmark
*Department of Physics and Astronomy, University of Amsterdam, The Netherlands
†Daresbury Laboratory, Warrington WA4 4AD, United Kingdom

Abstract. The absolute photoionization cross section for the astrophysically important C$^+$ ion has been measured for the first time thus allowing a test of the predicted cross section from the Opacity Project. The measurements are performed with a new ion - photon merged beam setup at the ASTRID storage ring utilizing an undulator beam line. In addition to the predicted 2s2p(^{3}P)np ^{2}D and ^{2}S autoionizing resonances, the ^{2}P states which are not included in the theoretical predictions also contribute significantly to the ionization yield below the 2s2p(^{3}P) limit. The cross section is determined with a precision of 10% and lies 5-25% below the theoretical prediction at 24-37 eV.

INTRODUCTION

Absolute photoionization cross sections for a large number of ions are essential for astrophysical modelling, which so far relies largely on theoretical predictions from in particular the Opacity Project [1–3], and its later extension, the Iron Project [4,5], and also the OPAL project [6]. Among the most wanted experimental cross sections is the photoionization cross section for the 2s^{2}2p ^{2}P ground state of the C$^+$ ion, which we have measured in this experiment [7].

EXPERIMENT

A merged photon-ion beam apparatus has been built at the ASTRID storage ring at University of Aarhus [8], based on a 2 keV ion accelerator and a photon beam from a newly established undulator beamline. The ion and photon beam are merged over a distance of 51 cm and the production of C^{++} ions is recorded by an electron multiplier [7,8]. The intensity of the primary ion beam is monitored by a Faraday cup and the higher charged ions by Johnston multipliers, which are calibrated absolutely with a Keithley 642 electrometer. The photon flux is measured

CP500, *The Physics of Electronic and Atomic Collisions*, edited by Y. Itikawa, et al.
© 2000 American Institute of Physics 1-56396-777-4/00/$17.00

by an Al photodiode. A double photoionization chamber containing He is used to calibrate the photodiode and to determine the amount of second order flux. The beam profiles in the interaction region are frequently scanned to secure a proper determination of the overlap between the two beams. We have 5 sets of x- and y-scanners, each with a width of 0.2 mm and separated by 12 cm. Overlap integrals of the ion- and the photon-beams are measured both one-dimensional for x and y and two-dimensional for xy to test variation with time. A precision of less than 2 to 4% is aimed at for all individual sources of error, resulting in a total precision of 10% for the absolute cross sections. Our photon flux is around 4×10^{11}/s and the typical ion current 55 nA. Compared to previous measurements on other ions, e.g. K$^+$ [9], we have gained a factor 1000 in peak to background ratio [7], primarily because of higher photon flux from the undulator. These improvements have made it possible to measure the low photoionization cross section for C$^+$.

RESULTS

To demonstrate the measurement of beam-overlaps Fig. 1 shows the x- and y-overlaps for the first, central and last scanners, separated by 24 cm, for the ion beam at the top and for the photon beam at the bottom for 24.6 eV photon energy.

Fig. 2 shows the two-dimensional xy-overlap at the same photon energy for the ion beam at the top and the photon beam at the bottom measured simultaneously. Fig. 2 shows left the raw result in 9×12 cells, used for getting the overlap integral by multiplication, and right a contour plot of the same data to get an impression of the distributions. An integration along the 51 cm long interaction length in most cases delivered a two-dimensional integral which was within 1% of the product of the two one-dimensional x- and y-overlaps integrated in the same way. Therefore one-dimensional scanning was used before and after measuring spectra changing the photon energy, taking often several hours, with 30-200 sec per point. It has been good to better than 1% to use only the first, central and last scanners instead of all 5 sets for integration over the interaction region.

The spectrum has been recorded from the threshold at 24 eV to 31 eV [7] and also higher energy ranges up to 105 eV have been measured. Two regions up to 37.5 eV are shown in Fig. 3. Above the first ionization threshold at 24.4 eV the photoionization cross section exhibits series of resonance lines converging to higher ionization limits. The first lines at 24.4 eV to 25.2 eV are due to a metastable beam fraction 2s2p^2 ^{4}P with excitation of a 2p electron to levels 7d, 8d, 9d and so on followed by autoionization. The relative intensity of these lines depends on how the ion source is operated and by running the ion source at a high pressure the fraction of metastable C$^+$ ions is kept down to 3% [7]. The dominant autoionization resonance lines between 26.5 and 30 eV come from the population of the 2s2p(^{3}P)np doublet levels. In addition weaker lines are observed leading to 2s2p(^{1}P)np doublet levels between 28.8 and 37 eV, with the excited electron in the states 4p, 5p, 6p, 7p or 8p according to a quantum defect calculation we have performed.

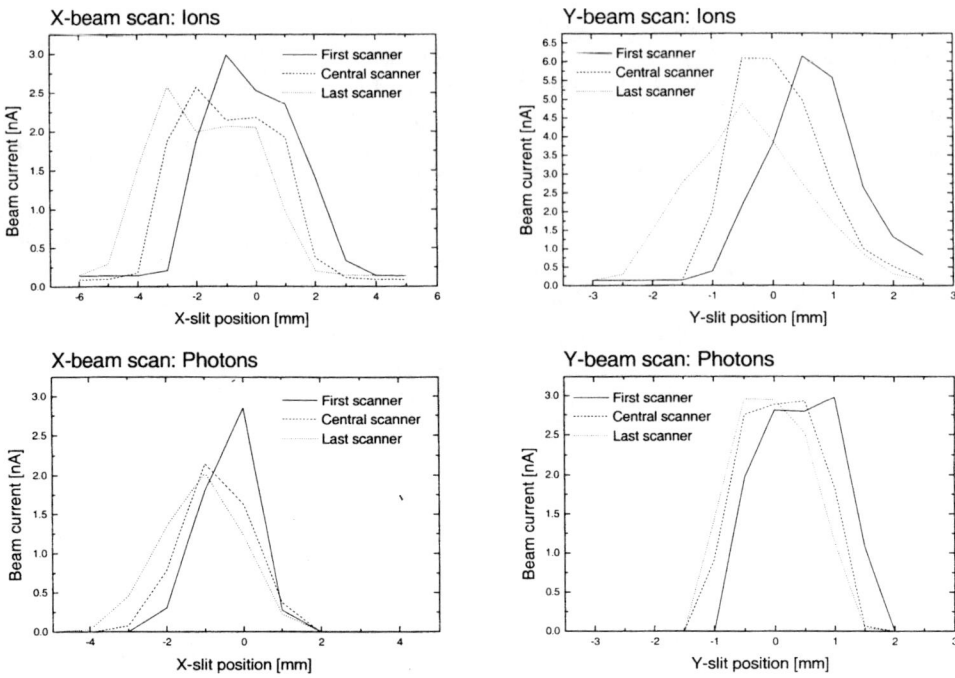

FIGURE 1. One-dimensional beam profiles for ions (top) and photons (bottom) in x- and y-direction measured with 3 sets of scanners, used for determination of the overlap of the two beams.

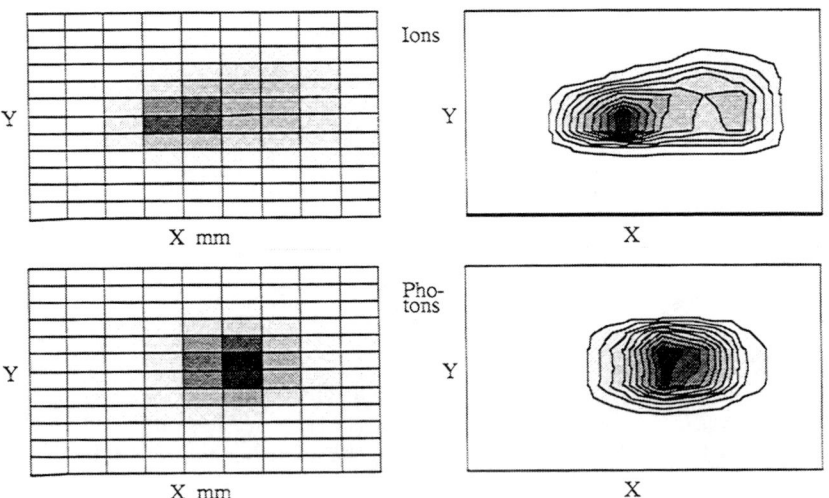

FIGURE 2. Two-dimensional xy-beam-profiles at the central scanner corresponding to Fig. 1. The overlap integral is determined from the two sets of 9×12 cells shown to the left. For X a cell is 1 mm and for Y 0.5 mm. Graphically derived contour plots are shown to the right.

DISCUSSION

Our experimental absolute photoionization cross section for C^+ has been compared with theoretical predictions. In the Opacity Project Yu Yan and Seaton [3] made calculations in the close-coupling approximation using the R-matrix method, and in the later Iron Project a larger basis set was used with the same technique [10,11]. There is good agreement between these two calculations except for the energy positions of some lines. In Fig. 3 our data are compared with the prediction from the Iron Project [10] and it is seen that there is good agreement in continuum level just after the threshold at 24 eV, but the experimental cross section falls grad-

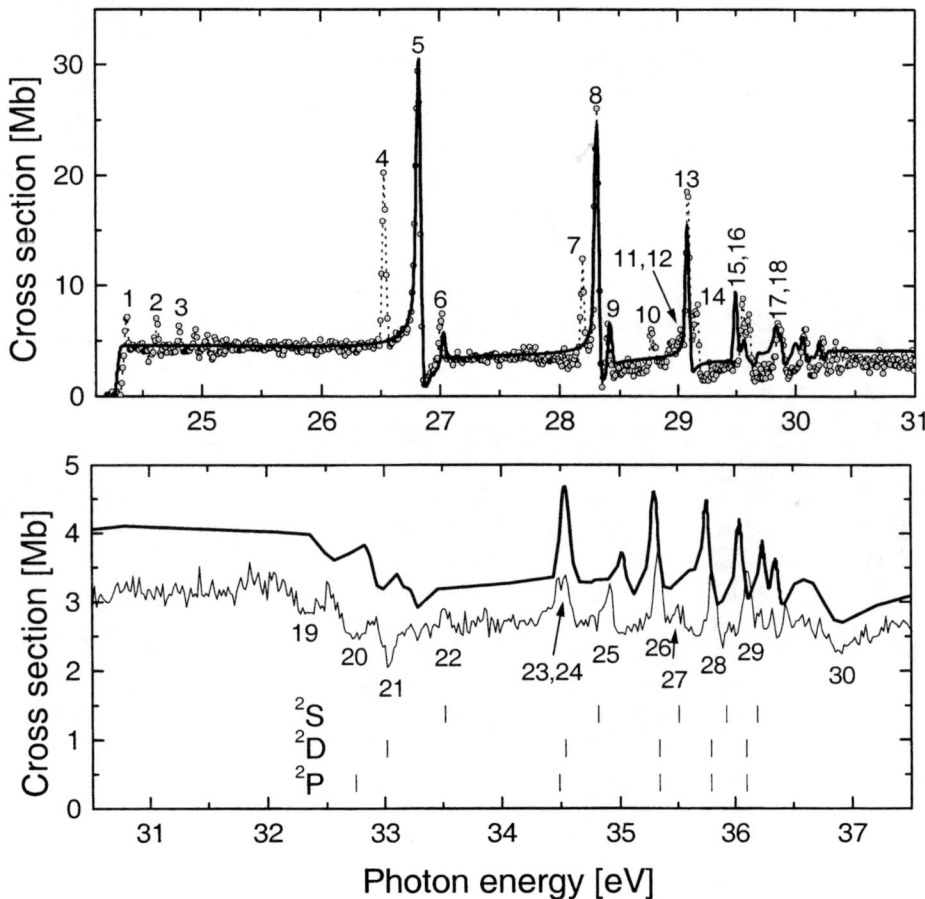

FIGURE 3. Photoionization cross section in Mb of C^+ ions. Present experimental data: dotted line and circles in upper and thin line in lower part. Theoretical data by Nahar [10]: thick line. The theoretical data are folded with a Gaussian of FWHM 35 meV in the upper part and 55 meV in the lower part.

ually to become 25% lower than the theoretical value at 31 eV and only 13% lower at 37 eV. In the region 26-30 eV transitions to $2s2p(^3P)np$ ^2D, ^2P and ^2S series are identified with n equal to 4 to 8. However the theoretical calculations miss the ^2P states (peaks No 4 and 7) which can not autoionize via Coulomb forces. Our explanation is that they may decay relativistically, e.g. via the spin-orbit interaction, or they may decay by radiation to a lower state, which can autoionize. ^2P states have also been observed in absorption measurements [12]. In the region around 29 eV lines are observed which are difficult to correlate with the published theories. We have made a configuration interaction calculation using [13], which identifies peaks No 10-14 by a strong mixing of $2s2p(^3P)6p$ and $2s2p(^1P)3p$ states [7].

In addition to our present C^+ results [7] we have measured the photoionization cross section of K^+ [8] which has a higher continuum value and corresponds well with earlier experimental data [9] but additional resonances are observed. Also Mg^+ has been measured and the study of other ions has been started and will be continued, so a reliable comparison can be made with theoretical calculations, especially for ions of primary astrophysical interest.

ACKNOWLEDGMENTS

The work was supported by the Aarhus Center for Atomic Physics, funded by the Danish National Research Foundation and by a Human Capital Mobility grant from the European Union.

REFERENCES

1. *The Opacity Project*, Vol. 1, Bristol: Institute of Physics Publishing, 1995.
2. Yu Yan, Taylor, K.T., and Seaton, M.J., *J. Phys. B* **20**, 6399–6408 (1987).
3. Yu Yan and Seaton, M.J. *J. Phys. B* **20**, 6409–6429 (1987).
4. Hummer, D.G., Berrington, K.A., Eissner, W., Pradhan, A.K., Saraph, H.E., and Tully, J.A., *Astron. Astrophys.* **279**, 298–309 (1993).
5. Nahar, S.N., and Pradhan, A.K., *Phys. Rev. A* **49**, 1816–1835 (1994).
6. Rogers, F.J., and Iglesias, C.A., *Science* **263**, 50–55 (1994).
7. Kjeldsen, H., Folkmann, F., Hansen, J.E., Knudsen, H., Rasmussen, M.S., West, J.B., and Andersen, T., *Astrophys. J. Lett.* (1999) in press.
8. Kjeldsen, H., Folkmann, F., Knudsen, H., Rasmussen, M.S., West, J.B., and Andersen, T., *J. Phys. B* (1999) in press.
9. Peart, B., and Lyon, I.C., *J. Phys. B.* **20**, L673–L675 (1987).
10. Nahar, S.N., *Astrophys. J. Suppl. Ser.* **101**, 423–434 (1995).
11. Nahar, S.N., and Pradhan, A.K., *Astrophys. J. Suppl. Ser.* **111**, 339–355 (1997).
12. Nicolosi, P., and Villoresi, P., *Phys. Rev. A* **58**, 4985–4988 (1998).
13. Cowan, R.D., *The Theory of Atomic Structure and Spectra*, Berkeley: University of California Press, 1981, Chapters 8 and 16.

Nuclear motion and symmetry breaking of core-excited polyatomic molecules — Anisotropic fragmentation of CF_4 following F 1s photoabsorption[*]

K. Ueda[†], Y. Muramatsu[†], Y. Shimizu[†], H. Chiba[†], K. Amano[†],
Y. Sato[†], and H. Nakamatsu[‡]

[†]*Research Institute for Scientific Measurements, Tohoku University, Sendai 980-8577, Japan*
[‡]*Institute for Chemical Research, Kyoto University, Uji, Kyoto 611-0011, Japan*

Abstract. Angle-resolved ion yield spectroscopy revealed that ionic fragmentation of a T_d symmetry molecule CF_4 is anisotropic following F 1s photoabsorption. The nuclear motion in the Auger-final state resulting in this anisotropic fragmentation is a consequence of asymmetric nuclear motion or *symmetry breaking* in the core-excited state, caused by vibronic coupling.

INTRODUCTION

When a core electron of a molecule is excited, the core hole relaxes *via* Auger electron emission and ionic fragmentation proceeds along the repulsive potential surface of the Auger-final state. The lifetime for the core-excited state of the light atom (C, N, O, F) is $\sim 10^{-14}$ s and thus nuclear motion in the molecular core-excited state can proceed before the Auger decay. A particularly interesting case is core excitation of the equivalent atom in a symmetric molecule, such as O 1s excitation in CO_2 [1–3] and F 1s excitation in SF_6 [4]. In this case vibronic coupling induces asymmetric nuclear motion in the core-excited state and thus the molecular symmetry is lowered. Then the question arises whether the asymmetric nuclear motion or the *symmetry breaking* in the core-excited state affects molecular dissociation, i.e., nuclear motion in the Auger-final state.

In our recent study on F 1s excitation of highly symmetric molecules SF_6 (O_h) [4] and CF_4 (T_d) [5], we found anisotropic angular distributions of F^+ fragment ions. In the present paper, we discuss this anisotropic fragmentation, due to asymmetric nuclear motion in the Auger-final state, in connection with the nuclear motion induced in the core-excited state, referring to our latest work on CF_4 [5]. The details of the experiment were described in our previous papers [4,5].

CP500, *The Physics of Electronic and Atomic Collisions*, edited by Y. Itikawa, et al.

RESULTS AND DISCUSSION

Figure 1 (a) shows ion yield spectra of CF_4 in the F $1s$ excitation region, where only the energetic fragment ions of kinetic energy ≥ 5 eV were detected in the directions parallel $(I(0°))$ and perpendicular $(I(90°))$ to the linear polarization axis of the incident light. It is clear that the ionic fragmentation is anisotropic at the resonance below the F $1s$ ionization threshold. The photon band-pass was ~ 1 eV. As we discussed previously [4,5], we consider that anisotropic fragmentation of this kind can be interpreted as preferential bond rupture at the F $1s$ core-hole site.

In order to confirm this localized bond-rupture picture, we have performed first-principles calculations for the F $1s$ absorption spectrum within the framework of the DV-Xα molecular-orbital method [6], using a package program SCAT with extensions for pseudo-states [7]. Here we have adopted the lower C_{3v} symmetry for the electronic structure, together with a ground state geometry of T_d symmetry (C–F bond length of 1.32 Å). Numerical basis functions consisting of core, valence, and continuum-like wave functions provide self-consistent charge distribution and square integrable (L^2) discretized wave functions [8] for pseudo-states. As a basis set we included atomic orbitals of $1s - 4f$ for the C atom, $1s - 4f$ for each F atom and furthermore s-, p-, d-, and f-type continuum wave functions with limited radii and energies between $0 - 1.2$ Hr for each atom. To obtain the spectra, the transition intensities to the discrete levels with the L^2 basis were replaced by Gaussian curves with a width of 3.2 eV. Dipole photoabsorption of the C_{3v} molecule in the A_1 ground state leads to the A_1 and E excited states. The dipole moment for the $A_1 \rightarrow A_1$ transition is parallel to the C_3 axis, whereas that for the $A_1 \rightarrow E$ transition is perpendicular to the C_3 axis. Figure 1 (b) shows symmetry-resolved absorption cross sections σ_{A_1} and $\sigma_E/2$, calculated separately for the $A_1 \rightarrow A_1$ and $A_1 \rightarrow E$ transitions. A small constant baseline, which corresponds to direct ionization from the valence orbitals, was added in such a way that the baseline contribution agreed reasonably well with the experimental spectrum.

The experimental spectra $I(0°)$ and $I(90°)$ clearly show some correlation with the calculated symmetry-resolved spectra σ_{A_1} and $\sigma_E/2$: $I(0°)$ is more intense and peaks at lower energy than $I(90°)$ and σ_{A_1} is more intense and peaks at lower energy than $\sigma_E/2$. This correlation is evidence for preferential bond rupture at the F $1s$ core-hole site, because the ion is preferentially detected in the 0° direction when the C–F bond with a F $1s$ hole is preferentially oriented in that direction.

In order to quantify this correlation, we consider the anisotropy parameter β for the fragment ions. If the incident light is completely linearly polarized, the angular distribution of the fragment ions relative to the incident light polarization axis can be described by the well-known formula:

$$I(\theta) = \frac{\sigma}{4\pi}[1 + \beta P_2(\cos\theta)] , \qquad (1)$$

where θ is the angle between the polarization axis and the direction of the ion detection and P_2 is the second-order Legendre polynomial. From Eq. (1), we have

$$\beta = \frac{2I(0°) - 2I(90°)}{I(0°) + 2I(90°)} \, . \tag{2}$$

The β values obtained using Eq. (2) are plotted in Fig. 1 (c).

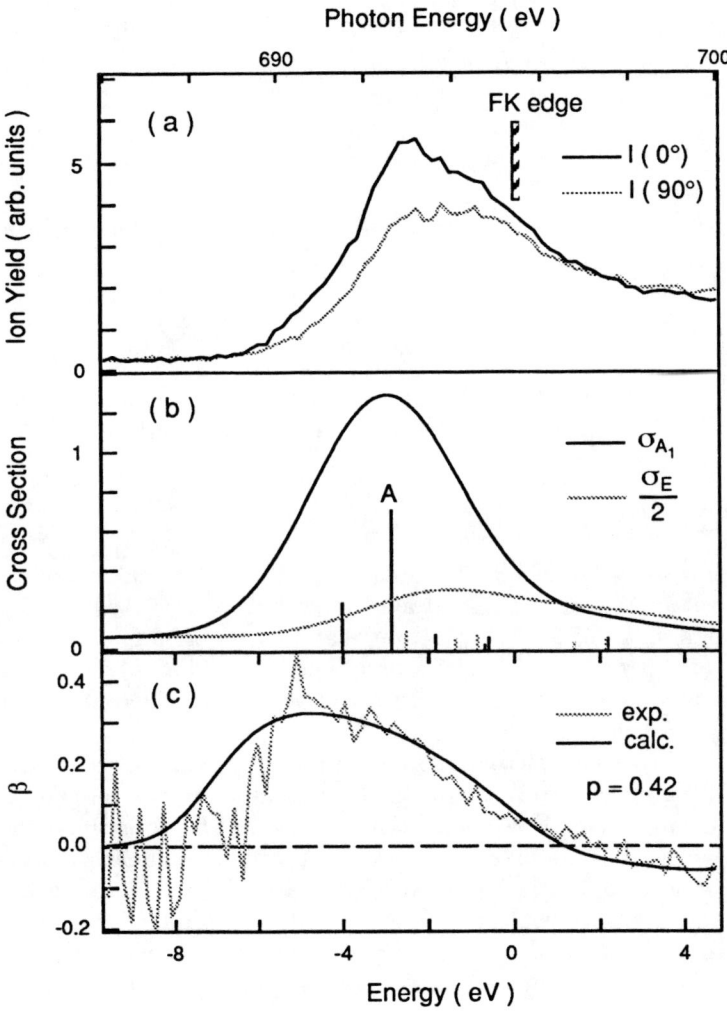

FIGURE 1. (a) Ion yield spectra of CF_4 in the F $1s$ excitation region. Ions with kinetic energy ≥ 5 eV are detected at $0°$ and $90°$ relative to the incident light polarization axis. (b) Calculated symmetry-resolved absorption cross sections σ_{A_1} and $\sigma_E/2$ of CF_4. (c) Measured and calculated ion anisotropy parameters β.

The anisotropy parameter β, on the other hand, is directly related to the angle χ between the dipole transition moment μ and the direction of ion detection:

$$\beta = 2P_2(\cos \chi) .\tag{3}$$

We denote here the probability that the ion is ejected along the C_3 axis on which F $1s$ hole was created as p and assume that the probability that the ion is ejected along one of the other three C–F axes is $(1 - p)/3$. Then, from Eq. (3), the ion anisotropy parameter β_{A_1} for the $A_1 \rightarrow A_1$ transition can be expressed as

$$\beta_{A_1} = \frac{8}{3}p - \frac{2}{3} ,\tag{4}$$

whereas the ion anisotropy parameter β_E for the $A_1 \rightarrow E$ transition can be expressed as

$$\beta_e = \frac{1}{3} - \frac{4}{3}p .\tag{5}$$

Where the $A_1 \rightarrow A_1$ and $A_1 \rightarrow E$ cross sections overlap, as seen in Fig. 1 (b), the anisotropy parameter β can be expressed as a weighted average:

$$\beta = \frac{\beta_{A_1}\sigma_{A_1} + \beta_E\sigma_E}{\sigma_{A_1} + \sigma_E} .\tag{6}$$

Substituting σ_{A_1} and σ_E of Fig. 1 (b) into Eq. (6) we have calculated the β curve and plotted it in Fig. 1 (c). Good agreement with the experimental β curve is obtained for a value of $p = 0.42$. Note that this probability is about twice the probability $(1 - p)/3$ for rupture of a bond not localized at a core-hole site. Thus the value of $p \sim 0.4$ clearly and quantitatively confirms preferential bond rupture at the F $1s$ core-hole site.

The localized core-hole picture adopted in the above analysis requires some clarification. Following core excitation of the equivalent atom, the core hole is spatially coherently created along the polarization direction of the incident light; the core hole is not localized at the time of excitation. Asymmetric nuclear motion resulting from the vibronic coupling in the core-excited state may however reduce this spatial coherence and lower the molecular symmetry, for example from T_d to C_{3v} in our CF_4 case. In this way dynamical core-hole localization can be realized [1–3] and the core-hole localization picture works well, as it does for O $1s$ excitation in the CO_2 molecule [1–3].

On the basis of this dynamical core-hole localization picture and our finding of preferential bond rupture at the localized F $1s$ core-hole site, we draw the following conclusion: asymmetric nuclear motion occurs *via* vibronic coupling in the F $1s$ core-excited states of CF_4. This nuclear motion is in the t_2 mode and thus the molecular symmetry is lowered; the direction of this nuclear motion is correlated to the direction of the polarization axis of the incident light. Then this nuclear motion is transferred to the Auger-final states, in accordance with the classical Franck-Condon principle, and results in anisotropic fragmentation. In other words, the anisotropic fragmentation (i.e., the asymmetric nuclear motion in the Auger-final states in a specific direction) is brought about by coupling with the asymmetric nuclear motion in the core-excited state.

It should however be noticed that, even though dynamical symmetry breaking does not occur in the core-excited state, the core-hole polarization can be transferred to the valence-hole polarization *via* Auger decay. Asymmetric nuclear motion, due to vibronic coupling in the Auger-final state, can then result in anisotropic ionic fragmentation. Strictly speaking, experimentally we cannot distinguish symmetry breaking in the core-excited state from that in the Auger-final state.

ACKNOWLEDGMENTS

This experiment was carried out with the approval of the Photon Factory Advisory Committee (Proposal Nos. 95G178) and supported in part by a Grant-in-aid for Scientific Research from the Japanese Ministry of Education, Science, Sports and Culture and by the Matsuo Foundation. The authors are grateful to J. B. West for critical reading of this manuscirpt, S. Tanaka for helpful discussion, and T. Hayaishi and the staff of the Photon Factory for their help in the course of the experiments.

REFERENCES

* Portions of the material presented in this paper have already appeared in J. Phys. B **32**, L213 (1999).

1. W. Domcke and L. S. Cederbaum, Chem. Phys. **25, 189** (1977).
2. L. S. Cederbaum, J. Chem. Phys. **103, 563** (1995).
3. A. Cesar, F. Gel'mukhanov, Y. Luo, H. Ågren, P. Skytt, P. Glans, J. Guo, K. Gunnelin, and J. Nordgren, J. Chem. Phys. **106, 3439** (1997).
4. K. Ueda, Y. Shimizu, M. Okunishi, K. Ohmori, J. B. West, Y. Sato, T. Hayaishi, H. Nakamatsu, T. Mukoyama, Phys. Rev. Lett. **79, 3371** (1997).
5. Y. Muramatsu, K. Ueda, Y. Shimizu, H. Chiba, K. Amano, Y. Sato, and H. Nakamatsu, J. Phys. B **32**, L213 (1999).
6. H. Adachi, M. Tsukada, and C. Satoko, J. Phys. Soc. Jpn. **45, 875** (1978).
7. H. Nakamatsu, Chem. Phys. **200**, 49 (1995).
8. W. P. Reinhardt, Comp. Phys. Commun. **17**, 1 (1979).

Three-Body Decay of Laser-Excited Triatomic Hydrogen Molecules[1]

U. Müller, M. Beckert, M. Braun, and H. Helm

Universität Freiburg, Fakultät für Physik,
Hermann-Herder-Str. 3, D-79104 Freiburg, Germany

Abstract. We investigate the photofragmentation of H_3 molecules in a fast-beam experiment. The H_3 molecules are laser-prepared in a single rovibrational state. Neutral fragments are detected in coincidence by a time- and position-sensitive multihit-detector. Triple coincidences from breakup into three H(1s) atoms have been detected for the first time. For each triple-hit, the momentum vectors of the three H(1s) fragments are determined. The 6-fold differential cross section of the fragment momentum components in the center-of-mass frame is then projected to distributions of the kinetic energy release, and the configuration of the fragment momenta relative to each other. The distributions of the fragment momenta are found to be highly structured and to depend sensitively on the H_3 initial electronic state.

INTRODUCTION

The triatomic hydrogen molecule has attracted the interest of physicists and chemists for many decades. Rydberg states of the H_3 molecule were first discovered by Herzberg's group [1]. They are embedded in the repulsive potential energy surface of the 2p $^2E'$ ground state. The lowest rotational level of the 2p $^2A_2''$ electronic state is the only metastable state of H_3 and can be produced in a fast beam by charge transfer of H_3^+ in Cs [2]. Starting from this platform, we laser-prepare the vibrationless 3s $^2A_1'$(N=1,K=0) or 3d $^2E''$(N=1,G=0,R=1) Rydberg states of H_3 which are known to decay in radiative and predissociative channels within 3 to 10 ns [2]. The two-body decay of these states has been studied experimentally [2,3] as well as theoretically [4,5]. To a certain extent, insight into the three-body decay of H_3 can be gained by measuring two of the three fragments in coincidence [6]. The objective of this paper is to detect triple-coincidences from photofragmentation into H(1s)+H(1s)+H(1s), and to achieve a kinematically complete analysis of the final state.

---

[1] This work is made possible by the generous support by Deutsche Forschungsgemeinschaft (SFB 276, TP C13).

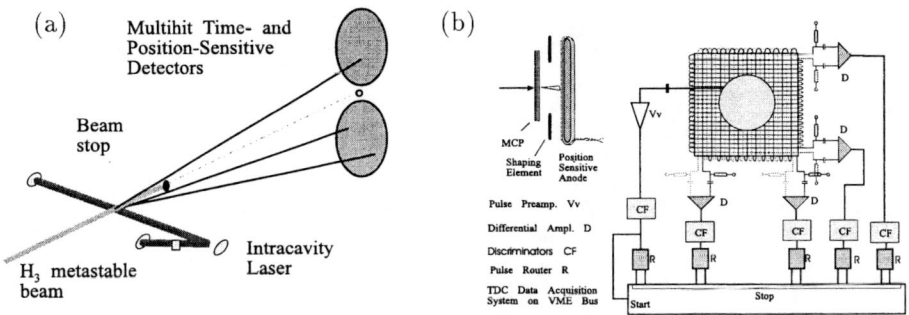

FIGURE 1. (a) Fast Neutral Beam Photofragment Spectrometer. (b) Time- and position sensitive delayline detector.

APPARATUS

The Freiburg photofragmentation spectrometer which is an advanced version of earlier forms [7,8] is shown in Fig. 1(a). H_3^+ ions are created in a hollow cathode discharge in H_2, accelerated to 3 keV, mass selected, and neutralized in cesium. The unreacted ions are removed by an electric field, and the products of dissociative charge transfer are stopped by an aperture. The well-collimated beam of neutral H_3 molecules is then crossed by an intracavity dye laser beam. After a free-flight of $L = 1.5$ m, the photofragments are detected in coincidence by a time- and position-sensitive multihit-detector which consists of two identical units shown in Fig. 1(b). The units consist of pairs of multichannel plates followed by position-sensitive delayline anodes [9]. The hit position on the detector surface is coded into the pulse arrival times at the ends of two waveguides, one for each cartesian coordinate. Pulse-routers distribute consecutive signals from each waveguide end into separate channels of time-to-digital converters (TDC). Each unit is able to detect consecutive fragments with a dead time as low as 10 ns. Additionally, the arrival time differences between events on both detector sides is determined. The readout and pre-analysis of the data is performed by a dedicated processor.

RESULTS

A double-hit on one detector unit and a single-hit on the other unit appearing within a 400 ns wide time window are recognized as a possible three-body event. From the binary data delivered by the TDC's, the six cartesian coordinates of the three impact positions in the detector plane and the two arrival time differences are determined. Redundant information on the arrival time differences is used to reject false coincidences. For each three-body event, the fragment momenta $m\,\mathbf{u}_1$, $m\,\mathbf{u}_2$, and $m\,\mathbf{u}_3$ in the c.m. frame (see Fig.2) are calculated making use of the well-known primary beam energy, the parent mass, and the flight distance L. Momentum conservation implies that the three vectors $\mathbf{u}_i$ have six independent components.

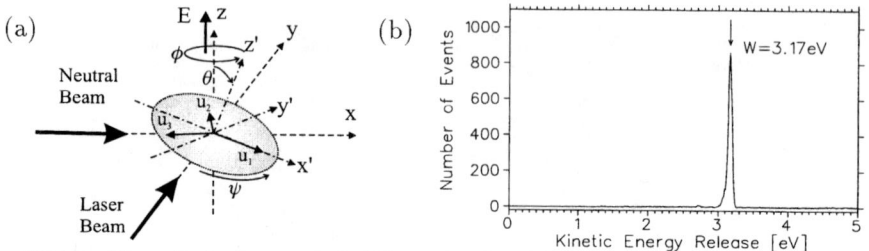

FIGURE 2. Three-body decay of the H$_3$ molecule. (a): Parameterization of the fragment momentum vectors. (b): Kinetic energy release spectrum of the 3s ^{2}A$'_1$ state.

In this way, we determine the 6-fold differential cross section, for the three-body photofragmentation.

To gain insight into the fragmentation pattern, we consider suitable projections of this high-dimensional set of information. We introduce six parameters which uniquely describe the three fragment momentum vectors. The $\mathbf{u}_i$ are contained in the (x', y')-plane of a new coordinate system (x', y', z') which is defined for each event, as shown Fig. 2a. Three Euler angles (ψ, θ, ϕ) describe the orientation of the (x', y', z')-system within the laboratory reference system (x, y, z). For the remaining three parameters describing the arrangement of the $\mathbf{u}_i$ in the (x', y')-plane we may choose the absolute momenta or the fragment energies $\epsilon_i = m\mathbf{u}_i^2/2$. We use the total c-m kinetic energy $W = \sum \epsilon_i$ and two parameters showing the correlation among the fragment momenta. The spectra of W pose a stringent test on our data-acquisition and -reduction procedures since the energies of the laser-excited initial states above the three-body limit are known. Figure 2(b) shows the kinetic energy release spectrum from triple coincidence data of the H$_3$ 3s ^{2}A$'_1$(N=1,K=0) initial state. A narrow peak appears at 3.17 eV. The excellent agreement between the measured value of W and the previously known state energy (3.155 eV above the three-body limit [2]) confirms the quality of the time- and position calibration of the detector. No continuous background is found in the W-spectra of the three-body decay. This shows, that radiative transitions from the 3s ^{2}A$'_1$ and 3d ^{2}E$''$ states to the repulsive ground state lead to H+H$_2$ fragment pairs as the only exit channel [3].

To show the correlation among the fragment momenta within the (x', y') plane, we use a Dalitz plot [10]. We plot for each event $(\epsilon_3/W - 1/3)$ vs. $((\epsilon_2 - \epsilon_1)/(W \cdot \sqrt{3}))$. Energy and momentum conservation require that the data points in this plot lie inside a circle with radius $1/3$, centered at the origin. In a Dalitz plot, the phase space density is conserved which means that a fragmentation process with a matrix element independent of the configuration leads to a homogeneous point density. Preferred fragmention pathways can be immediately recognized in such a plot. In Fig. 3a and b, the triple-coincident events following three-body breakup of the laser-excited H$_3$ 3s ^{2}A$'_1$(N=1,K=0) and 3d ^{2}E$''$(N=1,G=0,R=1) states are shown as Dalitz plots. In Fig. 3c, the correspondence between the configuration of the fragment momenta and the location in the plot is visualized. The six-fold sym-

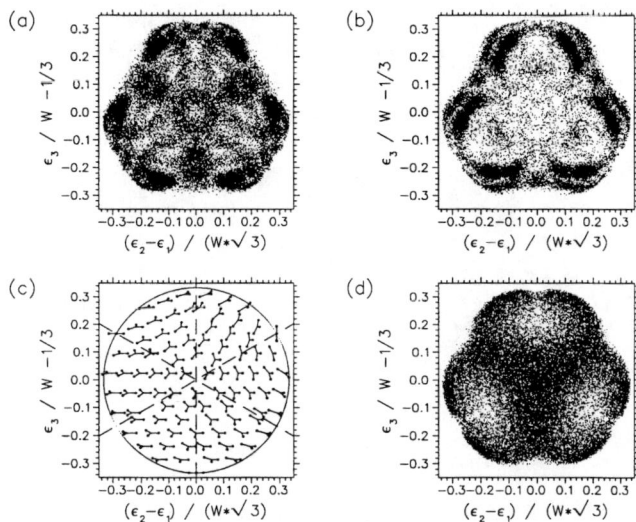

FIGURE 3. Dalitz plot of the raw data from three-body decay of the 3s $^2A_1'$ (a) and 3d $^2E''$ (b) initial states of H_3. In (c), the correspondence between the location in the plot and the fragmentation configuration is indicated. In (d) a Monte-Carlo simulation for a random distribution of fragmentation configuation in the phase space is shown, demonstrating the effect of the geometric detector collection efficiency.

metry in Fig. 3 result from the equal masses of the indistinguishable fragments. In order to understand the meaning of the very pronounced islands of correlation appearing in the experimental data in Fig. 3a and b, the limited collection efficiency of the detector has to be discussed first. Currently, detection of fragment triples where one of the momenta is close to zero (linear configuration) is excluded. Also, fragment hits which are close in time as well as in space ($H+H_2$ configuration) are suppressed due to the finite pulse pair resolution. The geometric and electronic detector collection efficiency was determined in a Monte Carlo simulation by generating a uniform distribution of fragmentation configurations and calculating the fragment propagation to the detector. A Monte-Carlo simulation of the detector response is shown in Fig. 3d. The collection efficiency vanishes only for the linear and the $H+H_2$ configurations on the circle boundary. The remainder of the Dalitz plot area shows a smooth variation of the detection efficiency.

As a consequence, the islands of high point density in Fig. 3 a and b are associated with the correlation of the fragment momenta produced by the dissociation process itself. Despite the high symmetry (D_{3h}) of the initial states, asymmetric fragmentation configurations are very much preferred in finding a path into the 3-particle continuum. Neither the totally symmetric configuration (center of the plot) nor isosceles configurations (dashed lines in Fig. 3 c) show emphasized population. It may surprise, that the preferred fragmentation configurations sensitively depend on the initial state, although the absolute energies of the states investigated here differ as little as 75 meV and the nuclear equilibrium configuration for each initial state is extremely close to that of the vibrationless H_3^+ ion in its ground electronic state. The striking difference of the final state distributions must reflect the different coupling mechanisms between the initial state and the two sheets of the repulsive ground state potential energy surface. The breakdown of the Born-Oppenheimer approximation is mediated by the zero-point motion in the degenerate vibration for the 3s $^2A_1'$ state, and by the rotational tumbling motion in case of the 3d $^2E''$ state [6]. While these processes mediate the first entry of the quasi-bound system into the continuum, a series of avoided crossings between the upper sheet of the repulsive ground state surface and the s- and d-Rydberg states of $^2\Sigma_g^+$ symmetry in linear geometry [11] governs the further evolution of the continuum state.

CONCLUSION

We have investigated the fragmentation of laser-prepared molecule H_3 molecules and achieved a kinematically complete analysis of the three-body decay into H(1s)+H(1s)+H(1s). The detailed maps of momentum partition among the three fragments obtained in our experiment pose a significant challenge to quantum-structure and quantum-dynamics calculations. The availability of such calculations might also help to understand the process of dissociative recombination of H_3^+ with slow electrons [12].

REFERENCES

1. G. Herzberg, J. T. Hougen, and J. K. G. Watson, Can. J. Phys. **60**, 1261 (1982)
2. P. C. Cosby and H. Helm, Phys. Rev. Lett **61**, 298, (1988)
3. U. Müller and P. C. Cosby, J. Chem. Phys. **105**, 3532, (1996)
4. A. E. Orel and K. C. Kulander, J. Chem. Phys. **91**, 6086 (1989)
5. I. F. Schneider and A. E. Orel, J. Chem. Phys., in press (1999)
6. U. Müller and P. C. Cosby, Phys. Rev. A **59**, 3632 (1999)
7. D. P. de Bruin and J. Los, Rev. Sci. Instr. **53**, 1020 (1982)
8. H. Helm and P. C. Cosby, J. Chem. Phys. **86**, 6813 (1987)
9. O. Jagutzki, V. Mergel, K. Ullmann-Pfleger, L. Spielberger, U. Meyer, and H. Schmidt-Böcking, SPIE Proc. "Imaging Spectroscopy IV", San Diego (1998), in print
10. R. H. Dalitz, Philos. Mag. 44, 1068 (1953); Ann. Rev. Nucl. Science **13**, 339 (1963)
11. I. D. Petsalakis, G. Theodorakopoulos, and J. S. Wright, J. Chem. Phys. **89**, 6850 (1988)
12. J. B. A. Mitchell, J. L. Forand, C. T. Ng, D. P. Levac, R. E. Mitchell, P. M. Mul, W. Claeys, A. Sen, and J. Wm. McGowan, Phys. Rev. Lett. **51**, 885 (1983); S. Datz, G. Sundström, Ch. Biedermann, L. Broström, H. Danared, S. Mannervik, J. R. Mowat, and M. Larsson, Phys. Rev. Lett. **74**, 896 (1995)

R-Matrix Study of Ionization in Barium
Via Two-Photon Interfering Routes

M. Aymar*, E. Luc-Koenig*, J.M. Lecomte*, M. Millet*, A. Lyras†

* *Laboratoire Aimé Cotton, CNRS II, Bât 505, 91405 Orsay Cedex, France*
† *Atomic and Molecular Physics Laboratory, University of Ioannina, 45110 Ioannina, Greece.*

Abstract. A quantitative analysis of part of the experimental data reported by Wang, Chen and Elliott [1,3] who studied in barium coherent control through two-color resonant interfering paths is reported. Dynamics of the two-color photoionization process, described as an adiabatic process in the rotating wave approximation, is governed by the coherent excitation of the 6s6p and 6s7p $^1P_1$ intermediate states. Interference effects are found to play a minor role. The required atomic parameters are obtained from a theoretical approach based on a combination of jj-coupled eigenchannel R-matrix and Multichannel Quantum Defect Theory.

Controlling branching ratios of various photofragmentation products by coupling through different laser excitation paths the same initial and final states has been extensively investigated [4,5]. Using a scheme, proposed by Pratt [6], insensitive to the laser phases, Wang *et al.* [1-3] studied coherent control of the branching ratios for photoionization in three different ionic states of atomic barium. The two-photon interactions used in that experiment are shown in Figure 1. Both routes correspond to absorption from the ground state of one visible laser beam of frequency ω_1 and intensity I_1 and one UV laser beam of frequency ω_2 and intensity I_2. Both lasers are pulsed and the linear polarizations of the two laser fields are either parallel or perpendicular. The two coherent interfering pathways are resonantly enhanced by either the 6s6p or the 6s7p $^1P_1$ level and populate final states with energy E_r, located above the Ba^+ $5d_{5/2}$ threshold, in an energy range including the 6p7p autoionizing resonances. The recorded photoionization spectra drawn as functions of the detuning Δ_1 of the visible laser exhibit very asymmetrical line shapes and display variations in the branching ratios for the produced Ba^+ 6s, $5d_{3/2}$ and $5d_{5/2}$ ions.

We report a quantitative analysis of the experimental data reported by Wang *et al.* [1-3] for the perpendicular polarization case.

The dynamical treatment of the coherent ionization process is carried out within the dipole approximation and the rotating wave approximation. We consider first

CP500, *The Physics of Electronic and Atomic Collisions*, edited by Y. Itikawa, et al.

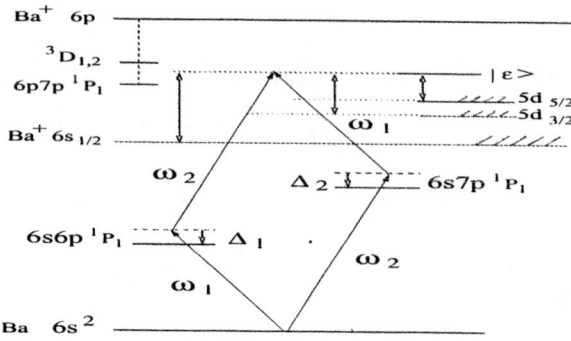

FIGURE 1. Energy level schematic diagram showing the two two-photon two-color ionizing pathways and the relevant final ionic states.

the couplings due to the laser field with ω_1 (and ω_2) frequency between the $6s^2$ ground state and the excited 6s6p $^1P_1$ (and 6s7p $^1P_1$) states. These couplings are characterized by the Rabi frequencies Ω_i with a slow time-dependence due to the pulse shape. We associate to the atomic system in the laser fields a Floquet space [7], where the three Floquet states corresponding to the discrete atomic states are denoted $|i>$. Thus, by diagonalizing, at each time t, the total Hamiltonian built on the $|i>$ basis we obtain three instantaneous adiabatic states $|\bar{i}(t)>$ which are connected to states $|i>$ at $t = 0$. In the adiabatic approximation, only the state $|\bar{0}>$ is populated. This state is unstable and acquires a width $\gamma_{\bar{0}}(t)$ due to its coupling with the continua. The photoionization process is characterized by electric dipole matrix elements D_{iE_rc} coupling the discrete states i to the continuum states c with energy E_r. By considering the different $Ba^+(N_c\ell_cj_c)$ ionic state we define threshold-resolved coherent ionization rates $\gamma_i^{coh}(N_c\ell_cj_c) = \sum_c 2\pi|D_{iE_rc}|^2$ and their sum γ_i^{coh}.

In the adiabatic approximation, the change in the population of the ionic state $(N_c\ell_cj_c)$ between t and $t + \delta t$ is:

$$\frac{\delta\sigma_{N_c\ell_cj_c}}{\delta t} = 2\pi \sum_c |\sum_{i=1}^2 D_{iE_rc}^* <\bar{0}|i>|^2 \exp(-\int_0^t dt'\gamma_{\bar{0}}(t')). \tag{1}$$

The squared term includes the interference effects between both ionization pathways. It can be noticed that the results do not depend on the exact form of the pulse except for pulses with very short duration of rise of time anf fall off.

In addition to the atomic parameters quoted above, the light shifts, the Raman coupling Ω_{21} and the threshold resolved interference term $\gamma_{12}^{coh}(N_c, \ell_c, j_c)$ between $|1>$ and $|2>$ are determined. All these atomic parameters are evaluated with an approach combining the eigenchannel R-matrix method with multichannel quantum defect theory (MQDT) [8]. This approach is also used to describe the odd- and even-parity barium spectra reached at each step of the two-photon ionization process [9]. A model Hamiltonian including spin-orbit terms is used for describing the two valence electrons outside the frozen Ba^{2+} core. Electron correlations are treated within a finite spherical volume V of radius $r_o = 50$ a.u. The wavefunctions of the $6s^2$ $^1S_0$ ground state and of the 6s6p and 6s7p $^1P_1$ intermediate states, which are assumed to be confined within the volume V, are obtained by diagonalization of two-electron Hamiltonian matrices built on large basis sets. The description of the final $J = 0^e, 1^e$ and 2^e states is obtained by using the eigenchannel jj-coupled R-matrix approach in combination with MQDT [9].

Theoretical calculations are performed with a pulse duration $\tau_p = 15$ ns and laser peak intensities $I_1^{max} = 8.15 \times 10^6$ W cm^{-2} and $I_2^{max} = 1 \times 10^6$ W cm^{-2} which correspond to the experimental conditions [1-3]. For such intensities, the light shifts and the Raman coupling Ω_{21} are negligible and one has:

$$\Omega_1 \gg \Omega_2 \gg \gamma_2^{coh} \gg \gamma_1^{coh} \tag{2}$$

The one-photon Rabi frequency $\Omega_1 \approx 5.6$ cm^{-1} strongly couples the $|0>$ and $|1>$ states and the coupling between the $|0>$ and $|2>$ states, associated with the one-photon Rabi frequency $\Omega_2 \approx 0.5$ cm^{-1}, can be considered as a perturbation. Ionization mainly occurs through the 6s7p state. The threshold-resolved ionization rates $\gamma_2^{coh}(N_c \ell_c j_c)$ prevail by a factor ~ 18 on $\gamma_1^{coh}(N_c \ell_c j_c)$.

The time-dependences of the energies and ionization rates for the three adiabatic states are reported in Figure 2, for $\Delta_2 = -0.7$ cm^{-1} and for $\Delta_1 = -5$ cm^{-1} (Fig. 2a and 2b) or $+5$ cm^{-1} (Fig. 2c and 2d). The strong coupling Ω_1 is responsible for the energy-gap between the two adiabatic states $|\bar{0}>$ and $|\bar{1}>$. This corresponds to the dynamical Stark effect. For negative Δ_1- values, the ionization rate of the

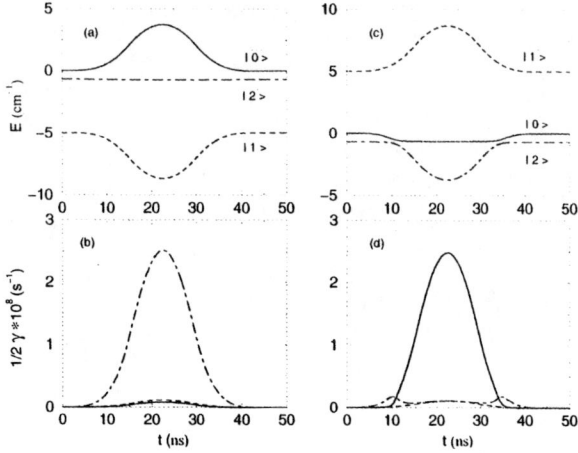

FIGURE 2. Time-dependence (in ns) of the energy positions (in cm^{-1}) and ionization rates (in s^{-1}) for the three adiabatic states for the detuning $\Delta_2 = -0.7$ cm^{-1}. (a) and (b): for $\Delta_1 = -5$ cm^{-1}. (c) and (d): for $\Delta_1 = +5$ cm^{-1}. $|\bar{0}>$: full line. $|\bar{1}>$: dashed line. $|\bar{2}>$: dot-dashed line.

adiabatic state $|\bar{0}>$ remains very small, $\sim \gamma_1^{coh}(t)$, during the pulse. For positive Δ_1- values, the adiabatic state $|\bar{0}>$ undergoes an anticrossing with the ionizing state $|\bar{2}>$ and thus acquires a large ionization rate $\sim \gamma_2^{coh}(t)$.

Figure 3 compares the experimental data to the partial ionization yields calculated in the adiabatic approximation.

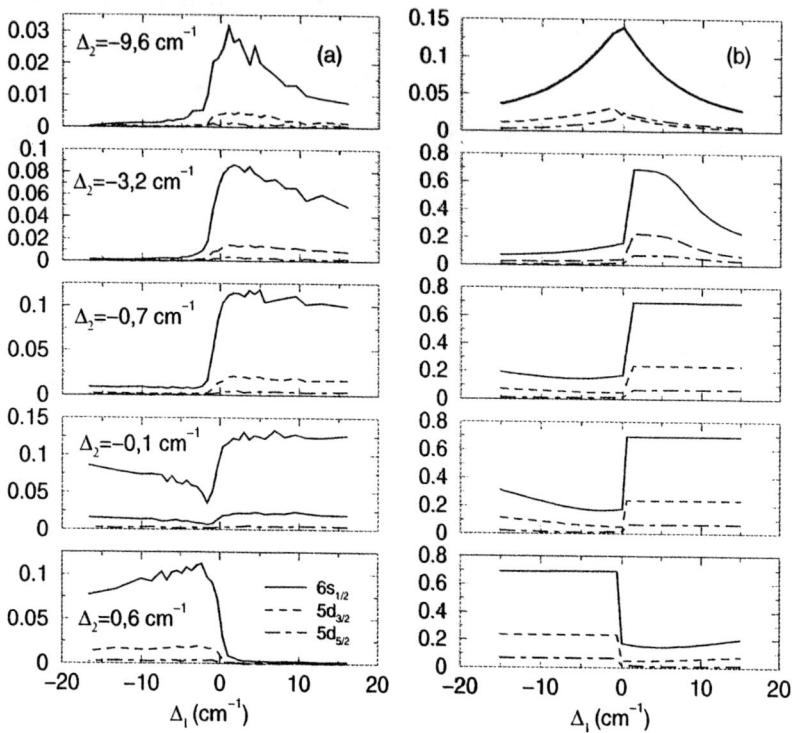

FIGURE 3. Experimental partial ionization yields (a) drawn as functions of the detuning Δ_1 for fixed values of the detuning Δ_2 are compared with results obtained with the adiabatic approximation (b). $6s_{1/2}$ ion yield: full line. $5d_{3/2}$ ion yield: dashed line. $5d_{5/2}$ ion yield: dot-dashed line.

The calculated spectra obtained in the adiabatic approximation are in rather good agreement with the experimental ones. The observed asymmetry of the ionization yields with respect to the different ionization thresholds are relatively well reproduced, except for large detuning Δ_2. For $\Delta_1 \times \Delta_2 < 0$, these yields do not vary significantly with Δ_1 as long as the adiabatic state $|\bar{0}>$ is almost identical to the unperturbed state $|2>$ during a long period.

The theoretical results obtained for $\Delta_2 = -3.2$ cm^{-1}, accounting for or disregarding interference effects are compared on Figure 4. The calculation disregarding the interference terms illustrates the weak role played by these terms. This results

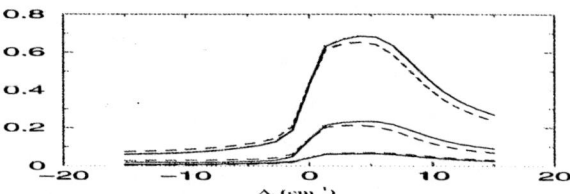

FIGURE 4. Partial ionization yields calculated for $\Delta_2 = -3.2$ cm^{-1} with (full line) or without (dashed line) interference effects.

from the fact that in (1) one has: $|D^*_{2Erc}| \gg |D^*_{1Erc}|$ and $< \bar{0}|i > \sim 1$ during the major part of the pulse.

We have shown, on the example of the perpendicular polarization case, that the main characteristics of the observed ion yields can be analyzed in the adiabatic approximation, following adiabatically during the pulse the evolution of the adiabatic state $|\bar{0} >$. The strong asymmetry observed in the spectra and persisting in a very large range of Δ_1-detunings is not related to interference effects which play a minor role. It is a manifestation of a coherent excitation of the intermediate states $|1 >$ and $|2 >$ through the adiabatic state $|\bar{0} >$. The ion yields result predominantly from a single ionization path. The branching ratios are mainly determined by the threshold resolved ionization rates $\gamma_2^{coh}(N_c, \ell_c, j_c)$ which removes the possibility of controlling the photoionization products.

As discussed in a forthcoming paper [10], our calculation gives a poorer description of the experimental data obtained for parallel polarizations. The weaker ionic yields are determined by the three contributions: ionization through the 6s6p or through the 6s7p and weak interference between the two ionization paths.

Acknowledgments. The authors would like to thank Professor D.S. Elliott and Doctor F. Wang for the discussions concerning the details of the experiment and for providing their experimental spectra.

References
1. F. Wang, D.S. Elliott, Phys. Rev. Lett. **77**, 4416 (1996).
2. F. Wang, Ph.D thesis, Purdue University, (1996).
3. F. Wang, Ce Chen, D.S. Elliott, Phys. Rev. A **56**, 3065 (1997).
4. M. Shapiro, P. Brumer, Int. Rev. Phys. Chem. **13**, 1987 (1986).
5. P. Brumer, M. Shapiro, Ann. Rev. Phys. Chem. **43**, 257 (1992).
6. S.T. Pratt, J. Chem. Phys. **104**, 5776 (1996).
7. H. Samble, Phys. Rev. A **7**, 2203 (1973).
8. E. Luc-Koenig, M. Aymar, J-M. Lecomte, A. Lyras, E.P.J.D, in press (1999).
9. M. Aymar, C.H.Greene, E. Luc-Koenig, Rev. Mod. Phys. **68**, 1015-1123 (1996).
10. E. Luc-Koenig, M. Aymar, J-M. Lecomte, M. Millet, A. Lyras, E.P.J.D, in preparation.

COLLISIONS INVOLVING
ELECTRONS AND POSITRONS

Recent Advances in our Understanding of (e,2e) Processes

Colm T. Whelan

Department of Applied Mathematics and Theoretical Physics, University of Cambridge, Silver Street, Cambridge, CB3 9EW, England

Abstract: A brief review is presented of the recent developments in the theory of electron impact ionization of atoms

1 Introduction

In this paper I am going to try to overview some of the highlights of research into electron impact ionization. In the last few years, revolutionary advances in experimental techniques and spectacular increases in computer power have offered unique opportunities to develop a much more profound understanding of the atomic few body problem. Part of the excitement of the area lies in the fact that modern measurements, because of their range and sophistication, constantly present new challenges to theory. The key technique is the coincidence measurement, where the fragments from a collision process are detected with their energies and angles resolved. The classic experiment of this kind is known as an (e,2e) one, where an electron is fired at a target ionizes it and the exiting electrons are detected in coincidence. (e,2e) experiments have now been performed on a remarkable range of targets, under an extraordinary range of kinematical conditions. Advances in experimental design mean that we now have at our disposal data over the full 3 dimensional space.[1]

It is, of course, impossible to present an overview of all aspects of coincidence studies in a few pages and in order to make my task almost manegable I have been forced to ignore some of the most exciting new developments. I am only going to deal with atomic targets thus excluding not only the massive literature on molecules, solids but also the recent first studies on surfaces[2]; I am, almost always going to assume the projectile is an electron and that only one electron is ejected. I will not discuss in any detail, $(e, 3e)$ or $(\gamma, 2e)$ processes or the very exciting new developments where full coincidence experiments are being performed with proton and heavier ion projectiles,[3]. This should not be taken to imply that I don't think this processes are important-quite the reverse they are too interesting to be constrained within a few paragraphs of this review. In the past year there have been a number of reviews of (e,2e) processes,[2],[4], [5], all of which have quite properly dwelt on the relation between theory and experiment. Here I will focus on some of the underlying theoretical developments. I plan to begin by giving a very brief overview of electron Hydrogen scattering and define some of the most

CP500, *The Physics of Electronic and Atomic Collisions*, edited by Y. Itikawa, et al.
© 2000 American Institute of Physics 1-56396-777-4/00/$17.00

commonly used approximations. Then I will spend just a little time talking about electron Helium scattering and the added interest of studying ionization processes with two active target electrons. Finally I will say a few words about the theory of relativistic (e,2e) processes.

2 Ionization with one active target electron

2.1 Basics

Suppose that we have an electron with momentum $\mathbf{k}_0$ and energy E_0 which collides with a hydrogen atom in the ground state and two electrons $(\mathbf{k}_s, E_s)$, $(\mathbf{k}_f, E_f)$ are detected in coincidence. The Hamiltonian for the system can be written

$$H = -\frac{\nabla_s^2}{2} - \frac{\nabla_f^2}{2} + V_s + V_f + V_{sf} \tag{1}$$

where $V_s = -1/r_S$, $V_f = -1/r_f$, $V_{sf} = 1/r_{sf}$. Let Ψ_a^+ be the exact solution of the Schrödinger equation, $H\Psi_a^+ = E\Psi_a^+$ with outgoing wave boundary conditions appropriate to the initial state

$$\Phi_a = \phi(\mathbf{r}_s)\frac{e^{i\mathbf{k}_0 \cdot \mathbf{r}_f}}{(2\pi)^{3/2}}. \tag{2}$$

Then following Whelan et. al. [6] and Walters et. al. [7] we may write[1]

$$f(\mathbf{k}_s, \mathbf{k}_f) = \left\langle \Phi_b \left| V_b \right| \Psi_a^+ \right\rangle \tag{3}$$

where

$$\Phi_b = \frac{e^{i\mathbf{k}_s \cdot \mathbf{r}_s} e^{i\mathbf{k}_f \cdot \mathbf{r}_f}}{(2\pi)^3}, \qquad V_b = V_s + V_f + V_{sf} \tag{4}$$

and the exchange amplitude is given by $g(\mathbf{k}_s, \mathbf{k}_f) = f(\mathbf{k}_f, \mathbf{k}_s)$. The spin averaged TDCS is defined as,

$$\frac{d^3\sigma}{d\Omega_s d\Omega_f dE_s} = \frac{(2\pi)^4 k_s k_f}{4k_0}\left(|f + g|^2 + 3|f - g|^2\right) \tag{5}$$

then, if we introduce some interaction W_b in to the left hand side we find that,

$$f(\mathbf{k}_s, \mathbf{k}_f) = \left\langle \psi_b^- \left| V_b - W_b \right| \Psi_a^+ \right\rangle - \left\langle \psi_b^- \left| V_a - V_b + W_b \right| \Phi_a \right\rangle \tag{6}$$

where $V_a = V_f + V_{sf}$. $\psi_b^-(\mathbf{r}_s, \mathbf{r}_f)$ is the wavefunction for the scattering of initially free electrons with momenta $\mathbf{k}_s, \mathbf{k}_f$ by W_b, that is,

$$\psi_b^- = \left[1 + \left(E + \frac{\nabla_s^2}{2} + \frac{\nabla_f^2}{2} - W_b - i\eta\right)^{-1} W_b\right]\Phi_b. \tag{7}$$

---

[1]In these papers it was assumed that the Coulomb potentials are cut off at some very large but finite distance.

Clearly if we take $W_b = V_b$ then the direct ionization amplitude can be written,

$$f(\mathbf{k}_s, \mathbf{k}_f) = \left\langle \Psi_b^- \left| V_a \right| \Phi_a \right\rangle. \tag{8}$$

Starting from equation (8) we could also go backwards and incorporate some interaction $W_a(\mathbf{r}_s, \mathbf{r}_f)$ in to the initial state, see Walters et. al. [7] for details.

It is clear from the analysis that once we treat the formalism with due respect we can move the interactions as we please from one side of the matrix element to the other. The equivalence between equations (3, 8) is the well known "post-prior" equivalence. Let us emphasize once again that Ψ_a^+, Ψ_b^- are exact and this freedom to move the interaction and the wavefunctions will be lost once we approximate either.

By approximating we lose information and the essence of a good theoretical approach is to retain as much of the dominant physics as possible. In order to exploit the formulation that we have developed let us assume there exists a wavefunction Ψ_{ansatz}^- which *explicitly* contains a great deal of the pertinent physics. Ψ_{ansatz}^- will satisfy the same boundary conditions as Ψ_b^- but generally it will *not* satisfy the same differential equation, that is, there exists a self adjoint operator $H_0 \neq H$ such that,

$$H_0 \Psi_{ansatz}^- = E \Psi_{ansatz}^-. \tag{9}$$

Then from equation (6), remembering that,

$$\Psi_a^+ = \left[1 + (E - H + i\eta)^{-1} V_a \right] \Phi_a \tag{10}$$

it follows that

$$\begin{aligned} f(\mathbf{k}_s, \mathbf{k}_f) &= \left\langle \Psi_{ansatz}^- \left| V_a \right| \Phi_a \right\rangle \\ &- \left\langle \Psi_{ansatz}^- \left| (H - H_0)(E - H + i\eta)^{-1} V_a \right| \Phi_a \right\rangle. \end{aligned} \tag{11}$$

Now if we neglect the second term in equation (11) we may write

$$f(\mathbf{k}_s, \mathbf{k}_f) \approx f(\mathbf{k}_s, \mathbf{k}_f)_{ansatz} = \left\langle \Psi_{ansatz}^- \left| V_a \right| \Phi_a \right\rangle. \tag{12}$$

We remark that equation (12) is exactly of the same form as equation (8) where we have approximated Ψ_b^- by Ψ_{ansatz}^-. Thus we can think of Ψ_{ansatz}^- as an ansatz for Ψ_b^-, hence the name. The advantage of the derivation is that we exhibit explicitly the term neglected. Clearly the validity of the ansatz approximation is determined solely by the relative importance of the second term on the right hand side of equation (11).

In the following discussion we will compare our results with a variant of the distorted wave Born approximation, (DWBA). The DWBA can be understood in terms of the general formalism, given above, as follows: returning to equation (6) and assuming that $W_b(\mathbf{r}_s, \mathbf{r}_f)$ is separable,

$$W_b(\mathbf{r}_s, \mathbf{r}_f) = V_1(\mathbf{r}_f) + V_2(\mathbf{r}_s) \tag{13}$$

243

where V_1, V_2 are effective potentials associated with the outgoing electrons, hence $\zeta_b(\mathbf{r}_s, \mathbf{r}_f) = \zeta_1(\mathbf{r}_f)\zeta_2(\mathbf{r}_s)$; then, [6], we see that the second term in equation (6) is zero. Quite generally we have that

$$f(\mathbf{k}_s, \mathbf{k}_f) = \left\langle \zeta_1^-(\mathbf{k}_f, \mathbf{r}_f)\zeta_2^-(\mathbf{k}_s, \mathbf{r}_s) \left| V_b - V_1 - V_2 \right| \Psi_a^+ \right\rangle. \tag{14}$$

As a first approximation to Ψ_a we could choose,

$$\Psi_a^+(\mathbf{r}_s, \mathbf{r}_f) = \zeta_o^+(\mathbf{k}_0, \mathbf{r}_f)\psi(\mathbf{r}_s) \pm \zeta_o^+(\mathbf{k}_0, \mathbf{r}_s)\psi(\mathbf{r}_f) \tag{15}$$

where $\pm$ denotes spin singlet$^+$ / triplet$^-$ static exchange wavefunction for electron scattering by the H atom in the state ψ_0.

This approximation is discussed in great detail elsewhere, Whelan et. al. [8, 6] Walters et. al. [7] and Rasch [9], here we will only make a few remarks. Firstly the wavefunctions, ζ_1^-, ζ_2^- are really quite general and once we have the exact wavefunction Ψ_a^+ on the left hand side of equation (14) we will get the exact scattering amplitude. However, when we make the approximation

$$\Psi_a^+ \approx \zeta_0^+(\mathbf{k}_0, \mathbf{r}_f)\psi(\mathbf{r}_s) \tag{16}$$

we have in effect produced a first order approximation in the $1/r_{sf}$ potential. Now as pointed out by Whelan et. al. [10], in this case we have recovered our post-prior equivalence which was characteristic of the full problem and one can interpret $\zeta_0^+(\mathbf{k}_0, \mathbf{r}_f)\psi(\mathbf{r}_s)$ as being associated with the incident channel and $\zeta_1^-\zeta_2^-$ as being associated with the final channel. The picture one has is of the incoming electron scattering in the incident channel distorting potential, in this case the static exchange potential of the hydrogen atom, the ionizing electron-electron interaction occurring once and the two exiting electrons elastically scattering in the potentials V_1 and V_2 on their way to the detectors. So after making the approximation, (equation (16)) the character of $\zeta_1^-\zeta_2^-$ changes. They are no longer "projectors" which act on Ψ_a^+ to pick out the exact scattering amplitude, they are now an intrinsic part of the approximation and it is important to include as much physics as possible in the choice of V_1, V_2.

For example, if $E_s \approx E_f$ then a reasonable choice would be,

$$V_1(\mathbf{r}_f) = -\frac{1}{r_f}, \qquad V_2(\mathbf{r}_s) = -\frac{1}{r_s}. \tag{17}$$

with this choice of DWBA direct and exchange scattering amplitudes become

$$f(\mathbf{k}_s, \mathbf{k}_f)^{DWBA} = \left\langle \zeta_1^-(\mathbf{k}_f, \mathbf{r}_f)\zeta_2^-(\mathbf{k}_s, \mathbf{r}_s) \left| \frac{1}{r_{sf}} \right| \zeta_0^+(\mathbf{k}_0, \mathbf{r}_f)\psi(\mathbf{r}_s) \right\rangle, \tag{18}$$

$$g(\mathbf{k}_s, \mathbf{k}_f)^{DWBA} = \left\langle \zeta_1^-(\mathbf{k}_s, \mathbf{r}_f)\zeta_2^-(\mathbf{k}_f, \mathbf{r}_s) \left| \frac{1}{r_{sf}} \right| \zeta_0^+(\mathbf{k}_0, \mathbf{r}_f)\psi(\mathbf{r}_s) \right\rangle. \tag{19}$$

We remark that a feature of this approximation is that ζ_0^+ will be different for singlet and triplet scattering, this will result in different direct and exchange amplitudes which we will denote by f^t, g^t, f^s, g^s with t and s denoting triplet and singlet. This is discussed in detail elsewhere, see Rasch [9]. The TDCS is the given by

$$\frac{d^3\sigma}{d\Omega_s d\Omega_f dE_s} = \frac{(2\pi)^4 k_s k_f}{4k_0} \left(|f^s + g^s|^2 + 3|f^t - g^t|^2 \right). \tag{20}$$

Despite its simplicity this approximation has been used to good effect to describe those (e,2e) measurements where interactions with the nucleus plays a significant role. For example in the study of multiple scattering effects in coplanar symmetric geometry, Whelan et. al. [6], the identification of strong interference effects in out of plane geometry, Rasch et. al. [11] and in the treatment of inner shell ionization processes, [12]. Indeed the relativistic generalization has proved spectacularly successful in the study of deep inner shell ionization of heavy metal targets, see Nakel & Whelan [5].

We can simplify still further and assume the incident and scattered electrons are sufficiently fast that one can neglect all distortion effects, i.e. use plane waves, then the direct amplitude becomes

$$f(\mathbf{k}_s, \mathbf{k}_f) = \left\langle e^{-i\mathbf{k}_f \cdot \mathbf{r}_f} \psi^-(\mathbf{k}_s : \mathbf{r}_s) \left| \frac{1}{|\mathbf{r}_f - \mathbf{r}_s|} \right| e^{i\mathbf{k}_0 \cdot \mathbf{r}_f} \psi_0 \right\rangle \tag{21}$$

where $\psi^-(\mathbf{k}_s, \mathbf{r}_s)$ is a continuum state of the atom. Equation (21) defines the first Born, it exhibits a symmetry about the direction of momentum transfer

$$\mathbf{q} = \mathbf{k}_0 - \mathbf{k}_f$$

and indeed is the exact solution when $q \to 0$. We have developed the theory for a hydrogenic ground state but the generalization to an arbitrary target state ψ_{nlm} is straight forward.

The DWBA as formulated above does not include any effect from the post-collisional electron-electron interaction, nor indeed from higher order effects such as polarization in the incident channel. One logical way to correct for these defects would be to expand Ψ_a^+ beyond the first order. However, even to take this expansion to second order presents serious technical difficulties, (see Rouet et. al. [13]; Madison et. al. [14]; Walters [15]). Of course the post-prior equivalence would be lost and with it the intuitively appealing idea of "incident" and "final" channels. One heuristic approach that has proved extremely useful has been to try to include higher order effects in the ζ_0 by means of a polarization potential and final sate interactions by means of a multiplicative factor either the Gamov factor or the Ward & Macek factor, see Whelan et. al. [16, 8, 6] and Rasch [9]. This approach contains something of the philosophy of the DWBA and allows one to retain the concept of initial and final channel effects. In this approximation the TDCS is given by

$$\frac{d^3\sigma}{d\Omega_s d\Omega_f dE_s} = N_{ee} \frac{d^3\sigma^{DWBA}}{d\Omega_s d\Omega_f dE_s} \tag{22}$$

where N_{ee} is either the Gamov factor, $N_{ee} = \gamma/(e^\gamma - 1)$, with $\gamma = 2\pi/|\mathbf{k}_s - \mathbf{k}_f|$ or the Ward & Macek factor [17]. These factors act to take account of some of the final channel electron-electron repulsion, in that they induce the correct kinematical character on the TDCS. For example, $N_{ee} = 0$ when $\mathbf{k}_s = \mathbf{k}_f$. This is discussed in great detail elsewhere Whelan et. al. [16] and Röder et. al. [18].

A much better way of going beyond the first Born is to solve the close-coupling equations for the exact solution Ψ^+,(or equivalent Ψ^-) using a discrete basis set of real and psedudo-states. This approach was pioneered by Curran and Walters, [19],[20] and is at the basis of the convergent close coupling approach [21].

3   Comparison between Theory and Experiment

Consider the ionization of helium in coplanar symmetric (Pochat) geometry. In this geometry the two outgoing electrons have equal energies and their momenta $\mathbf{k_f}, \mathbf{k_s}$ lie in the same plane as the incident momentum $\mathbf{k_0}$ and make the same angle, θ, with the beam direction. This can be described as a very "hard" collision, with the incident electron losing over half its initial momentum. If this were a collision between two free particles, conservation of energy and momentum would mean that the two electrons would be detected at 90° to each other, i.e at $\theta = 45^o$. In our case, the target electron is not free but is in a bound state of the Helium atom-thus it has a momentum distribution before the collision and the nucleus may take an active part in the process. Let us assume that ionization occurs as the result of a single electron-electron interaction then,intuitively, one would expect the cross section to be greatest when the angle between the two particles is approximately 90°. Now in this geometry this can happen when $\theta = 45^o$ and $\theta = 135^o$. In the original experiments, of Pochat et al,[22], the TDCSs were measured for Helium targets at impact energies of 100, 150 and 200 eV. The cross sections were found to be peaked around the 45° point and there was a clear indication of a rise at the larger angle. This prompted Whelan and Walters,[23] to speculate that a multiple scattering mechanism might be at work. They argued that the large angle structure could be interpreted in terms of the incident electron first elastically back scattering from the atom, essentially the nucleus, and then colliding with the target electron. Motivated by this simple intuitive model, a distorted wave Born approximation calculation was performed,[24]. This calculation turned out to be in excellent agreement with the measured cross sections,[25] There have been a number of experiments for a range of targets in this geometry and for the higher energies and heavier targets agreement with the DWBA was good, see [10] but as the energies were decreased things went badly wrong. Whelan et al [9] suggested that polarization effects needed to be taken into account in the incident channel and post-collisional interactions in the final and were able to come up with a model that had the character of the TDCS right and successfully predicted the shapes of the low energy Hydrogen cross sections in Pochat geometry,[8]. The experimental situation for hydrogen is, now,largely resolved we have high quality absolute data over the entire energy range- but theory is still lagging behind. Of the available models the close-coupling approach does best but is not able to reproduce the correct absolute size for the cross-section. I will not go into detail on this approach here- rather I would like to say a few words about the various ansatz type approximations that have been used. The ansatz method was first applied to the calculation of hydrogen cross sections by Brauner, Briggs and Klar, [26] and since then a number of different variants have been used. These authors originally represented the wave function as the product of 3 coulomb waves with fixed charges. It rapidly became apparent that this approach lead to cross sections which were far too small at low impact energies. Berakdar and Briggs [27], [28] attempted to improve the situation by introducing effective charges in the Coulomb waves. At first these charges were constant but in the most sophisticated version, known as DS3C, they were positional dependent. Using these wavefunctions presented serious difficulties in actually calculating the cross section. At this point it is important to point out that one can only calculate TDCS involving Ψ_f^{DS3C} using a fully numerical method. However, calculations have been performed for (e,2e) on H using a "momentum" version or asymptotic version of Ψ_f^{DS3C}. In this method one assumes that classically in the asymptotic region the three particles can be converted in to velocity coordinates in which $r_{s,f,sf} = k_{s,f,sf}\, t$, where t is time. In Lucey et al.[29] a new numerical method was employed and a full set of cross sections were calculated. These were compared with the DWBA with polarization and Gamov factor. It was found that the ansatz type approximations gave only poor agreement, especially for situations where polarization was found to be important.

4   Two Active Target Electrons

Consider the excitation-ionization of helium in which the final helium ion is $2s$ or $2p$, ie,

$$e^-(\mathbf{k}_0) + He(1\ ^1S) \rightarrow e^-(\mathbf{k}_f) + e^-(\boldsymbol{\kappa}) + He^+(2s\ \text{or}\ 2p) \tag{23}$$

This is characteristic of an ionization process in which both electrons in the target have to be active. To see this it is instructive to follow the argument given in Marchalant et al,[30], who considered a perturbative approach. The scattering amplitude in the first Born approximation, for helium target, is given by:

$$f^{B1} = -\frac{4}{q^2}\langle \Psi_f(\mathbf{r}_2, \mathbf{r}_3)\, |-1 + e^{i\mathbf{q}\cdot\mathbf{r}_2}\, |\, \Psi_0(\mathbf{r}_2, \mathbf{r}_3)\rangle \tag{24}$$

where $\Psi_0(\mathbf{r}_2, \mathbf{r}_3)$ is the initial state of the atom, and $\Psi_f(\mathbf{r}_2, \mathbf{r}_3)$ is the final doubly excited state of the atom and

$$\mathbf{q} = \mathbf{k}_0 - \mathbf{k}_f \tag{25}$$

is the momentum transfer in the collision.

Following Marchalant et al[30] it is of interest to examine the first Born approximation in a model in which uncorrelated wave functions are used for the helium states. In this model the target wave functions are constructed out of orthonormal orbitals $\phi_n(\mathbf{r})$ with any helium state being represented as a product of two

247

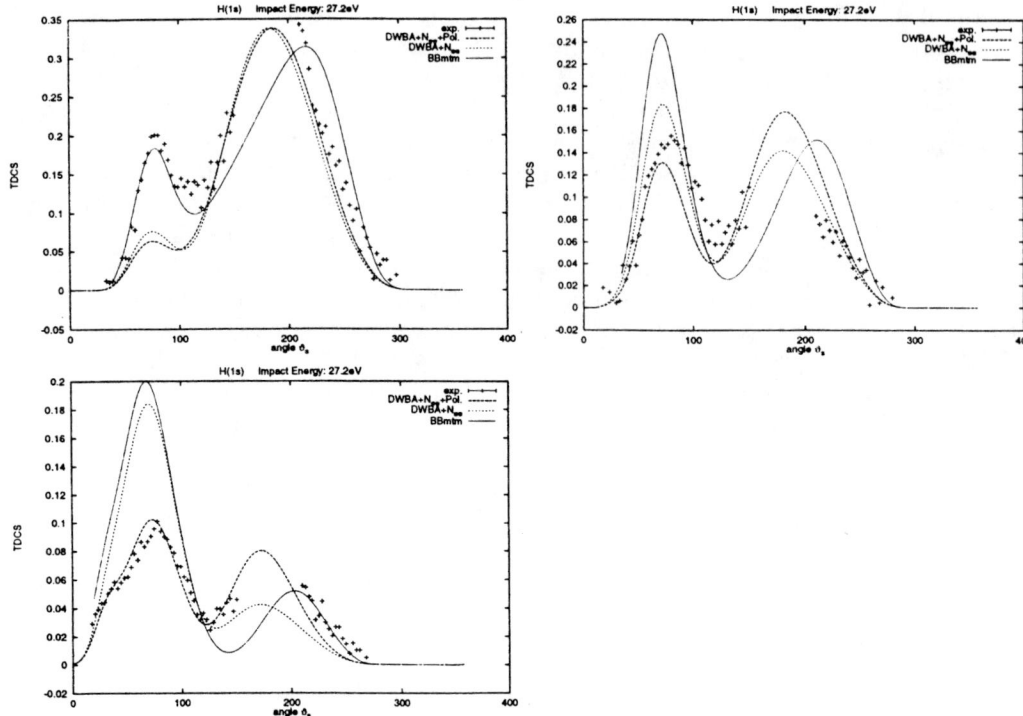

FIGURE 1. TDCS in coplanar energy sharing geometry at an impact energy of $E_0 = 27.2\text{eV}$, $E_s = E_f = 6.8\text{eV}$, experimental data is inter normalized but relative therefore to all exp. data is scaled by the same overall factor to give the best fit to theory. Curves in all figures: (a) DWBA with polarization and PCI effects included, (long dashed); (b) standard DWBA with PCI effects included (short dashed): (c) Berakdar and Briggs effective charge calculation (dotted). Angle of one of the two outgoing electrons is fixed at 345° (top left), 330° (top right), 315° (bottom left).

one-electron orbitals ϕ_n and ϕ_m; we write

$$\Psi_{nm}^S(\mathbf{r}_2, \mathbf{r}_3) = N_{nm}\{\phi_n(\mathbf{r}_2)\phi_m(\mathbf{r}_3) + (-1)^S\phi_n(\mathbf{r}_3)\phi_m(\mathbf{r}_2)\} \tag{26}$$

where N_{nm} is the normalization constant and S is the total spin. Consider a transition from the ground state:

$$\begin{aligned} \Psi_0(\mathbf{r}_2, \mathbf{r}_3) &= \Psi_{00}^{S=0}(\mathbf{r}_2, \mathbf{r}_3) \\ &= \Phi_0(\mathbf{r}_2)\phi_0(\mathbf{r}_3) \end{aligned} \tag{27}$$

to the final state

$$\Psi_{fg}^{S=0}(\mathbf{r}_2, \mathbf{r}_3) = N_{fg}(\phi_f(\mathbf{r}_2)\phi_g(\mathbf{r}_3) + \phi_f(\mathbf{r}_3)\phi_g(\mathbf{r}_2)) \tag{28}$$

the first Born amplitude (24) then becomes

$$\begin{aligned} f^{B1} &= -\frac{4N_{fg}}{q^2}\left\{\langle\phi_f(\mathbf{r})\,|-1+e^{i\mathbf{q}\cdot\mathbf{r}}\,|\,\phi_0(\mathbf{r})\rangle\langle\phi_g\,|\,\phi_0\rangle \right. \\ &\quad \left. + \langle\phi_g(\mathbf{r})\,|-1+e^{i\mathbf{q}\cdot\mathbf{r}}\,|\,\phi_.(\mathbf{r})\rangle\langle\phi_f\,|\,\phi_0\rangle\right\} \end{aligned} \tag{29}$$

If $\Psi_{fg}^{S=0}$ is a singly excited state, so that $\phi_g = \phi_0$ and $\phi_f \neq \phi_0$, 29 gives

$$f^{B1} = -\frac{4N_{fg}}{q^2} \langle \phi_f(\mathbf{r}) \mid e^{i\mathbf{q}\cdot\mathbf{r}} \mid \phi_0(\mathbf{r}) \rangle \qquad (30)$$

However, if $\Psi_{fg}^{S=0}$ is a doubly excited state, $\phi_g \neq \phi_0$ and $\phi_f \neq \phi_0$, (29) gives

$$f^{B1} = 0 \qquad (31)$$

Thus, whereas this model leads to a non-zero value for the first Born approximation to single excitation processes, it gives a zero first Born amplitude for double excitation reactions. Of course in practice no one should ever use helium wave functions as crude as those of the uncorrelated model described above. However, the uncorrelated result, (31), warns that correlation effects in the initial and final states are going to be much more important for double excitation processes, such as excitation–ionization, than for single excitation processes, such as ionization to the ground state ion. It should also be noted that the second Born amplitude f^{B2} remains non-zero even in the simple minded uncorrelated approximation -which suggests that the 2nd Born term will be more important than in the equivalent simple ionization process.

In Marchalant et al[30],[31] the first Born element is evaluated with a close coupled wave function on the left hand side and various highly correlated target wavefunctions on the right hand side. The second Born term is evaluated using closure. It is clear that the 2nd born term remains important even up to quite high energies.

A similar analysis was applied by Marchalant et al, [30], [31],[32] to the study of other two active target electron processes in particular excitation-autoionization and (e,3e). In each case the 2nd Born term was found to be very important.

There is an urgent need for new experimental data. It is clear that the cross sections will depend on target correlation, final state interactions between the ejected electrons and the nucleus as well as higher order interactions with the fast incident/scattered electron. It will be a major theoretical task to isolate the different effects- it would exceedingly nice if the effect of target correlation could be seen, even implicitly in the cross-section. (It should be remarked that in [32] one can see measurable differences in the computed TDCS between different high quality Helium target wavefunctions). We need to exploit the kinematical freedom inherent in an (e,2e) experiment to isolate regions of phase space where one effect is stronger than the other. We need a systematic study of the 1st and 2nd Born terms, we need to know their regions of applicability. In this regard the new generation of coincidence experiments with differently charged projectiles will be very valuable, e.g. if we use a positron rather than an electron then the relative sign of the terms in the sum

$$|f^{B1} + f^{B2}|^2$$

changes. If we increase the charge on the projectile it will emphaise the second Born part. We can also exploit complementary $(\gamma, 2e)$ experiments to give us information on the the 1st Born term.

5   Relativistic (e,2e)

In the previous sections we have concentrated our attention on electron impact
ionization at low or indeed very low impact energies, where relativistic effects will
be entirely negligible. There have been, however, studies at much higher energies
on the deep inner shells of heavy targets; these began in 1982 with (e,2e) experiments by Schüle and Nakel,[33] at an incident energy of 500 keV on the K shell of
silver. The description of such processes opens up a whole new range of problems.
Relativistic effects are globally important the target electron is in a deep inner
shell, the incident and final electrons have velocities that are a significant proportion of the speed of light, retardation and spin dependent interactions need to be
considered. We have seen above that at low energies the DWBA is the simplest
possible approximation that we can use to include multiple scattering effects in
both the incident and final channels. It works well for a range of geometries where
higher order effects - for example polarization of the target in the incident channel
or electron-electron repulsion terms in the final are weak. It gave excellent agreement with inner shell ionization experiments on Argon, [12]. Clearly therefore it
is a prime candidate as a theoretical model for the relativistic inner shell problem.
Indeed, there are some simplifications at the energies we are working: exchange in
the elastic channels is likely to be negligible, final state $e-e$ repulsion will certainly
play no role. However, as we mentioned earlier, we are now dealing with a fully
relativistic problem. This means that we will have to solve for Dirac spinors rather
than Schrödinger wave functions and we will need to include the full QED photon
propagator.

In the earliest attempts to model these processes a number of assumptions were
made, which subsequently turned out to false and it is perhaps valuable to list these
here. The simplest approximation one could use is the plane wave Born approximation, (PWBA) in which all the electron wavefunctions are represented by plane
waves. At non-relativistic energies this approximation has been used extensively
in impulsive experimental arrangements and when it is valid it allows the use of
the (e,2e) method as a means of mapping the momentum distribution of the target
electron. A feature of the, low energy, PWBA approximation is that it factorises
into two terms, one which is independent of the target wavefunction and the other
is in essence the wavefunction of the target in momentum space. In Bell,[34], a
relativistic version was proposed in which the TDCS was taken to be the product
of the free first order electron-electron (Møller) cross section and the momentum
profile of the bound state. Keller and Whelan,[35], analysed the relativistic plane
wave approximation and concluded that in fact it did not factorize and that for
any given kinematical arrangement and bound state energy, the "cross section function" part of the TDCS depended on the spinor structure of the bound state. The
results obtained in the relativistic plane wave Born approximation are substantially

different from those found in the impulsive treatment of Bell,[34], though both theories share the same, factorized, form in the non-relativistic limit. A number of semi-relativistic variants on the Born approximation were tried. Das and Konar, [36], employed a semi-relativistic Sommerfeld Maue function,[37], for one of the outgoing electrons and Jakubaßa-Amundsen, in a series, of calculations, studied the influence of different approximate semi-relativistic scattering and bound state wave functions on the TDCS,[38, 39, 40]. We will not discuss the validity of using semi-relativistic wavefunctions here and only remark that they can easily lead one to error, [41]. Another common feature of all these approximations, was that they only included those spin channels which would contribute at low energies, the other relativistic "spin flip" channels were not included. In Walters et al,[42], it was shown that this was an invalid assumption and that, especially in the symmetric case, they made a very large contribution to the TDCS.

It is clear that only a fully relativistic approximation would stand any chance of describing the TDCS, further it is also evident that distortion effects needed to be included ie the strong nuclear field could not be neglected. In [43], [44] a relativistic version of the DWBA was developed (rDWBA). An approximation was devised which assumed that the ionizing interaction between the two electrons could be treated in first order perturbation theory but the strong field of the atom was retained to all orders. It was considered absolutely vital that no unwarranted non-relativistic assumption be made. In [44] the rDWBA is defined with great care, here we will only give a brief summary of the assumptions made and a few key points about the calculations. It was assumed that the electromagnetic field of the nucleus and the atomic electrons could be incorporated in the form of a classical field. In all calculations, to date, this field was assumed to be purely electrostatic and radially symmetric in the reference frame of the calculations(which was taken to be the rest frame of the nucleus). This meant that the Dirac equation, that had to be solved in the elastic channels, could be separated in spherical co-ordinates. In all the calculations reported a self consistent relativistic Kohn-Sham local density approximation potential was used, [45], to represent the atomic potential.

The full free QED photon propagator,in the Feynman gauge, was included. In relativistic physics the interaction between the two electrons is represented by photon exchange between the four currents of the two electrons,in contrast the non relativistic Coulomb interaction can be thought of as an instantaneous interaction between point charges. By using the full propagator one automatically includes the magnetic interactions between the currents of the particles and retardation effects. As I have stressed this is a first order theory so the propagator acts only once.

The photon propagator can be represented in polar coordinates and then expanded in multi-poles. Keller et al,[44], used the program of Salvat and Mayol,[46], to numerically evaluate the radial functions, in the elastic channels, and the Clebsch Gordan coefficients were extracted from Burgess and Whelan, [47]. They approximated the K shell electron by a relativistic hydrogenic 1s state. The success of this approximation has been well documented elsewhere, [5],[4], and it is now clear and that we understand the role of the q.e.d photon propogator, Mott scattering,

Pauli blocking, spin-flip, fine-structure and multiple scattering effects. Agreement with experiment is excellent.

ACKNOWLEDGMENTS

I have been very fortunate to be involved in a collaborative project to study relativistic (e,2e) processes. This joint effort has involved the Queen's University of Belfast, the University of Frankfurt am Main as well as Cambridge. I am grateful to Professor Dreizler, Drs Ancarani, Ast, Keller and Rasch for very many useful discussions. A special word of thanks is due to Professor Walters with whom I have the privilege of collaborating on a whole range of different (e,2e) problems. Support from EPSRC and the British Council is gratefully acknowledged. The numerical calculations shown in this work were performed on the Hitachi at Cambridge.

References

[1] R. Dörner, T. Weber, Kh. Khayyat, V. Mergel, H. Bräuning, M. Achler, O. Jagutzki, L. Speilberger, J. Ullrich, R. Moshammer, W. Schmitt, R. E. Olson, C. Woods, and H. Schmidt-Böcking. Recoil Ion Momentum Spectroscopy-Momentum Space Images of Atomic Reactions . In Colm T. Whelan, R.M. Dreizler, J.H. Macek, and H. R. J. Walters, editors, *New Directions in Atomic Physics*, pages 33–46, New York, 1999. Plenum/Kluwer.

[2] Stefani G. Recent advances in electron-electron coincidence experiments. In Colm T. Whelan, R.M. Dreizler, J.H. Macek, and H. R. J. Walters, editors, *New Directions in Atomic Physics*, pages 17–32, New York, 1999. Plenum/Kluwer.

[3] B. Bapat, R. Moshammer, S. Keller, W. Schmitt, A. Cassimi, L. Adoui, H. Kollmus, R. Dörner, Th. Weber, K. Khayyat, R. Mann, J.P. Grandin, and Ullrich J. Double ionization of helium in fast ion collisions: the role of momentum transfer. *J. Phys. B*, 32:1859–1872, 1999.

[4] Colm T. Whelan. (e,2e) processes. In Colm T. Whelan, R.M. Dreizler, J.H. Macek, and H. R. J. Walters, editors, *New Directions in Atomic Physics*, pages 87–104, New York, 1999. Plenum/Kluwer.

[5] W. Nakel and Colm T. Whelan. Relativistic (e,2e) processes. *Phys. Rep.*, 315:409–471, 1999.

[6] Colm T. Whelan, H. R. J. Walters, and X. Zhang. (e,2e) and All That ! In Colm T. et. al.Whelan, editor, *(e,2e) and Related Processes*, pages 1–32. Kluwer academic publishers, Netherlands, 1993.

[7] H. R. J. Walters, X. Zhang, and Colm T. Whelan. Directions in (e,2e) and related processes. In Colm T. et. al.Whelan, editor, *(e,2e) and related processes*. Kluwer academic publishers, Netherlands, 1993.

[8] Colm T. Whelan, R. J. Allan, and H. R. J. Walters. PCI, polarisation and exchange effects in (e2e) collisions. *in "Journal de Physique", Volume 3, Colloque 6*, pages 39–49, 1993.

[9] J. Rasch. *(e,2e) processes with neutral atom targets*. PhD thesis, Cambridge University, 1996.

[10] Colm T. Whelan, H. R. J. Walters, J. Hannsen, and R. M. Dreizler. High-energy electron-impact ionisation of H(1s in coplanar asymmetric geometry). *Aust. J. Phys.*, 44(1):39–58, 1991.

[11] J. Rasch, Colm T. Whelan, R. J. Allan, S. P. Lucey, and H. R. J. Walters. Strong interference effects in the triple differential cross section of neutral atom targets. *Phys. Rev. A*, 56(2):1379–83, 197.

[12] Zhang X., Colm T. Whelan, H. R. J. Walters, R. J. Allan, P. Bickert, W. Hink, and S. Schönberger. (e,2e) cross-sections for inner-shell ionization of argon and neon. *J. Phys. B*, 25(20):4325–35, 1992.

[13] F. Rouet, R. J. Tweed, and J. Langlois. The effect of target atom polarisation and wavefunction distortion in (e,2e) ionisation of Hydrogen. *J. Phys. B*, 29(9):1767–83, 1996.

[14] D. H. Madison, I. Bray, and I. E. McCarthy. Exact Second Order Distorted Wave Calculation for Hydrogen Including Second Order Exchange. *J. Phys. B*, 24:3861–88, 1991.

[15] H. R. J. Walters. Perturbative methods in electron – and positron – atom scattering. *Physics Reports*, 116(1-2):1–102, 1984.

[16] Colm T. Whelan, R. J. Allan, J. Rasch, H. R. J. Walters, X. Zhang, J. Röder, K. Jung, and H. Ehrhardt. Coulomb three-body effects in (e,2e) collisions: The ionization of H in coplanar symmetric geometry. *Phys. Rev. A*, 50:4394–4396, 1994.

[17] S. J. Ward and J. H. Macek. Wave-functions for continuum states of charged particles. *Phys. Rev. A*, 49(2):1049–56, 1994.

[18] J. Röder, J. Rasch, K. Jung, Colm T. Whelan, H. Ehrhardt, R. J. Allan, and H. R. J. Walters. "On the role of Coulomb 3 body effects in low energy impact ionisation of H(1s)". *Phys. Rev. A*, 53(1):225–33, 1996.

[19] E. P. Curran and H. R. J. Walters. Triple differential cross sections for the electron impact ionisation of atomic hydrogen-a coupled pseudo state approach. *J. Phys. B*, 20:337–365, 1987.

[20] E. P. Curran, Colm T. Whelan, and H. R. J. Walters. On the electron impact ionisation of H(1s) in coplanar asymmetric geometry. *J. Phys. B*, 24:L19–25, 1991.

[21] I. Bray and Andris T. Stelbovics. The convergent clse coupling method for a coulomb three-body problem. *Comp. Phys. Commun.*, 85:1, 1995.

[22] A. Pochat, R. J. Tweed, J. Peresse, C. J. Joachain, B. Piraux, and F. W. Byron. Second-order effects in large-angle coplanar symmetric (e,2e) processes. *J. Phys. B*, 16:L775–9, 1983.

[23] Colm T. Whelan and H. R. J. Walters. A new version of the distorted wave impulse approximation – applications to large angle scattering. *J. Phys. B*, 23:2989–2995, 1990.

[24] Zhang X., Colm T. Whelan, and H. R. J. Walters. (e,2e) cross-sections for ionsiation of helium in coplanar symmetric geometry. *J. Phys. B*, 23(25):L509–16, 1990.

[25] L. Frost, P. Freinstein, and M. Wagner. 200ev coplanar symmetric (e,2e) on helium : a sensitive test of reaction models. *J. Phys. B*, 24:657–73, 1990.

[26] M. Brauner, J. S. Briggs, and H. Klar. Triply differential cross sections for ionisation of hydrogen atoms by electrons and positrons. *J. Phys. B*, 22(14):2265–87, 1989.

[27] J. Berakdar and J. S. Briggs. Three-body continuum problem. *Phys. Rev. Lett.*, 72(24):3799–82, 1994.

[28] J. Berakdar. Parabolic-hypersherical approach to the fragmentation of three-particle Coulomb systems. *Phys. Rev. A*, 54(2):1480–6, 1996.

[29] S. P. Lucey, J. Rasch, Colm T. Whelan, and H. R. J. Walters. Gauge discrepancies in calculations of (γ,2e) on He. *J. Phys. B*, 31:1237–1258, 1998.

[30] P. J. Marchalant, Colm T. Whelan, and H. R. J. Walters. Excitation-ionization and excitation-autoionization of helium. In Colm T. Whelan and H. R. J. Walters, editors, *Coincidence Studies of Electron and Photon Impact Ionization*, pages 21–44. Plenum, 1997.

[31] P. J. Marchalant, Colm T. Whelan, and H. R. J. Walters. Second order effects in (e,2e) excitation-ionization of helium to $he^+(n = 2)$". *J. Phys. B*, 31:1141–1178, 1999.

[32] P. J. Marchalant, J. Rasch, Colm T. Whelan, and H. R. J. Walters. New results for double excitation processes with helium targets. In Colm T. Whelan, R.M. Dreizler, J.H. Macek, and H. R. J. Walters, editors, *New Directions in Atomic Physics*, pages 355–362. Plenum, 1997.

[33] E. Schüle and W Nakel. *J. Phys. B.*, 15:L639–41, 1982.

[34] F. Bell. *J. Phys. B*, 22, 1989.

[35] S. Keller and Colm T. Whelan. On the plane wave born approximation for relativistic (e,2e) processes. *J. Phys. B*, 27:L771–776, 1994.

[36] J.N. Das and A.N. Konar. *J. Phys. B*, 7:2417, 1974.

[37] M. E. Rose. *Relativistic Electron Theory*. John Wiley, New York, second edition, 1961.

[38] D.H. Jakubaßa-Amundsen. *Z. Phys. D.*, 11:305, 1989.

[39] D.H. Jakubaßa-Amundsen. *J. Phys. B.*, 25:1297, 1992.

[40] D.H. Jakubaßa-Amundsen. *J. Phys. B.*, 28:259–73, 1995.

[41] L.U. Ancarani, S. Keller, H. Ast, Colm T. Whelan, H. R. J. Walters, and R.M. Dreizler. Coulomb boundary conditions for relativistic (e,2e) processes. In Colm T. Whelan and H. R. J. Walters, editors, *Coincidence Studies of Electron and Photon Impact Ionization*, pages 215–220. Plenum, 1997.

[42] H. R. J. Walters, H. Ast, Colm T. Whelan, R.M. Dreizler, H. Graf, C. D. Schröter, J. Bonfert, and W. Nakel. Relativistic (e,2e) collisions on atomic inner shells in symmetric geometry. *Z. Phys, D*, 23:353–357, 1992.

[43] H. Ast, S. Keller, Colm T. Whelan, H. R. J Walters, and R.M. Dreizler. Electron-impact ionization of the k shell of silver and gold in coplanar asymmetric geometry. *Phys. Rev. A*, 50:R1–R3, 1994.

[44] S. Keller, Colm T. Whelan, H. Ast, H.R.J. Walters, and R.M. Dreizler. Relativistic distorted-wave Born calculations for (e,2e) processes on inner shells of heavy atoms. *Phys. Rev. A*, 50:3865–3877, 1994.

[45] R.M. Dreizler and E.K.U. Gross. *Density Functional Theory*. Springer, Berlin, 1990.

[46] F. Salvat and R. Mayol. *Comput. Phys. Commun.*, 62:65, 1991.

[47] A. Burgess and Colm T. Whelan. Betrt - a procedure to evaluate the cross section for electron-Hydrogen collisions in the Bethe approximation to the reactance matrix. *Comp. Phys. Commun.*, 47:295–304, 1987.

Electron Scattering from Laser Excited Atoms

P. J. O. Teubner

School of Chemistry, Physics and Earth Sciences,
The Flinders University of South Australia,
Adelaide, Australia 5001.

Abstract. This paper reviews advances in the area of electron scattering from laser excited atoms since the XX International Conference on the Physics of Electronic and Atomic Collisions (ICPEAC) in Vienna. It discusses superelastic scattering of electrons from potassium and from rubidium. The excitation of higher lying states from excited P states in both sodium and barium is discussed. Elastic scattering from the (... 6s 6p) ^{1}P state and (... 6s 5d) ^{1,3}D states of barium is discussed as are experiments that observe elastic scattering from the 3 ^{2}P state in sodium.

This paper reviews the very significant activity in the field of electron scattering from laser excited states since the last ICPEAC. Although it focuses on the progress of experiments in the field much of the published work includes collaboration with theoretical groups so the comparison between experiment and theory has often been easy to make. Nevertheless it should be remembered that this review is given by an experimentalist and is primarily about experiments. The scope of the review is governed by what can be included in a 40 minute talk so it is concerned only with discrete inelastic and elastic scattering of electrons from excited states. Thus it does no more than mention the experiments of Weigold's group of polarised electron atom (e,2e) experiments in sodium (1) in which evidence of dichroism has been identified.

The review can be separated in to a discussion of three different types of experiments. The superelastic scattering experiments of electrons from the alkalis, the measurement of orientation and alignment parameters for transitions from excited P states to states other than the ground state for example 3P to 4S in sodium and (... 6s 6p) ^{1}P to (... 6s 5d) ^{1}D states in Barium and finally the elastic scattering of electrons from excited states in barium and in sodium.

SUPERELASTIC ELECTRON EXPERIMENTS

In a typical superelastic scattering experiment a well collimated beam of ground state atoms lies in the scattering plane and is optically pumped to the first excited state. The resonance laser radiation is introduced normal to the scattering plane. A beam of electrons is directed onto the region where the laser and atomic beams

CP500, *The Physics of Electronic and Atomic Collisions*, edited by Y. Itikawa, et al.
© 2000 American Institute of Physics 1-56396-777-4/00/$17.00

overlap; the overlap of the three beams defines the interaction region. An electron spectrometer views the interaction region and analyses the energy of the electrons that have been scattered through an angle θ. Electrons that have gained kinetic energy equal to the energy difference between the atomic ground and excited state are detected and counted as a function of the scattering angle θ for six different states of laser polarization.

The three equivalent Stokes parameters P_i can be deduced from these measurements. Specifically

$$P_1 = \frac{I_0 - I_{90}}{I_0 + I_{90}}$$

$$P_2 = \frac{I_{45} - I_{135}}{I_{45} + I_{135}} \tag{1}$$

$$P_3 = \frac{I_{RHC} - I_{LHC}}{I_{RHC} + I_{LHC}}$$

Here I_0 represents the superelastic count rate when the laser light is polarised at an angle ϕ with respect to the scattered electron direction. The terms I_{RHC} and I_{LHC} denote the count rates for right and left hand circular polarization.

The optical pumping process in superelastic scattering experiments that involve the alkalis is complicated by the influence of hyperfine structure on the linear and circular polarization. This polarization influence can be accounted for by the coefficients K and K'. These coefficients are related to the equivalent Stokes parameters (2) by

$$\overline{P}_1 = \frac{1}{K} P_1$$

$$\overline{P}_2 = \frac{1}{K} P_2 \tag{2}$$

$$\overline{P}_3 = \frac{1}{K'} P_3$$

where the components $\overline{P}_i$ are the reduced or equivalent Stokes parameters.

The coefficient K can be determined by measuring the linear polarization of the fluorescent radiation emitted in the scattering plane by the optically pumped target atoms. The parameter K' is much more difficult to determine empirically and there are strong reasons to suggest that in most projects K' = 1 (2).

In the $| LM_L \rangle$ basis the final state wavefunction can be represented by

$$\Psi = a_1 | 1 \ 1 \rangle + a_{-1} | 1 - 1 \rangle \tag{3}$$

where we have taken the natural coordinate frame that chooses the z axis as that of the laser beam. Andersen *et al.* (3) have demonstrated the relationship between the reduced Stokes parameters and the scattering amplitudes a_1 and a_{-1}.

These are

$$\overline{P}_1 = 2 \, \text{Re} \, (a_1 \, a_{-1}^*)$$
$$\overline{P}_2 = 2 \, \text{Im} \, (a_1 \, a_{-1}^*) \tag{4}$$
$$\overline{P}_3 = |a_1|^2 - |a_{-1}|^2$$

It has also been shown by Hermann and Hertel (4) that the angular distribution of the total scattered intensities for linear polarization I_{lin} is proportional to the differential cross section. That is

$$I_{lin} = I_0 + I_{90} = I_{45} + I_{135} \tag{5}$$

$$= \sigma(\theta) \tag{6}$$

$$= |a_1|^2 + |a_{-1}|^2 \, . \tag{7}$$

These simple relationships hold for the case in which there is positive reflection symmetry in the scattering plane. This implies that the spin flip cross section is zero which is a reasonable assumption for the lighter alkalis up to potassium. Equations (5) and (6) imply that it is possible to determine the differential cross section simultaneously with the Stokes parameters in a superelastic scattering experiment.

Andersen *et al.* (3) introduced an alternative way of describing the scattering parameters. These are the orientation and alignment parameters. The orientation parameter $L_\perp = -\overline{P}_3$ measures the angular momentum transferred to the atom normal to the scattering plane. The alignment parameter γ measures the alignment angle of the final state wavefunction relative to the incident beam direction.

$$\gamma = \frac{1}{2} \, \text{arg} \, (\overline{P}_1 + i \overline{P}_2) \tag{8}$$

Superelastic scattering experiments provide a description not only of the magnitude of the scattering amplitudes but also of their relative phase. In all of the work described in this review incident unpolarized electrons were used and there was no attempt made to determine the polarization of the electrons after the collision. Thus the data represent spin averaged amplitudes.

The extension of the superelastic scattering technique to alkali targets other than sodium has coincided with the development by Bray of the convergent close coupling (CCC) approximaton to describe electron scattering from the alkalis (5). At the last

ICPEAC we reported a series of orientation and alignment parameters for electron scattering from lithium at several energies between 5 and 20 eV [Karaganov and Teubner (6)]. When these parameters were compared with the predictions of the CCC theory of Bray it was found that there was excellent agreement between experiment and theory over the whole energy range and for scattering angles between 0° and 140°. This congruence tempted us to speculate that the electron lithium problem was solved.

The preparation of the excited target state by laser light follows a technique that was first demonstrated by Hertel and coworkers (7) twenty five years ago. The basic principles can be seen in Figure 1.

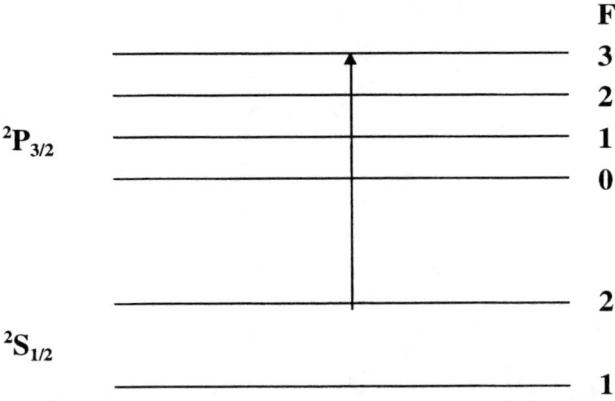

FIGURE 1. A Schematic Energy level diagram shown HFS splitting for an atom with nuclear spin $^3/_2$. The arrow indicates the frequency of the pump laser.

In this case we assume that we are dealing with alkalis with nuclear spin $I = ^3/_2$ so that the discussion applies to ^{23}Na, ^{7}Li and ^{39}K.

Laser light is tuned to the transition between the $\overline{F} = 2$ level in the ground state to the F = 3 level in the J = $^3/_2$ excited state. Stimulated and spontaneous emission from the upper to the lower level results in an aligned and oriented excited state which can serve as the target in a series of superelastic scattering experiments.

In practice the optical pumping process is complicated by the fact that the hyperfine structure levels have a finite width of between 5 and 10 MHz, the atomic beam has a Doppler width which can be as small as 15 MHz or a large as 150 MHz. The laser line is power broadened with a width greater than or equal to 40 MHz. Consequently all of the HFS levels in the excited state are very strongly mixed. So pumping with single frequency laser light pumps all of the atoms into the $\overline{F} = 1$ ground state. These atoms no longer can be accessed by the laser light and they are of no use to the experiment. This phenomenon is known as hyperfine structure trapping.

The problem can be removed either by pumping the F = 3 level from both ground state levels with two separate lasers which differ by the frequency separation of the HFS splitting in the ground state. This approach has been used by the Griffith group (8). A less expensive solution which can be used for lithium and potassium where the HFS splitting is less than it is in sodium is to pass the laser light through an electro optic modulator crystal. This crystal acts as the capacitor in an LC circuit that is tuned to oscillate at half the frequency separation of the HFS splitting of the ground state. The oscillating field phase modulates the laser field and converts single frequency light into light with frequency shifted side bands. This approach has been used in the Flinders experiments on lithium and potassium. In the case of potassium the dual frequency pumping technique increased the excited state population by a factor of about 4 when compared to single frequency pumping.

It was surprising therefore that the CCC theory which had proven so successful in describing the excitation of the resonance transition in lithium and in sodium seemed unable to predict the differential cross section for the excitation of the 4 ^{2}P state in potassium. There has been considerable interest in the theoretical description of electron scattering from potassium. The theories are reviewed in Stockman et al. (9). The agreement between all of the theories at an incident energy of 54.4 eV and the differential cross section measurements of Buckman et al. (10) and Vuskovic and Srivastava (11) is poor. These two sets of experimental data essentially agree out to scattering angles of 110° whereas there is a difference of about a factor of 10 between experiment and most theories at middle angles. This lack of agreement between the CCC theory and the previous experiments prompted the Flinders group to study potassium using a superelastic scattering technique. This technique not only provided experimental data that offered in principle a more sensitive test of the theory than did the differential cross section but as can be seen from equation (6), the differential cross section could be measured at the same time as the orientation and alignment parameters.

Stockman et al. (12) reported results on potassium at an incident energy of 10 eV which showed that there was excellent agreement between the predictions of the CCC theory and experiment. This was consistent with the results in the other alkalis but it did not solve the problem at 54.4 eV. Doubt was cast on the earlier measurements when the orientation and alignment parameters were measured at 54.4 eV. There was excellent agreement between the CCC theory and experiment for the linear component of the pseudo Stokes vector. Apart from some minor departures between 40° and 60° the agreement with the orientation parameter was also excellent.

Equations (4) show that the orientation parameter is given by the difference of the squares of the moduli of the scattering amplitudes, whereas the differential cross section is given by the sum of the squares of these amplitudes. It was difficult to reconcile how a theory could correctly predict the difference between the squares of two amplitudes and not at the same time predict the sum. Consequently it is not surprising that the differential cross sections reported by Stockman et al. (9) for the excitation of the 4 ^{2}P state in potassium essentially agree with those predicted by the CCC theory.

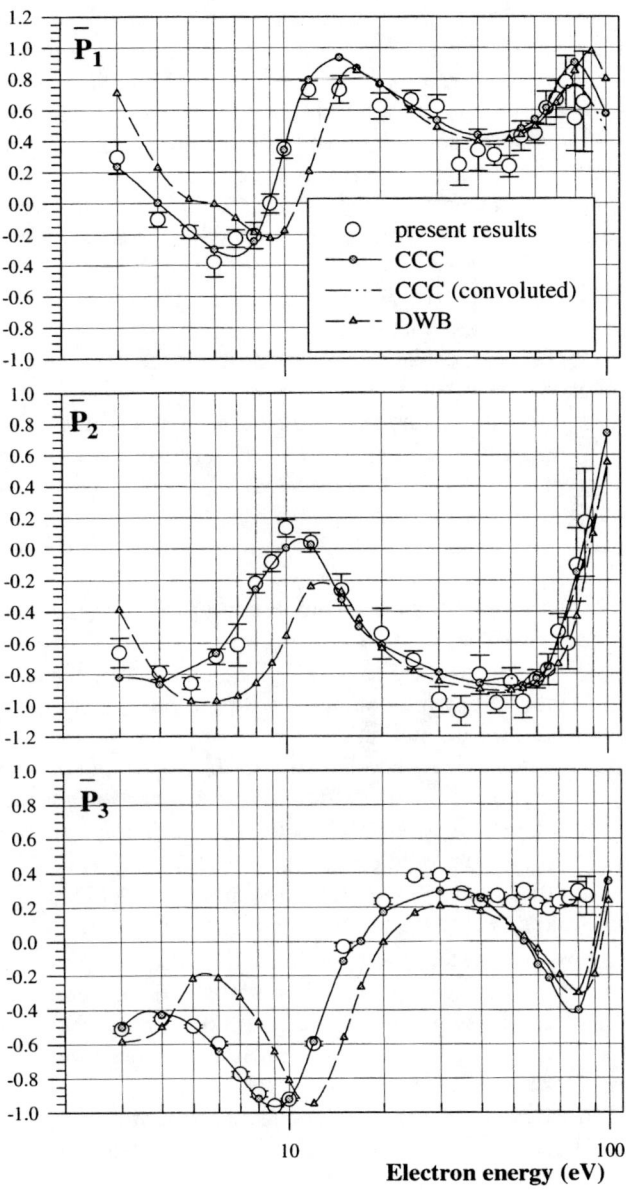

FIGURE 2. The equivalent Stokes parameters measured at a scattering angle of 60° as a function of the incident energy compared with a CCC calculation and a distorted wave Born calculation (DWB).

In this case angular distributions were measured using equation (6). The measurements were normally taken in several parts and were repeated several times to check for consistency. Different parts of the measured angular distributions were matched in overlapping angular regions and particular care was taken to ensure that there were large overlaps between each range. The resulting angular distribution was integrated numerically using Simpson's rule and normalised to the absolute value of the integral cross section measured by Phelps et al. (13). This value had been corrected for cascades and was in good agreement with the measurements of Chen and Gallagher (14). Stockman et al. (9) compare their differential cross sections at 54.4 eV with the predictions of five different theories. The theory that best predicts the observed behaviour is the six state unitarized Born approach of Mitroy (15). This theory uses a more sophisticated wavefunction for the target than does that used by Bray in the CCC theory.

Bray treats potassium as a quasi one electron atom. The eighteen electrons in the core are modelled by the Hartree Fock frozen core approximation together with a small phenomenological core polarization potential [Bray et al. (16)]. At an incident energy of 54.4 eV it is possible to eject two electrons from the core 3p orbital and as the energy is increased the validity of the frozen core model should be increasingly tenuous.

In an effort to explore this phenomenon, Stockman et al. (17) measured the three pseudo Stokes parameters at fixed scattering angles of 45°, 60° and 90° whilst the electron energy was varied from 2 eV to 90 eV. The results of such an investigation are shown in Figure 2 where they are compared with the predictions of a CCC theory and a distorted wave Born approximation theory.

It is clear that the excellent agreement between CCC theory and experiment for the linear components $\bar{P}_1$ and $\bar{P}_2$ is maintained over a very large energy range. Nevertheless at an incident energy of about 60 eV significant departures are seen between the experiments and the predictions of both of the theories. This behaviour was reproduced at the other scattering angles. Stockman et al. (17) measured all three components of the Stokes vector during the same run. Thus the discrepancy in $\bar{P}_3$ cannot be explained by any systematic error in the angular position of experiments. We suspect that the disagreement between theory and experiment in this case is reflected by inadequacies in the target wavefunction. This behaviour is the first example where the CCC theory is in serious conflict with observations for discrete inelastic scattering.

In the course of an extensive study of superelastic scattering from sodium, lithium and potassium it has become clear that the CCC theory is outstandingly successful in predicting orientation and alignment parameters over a wide range of energy and scattering angles. Indeed this theory is so successful that the Flinders group has uncovered rather esoteric sources of systematic errors that at first sight obscured the validity of the experiments. In other words the existence of a reliable theory places very stringent conditions on acceptable experimental practice in the field.

The lighter alkalis can clearly be described successfully with non relativistic theories both in the structure and scattering part of the calculations. The next step in superelastic scattering experiments lies with targets that require relativistic treatments to describe the scattering process. Hall *et al.* (18) report measurements of the orientation parameter in a series of superelastic scattering experiments from rubidium at an energy of 20 eV referred to the ground state. The technique that was used in these experiments was essentially the same as that which has been described above. The major difference arises from the premise that LS coupling may not apply in the optical pumping phase so that an additional parameter K″ has to be introduced to account for this. K″ can be calculated and it can be measured. Hall *et al.* find excellent agreement between theory and experiment for this parameter.

They compare their measurements of the orientation parameter with those predicted by the CCC theory of Bray. They find that there is excellent agreement out to an angle of about 90° where significant departures are observed. This disagreement between theory and experiment may reflect inadequacies in the description of the target wavefunction or that the CCC theory is non relativistic or both.

P – S AND P – D EXCITATION IN SODIUM

Having prepared an excited state, it is possible in principle to scatter electrons from that state to higher lying states. Such processes are characterised in the electron energy spectrum by specific energy loss. For example, the excitation of the $4\,^2S$ state from the $3\,^2P$ state in sodium yields electrons with an energy loss of 1.08 eV whilst excitation of the $3\,^2D$ state gives rise to electrons that have lost 1.5 eV.

The experimental technique is very similar to that used in superelastic scattering experiments except that in this case the polarisation of the laser light is defined with respect to the incident beam direction rather than the scattered beam. The overall energy resolving power of the apparatus needs to be much greater than in standard superelastic experiments so that the individual excitation processes can be resolved. Shurgalin *et al.* (19) report an energy resolution of 140 meV compared with a typical energy resolution of the Flinders group of 600 meV. Energy resolutions of 140 meV can only be attained by using a monochromator on the incident beam and this necessarily reduces the incident electron current. Consequently the results of such experiments are limited to scattering angles less than 30°. Pseudo Stokes parameters as defined in equations (1) are measured for the P – S transitions and in the case of P - – D transitions it is also necessary to measure P_4 the linear component of the radiation emitted in the scattering plane.

The group from Griffith university has done extensive measurements on sodium. Shurgalin *et al.* (8), (20) have reported measurements of the 3P – 4S transition at 22, 30 and 50 eV over the angular range 2° to 25°. The first account focussed on the behaviour of the parameter $L_\perp$ which measured the angular momentum transferred normal to the scattering plane. Madison and Winters (21) had established a propensity rule for $L_\perp$ for the excitation of a P state from a ground S state. Their analysis was based on a Born series expansion for the transition matrix up to second order and they concluded that $L_\perp$ should be positive at small scattering angles and negative at larger angles. This rule has been verified by the experiments of Scholten et al. (22) at least out to scattering angles where the Born series expansion was valid. Extension of this model to positron scattering led Andersen and Hertel (23) to propose that $L_\perp$ should be negative at all scattering angles for electron impact deexcitation of an S state to a P state. This hypothesis was tested by Shurgalin et al. (8) in the time reversed equivalent experiment by studying the excitation of the 4S state from the 3P state.

The results at an incident energy of 23.1 eV showed that $L_\perp$ is almost zero at scattering angles less than 8°, then decreases to a negative value of 0.25 at a scattering angle of 15° in accordance with the suggestion of Andersen and Hertel (23). However $L_\perp$ increases at larger scattering angles and becomes positive at about 20° then continues to increase with the scattering angle. The behaviour of $L_\perp$ is in excellent agreement with that predicted by the CCC calculation. The second order Born calculation was demonstrably in serious disagreement with the results of the experiments over the whole angular range. The failure of the propensity rule in this case then reflects the failure of the theory on which it was based.

These results also show a benefit in studying electron scattering from excited states over a standard superelastic scattering experiment. Shurgalin et al. (8) also measured $L_\perp$ for the superelastic scattering process at an equivalent energy of 19.9 eV. These results are in excellent agreement with the predictions of the CCC theory and this is consistent with what Bray (5) had found when he compared his CCC predictions with the results of Scholten et al. (22) at an equivalent energy of 22.1 eV. There is however only a marginal difference between the CCC predictions and those of the distorted wave Born approximation of Madison for $L_\perp$ for the superelastic case. Indeed Shurgalin et al. (8) cannot distinguish between the two theories for this process. On the other hand there is a very large difference between the two theories for the excitation of the higher states.

This difference is also present when the theories are used to predict the linear components P_1 and P_2. However, at an incident energy of 23.1 eV the position is not as transparent as it was in the case of $L_\perp$. Neither theory predicts the observed behaviour of P_1 whereas the CCC theory is superior to the DWB2 theory in predicting P_2. The uncertainties in the experimental data are considerably greater than is the case for $L_\perp$ and this reflects the more significant role of the hyperfine structure in depolarising the linear components rather than the circular components.

These general features are observed at the higher energies. That is, there is excellent agreement between the experiment and the CCC theory for the superelastic scattering case and for $L_\perp$ and P_2 for 3P – 4S excitation but not for P_1.

Experiments such as those reported by the Griffith group (19) also yield data for the process of excitation of the $3\,^2D$ state in sodium from the $3\,^2P$ excited state. The study of such processes has been proposed by Andersen and Bartschat (24). They point out that increasing the angular momentum of the states involved in the electron scattering process leads to a substantial increase in the number of individual scattering amplitudes that characterise the scattering process. In the case of a P – D transition in sodium Andersen and Bartschat (24) suggest that 8 scattering amplitudes are required for each spin channel compared with only 2 scattering amplitudes for an S – P transition for the case where the total spin is conserved. Given the level of agreement that now exists between experiment and the CCC theory for S – P transitions there clearly is justification to test this theory for more complex P – D transitions not only in sodium but also in barium. These latter collisions are discussed below.

Shurgalin *et al.* (19) report on a study of 3P – 3D transitions in sodium at an incident energy of 30 eV over the angular range from – 10° to + 22°. In this study they measured the pseudo Stokes parameters P_1, P_2, P_3 as defined in equation (1) as well as the parameter P_4. P_4 is related to the height of the final state charge cloud which is in general non zero for a D state. The results of these experiments are used to derive either density matrix elements or the standard atomic collision parameters $L_\perp$, ρ_{00}, P_ℓ and γ. When these parameters are compared to the predictions of the CCC theory of Bray it is found that there is excellent agreement between theory and experiment.

The technique that has now been used successfully to test and prove the CCC theory was first described by Hertel and his collaborators over twenty years ago (7). These original experiments were performed with an electron monochromator on the incident electron beam so this group was able to observe electron scattering from excited states. In fact the first observation of the excitation of a D state was made by Hermann (25) at an incident energy of 5 eV.

INELASTIC SCATTERING FROM BARIUM

There is a long and comprehensive literature on the study of electron scattering from excited states in barium. Initial interest in barium as a target [Register *et al.* (26)] was stimulated partly by the fact that it could be efficiently pumped by a dye laser operating with a high gain dye. The most common isotope ^{138}Ba has zero nuclear spin so the optical pumping process is not complicated by fine and hyperfine structure as is the case in the alkalis. The equivalent Stokes parameters are then unhampered by depolarising influences and this has important consequences to the signal to noise ratios of barium excited state experiments. There is growing theoretical interest in the barium atom .

Following the success of Fursa and Bray (27) in applying the CCC theory to electron scattering from helium, these workers have attempted to apply the CCC theory to barium by considering that barium can be described as two valence electrons bound to a relatively inert core. Whilst this approximation may be considered drastic for an atom with an atomic number of Z = 56, it is clear that at least at forward angles it adequately describes the excitation of the P state from the ground state (28).

The experiments of Zetner's group at the University of Manitoba have concentrated on two separate processes; the measurement of differential cross sections and on the measurement of the Stokes parameters P_1, P_2 and P_3. These experiments have been done for the excitation of a variety of states and it is not possible to review all of these studies in this paper. Instead we focus on the excitation of the (… 6s 5d) $^1D_2$ state from the (… 6s 6p) $^1P_1$ laser excited state.

The excitation of the $^1D_2$ state involves an energy loss of 0.826 eV and one can measure differential cross sections for the excitation of this state using the approach implied by equation (6). Johnson et al. (29) have measured differential cross sections and Stokes parameters for this process at energies between 10 eV and 40 eV over the angular range from 0° to 35°. These data complement the work of Li and Zetner (30) at 20 eV for the same process. There is reasonable agreement between the measured differential cross sections and the CCC theory over the whole energy range but it is clear that the theory is better at 10 eV than at 40 eV. This level of agreement is also reflected in the Stokes parameters. At 10 eV there is excellent agreement between the CCC theory and experiment for the Stokes parameters but at 40 eV the theory bears little relationship to the measurements. Johnson et al. (29) show the results of two CCC calculations, one with 115 states and one containing 55 states. The differences between the two calculations are much more pronounced at 40 eV than they are at 10 eV.

ELASTIC SCATTERING FROM EXCITED STATES

Barium

Elastic scattering experiments in general are characterised by extremely small signal to noise ratios. These arise because electrons scattered from the background gas in the apparatus have essentially the same energy as those scattered from the target. Usually the partial pressure of the background gas is greater than that in the beam so elastic scattering experiments are inherently difficult. These problems are further compounded at forward scattering angles by the incident electron beam which also has the same energy as that of the scattered electrons.

In the case of elastic scattering from excited states the problem is further complicated by elastic scattering from ground state atoms in the beam. Nevertheless by chopping both the target barium beam and the laser beam and by using two laser beam positions the group at Jet Propulsion Laboratory (JPL) have been able to measure differential cross sections for elastic scattering from both the (… 6s 6p) $^1P^1$ and (… 6s 5d) ^1D, ^3D states of barium. The complexity of these experiments was described by Trajmar (31) in an invited talk at the last ICPEAC.

The experiments on the $^1P_1$ state involved 116 separated measurements for each fixed (E_0, θ) case so it is not surprising that differential cross sections have only been reported at 20 eV for scattering angles of 10°, 15° and 20°. Nevertheless the experiments are in excellent agreement in absolute value with the predictions of the CCC theory [Trajmar et al. (32)].

The (... 6s 6p) $^1P_1$ state in barium can decay into the (... 6s 5d) $^1D_2$ and the $^3D_2$ state. Consequently pumping the $^1P_1$ state with laser light of wavelength 553.71 nm from the ground (...$6s^2$) S_0 state yields a significant population of $^1D_2$ and $^3D_2$ atoms. The decay of these states to the ground state involves a $\Delta J = 2$ transition which is strictly forbidden under electric dipole selection rules. These states are therefore metastable and their influence on the elastically scattered signal can be investigated by adjusting the position of the laser beam in the experiments. Trajmar et al. (33) have reported differential cross sections for elastic scattering from the ^1D and ^3D states of barium at 20 eV; the mixture comprised 70% singlet and 30% triplet. The experiments are confined to the angles 10°, 15° and 20° and they are in excellent agreement with the predictions of a CCC theory. In this case the absolute values were assigned by comparing the measured D state signal with the known differential cross section for the S – P transition measured by Wang et al. (28).

Sodium

Vuscovic and her collaborators have studied elastic scattering from the 3 ^2P state in sodium using an atomic recoil technique that was first developed by Bederson and coworkers in the 1950s. When an electron is scattered elastically from a target its momentum changes. Consequently the target must recoil and the recoil momentum is such that the overall momentum of the system is unchanged in the collision. In principle one can gain as much information by looking at the atom as looking at the electrons. In this case the 3 $^2P_{3/2}$ state was prepared using optical pumping techniques (34) that have been described above. This group has also done a series of experiments on elastic scattering from ground state sodium atoms at a range of energies from 1.25 eV to 3.0 eV and for scattering angles from 20° to 60°. These results are in excellent agreement both in absolute value and shape with the predictions of a CCC calculation of Bray (5). It is surprising then that the CCC is about a factor of 3 less than the observations over the whole energy and angular range for elastic scattering from the excited state. Given the excellent agreement in the shape it is tempting to suggest that the discrepancy can be explained by a simple numerical error in translating a_0^2 to square angstroms or indeed a factor of $(2L + 1)$ may have been ignored.

CONCLUSIONS

There have been significant advances in the field of electron scattering from excited states since the last ICPEAC. There is now a comprehensive data set on superelastic scattering from the light alkalis. These data in general reinforce the outstanding role that the CCC theory plays in predicting the orientation and alignment parameters in electron scattering from the alkalis.

Results have been reported from the first experiments on an alkali target where one may expect the influence of relativistic effects to be evident. The group at Griffith University have also reported on a wide range of experiments in sodium in which higher lying states have been excited from the 3 ^{2}P state. The D state results are accurately described by the CCC theory whereas there is not such good agreement between theory and the experiments on the excitation of the S states.

There are now a wealth of results for electron scattering from excited states in barium. There is excellent agreement between CCC theory and experiment for elastic scattering from excited P and D states. With respect to the excitation of the D state from the P state, the orientation and alignment parameters predicted by the CCC theory do not agree with the observations as well in barium as they did in the D state in sodium. This may reflect inadequacies in the target wavefunctions used in the barium case.

ACKNOWLEDGEMENTS

Presenting a review of an active field requires the cooperation of a large number of people in the field. I am delighted to acknowledge the assistance of Ms. K. Stockman in the preparation of the results on potassium and I thank Drs. V. Karaganov and I. Bray for permitting me to discuss some of our potassium results prior to publication. Thank you also to Professor W. MacGillivray for providing data from the Griffith group and to Dr. D. Fursa, Dr. P. Zetner and Dr. L. Vuscovic for their data.

REFERENCES

1. Berakdar, J., Dorn, A., Elliot, A., Lower, J. and Weigold, E., in *Photonic, Electronic and Atomic Collisions*, Aumayr, F. and Winter, H.P. eds., World Scientific, Singapore p.295 (1998).
2. Farrell, P.M., MacGillivray, W.R. and Standage, M.C., *Phys. Rev. A* **44**, 1828 (1991).
3. Andersen, N., Gallagher, J.W. and Hertel, I.V., *Phys. Rep.* **165**, 1 (1988).
4. Hermann, H.W. and Hertel, I.V., *Comm. At. Mol. Phys.* **12**, 61 (1982).
5. Bray, I., *Phys. Rev. A* **49**, 1066 (1994).
6. Karaganov, V. and Teubner, P.J.O., in *Photonic, Electronic and Atomic Collisions*, Aumayr, F. and Winter, H.P. eds., World Scientific, Singapore p.291 (1998).
7. Hertel, I.V. and Stoll, W., *J. Phys. B: At. Mol. Phys.* **7**, 583 (1974).
8. Shurgalin, M., Murray, A.J., MacGillivray, W.R., Standage, M.C., Madison, D.H., Winkler, K.D. and Bray, I., *Phys. Rev. Lett.* **81**, 4604 (1998).
9. Stockman, K.A., Karaganov, V., Bray, I. and Teubner, P.J.O., *J. Phys. B: At. Mol. Opt. Phys.* **32**, 3003 (1999).
10. Buckman, S.J., Noble, C.J. and Teubner, P.J.O., *J. Phys. B: At. Mol. Phys.* **12**, 3077 (1979).
11. Vuskovic, L. and Srivastava, S.K., *J. Phys. B: At. Mol. Phys.* **13**, 4849 (1980).
12. Stockman, K.A., Karaganov, V., Bray, I. and Teubner, P.J.O., *J. Phys. B: At. Mol. Opt. Phys.* **31**, L867 (1998).
13. Phelps, J.O., Solomon, J.E., Korff, D.F. and Lin, C.C., *Phys. Rev. A* **20**, 1418 (1979).
14. Chen, S.T. and Gallagher, A.C., *Phys. Rev. A* **17**, 551 (1978).
15. Mitroy, J., *J. Phys. B: At. Mol. Opt. Phys.* **26**, 2201 (1993).
16. Bray, I., Fursa, D.V. and McCarthy, I.E., *Phys. Rev. A* **47**, 3951 (1993).

17. Stockman, K.A., Karaganov, V., Bray, I. and Teubner, P.J.O., to be submitted J. Phys. B.

18. Hall, B.V., Murray, A.J., MacGillivray, W.R. Standage, M.C. and Bray, I., *Abstracts of Contributed Papers 21st International Conference on the Physics of Electronic and Atomic Collisions*, eds. Itikawa, Y., Okuno, K., Tanaka, H., Yagishita, A. and Matsuzawa, M., University of Electro-Communications, Tokyo, p.190.

19. Shurgalin, M., Murray, A.J., MacGillivray, W.R. and Standage, M.C., *J. Phys. B: At. Mol. Opt. Phys.* **31**, 4205 (1998).

20. Shurgalin, M., Murray, A.J., MacGillivray, W.R., Standage, M.C., Madison, D.H., Winkler, K.D. and Bray, I., *J. Phys. B: At. Mol. Opt. Phys.* **32**, 2439 (1999).

21. Madison, D.H. and Winters, K.H., *Phys. Rev. Lett.* **47**, 1885 (1981).

22. Scholten, R.E., Shen, G.F. and Teubner, P.J.O., *J. Phys. B: At. Mol. Opt. Phys.* **26**, 987 (1993).

23. Andersen, N. and Hertel, I.V., *Comment. At. Mol. Phys.* **19**, 1 (1986).

24. Andersen, N. and Bartschat, K. *Adv. At. Mol. Opt. Phys.* **36**, 1 (1996).

25. Hermann, H.W., Ph.D. Thesis, University of Kaiserslautern, Kaiserslautern (1979).

26. Register, D.F., Trajmar, S., Jensen, S.W. and Poe, R.T., *Phys. Rev. Lett.* **41**, 749 (1978).

27. Fursa, D.V. and Bray, I., *Phys. Rev. A* **52**, 1279 (1995).

28. Wang, S., Trajmar, S. and Zetner, P., *J. Phys. B: At. Mol. Opt. Phys.* **27**, 1613 (1994).

29. Johnson, P.V., Eves, B., Zetner, P., Fursa, D. and Bray, I., *Phys. Rev. A* **59**, 439 (1999).

30. Li, Y. and Zetner, P., *J. Phys. B: At. Mol. Opt. Phys.* **29**, 1803 (1996).

31. Trajmar, S., Kanik, I., Khakoo, M.A., LeClair, L.R., Bray, I, Fursa, D. and Csanak, G., in *Photonic and Atomic Collisions*, eds. Auymayr, F. and Winter, H.P., World Scientific, Singapore p.187 (1998).

32. Trajmar, S., Kanik, I., Khakoo, M.A., LeClair, L.R., Bray, I., Fursa, D.V. and Csanak, G., *J. Phys. B. At. Mol. Opt. Phys.* **31**, L393 (1998).

33. Trajmar, S., Kanik, I., Khakoo, M.A., LeClair, L.R., Bray, I., Fursa, D.V. and Csanak, G., *J. Phys. B: Atl. Mol. Opt. Phys.* **32**, 2801 (1999).

34. Jiang, T.Y., Zuo, M., Vuskovic, L. and Bederson, B., *Phys. Rev. Letts.* **68**, 915 (1992).

Studies of Dissociative Attachment Reactions:
from Gas Phase to Condensed Phase

Ilya I. Fabrikant

Department of Physics and Astronomy
University of Nebraska, Lincoln, NE 68588, USA

Abstract. The resonance theory of dissociative electron attachment developed by O'Malley and Bardsley more than thirty years ago has become an efficient tool for *ab initio* calculations of dissociative attachment to diatomic molecules. We illustrate this by discussing application of non-local complex potential theory to the process of dissociative attachment to the HF molecule. For polyatomics we have developed a semiempirical resonance method based on the one-pole approximation for the R matrix. We demonstrate results of application of this theory to methyl halides in the gas and condensed phases. Many interesting observations, particularly a strong temperature effect for methyl chloride, vibrational Feshbach resonances for methyl iodide, and condensed-matter effects in dissociative attachment to methyl- and perfluoromethyl chloride are explained by our theoretical calculations.

INTRODUCTION

The processes of dissociative attachment (DA) of electrons to molecules are important for the description of a broad variety of phenomena including gas discharge, controlled thermonuclear fusion, astrophysical processes and environmental control [1]. In particular, dissociative attachment increases the breakdown threshold and decreases conductivity. The conductivity can even become negative, temporarily [1], due to the so-called attachment heating [2]. In fusion plasmas the attachment at the walls is the major source of H^- and D^- ions. In environmental applications, the chain of reactions, which remove SO_2 and NO_x pollutants and freons, involves DA.

In spite of the importance of DA reactions, they are relatively poorly studied theoretically. Recent reviews concentrate on the theoretical formulation and some sample results for diatomics [3], and experimental results [4,5]. Theoretical calculations still cannot be performed routinely like in the area of electron–atom and electron–molecule (non-reactive) collisions, and there are many examples where we have not achieved even qualitative understanding of the DA processes. This can be explained

CP500, *The Physics of Electronic and Atomic Collisions*, edited by Y. Itikawa, et al.
© 2000 American Institute of Physics 1-56396-777-4/00/$17.00

by fundamental theoretical difficulties: the DA process involves motion of two very different entities: electrons which should be described quantum-mechanically and nuclei whose motion is typically quasiclassical. The exact quantum-mechanical formalism is not well suited for a description of the nuclear motion, although it is possible in principle to describe the whole process fully quantum-mechanically [3]. Early quasiclassical ideas, which were introduced in fundamental DA papers [6,7], were recently advanced further [8], and it has become possible to develop more or less standard methods for diatomic systems. However, polyatomic systems still present a big challenge to the theory.

The situation becomes even more complicated when the DA process is affected by environment, like in the case of DA to clusters and adsorbates, and molecules embedded in condensed medium. Nevertheless, several successful attempts to describe the DA process in these cases have recently been made.

The present paper starts with reviewing relatively simple examples of attachment to diatomic molecules in the gas phase when DA cross section can be calculated *ab initio*. Then we will discuss a more complicated situation involving polyatomic molecules when semiempirical methods have been used. Finally we will discuss generalization of semiempirical methods allowing us to incorporate the evironmental effects.

METHODS

DA is essentially a resonance process going through formation of an intermediate molecular negative-ion state. It is very unlikely to transfer energy from electron to the nuclear motion without a resonance mechanism due to the large difference between the atomic mass and the electron mass. However not all methods use the resonance concept explicitly. In particular the DA process can be described by the three-body Faddeev equation method [9] and the effective range theory [10] which do not incorporate explicitly the resonance concept. However, both methods make certain model assumptions about the interaction between the electron and the molecule. On the other hand, the R-matrix approach [11] is able, in principle, to describe the DA process in an *ab initio* manner, although in practical realizations several R-matrix poles are necessary to describe one resonance, therefore this method is also not explicitly a resonance method.

The Feshbach projection-operator method, or its equivalent, Fano resonance theory, were used as long as thirty years ago [6,7] to formulate the resonance theory of DA. It was shown later [12] that an equivalent description can be given in terms of the one-pole approximation of the R-matrix theory. The present paper concentrates on these two approaches.

NON-LOCAL COMPLEX POTENTIAL THEORY

The DA process goes through two stages. At the first stage the electron is captured into (typically) an antibonding, unoccupied orbital, and at the second the atomic fragments move apart along the repulsive potential curve until they reach the stabilization point beyond which the autodetachment of the electron is not possible. These two stages can be described by a simple formula for the DA cross section [6,7]

$$\sigma_{DA} = \sigma_{cap} S \tag{1}$$

where σ_{cap} is the capture cross section and S is the survival factor. The capture occurs in accordance with the Franck-Condon principle at a fixed internuclear separation which is determined from the Franck-Condon condition

$$\epsilon_v - V_0(R) = E - V_-(R), \tag{2}$$

where E is the total energy of the electron+molecule system, ϵ_v is the vibrational energy of the molecule in the initial state, $V_0(R)$ is the potential energy curve for the molecule, and $V_-(R)$ is the potential energy curve for the negative ion.

Eq. (1) works quite well for narrow resonances. However, for broad shape resonances it usually fails: due to the uncertainty in the resonance energy, the transition region is not localized. More rigorously, the nuclear motion in the resonance state is described by the Schrödinger equation with a non-local complex potential [3]. This potential can be constructed by using the probability amplitude V_{bk} for the electron capture from the initial continuum state $|k\rangle$ into the discrete resonance state $|b\rangle$. However, there are several principal and computational difficulties in solving this equation. The simplest method to circumvent these difficulties is to use a local approximation [6,7,13] for the non-local potential. There are several versions of the local approximation which include the so-called semi-local approximation where the imaginary part of the non-local complex potential is taken into account exactly [14,15]. This is easy to do below the dissociation threshold since in this case the imaginary part is reduced to a finite sum of separable terms. One can try the same approach for the real part. This corresponds to the separable approximation used in several calculations [16,17]. An accurate approach which fully incorporates the non-local effects is based on the discretization of the vibrational continuum using the Lanczos [18] or Schwinger-Lanczos [19] discretization method.

An alternative approach [8] employs the quasiclassical representation of the Green's operator for the negative-ion state. This representation simplifies substantially the structure of the equation for the negative-ion wavefunction and allows us to obtain the solution through simple recursive equations containing Franck-Condon factors for transitions between neutral and negative-ion states. The Franck-Condon factors are calculated using a uniform Airy-function approximation [20]. Although this approach is not exact from the point of view of quantum mechanics, it is natural in this problem because the nuclear motion is essentially quasiclassical, and its accuracy was shown to be very high.

The quasiclassical approach was recently applied to *ab initio* calculations of DA to H_2 and HF molecules [21]. Both targets attach electrons at low energies but with very small cross sections, of the order of 10^{-20} cm^2. This occurs because of a very large width of the shape resonance responsible for the process. We illustrate the main features of the cross sections by presenting results for the HF molecule in Fig. 1.

The vertical onset at threshold is typical for DA processes going through a low-energy shape resonance, particularly for hydrogen halides [13]. Although the Wigner threshold law [22] predicts the square-root of the excess energy behavior, in practice the cross section becomes finite very close to the threshold because of the qasiclassical character of the nuclear motion. The measurements of the DA cross sections for HF [23] confirm these features. However, agreement with theory should be considered more qualitative than quantitative at the present stage.

Another feature which can be noticed from Fig. 1 is a strong enhancement of the cross section if the initial state is vibrationally excited. This effect appears due to a substantial increase of the Franck-Condon overlap and the survival factor for larger

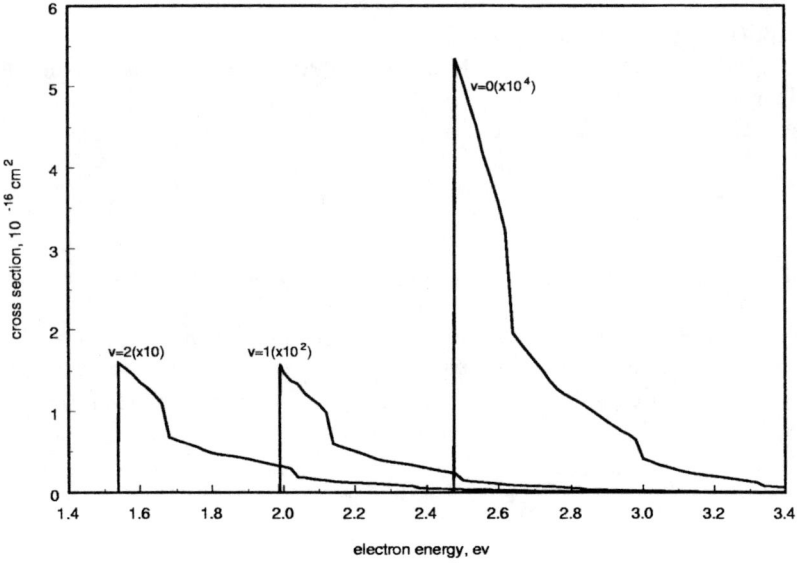

FIGURE 1. Dissociative attachment for various vibrational states of HF; note different scale for each v.

initial vibrational quantum number v. It can be usually measured by increasing the vibrational temperature of the target. The calculated results for vibrational enhancement in HF agree very well with the measurements of Allan and Wong [24].

RESONANCE R-MATRIX METHOD

In the R-matrix method the whole space for electron motion is separated into two regions: outside and inside the R-matrix sphere of radius r_0. In the outer region only the long-range part of the electron-molecule interaction (typically dipolar and polarization) is included. The wavefunction obtained in the outer region is then matched with the internal wavefunction using the R-matrix boundary condition. In the resonance approximation the R matrix can be written as [25] (for simplicity we consider the single-channel case)

$$R = \frac{\gamma^2}{W - E_e} + R_r,\tag{3}$$

where γ is the surface amplitude, W is the lowest pole of the R-matrix as a function of the electron energy E_e, and R_r is a residual term whose energy dependence can typically be ignored.

Expression (3) is completely equivalent to the Breit-Wigner formula for the scattering amplitude, therefore it is possible to find a one to one correspondence between the parameters of the Fano-Feshbach and R-matrix theories of the resonance scattering. In particular the complex shift F of the resonance (whose imaginary part gives the half-width) can be written as

$$F = \Delta - i\frac{\Gamma}{2} = -\gamma^2 L_E,\tag{4}$$

where $L_E = (u_E^+)'/[u_E^+ - R_r(u_E^+)']$ is the generalized logarithmic derivative of the electron wavefunction u_E^+ on the R-matrix sphere.

All the above equations are written for a fixed internuclear distance. However, it is easy to generalize them if we want to incorporate nuclear dynamics [12]. In particular the complex shift, Eq. (4), turns into the non-local complex potential whose explicit form is

$$F = -\sum_v \!\!\!\!\!\!\int \gamma|v\rangle L_v \langle v|\gamma,\tag{5}$$

where L_v represents now the logarithmic derivative in different vibrational channels, and the ket and bra represent the vibrational states of the target molecule. Both the summation over the discrete states and the integration over the vibrational continuum should generally be carried out.

The resonance R-matrix method can be easily employed for semiempirical calculations. The R-matrix pole W is basically equivalent to the resonance state of the

Fano-Feshbach theory. In the region of internuclear distances where the negative ion is stable it can be calculated using the quantum chemistry codes. Calculation of W for lower internuclear distances and calculation of surface amplitude often represent certain difficulties. In this case relevant information can be obtained from experimental data on resonance scattering (e.g., vibrational excitation) or from *ab initio* calculations of scattering phase shifts. The advantage of the R-matrix method is that all of the energy dependence comes from the logarithmic derivatives, L_E, and can be easily calculated, whereas in the similar applications of the projection-operator approach [26,27] the energy dependence of the width should be parametrized in a form consistent with the threshold behavior.

POLYATOMIC TARGETS

The semiempirical R-matrix method has been applied to several polyatomic molecules [28–31]. We will start our discussion with DA to methyl chloride. The C–Cl symmetric stretching mode ν_3, which is responsible for DA at low energies, is essentially uncoupled from other vibrational modes, therefore we treat the DA process assuming a fixed nuclear configuration in the CH_3 complex. According to the electron transmission experiment [32] electron capture into the lowest unoccupied molecular orbital of symmetry a_1 leads to a resonance whose position is 3.4 eV and width 1.3 eV. Experimental data on vibrational excitation [33] were used to extrapolate the negative-ion curve into the Franck-Condon region and to obtain the R-matrix parameter γ which determines the resonance width. The parameters of our earlier calculations [28,29] were substantially improved recently [34] and the experimental vibrational excitation cross sections are reproduced very well by our calculations.

In Fig. 2 we present the energy dependence of the updated DA cross sections at several vibrational temperatures. Because of the large resonance width, the negative-ion survival probability is extremely small if the process starts with the ground vibrational state. This explains why the cross sections are so small at room temperature. With growing temperature the population of higher vibrational states increases exponentially. This leads to a very rapid growth of DA cross sections [29]. For a fixed electron energy, the temperature dependence is well described by the Arrhenius law and agrees very well with measurements [35].

Another interesting feature which can be seen from Fig. 2 is sharp peaks near each vibrational excitation threshold. For comparison with the experimental data available for methyl chloride, the cross sections were averaged [35] over the experimental electron energy distribution of about 0.07 eV. Although the agreement is good, the peaks disappear after the averaging. More recent measurements [36,37] for methyl iodide performed with very high energy resolution confirmed the existence of the near-threshold peaks. Our semiempirical R-matrix calculations [37] allowed us to interpret these peaks as vibrational Feshbach resonances associated with the temporary capture of the incident electron into the long-range field of the vibrationally excited molecule. It was shown that both the dipolar and polarization interactions are responsible for the vibrational Feshbach resonances.

In Fig. 3 we present cross sections for attachment to CH_3I [37]. The two sets of experimental data correspond to two different degrees of initial vibrational excitation: in the nozzle-beam measurements vibrations are cooled down to essentially zero temperature. In the room-temperature experiment there is a substantial contribution from the first excited state. The theory is able to reproduce both sets of experimental data using one set of R-matrix parameters. In the case of the CH_3I molecule only one vibrational Feshbach resonance is observed. The cross section near the next vibrational excitation threshold exhibits a sharp cusp structure. Note that in contrast to the cusps observed in DA to the SF_6 molecule [38], the present cusp leads to a sharp drop of the cross section *below* rather than *above* the threshold. Therefore, a common interpretation in terms of loss of flux due to opening of new channels should be used with caution. Of course, the cusp structure is associated with unitarity, but the unitarity condition cannot be simply expressed in terms of probabilities.

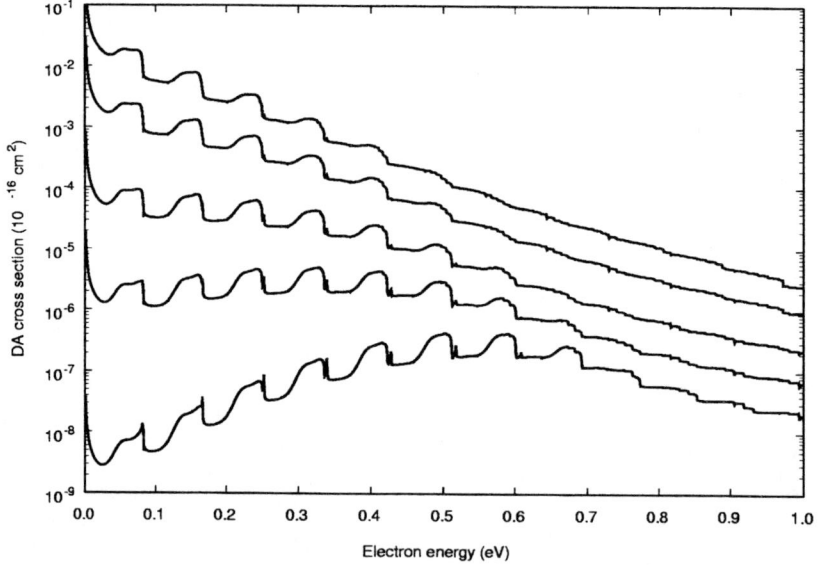

FIGURE 2. Dissociative attachment cross sections for CH_3Cl. Curves from bottom to top correspond to vibrational temperatures 300, 400, 500, 650, and 800 K respectively.

Comparison of the DA cross sections for CH_3Cl and CH_3I show enormous difference in absolute values and temperature dependencie. This can be explained by the different behavior of the anion curves: both are repulsive but the CH_3I curve crosses the neutral curve at much smaller internuclear distances which are closer to the Franck-Condon region. It is therefore interesting to compare the results for three methyl halides.

In Fig. 4 we present the potential curves plotted as functions of ρ, the internuclear distance relative to the equilibrium separation (which is different for each molecule). The anion curve for methyl chloride was computed [35], whereas those for methyl bromide and iodide where obtained semiempirically from the data on attachment rates. A small change in the position of crossing between neutral and anion curves leads to very sharp change in the magnitude of DA rate. In Fig. 5 we compare DA rate coefficients for the three targets.

In addition to very large difference in absolute values we see very different variation of rates with temperature. Whereas for methyl chloride the DA rate changes

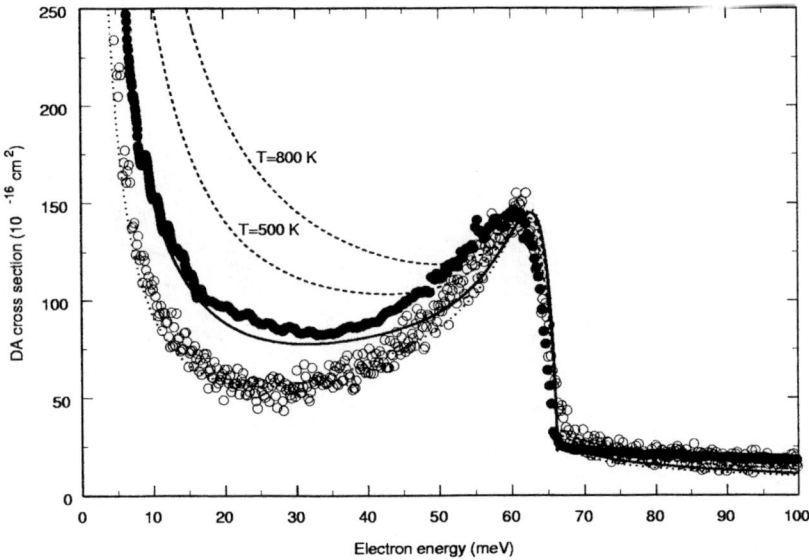

FIGURE 3. Dissociative attachment to CH_3I molecules. Full circles: experimental results for gas temperature $T_G = 300$ K; open circles: experimental results obtained for a supersonic beam target; full curve: calculation for $T_G = 300$ K; dotted curve: calculation for $T = 0$ K; dashed curves: calculations for T=500 and 800 K. Experimental cross sections are normalized to 145×10^{-16} cm^2 at the peak of the vibrational Feshbach resonance. (From Ref. [31])

277

substantially, it is almost flat for methyl iodide. Moreover, the methyl iodide data exhibit a rate increase at temperatures below 200 K. This occurs because of contribution of low-energy electrons. Due to the permanent dipole moment of methyl iodide, the DA cross section at low electron energy E increases as $1/E$ instead of the more common Wigner-law behavior, $1/E^{1/2}$. This leads to a $T^{-1/2}$ dependence on temperature which is consistent with the experimental data.

CONDENSED-MATTER EFFECTS

We have already seen that variation of the gas temperature can drastically change the absolute magnitude of the DA cross section. Similar effects occur if a molecule is placed near or in a polarizable medium. The polarization interaction between the temporary negative ion and the medium lowers the anion curves and changes the crossing point [42]. This effect was observed experimentally [43,44] by depositing

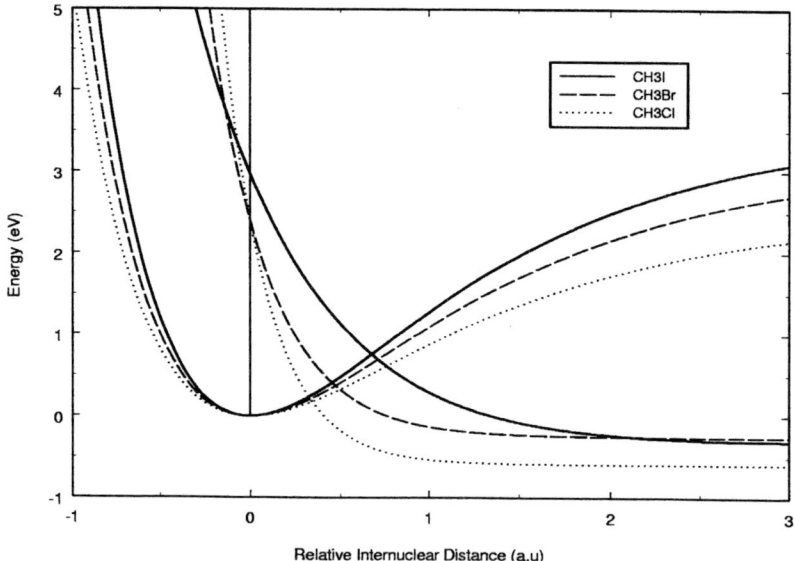

FIGURE 4. Neutral and negative-ion curves for three methyl halides plotted as functions of the internuclear distance relative to the equilibrium. Crossing of the solid vertical line with a negative-ion curve gives the vertical attachment energy.

methyl halide molecules on a film of rare-gas atoms with a Pt substrate or by sandwiching [45,46] a layer of molecules between two films. The second case is easier to treat theoretically. Indeed, if the film thickness is large enough, the electron wavefunction in the continuum can be calculated in the approximation of infinite medium. At very low electron energies, when the electron wavelength is large compared to the distance between atoms in the lattice, the Bloch wave can be approximated by a plane wave with a modified relation between the electron energy E and its momentum k

$$ E = E_0 + \frac{k^2}{2m^*}, \tag{6} $$

where E_0 is the bottom of the conduction band, and m^* is the electron effective mass in the lattice.

The result of the application of this model to DA from CH_3Cl and CF_3Cl molecules embedded in a Kr crystal is shown in Fig. 6. Calculations [47] using an electrostatic model give the polarization energy of a negative ion embedded in solid Kr to be $E_{pol} = 1.26$ eV. DA calculations [46] with this polarization energy satisfactorily reproduce the energy dependence of the DA cross section and its

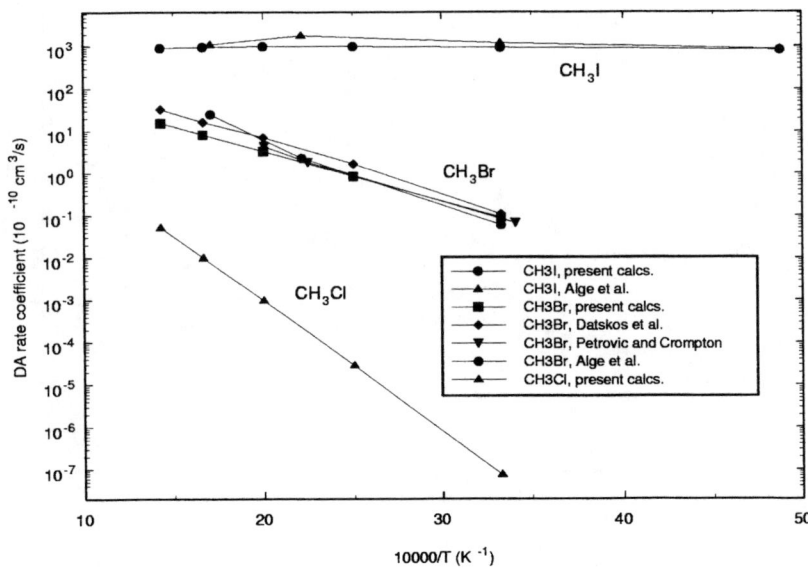

FIGURE 5. DA rates for three methyl halides. Experimental data are taken from Alge *et. al.* [39], Datskos *et. al.* [40], and Petrovic and Crompton [41].

enormous increase (by almost six orders of magnitude) due to condensed-matter effects. However, it is lower than the measured value by a factor of five. Assuming that the simple electrostatic model for an infinite medium does not reproduce the polarization energy quite correctly, we have adjusted E_{pol} and have obtained good agreement for $E_{pol} = 1.52$ eV. The situation with the CF_3Cl molecule is somewhat different. The cross section enhancement due to the presence of the polarizable medium is not as big, less than three orders of magnitude. The theoretical result (dashed curve in Fig. 6) employing the polarization energy $E_p = 1.26$ eV agrees with the experiment [45] for the position of the peak of the cross section, however the absolute magnitude of the theoretical cross section is higher than the experiment [45] by a factor of 8. To bring the theoretical value in accord with the experimental result, it is necessary to *decrease* the polarization energy which is not consistent with the adjustment procedure for methyl chloride. In addition, this adjustment moves the position of the peak of the cross section towards higher energies.

A possible explanation for this disagreement might be a suppression of the efficiency of the dissociative attachment due to the cage effect [48]. The dissociating

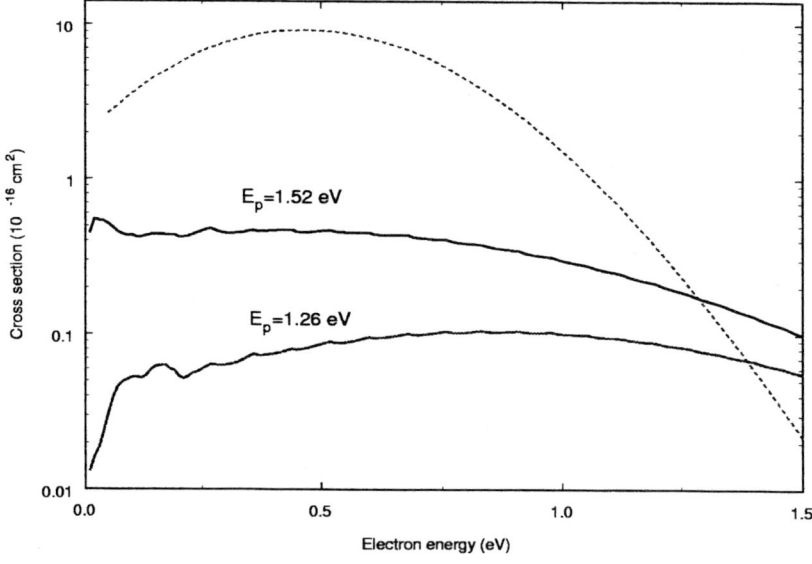

FIGURE 6. Dissociative attachment cross section for molecules embedded within a lattice of Kr film. Solid lines: results for CH_3Cl for two polarization energies; dashed line: result for CF_3Cl for $E_p = 1.26$ eV.

fragments, CF_3 radicals, interact repulsively with the Kr atoms and can be returned back to the reaction zone where the negative ion becomes unstable again. Due to the large decay width of CF_3Cl this leads to the large probability of autodetachment. The effect might be more important for the CF_3 rather than the CH_3 fragment because the velocity of the former after reaction is lower due to a larger mass.

For the molecules physisorbed on surfaces the problem of calculation of the electronic wavefunction becomes more complicated since the spherical symmetry is broken. In addition, in the experiments of the University of Sherbrooke group the thickness of the Kr film was varied. At low thicknesses the effects of the metal substrate should be included. Our preliminary analysis [43,44] ignores non-spherical effects by averaging the electron–surface interaction over the orientation of the electron momentum. Although this approach is rather crude, it is able to reproduce the major features of the DA cross sections: a strong increase in absolute value as compared to the gas phase and a non-monotonic dependence on the film thickness: for a thick film the polarization energy increases with decreasing thickness due to the metal substrate, and the cross section grows. However, for a thin film the resonance width starts to increase, and the cross section drops. For a more quantitative analysis we should use the low-energy electron diffraction (LEED) methods.

CONCLUSIONS

The resonance theory of electron–molecule collisions developed by O'Malley and Bardsley thirty years ago [6,7] has become an efficient tool for *ab initio* calculation of DA processes for diatomic molecules. Several technical difficulties associated with the treatment of nuclear motion in the non-local complex potential have recently been resolved by developing the Lanczos and Schwinger-Lanczos approach, and the quasiclassical approach.

The situation with polyatomic targets remains more complicated. In addition to much more complicated vibrational dynamics, we still do not have reliable methods for calculation of the capture amplitudes. In this case the semiempirical R-matrix method becomes very useful. In this approach the surface amplitudes of the R-matrix theory are adjusted to reproduce experimental data on other processes, e.g. vibrational excitation or resonant elastic scattering. Then these amplitudes can be used for the DA calculations with an additional assumption of the dominance of one vibrational mode for a given DA channel. This approach has allowed us to describe the observed energy and temperature dependence of DA cross sections for methyl halides and to find sharp vibrational Feshbach resonances in DA below vibrational excitation thresholds.

The semiempirical R-matrix approach is also useful in extending the theory to the description of condensed-matter effects in DA. The position of the negative-ion resonance is affected in this case by the polarization of the environment by the negative ion. This effect leads to a massive enhancement (by almost six orders of magnitude) of the DA cross section for the methyl chloride molecules adsorbed on surfaces or embedded in a bulk medium.

Future theoretical work should take into account interaction between different vibrational modes in polyatomic molecules and pursue the development of reliable methods of calculation of the capture amplitudes. For a more accurate account of condensed-matter effects we should use the electron wave function which incorporates the band structure of solids.

ACKNOWLEDGMENTS

This work became possible due to the contribution of graduate students, R. S. Wilde and Y. Xu, and due to a fruitful collaboration with P. D. Burrow, G. A. Gallup, H. Hotop, A. K. Kazansky and L. Sanche. Support from the US National Science Foundation through Grant No. PHY-9801871 is gratefully acknowledged.

REFERENCES

1. Aleksandrov, N. L., and Napartovich, A. P., *Usp. Fiz. Nauk* **163**, 1 (1993) [*Physics - Uspekhi* **36**, 107 (1993)].
2. Ness, K. F., and Robson, R. E., *Phys. Rev. A* **34**, 2185 (1986).
3. Domcke, W., *Phys. Rep.* **208**, 97 (1991).
4. Chutjian, A., Garscadden, A., and Wadehra, J. M., *Phys. Rep.* **264**, 393 (1996).
5. Burrow, P. D., Gallup, G. A., Fabrikant, I. I., and Jordan, K. D., *Aust. J. Phys.* **49**, 403 (1996).
6. O'Malley, T. F., *Phys. Rev.* **150**, 14 (1966).
7. Bardsley, J. N., *J. Phys. B* **1**, 349 (1968).
8. Kalin, S. A., and Kazansky, A. K., *J. Phys. B* **23**, 4377 (1990).
9. Drukarev G., and Pozdneev, S., *J. Phys. B* **13**, 2611 (1980).
10. Gauyacq, J. P., *J. Phys. B* **18**, 1859 (1985).
11. Schneider, B. I., LeDourneuf, M., and Burke, P. G., *J. Phys. B* **12**, L365 (1979).
12. Fabrikant, I. I., *Comments At. Mol. Phys.* **24**, 37 (1990).
13. O'Malley, T. F., *Phys. Rev.* **162**, 98 (1967).
14. Fiquet-Fayard, F., *J. Phys. B*, **7**, 810 (1974).
15. Bardsley, J. N., and Wadehra, J. M., *J. Chem. Phys.* **78**, 2227 (1983).
16. Atems, D. E., and Wadehra, J. M., *Phys. Rev. A* **42**, 5201 (1990).
17. Hickman, A. P., *Phys. Rev. A* **43**, 3495 (1991).
18. Mündel C., and Domcke, W., *J. Phys. B* **17**, 3593 (1984).
19. Meyer, H.-D., Horacek, J., and Cederbaum, L. S., *Phys. Rev. A* **43**, 3587 (1991).
20. Kazanskii, A. K., and Elets, I. S., *Zh. Eksp. Teor. Fiz.* **80**, 982 (1981) [*Sov. Phys. JETP* **53**, 499 (1981)].
21. Gallup, G. A., Xu, Y., and Fabrikant, I. I., *Phys. Rev. A* **57**, 2596 (1998).
22. Wigner, E. P., *Phys. Rev.* **73**, 1002 (1948).
23. Abouaf, R., and Teillet-Billy, D., *Chem. Phys. Lett.* **73**, 106 (1980).

24. Allan, M., and Wong, S. F., *J. Chem. Phys.* **74**, 1687 (1981).
25. Lane, A. M., and Thomas, R. G., *Rev. Mod. Phys.* **30**, 257 (1958).
26. Domcke W., and Mündel, C., *J. Phys. B* **18**, 4491 (1985).
27. Horacek, J., Domcke, W., and Nakamura, H., *Z. Phys. D* **42**, 181 (1997).
28. Fabrikant, I. I., *J. Phys. B* **24**, 2213 (1991).
29. Fabrikant, I. I., *J. Phys. B* **27**, 4325 (1994).
30. Fabrikant, I. I., Pearl, D. M., Burrow, P. D., and Gallup, G. A., in *Electron Collisions with Molecules, Clusters, and Surfaces*, ed. H. Ehrhardt and L. A. Morgan, (New York: Plenum), 1994, p. 119.
31. Wilde, R. S., Gallup, G. A., and Fabrikant, I. I., *J. Phys. B* **32**, 663 (1999).
32. Burrow, P. D., Modelli, A., and Chiu, N. S., *J. Chem. Phys.* **77**, 2699 (1982).
33. Shi, X., Chan, V. K., Gallup, G. A., and Burrow, P. D., *J. Chem. Phys.* **104**, 1855 (1996).
34. Fabrikant, I. I., and Wilde, R. S., *J. Phys. B* **32**, 235 (1999).
35. Pearl, D. M., Burrow, P. D., Fabrikant, I. I., and Gallup, G. A., *J. Chem. Phys.* **102**, 2737 (1995).
36. Hotop, H., Klar, D., Kreil, J., Ruf, M.-W., Schramm, A., and Weber, J. M., in *The Physics of Electronic and Atomic Collisions* ed. L. J. Dubé *et al* (New York: AIP), 1995, p. 267.
37. Schramm, A., Fabrikant, I. I., Weber, J. M., Leber, E., Ruf, M.-W., and Hotop, H., *J. Phys. B* **32**, 2153 (1999).
38. Klar, D., Ruf, M.-W., and Hotop, H., *Aust. J. Phys.* **45**, 263 (1992).
39. Alge, E., Adams, N. G., and Smith, D., *J. Phys. B* **17**, 3827 (1984).
40. Datskos, P. G., Christophorou, L. G., and Carter, J. G., *J. Chem. Phys.* **97**, 9031 (1992).
41. Petrovic, Z. Lj., and Crompton, R. W., *J. Phys. B* **20**, 5557 (1987).
42. Sambe, H., Ramaker, D. E., Deschenes, M., Bass, A. D., and Sanche, L., *Phys. Rev. Lett.* **64**, 523 (1990).
43. Sanche, L., Bass, A. D., Ayotte, P., and Fabrikant, I. I., *Phys. Rev. Lett.* **75**, 3568 (1995).
44. Ayotte, P., Gamache, J., Bass, A. D., Fabrikant, I. I., and Sanche, L., *J. Chem. Phys.* **106**, 749 (1997).
45. Nagesha, K., and Sanche, L., *Phys. Rev. Lett.* **78**, 4725 (1997).
46. Fabrikant, I. I., Nagesha, K., Wilde, R., and Sanche, L., *Phys. Rev. B* **56**, R5725 (1997).
47. Michaud, M., and Sanche, L., *J. Electron Spectrosc. Relat. Phenom.* **51**, 237 (1990).
48. Schriever, R., Chergui, M., Ünal, Ö., Schwentner, N., and Stepanenko, V., *J. Chem. Phys.* **93**, 3245 (1990).

Positron—Polyatomic Molecule Scattering: a Comparative Study with Electron Scattering

M. Kimura and O. Sueoka

Graduate School of Science and Engineering, Yamaguchi University, Ube, Japan

Y. Itikawa

Institute of Space and Astronautical Science, Sagamihara, Japan

Abstract. A comparable study on various processes in positron- and electron-molecule collisions is carried out. Positron-molecule interactions are known to give significantly different characteristics than those of electron-molecule are, consequently resulting in quite different scattering dynamics. We specifically examine some interesting cases: for ionization, it appears that positron impact gives slightly larger cross sections than those by electron impact, while for vibrational excitation, electron impact gives much larger cross sections by a few orders of magnitude for certain excitation modes than positron impact does. We also examine a few different experimental evidences for a possible resonance state arising from positronium annihilation, fragmentation, and positronium formation rate in positron-molecule collisions and discuss relations and implications among these experiments.

INTRODUCTION

A study of positron scattering from molecules contains a very interesting physics which is warrant for careful examination for research in atomic and molecular physics, condensed matter physics and high-energy physics experimentally and theoretically [1]. Yet, no systematic and comprehensive study exists for molecular targets. Although a great deal of applications based on positron impact on materials has been developed and indeed employed in recent years notably such as the positron emission tomography (PET), and positron-microscope, very little basic knowledge and information for positron scattering dynamics from gaseous targets has been known. For example, it has been speculated over decades that positron can form a resonance, or attach to atoms or molecules when it interacts with these under certain circumstances. However, there has been observed no firm experimental evidence for the resonance/attachment. For theoretical side, rigorous investigation whether positron can really form a bound state with a molecule has not been conducted because the positron interaction becomes relatively weaker compared to electron counterpart, and therefor, any theoretical calculation depends very sensitively on any model interaction proposed. Hence high precision numerical calculation is exceedingly difficult for molecules.

By comparing various processes in positron and electron scattering from atoms and molecules, one can notice a quite different feature in their behavior and

CP500, *The Physics of Electronic and Atomic Collisions*, edited by Y. Itikawa, et al.
© 2000 American Institute of Physics 1-56396-777-4/00/$17.00

characteristics in dynamics. For example, based on a simple Born argument, Takayanagi and Inokuti [2] estimated rotational excitation cross sections for positron collisions with molecules. They demonstrated a notable difference between positron and electron collision in the process. Raith's group at Bielefeld [3-4] measured ionization cross section from rare gas atoms such as He and Ne and H_2 by both electron and positron impact, and found that ionization cross section by positron impact appears to be larger than that of electron. They suspected that the absence of exchange interaction is in part responsible. Through careful and thorough investigation, essential knowledge of interactions and scattering dynamics is attained at much deeper level than those studied for each projectile alone individually [5]. Therefore we believe that since vigorous research activity on positron scattering from molecular targets has been performed at various institutions in the world last a few years, it is extremely timely to review the current status of various aspects of positron scattering from large molecules. We place the special emphasis on the close comparison with electron scattering.

Although the positron-molecule collision research group is small in number, all members are extremely active and productive, and very important studies have recently come out from this group as exemplified in references (see experimental new findings [6 – 28] and theoretical results [29 – 42]. We also intend to review some of these recent results.

BASIC IDEAS

Interactions: The interaction potential formed by a molecule is always anisotropic. Molecules have electric multipole (i.e., dipole, quadrupole and so on) moments. Interactions of an incident positron or electron with those multipoles are usually anisotropic and of long range. This feature is one of the noticeable differences between the positron/electron-atom and -molecule collisions. The interaction potential between the positron or electron and a molecule consists of three components: *electrostatic, exchange* and *polarization*. The static potential arises from the interaction between the projectile and the electrostatic field of the molecule. Thus the static potential for the positron collision is the same as that for the electron collision except for its sign, i.e., *attractive for electron* and *repulsive for positron*. There is no exchange interaction in the case of positron-molecule collision. Induced polarization plays a special role in the positron and electron collision with a molecule. In the asymptotic (long-range) region, the polarization interaction for the positron-molecule collision is exactly the same as that for the electron-molecule collision, which is attractive for both projectiles. As the projectile comes closer to the molecule, the distortion of the molecular charge cloud becomes different for different projectiles. The polarization potential at short to intermediate range, particularly its anisotropic part, should be quite different for the two projectiles because the electron-electron correlation is strongly repulsive while the positron-electron correlation is strong attraction. As mentioned above, the long-range interaction due to the molecular multipoles is important in the positron-molecule collision. The polarization interaction is another example of long-range interaction, but the cumulative effects of the two types of interactions are totally different for the two projectiles (additive for electron and canceling for positron).

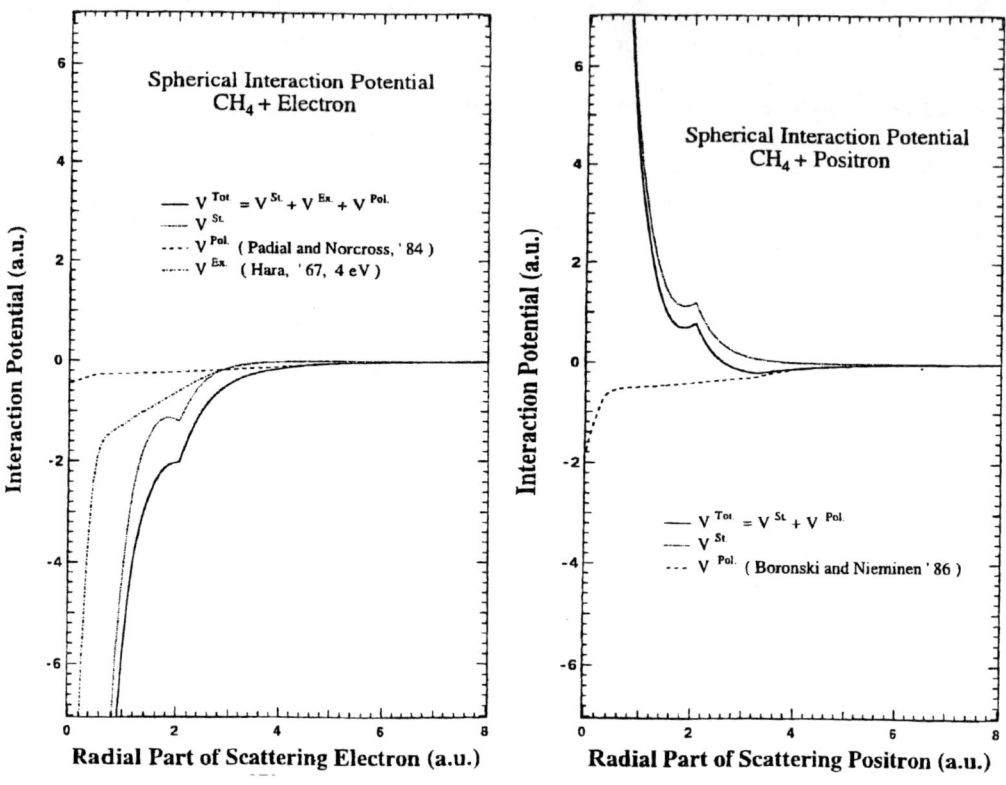

Figs. 1-a and **1-b**   Interaction potentials for electron-CH_4 and positron-CH_4, respectively.

For clear visualization, the interaction potentials arising from three terms above are plotted in Figs.1-a and 1-b for electron and positron, respectively, interacting with CH_4 molecule as a function of electron/positron-molecule distance.   For electron interaction, the static, correlation-polarization and exchange are all attractive and hence, total interaction is the sum of all three contributions, while for positron interaction, the static interaction and correlation-polarization interaction are in the different sign and hence, they cancel.   Therefore, as seen for positron, at smaller r within about 1.4 a_0, the total interaction becomes strongly repulsive, while at large r beyond 1.4 a_0, it becomes attractive.   In the region in-between, one can observe a shallow well around 1.5 a_0, and this type of the well can hold the incoming positron temporarily, causing a resonance state, i.e., positronium-in-molecule. And this resonance may be significant from dynamical point of view. Figure 1-a clearly shows that the total interaction for electron is uniformly attractive and stronger as a whole. Although very little information is available for the correlation, within the present model, the positron-electron correlation is stronger than the electron-electron correlation, and this should be reasonable. The kink seen at 2.05 a_0 is due to charge distribution from H atom in methane. At much larger r, the total interaction becomes equivalent to that of the polarization. In general, because of the cancellation of the interaction for positron, the total interaction becomes highly sensitive, as exemplified in the figure, to the choice of the polarization potential since it governs the point where the total changes from attractive to repulsive and the slope the potential. These factors in turn reflect on scattering calculation suggesting that the *ab initio* high-precision calculation for the case of positron scattering is more difficult to achieve.

Positronium formation is one of the most unique features observed in positron scattering. Despite of a relatively large volume of experimental attempts, very few accurate measurements of absolute cross sections for positronium formation from atomic and molecular targets has been reported. This process is regarded to be equivalent to *charge transfer* or electron capture in ion-atom collisions. If an electron from the target fails to be captured by an incoming positron, then it ends up as an ionization event, or in ion-atom terminology, *"charge transfer-to-continuum."*. Another conspicuous phenomenon in electron and positron scattering is a **resonance** [44]. Due to the complexity of the potential, an electron is known that often it is temporarily captured by a molecule, and consequently, this temporal negative ion contributes to enhancement for molecular dissociation. This subject has been well studied, and gives rise to a so-called "shape resonance" in the electron scattering. For positron impact, although there have been some indirect experimental evidences which suggest the existence of the resonance state of molecules, nevertheless it is of extremely interest to see if more solid and complete finding of a resonance can be made for positron impact. The shape resonance affects also the rotational and vibrational excitation processes. Even if no shape resonance occurs, the dependence on the projectile (i.e., positron vs. electron) can appear differently in the processes of rotational and vibrational excitations. A comparative study of the positron- and electron-molecule collisions, particularly that of the rotational and vibrational excitation (and dissociation) would be fruitful in understanding the dynamics of the interaction of both the positron and electron with molecules. A polyatomic molecule has a multiple mode of vibration and the dependence on the projectile may be different for different modes. In this sense, vibrational excitation is more sensitive to the short-range part (i.e., near the molecular nuclei) of the interaction. For the same reason, preferable dissociation channels in the positron molecule collision may differ significantly from those in the electron molecule collision.

In what follows, we discuss the difference and similarity for a specific process from ionization and rovibrational excitation to positronium formation by positron and electron impacts on polyatomic molecules.

EXAMPLES OF POSITRON AND ELECTRON SCATTERING

A. Ionization and Electronic Excitation Processes

A comparison of the electron-impact and positron-impact ionization of atomic targets has been studied several times. Raith's group at Bielefeld measured ionization cross section for both projectiles for rare gases and H_2 targets [3 – 4]. Figure 2 summarizes their results for He in the energy from threshold to a few 100 eV region as an example. Although error bars are relatively large, they appear that ionization (without Ps formation) cross section by positron impact is somewhat larger than that of electron impact. Although the behavior of the ionization cross section of H_2 is slightly different from that of rare-gas atoms, the larger ionization cross section for positron impact is also seen. For the interpretation of this feature of larger cross section in positron impact, although the absence of exchange interaction for positron is expected to play some roles, the *charge transfer-to-continuum* should also be important and contribute significantly. If one includes the Ps formation as an ionization, then clearly total ionization (direct ionization + Ps formation) by

Fig. 2 e⁻ and e⁺ impact ionization and Ps formation from He

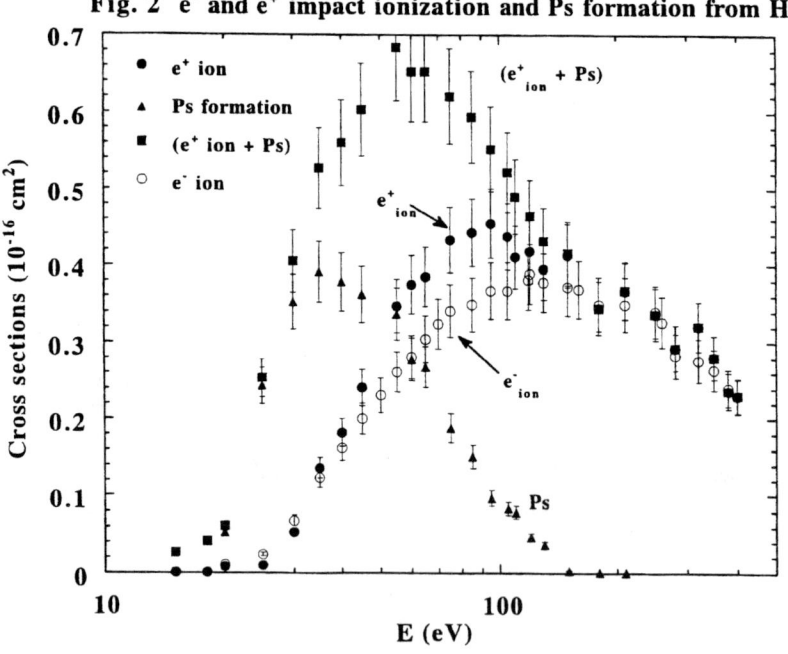

positron impact fur exceeds that of electron impact. To the best of our knowledge, there is no systematic and rigorous study for ionization process experimentally and theoretically. In addition, Raith's group is among the first who observed the Wigner cusp in elastic cross section in positron-He scattering.

Only example of the calculation of the positron-impact excitation of electronic state of a molecule is the excitation of B Σ state of H_2. Two groups used two different theoretical approaches to obtain the differential, as well as integral, cross sections for the excitation of the B state [35, 45]. However, there found a large disagreement between the integral cross sections obtained by these two groups. The difference may be ascribed to the description of the excited state in the two calculations. It should be noted that differential cross sections for positron collisions obtained by Lino et al. [35] at 20 and 30 eV are similar to those for the excitation by electron collisions. The authors explained this by the fact that the excitation of the B Σ state is governed mainly by a dipole transition, which is strong and long-range interaction.

B. Rovibrational Excitation Processes

Vibrational excitation of molecule by electron impact is often enhanced by the formation of a temporal negative ion state (resonance). Vibrational excitation through the resonance causes a variety of structures in addition to a large magnitude of the cross section. For positron impact, no such resonance state is known to exist at the present, and hence a cross section is expected to be small and rather smooth in general. However, as is shown in Fig. 3 the total cross sections for CO_2 by positron and electron impacts, the cross section

Fig. 3 e⁻- and e⁺-CO₂ scattering total cross sections

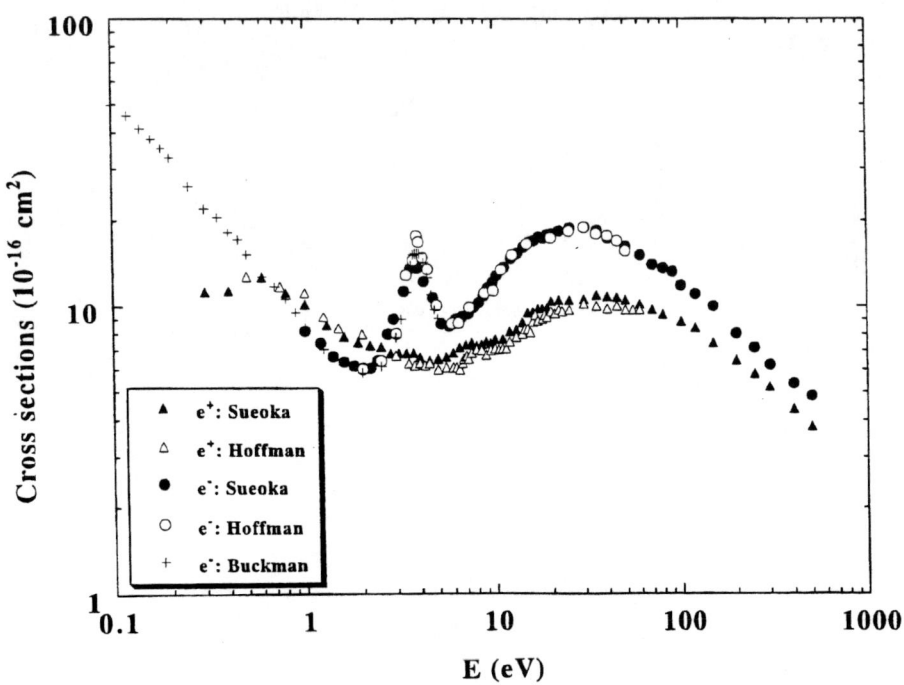

for positron becomes larger than that of electron below 2 eV, and it reverses the order at around 0.7 eV [41]. As stated above, generally the cross section for positron is expected to be smaller than that of electron since the interaction is weaker. So what causes this peculiar behavior in the cross section? In order to answer to this question, we have recently carried out a joint theoretical and experimental study for CO_2 molecular vibrational excitation by using the two-channel close coupling method for electron and positron impacts at 2 - 6 eV as illustrated in Fig. 4 [42]. For the (100) symmetric stretching excitation, we have found that the cross section by electron impact is larger by almost three orders of magnitude than that by positron impact, while the cross sections for (010) bending and (001) asymmetric stretching excitation are found to be nearly identical for both projectiles. This theoretical finding is substantiated by the experiment that utilizes the energy-loss technique. The result is surprising and unexpected. But the interpretation for this finding is that for the (100) symmetric excitation, the weaker polarization interaction is short range and whether positron or electron can penetrate inside the molecule is important to yield sufficient overlap with the interaction. On the other hand, for (010) bending and (001) asymmetric stretching excitation, the induced dipole interaction is long range, and hence, whether positron or electron can penetrate inside molecule would not affect the coupling so much. As a summary, for the CO_2 case, the sum of all vibrational excitation cross sections for electron impact is found to be much larger than that of positron impact.

There have been a few calculations on the rotational excitation of molecules by positron impact by using the perturbative and close-coupling methods [30, 38 – 40].

Fig. 4 e⁻- and e⁺-CO₂ vibrational excitation
for (100), (010) and (001) modes

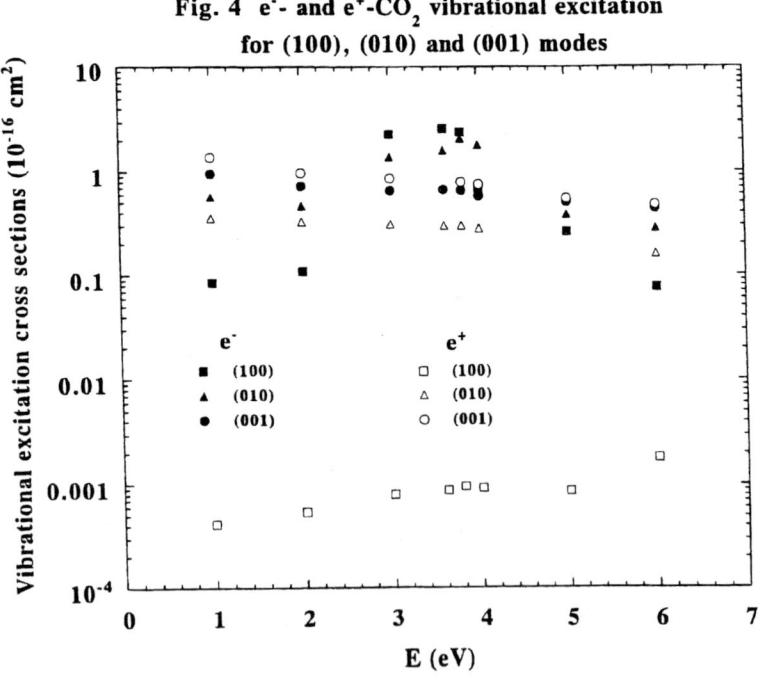

Earlier, Takayanagi and Inokuti [1] investigated rotational excitation by electron and positron impact by using the Born approximation, and found that for diatomic molecules with a negative quadrupole, the rotational excitation cross section by positron should be larger than that by electron. Contrarily, the situation reverses when the molecule has a positive quadrupole moment. For the CO_2 case, its quadrupole has a value of -3.802 a.u . and hence, the rotational excitation by positron impact becomes larger than that by electron impact even at threshold, and the difference widens as the impact energy increases. Although the validity of this theory is questionable for providing accurate numerical values particularly at higher energies, it still may be correct qualitatively. Hence, tentatively, the reverse of the cross section below 2eV for CO_2 may be due to the rotational excitation. The representative example for the rotational excitation for J=16 → 18 obtained by the Born theory is displayed in Fig. 5. As discussed, the rotational cross section by positron is apparently larger, and the difference widens as the incident energy increases.

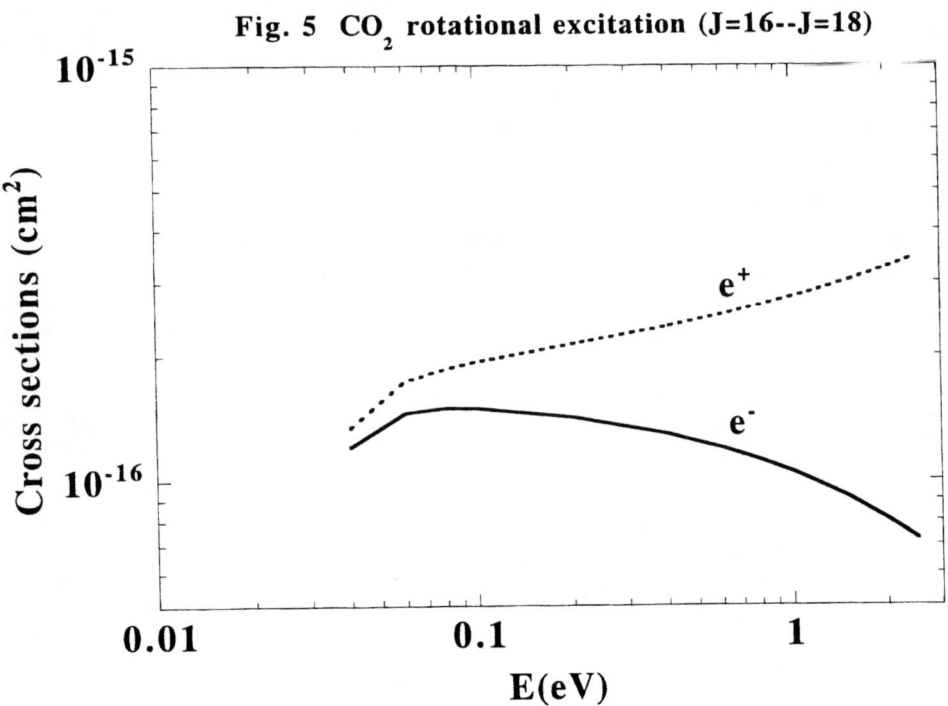

Fig. 5 CO_2 rotational excitation (J=16--J=18)

C. Positronium Formation

Positronium formation is a unique channel, which is possible only for positron impact. This process is considered to be equivalent to charge transfer in ion-atom collisions. When the speed of the incoming positron slows, and the molecular potential is sufficiently deep to support a bound state for the positron, then this positron-molecule complex system might have a bound state, i.e., positronium-in-molecule. If the size of the molecule becomes large, then the number of degree of freedoms of internal motions increases, in which they can easily absorb the extra kinetic energy of the incoming positron and trap it. Since the static interaction is negative for positron, the trapping should occur at the tail, or, near the tail, of the electronic wavefunction of outer most atoms in a molecule. In table I., the ratio of the Ps formation to the total cross-section is tabulated for some simple hydrocarbons, and their fluorine-substituted hydrocarbons at 2 eV above Ps formation threshold [43]. Interestingly, the ratios for fluorine-substituted hydrocarbons are much smaller by a factor of four than those for hydrocarbons. For F-substituted hydrocarbons, the electron charge distribution concentrates more on F atoms due to their higher electronegativity, while for hydrocarbons, the charge distribution is on C atoms in the center of molecules. Therefore, the incoming positron is likely to be attracted to F-substituted hydrocarbons and with high possibility, it may be captured and form a bound state, thus reducing the Ps formation. This may be an interpretation for smaller ratios for F-substituted hydrocarbon, and is indicative to a possible bound state or resonance.

Table I. Ps formation cross sections (Q_{Ps}) and total cross sections (Q_T) for hydrocarbons and fluorine-substituted hydrocarbons and their ratios at 2 eV above each threshold.

Molecules	$Q_{Ps}(\text{Å}^2)$	$Q_T(\text{Å}^2)$	Q_{Ps}/Q_T (%)
CH_4	1.8	7.2	25
C_2H_6	3.2	13.9	23
C_3H_8	4.6	22.2	21
CF_4	0.45	9.0	5.0
C_2F_6	0.73	15.2	4.8
C_3F_8	0.90	18.6	4.9

Two sets of circumstantial evidences for possible bound states were also obtained by Surko et al. [24, 25], and Xu et al.[27], who measured the positronium annihilation rates, and fragmentation products, respectively, for a variety of large molecules. In measurements by Surko and his group, they found unusually large annihilation rates for larger hydrocarbons, and smaller rates for F-substituted hydrocarbons. The origin of these phenomena of their experimental results is expected to share the same dynamics with our measurements in which we looked at the ratios of the total and Ps formation for a number of molecules. Hence we believe that these two different experimental results should be consistent. The experimental results by Xu et al show a large fragmentation probability and a variety of fragment products, i.e., from diatomic to relatively larger fragments, when low energy positron is impinged upon hydrocarbons below a few eV. These fragmentation processes may be proceeded through resonances since the collision energy is well below Ps formation threshold. These experimental findings described are very important and significant, because these experiments may open up new area in atomic physic [46]. Furthermore, these processes closely relate to similar phenomena observed in condensed phases such as positron trapping in solid and liquid phases, and positron stowing down in these phases. However these very interesting experimental observations still wait for rigorous theoretical interpretations.

D. The Ramsauer-Townsend effect

As well studied for rare gases, election impact gives the Ramsauer-Townsend (RT) minimum in elastic cross section for Ar, Kr, and Xe at around 0.1 – 0.3 eV, while it does not give to He and Ne because of the stronger repulsive potential for these lighter rare gases. Contrary, for positron impact, however, He and Ne are found to show the RT minimum while Ar, Kr, and Xe do not. For molecular targets, CH_4 appears to be only one case studied rather carefully. For electron impact, the eigenphase sum for s-wave crosses zero at 0.2 eV region, thus indicating the RT minimum in elastic cross section, while for positron impact, eigenphase sums for some waves cross zero at rather higher energy above 1 eV. In this energy domain, many partial waves contribute to a scattering event, and hence, even minima in certain partial-wave phases are likely to be washed out hence causing the RT effect disappeared entirely. These are another clear and marked example for the difference of dynamics for electron and positron impacts. The RT minimum is important for understanding the scheme for positron- and electron-molecule interactions, and also for applications such as mobility in condensed matter.

E. Positron in Condensed Phases

Brief comments on positron in condensed matter would be interesting in comparison with positrons in gaseous phase. There have been wide and resurgent interests in the use of positrons as a probe for studying condensed phases. Important physical quantities in describing the interaction of positrons and electrons in condensed phases are the inelastic mean free-path and energy-loss per unit pathlength. Theoretical calculations for these quantities, based on the dielectric response function, in various types of condensed matter show the larger energy-loss for positron in comparison with electron case [47]. This has been interpreted that as a positron approaches other atoms and molecules in the condensed medium, electrons from these atoms and molecules "swarm" toward the incoming positron. Hence, the positron undergoes more collisions thus loosing more energy, while as an electron approaches to these atoms and molecules, all electrons are repelled from the incoming electron thus reducing interactions. This difference in feature causes the larger stopping power for positron than that for electron, and also the smaller mean free path for positron. However, very little study has been carried out in this field, and much less information is available.

SUMMARY

Reflecting the strong contrast in the interaction scheme for two projectiles, there have been seen significant differences in dynamics for various collision processes, and some of these significant points can be summarized:
- (i) For ionization, positron impact appears to give larger cross sections than those of electron impact.
- (ii) For electronic excitation, because of no exchange interaction for positron impact, specific excitation processes such as singlet-triplet transitions through the exchange are forbidden for positron impact.
- (iii) For vibrational excitation, electron impact gives much larger cross sections than those of positron impact for certain types of excitation modes,
- (iv) For positron attachment or resonance, there are some indirect evidences to support their existence such as annihilation rates, Ps formation rates and measurements of fragment products.

Although some exploratory studies for positron and electron impact on large molecules exist, a systematic and comprehensive study has not been carried out yet for better understanding the interaction and scattering dynamics to provide rationales to these experimental findings. Much concentrated and joint efforts are desirable. The knowledge for these for molecular targets closely relates with that in condensed phases at much fundamental level and bridges among different research fields. Therefore, a close collaboration in positron research community from all sub-fields of physics and chemistry would be essential for further progress of our understanding.

ACKNOWLEDGEMENT

This project was supported in part by the Grant-in-Aid from the Ministry of Education, Science and Culture, Japan. The authors thank Tamio Nishimura, Bill Raith, Cliff Surko, and Walter Kauppila for providing us numerous suggestions, comments and data.

REFERENCES

[1] M. Kimura, O. Sueoka, A. Hamada and Y. Itikawa, Adv. Chem Phys. **111**, 537 (1999).

[2] K. Takayanagi and M. Inokuti, J. Phys. Soc. Jpn. **23**, 1412 (1967).

[3] D. Fromme, G. Kruse, W. Raith and G. Sinapius J. Phys. B: At. Mol. Opt. Phys. **21**, L261 (1988).

[4] A. Deuring, K. Floeder, D. Fromme, W. Raith, A. Schwab, G. Sinapius, P. W. Zitzewitz and J. Krug, J. Phys. B**16**, 1633 (1983).

[5] W. Kauppila and T. S. Stein, Adv. At. Mol. Opt. Phys. **26**, 1 (1989).

[6] M. Charlton, T.C. Griffith, G.R. Heyland, K.S. Lines and G.L. Wrigh, J. Phys. B**13**, L757 (1980).

[7] G. Laricchia, J. Moxom and M. Charlton, Phys. Rev. Lett. **70**, 3229 (1993).

[8] P. Ashley, J. Moxom and G. Laricchia, Phys. Rev. Lett.**77**, 1250 (1996).

[9] G. Laricchia, J. Moxom and M. Charlton, Phys. Rev. Lett. **70**, 3229 (1993).

[10] A. Kover and G. Laricchia, Phys. Rev. Lett. **80**, 5309 (1998).

[11] L. S. Fornari, L.M. Diana and P. G. Coleman, Phys. Rev. Lett. **51**, 2276 (1983).

[12] L. M. Diana, P. G. Coleman, D. L. Brooks, P. K. Pendleton and D. M. Norman, Phys. Rev. A **34**, 2731 (1986).

[13] T. Falke, W. Raith and M. Weber, Phys. Rev. Lett. **75**, 3418 (1995).

[14] A. Schmitt, U. Cerny, H. Moller, W. Raith and M. Weber, Phys. Rev. A**49**, R5 (1994).

[15] C. K. Kwan, Y. F. Hsieh, W. E. Kauppila, S. J. Smith, T. S. Stein, M. N.Uddin and M. S. Dababneh, Phys. Rev. A **27**, 1328 (1983).

[16] C. K. Kwan, Y. F. Hsieh, W. E. Kauppila, S. J. Smith, T. S. Stein, M. N. Uddin and M. S. Dababneh, Phys. Rev. Lett. **52**, 1417 (1984).

[17] S. Zhou, W. E. Kauppila, C. K. Kwan and T. S. Stein, Phys. Rev. Lett. **72**, 1443 (1994).

[18] S. Zhou, H. Li, W. E. Kauppila, C. K. Kwan and T. S. Stein, Phys. Rev. A **55**, 361 (1997).

[19] O. Sueoka, S. Mori and A. Hamada, J. Phys. B **27**, 1453 (1994).

[20] O. Sueoka and S. Mori, J. Phys. B**19**, 4035 (1986).

[21] O. Sueoka, S. Mori and Y. Katayama, J. Phys. B**19,** L373 (1986).

[22] D. M. Schrader, F. M. Jacobsen, N. P. Frandsen and U. Mikkelsen, Phys. Rev. Lett. **69**, 57 (1992).

[23] N Jiang and D. M. Schrader, J. Chem. Phys. **109**, 9430 (1998).

[24] C. M. Surco, A. Passner, M. Leventhal and F. J. Wysocki, Phys. Rev. Lett. **61**, 1831 (1988).

[25] T. J. Murphy and C. M. Surco, Phys. Rev. Lett. **67**, 2954 (1991).

[26] S. J. Gilbert, R. G. Greaves and C. M. Surco, Phys. Rev. Lett. **82**, 5032 (1999).

[27] J. Xu, L. D. Hulett, Jr., T. A. Lewis, D. L. Donahue, S. A. McLuckey and O. H. Crawford, Phys. Rev. A**49**, R3151 (1994).

[28] H. Bluhme, H. Knudsen, J. P. Merrison, and M. R. Poulsen, Phys. Rev. Lett. **81**, 73 (1998).

[29] A. Jain, J. Phys. B**19**, L105 (1986).

[30] A. Jain and D.G. Thompson, J. Phys. B**16**, 1113 (1983).

[31] A. Jain, J. Phys. B**19**, L807 (1986).

[32] J. Tennyson, J. Phys. B**19**, 4255 (1986).

[33] J. Tennyson and L. Morgan, J. Phys. B**20**, L641 (1987).

[34] G. Danby and J. Tennyson, J. Phys. B**24**, 3517 (1991).

[35] J. L. S. Lino, J. S. E. Germano and M. A. P. Lima, J. Phys. B**27**, 1881 (1994).

[36] E. P. da Silva, J. S. E. Germane, and M. A. P. Lima, Phys. Rev Lett. **77**, 1028 (1996).

[37] F. A. Gianturco, P. Paioletti and J. A. Rodriguez-Ruiz, Z. Phys. D **36**, 51 (1996).

[38] F. A. Gianturco and T. Mukherjee, J. Phys. B**30**, 3567 (1997).

[39] F.A. Gianturco, T. Mukherjee and P. Paioletti, Phys. Rev. A **56**, 3638 (1997).

[40] F.A. Gianturco and P. Paioletti, Phys. Rev. A **55**, 3491 (1997).

[41] M. Kimura, O. Sueoka, A. Hamada, M. Takekawa, Y. Itikawa, H. Tanaka and L. Boesten, J. Chem. Phys. **107**, 6616 (1997).

[42] M. Kimura, M. Takekawa, Y. Itikawa, M. Takaki and O. Sueoka, Phys. Rev. Lett. **80**, 3936 (1998).

[43] O. Sueoka and M. Kimura, Phys. Rev. Lett.**xx**, xxx (2000).

[44] Sir H. Massey, Physics Today, March (1976).

[45] T. Mukherjee and A. S. Ghosh, J. Phys. B**24**, 1449 (1991).

[46] G. G. Ryzhikh and J. Mitroy, Phys. Rev. Lett. **79**, 4124 (1997).

[47] J. C. Ashley, J. Electr. Spect. and Related Phenom. **50**, 323 (1990).

Quantum mechanically complete measurements in electron impact excitation of helium

Andrew G. Mikosza

Centre for Atomic, Molecular and Surface Physics,
Physics Department, The University of Western Australia,
Nedlands, Perth. 6907, Australia.

Abstract. A complete quantum description of the 3^1D state of helium, that is the amplitudes and relative phases for the m = 0, ±2 magnetic sublevels are reported. All state multipoles $\langle T(2)_{kq} \rangle$ ($k = 0-4, k \geq q \geq -k$) that describe the population and the anisotropy of the excited He(3^1D) state are presented. The data were determined for 60eV incident electrons and 40° electron scattering angle.

The experimental procedure was to measure the triple-coincidence, polarisation correlations of the sequential cascading photons (667.8 nm $3^1D \rightarrow 2^1P$ and 58.4 nm $2^1P \rightarrow 1^1S$ transitions) and the scattered n = 3 energy loss electrons. Simultaneous measurements of the three two-particle coincidence pairs determined accurate magnitudes of the scattering amplitudes and phases as well as enabling consistency checks with earlier data. The triple coincidence data determined the sign of the relative m = ±2 phases. The results are presented in a simplifying diagrammatic form. As yet there are no other complete experimental determinations of the 3^1D state.

The measurements are in agreement, within the experimental uncertainty, with the Convergent Close Coupling (CCC) calculations.

Previous data for helium at 40eV incident electrons are discussed in the context of the present triple-coincidence results. The desirability of a recently introduced parameter notation for the helium $3^{1,3}D$ states is questioned with respect to physical insight.

INTRODUCTION

The experimental and theoretical investigations of electron collisions with atoms leads to a better understanding of atomic structure, particle interactions and collision processes. Here we are concerned with electron impact excitation of helium for which the structure and interactions are well known and the fundamental information to be obtained is the scattering amplitudes and their relative phases. From those quantities the data of differential and total scattering cross sections can be calculated for applications in, for example, astrophysics, stellar and planetary atmospheres, laser physics and plasma physics.

Although there is at present a considerable interest in quantum mechanically complete description of an atomic states, experimentally the goal has only been

CP500, *The Physics of Electronic and Atomic Collisions*, edited by Y. Itikawa, et al.
© 2000 American Institute of Physics 1-56396-777-4/00/$17.00

attained for the 2^1P and 3^1P states of helium. Reviews have been given by, for example, Andersen and Bartschat (1) and Andersen (2,3).

The necessary set of measurements for these P states has been obtained for electron impact through measurements of coincident pairs of scattered electrons and radiated photons emitted after the excitation process.

Earlier studies of the 3^1D excitation process made observations in the scattering plane, defined by the incident and scattered electron momentum vectors, of scattered pairs of particles and observed either angular or polarisation correlations. Andersen and Bartschat (4) have also reviewed electron impact excitation of D states of atoms, pointing out that after the work in the early 1980's (5-7) on the D state excitation, it became clear that a scattered particle and one photon coincidence experiments would not allow a unique determination of all the complex scattering amplitudes.

In this paper we extend the complete quantum mechanical description to the 3^1D state of helium with measurements of triple coincidence observations of the scattered electron and two photons (D $\rightarrow$ P followed by P $\rightarrow$ S). The most recent studies on the D state are discussed first.

THE HELIUM 3^1D STATE AND COMPLETE EXPERIMENTS

The coincidence detection of the energy loss electron and the cascade 58.4 nm $2^1P \rightarrow 1^1S$ photon enabled the determination of the polarisation anisotropy parameters λ, $|\chi|$, $|\mu|$ and η defined by van Linden van Den Heuvell et al. (6). The only other angular correlation measurement, by Perera and Burns (8), observed the 667.8 nm $3^1D \rightarrow 2^1P$ photons in coincidence with the energy loss electron. They expressed their results in terms of γ_a, P_{lin}, ρ_{oo}

Further coincidence measurements of the scattered energy loss electron and the 667.8 nm photon used polarisation correlations, in the direction perpendicular to the scattering plane, to obtain the Stokes parameters P_i (i=1-3) and hence the χ, P_ℓ and $L_\perp$ parameters (9-11).

More detailed measurements included the determination of Stokes parameter P_4 by measurements in-the-scattering plane of polarisation correlations and hence the electron charge cloud height parameter ρ_{oo} (12-14).

Coincidences between the sequential cascading photons, with polarisation analysis of the first photon (15,16), enabled the first determination of a rank four, zero order, state multipole for the 3^1D state, and of the total (integrated over scattered electron angle) magnetic sublevel cross sections for m = 0, 1 and 2 separately.

The above experimental results for the 3^1D state may be compared in terms of the fundamental scattering amplitudes and phases, the Stokes parameter measurements, and also in terms of the state multipoles. These comparisons, shown in table 1, highlight the missing information required for a complete description of the helium D state.

TABLE 1. Selected experimental work leading up to the full description of the He(3^1D).

Selected experimental work on He(3^1D)	Scattering Amplitudes and Phases [a]	Target / coherence parameters [a]	Stokes Parameters	State Multipoles $\langle T(L_2)_{kq}\rangle$
van Linden van Den Heuvell et al. (1983)	$\lvert\alpha_0\rvert^2$, $\alpha_{+2}^{*}+\alpha_{-2}$	γ_a, P_{lin}, ρ_{00}	P_1, P_2, P_4	$k = 0, 2$ $q = 0$
Beijers et al. 1987	$\alpha_{+2}^{*}+\alpha_{-2}$	γ_a, P_{lin}, $(L_\perp = -\sqrt{1-P_{lin}^2}\,)$	P_1, P_2 $(P_3 = \sqrt{1-P_{lin}^2}\,)$	$k = 0, 2$ $q = 0$
Batelaan et al. 1988	α_{+2}, α_{-2}, $\alpha_{+2}^{*}+\alpha_{-2}$	γ_a, $L_\perp$, P_{lin}	P_1, P_2, P_3	$k = 0, 1$ $q = 0$
Donnelly and Crowe 1988	α_{+2}, α_{-2}, $\alpha_{+2}^{*}+\alpha_{-2}$	γ_a, $L_\perp$, P_{lin}	P_1, P_2, P_3	$k = 0, 1$ $q = 0$
Perera and Burns 1990	$\lvert\alpha_0\rvert^2$, $\alpha_{+2}^{*}+\alpha_{-2}$	γ_a, P_{lin}, ρ_{00}	P_1, P_2, P_4	$k = 0, 2$ $q = 0$
Batelaan et al., 1991	$\alpha_0, \alpha_{+2}, \alpha_{-2}$, $\alpha_{+2}^{*}+\alpha_{-2}$	γ_a, $L_\perp$, P_{lin}, ρ_{00}	P_1, P_2, P_3, P_4	$k = 0-2$ $q = 0-2$
Mikosza et al., 1994	$\alpha_0, \alpha_{+2}, \alpha_{-2}$, $\alpha_{+2}^{*}+\alpha_{-2}$	γ_a, $L_\perp$, P_{lin}, ρ_{00}	P_1, P_2, P_3, P_4	$k = 0-2$ $q = 0-2$
Donnelly et al., 1994	$\alpha_0, \alpha_{+2}, \alpha_{-2}$, $\alpha_{+2}^{*}+\alpha_{-2}$	γ_a, $L_\perp$, P_{lin}, ρ_{00}	P_1, P_2, P_3, P_4	$k = 0-2$ $q = 0-2$
Mikosza et al., 1995	$\lvert\alpha_0\rvert^2$, $\lvert\alpha_{+2}\rvert^2$, $\lvert\alpha_{-2}\rvert^2$	ρ_{00}, $\sigma_1, \sigma_2, \sigma_3$	P_1, P_4 $(P_2 = P_3 = 0)$	$k = 0-4$ $q = 0$
Complete description (present work)	$\alpha_0, \alpha_{+2}, \alpha_{-2}$ $(\beta_{+2})_0$, $(\beta_{-2})_0$	γ_a, $L_\perp$, P_{lin}, ρ_{00} $\gamma_a^{\pm}$, $L_\perp^{\pm}$, $P_{lin}^{\pm}$, $\rho_{00}^{\pm}$ [b]	P_1, P_2, P_3, P_4 $(P_1)^{triple}$, $(P_2)^{triple}$	$k = 0-4$ $q = 0-4$

[a] The parameters γ_a, $L_\perp$, P_{lin}, ρ_{00} or α_0, α_{+2}, α_{-2}, and $\alpha_{+2}^{*}+\alpha_{-2}$ do not allow a unique determination of the 3^1D state.

[b] Parameters with the indicated (±) separate values are obtained from the D state decay channels via the m = ± 1 magnetic sublevels.

An analysis of the decay of excited D-state atoms is based on Fano and Macek (17), Blum and Kleinpoppen (18), Blum (19), Heck and Gauntlett (20) and Wang et al. (21). The present work (last row in the above table), is the first complete description of the 3^1D state obtained from the triple coincidence observation of the scattered energy loss electron and the two sequential cascading photons. In the experimental setup, where the first photon polarisation and the second photon intensity (without polarisation) were measured in coincidence, the Stokes parameters are given by, (21)

$$IP_1 = -\sum_{\lambda_1 = \dot\lambda_1 = \pm 1}[\rho(\lambda_2 = -1; \dot\lambda_2 = 1) + \rho(\lambda_2 = 1; \dot\lambda_2 = -1)],$$

$$IP_2 = i \sum_{\lambda_1 = \dot{\lambda}_1 = \pm 1} [\rho(\lambda_2 = -1; \dot{\lambda}_2 = 1) - \rho(\lambda_2 = 1; \dot{\lambda}_2 = -1)] \quad \text{and}$$

$$IP_3 = \sum_{\lambda_1 = \dot{\lambda}_1 = \pm 1} [\rho(\lambda_2 = 1; \dot{\lambda}_2 = 1) - \rho(\lambda_2 = -1; \dot{\lambda}_2 = -1)] \tag{1}$$

The Stokes parameters obtained from these three particle polarisation correlations measurements are related to the D state multipoles $\langle T(L_2)^\dagger_{kq} \rangle$, with rank k and order q up to 4 by, (22)

$$IP_i = C \sum_{kq} IPi[k,q]\langle T(L_2)^\dagger_{kq} \rangle_{lab} \qquad (\text{for } i = 1,2,3)$$

The rank 4 state multipoles describe the population and the anisotropy of the excited He(3^1D) state fully.

A description of the experimental method follows, together with the information that can be obtained obtained, firstly from the simultaneous observation of two-particle coincidences and then from observations of three-particle coincidences.

EXPERIMENTAL ASPECTS

The experimental method, schematically indicated in figure 1, involves electron impact excitation of helium atoms and the triple coincident detection of the energy loss electron and the two sequential cascade photons following the decay of the excited 3^1D state via the intermediate 2^1P to the 1^1S ground state. The apparatus is a crossed electron and atomic beams system, which has been described in detail previously (23).

Briefly it consists of a cylindrical vacuum chamber with rotatable electron gun, electrostatic $180°$ electron energy analyser, and an in-the-scattering-plane vacuum ultraviolet (VUV) photon detector. The helium target gas effused from a single capillary with 0.3 mm diameter and 5 mm length located on the central axis of the rotary tables, with typical driving pressures behind the capillary were 0.2–0.5 Torr resulting in a background pressure of about 1×10^{-7} Torr. This pressure was low enough for pressure-dependent effects, due to resonance trapping arising from the absorption and subsequent re-emission of the emitted 2^1P $\rightarrow$ 1^1S (58.4 nm) photons, to be negligible for the present measurements.

The first (visible) photon (γ_1) originates from the 3^1D $\rightarrow$ 2^1P transition at 667.8 nm. These photons were selected by a cone with entrance solid angle of 85 degrees, followed by a lens to form a parallel beam which was subsequently passed through the appropriate 'retarder and linear polarizer combination' to permit a full polarization analysis. Two different visible polarizer/detector systems were used to view the

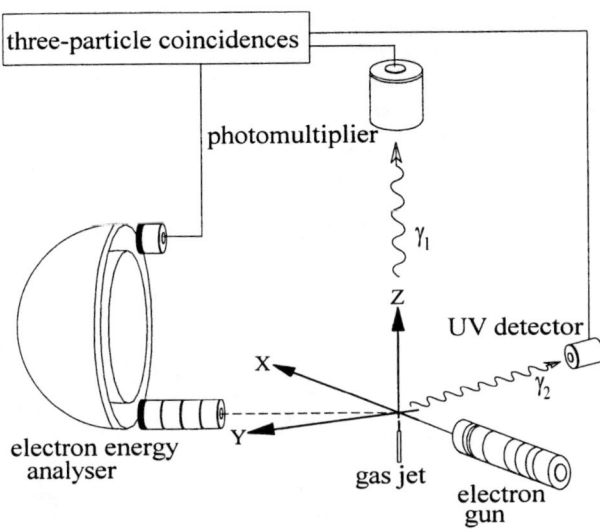

FIGURE 1. A schematic of the apparatuss and geometry for three-particle coincidence measurements.

collision region perpendicular to the electron beam direction in either perpendicular (vertical) or parallel (horizontal), respectively, to the collision plane.

All (visible and VUV) photons were detected perpendicular to the electron beam ($\theta_1 = \theta_2 = 90°$). The relative (azimuthal) angle $\Delta\phi = \phi_1 - \phi_2$ between the visible (γ_1) and the VUV (γ_2) photon was chosen as $\Delta\phi = 90°$ and $\Delta\phi = 180°$ for the vertical and horizontal visible photon detection systems, respectively.

The signals from the scattered electron (e⁻), the visible photon (γ_1) and the cascade UV photon (γ_2) were used to start and stop two TACs resulting in three types of output. The first type are the three-particle coincidence counts with the scattered electron and the two photons originating from the same helium atom (e⁻, γ_1, γ_2). The second type originates from the three possible ways in which two of the three signals are correlated, (e⁻, γ_1), (e⁻, γ_2) and (γ_1, γ_2), with the third signal γ_2 , γ_1 and e⁻ respectively, accidental. These three two-particle correlated contributions to the overall counts represent the accidental coincidences in triple coincidence measurements. The third type of output is the totally accidental contribution with all three, e⁻, γ_1 and γ_2 , uncorrelated.

These types of output signals, and their low coincidence count rates, required the development and optimisation of a computer controlled interactive real-time data acquisition, analysis and display system (24). The system is capable of uniquely identifying the three-particle, the two-particle coincidences and the totally accidental counts from the three detectors.

To describe the excited 3^1D state fully, we employed our three-particle coincidences detection simultaneously with the most common experimental approach used on the 3^1D state, the two-particle polarization correlations detection. From these polarization correlations measurements of the first cascade photon and the scattered electron, with the visible light analysis perpendicular to the scattering plane, we obtained the Stokes parameters P_i (i = 1-3). Further polarization correlations measurements with in-the-scattering-plane analysis of the visible light provided the P_4 Stokes parameter. The Stokes parameter values for P_i (i = 1, 2 and 4) were obtained concurrently with the three-particle, P_1^t and P_2^t, coincidences Stokes parameter measurements, and the Stokes parameter P_3 subsequently, using a suitable retarder.

The three and two-particle coincidences were accumulated simultaneously for other reasons also. The accumulated data enabled consistency checks to be made of the deduced scattering amplitudes and phases of earlier data and to obtain the present amplitude data with decreased experimental uncertainties. The latter was made possible by the long three-particle coincidences accumulation times.

RESULTS

Two-particle coincidence experiments

In the natural coordinate frame and with negative reflection symmetry, the m = $\pm$ 1 states are not populated. With one arbitrary phase and normalisation to the differential cross section for exciting the 3^1D state, there are only four independent parameters to be determined, namely the two m = $\pm$ 2 amplitudes, $\alpha_{\pm2}$, and their phases, $\beta_{\pm2}$, with respect to the α_0 amplitude (25). In the present study using the double coincidence electron-photon polarization correlation experimental method, in which polarization analysis is made only of the $3^1D \rightarrow 2^1P$ 667.8 nm photons, the scattering amplitudes are related to the Stokes parameters by

$$I_{in-plane} = \tfrac{2}{3}\alpha_0^2 = (1-P_1)(1-P_4)/[4-(1-P_1)(1-P_4)] \tag{2}$$

$$I_z P_3 = -(\alpha_{+2}^2 - \alpha_{-2}^2) \tag{3}$$

$$I_z = \alpha_{+2}^2 + \alpha_{-2}^2 + \tfrac{1}{3}\alpha_0^2 \tag{4}$$

$$I_z P_1 = -\sqrt{\tfrac{2}{3}}\,\alpha_0(\alpha_{+2}\cos\beta_{+2} + \alpha_{-2}\cos\beta_{-2}) \tag{5}$$

$$I_z P_2 = -\sqrt{\tfrac{2}{3}}\,\alpha_0(-\alpha_{+2}\sin\beta_{+2} + \alpha_{-2}\sin\beta_{-2}), \tag{6}$$

where polarization analysis of the visible light intensity in the z-direction, I_z, perpendicular to the scattering plane determines P_1, P_2 and P_3. In-the-scattering plane linear polarization analysis determines P_4. The values of α_0, α_{+2} and α_{-2} are then obtained from equations (2) to (4). Polarization correlations measurements of P_1 and P_2, equations (5) and (6) determine the phases β_{+2} and β_{-2}, but not the relative phase between β_{+2} and β_{-2}, with the consequence that the corresponding amplitude sum can be obtained by two alternative phase combinations; the 'real' and the 'ghost' solutions, that are indistinguishable in two particle-coincidence experiments.

The two solutions for the amplitude (vector) sum $\alpha_{+2} + \alpha_{-2}$ are best described by the alternative phases (β^I_{+2} and β^{II}_{+2}) and (β^I_{-2} and β^{II}_{-2}), and our present results are shown in table 2. With the stated aim of describing the 3^1D state fully, table 2 also lists all the other results that can be obtained using the two-particle coincidence, polarization correlation experimental method, that does not identify the 'real' solution. For example, the derived atomic properties of the charge cloud, namely the linear polarization P_{lin}, the alignment of the charge cloud γ, the angular momentum L and the height of the charge cloud (ρ_{oo}), are identical for the two solutions.

The two-particle coincidence results, obtained with 60eV electron impact energy and 40° electron scattering angle, represent the 'first step' towards the full description of the 3^1D state. The final and 'second step', which can only be provided by the more difficult three-particle coincidence measurements, allows the identification of the 'real' solution.

However before the three-particle coincidence results are presented and the identification of the real solution is made, it is instructive to illustrate diagramatically the alternative solutions provided by the two-particle coincidence polarization

TABLE 2. The 3^1D parameters obtained from polarization correlations (e⁻, γ, e⁻) experiments, with 60eV energy electron impact electrons and 40° scattering angle. The right hand side columns show the alternative phases β^I_{+2} or β^{II}_{+2}, and β^I_{-2} or β^{II}_{-2}, for the α_{+2} and α_{-2} amplitudes, respectively. The values of P_i (i = 1, 2, 3 and 4) agree well with our previous polarization correlations experiments (13).

Stokes Parameters		Charge Cloud Parameters		Scattering Amplitudes		Alternative Phases	
P_1	0.424 ± 0.049	P_{lin}	0.427 ± 0.049	α_{+2}	0.513	β^I_{+2}	-150.0° $\pm 14.2^\circ$
P_2	-0.047 ± 0.060	γ_a	$-3.16^\circ \pm 4.02^\circ$		± 0.043	β^{II}_{+2}	-197.4° $\pm 14.2^\circ$
P_3	-0.375 ± 0.040	L	0.410 ± 0.101	α_{-2}	0.240	β^I_{-2}	294.4° $\pm 37.8^\circ$
P_4	-0.075 ± 0.086	ρ_{oo}	0.679 ± 0.066		± 0.027	β^{II}_{-2}	52.9° $\pm 37.8^\circ$
				α_0	0.824 ± 0.066		

correlations experiments. Briefly, for the selected geometry and polarisation correlation approach mentioned above, the measured two and three-particle coincidence Stokes parameters are directly related to the excitation amplitudes and their phases by equation (1). The limitations of the amplitude information obtainable in the two-particle coincidence case (equations (5) and (6)), the result of summing equation (1) over all angles of the undetected second cascade photon, are most apparent from the alternative phases for α_{+2}, and α_{-2} in table 2, and from figure 2. Figure 2 clearly shows the amplitude vector sum $\alpha_{+2}^{*} + \alpha_{-2}$ together with the two possible solutions, and the angular values of the alternative phases $-\beta_{+2}^{I}$ or $-\beta_{+2}^{II}$ and β_{-2}^{I} or β_{-2}^{II}, for α_{+2}^{*} and α_{-2}, respectively. The diagram also shows the electron space charge alignment angle γ_a, for our chosen electron scattering angle. The choice of the 40° scattering angle is crucial, and can be seen from our previous measurements (13) of two-particle coincidences at the scattering angles of 15° and 25°, shown in figure 3. The experimental uncertainties for these solutions overlap and are significantly larger than the uncertainties of the present results at the electron scattering angle of 40°. The reduced experimental uncertainties for the 40° data reflect that a much larger data accumulation time, as well as apparatus improvements, for these measurements. The symbols at 15° and 25° indicate paired real and ghost solutions since they have not yet been identified. It can be seen that at 40° electron scattering angle the real and ghost solutions diverge, allowing easier identification of the real solution from the three particle coincidence results. Also, it is easy to deduce the statistical accuracy needed for both two and three-particle coincidence measurements of specific relative phases in order for the present approach to be successful in separating the real and ghost phases. The more difficult three-particle coincidence measurements were used only to identify the real solution. The statistically more accurate two-particle coincidence measurements provided all of the necessary excitation amplitudes data apart from the relative phase of $\beta_{\pm2}$.

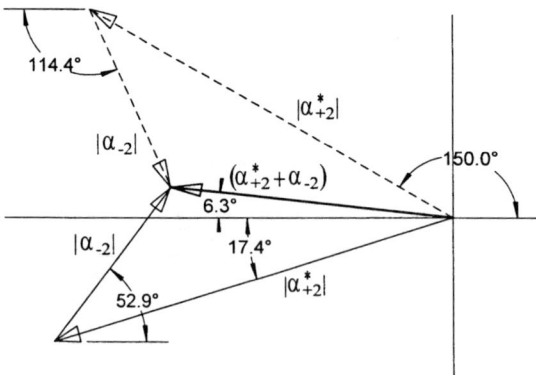

FIGURE 2. The vector sum $\alpha_{+2}^{*} + \alpha_{-2}$ together with the two possible solutions (see text).

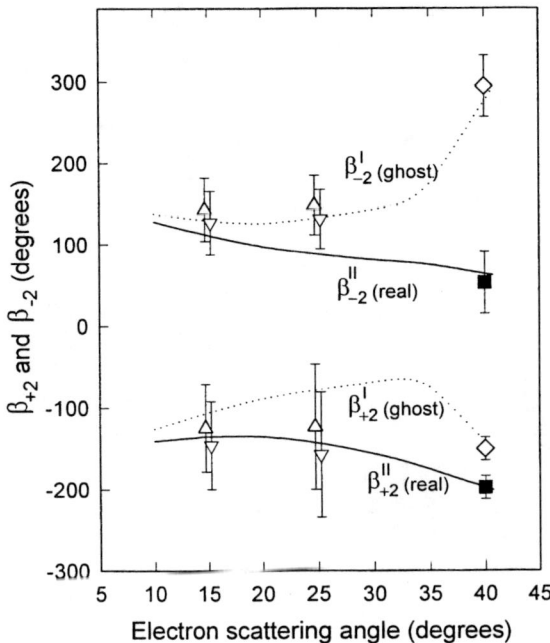

FIGURE 3. The relative phases, β_{+2} and β_{-2}, of the amplitudes α_{+2} and α_{-2}, respectively, as a function of the electron scattering angle. The two-particle (e^-, γ_1) coincidence measurements (13) provide the possible real β^{I} (∇) and ghost β ($\triangle$) values. At 40^0 scattering angle, the two-particle coincidence data, obtained concurrently with three-particle coincidence measurements, are identified as the real β^{I} ($\blacksquare$) and as the ghost β ($\lozenge$) solutions (see Three-partile Coincidence Experiments section). The convergent close coupling (CCC) phases (27) are shown as the full lines, while the inferred CCC ghost values are the dotted lines.

Figure 3 showing the alternative $(\beta_{\pm2})$ phases is preferred to the equivalent $\gamma^{\pm}$ plots for the excited D state (4), where the $\pm$ separate values are obtained from decay channels via the $m = \pm 1$ magnetic sublevels. Also the real and ghost solutions are more clearly represented by vector diagrams as in figure 2 than by $\gamma^{\pm}$ triangles (25), in particular when errors determine triangle closures.

The real and the ghost solutions can also be shown schematically, in the atomic scattering reference frame ($\gamma_a = 0$), by the two 3^1D electron charge cloud distributions (figure 4 (top)), given by the angular part of the electron charge density equation (26),

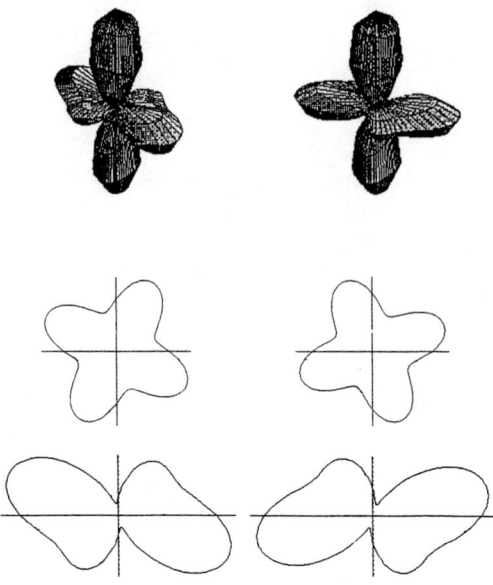

FIGURE 4. The real and the ghost solutions of the He(3^1D) state excited by 60eV energy electron impact electrons, at 40^0 electron scattering angle (see text).

$$Y^a(\theta,\phi) = (\alpha_2^2 + \alpha_{-2}^2)\tfrac{3}{8}\sin^4\theta + \alpha_0^2\tfrac{1}{4}(3\cos^2\theta - 1)^2$$
$$- P_{lin}(\alpha_2^2 + \alpha_{-2}^2 + \tfrac{1}{3}\alpha_0^2)\tfrac{3}{4}\sin^2\theta(3\cos^2\theta - 1)\cos 2\phi$$
$$+ \alpha_2\alpha_{-2}\tfrac{3}{4}\sin^4\theta\cos 4(\phi - \eta)$$
$$= A_0(\theta) + A_2(\theta)\cos 2\phi + A_4\cos 4(\phi - \eta),$$

where $\eta = (\beta_{-2} - \beta_2)/4$. The difference between the real and ghost charge clouds is the sign of η. The two possible solutions for the sum $\alpha_{+2}^{\bullet} + \alpha_{-2}$, are now represented by the undetermined sign of the third term of the equation. If the charge cloud is separated into its components, the asymmetric contribution of the third term, together with the isotropic first term, is shown in figure 4 (middle) projection onto the X-Y plane. As a result the complete ghost and the real charge clouds display a reflection symmetry about the X-Z plane, in the atomic scattering reference frame, as shown in figure 4 (bottom) projections onto the X-Y plane. The two charge clouds are indistinguishable by any of the parameters, α_{-2}, α_0, α_{+2}, γ_a, L, P_{lin}, ρ_{00}, P_1, P_2, P_3, P_4; only the relative phase difference $(\beta_{+2} - \beta_{-2})$, if known, will distinguish the 'real' from the 'ghost' distribution.

Three-particle coincidence experiments

The double coincidence correlations measurements of P_1 and P_2, have determined the phases β_{+2} and β_{-2} but not their relative phase as indicated by the two possible solutions to the vector sum $\alpha_{+2}^* + \alpha$.

The determination of the relative phase requires interference between the scattering amplitudes α_{+2} and α_{-2} and necessitates triple coincidence measurements. In our experimental geometry the visible photon detector is located perpendicular to the scattering plane, the UV detector is in the scattering plane and both detectors are at 90 degrees to the incident electron beam. Then the triple coincidence intensity and polarisations of the visible photon perpendicular to the scattering plane in the natural reference frame are given by

$$I_z^t = \frac{1}{6}\alpha_0^2 + \frac{1}{2}(\alpha_{+2}^2 + \alpha_{-2}^2) - \frac{1}{\sqrt{6}}\alpha_0(\alpha_{+2}\cos\beta_{+2} + \alpha_{-2}\cos\beta_{-2}) \tag{7}$$

$$I_z^t P_1^t = \frac{1}{6}\alpha_0^2 - \frac{1}{\sqrt{6}}\alpha_0(\alpha_{+2}\cos\beta_{+2} + \alpha_{-2}\cos\beta_{-2}) + \alpha_{+2}\alpha_{-2}\cos(\beta_{+2} - \beta_{-2}) \tag{8}$$

$$I_z^t P_2^t = -\frac{1}{\sqrt{6}}\alpha_0(\alpha_{+2}\sin\beta_{+2} - \alpha_{-2}\sin\beta_{-2}) + \alpha_{+2}\alpha_{-2}\sin(\beta_{+2} - \beta_{-2}) \tag{9}$$

where the superscript indicates triple coincidence values. The relative phase difference $(\beta_{+2} - \beta_{-2})$, for the α_{+2} and α_{-2} amplitudes, is apparent in the cos and sin functions in equations 8 and 9, and hence the ambiguity inherent in the two particle coincidence polarization correlations experiments is resolved by measuring I_z^t, P_1^t and P_2^t. It is noted that there is no requirement to measure *both* of the three particle coincidence Stokes parameters P_1^t and P_2^t for the relative phase difference to be determined.

The measured values for the triple coincidence Stokes parameters are $P_1^t = 0.436 \pm 0.349$ and $P_2^t = 0.285 \pm 0.376$. Now, rather than solve equations 8 and 9 to determine β_{+2} and β_{-2}, we choose to (i) use the two phases determined from double coincidence measurements since they are considerably more accurate because the double coincidence count rates are about 10^3 times higher than the triple coincidence count rates and (ii) use the triple coincidence values of the Stokes parameters as limits for the true phases. The triple coincidence Stokes parameters which are implied by the two alternative values of the scattering amplitudes and phases of our $(e^-\gamma_1)$ polarization correlations experiments (table 2) are $P_1^t = 0.619$ and 0.482, and $P_2^t = -0.282$ and 0.330. The measured triple coincidence value of P_1^t of 0.436 ± 0.349 cannot separate the double coincidence values which both fall within the experimental uncertainty. However for P_2^t the determined $(e^-\gamma_1)$ value of -0.282 is outside the experimental uncertainty of the triple coincidence measured value of 0.285 ± 0.376 hence it is the ghost solution. Therefore 0.330 is the *real* solution. This double coincidence real value of P_2^t was obtained using the β_{+2}^{II} ($= -197.4°$) and β_{-2}^{II} ($= 52.9°$) phases from the two possible solutions (table 2) scattering amplitudes α_{+2} and α_{-2}, respectively. Both the real and ghost double coincidence phases are shown in figure 2. The convergent close coupling (CCC) results (27) at 64.6 eV are $P_1^t = 0.616$ and $P_2^t = 0.280$ which are in agreement with the measured values within the experimental uncertainties.

The real and ghost value obtained with 60eV electron impact energy and 40° electron scattering angle, can be extended to nearby scattering angles and energies because of the monotonic variation of the phases with scattering angle and the similarity of figure 3 plots at energies close to 60eV.

Finally the state multipoles obtained from the two and three-particle results are shown, both in the natural and collision scattering frames.

TABLE 3. The state multipoles $\langle T(2)_{kq} \rangle$ (L=2) for the 3^1D state of helium excited by 60eV energy electrons and 40° scattering angle.

State Multipole	Collision frame	Natural frame
T_{00}	0.4471	0.4471
T_{1-1}	0.0919 i	0
T_{10}	0	0.1300
T_{11}	0.0919 i	0
T_{2-2}	0.1932	$-0.1518 + 0.0167$ i
T_{2-1}	0.0167	0
T_{20}	-0.0902	-0.1915
T_{21}	-0.0167	0
T_{22}	0.1932	$-0.1518 - 0.0167$ i
T_{3-3}	**0.1900 i**	**0**
T_{3-2}	**-0.2009 i**	**$-0.3696 - 0.2009$ i**
T_{3-1}	**-0.3203 i**	**0**
T_{30}	**0**	**0.0650**
T_{31}	**-0.3203 i**	**0**
T_{32}	**0.2009 i**	**$-0.3696 + 0.2009$ i**
T_{33}	**0.1900 i**	**0**
T_{4-4}	**0.1825**	**$-0.0415 - 0.1159$ i**
T_{4-3}	**0.0274**	**0**
T_{4-2}	**0.3008**	**$-0.1315 + 0.0145$ i**
T_{4-1}	**-0.1136**	**0**
T_{40}	**0.2575**	**0.5253**
T_{41}	**0.1136**	**0**
T_{42}	**0.3008**	**$-0.1315 - 0.0145$ i**
T_{43}	**-0.0274**	**0**
T_{44}	**0.1825**	**$-0.0415 + 0.1159$ i**

The additional state multipoles provided by the three-particle coincidence results are shown in bold.

CONCLUSIONS

In conclusion, the present the three-particle coincidence measurements have resolved the phase ambiguity of the less difficult two-particle coincidence (scattered electron and radiated photon) measurements of the excitation amplitudes and their phases for the 3^1D state of helium. A complete experiment for the 3^1D state of helium has been achieved. The real and ghost phases for other nearby energies and scattering angles can be inferred from the experimental results. The principle can be extended to higher-order coincidence measurements for higher angular momentum states.

REFERENCES

1. Andersen, N. and Bartschat, K., *Adv. At. Mol. Opt. Phys.* **36**, 1-85 (1996).
2. Andersen, N., Broad, J. T., Campbell, E. E. B., Gallagher, J. W., and Hertel, I. V., *Phys. Rep.* **278**, 107 (1997).
3. Andersen, N., Bartschat, K., Broad, J. T., and Hertel, I. V., *Phys. Rep.* **279**, 251 (1997).
4. Andersen, N. and Bartschat, K., *Adv. At. Mol. Opt. Phys.* **36**, 1 (1996).
5. Nienhus, G., *Coherence and Correlation in Atomic Collisions*, New York: Plenum, 1980, pp.121-32.
6. van Linden van Den Heuvell, H.B., van Gasteren, E.M., van Eck, J., and Heideman, H.G.M., *J. Phys B* **16**, 1619 (1983)..
7. Andersen, N., Gallagher, J. W., and Hertel, I. V., *Phys. Rep.* **165**, 1-188 (1988).
8. Perera, N.W.P.H., and Burns, D.J.J., *J. Phys B: At. Mol. Opt. Phys.* **23**, 3007 (1990).
9. Beijers, J.P.M., Doornenbal, S.J., van Eck and H.G.M. Heideman, J., *J. Phys B: At. Mol. Opt.* Phys. **20**, 6617 (1987).
10. Batelaan, H., van Eck, J., and Heideman, H.G.M., *J. Phys B* **21**, L741 (1988).
11. Donnelly, B.P., and Crowe, A., *J. Phys B* **21**, L637 (1988).
12. Batelaan, H., van Eck, J., and Heideman, H.G.M., *J. Phys B* **24**, L397 (1991).
13. Mikosza, A.G., Hippler, R., Wang, J.B., and Williams, J.F., *Zeit. Phys.* D **30**, 129 (1994).
14. Donnelly, B.P., McLaughlin, D.T., and Crowe, A., *J. Phys B: At. Mol. Opt. Phys.* **27**, 319 (1994).
15. Mikosza, A.G., Hippler, R., Wang, J.B., and Williams, J.F., *Phys. Rev. Lett.* **71** 235 (1993).
16. Mikosza, A.G., Hippler, R., Wang, J.B., and Williams, J.F. *Phys Rev. A* **53**, 3287-94 (1996).
17. Fano, U., and Macek, J.H., *Rev. Mod. Phys.* **45** 553 (1973).
18. Blum, K., and Kleinpoppen, H., *Phys. Rep.* **52** 203 (1979).
19. Blum, K., *Density Matrix Theory and its Application*, New York: Plenum Press, (1981).
20. Heck, E.I., and Gauntlett, J., *J. Phys. B* **19** 3633 (1986).
21. Wang, J.B., Williams, J.F., Stelbovics, A.T., Furst, J.E., and Madison, D.H., 1995 *Phys Rev A* **52**, 2885 (1995).
22. Wang, J.B., Stelbovics, A.T., and Williams J. F., *Zeit. Phys.* D **30**, 119-127 (1994).
23. Mikosza, A.G., Hippler, R., Wang, J.B., Williams, J.F., and Wedding, A.B., *J Phys B* **27**, 1429 (1994).
24. Mikosza, A.G., Williams, J.F., and Wang, J.B., *Phys. Rev. Lett.* **79** 3375-78 (1993).
25. Fursa, D.V., Bray, I., Donnelly, B.P., McLaughlin, D.T., and Crowe, A., *J Phys B* **30**, 3459 (1997).
26. Andersen, N., Gallagher, J.W., and Hertel, IV., *Phys. Rep.* **165**, 1 (1988).
27. Fursa, D.V., and Bray, I., *Phys. Rev. A* **52**, 2885 (1995), and private communication.

Calculation of electron-barium scattering

D. V. Fursa and I. Bray

The Flinders University of South Australia, G.P.O. Box 2100, Adelaide 5001, Australia

Abstract. Convergent close-coupling method is applied to investigation of electron-barium scattering. Scattering from the ground state of barium, laser prepared $6s6p\,^1P_1$ state, and cascade populated metastable $6s5d\,^{1,3}D_2$ states has been studied. Calculated integral and differential cross sections and electron-impact coherence parameters have been found to be in a good agreement with available experimental data. For many transitions the break-down of the nonrelativistic approximation is substantial and has to be accounted for.

INTRODUCTION

The convergent close-coupling method (CCC) has been applied with considerable success to the investigation of electron scattering from relatively light targets (H [1], Li [2], Na [3], He [4], Be [5]). The comparison with accurate experimental data available for electron scattering from these atoms has allowed us to test many aspects of the underlying scattering theory. It is with this confidence in the basics of the CCC method that we turn our attention to electron scattering from heavy targets. The latter poses an array of unexplored problems associated with considerably more complex target wave functions and the break-down of the nonrelativistic approximation both in target structure and electron scattering.

The present formulation of the CCC method allows us to calculate electron scattering from atoms and ions that are well described by a model with one or two valence electrons. No relativistic effects are accounted for in CCC scattering or target structure calculations. Calculation of electron scattering from barium allows us to test these aspects of the CCC method due to availability of the detailed experimental data. Modifications to the CCC method which will overcome the current limitations can be also tested.

There has been an extensive e-Ba scattering experimental program over the last two decades. For scattering from the ground state of barium these include measurements of the $6s6p\,^1P_1$ state apparent cross section [6], total ionization cross section [7,8], total cross section [8], differential cross sections for elastic scattering and excitations of the $(6s6p)^1P_1^o$ and $(6s5d)^1D_2^e$ states [9,10]. These measurements have been recently complemented by an experimental study of electron scattering from

CP500, *The Physics of Electronic and Atomic Collisions*, edited by Y. Itikawa, et al.

the laser prepared $6s6p\ ^1P_1$ state [11–16] and from cascade populated metastable $6s5d\ ^{1,3}D_2$ states [17].

Probably the most interesting and welcome development in e-Ba scattering is the interplay between fundamental science and industrial applications. A large number of excitation cross sections for transitions between ground and excited states and for transitions between excited states are required for modelling of Ba vapor lasers [18,19], discharge lamps [20], plasma switches [21], and various planetary iono-spheres [22–25]. Experimental determination of all these cross sections is unlikely and therefore large scale theoretical calculations are required. The CCC method is well placed to perform such calculations.

THEORY

The CCC method uses a set of (barium) target states $\{\Phi_n\}$, $n = 1, \ldots, N$ to perform a multichannel expansion of the total (projectile and target electrons) wave function $\Psi^{(+)}$

$$|\Psi_N^{(+)}\rangle = (1 - P_{rs}) \sum_{n=1}^{N} |\Phi_n\rangle\langle\Phi_n|\psi_i^{(+)}\rangle, \tag{1}$$

here the index N indicates the approximation of including only N states in the close-coupling expansion, the superscript '+' indicates outgoing spherical wave boundary conditions, and the space and spin exchange operator P_{rs} ensures the antisymmetry of the total wave function and allows us to work with a nonsymmetrized function $\psi_i^{(+)}$.

The scattering information is obtained from the calculation of the T matrix defined as

$$\langle \mathbf{k}_f\Phi_f|T^N|\mathbf{k}_i\Phi_i\rangle = \langle\mathbf{k}_f\Phi_f|(H - E)|\Psi^{(+)}\rangle, \tag{2}$$

where $|\mathbf{k}\rangle$ is a plane wave, and H and E are the total Hamiltonian and energy of the scattering system, respectively. In order to find the T-matrix we solve the coupled Lippmann-Schwinger equations in momentum space,

$$\langle\mathbf{k}_f\Phi_f|T^N|\Phi_i\mathbf{k}_i\rangle = \langle\mathbf{k}_f\Phi_f|V|\Phi_i\mathbf{k}_i\rangle$$
$$+ \sum_{n=1}^{N} \sum_{\mathbf{k}} \frac{\langle\mathbf{k}_f\Phi_f|V|\Phi_n\mathbf{k}\rangle\langle\mathbf{k}\Phi_n|T^N|\Phi_i\mathbf{k}_i\rangle}{E^{(+)} - \varepsilon_k - \epsilon_n}, \tag{3}$$

where $V = H - H_T - K_0 - (H - E)P_{rs}$, and K_0 is the projectile kinetic energy operator.

We use a model of two valence electrons above an inert Hartree-Fock core $(1s^22s^22p^63s^23p^63d^{10}4s^24p^64d^{10}5s^25p^6)$ to describe barium wave functions. The set of barium target states $\{\Phi_n\}$ is obtained by performing a configuration-interaction expansion for the valence electrons. The barium Hamiltonian

$$H_T = H_1 + H_2 + V_{12} \tag{4}$$

is diagonalized in a basis of antisymmetric two-electron configurations. These configurations are built from one-electron functions obtained by diagonalizing the one-electron Hamiltonian (H_1) of the Ba^+ ion in a Sturmian (Laguerre) basis. One- and two-electron polarization potentials have been used to account for the inert core polarization. We refer to Ref. [26] for details of the structure calculations and comparison with experimental energy level and oscillator strength data. The resulting target states Φ_n of the barium atom satisfy

$$\langle \Phi_{n'} | H_T | \Phi_n \rangle = \epsilon_n \delta_{n'n}, \quad n = 1, ..., N, \tag{5}$$

where ϵ_n is the energy associated with Φ_n. The use of the Sturmian basis in the structure calculation leads to a square-integrable discretization of the barium atom continuum and allows for coupling to the ionization channels in the scattering calculations.

The CCC method is formulated as a purely nonrelativistic theory. However, it is well known that the nonrelativistic approximation breaks down for heavy targets, including barium. The major relativistic effect is a substantial singlet-triplet mixing in the barium spectrum [27–29]. It affects most scattering cross sections for weak transitions, especially those transitions which can only occur in the nonrelativistic approximation via exchange scattering. They change substantially if the direct scattering becomes possible due to the singlet-triplet mixing.

While a consistent relativistic formulation of the CCC method is some time away, we can account for major relativistic effects using a technique essentially identical with the transformation scheme described by Saraph [30]. The nonrelativistic CCC scattering amplitudes $f^S_{\pi_f s_f l_f m_f, \pi_i s_i l_i m_i}$, describing transition between state Φ_f and Φ_i with parity π_f (π_i), spin s_f (s_i), orbital angular momentum l_f(l_i) and m_f (m_i) its projection on the Z-axis of the collision frame, are transformed to the amplitudes describing transitions between fine-structure levels J_f and J_i,

$$f^{\sigma_f, \sigma_i}_{\pi_f J_f M_f, \pi_i J_i M_i}(s_f l_f \gamma_f, s_i l_i \gamma_i) = \sum_{m_f, q_f, m_f, q_f, S} C^{J_f M_f}_{l_f m_f, s_f q_f} C^{S M_S}_{\frac{1}{2} \sigma_f, s_f q_f} C^{J_i M_i}_{l_i m_i, s_i q_i} C^{S M_S}_{\frac{1}{2} \sigma_i, s_i q_i}$$

$$f^S_{\pi_f s_f l_f m_f, \pi_i s_i l_i m_i}(\gamma_f, \gamma_i), \tag{6}$$

where S is total spin and final (initial) projectile spin projection on the Z-axis of the collision frame is indicated as σ_f (σ_i). The index γ distinguishes states with the same orbital angular momentum, spin and parity. The above amplitudes are used to form amplitudes in the intermediate coupling scheme

$$F^{\sigma_f, \sigma_i}_{\pi_f J_f M_f, \pi_i J_i M_i}(\beta_f, \beta_i) = \sum_{s_f, l_f, s_i, l_i} \sum_{\gamma_f, \gamma_i} C^{\beta_f}_{\gamma_f} C^{\beta_i}_{\gamma_i} f^{\sigma_f, \sigma_i}_{\pi_f J_f M_f, \pi_i J_i M_i}(s_f l_f \gamma_f, s_i l_i \gamma_i), \tag{7}$$

where the index β distinguishes target states with the same total angular momentum J and parity π. The mixing coefficients C^β_γ are obtained by diagonalizing the one-body spin-orbit term of the Breit-Pauli Hamiltonian in the basis of the non-relativistic barium target states. We have found that typically only a few terms in (7) have large mixing coefficients.

RESULTS

We present results of 55 state close-coupling (CC(55)) and 115 state CCC calculations. The CC(55) calculations include only negative energy states (relative to Ba^+ ground state) and do not account for any flux to ionization channels. The CCC calculations include a large number of positive energy states and therefor allow us to model coupling to the Ba ionization continuum (see Ref. [26] for details).

Scattering by the ground state

Accurate measurement of the $6s6p\ ^1P_1$ state apparent cross section by Chen and Gallagher [6] allows for a stringent test of theoretical methods. This cross section is a sum of the direct $6s6p\ ^1P_1$ state excitation cross section and cascade contributions from the higher lying levels. In Fig. 1 we present experimental results (marginally renormalized by a factor of 1.03 to account for more accurately known now normalization at high energy) together with results of CCC and CC(55) calculations. Results from earlier calculations in the unitarized distorted-wave approximation (UDWA) [31], relativistic distorted-wave approximation (RDWA) [32,33], and two-state close-coupling calculations (CC(2)) [34] are also given for the direct excitation cross section. We find that all earlier calculations are larger than experimental data even without addition of the cascades. The CCC(55) apparent cross section results also overestimate the experimental data. This happens in the energy region where flux to ionization channels is substantial [7]. The CCC calculations account for a large part of the flux to ionization channels [26] and are in very good quantitative agreement with the experimental data of Chen and Gallagher.

The CCC method and the unitarized first-order many-body theory (UFOMBT) [31] method have been used to calculate integrated cross sections for a large number of transitions in barium [35]. Excitation of the $6p5d\ ^1D_2$ level by electron scattering from the Ba ground state is of special interest as it offers an interesting interplay between the complex target wave function structure in barium and theoretical methods used to calculate e-Ba scattering. The results of CCC, CC(55) and UFOMBT calculations for this transition are presented in Fig. 2. The negative parity $6p5d\ ^1D_2$ level is one of many double excited states that are present in the barium discrete spectrum. In a first-order theory an excitation of this state can occur only by exchange scattering if the nonrelativistic approximation holds. However, in a close-coupling method such a transition can occur via direct scattering as a two-step process, $6s^2\ ^1S \rightarrow 6s5d\ ^1D_2 \rightarrow 6p5d\ ^1D_2$, and leads to significantly larger cross sections. Accordingly, we find that the close-coupling results are much larger than the UFOMBT results, particularly at large energies. Account of the relativistic corrections does not lead to any significant changes in the behaviour of the cross section.

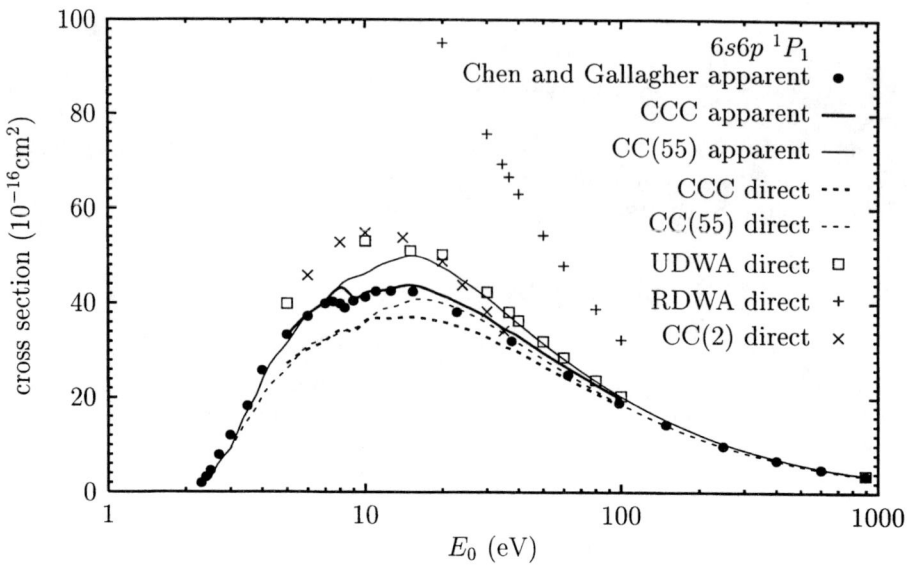

FIGURE 1. Apparent and direct $6s6p\ ^1P_1$ excitation integrated cross sections for electron scattering from the Ba ground state. The CCC(115) and CC(55) calculations are described in the text. The UDWA calculations are due to Clark *et al.* [31], RDWA calculations are due to Srivastava *et al.* [32] and CC(2) calculations are due to Fabrikant [34]. The experiment is by Chen and Gallagher [6].

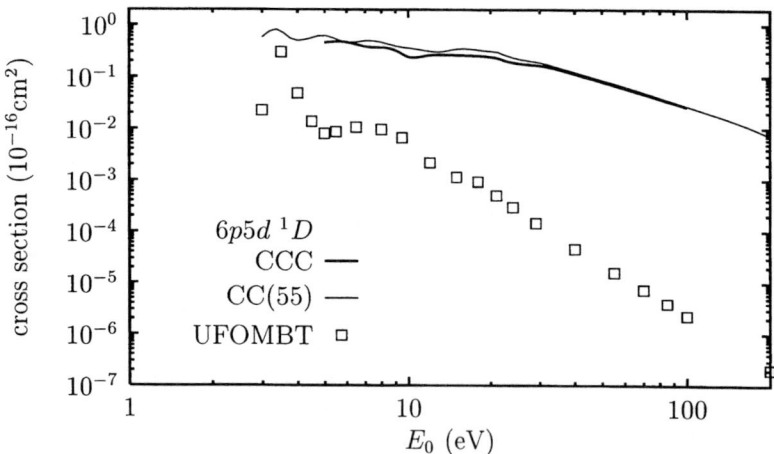

FIGURE 2. Integrated cross sections for excitation of the $6p5d\ ^1D_2$ level by electron scattering from the Ba ground state [35].

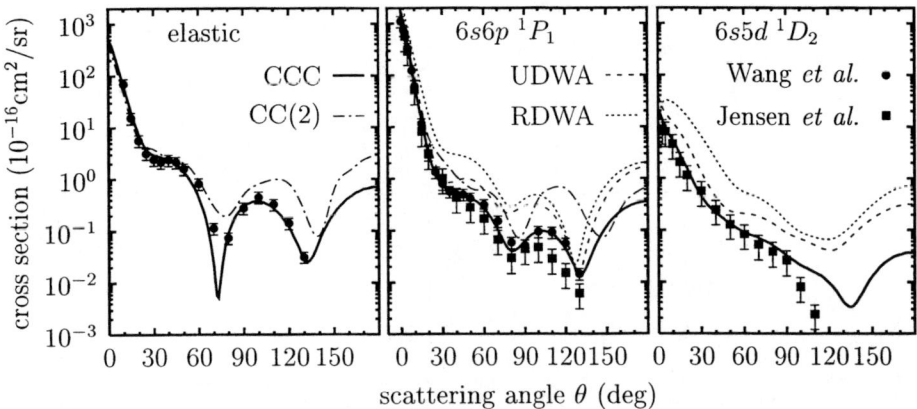

FIGURE 3. The elastic, $6s6p\,^1P_1$, and $6s5d\,^1D_2$ excitation differential cross section for electron scattering on the Ba ground state at 20 eV. Calculations as in Fig 1 and in addition for the $6s5d\,^1D_2$ level RDWA calculations are due to Srivastava *et al.* [33]. Measurements are by Wang *et al.* [10] and Jensen *et al.* [9].

Comparison between theoretical and experimental differential cross section (DCS) results for the elastic scattering and excitations of the $(6s6p)^1P_1^o$ and $(6s5d)^1D_2^e$ states at 20 eV is given in Fig. 3. We find that measurements of Jensen *et al.* [9] and Wang *et al.* [10] are in a very good agreement with our results, but this is not the case for other (CC(2), RDWA, UDWA) theoretical methods. Similarly good agreement between our results and experiment is found at other incident electron energies, while the RDWA and UDWA results is in much better agreement with our results and experiment as incident electron energy increases to 100 eV [26].

Scattering by excited states

Electron scattering from barium excited states has recently attracted a lot of interest. This is primarily due to the ability to produce significant populations of the $6s6p\,^1P_1$ state by applying readily available lasers to the ground state of barium. Performing scattering from the laser prepared $6s6p\,^1P_1$ state allows us to probe fine details of scattering process. The superelastic scattering technique has been used to determine DCS and electron-impact coherence parameters (EICPs) [36] for $6s6p\,^1P_1$-$6s\,^1S$ [11–13] and $6s6p\,^1P_1$-$6s5d\,^1D_2$ [16] transitions. A similar technique has been used to measure DCS and EICPs for elastic scattering from laser prepared $6s6p\,^1P_1$ state [15]. In Fig. 4 we present a comparison between theoretical and experimental results for EICP for the $6s6p\,^1P_1$-$6s^2\,^1S$ transition. The CCC method is clearly superior to other calculations though some discrepancies with experiment remain.

Finally, in Fig. 5 and 6 we present theoretical and experimental DCS results

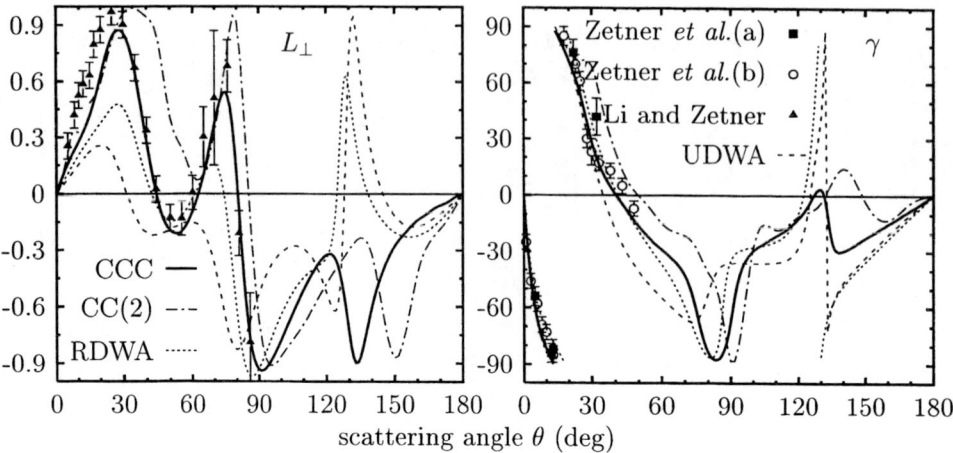

FIGURE 4. EICPs $L_\perp$ and γ of the $6s6p\ ^1P_1$ state for electron scattering on the Ba ground state at 20 eV. Theoretical calculations are as for Fig. 3. Measurements are by Zetner *et al.*(a) [11], (b) [12] and Li and Zetner [13].

for selected excitations for electron impact on laser prepared $6s6p\ ^1P_1$ state and on the cascade populated $6s5d\ ^{1,3}D_2$ metastable states at 20 eV incident electron energies. Results for other transitions and incident electron energies can be found in Refs. [14] and [17]. Singlet-triplet mixing is substantial for many excited states in barium. In electron scattering it is responsible for strong forward peaking in DCS for transitions which in nonrelativistic approximation can occur via exchange scattering only, see $5d^2\ ^3P_2 - 6s6p\ ^1P_1$ and $6p5d\ ^3F_2 - 6s5d\ ^1D_2$ transitions. We find that CCC and CC(55) results are very similar, which demonstrates good convergence of the close-coupling calculations. Good agreement between the CCC results and experiment indicates a correct account of the singlet-triplet mixing in the barium spectrum by the CCC calculations. It also gives credence to CCC results for integrated cross section for transitions between excited states which are dominated by the small angle region of the corresponding DCS and are important for industrial applications.

CONCLUSIONS

Over recent years the CCC method has been successfully applied to study a large number of transitions in e-Ba scattering. Limitations of the present formulation of the CCC method have been tested. We found that the absence of inner core excitation in the CCC method is responsible for substantial disagreement with measurements of total ionization cross sections (see Ref. [26] for details) but seems to be of little importance for other transitions. We have also demonstrated that the break-down of the nonrelativistic approximation for barium can be well accounted

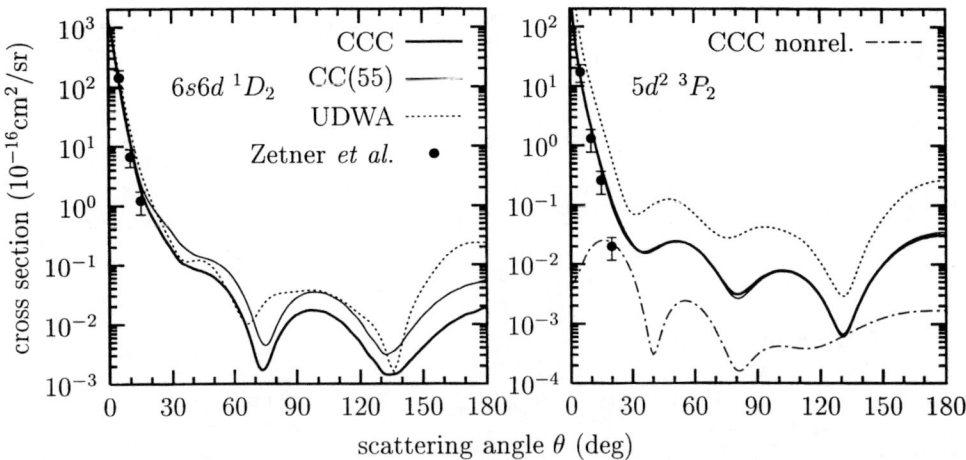

FIGURE 5. Differential cross section for electron scattering from the $6s6p\,{}^1P_1$ state of barium to $6s6d\,{}^1D_2$ and $5d^2\,{}^3P_2$ states at 20 eV. Measurements and UDWA results are from Zetner *et al.* [14].

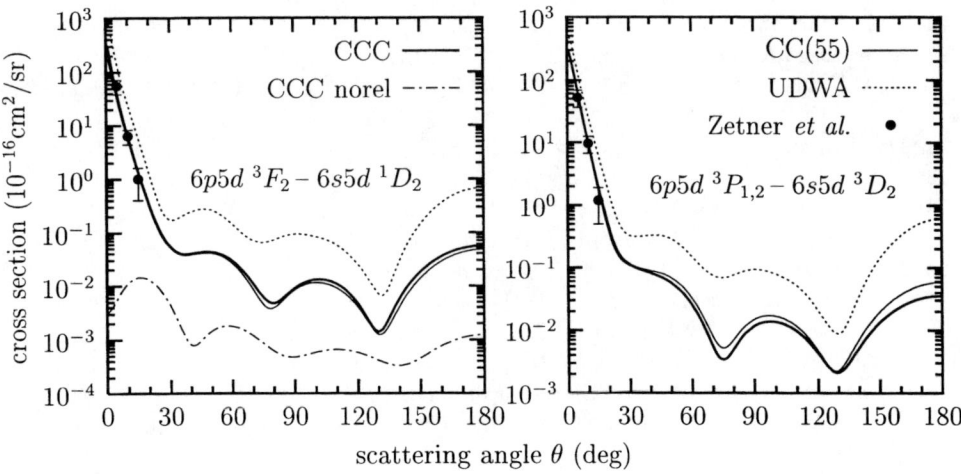

FIGURE 6. Differential cross section for $6p5d\,{}^3F_2 - 6s5d\,{}^1D_2$ and $6p5d\,{}^3P_{1,2} - 6s5d\,{}^3D_2$ transitions at 20 eV. Measurements and theoretical results are from Zetner *et al.* [17].

for by transformation of the CCC nonrelativistic amplitudes to the intermediate coupling scheme.

REFERENCES

1. I. Bray and A. T. Stelbovics, Phys. Rev. A **46**, 6995 (1992).
2. V. Karaganov, I. Bray, P. J. O. Teubner, and P. Farrell, Phys. Rev. A **54**, R9 (1996).
3. I. Bray, Phys. Rev. A **49**, 1066 (1994).
4. D. V. Fursa and I. Bray, Phys. Rev. A **52**, 1279 (1995).
5. D. V. Fursa and I. Bray, J. Phys. B **30**, 5895 (1997).
6. S. T. Chen and A. Gallagher, Phys. Rev. A **14**, 593 (1976).
7. J. Dettmann and F. Karstensen, J. Phys. B **15**, 287 (1982).
8. N. I. Romanyuk, O. B. Shpenik, and I. P. Zapesochny, JETP Lett. **32**, 452 (1980).
9. S. Jensen, D. Register, and S. Trajmar, J. Phys. B **11**, 2367 (1978).
10. S. Wang, S. Trajmar, and P. W. Zetner, J. Phys. B **27**, 1613 (1994).
11. P. W. Zetner, Y. Li, and S. Trajmar, J. Phys. B **25**, 3187 (1992).
12. P. W. Zetner, Y. Li, and S. Trajmar, Phys. Rev. A **48**, 495 (1993).
13. Y. Li and P. W. Zetner, Phys. Rev. A **49**, 950 (1994).
14. P. W. Zetner *et al.*, J. Phys. B **30**, 5317 (1997).
15. S. Trajmar *et al.*, J. Phys. B **31**, L393 (1998).
16. P. V. Johnson *et al.*, Phys. Rev. A **59**, 439 (1999).
17. P. W. Zetner *et al.*, submitted to J.Phys.B .
18. R. P. Mildren, D. J. W. Brown, and J. A. Piper, J. Appl. Phys. **82**, 2039 (1997).
19. R. P. Mildren, D. J. W. Brown, and J. A. Piper, Optics Comm. **137**, 299 (1997).
20. A. K. Bhattacharya, J. Appl. Phys. **137**, 299 (1997).
21. C. M. Yang and A. E. Rodrigez, Wright Laboratory, Report TR-92-006,Wright Patterson AFB, Ohio, March 1992.
22. D. Winske, J. Geophys. Res. **93**, 2539 (1988).
23. S. C. Chapman, J. Geophys. Res. **94**, 227 (1989).
24. R. W. Shuk and E. P. Szuszgzewiez, J. Geophys. Res. **96**, 1337 (1991).
25. E. M. Wescott *et al.*, J. Geophys. Res. **98**, 3711 (1993).
26. D. V. Fursa and I. Bray, Phys. Rev. A **59**, 282 (1999).
27. E. Trefftz, J. Phys. B **7**, L342 (1974).
28. S. J. Rose, N. C. Pyper, and I. P. Grant, J. Phys. B **5**, 755 (1978).
29. C. W. Bauschlicher Jr *et al.*, J. Phys. B **18**, 2147 (1985).
30. H. E. Saraph, Comp. Phys. Comm. **3**, 256 (1972).
31. R. E. Clark, J. Abdallah Jr., G. Csanak, and S. P. Kramer, Phys. Rev. A **40**, 2935 (1989).
32. R. Srivastava, T. Zuo, R. P. McEachran, and A. D. Stauffer, J. Phys. B **25**, 3709 (1992).
33. R. Srivastava, R. P. McEachran, and A. D. Stauffer, J. Phys. B **25**, 4033 (1992).
34. I. I. Fabrikant, J. Phys. B **13**, 603 (1980).
35. D. V. Fursa *et al.*, submitted to PRA .
36. N. Andersen, J. W. Gallagher, and I. V. Hertel, Phys. Rep. **165**, 1 (1988).

Polarization and correlations in electron-impact autoionization studies

V.V.Balashov

Institute of Nuclear Physics, Moscow State University,
Moscow 119899, Russia

Abstract. Theoretical arguments for coincidence and polarization measurements on electron-impact excitation of atomic autoionizing states are analysed in the light of modern trends in experimental studies in the field.

Introduction

The wide scope of problems on excitation and decay of atomic autoionizing states in electron-atom collisions is concentrated around two fundamental points: a)the interference between the resonant and direct mechanisms of the ionization process; b)polarization (alignmen) of autoionizing states induced by the incoming electron beam in the excited atom. The first one leads to the *profile problem*: contrary to the resonant photoabsorption phenomena, the Fano profile index in the electron-impact processes is not a fixed structural parameter of each autoionization resonance and can differ considerably when the same resonance is observed in the energy-loss spectra, in spectra of the ejected electrons or by the coincidence (e,2e) method. As for the collisionally induced polarization of the autoionizing states, it results in a wide number of various polarization and correlation phenomena when one detects the autoionization products.

The first autoionization (e,2e) experiment was performed with helium atom [1]. Now helium atom continues to serve as a good laboratory for experimental and theoretical investigations in physics of autoionization phenomena. It is interesting to note that some very old problems in the field show a tremendous vitality and take an outstanding place in current investigations. The problem of mechanism of electron-impact excitation of both two electrons in helium atom from its ground state to autoionizing states is one of them.

One-step and two-step excitations of autoionizing states

Consider the doublet of $2s2p$ $:^1P$ and $2p^2$ $:^1D$ autoionizing states of heliun atom lying between the first and second ionization threshold. Investigated in detail in a number of non-coincidence experiments in 60-s and early 70-s and, also, in recent very careful coincidence (e,2e) experiments [2,3] it remains a difficult point for the

CP500, *The Physics of Electronic and Atomic Collisions*, edited by Y. Itikawa, et al.
© 2000 American Institute of Physics 1-56396-777-4/00/$17.00

theory using the first order PWBA and DWBA approximations to describe the scattering process [4-7]. The question is *what is the role of the two-step mechanism of excitation of $2p^2$:^{1}D state via virtually excited intermediate states?*

TABLE 1. Calculated integrated cross section $\sigma_r(E_0)$ for electron-impact excitation of $2p^2$:^{1}D autoionizing state in helium atom with and without taking into account the two-step transition $1s^2$:^{1}S$\rightarrow 1s2p$:^{1}P$\rightarrow 2p^2$:^{1}D [9].

	$\sigma_r(E_0)$, $a_0^2 \cdot 10^{-4}$	
$E_0(eV)$	without two-step excitation (PWBA)	with two-step excitation (MCDA)
200	0.646	1.706
500	0.471	0.742
1000	0.275	0.352

Earlier this question was investigated within the multichannel diffraction approximation (MCDA) [8] in a schematical way [9] where excitation of $2p^2$:^{1}D autoionizing state was considered as if one deals with a discrete level disregarding complitely its interaction with the adjacent continuum (Table 1). Now we incorporate this aspect into the unified description of the direct and resonant ionization processes. Calculations show serious influence of the two-step mechanism $1s^2$:S$\rightarrow 1s2p$:^{1}P$\rightarrow 2p^2$:^{1}D of excitation of the autoionizing state $2p^2$:^{1}D on the profile of the (e,2e) triple differential cross section $\frac{d^3\sigma}{d\Omega_{sc}d\Omega_{ej}d\omega}$ in the vicinity of the two resonances (Figure 1). Agreement with experiment becomes better. Wider calculations with realistic wave functions of the states involved into the transition should be done to have a more definite conclusion. It concerns, in particular, the wave function of the ground state of the target atom.

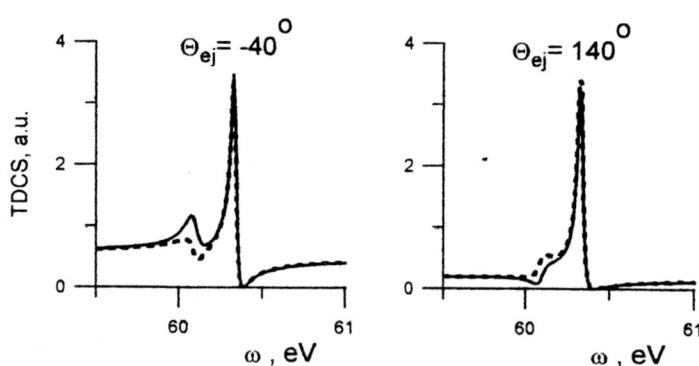

FIGURE 1. TDCS for reaction He$(e, 2e)$He$^+$ in the vicinity of $2s2p$:^{1}P and $2p^2$:^{1}D autoionizing states as a function of excitation energy ω calculated at $E_0=400$ eV and $\theta_{sc} = 3^o$ with (solid lines) and without (dashed lines) taking into account the two-step transition $1s^2$:S$\rightarrow 1s2p$:^{1}P $\rightarrow 2p^2$:^{1}D.

Autoionization in the ionization-excitation processes

The old problem how to construct a realistic wave function of helium atom in the ground state seems to become important again in connection with growing experimental and theoretical studies on the electron-impact ionization-excitation process $He(e, 2e)He^*(n = 2)$ [10-13]. Double coincidence $(e, e'\gamma)$ measurements performed in Perth and Newcastle [14,15] with intermediate-energy electron beams have revealed the resonance structure in the excitation cross section of helium ion $He^*(2p)$ in the $3ln'l'$ autoionization resonances region. It was interesting to consider this question at higher energy, under geometry and kinematical conditions of the well-known series of asymmetric coincidence experiments performed by the Roma and Orsay groups [10-12]. Our theoretical analysis has shown [16] that the resonance structure in the corresponding cross sections in the $3ln'l'$ states region could be hardly *washed out* under typical conditions of energy resolution in modern coincidence experiments. Figure 2 gives an example of these calculations. We emphasize a strong cooperative effect of a great number of non-resolved near-lying autoionization resonances in the region of $E_{ex} = 71.0 - 72.5eV$.

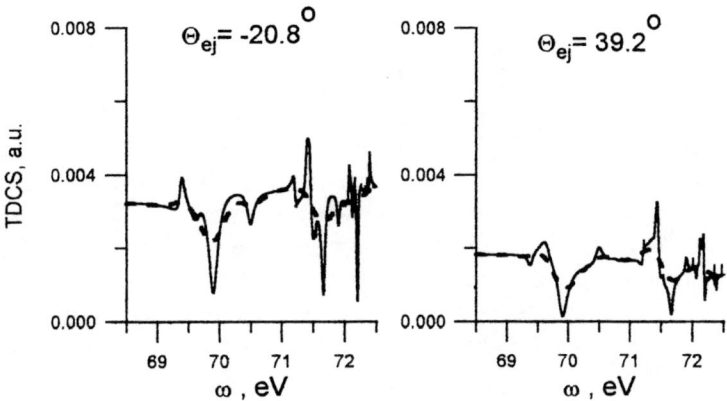

FIGURE 2. TDCS for reaction $He(e, 2e)He^+(n = 2)$ in the scattering plane at $E_0 = 3$ keV and $\theta_{sc} = 3°$; dashed lines - after averaging with FWHM=0.35 eV.

To consider influence of the electron-electron correlations in the ground state of the target atom on various characteristics of the ionization- excitation process we use the 41-parameter wave function by Tweed [17] and the 6-parameter Hilleraas wave function by Steward and Webb [18] which reproduce the two-body radial and angular correlation effects in different manner and whose multipole expansions differ considerably as it concerns higher L-multipoles. The difference between the results obtained within these two approaches is considerable. These calculations bring also interesting results concerning the direct (e,2e) ionization process beyond the autoionization resonances region [19]. One can compare them with experimental data (Figure 3). The corridor between two versions of our first order PWBA calculations is of about the same width than that between the first order and second order PWBA calculations performed recently in [13].

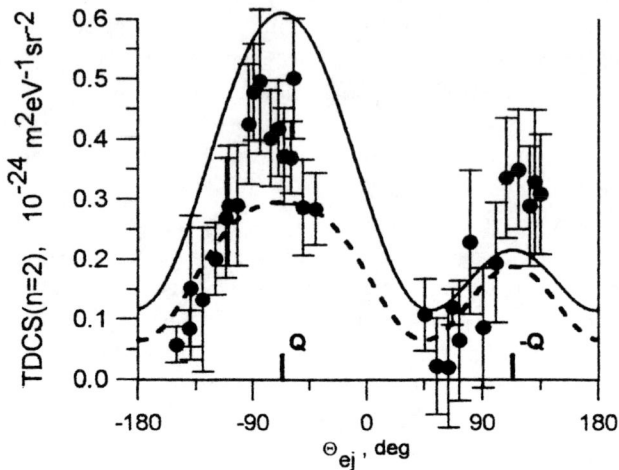

FIGURE 3. TDCS for reaction $He(e, 2e)He^+ (n = 2)$ at $E_0 = 1584.5$ eV, $\theta_{sc} = 4^\circ$ and $E_{ej} = 20$ eV: calculations with the ground state wave function by Steward and Webb [18] (solid line) and by Tweed [17] (dashed line); experimental data - from [12].

Calculations show that partial TDCS's for reaction $He(e, 2e)He^*(n = 2)$ corresponding to the 2s and 2p ionization channels, including the resonance structure of these cross sections, differ from each other considerably. To separate their contribution by triple coincidence $(e, 2e\gamma)$ measurements one must take into account possible angular anisotropy of the 2p→1s fluorescence radiation caused by the polarization (alignment) of the produced excited ion.

The origin of the alignment of excited states of the residual ion, when observed by the coincidence $(e, 2e\gamma)$ method, is complicated. As concerns its resonant component, it is the alignment of the atomic autoionizing state induced by the electron-atom collision and transferred to the residual ion and also the alignment connected with angular anisotropy of the ejected electron and created at the very process of the autoionization. This resonant mechanism of the alignment acts together with the direct one where the excited ion is formed by direct knock-out of an electron from the ground state of the target atom.

Starting from [20] we have elaborated an unified theoretical description to take into consideration all above mentioned aspects of this process together [21]. The start point of the approach is that the amplitude $< \gamma_1 L_1 l L M \mid \hat{T}(\vec{k}_0, \vec{k}_{sc}) \mid \gamma_0 >$ of the ionization process

$$A(\gamma_0 L_0) + e_0 \rightarrow \left(A^+(\gamma_1 L_1) + e_{ej} \right) + e_{sc} \qquad (1)$$

in the vicinity of a number of near-lying autoionization resonances $\mid \gamma_r L_r >$ is a sum of direct and resonant terms. The direct one $< \gamma_1 L_1 l L_r M \mid \hat{T}(\vec{k}_0, \vec{k}_{sc}) \mid \gamma_0 >_{dir}$

is calculated within standard channel-coupling procedures. Each resonant term

$$\sum_{\gamma_r L_r} \frac{V_r(\gamma_1 L_1 lL)}{V_r} \cdot < \nu_r L_r M \mid \hat{T}(\vec{k}_0, \vec{k}_{sc}) \mid \gamma_0 >_{dir} \left(\frac{q_r - i}{\epsilon_r + i}\right). \tag{2}$$

contains the generalized Fano profile index $q_r \equiv q_r(\vec{k}_0, \vec{k}_{sc})$ which depends on the target atom internal structure and the dynamics of the ionization process.

All observables of the process are obtained from the density matrix or, equivalently, statistical tensors of the final system $(A^+ + e_{ej})$ which are built up from the amplitudes $< \gamma_1 L_1 l L_r M \mid \hat{T}(\vec{k}_0, \vec{k}_{sc}) \mid \gamma_0 >$ in a standard way [22]. Among them are angular distribution and polarization parameters of the fluorescence radiation emitted by the residual ion.

Excitation of atomic autoionizing states
via formation of negative-ion resonances

This problem became topical when detailed measurements of alignment of autoionizing state $3p^5 4s^2 :^2 P_{3/2}$ in potassium atom performed by the Freiburg group from 0.5 keV down to about 30 eV [23] revealed an unusual behaviour of the alignment parameter $A_{20}(E_0)$ in the lowest part of this range. The effect was suggested to be caused by formation of negative-ion resonances and sharing of their angular momentum between the excited target atom and the scattered electron. Indeed, such negative-ion resonances were observed later by tracing the alignment of the autoionizing state as a function of the incoming electron energy with a higher energy resolution [24]. The resonance structure is very complicated. The problem of identification of the observed resonances came to the stage.

Similar negative-ion resonances effects are known for a long time in experiments on near-threshold excitation of discrete atomic levels [25] including a series of measurements of the polarization degree of the radiation emitted by the excited atom in coincidence with the low energy scattered electrons [26]. Exploring this analogy we have considered a new version of (e,2e) measurements [27] opposite to the widely used scheme of the asymmetric (e,2e) experiment: the energy of the scattered electron in our case (after formation and decay of the negative-ion resonance) is very small while that of the second electron (produced at the decay of the autoionizing state of the target atom) can be high enough.

We have applied this scheme to theoretical analysis of alignment of autoionizing state $2p^5 3s^2 :^2 P_{3/2}$ state in sodium atom ($E_{ex} = 30.77$ eV) in a narrow range of the incoming electron energy which is only few eV above the excitation threshold of this state [28] (the excitation amplitudes were calculated within the 26-parameter R-matrix approach [29]). In one-arm experiments, when the scattered electron is not detected, alignment $A_{20}(E_0)$ is the only parameter characterizing polarization properties of the produced autoionizing state and, hence, there is only one combination of the excitation amplitudes which can be extracted from angular distribution of the autoionization electrons. The angular correlation function between the ejected and scattered electrons is more informative. It is determined by a set of statistical tensors $A_{2q}(E_0, \theta_{sc})$ of the autoionizing state with q=0;1 and 2. As an example, we demonstrate them for two values of the incoming electron energy corresponding to very close values of the integral parameter $A_{20}(E_0)$ (Figure 4).

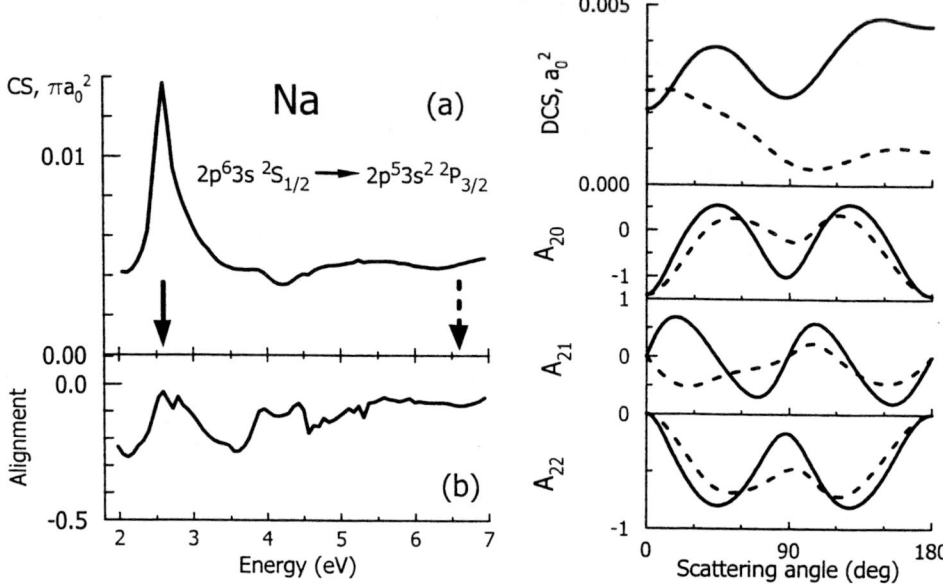

FIGURE 4. *Left*: integral excitation cross section (a) and alignment (b) of the autoionizing state $2p^5 3s^2\,^2P_{3/2}$ as functions of the incident electron energy counted from the excitation threshold; *right*: differential excitation cross section and statistical tensors of this state for two values of the incident electron energy marked in (a).

Autoionization with polarized electron beams and polarized targets

Earlier we made a proposal for coincidence $(\vec{e}, 2e)$ measurements on excitation of autoionizing states by polarized electron beam [30]. The interest was in a specific interrelation between the left-right asymmetry effects expected in the differential cross section of the scattered electron and, on the other hand, in the (e,2e) angular correlation function measured when the incoming beam is polarized **up** or **down** relative to the scattering plane. No measurements have been made yet on this line. It is worthwhile to confirm that the experiment under discussion could be an important contribution to better understanding the role of spin-dependent interactions in electron-atom collisions.

Another line of our investigations in this field concerns experiments with polarized (laser excited) target atoms. We began with the alignment aspect of the problem [31-34] related to experiments with linear polarized laser. To extend them to oriented (vector-polarized) targets excited by circular polarized laser beam we investigated [35] excitation of sodium autoionizing states $2p^53s3p\,^2D_{3/2,5/2}$ from laser excited totally polarized state $2p^63p^2P_{3/2}$ with the atomic angular momentum $<\vec{J}_0>$ directed **up** and **down** along the normal to the scattering plane. The strong left-right (up-down) asymmetry effect in the differential cross section of this process is predicted as a kind of the *circular pumping dichroism in the angular distributions* effect (CPDAD) [34]. Similar orientation effects are under current theoretical and experimental investigations of other processes [36-40]. Looking similar to the left-right asymmetry effect in scattering of polarized electrons from nonpolarized atoms, CPDAD originates, mainly, not in spin-dependent electron-atom interactions and does not vanish when all interactions of this sort including exchange electron-atom interactions are switched off [41].

Triple coincidence $(e, 2e\gamma)$ measurements as a 'perfect experiment' instrument for ionization-excitation studies

Autoionization (e,2e) measurements are similar, in a sence, to the electron-photon correlation $(e, e'\gamma)$ experiments on excitation of atomic discrete levels serving as a sourse of 'perfect' and model independent information on the m-resolved excitation amplitudes corresponding to magnetic substates of the levels. Can one suggest a similar program in the autoionization case taking into account that the analogy between the (e,e'γ) method in the case of excitation of discrete levels and the (e,2e) method for autoionization studies is not perfect because of competing direct ionization processes in the latter case? As the first step, we considered the case of an isolated singlet autoionizing state decaying to an S-state of the residual ion [42] and have found a model-independent way to extract the direct and resonant amplitudes of the ionization process from Shore parameters used to parameterize autoionization resonances in (e,2e) experiments. We considered also excitation of autoionizing states in "one-electron" atoms Li, Na, K, ... in a special case of negligibly small contribution of the competing direct ionization mechanism. A complex of experiments with polarized electron beam and polarized target was suggested to obtain model independent information on the excitation amplitudes [35].

To estimate perspectives of triple coincidence $(e, 2e\gamma)$ measurements as a 'perfect experiment' instrument in ionization-excitation studies we consider reaction $He(e, 2e\gamma)He^+(2p)$. A fundamental difference of $(e, 2e\gamma)$ measurements in comparison with double coincidence measurements $(e, e'\gamma)$ (where the ejected electron is not detected) is that in the first case the final state $2p$ is produced coherently as a *pure state* characterized with a definite state vector. According to general symmetry arguments there are three independent real parameters for three complex (e,2e) amplitudes $F_{m=0;\pm1}(\vec{k_0}, \vec{k_{sc}}, \vec{k_{ej}}) \equiv F_{0;\pm1}$ describing population of the 2p-state. In the natural frame we take: $F_0 = 0$, $F_{+1} = a \cdot \sin\psi \cdot e^{-i\delta}$ and $F_{-1} = a \cdot \cos\psi \cdot e^{i\delta}$ $(0 \le \psi \le \pi/2)$. Parameters ψ and δ can be obtained from angular distribution of the 2p→1s fluorescence radiation in the reaction plane detected in coincidence with the scattered and ejected electrons:

$$W_\gamma(\phi_\gamma) \sim 1 - A \cdot \cos 2(\phi_\gamma - \gamma). \tag{3}$$

The anisotropy magnitude parameter $A = \frac{3}{11}\sin 2\psi$ shows the ratio of absolute values of amplitudes F_{+1} and F_{-1} while parameter $\gamma = \delta + \pi/2$ (angle between the short axis of symmetry of the correlation function $W_\gamma(\phi_\gamma)$ and the incoming electron beam) shows the phase relation between these amplitudes (these equations are given with the fine structure disalignment corrections taken into account).

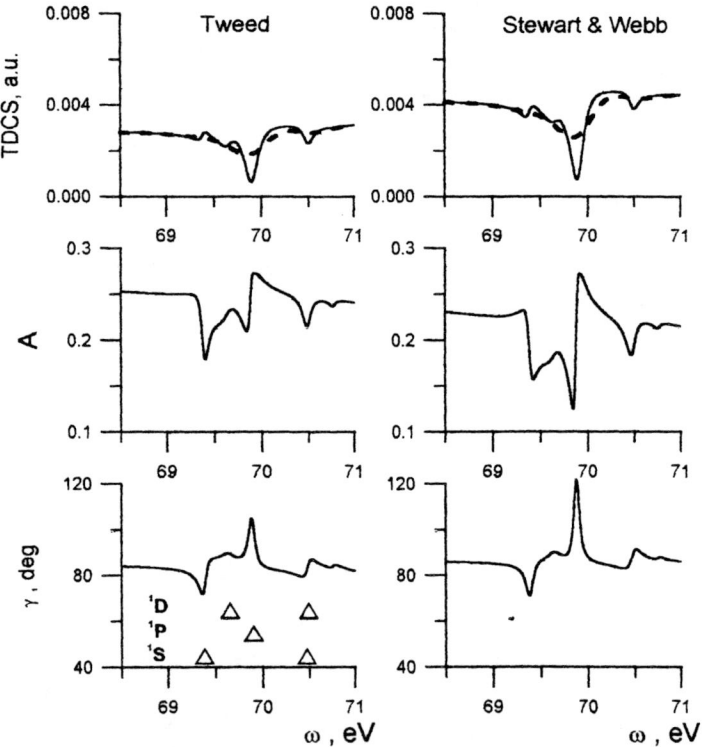

FIGURE 5. TDCS for reaction He$(e, 2e)$He$^+(2p)$ in the scattering plane at $E_0 = 3$ keV, $\theta_{sc} = 3°$, $\theta_{ej} = -20.8°$ and corresponding angular distribution parameters A and γ for the 2p→1s fluorescence radiation in the reaction $(e, 2e\gamma)$.

So, measurements of angular distribution of the fluorescence radiation in the reaction plane together with the absolute fluorescence radiation yield provide the *perfect* information on the process He$(e, 2e\gamma)$He$^*(2p)$ and, by the same way, on the electron cloud distribution in the excited state $2p$ of the residual ion He$^+$. All other observables including polarization parameters of the 2p→1s fluorescence radiation can be calculated, when necessary, using parameters ψ and δ extracted from angular distribution of the fluorescence radiation in the $(e, 2e\gamma)$ measurements.

Figure 5 presents calculated parameters A and γ for reaction He$(e, 2e\gamma)$He$^*(2p)$ in the excitation energy region between the second and third ionization thresholds [43]. They demonstrate strong influence of the autoionization resonances on the angular distribution of the 2p→1s fluorescence radiation.

Conclusion

Two categories of theoretical procedures are known to calculate autoionization resonances in atomic collisions. It is typical for one of them (the *non-parameterized approach*) to proceed without any form taken *a-priori* for energy dependence of ionization amplitudes and other characteristics of resonance processes. Calculated cross sections are analyzed using the Shore of Fano formulas to extract resonance parameters of interest as if these curves are obtained experimentally. Various close-coupling methods such as the R-matrix method are used here.

On the contrary, when (starting from the Feshbach unified theory of nuclear reactions and the Fano theory in atomic physics) one uses the *parameterized approach*, the aim is to calculate first just these resonance parameters provided that various energy dependent characteristics, including those related to correlation and polarization measurements, can be built up, when necessary, using these parameters in accordance with corresponding theoretical formulas.

We use both of these approaches exploring specific adventages of each of them depending of the concrete problem under consideration. Combining them one gets, in particular, a solid base for productive contacts with experimental research.

Acknowledgments

The work has been supported by the Program "University of Russia - Basic Research" (grant N 5340).

References

1. Weigold E., Urgabe A., Teubner R.J.O., *Phys.Rev.Lett.* **35**, 209 (1975).
2. Lower J. and Weigold E., *J.Phys.* B**23**, 2819 (1990).
3. McDonald D.G. and Crowe A., *Z.Phys.* B**23**, 371 (1992).
4. Balashov V.V., Lipovetsky S.S., Senashenko V.S., *Phys.Lett* **39**A, 103 (1972).
5. Pochat A. et al., *J.Phys.* B**15**, 2269 (1982).
6. Kheifets A., *J.Phys.* B**26**, 1993 (1993).
7. McCarthy I. and Shang B., *Phys.Rev.A* **47**, 4807 (1993).
8. Balashov V.V., Kozhevnikov I.V., Magunov A.I., *J.Phys.* B**14**, 2059(1981).

9. Balashov V.V., Crowe A., Gorelenkova M.V., *Two-step contribution to electron-impact excitation of atomic autoionizing states*, 5th ECAMP (Edinburg, UK, 3-7 april, 1995); Contributed Papers, Part II, p.675.

10. Stefani G., Avaldi L., Camilloni R., *J.Phys.* B**23**, L227(1990).

11. Dupre C. et al., *J.Phys.* B**25**, 259(1992).

12. Avaldi L. et al., *J.Phys.* B**31**,2981(1998).

13. Marchalant P.J., Whelan C.T., Walters H.R.J., *J.Phys.* B**31**, 1141(1998).

14. Hayes P.A. and Williams J.F., *Phys.Rev.Lett.* **27**,955(1996).

15. Dogan M., Crowe A., Bartschat K., Marchalant P., *J.Phys.* B**31**, 1611(1998).

16. V.V.Balashov, Proc. Int. Conf. on Coincidence Spectroscopy, (Brest, France; Sept. 23-26, 1998). J. de Physique (to be published).

17. Tweed R.J., *J.Phys.* B**5**, 810(1972).

18. Stewart A.L. and Webb T.G., *Proc.Phys.Soc.* **82**, 532(1963).

19. Balashov V.V. and Bodrenko I.V., *J.Phys.* B (submitted).

20. Balashov V.V., Lipovetsky S.S., Senashenko V.S., *Sov.Phys.JETP* **36**, 853(1973).

21. Balashov V.V., *Profile of autoionization resonances in electron-impact ionization processes*, INP MSU Preprint 97-3/454, Moscow, 1997.

22. Balashov V.V., Grum-Grzhimailo A.N., Dolinov V.K., Korenman G.Ya., Krementsova Yu.N., Smirnov Yu.F., Yudin N.P., *Theoretical Practicum in Nuclear and Atomic Physics* (in Russian), Moscow, Energoatomizdat, 1984.

23. Matterstock B., Huster R., Paripas B. et al. , *J.Phys.* B**28**, 4301(1995).

24. Grum-Grzhimailo A.N., Bartshat K., Feierstein B., Mehlhorn W., *Phys.Rev.* A (in press).

25. Defrance A., *J.Phys.* B**13**, 1229(1980).

26. Batelaan H., van Eck J., Heideman H.G.M., *J.Phys.* B**24**, 5151(1991).

27. Balashov V.V., *Two-step mechanism of electron-impact excitation of atomic autoionizing states via formation of negative-ion resonances*, INP MSU Preprint 97-50/501, Moscow, 1997.

28. Balashov V.V. and Grum-Grzhimailo A.N., XXI ICPEAC (Sendai, 1999), Abstracts of contributed papers, v.1, p.197.

29. Feierstein B., Grum-Grzhimailo A.N., Mehlhorn W., *J.Phys.* B**31**, 593(1998).

30. Balashov V.V., Bodrenko I.V., Grum-Grzhimailo A.N., *Left-right asymmetry in excitation of atomic autoionizing states by polarized electrons*, in AUTOIONIZATION PHENOMENA IN ATOMS (5th Int. Workshop, Dubna, Russia, December 12-14, 1995), Moscow, Moscow University Press, 1996; p.72.

31. Balashov V.V. and Grum-Grzhimailo A.N., 1986 *Proc. of 3rd All-Union Workshop on the Autoionizing Phenomena in Atoms (Moscow, 1995)*; Moscow University Press, p 46.

32. Dorn A., Nienhaus J, Wetzstein M. et al., *J.Phys.* B**27**, L529(1994).

33. Balashov V.V., Golokhov E.I., Grum-Grzhimailo A.N., *Phys.Lett.* **222**A, 81(1996).

34. Balashov V.V., Grum-Grzhimailo A.N., Kabachnik N.M., *J.Phys.* B**30**, 1269(1997).

35. Balashov V.V., *Perspectives in autoionization (e,2e) studies with polarized electron beams and polarized targets*, Int. Symposium on (e,2e) and related processes; Frascati, Roma, 1997 (unpublished).

36. Berakdar J., Engelis A., Klar H., *J.Phys.*, B**29**, 1109(1996).

37. Dorn A., Elliott A., Lower J. et al., *Phys.Rev.Lett.* **80**, 257(1998).

38. Berakdar J., Huetz A., Klar H., Selles P., *J.Phys.* B**26**, 1403(1993).

39. Viefhaus J., Avaldi L., Snell G. et al., *Phys.Rev.Lett.* **77**, 3975(1996).

40. Hansen J.P., Kocbach L., Jun Lu, Dubois A., XX ICPEAC (July 22-27, 1997, Vienna), Abstracts of contributed papers, v.1, TH 076.

41. Balashov V.V., Golokhov E.I., *Phys.Lett.* **257**A, 70(1999).

42. Balashov V.V., Martin S.E., Crowe A., *J.Phys.* B**29**, L337(1996).

43. Balashov V.V., Bodrenko I.V., XXI ICPEAC (Sendai, Japan, 1999), Abstracts of contributed papers, v.1, p.245.

Inelastic low-energy electron collisions with hydrogen halides

J. Horáček

Department of Theoretical Physics,
Faculty of Mathematics and Physics, Charles University Prague,
V Holešovičkách 2, 180 00 Praha 8, Czech Republic
horacek@indigo-1.troja.mff.cuni.cz

Abstract. Inelastic low-energy electron collisions with hydrogen halides HCl, HBr, DBr and HI are studied theoretically on the basis of the nonlocal resonance model of Domcke and Mündel. The model takes account of the dependence of the dipole-modified threshold exponent on the internuclear distance and of the precise form of the long-range part of the negative ion potential. Cross sections for vibrational excitation, dissociative attachment and associative detachment have been calculated. For all three collision processes, the cross sections calculated are in better agreement with experiment than previous calculations.

I   INTRODUCTION

The collision of low-energy electrons with hydrogen halides represents a typical example of a molecular resonance process in which the polar nature of the molecular target plays an important role. The resonant character of the collision leads to large cross sections for inelastic and reactive processes, i.e., vibrational excitation (VE), dissociative electron attachment (DA) and the inverse process of associative detachment (AD).

The most interesting feature in hydrogen halides is the existence of pronounced threshold peaks in the VE cross sections observed first by Rohr and Linder [1,2]. The cross sections in the threshold region attain very high values and represent the highest vibrationally inelastic cross sections measured. The step-like structures in the DA cross section (Wigner cusps) first observed by Abouaf and Teillet-Billy [3] represent another striking feature of low-energy electron-polar molecule collisions. More recently, additional fine structure has been discovered in the $0 \to 1$ and $0 \to 2$ VE functions of HCl, which appears in the form of oscillations converging towards the DA threshold [4,5].

Several theoretical models have been put forward to provide an explanation of

CP500, *The Physics of Electronic and Atomic Collisions*, edited by Y. Itikawa, et al.
© 2000 American Institute of Physics 1-56396-777-4/00/$17.00

these unusual phenomena [6–15]. They are based on a variety of theoretical concepts such as the projection-operator formalism [6,7,10,11,15], the R-matrix formalism [12–14], or the zero-range potential model [8,9]. The nonlocal resonance model (NRM) developed by Domcke and Mündel [11] has so far provided the most complete description of the resonance and threshold features observed in low-energy electron-HCl collisions [4,16,17]. This model describes qualitatively all the observed features: threshold peaks and broad resonances in the VE cross sections, Wigner cusps in the DA cross sections, as well as oscillatory structures in the VE cross sections below the threshold of the DA channel.

It is the purpose of this paper to give a brief review of the work which has been done at the Department of Theoretical Physics, Charles University, Prague. During several years we have developed efficient numerical techniques for solving integral equations of NRM theory, generalized the original model of Domcke and Mündel (DM), obtained the parameters for several molecules and calculated cross sections for all three processes.

II DESCRIPTION OF THE NRM MODEL

The nonlocal resonance model of Domcke and Mündel is given in terms of the functions $V_d(R)$, $V_0(R)$ and $V_{dE}(R)$ describing the discrete state of the negative ion, the neutral ground state and the discrete-state-continuum coupling. There is no restriction on the functional dependence of the potential functions $V_d(R)$ and $V_0(R)$. For the discrete-state-continuum coupling we assume the following very general form which conforms to the Wigner threshold law

$$V_{dE}(R) = A \left(\frac{E}{B}\right)^{\alpha(R)/2} e^{-E/2\beta(R)} g(R). \tag{1}$$

Note that the coupling $V_{dE}(R)$ is not a separable function of the variables E and R. For the calculation of the AD cross sections the long-range part of the negative ion-atom interaction must be known with a high precision. Several accurate ab initio studies of the bound part of the $^2\Sigma^+$ potential of HX$^-$ are now available in the literature [21,22]. This makes it possible to construct an improved model exhibiting the correct long-range behaviour of the HX$^-$ potential function by modifying the function $V_d(R)$ while retaining the functions $V_0(R)$ and $V_{dE}(R)$.

III RESULTS

To apply the nonlocal resonance model to a specific molecule the parameters defining the functions $V_d(R)$ and $V_{dE}(R)$ must be obtained. These parameters have been obtained by fitting the ab initio data (phaseshifts) in the case of HCl and HBr. Since for HI no such data are available the parameters were obtained by fitting the very accurate DA data of Klar et al. [23]. Here we give a brief overview of our results.

A   Vibrational excitation

Let us start with the discussion of the vibrational excitation process. Previous calculations [11,14,15] have described qualitatively the observed features, i.e., the threshold peaks and the broad resonance region, but the detailed shapes of the cross sections were not reproduced correctly. Here we demonstrate that for energies close to the threshold the modifications of the original DM model mentioned above improve the agreement between theory and experiment to such an extent that not only the gross features, but also very minute details of the calculated cross sections can be compared with the measured data.

It has been shown in [20] that taking the variation of the threshold exponent α with R into account changes the cross section near threshold significantly for molecules with larger dipole moments (HCl), but has only a marginal effect for molecules with small dipole moments (HBr and HI). The agreement between theory and experiment is further improved when the long-range behaviour of the HCl^- potential function is corrected according to ab initio calculations. In figure 1, the $0 \rightarrow 1, ..., 3$ calculated VE cross sections (left) are compared with the data measured by Schaffer and Allan [5] (right).

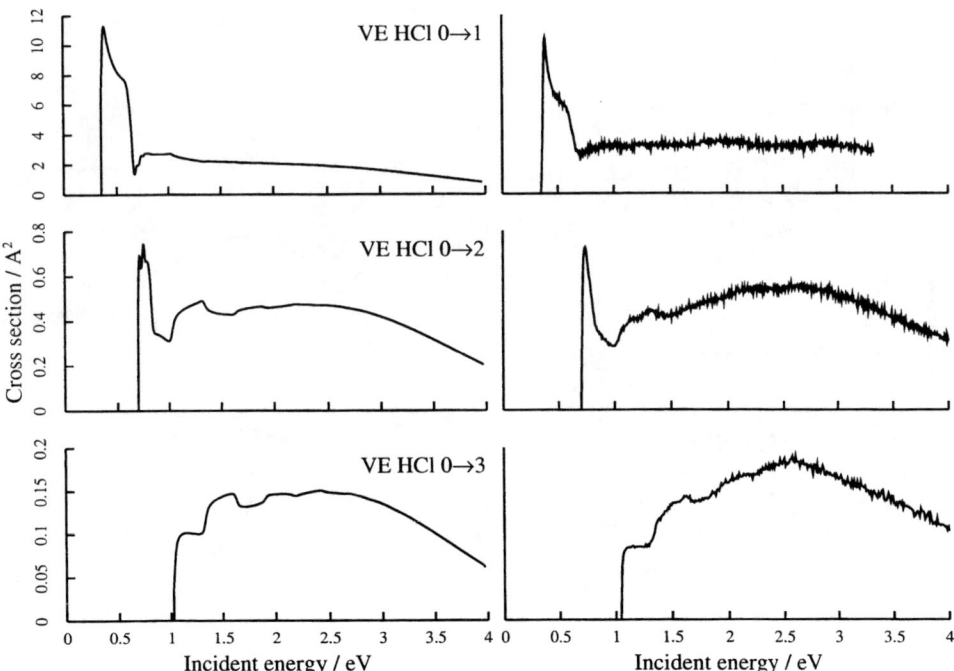

Figure 1. VE $0 \rightarrow 1,..,3$ cross sections for HCl. Left: NRM theory, right: experiment of Schaffer and Allan.

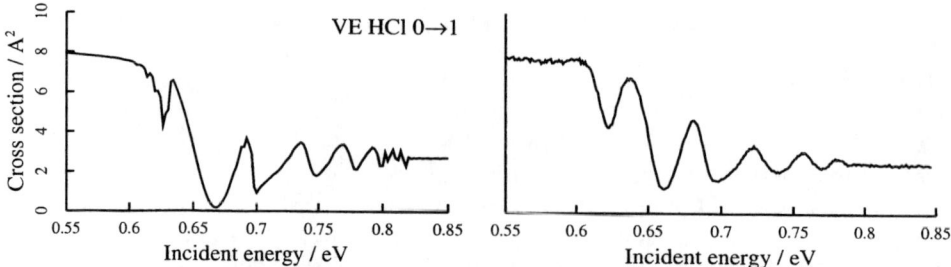

Figure 2. Details of the VE $0 \rightarrow 1$ cross sections for HCl below the opening of the DA channel. Left: NRM theory, right: experiment of Allan et al. [32].

Our calculation predicts oscillatory structures in the VE cross sections, see figure 2, in an energy range which extends from the bottom of the negative ion potential curve to the DA threshold. The oscillations reflect interference with quasi-bound levels supported by the attractive well at intermediate internuclear distances. Such oscillations have been predicted also by some other HCl models [15] and observed experimentally [4,5]. The present model constructed from the most sophisticated ab initio data predicts oscillations in an excellent agreement with the most recent high resolution experiments of Allan et al. [32].

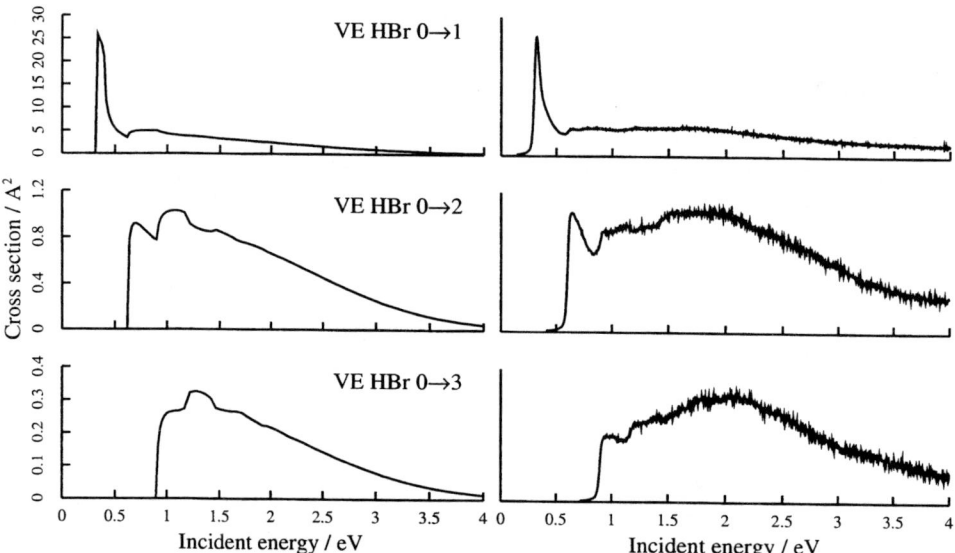

Figure 3. VE $0 \rightarrow 1, .., 3$ cross sections for HBr. Left: NRM theory, right: experiment of Allan et al. [32].

The results of our calculations of the VE cross sections for HBr and DBr molecules are compared with the recent preliminary data of Allan et al. [32] in figures 3 and 4. It is clearly seen that all the major features of the cross sections are very well reproduced with the present model.

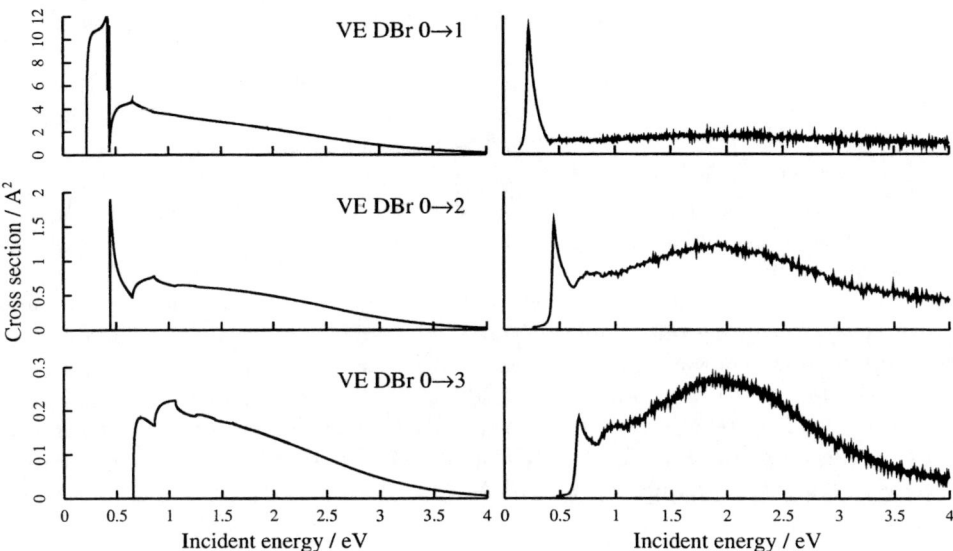

Figure 4. VE $0 \to 1, .., 3$ cross sections for DBr. Left: NRM theory, right: experiment of Allan et al. [32].

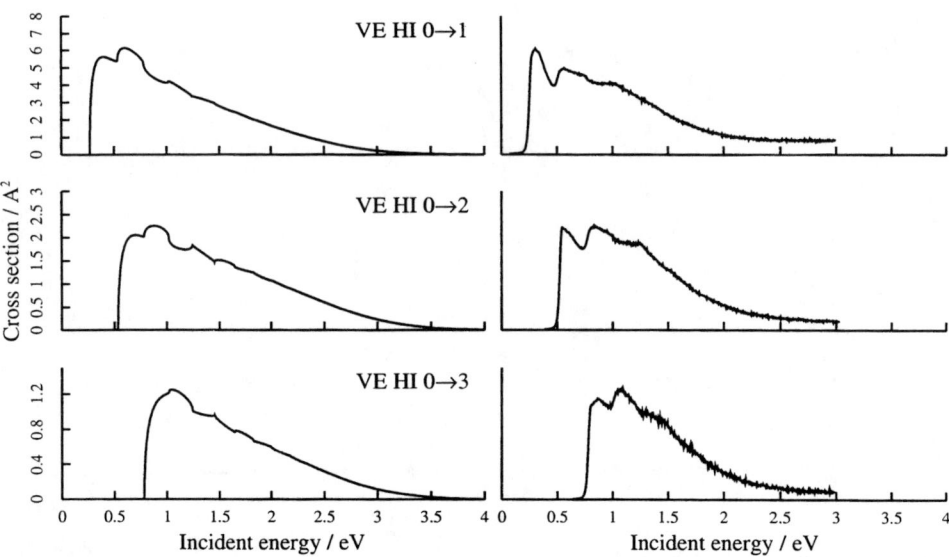

Figure 5. VE $0 \to 1, .., 3$ cross sections for HI. Left: NRM theory, right: experiment of Allan et al. [32].

The calculated VE cross sections for HI molecule are shown together with the preliminary data of Allan et al. [32] in figure 5. In agreement with the calculation the measured VE cross section do not show threshold peaks in any channel.

B   Dissociative attachment

Calculations of the DA cross section for HCl have been reported by several authors, see, e.g. [6,9,11–15]. The majority of the calculations provides a qualitatively correct description of the observed features, i.e., an essentially vertical onset at threshold, step-like Wigner cusps at the openings of the VE channels, and a rapid increase of the cross section with increasing rovibrational energy of the target. The existing calculations taking full account of the nonlocality of the problem have been performed without inclusion of the rotational degrees of freedom ($l = 0$) [11,14,15,30]. Since in the real experiments the target gas is at nonzero temperature T, the target molecules may be excited either rotationally or vibrationally. To describe this situation we have calculated the DA cross section for target molecules in their four lowest vibrational states and for angular momenta $l = 0, \ldots, 30$ and then averaged the results over the Maxwell-Boltzmann distribution of the target states. The cross sections for the temperature $T = 1000K$ is shown in figure 6 together with the experimental data of Allan and Wong [31].

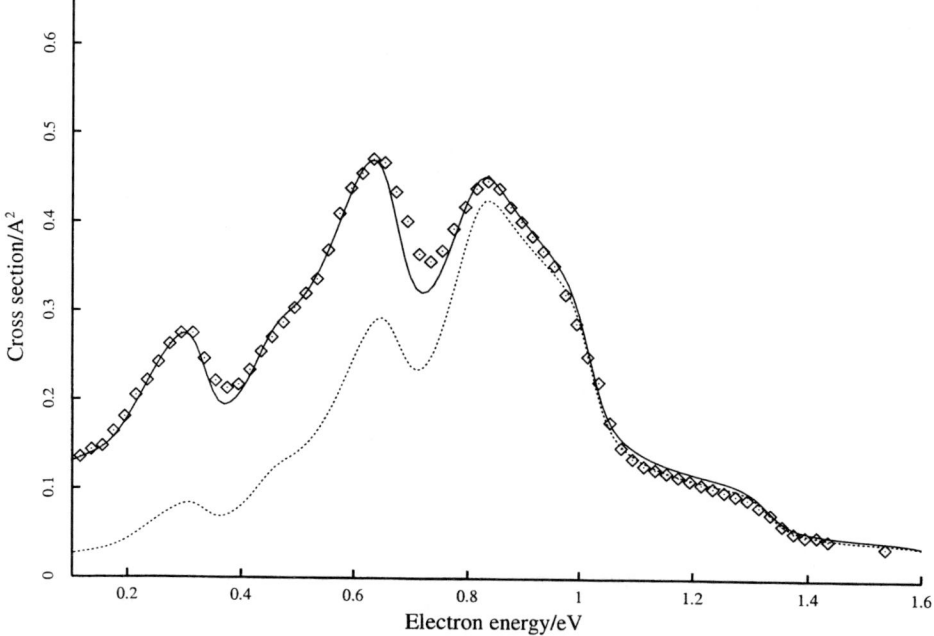

Figure 6. DA cross section for HCl; solid line - the NRM results calculated at the temperature 1230K, dotted line - the same calculation at the temperature 1000K. Diamonds are experimental values normalised to the maximum of the solid line curve.

To compare our results with the experimental data we convoluted the theoretical data with a Gaussian distribution with the full width at half maximum (FWHM) equal to the energy spread of the electrons in the experiment (50meV). Comparing the theoretical data with the experiment of Allan and Wong, we see that the position of the structures agrees quite well but their relative height differs significantly. It is worth to point out that perfect agreement between theory and the measured data may be obtained by assuming a higher temperature of the target gas. This is demonstrated in figure 6. If, for example, the cross section calculated at the temperature $T = 1230K$, solid line, is compared with the data measured at $T = 1000K$ we observe perfect agreement. The same holds also for other temperatures.

C   Associative detachment

The AD cross section is a highly integrated quantity providing little insight into the details of the dynamics. The total AD cross sections for HCl are shown in figure 7a together with the experimental data [28]. The AD cross section exhibits an extended series of threshold peaks, each at the opening of a rovibrational channel of the HCl molecule. Detailed plots of the threshold peaks in the AD cross section are shown in figure 7b.

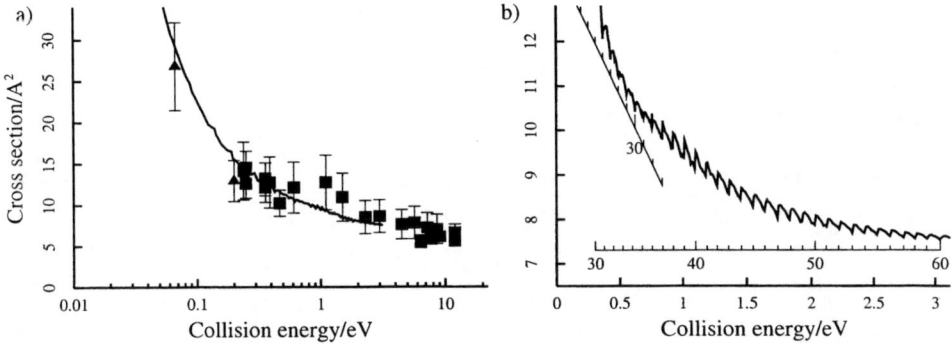

Figure 7. a) The total associative detachment cross section (full curve), experimental data (squares) and data derived from experimental rates (triangles) are taken from Huels et al. [28].
b) Details of the threshold peak structures in the AD cross section.

The spectrum of electrons ejected in the process of the associative detachment depends strongly on the energy distribution in the entrance channel. In figure 8 we plot the convoluted electron spectrum for H + Cl⁻ AD process. The distribution of energy in the incident beam is shown in the inset of figure 8.

Contrary to the AD process in HCl which is exothermic the associative detachment in HI is endothermic with a very small threshold energy. The calculated cross section is shown together with the experimental data of [28] in figure 9. Although the data for the HI model were obtained by fitting the low energy DA cross section and the model needs improvement we observe a very good agreement of the calculated AD cross section with the experimental data.

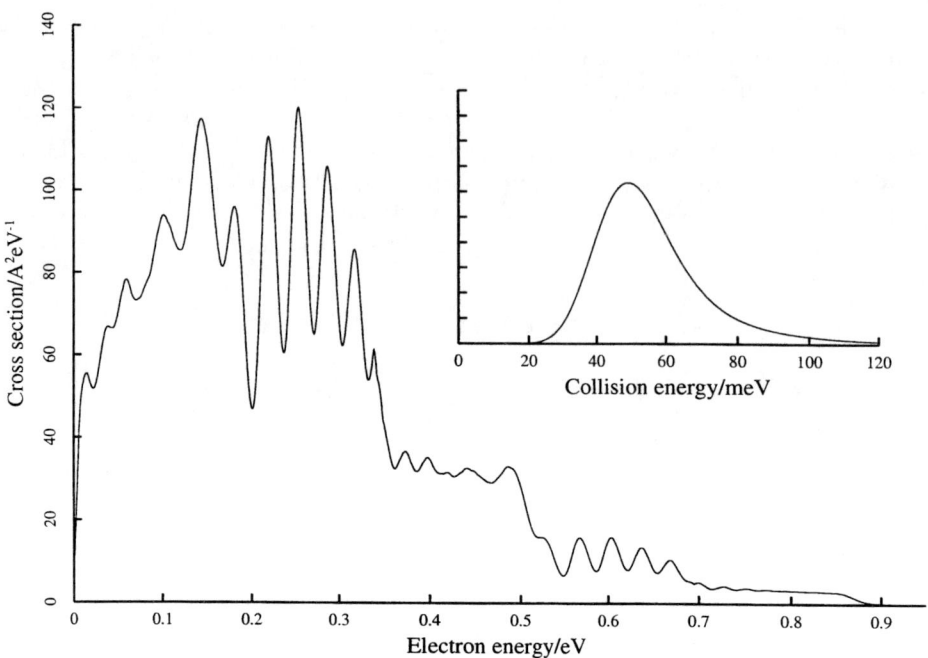

Figure 8. The spectrum of electrons ejected in the AD process for HCl.

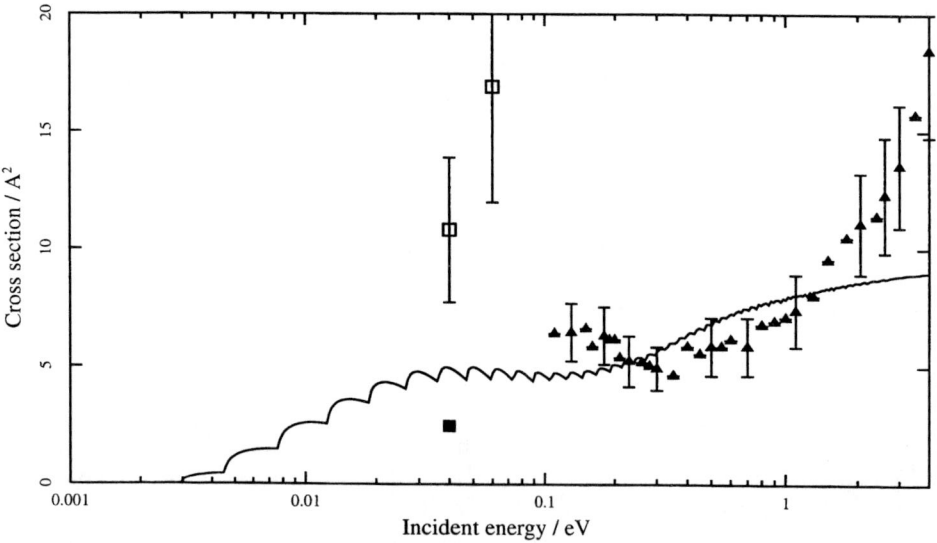

Figure 9. The AD cross section for HI. Solid line - the NRM results, experiment of Huels et al. [28]

IV CONCLUSIONS

In this paper we report results of our study of three inelastic processes, namely: vibrational excitation, dissociative attachment and associative detachment in the system of electron-hydrogen halide molecule based on the generalized nonlocal resonance model. We have constructed improved nonlocal resonance models for electron-HX collisions and have performed calculations of VE, DA and AD cross sections. It can be summarized that the cross sections calculated with the present model are in much better agreement with the experimental data for all processes considered here than any other calculation published so far. The generalization and improvement of the nonlocal resonance model leads to almost perfect agreement between theory and experiment in the low-energy range. For VE at higher energies further improvement of the model is needed.

ACKNOWLEDGEMENTS

This paper represents an effort of many colleagues and collaborators to whome I would like to express my gratitude. In particular I would like to thank Wolfgang Domcke, Hiroki Nakamura, Michael Allan, Hartmut Hotop, Ilya Fabrikant, S. L. Cederbaum, Hans-Dieter Meyer and my students Martin Čížek, František Gemperle and Karel Houfek.

REFERENCES

1. K. Rohr and F. Linder, *J. Phys.* B**8**, L200 (1975)
2. K. Rohr and F. Linder, *J. Phys.* B**9**, 2521 (1976)
3. R. Abouaf and D. Teillet-Billy, *J. Phys.* B**10**, 2261 (1977)
4. S. Cvejanović, in: 18th Int. Cont. on Electronic and Atomic Collisions (ICPEAC), bool of invited papers, Aarhus (1993).
5. O. Schafer and M. Allan, *J. Phys.* B**24**, 3069 (1991)
6. F. Fiquet-Fayard, *J. Phys.* B**7**, 810 (1974)
7. W. Domcke and L. S. Cederbaum, *J. Phys.* B**14**, 149 (1980)
8. J. P. Gauyacq and A. Herzenberg, *Phys. Rev.* A**25**, 2959 (1982)
9. D. Teillet-Billy and J. P. Gauyacq, *J. Phys.* B**17**, 4041 (1984)
10. N. Bardsley and J. M. Wadehra, *J. Chem. Phys.* **78**, 7227 (1983)
11. W. Domcke and C. Mündel, *J. Phys.* B**18**, 4491 (1985)
12. I. I. Fabrikant, *Z. Phys.* D**3**, 401 (1986)
13. L. A. Morgan, P. G. Burke, and C. J. Gillan, *J. Phys.* B**23**, 99 (1990)
14. I. I. Fabrikant, S. A. Kalin, and A. K. Kazansky, *J. Chem. Phys.* **95**, 4966 (1991)
15. P. L. Gertitschke and W. Domcke, *Z. Phys.* D**31**, 171 (1994)
16. M. A. Morrison, *Adv. At. Mol. Phys.* **24**, 51 (1988)
17. W. Domcke, *Phys. Rep.* **208**, 97 (1991)
18. J. Horáček and W. Domcke, *Phys. Rev.* A**53**, 2262 (1996)

19. J. Horáček, W. Domcke, and H. Nakamura, *Z. Phys.* D**42**, 181 (1997)
20. J. Horáček, M. Čížek, and W. Domcke, *Theor. Chem. Acc.* **100**, 31 (1998).
21. S. V. O'Neil, P. Rosmus, D. W. Norcross, and H. Werner, *J. Chem. Phys.* **85**, 7232 (1986)
22. P. Åstrand and G. Karlström, *Chem. Phys. Lett.* **175**, 624 (1990)
23. D. Klar, B. Mirbach, H.J. Korsch, M.-W. Ruf, H. Hotop, *Z. Phys.* D**31**, 235 (1994)
24. J. P. Gauyacq, *J. Phys.* B**15**, 2721 (1982)
25. M. Čížek, J. Horáček, and W. Domcke, *J. Phys.* B**31**, 2571 (1998)
26. H.-D. Meyer, J. Horáček, and L. S. Cederbaum, *Phys. Rev.* A**43**, 3587 (1991)
27. H. J. Werner and P. Rosmus, *J. Chem. Phys.* **73**, 2319 (1980)
28. M. A. Huels, J. A. Fedchak, R. L. Champion, L. D. Doverspike, J. P. Gauyacq, and D. Teillet-Billy, *Phys. Rev.* A**49**, 255 (1994)
29. C. J. Howard, F. C. Fehsenfeld, and M. McFarland, *J. Chem. Phys.* **60**, 5068 (1974)
30. V. Pless, B. M. Nestmann, and S. D. Peyerimhoff, *J. Phys.* B**25**, 4649 (1992)
31. M. Allan and S. F. Wong, *J. Chem. Phys.* **74**, 1687 (1981)
32. M. Allan, preliminary results (to be published)

Electron Impact Dissociation of Molecules and Molecular Ions

Ann E. Orel

Department of Applied Science
University of California, Davis
Livermore, CA 94550

Abstract. In most collisions between electrons and molecules or molecular ions, there is little transfer of energy from the electron to the nuclear motion of the system. However, in resonant cases, the electron forms a temporary state, which changes the forces between the atoms, leading to vibrational excitation and in some cases dissociation. We have developed a completely *ab initio* approach to these problems. We use the Complex Kohn variational principle to determine the resonance parameters and a wave packet method to describe the nuclear dynamics. We will outline the basics of the method and show examples in dissociative recombination and dissociative attachment.

INTRODUCTION

One of the most fundamental (and one of the more useful) approximations in molecular physics is the Born-Oppenheimer approximation. The simple observation that the mass of the electron is at least three orders of magnitude less than the nuclei, and hence that the electron can rapidly adjust to the nuclear motion, allows the dynamics to be separated into distinct electronic and nuclear parts. This allows, for example, the calculation of the electronic spectra of a molecule at fixed internuclear geometry, with vibrational and rotational effects added later (if at all) or calculations of the elastic scattering of electrons from neutral molecular targets in the fixed nuclei approximation. This would lead to the general prediction that in electron-molecule or molecular ion collisions, there is little vibrational energy transfer. However, there is an important exception to this rule, systems where there is an intimate coupling between the electron interaction with the target and the nuclear dynamics of the target, causing dramatic effects. In these cases, the electron can temporarily attach to the molecule and change the forces felt between its atoms for a period of time comparable to a vibrational period. In dissociative recombination (DR)

$$ABC^+ + e^- \rightarrow A + BC \tag{1}$$

and in dissociative attachment (DA)

CP500, *The Physics of Electronic and Atomic Collisions*, edited by Y. Itikawa, et al.
© 2000 American Institute of Physics 1-56396-777-4/00/$17.00

$$ABC + e^- \rightarrow A^- + BC \qquad (2)$$

the 'light' low-energy electron can even break apart the molecular system.

Despite the crucial importance of DA and resonant vibrational excitation in modeling plasmas, and the pivotal role of DR in the formation of molecules in the interstellar media and planetary atmospheres, most calculations have been limited to diatomics, or one-dimensional models of polyatomic systems. In most cases, especially for DA, the resonance parameters used for these systems have been approximated, or based on input from experiments. We have used a combination of the Complex Kohn variational method to determine the resonance surfaces and autoionization widths, coupled with a wave packet method to determine the dynamics to treat DR in a number of systems. We will review this methodology in the next section. We will then discuss two examples, first application of this method to the DR of HeH^+ and and second, our progress towards a wave packet calculation of the DA cross section of $ClCN$.

METHODOLOGY

Both DR and DA are viewed as three-step processes. First, the electron is captured by the ion (in the case of DR) or the neutral (in the case of DA) into a resonant dissociative state. Second, the molecule begins to fragment, moving on the excited state (resonant) potential energy surface. During this process, the molecule can re-emit the electron (autoionize) possibly leaving the molecule in an excited vibrational state, or even dissociating it. (This process is referred to as resonant vibrational excitation (VE) or dissociative excitation (DE).) Finally, if no autoionization occurs, the molecule fragments into products, during which the probability may be distributed into various product states due to non-adiabatic coupling. Therefore, a complete study of a DR or DA process must include:

1. The determination of the resonant potential energy surface.

2. The determination of the resonant lifetime on each point of the potential energy surface.

3. The determination of the coupling between the relevant surfaces.

4. The calculation of the dynamics of the dissociation.

Dynamics

Wave packet methods, that is, numerical solutions of the time-dependent Schrödinger equation, are standard techniques in the study of the dynamics of chemical reactions, dating back to their introduction in this area by Heller in the 1970's [1]. The wave packet picture is physically intuitive, allowing one to watch the evolution of the system dynamics in time, as energy flows through the available modes of the molecule leading to bond breaking and formation, and in a coupled surface calculation, the flow of flux between the various states of the system. Since the wave packet is described by its value on a numerical grid of points, there are no serious problems with coupled potential energy surfaces [2,3] or a three-body continuum [4], which can be difficult to handle with basis set techniques. Wave packet methods have the additional advantage that the cross section is obtained at all energies from a single calculation.

The wave packet method proceeds by the direct integration of the time-dependent Schrödinger equation:

$$i\frac{\partial}{\partial t}\Psi(\mathbf{R}, t) = H(\mathbf{R}, t)\Psi(\mathbf{R}, t) \tag{3}$$

where, as throughout this paper, atomic units are used.

$$H = T + V \tag{4}$$

where T is the kinetic energy operator and $V(\mathbf{R})$ is a complex potential energy surface:

$$V(\mathbf{R}) = V_0(\mathbf{R}) + i\frac{\Gamma(\mathbf{R})}{2} \tag{5}$$

$V_0(\mathbf{R})$ is the real part of the resonance energy surface and $\Gamma(\mathbf{R})$ is the autoionization width. The complex energy surface reflects the fact that the resonant state can release the electron as long as the resonance energy lies within the continuum [5].

Our treatment of dissociative recombination and dissociative attachment is a generalization of the time dependent treatment of photodissociation given by Heller [6]. The initial wave packet at $t = 0$ is given by:

$$\Psi(\mathbf{R}, t = 0) = \sqrt{\frac{\Gamma(\mathbf{R})}{2\pi}}\chi_{v_i}(\mathbf{R}) \tag{6}$$

where $\chi_{v_i}(\mathbf{R})$ is the initial vibrational wave function of the molecule. From the time propagation on the dissociative surfaces, we can determine the total capture probability and the resonant DA or DR cross section. The total capture probability is proportional to the overlap integral:

$$S(E) = \int_{-\infty}^{\infty} e^{iEt}\langle\Psi(\mathbf{R}, t = 0)|\Psi(\mathbf{R}, t)\rangle dt \tag{7}$$

where $\Psi(\mathbf{R}, t)$ is the wave packet at time t, determined by direct solution of Eq. 3 for motion on the real part of the resonant potential surface. This is equivalent to the expression for the absorption profile in photodissociation calculations.

The cross section is given by:

$$\sigma(E) = \frac{2\pi^3}{E}|T(E)|^2 \tag{8}$$

where $T(E)$ is given by the projection of the wave packet $\Psi(\mathbf{R})$ for long times when the wave packet has reached the asymptotic region of the potential and the autoionization loss has gone to zero. For a diatomic this is given by:[1]

$$T(E) = \lim_{t\to\infty}\left(-i\int_0^\infty \sqrt{\frac{\mu}{2\pi k}}e^{ikR}\Psi(R)dR\right) \tag{9}$$

where k is the wave number. For a triatomic which has fragmented to an atom and a diatomic it is given by:

$$T_{v,j}(E) = \lim_{t\to\infty}|\int \phi_{v,j}(r_1)\phi_t(r_2)\phi_j(\theta)\Psi(r_1,r_2,\theta,t)dr_1dr_2d\theta|^2 \tag{10}$$

$\phi_{v,j}$ is a vibrational wave function of the diatomic fragment, ϕ_j the rotational wave function, and ϕ_t is a translational wave function describing the relative motion between the atom and the diatomic. The generalization of these equations to polyatomic systems is obvious.

In the case of vibrational excitation, using the time-dependent formalism, for a diatomic system, we can define [11]:

$$T_{v_f v_i}(E) = -i\int_0^\infty e^{iEt}\langle\chi_{v_f}(R)|\sqrt{\frac{\Gamma(R)}{2\pi}}|\Psi_{v_i}(R,t)\rangle dt \tag{11}$$

where $\Psi_{v_i}(R)$ is the time-dependent wave function where the ion was initially in the state v_i and $\chi_{v_f}(R)$ is the wave function for the ion in the state v_f. Note that this state can either be a bound or free vibrational state of the ion.
The cross section is defined as in Eq. (8)

This method, where the T-matrix is accumulated as the Fourier transform of a time-dependent overlap of the wave function with vibrational states is not practical for a polyatomic. The calculation and storage of the vibrational states in multi-dimensions, and the computation time involved in the overlap is prohibitively expensive. Instead one returns to the original time-dependent Schrödinger equation, Eq. 1. However, H now becomes a matrix:

$$H = \begin{pmatrix} T + V_{ion} & V_E(\mathbf{R}) \\ V_E(\mathbf{R}) & T + V_{res} \end{pmatrix} \tag{12}$$

Where T is the kinetic energy operator, V_{ion} is the potential energy surface for the ion, and V_{res} is the potential energy surface for the resonant state. These surfaces are coupled by the electronic coupling, $V_E(\mathbf{R})$. Ψ is now a vector:

---

[1] Note that this definition of the cross section differs by a factor of two from that used previously [7–11]. This is the correct definition.

$$\left(\begin{array}{c} \Psi_{ion} \\ \Psi_{res} \end{array} \right)$$ (13)

where Ψ_{ion} is the time dependent wave function on the ion surface and Ψ_{res} is the time dependent wave function on the resonance surface. Now flux can flow from the resonant state to the ion state. At the end of the calculation, when the wave packet has passed the region where $V_E(\mathbf{R})$ is non-zero, the cross section for vibrational and dissociative excitation can be extracted, using Eq. 10.

Resonance Parameters

The cross sections resulting from the dynamics calculations are only as accurate as the resonance parameters used. We have done a number of calculations on diatomic systems, studying how to determine these parameters accurately. The electronic resonance parameters, $V_0(\mathbf{R})$ and $\Gamma(\mathbf{R})$ are determined by analyzing the S-matrix from state-of-the-art electron-molecule or electron-molecular ion scattering calculations. For single, isolated resonances, the eigenphase sum is fit to a Breit-Wigner form. For more complicated cases, a multichannel generalization can be used [8]. We used the Complex Kohn variational method to calculate these parameters. The Complex Kohn variational method is an algebraic variational technique which has been developed to study both heavy-particle (reactive) collisions [12] and electron scattering problems [13]. We extended it to the scattering of electrons from positively charged molecular ions, and demonstrated the accuracy of the method on a number of systems, including electron-impact excitation [14] and dissociation [15,16]. The complex Kohn method allows us to use the highly accurate configuration interaction wave function that are necessary to study these resonant systems. These calculations extend beyond static-exchange and include the effects of polarization, producing accurate reliable cross sections and resonance parameters. This aspect of our treatment puts us in the unique position of developing an entirely *ab initio* approach to the problem, and removed a level of empiricism that characterizes most other methods.

DISSOCIATION RECOMBINATION OF HeH^+

We will first illustrate the application of this method to the DR of HeH^+. This molecule was predicted to exist in observable abundance in the interstellar media [17]. However, no evidence of HeH^+ has been observe [18], which has been attributed to a high DR rate at low electron energy. Several storage rings have measured the DR rate over a range of energies, confirming a large recombination rate $10^{-8}cm^3/sec$ at 300K [19–21]. These experiments show a monotonic decrease as a function of increasing energy followed by a large peak centered at about 17 eV. It is this resonance peak we will describe.

The ground state of this system is $(1\sigma)^2$ which correlates asymptotically to $He + H^+$. The next excited states are $(1\sigma)(2\sigma)$, either singlet or triplet coupled, which dissociate to $He^+ + H$. There exist a number of resonances, either $(1\sigma2\sigma)^3n\sigma$, $(1\sigma2\sigma)^3n\pi$ converging to the first excited state of the ion, or $(1\sigma2\sigma)^1n\sigma$, $(1\sigma2\sigma)^1n\pi$ converging to the second excited state of the ion. We carried out a series of Complex Kohn variational calculations at fixed nuclear geometries. The resonance parameters were obtained in the most part by Breit-Wigner fits to the eigenphase sum, although for some resonances at small internuclear geometry, analysis of the individual T-matrix elements was required to obtain the partial width into the two open electronically excited ion states. Details can be found in Ref. 8. These resonance parameters were used as input to a wave packet dynamics with the modification that the initial state becomes:

$$\Psi(R, t = 0) = \frac{\mid \gamma(R) \mid}{\sqrt{2\pi}} \chi_{v_i}(R) \tag{14}$$

where $\mid \gamma(R) \mid^2$ is the *partial* resonance width into the initial electronic state and $\chi_{v_i}(R)$ is the initial vibrational wave function. If only one channel is open,

$$\mid \gamma(R) \mid^2 = \Gamma(R) \tag{15}$$

The results of the calculation for DR of $^3HeH^+$ are shown in Figure 1 compared to the results of Sundström *et al* [21]. The lowest six resonance states were included

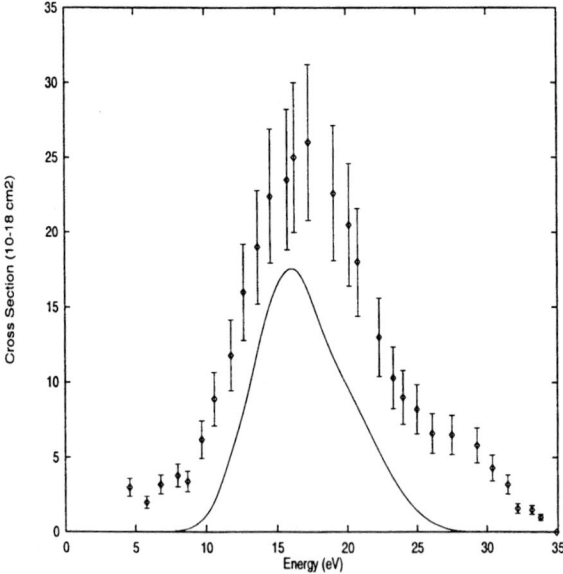

FIGURE 1. Dissociative Recombination of $^3HeH^+$. Solid line theory. Crosses experiment. Experimental results are those of Sundström *et al* [19]. Theoretical results are those of Larson *et al* [9] corrected by a factor of 2.

in the calculation. The calculation is in good agreement with experiment. The high energy shoulder in the experimental cross section, centered at about 27 eV is due to higher resonance states which were not included in the calculation.

DISSOCIATION ATTACHMENT OF ClCN

Another interesting system is ClCN. Dissociative attachment in this system has been studied by Illenberger [22]. This system (and the analogous pseudobihalogens, such as HCN, BrCl...) has the interesting property that since both fragments have positive electron affinities, two fragmentation channels are open. These are:

$$e^- + ClCN \rightarrow Cl^- + CN \tag{16}$$

or

$$e^- + ClCN \rightarrow Cl + CN^- \tag{17}$$

The first channel, producing Cl^-, is believed to proceed through a $ClCN^-$ $^2\Sigma$ resonance state, formed by the addition of an electron to a σ^* antibonding orbital. The second channel, producing CN^-, has been explained as a two-step mechanism. The first step is the transition to the $^2\Pi$ resonance, which is the addition of the electron to a π^* antibonding orbital. However, the dissociation is not believed to

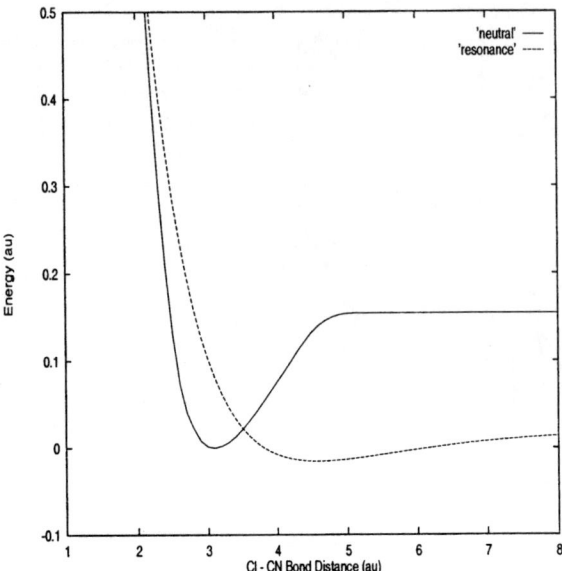

FIGURE 2. Potential Energy Curves for ClCN. The solid curve is the neutral potential energy, the dashed curve shows the dissociative negative ion $^2\Sigma$ resonance state.

be direct, but occurs after energy transfer from the $C - N$ stretching mode into the dissociative $Cl - CN$ mode [22]. If this is true, then at least two degrees of freedom must be include in the calculation, the symmetric and asymmetric stretch modes. The experiment studied the relative cross section as a function of energy from 0 to 15 eV for Cl^- and CN^- production, the kinetic energy of the dissociating fragments, and the competition between the two channels, that is, the branching ratio as a function of temperature.

No theory exists for the DA in this system. We have begun a series of scattering calculation using the Complex Kohn variational method to determine the resonance energy surface and autoionization width as a function of the symmetric and asymmetric stretch, for both resonance symmetries. The ground state of $ClCN$ is:

$$[1\sigma^2 2\sigma^2 3\sigma^2 4\sigma^2 5\sigma^2 1\pi^4]6\sigma^2 7\sigma^2 8\sigma^2 2\pi^4 9\sigma^2 3\pi^4 \qquad (18)$$

The $^2\Sigma$ of $ClCN^-$ is formed by the addition of an electron into a σ^* antibonding orbital. We have carried out some preliminary calculations on this $^2\Sigma$ resonance channel as a function of the $Cl - CN$ and $C - N$ bond distances. The calculations started with a SCF in a 'small' basis consisting of a McLean [23] on the chlorine and triple-zeta plus polarization on the carbon and nitrogen. This was then expanded by the addition of diffuse functions to 104 basis functions. A polarized-SCF calculation polarizing the eight highest occupied orbitals was carried out. Details on the polarized-SCF method can be found in Ref. 23. The lowest seven occupied

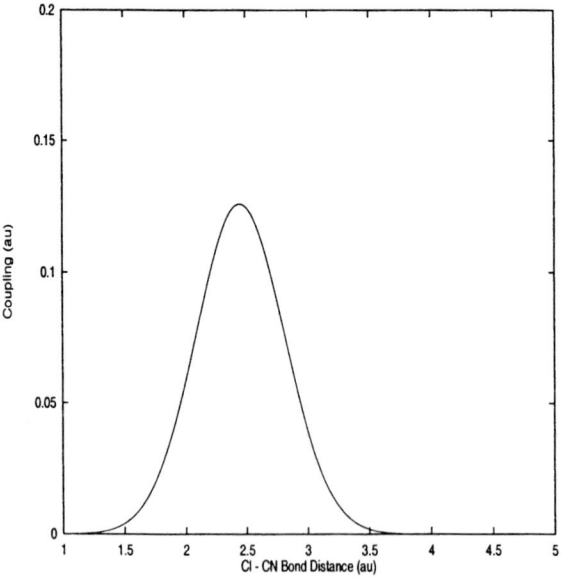

FIGURE 3. Autoionization Width for ClCN $^2\Sigma$ resonance state.

orbitals were frozen, leading to an active space of 22 polarized orbitals and 26 virtual orbitals. The eigenphase sum was fit to a Breit-Wigner form. The results for the resonance energy and autoionization width as a function of the $Cl - CN$ bond distance is shown in figure 2 and 3. The resonance energy as a function of the $C - N$ bond distance was found to be harmonic, with little change in the autoionization width. These calculations confirm the mechanism for DA through this resonance proposed by Illenberger [22]. We plan to continue these calculations.

CONCLUSIONS

There have been dramatic advances in the area of chemical reaction dynamics, allowing the study of three- and even four-atom systems treating all degrees of freedom. There has been a parallel development in the area of electron collisions with molecules and molecular ions, resulting in the ability to calculate accurate cross sections for low-energy collisions in polyatomic systems. Combining this capabilities, we can now begin to treat at a fully *ab initio* level, resonant vibrational excitation, dissociative attachment, and dissociative recombination, in systems where more than one degree of freedom is important. This area has just begun to be explored and holds great promise for the future.

ACKNOWLEDGMENTS

The author would like to thank W. H. Woodin for TEXnical support, Åsa Larson for supplying some of the data shown and K. C. Kulander and T. N. Rescigno for discussions leading to the development of this work. The author acknowledges support from the National Science Foundation, Grant No. PHY-97-22136, and the hospitality of Prof. J. Manz and his group during her sabbatical year at the Freie Universität Berlin, Germany. Part of this work was performed under the auspices of the U.S. Department of Energy at Lawrence Livermore National Laboratory under contract number W-7405-Eng-48.

REFERENCES

1. For a review, see Heller, E. J., *Acc. Chem. Res.*, **74**, 368 (1981).
2. Orel, A. E. and Kulander, K. C., *Chem. Phys. Lett.*, **146**, 428 (1988).
3. Heather, R., Jiang, X.-P., Metiu, H., Bjorken, J. D. and Dunietz, I., *J. Chem. Phys.*, **90**, 2555 (1988).
4. C. Leforestier, C., Bisseling, R. H., Cerjan, C., Feit, M. D., Freisner, R., Goldberg, A., Hammerich, A., Jolicard, G., Karrlein, W., Meyer, H. -D., Lipkin, N., Roncero, O. and Kosloff, R., *J. Comput. Phys.*, **94**, 59 (1991).
5. Herzenberg, A., *J. Phys. B*, **1**, 548 (1968), Birtwistle D. T., and Herzenberg, A., *J. Phys. B*, **4**, 53 (1971), Dube, L., and Herzenberg, A., *Phys. Rev. A*, **11**, 1314 (1975).

6. Heller, E. J., *J. Chem. Phys.*, **68**, 3891 (1978).

7. Orel, A. E. and Kulander, K. C., *Phys. Rev. Letts.*, **71**, 4315 (1993).

8. Orel, A. E., Kulander, K. C. and Rescigno, T. N., *Phys. Rev. Letts.*, **74**, 4807 (1995).

9. Larson, A., and Orel, A. E., *Phys. Rev. A.*, **59**, 3601 (1999).

10. Orel, A. E. and Kulander, K. C., *Phys. Rev. A*, **54**, 4992 (1996).

11. McCurdy, C. W. and Turner, J. L., *J. Chem. Phys.*, **78**, 6773 (1983).

12. Zhang, J. Z. H., and Miller, W. H., *Chem. Phys. Lett.*, **140**, 329 (1987).

13. Rescigno, T. N., McCurdy, C. W., Orel, A. E., and Lengsfield III, B. H., *Computational Methods for Electron-Molecule Collisions*, New York: Plenum, 1995, pp.1-44.

14. Orel, A. E., Rescigno, T. N. and Lengsfield III, B. H. , *Phys. Rev. A*, **42**, 5292 (1990).

15. Orel, A. E., Rescigno, T. N. and Lengsfield III, B. H. *Phys. Rev. A*, **44**, 4328 (1991).

16. Orel, A. E. *Phys. Rev. A*, **46**, 1333 (1992).

17. Roberge, W., and Dalgarno, A., *Astrophys. J.*, **255**, 489 (1982).

18. Moorhead, J. M., Lowe, R. P., Maillard, J. P., Whelau, W. H., and Bernath, P. F., *Astrophys. J.*, **326**, 899 (1988).

19. Sundström, G., Datz, S., Mowat, J. R., Mannervik, S., Broström, L., Carlsson, M., Danared H., and Larsson, M., *Phys.Rev. A*, **50**, R2806 (1994).

20. Mowat, J. R., Danared, H., Sundström, G., Carlsson, M., Andersson, L. H., Vejby-Christensen, L., Ugglas, M. af,and Larsson, M., *Phys. Rev. Lett.*, **74**, 50 (1995).

21. Strömholm, C., Semaniak, J., Rosén, S., Danared, H., Datz, S., van der Zande, W., and Larsson, M., *Phys. Rev. A*, **50**, 3086 (1996).

22. Brüning, F., Hahndorf, I., Stamatovic, A., and Illenberger, E., *J. Phys. Chem.*, **100**, 19740 (1996).

23. McLean, A. D. and Chandler, G. S., *J. Chem. Phys.*, **72**, 5639 (1980).

24. Lengsfield III, B. H., Rescigno, T. N., and McCurdy, C. W., *Phys. Rev. A*, **44**, 4296 (1991).

Measurements of Electron-Impact-Dissociation Cross Section for Neutral Products

Hideo Sugai and Hirotaka Toyoda

Department of Electrical Engineering, Nagoya University
Furo-cho, Chikusa-ku, Nagoya 464-8603, Japan

Abstract. Neutral radicals play key roles in plasma CVD and etching, and a comprehensive data base is needed on electron-impact-dissociation cross section for neutral radical production. This paper reviews the direct measurement of the neutral dissociation cross section accomplished by using appearance mass spectrometry in a dual-electron-beam system. The collision energy dependence of neutral dissociation cross sections is investigated for typical molecules used for plasma processing; CH_4, CF_4, CHF_3, C_4F_8, SF_6, SiF_4 and so on.

INTRODUCTION

Recently, much attention has been given to studies on cross sections for electron impact dissociation of molecule into ions and neutral radicals, from plasma processing point of view. The ions and radicals produced in the plasma are transported to a substrate and chemical reactions on the surface give rise to deposition and etching of thin films. Many kinds of feed gases have been used in actual materials processing in industries. In order to model and control the plasma processing, there has been a great need for data base of the cross section set for each gas.

In a plasma, feed gas (XY) is decomposed by electrons by the following reactions:

$$XY + e \rightarrow X + Y + e \qquad (\varepsilon > \varepsilon_n) \qquad (1)$$

$$XY + e \rightarrow X + Y^+ + 2e \qquad (\varepsilon > \varepsilon_{di}) \qquad (2)$$

where X and Y denote neutral (molecular or atomic) radicals and Y^+ does an ionic radical. The reaction (1) is "pure" neutral dissociation and the reaction (2) is dissociative ionization which produces neutral X as well. For simplicity, let the energy dependence of the collision cross sections $\sigma_n(\varepsilon)$ and $\sigma_{di}(\varepsilon)$ be a linear function of slope α_n and α_{di} above the threshold of ε_n and ε_{di} for the reaction (1) and (2), respectively. Then, if electron energy distribution function is given by a Maxwell distribution of temperature T_e, the reaction rate constant $k_n(T_e)$ of the radical X production by the reaction (1) is expressed as

$$k_n(T_e) = \alpha_n(1 + \frac{2T_e}{\varepsilon_n})\exp(-\frac{\varepsilon_n}{T_e}) \qquad (3)$$

CP500, *The Physics of Electronic and Atomic Collisions*, edited by Y. Itikawa, et al.
© 2000 American Institute of Physics 1-56396-777-4/00/$17.00

Similar expression is obtained for the reaction (2). As the rate constant is mainly determined by the exponential factor, the key parameter is the ratio of the threshold energy to electron temperature. In conventional processing plasmas, ε_n and ε_{di} are about 10 - 20 eV, while T_e is 1-3 eV. Thus, the important range of electron impact energy for the cross section measurement is from the threshold to 40 eV in view of plasma processing.

Earlier electron impact studies have given the extensive data of ionization cross sections, so that most of the data are available for parent molecule ionization, except for the radical ionization such as

$$X + e \rightarrow X^+ + 2e \qquad (\varepsilon > \varepsilon_{ri}) \qquad (4)$$

In plasma processing, however, neutral radicals rather than ions play major roles in most cases since the neutral radical flux to the substrate is usually much higher than the ion flux. In present day, it is still almost impossible to theoretically find the cross section for neutral dissociation. Experimentally, however, little is known about dissociation into neutral radicals since detection of neutral radicals is extremely difficult.

Recently, we have developed a highly sensitive technique for radical detection, i.e., *appearance (potential) mass spectrometry* [1], and succeeded in measuring cross sections for neutral radical yield in electron-impact dissociation of the following species of molecules used in CVD and etching:

methane; CH_4 [2, 3], carbontetrafluoride; CF_4 [4, 8],
tetrafluorosilane; SiF_4 [5], trifluoromethane; CHF_3 [6],
sulfur hexafluoride; SF_6 [7], octafluorocyclobutane; c-C_4F_8 [9],
1,2,2-tetrafluoroethyl-trifluoromethyl ether; C_3HF_7O [10].

Here the radical species X produced by the two reactions (1) and (2) is directly measured by appearance mass technique. Thus, for $\varepsilon_n < \varepsilon < \varepsilon_{di}$, the measured cross section solely corresponds to $\sigma_n(\varepsilon)$, that is, "pure" neutral dissociation. The cross section for $\varepsilon_{di} < \varepsilon$ is a sum of σ_n and σ_{di}. In this paper, we review the experimental method for measuring the cross section for neutral radical yield, some examples of the obtained data, and finally the challenges in future.

Appearance Mass Method for Cross Section Measurement

Cross sections for electron-impact neutral production are measured in a *dual-electron-beam device* combined with a quadrupole mass spectrometer (QMS), as schematically shown in Fig. 1 [1, 2]. Let us take an example of electron collision with CHF_3 molecule which is widely used in plasma etching of silicon dioxide in semiconductor industry. The first electron beam (5-300 eV, 0.1-60 μA) injected through a 3-mm-diam orifice along magnetic field of ~0.06 T dissociates CHF_3 into CF_3 for example, in a dissociation cell at a CHF_3 pressure of 10^{-3} - 10^{-1} Pa. The resultant CF_3 radicals effuse through a 4-mm-diam orifice into an ionization cell at a pressure of 10^{-5}-10^{-4} Pa where the second electron beam ionizes CF_3 radicals, producing CF_3^+ ions and these ions are detected by the QMS.

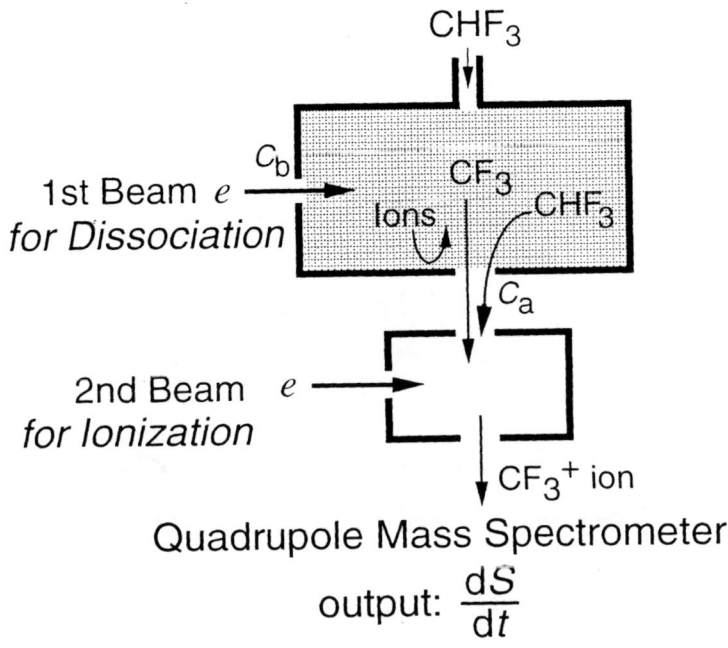

CHF$_3$

C_b

1st Beam e
for Dissociation

Ions

CF$_3$

CHF$_3$

C_a

2nd Beam e
for Ionization

CF$_3^+$ ion

Quadrupole Mass Spectrometer

output: $\dfrac{dS}{dt}$

FIGURE 1. Schematic diagram of experimental setup.

In order to measure the cross section for CF$_3$ production from CHF$_3$ in the dissociation cell, one should selectively detect the CF$_3^+$ ions produced from CF$_3$ radicals effusing from the dissociation cell. To do this, firstly the ionization cell and the QMS are biased positively ($\sim$5 eV) with respect to the dissociation cell to prevent the CF$_3^+$ ions produced in the dissociation cell from effusing into the ionization cell. Secondly, the *appearance mass spectrometry* [1] is used in the ionization cell to discriminate CF$_3$ from CHF$_3$. The appearance mass technique is also called a threshold ionization technique which is based on about 5 eV difference in ionization threshold (appearance potential of ion) for the parent molecule dissociation expressed by the reaction (2) and the radical ionization expressed by the reaction (4): (ε_{di} - ε_{ri}) $\sim$ the binding energy between X and Y, typically 5 eV. For example, Fig. 2 shows the threshold energy of ε_{di} =15.2 eV for the dissociative ionization of CHF$_3$ into CF$_3^+$, and ε_{ri} =10.6 eV for the direct ionization of CF$_3$ into CF$_3^+$, where the absolute value of electron energy is calibrated by measuring the known ionization threshold of argon. Thus, only CF$_3$ radical is detected when the second beam energy is set at the value between ε_{ri} and ε_{di}, say 12 eV in the present case.

In the actual measurement, we subtract the QMS output signal with the first beam turned off (S_{OFF}) from the output signal with the beam turned on (S_{ON}). Furthermore, we use a pulse counting technique in the QMS to improve signal-to-noise ratio as the radical signal is very weak, especially near the threshold energy. In addition, the first electron beam is pulsed at a repetition frequency of 1/60 Hz with square-wave-form voltage (duration 30 s), and the signal difference ($S = S_{ON} - S_{OFF}$) is integrated in time [2] whose growth rate in time, dS/dt, is proportional to the collision cross section.

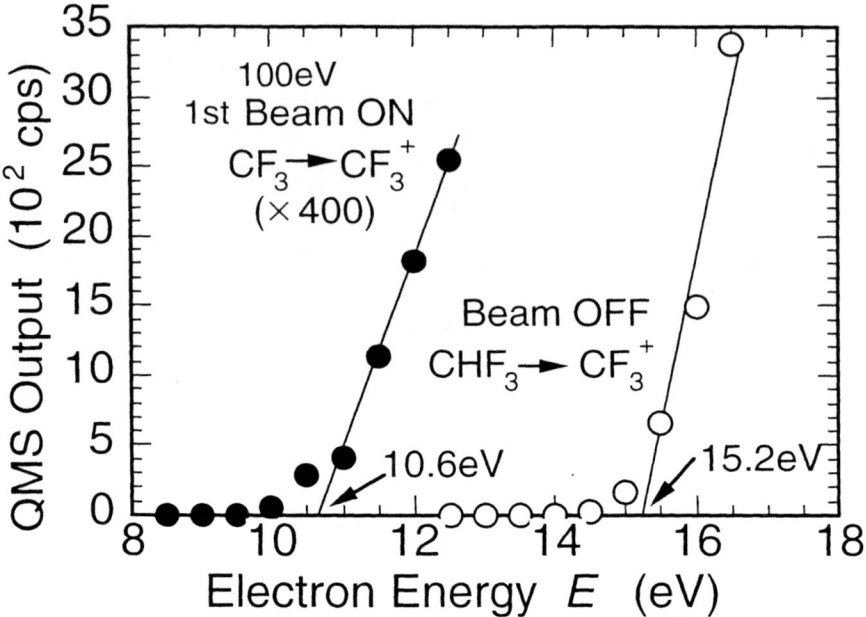

FIGURE 2. Quadrupole mass spectrometer output for CF_3^+ as a function of the second electron beam energy E with the first electron beam at 100 eV turned on (filled circles) and off (open circles).

The threshold energy ε_{di} for dissociative ionization can be easily measured as shown above, however the threshold ε_{ri} for ionization from neutral radical (CF_3, CF_2, CHF_2, etc) are usually unknown. For limited species, one may find the experimental values in literatures, and also calculate them, combining a few related reactions if the bond dissociation energies are known [4,5,6]. When setting the energy of the second beam to detect radicals, special care should be paid to eliminate contributions from other radicals produced in the same conditions. For instance, CF^+ ion may be produced not only from CF but from CF_2 and CF_3 as well, so that the second beam energy should be set at the value very close to the ionization potential of CF radical.

If neutral radical could be detected in a way described above, the QMS output will be proportional to the radical production rate in the dissociation cell, which is proportional to the cross section for electron impact dissociation to the detected radical. Before starting the cross section measurement, one should confirm a linearity of the QMS signal against both the first beam current and the feed gas pressure. The linearity implies that the secondary reaction and multiple scattering processes are negligible, thus supporting the observation of single collision event. Figure 3 shows examples of such linearities confirmed in the case of CHF_3 dissociation into CF radical: the upper figure shows the CF signal vs CHF_3 pressure and the lower figure shows the CF signal vs electron current.

As descibed by Eq.(3), the radical production rate in a plasma strongly depends on the threshold energy for electron-impact dissociation. This threshold is determined by "pure" neutral dissociation of Eq. (1) which has the lower threshold than the dissociative ionization shown by Eq. (2). The value of the threshold energy for neutral dissociation

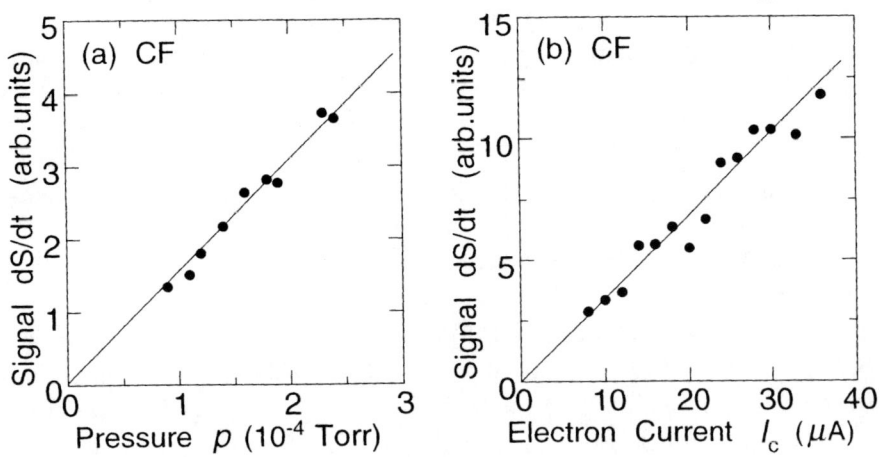

FIGURE 3. (a) CF signal vs pressure at I_c=20 μA; (b) CF signal vs electron current at p=1.8x10^{-2} Pa; electron impact energy = 80 eV.

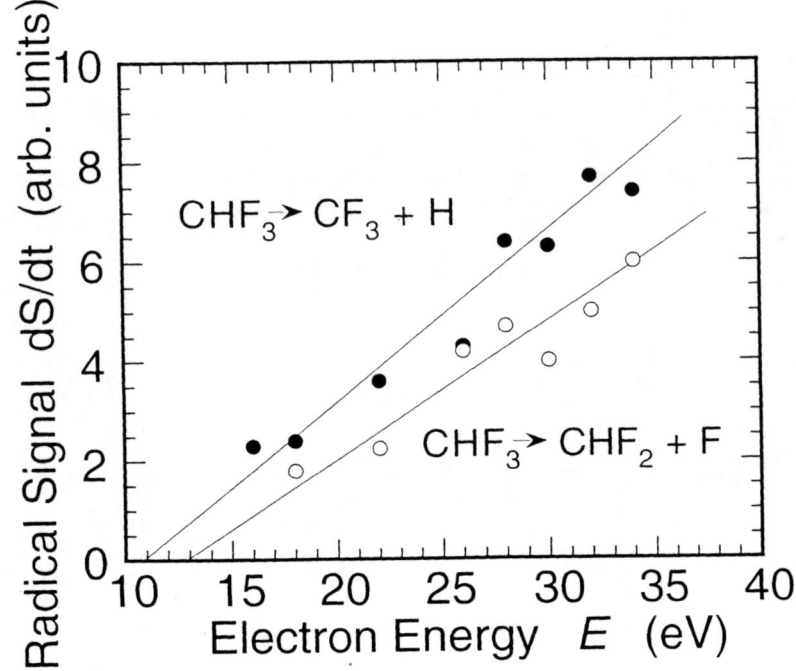

FIGURE 4. Threshold energy for neutral dissociation of CHF$_3$ into CF$_3$ (filled circles) and CHF$_2$ (open circles).

can be found, within the error of 1 eV, by fitting a straight line on the data in a range of low energies in a limit of the radical signal detection. In case of CHF_3, for example, Fig. 4 shows the electron impact energy dependence of CHF_2 and CF_3 radical signals near the dissociation thresholds. According to these data, the threshold energies are determined to be 11.0 eV for the neutral dissociation of $CHF_3 \rightarrow CF_3 + H$, and 13.0 eV for the neutral dissociation of $CHF_3 \rightarrow CHF_2 + F$, respectively. These thresholds agree with the values expected from bond dissociation energies [6].

As mentioned above, the QMS output dS/dt, is proportional to the cross section $\sigma(E)$ at the electron impact energy E. The energy dependence of the cross section can be obtained in a straightforward way by plotting the QMS output as a function of the first electron beam energy E. However, one should carefully calibrate such *relative* cross section, to obtain the *absolute* cross section. Referring to Fig. 1, let us consider the calibration procedure. In the dissociation cell of volume V, the dissociation of molecule XY by the first electron beam gives rise to the radical X at the rate

$$G = [XY] (l_1 I_1/e)\, \sigma_{XY \rightarrow X}$$

where [XY] represents the parent molecule density, I_1 and l_1 are the current and the path length of the first electron beam, $\sigma_{XY \rightarrow X}$ is the cross section for dissociation from XY to X, respectively. In general, a time variation of the radical density [X] can be written as

$$V\frac{d[X]}{dt} = G - L - k[X] - (C_a + C_b)[X] \tag{5}$$

where the second term L on the right hand side (RHS) represents the gas-phase-loss caused by such secondary reactions as radical-radical reactions, and this term is negligible in the present low-pressure experiment. The third term corresponds to the surface loss expressed by a rate constant $k=svA/4$, where s is the sticking coefficient (probability)

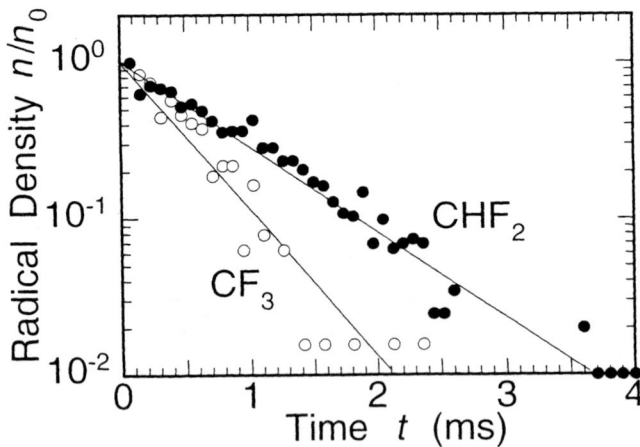

FIGURE 5. Density decay after turning off the first electron beam for CHF_2 (filled circles) and CF_3 (open circles).

and v is the radical thermal speed. The last term stands for the loss of radicals evacuated through two orifices (see Fig. 1) with the vacuum conductance C_a and C_b. When the first electron beam is turned off ($G=0$ at $t=0$), the radical density exponentially decreases in time due to the wall loss and the pumping loss: the density changes as

$$[X] = [X]_0 \exp(-t/\tau) \tag{6}$$

where $\tau = V/(k + C_a + C_b)$. Fig. 5 shows examples of the time decay of CHF_2 and CF_3 densities after the electron beam turned off where $\tau = 0.80$ ms for CHF_2 and $\tau = 0.48$ ms for CF_3.

In steady state ($d[X]/dt \to 0$), Eq. (5) gives the radical density in the dissociation cell as $[X]_0 = G/(k + C_a + C_b) = G\tau/V$. Here the radical flux effusing into the ionization cell is given by $C_a[X]_0$, and the radical is ionized by the second electron beam of the current I_2 and the path length l_2, giving the ion density as $[X^+] \propto C_a[X]_0 (l_2 I_2/e) \sigma_{X \to X+}$ where $\sigma_{X \to X+}$ represents the cross section for radical ionization. The ions are transferred to the QMS to select the X^+ species, and a secondary electron multiplier gives a current pulse per each detected ion. Since the radical signal dS/dt measured by the QMS is proportional to $[X^+]$, one can express the unknown cross section $\sigma_{XY \to X}$, using the known or measurable parameters, as

$$\sigma_{XY \to X} = \frac{\alpha \beta V}{\mu C_a \tau [XY] \sigma_{X \to X+}} \frac{dS}{dt} \tag{7}$$

where α is the constant independent of radical species, $\beta = e^2/l_1 l_2 I_1 I_2$, μ represents the QMS sensitivity depending on the ion mass.

According to Eq. (7), we can determine the absolute cross section as follows. The parent gas pressure was measured by a B-A gauge and a spinning rotor gauge which enable us to determine $[XY]$ and the vacuum conductance C_a. In case of radicals presented here, we used the radical ionization cross section $\sigma_{X \to X+}$ reported by Freund and co-workers [11, 12] and Becker and co-workers[13, 14]. The mass discrimination effect expressed by μ was investigated in a mass range from m/e=28 to 117 by measuring the ion signal and comparing the known ionization cross sections of rare gases, CF_4 [15] and SF_6 [16].

In order to eliminate α, β and V in Eq. (7), we measured the relative cross section for electron impact dissociation of N_2 molecule into N atom in the same condition of β and V [2]. Here Eq. (7) stands also for the nitrogen dissociation cross section $\sigma_{N2 \to N}$, and hence we obtain the following ratio

$$\frac{\sigma_{XY \to X}}{\sigma_{N2 \to N}} = \frac{[N_2]}{[XY]} \frac{\mu}{\mu_N} \frac{C_{aN}}{C_a} \frac{\tau}{\tau_N} \frac{\sigma_{N \to N+}}{\sigma_{X \to X+}} \frac{(\frac{dS}{dt})}{(\frac{dS}{dt})_N} \tag{8}$$

where the suffix "N" indicates the nitrogen dissociation and the cross sections $\sigma_{N2 \to N}$

can be derived using the value of $\sigma_{N \to N^+}$ reported by Cosby [17] and the total dissociation cross section of N_2 reported by Winters [18]. In this way, the absolute cross section has been determined for various molecules used in plasma processing [2-10].

Examples of Cross Sections for Neutral Products

Seven species of molecules conventionally used in materials processing by discharge plasma have been measured in the appearance mass method. For example, tetrafluorosilane (SiF_4) is used as a feed gas in silicon thin film deposition while SiF_4 molecules are desorbed as etching product from the silicon and silicon dioxide surface, and they are re-ionized in F-atom containing etching plasma. Fig. 6 shows the absolute values of partial cross sections for electron impact dissociation of SiF_4 into SiF_3, SiF_2, SiF, and Si radicals. The relative uncertainty of the cross section is estimated to be about ±20%, while the absolute cross sections derived here will be accurate to ±100% in the worst estimation as discussed previously [5].

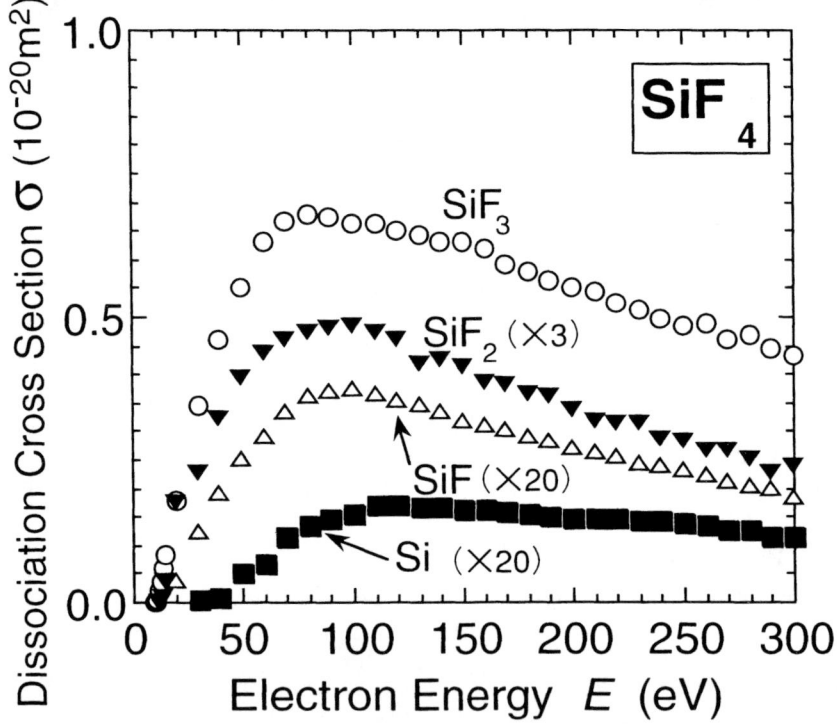

FIGURE 6. Absolute cross sections for dissociation of SiF_4 into SiF_x radicals (x=0-3).

Most of molecules we measured showed the energy dependence of the cross section similar to Fig. 6. Namely, the cross section increases from the threshold energy around 15 eV and reaches the peak value around 100 eV, above which it gently decreases with

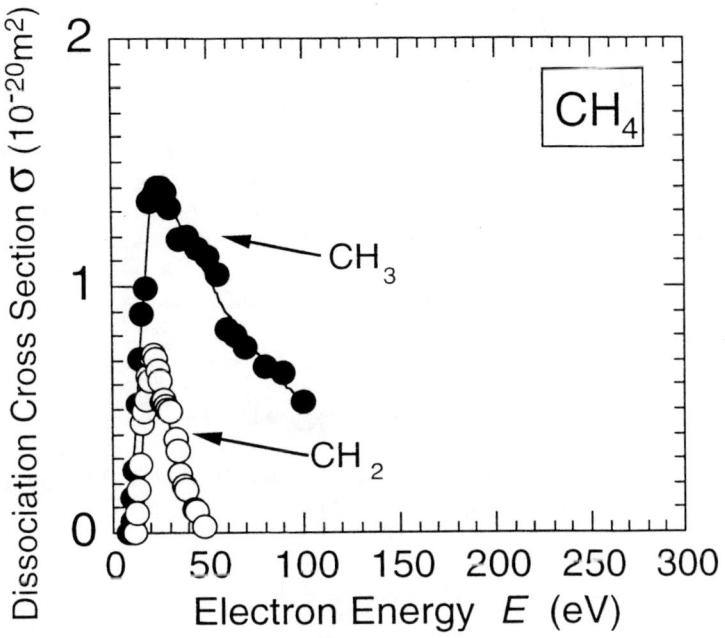

FIGURE 7. Absolute cross sections for dissociation of CH_4 into CH_3 and CH_2 radicals.

impact energy. However, methane is exceptional as shown in Fig. 7. The cross section quickly rises up from the threshold, shows the peak around 25 eV, and then sharply drops, especially in case of dissociation into CH_2 radical. Such resonance-like behavior is also observed for the dissociation from H_2 into H [19] and explained by spin exchange collisions from singlet to triplet.

Challenges in Future

To date, information on production and loss processes of neutral radicals in plasma is very poor, which hinders development of plasma processing for the next generation electronic devices. The primary reaction in plasma is the electron-impact dissociation of feed gas into neutral radicals, so that the basic study of the dissociation cross section is crucial to understand and control the reactive plasma. We have developed *appearance mass method* for measuring the cross section for electron-impact dissociation into neutral radical. This method enables one to disclose many unknown cross sections of parent molecules used in plasma processing.

Although the relative cross sections are readily obtained using this method, determination of the absolute cross sections necessitates the ionization cross sections of radicals. At present, such data are available only for several restricted species of radicals [11-14]. It is notable here that Kim and co-workers recently developed a new analytical model to obtain ionization cross sections [20] which is applicable to radical ionization as well.

Accuracy of the cross sections obtained by the appearance mass method should be cross-checked by another method. Recently, Motlagh and Moore investigated electron-impact dissociation of methane and a variety of fluoroalkanes, based on the surface reaction of the radicals at a tellurium mirror. In this approach, some discrepancy was found between the neutral dissociation cross sections obtained by the two different methods. Thus, both the appearance mass and the surfave reaction method should be carefully evaluated in reliability and be successively improved. The appearance mass method allows direct detection of radicals and has wider applications than the surface reaction method. Anyhow, development of innovative techniques is desired for precise cross-section measurement since there still remain many unknown cross sections of feed gases important in plasma processing.

ACKNOWLEDGMENTS

This work was supported by a grant-in-Aid for Scientific Research from the Ministry of Education, Science, Sports and Culture in Japan.

REFERENCES

1. Sugai, H. and Toyoda, H., *J. Vac. Sci. Technol.* A**10**, 1193- 1200 (1992).
2. Nakano, T., Toyoda, H. and Sugai, H. *Jpn. J. Appl. Phys.* **30**, 2908-2911 (1991).
3. Nakano, T., Toyoda, H. and Sugai, H. *Jpn. J. Appl. Phys.* **30**, 2912-2915 (1991).
4. Nakano, T. and Sugai, H., *Jpn. J. Appl. Phys.* **31**, 2919-2924 (1992). Subsequently, corrected by Reference 8.
5. Nakano, T. and Sugai, H., *J. Phys.* D**26**, 1909-1915 (1993).
6. Goto, M., Nakamura, K., Toyoda, H., Sugai, H., *Jpn. J. Appl. Phys.* **33**, 3602-3607 (1994). Subsequently, corrected by Reference 8.
7. Iio, M., Goto, M., Toyoda, H. and Sugai, H., *Contrib. Plasma Phys.* **35**, 405-413 (1995) .
8. Sugai, H., Toyoda, H., Nakano, T. and Goto, M., *Contrib. Plasma Phys.* **35**, 415-420 (1995).
9. Toyoda, H., Iio, M. and Sugai, H., *Jpn. J. Appl. Phys.* **36**, 3730-3735 (1997).
10. Tanaka, H., Toyoda, H. and Sugai, H., *Jpn. J. Appl. Phys.* **37**, 5053-5059 (1998).
11. Baiocchi, F.A., Wetzel, R.C. and Freund, R.S., *Phys. Rev. Lett.* **53**, 771 (1984).
12. Hayes, T.R., Shule, R.J., Baiocchi, F.A., Wetzel, R.C. and Freund, R.S., *J. Chem. Phys.* **88**, 823-829 (1987); **89**, 4035-4041 (1988); **89**, 4042-4047 (1988).
13. Tarnovsky, V., Becker, K., *J. Chem Phys.* **98**, 7968 (1993).
14. Tarnovsky, V., Kurunczi, P., Rogozhnikov, D., Becker, K., *Int. J. Mas. Spectrom. Ion Processes.* **128**, 181 (1993).
15. Ma, Ce., Bruce, M.R., Bonham, R.A., *Phys. Rev,* A**44**, 2921 (1991).
16. Margreiter, D., Walder, G., Deutsch, H., Poll, H.U., Winkler, C., Stephan, K., Maerk, T.D., *Int. J. Mas. Spectrom. Ion Processes.* **100**, 143 (1990).
17. Cosby, P. C., in *Abstracts of Contributed Papers of 16th Int. Conf. Phys. of Electronic and Atomic Collisions,* 1989, p.348.
18. Winters, H.F., *J. Chem. Phys.* **44**, 1472 (1966).
19. Corrigan, S.J.B., *J. Chem Phys.* **43**, 4381 (1965).
20. Kim, Y.-K., Rudd, M.E., *Phys. Rev.,* A**50**, 3954 (1996); *J. Chem Phys.* **104**, 2956 (1996); *J. Chem Phys.* **106**, 1026 (1997); *J. Chem Phys.* **106**, 9602 (1997).
21. Motlagh, S., Moore, J.H.., *J. Chem Phys.* **109**, 432 (1998).

Dissociative Attachment to Excited Molecules

S. V. K. Kumar[1], E. Krishnakumar, S. A. Rangwala and
V. S. Ashoka

Tata Institute of Fundamental Research, Homi Bhabha Road, Colaba, Mumbai 400 005, India.

Abstract. Cross-sections for dissociative attachment of electrons to electronically excited molecules produced by optical pumping has been studied. Negative ion resonance states could be identified based on initial state selection of the neutral. Couplings between various states of the neutral molecule could be unraveled which has been difficult using high resolution optical spectroscopy.

INTRODUCTION

Dissociative attachment (DA) is one of the important processes in the 0 to 15eV energy region of electron-molecule interactions and has been studied for several decades. The first step of the DA process is electron attachment to the neutral molecule forming a Negative Ion Resonance State (NIRS). This NIRS can decay by either ejecting the additional electron to go back to the neutral state or by fragmenting to form a negative ion and neutral fragment/s - the DA channel. Therefore, the DA channel has been used to study NIRS. The formation of NIRS is characterized by selection rules which are based on the conservation of the initial symmetry of the electron plus neutral molecule system and the symmetry of the NIRS formed after the electron attachment. Thus for a given orientation of a molecule with respect to the incident electron momentum vector, only specific negative ion resonances can be formed depending on the electronic wave function of the initial neutral state [1]. Therefore, by changing the initial symmetry of the neutral molecule through an electronic excitation, new channels may open up for the negative ion resonances or close the existing channels, which is depicted schematically in Figure 1. The manifestation of this phenomenon was observed in DA experiments carried out on the metastable molecule, $O_2(\tilde{a}\ ^1\Delta_g)$ produced in a microwave discharge. Two new resonances were observed from the metastable molecule as compared to that seen from the ground $\tilde{X}\ ^3\Sigma_g^-$ state in these experiments [2].

---

[1] svkk@tifrc4.tifr.res.in

CP500, *The Physics of Electronic and Atomic Collisions*, edited by Y. Itikawa, et al.
© 2000 American Institute of Physics 1-56396-777-4/00/$17.00

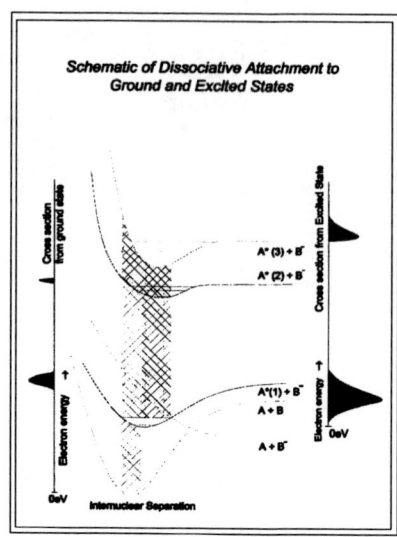

FIGURE 1. Schematic representation of Dissociative Attachment to ground and excited states of a molecule. The appropriate Franck-Condon overlaps of the two neutral states with the negative ion states are shown as hatched regions. The cross-section functions for attachment to the ground state are shown on the left side and for the electronically excited state on the right side.

It is well known that a large number of molecules are in various excited states in many kinds of plasmas, ionosphere, etc. and very little data is available on electron collisions with excited molecules and particularly on DA. From a practical point of view to model any plasma where negative ions are present the cross-sections for the formation of negative ions by electron attachment is required not only from the ground state but also from excited states. In addition, DA cross-section data for excited molecules is needed to understand, model and exploit various kinds of plasmas and discharges, upper atmosphere, etc. This is particularly the case for those molecules which are used in the dry etch process, used as electrical insulants and atmospheric pollutants. The study of DA from excited molecules can also provide information on the NIRS as well as dynamics of the DA process (in some cases). Though, the need for these numbers and other information was realized a long time ago, the formidable difficulty of creating excited molecules in sufficient number densities has resulted in very few experimental measurements.

Measurements on vibrationally excited SF_6 [3], HCl and HF produced by laser photodissociation [4] were reported way back in the eighties. Optical pumping techniques have been used to carry out dissociative attachment experiments on vibrationally excited Li_2 [5]. The most sophisticated experiments so far have been on vibrationally and rotationally state selected Na_2 from Bergmann's group [6,7].

They have used the STIRAP population transfer method to produce Na_2 in a specific vibrational and rotational state and carried out DA measurements over the vibrational ladder . The measurements reported so far on the dissociative attachment from electronically excited states have been on the long lived excited state of O_2 ($^1\Delta_g$) produced by microwave discharge [2] and by using laser pumping to NO [8,10], C_6H_5SH [9], H_2 [10,12], D_2 [10], SO_2 [11] and Triethylamine [13]. Most of these measurements in which lasers have been used to excite the molecules, the term 'Laser Enhanced Dissociative Attachment' has been used and these have focused on the qualitative measurements in the DA cross-sections.

The results on DA presented here is part of our efforts to study interaction of electrons, ions and photons with excited molecules. Molecules are excited to a specific electronic state using laser photons and DA to excited molecules are studied. The measured cross-sections for DA to excited SO_2 and CS_2, identification of NIRS and resolving the couplings of the neutral electronic states using the DA data from ground and excited states are presented here.

EXPERIMENTAL

The experimental arrangement is shown in Figure 2. The measurements were carried out in a triple crossed beams geometry in which a pulsed beam of photons of appropriate wavelength from either an excimer laser or an excimer pumped dye laser (LPX240i - LPD3002, Lambda Physik) excited the molecules in an effusive beam. The laser pulse interacted with the molecular beam, exciting a fraction of the molecules to a specific excited state. Immediately after the laser pulse (a few tens of nanoseconds) a pulsed and magnetically collimated electron beam of 300 ns duration intersected this effusive beam. Following this, the ions produced were extracted by the application of a 200 V/cm electric field pulse of 1 μsec duration, to the pusher plate of the ion extraction assembly of a Time-of-Flight Mass Spectrometer (TOFMS). The ions entering the TOF tube assembly were focused at the exit of the assembly and detected by the channel electron multiplier mounted off-axis and operated in the pulse counting mode. The time jitter and drift in the laser pulsing was minimized by using a synchronizing unit. It may be noted that the ion extraction pulse comes into effect only after the electron pulse has left the interaction region. In this way, a high ion extraction field could be employed for extracting all the ions irrespective of their initial kinetic energies and angular distributions without affecting the electron beam. The flight tube consists of segments which are appropriately biased to act as a lens to focus all the ions entering it into the cone of the channel electron multiplier. This arrangement makes sure that all the ions formed in the interaction region reaches the detector, thus making accurate measurements of cross sections possible. However, there is a trade off in the mass resolution of the TOFMS. This trade off in the mass resolution does not affect the results reported here as the resolution is sufficient to resolve the mass peaks of interest. The TOF spectra were recorded using a time-to-

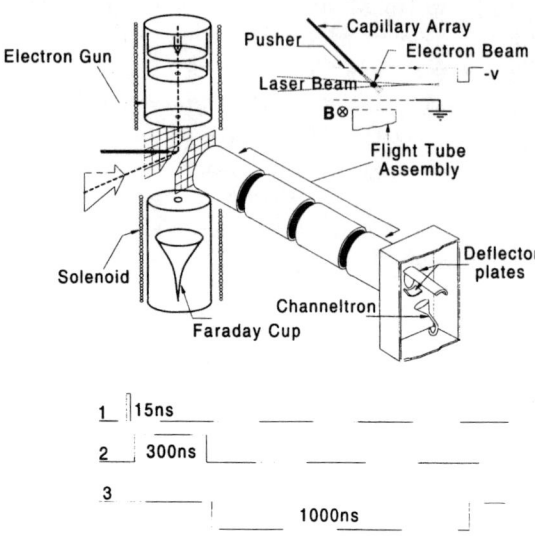

FIGURE 2. Schematic of the Experimental Setup.

amplitude converter (TAC) and a pulse height analyzer (PHA). The count rate of a particular ion versus the incident electron energy (excitation function) was obtained by selecting the appropriate time window in the TAC. The whole experiment is controlled by a PC and a GPIB based data acquisition and control system [14]. The electron energy resolution was about 0.5 eV in the present measurements. The performance of this apparatus was evaluated as discussed in our recent work on excited SO_2 molecules [15].

This data acquisition system ensured simultaneous monitoring of the electron beam current and the intensities of all the ionic species of interest. The above arrangement provides accurate compensation of the excitation functions for the variation of electron current with energy, reducing the data acquisition time considerably and also improving the reliability.

In order to obtain the cross sections for the excited state molecules, it is necessary to obtain their number density relative to that of the ground state molecules. This was achieved by taking the mass spectra at electron energies corresponding to the peaks in the dissociative attachment spectrum from the ground state with and without the laser beam, keeping all other parameters like the gas pressure and the electron beam current constant. It is noticed that the the TOF spectra taken in the presence of the laser has a lower count rate as compared to that taken without the laser. This could only be due to a reduction in the number of molecules in their ground state due to the laser excitation of a fraction of these molecules to the upper state, since at the electron energy used, attachment occurs only to the ground state molecules. Using the difference in the intensities for the two cases the percentage depletion of the target molecules due to laser excitation was calculated. Checks were made for signatures of multi-photon excitation and dissociation processes by varying the laser intensity and they did not show significant effects to indicate presence multiphoton processes. Thus, for all practical purposes, the experiments were done in single photon absorption regime. The lifetimes of the excited state involved are much larger than the time duration of the laser pulse and the subsequent electron pulse. Since multiphoton processes are found to be negligible, the observed depletion corresponds to the fraction of molecules in the excited state. Using this fraction, the cross sections for the excited states were determined. The excited state fraction determination is discussed in greater detail in our recent work on SO_2 molecules [15].

RESULTS AND DISCUSSION

Results of SO_2

SO_2 is a bent molecule with C_{2v} symmetry. The products of DA to SO_2 in its ground state are O^-, S^- and SO^-. Of these O^- and SO^- channels have relatively large cross sections as compared to the S^- channel [19]. DA from the ground state of SO_2 results in two peaks at 4.6 eV and 7.3 eV respectively in the O^- cross sections and at 4.8 eV and 7.2 eV respectively in the SO^- cross sections. The electron energy scale was calibrated against the O^- peak from CO at 10.6 eV by flowing SO_2 and CO simultaneously. It is seen that both in the case of O^- and SO^-, a new peak appears at very low energies when laser excitation is present. In the case of S^- the overall count rate was very low and there was no discernible new peak. Spurious effects due to photoelectrons from surfaces produced by scattered light were investigated by making the measurements without the electron beam and was found to be absent and these new peaks were solely due to dissociative attachment from excited molecules. Several checks were carried out for possible contributions from multiphoton processes, effect of laser intensity, gas pressure, time delay between laser and electron pulses, etc. which has been described in detail in earlier publications [16,15]. We find that the O^- and S^- intensities due to the excited state peak at about 0.4 and 0.6 eV respectively. The position of these peaks are separated from the lower energy peaks due to the ground state by 4.2 eV which is close to the photon energy used in producing the excited state. Thus based on energetics, we may conclude that the NIRS accessed from the excited state is the same as that which is giving rise to the peak at 4.6 eV from the ground state. The measured cross-sections for the formation of O^- by DA to ground and excited states of SO_2 is shown in Figure 3.

The most noteworthy part of the results is the absence of a second peak due to DA from the excited state. One would expect to see additional peaks at about 3 eV in both the O^- and SO^- data due to DA to the excited molecules corresponding to the

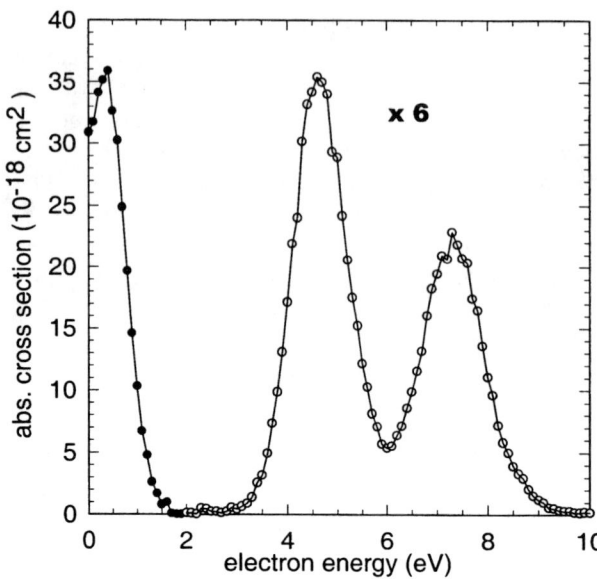

FIGURE 3. Cross-sections for the formation of O^- by Dissociative Attachment to SO_2 ground state and excited state created by 308nm pumping. Filled circles are cross-sections from the excited state and open circles are cross-sections from the ground state.

peaks seen at about 7 eV from the ground state. It may be argued that the absence of the above peak from the excited state is due to small 'survival probability' of the NIRS. However, as the excited neutral states are likely to be in higher vibrational levels, the NIRS formed from them should have larger survival probability, leading to larger dissociative attachment cross sections. Thus the absence of the peak has to be attributed to selection rules governing the electron capture process. An analysis of this based on the selection rules helps not only to identify the NIRS but also to characterize the excited neutral state itself, which has not been possible by high resolution optical spectroscopy.

Using the general rule laid out by Dunn [1], the selection rules have been worked out for DA to molecules with C_{2v} symmetry, which can be applied to SO_2, taking into account all possible orientations of the molecule with respect to the momentum vector of the incident electron as discussed below. In the laboratory frame, the molecules are randomly oriented and the incident electron momentum vector is fixed. In order to get the selection rules we consider three mutually perpendicular orientations of the molecule with respect to the electron momentum vector **k**. These are shown in Figure 4. In order to have a nonzero transition probability to the negative ion state, the initial molecular state plus electron and final NIRS should have the same symmetry. The list of allowed and not allowed transitions for each orientation is given in Table 1.

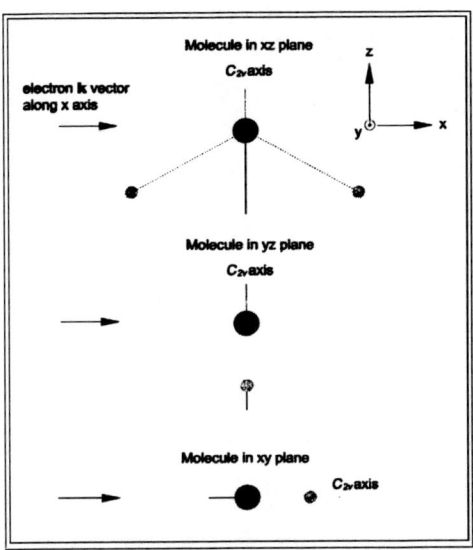

FIGURE 4. Selection rules for Negative Ion Resonances of molecules with C_{2v} symmetry.

TABLE 1. Selection rules for resonant attachment of electrons for the three mutually perpendicular orientations of the molecules of C_{2v} symmetry

Neutral state	NIRS			
	A_1	A_2	B_1	B_2
A_1	a—a—a	n—n—n	n—n—a	n—a—n
A_2	n—n—n	a—a—a	n—a—n	n—n—a
B_1	n—n—a	n—a—n	a—a—a	n—n—n
B_2	n—a—n	n—n—a	n—n—n	a—a—a

a : Allowed, n : Not allowed. Case 1, case 2 and case 3 correspond to the top, middle and bottom shown if Figure 4.

In the case where molecules are randomly oriented and measurements made without any angular discrimination as in our case, the selection rules will be the union of all the three cases and can be summarized as $A_1 \leftrightarrow A_2$ and $B_1 \leftrightarrow B_2$ **not allowed** and rest are allowed.

From the absorption spectrum of SO_2 shown in Figure 5, also known as Clements bands [18], the broad underlying continuum is ascribed to the allowed $\tilde{B} \, ^1B_1$ state and the finger like structure to the $\tilde{A} \, ^1A_2$ state. At the photon wavelength of 308nm which is used to pump SO_2, the absorption is to the allowed $\tilde{B} \, ^1B_1$ state. Selection rules does not allow access to the B_2 negative ion state from B_1. As one does not observe a DA peak from the excited state corresponding to the 7.3eV peak, the

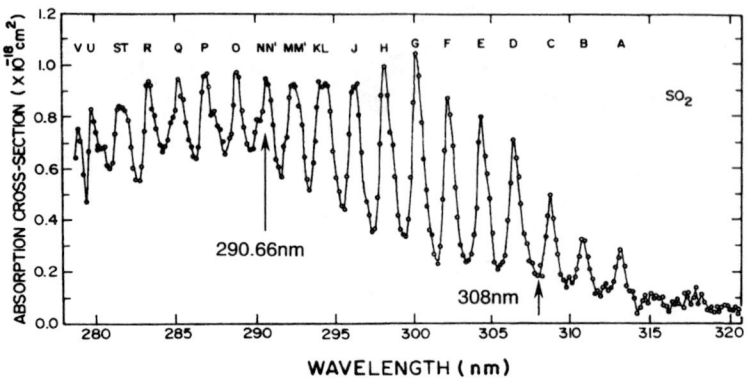

FIGURE 5. Absorption spectrum of SO_2. Adapted from [17].

negative ion state corresponding to the 7.3eV peak can be identified to be a $^2B_2$ state.

Ab initio molecular orbital calculations have been carried out for ground and first three excited states of SO_2 and SO_2^-. This was done using the Gaussian 94 package [21] at fourth order Møller-Plesset correlation correction (MP4) level with aug-cc-pVDZ basis set. The ground state Molecular Orbital (MO) configuration of SO_2 is ... $(7a_1)^2$, $(1a_2)^2$, $(4b_2)^2$, $(8a_1)^2$ giving an overall $\tilde{X}\ ^1A_1$. Adding an electron to the next unoccupied orbital $3b_1$ gives rise to $SO_2^-(\tilde{X}\ ^2B_1)$. Promotion of $8a_1$, $4b_2$ and $1a_2$ electron to $3b_1$ orbital of neutral SO_2 results in the first three excited states of $\tilde{B}\ ^1B_1$, $\tilde{A}\ ^1A_2$ and $\tilde{C}\ ^1B_2$ respectively. Calculations show that adding the extra electron to the $3b_1$ orbital of these excited molecules can produce $^2A_1$, $^2B_2$ and $^2A_2$ NIRS in increasing order of energy from the ground state of SO_2^- for the Franck - Condon overlap region of SO_2 neutral ground state. Based on this, the first peak at 4.6 eV seen in the dissociative attachment from the ground state may be due to a $^2A_1$ state and the second one at 7.3 eV due to a $^2B_2$ state. Our experimental identification of the 7.3 eV peak as $^2B_2$ is in agreement with these calculations.

In addition to identifying the NIRS, the present results have important implications in characterizing the excited neutral state itself. The Clements' bands [18] has been characterized by some very unusual properties. Dipole selection rules allow excitation only to the $\tilde{B}\ ^1B_1$ state. However, no signature of any vibrational bands of $\tilde{B}\ ^1B_1$ has been observed yet. What is observed are the finger like bands with anomalous intensity pattern (Clement's bands). This band structure has been explained as due to a $\tilde{A}\ ^1A_2$ state which vibronically couples strongly with $\tilde{B}\ ^1B_1$ [22,23]. The only signature of the $\tilde{B}\ ^1B_1$ state is the quasi continuum like structure underlying the Clements' bands. Also fluorescence measurements from this state have shown anomalously large lifetime [20]. This and the absence of any clear vibrational bands due to $\tilde{B}\ ^1B_1$ has been explained as due to strong mixing with closely spaced high lying states of the ground state $\tilde{X}\ ^1A_1$ through vibronic

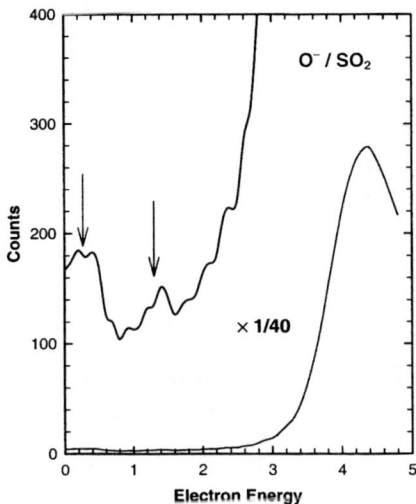

FIGURE 6. Dissociative attachment excitation function for SO_2 from ground and excited state.

interaction and with $\tilde{a}\ ^3B_1$ state through spin orbit interaction [22,24]. Thus based on the existing spectroscopic data, excitation by 308 nm photons leads to a state which is a mixture of $\tilde{B}\ ^1B_1$, $\tilde{A}\ ^1A_2$, $\tilde{X}\ ^1A_1$ and possibly $\tilde{a}\ ^3B_1$ states. However, if $\tilde{A}\ ^1A_2$ and $\tilde{X}\ ^1A_1$ were present in the admixture, we would expect to see a second peak in the resonance attachment from the excited molecule as the selection rules do not overrule the attachment process from these states to form a $^2B_2$ NIRS. Since we do not see such a process, we have to conclude that the state excited by the 308 nm radiation does not have any $^1A_2$ or $^1A_1$ characteristics but is only B_1 in nature. The absence of vibronic mixing between the $^1A_2$ and the $^1B_1$ states at 308 nm excitation may be due to the fact that the rovibrational levels of the $^1B_1$ state is far removed from the rovibrational levels of the $^1A_2$ state. This is supported by the fact that the absorption at 308 nm lies right in the valley between the Clements' C and D bands [18,22]. The observation, that the excited state does not have any $^1A_1$ characteristic leads us to the conclusion that the quasi continuum nature seen for the $\tilde{B}\ ^1B_1$ is largely due to mixing with the $\tilde{a}\ ^3B_1$ state.

The next obvious attempt would be to prepare the excited state corresponding to the peak of one of the $\tilde{A}\ ^1A_2$ band and check for the validity of the arguments discussed above. SO_2 was pumped at the NN' peak using 290.66nm photons generated by pumping Rhodamine 6G dye and doubling using a KDP crystal. The excitation function for DA to excited SO_2 pumped at 290.66nm is shown in Figure 6. Two DA peaks from the excited state are observed at 0.3eV and 1.4eV. The peak at 0.3eV corresponds to a shift towards lower energy corresponding to the photon energy of 4.265eV of the first DA peak from the ground state, which is in accordance with the previously observed results for 308nm pumping. However, the 1.4eV peak does not correspond to a shift of the 7.3eV peak by the photon energy. It could be due to the excited neutral accessing a different part of the potential energy surface of the negative ion state compared to the ground neutral state. Such differences are observed in the case of CS_2 and is discussed in the next section on CS_2.

The DA processes in SO_2 can be better understood when the potential energy surfaces becomes available and the final states of the dissociating products can be identified.

CS_2 results

CS_2 is a linear molecule which, belongs to the $D_{\infty h}$ point group, and has a $\tilde{X}\,^1\Sigma_g^+$ symmetry in its ground state. The products of DA to the ground state are S^-, CS^-, S_2^-, and relatively small quantity of C^-. The S^- cross sections exhibit resonant peaks at 3.6 eV and 6.2 eV, where as CS^- and S_2^- are found to have a major resonant peak at 6.2 eV. The excited states of CS_2 are bent [25]. When CS_2 is pumped by 308 nm (4.03 eV) photons to the $^1B_2$ state, a new peak was observed in each of the S^- and CS^- channels. These are shown in Figure 7. The peak for S^- is centered at 0.5 eV and and that for CS^- is centered at 0.7 eV. Both peaks have finite cross section even at 0 eV. The CS^- peak seem to be broader than the S^- peak. The ground state counts in the present measurement were normalized with the values of Krishnakumar and Nagesha [26]. The excited state cross sections were quantified using the percentage depletion of the ground state ions.

If the precursor NIRS accessed from the ground state and the excited state is the same, one would expect a preferentially large cross section for the formation of S_2^- from the $^1B_2$ excited state which has a bent geometry. We do not observe the formation of S_2^- from the excited state. Two plausible explanations can be put forward to interpret this observation: (1) the selection rules for electron capture may imply that the NIRS, accessed from the ground state and the laser excited state may be completely different and (2) the electron capture may be occurring to the same molecular negative ion state, from both the ground $^1\Sigma_g^+$ state as well as from the laser excited $^1B_2$ state, but access very different parts of the potential energy surface due to differing Franck-Condon overlaps, as the ground state has a linear geometry and the excited state has a bent geometry. Both these possibilites will reflect in the cross-section of DA peaks from the excited state being observed at energies not translated downward by the photon excitation energy.

Additional arguments that can be put forth to support the first possibility by invoking the selection rules for the formation of NIRS [1]. From the ground neutral state $^1\Sigma_g^+$ of CS_2 NIRS with symmetries $^2\Sigma_g^+$, $^2\Sigma_u^+$, $^2\Pi_u$, $^2\Delta_g$ etc can be accessed. The formation of S_2^- is signature of bending of the molecular negative ion. These linear symmetries map to $^2\Sigma_g^+ \rightarrow {}^2A_1$; $^2\Sigma_u^+ \rightarrow {}^2B_2$; $^2\Pi_u \rightarrow {}^2A_1$, $^2B_1$ and $^1\Delta_g \rightarrow$

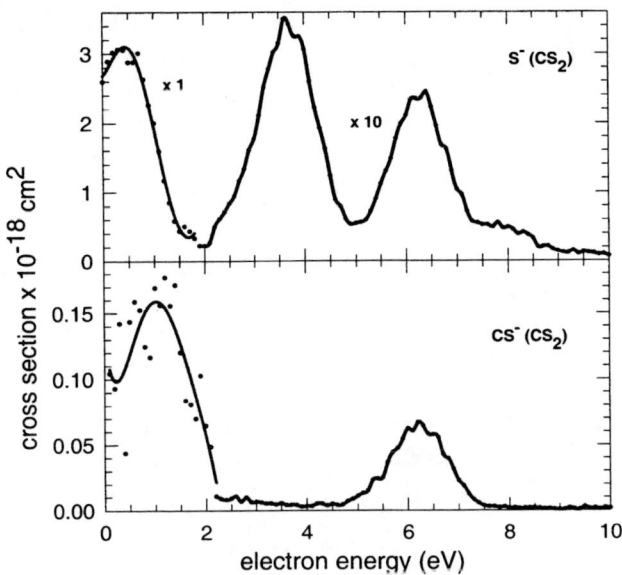

FIGURE 7. Cross-sections for Dissociative Attachment to CS_2 ground and excited states (pumped by 308nm photons). Gray colored curves represent the cross-sections from the excited state.

$^2A_1$, $^2B_1$ symmetries in the bent geometry (C_{2v} point group). The symmetry of the laser excited state by absorption of a single 308nm photon is $^1B_2$ in its equilibrium configuration which correlates to the $^2\Delta_u$ state via a Renner-Teller mechanism [27]. The absence of the S_2^- negative ion from the laser excited state and the fact that $B_2 \leftrightarrow B_1$ is forbidden by selection rules [16] for electron capture to C_{2v} molecules implies, that the symmetry of the NIRS that results in the formation S_2^- ion fragment from the ground state of the neutral, should correlate to either $^2\Pi_u$ or $^2\Delta_g$ states. So far, there exists no experimental or theoretical data to identify the symmetry of the negative ion state seen at 6.2 eV, from the ground electronic state of CS_2. The above analysis, of the excited state measurement, indicates that this symmetry could well be B_1. Further, given that the excited state has B_2 equilibrium symmetry, the NIRS seen in the S^- and CS^- channels can have A_1, A_2 and B_2 symmetries consistent with the selection rules. A detailed discussion will be presented elsewhere [28].

It must however be kept in mind that the excited $^1B_2$ neutral state is highly perturbed by the rotational and vibrational manifolds of other electronic states at the same energy [29], which are dipole forbidden from the ground electronic state [27]. In addition, some evidence suggests a significant perturbation of the $^1B_2$ neutral state by close lying triplet states by a spin orbit interaction [27]. The above selection-rule identification of the ground NIRS does not consider the perturbations mentioned above, and so appropriate caution must be exercised in taking the above identification at face value. Theoretical calculations on the potential energy surfaces of negative ion states and electron scattering calculations are needed to unambiguously identify the NIRS and the dissociation pathways.

CONCLUSIONS

Dissociative attachment from ground and electronically excited states have been studied. Using the selectivity of the DA process, the negative ion resonances have been characterized utilizing the DA data from neutral and excited states. Cross-sections for DA from excited state prepared by laser pumping have been measured in favorable cases. Dissociative attachment experiments from excited states have been shown to be a new tool in providing information on the excited neutrals that could not be obtained using high resolution spectroscopy. For a better understanding of the DA process, potential energy surfaces of negative ions are needed so that the dynamics of dissociation can be studied. Electron scattering calculations from ground and excited states would help in identifying the NIRS and correlating with the correct DA channels.

ACKNOWLEDGMENTS

Thanks are due to S. T. Tare and Y. V. Upalekar for the timely and highly useful technical support. It is a pleasure to thank T. S. Ananthakrishnan, B. A. R. C., Mumbai for the development of the software for Data Acquisition and Control of the experiment.

REFERENCES

1. G. H. Dunn, Phys. Rev. Lett. **8**, 64 (1962).
2. D. S. Belić and R. I. Hall, J. Phys. B, **14**, 365 (1981).
3. C. L. Chen and P. J. Chantry, J. Chem. Phys. **71**, 3897 (1979).
4. M. J. Rossi, H. Helm, and D. C. Lorents, Appl. Phys. Lett. **47**, 576 (1985).
5. M. W. McGeoch and R. E. Schlier, Phys. Rev. A, **33**, 1708 (1986).
6. M. Külz, M. Keil, A. Kortyna, B. Schellhaß, J. Hauck, K. Bergmann, W. Meyer and D. Weyh, Phys. Rev. A, **53**, 3324 (1996).
7. M. Keil, C. Gebauer-Rochholz, O. Kaufmann, W. Meyer and K. Bergmann, in Book of Abstracts, XXI International Conference on the Physics of Electronic and Atomic Collisions, Sendai, Japan, p.325 (1999).
8. C. T. Kuo, Y. Ono, J. L. Hardwick, and J. T. Moseley, J. Phys. Chem. **92**, 5072 (1988)
9. L. G. Christophorou, S. R. Hunter, L. A. Pinnaduwage, J. G. Carter, A. A. Christodulides, and S. M. Spyrou, Phys. Rev. Lett. **58**, 1316 (1987).

10. L. A. Pinnaduwage and L. G. Christophorou, Chem. Phys. Lett., **186**, 4 (1991); L. A. Pinnaduwage, and L. G. Christophorou, Phys. Rev. Lett., **70**, 754 (1993).

11. T. Jaffke, R. Hashemi, L. G. Christophorou, E. Illenberger, and H. Baumgartel, Chem. Phys. Lett. **203**, 21 (1993).

12. P. G. Datskos, L. A. Pinnaduwage, and J. F. Kielkopf, Phys. Rev. A, **55**, 4131 (1997).

13. L. A. Pinnaduwage and D. L. McCorkle, Chem. Phys. Lett. **255**, 410 (1996).

14. S. V. K. Kumar, T. S. Ananthakrishnan, E. Krishnakumar, S. A. Rangwala, in Book of Abstracts, XX International Conference on the Physics of Electronic and Atomic Collisions, Vienna, Austria, FR201 (1997).

15. E. Krishnakumar, S. V. K. Kumar, S. A. Rangwala and S. K. Mitra, Phys. Rev. A, **56**, 1945 (1997).

16. E. Krishnakumar, S. V. K. Kumar, S. Rangwala, and S. K. Mitra, J. Phys. B, **29** L657 (1996).

17. S. M. Ahmed and Vijay Kumar, J. Quant. Spectrosc. Radiat. Transfer, **47**, 359 (1992).

18. J. H. Clements, Phys. Rev. **47**, 224 (1935).

19. H. -X. Wan, J. H. Moore, J. K. Olthoff and R. J. Van Brunt, Plasma Chem. Plasma Process. **13**, 1 (1993).

20. L. E. Brus and J. R. McDonald, Chem. Phys. Lett. **21**, 283 (1973).

21. Gaussian 94, Revision C.2, M. J. Frisch, G. W. Trucks, H. B. Schlegel, P. M. W. Gill, B. G. Johnson, M. A. Robb, J. R. Cheeseman, T. Keith, G. A. Petersson, J. A. Montgomery, K. Raghavachari, M. A. Al-Laham, V. G. Zakrzewski, J. V. Ortiz, J. B. Foresman, J. Cioslowski, B. B. Stefanov, A. Nanayakkara, M. Challacombe, C. Y. Peng, P. Y. Ayala, W. Chen, M. W. Wong, J. L. Andres, E. S. Replogle, R. Gomperts, R. L. Martin, D. J. Fox, J. S. Binkley, D. J. Defrees, J. Baker, J. P. Stewart, M. Head-Gordon, C. Gonzalez, and J. A. Pople, Gaussian, Inc., Pittsburgh PA, 1995.

22. Y. Hamada and A. J. Merer, Can. J. Phys. **53**, 2555 (1975); J. C. D. Brand, J. L. Hardwick, D. R. Humphrey, Y. Hamada and A. J. Merer, Can. J. Phys. **54**, 186 (1976).

23. R. J. Shaw, J. E. Kent, M. F. O'Dwyer, J. Mol. Spectrosc. **82**, 1 (1980).

24. R. Kullmer and W. Demtröder, J. Chem. Phys, **83**, 2712 (1985); J. S. Baskin, F. Al-Adel and A. Hamdan, Chem. Phys., **200**, 181 (1995).

25. Q. Zhang and P. H. Vaccaro, J. Phys. Chem. **99**, 1799 (1995); G. Brasen, M. Leidecker, and W. Demtröder, T. Shimàmoto and H. Kâto, J. Chem. Phys., **109**, 2779 (1998).

26. E. Krishnakumar and K. Nagesha, J. Phys. B, **25**, 1645 (1993).

27. Ch. Jungen., D. N. Malm and A. J. Merer, Can. J. Phys., **51**, 1471 (1973).

28. S. A. Rangwala, S. V. K. Kumar and E. Krishnakumar *(to be submitted)*.

29. A. E. Douglas, J. Chem. Phys., **45**, 1007 (1966).

Low-energy electron and positron scattering from C_{60}: A Progress report on calculations

F. A. Gianturco[*], Robert R. Lucchese[†], and N. Sanna[‡]

[*] Department of Chemistry, The University of Rome,
Città Universitaria, 00185 Rome, Italy
[†] Department of Chemistry, Texas A&M University,
College Station, Texas, 77843-3255, USA
[‡] Center for Supercomputing Applications to University and Research,
P.le A. Moro, 00185 Rome, Italy

Abstract. The scattering of low-energy electrons and positrons from gaseous C_{60} molecules provides much interesting information on the behavior of this molecule and on the relationship between its structural features and its response to slow electrons and positrons. Our model for these processes has no adjustable parameters and includes both interaction potentials to describe polarization effects and a non-local exchange potential in the case of electron scattering. These theoretical investigations have allowed us to give a microscopic interpretation of observed experimental behavior and has allowed us to propose additional experiments which should exhibit features in both electron and positron low-energy collisions which are unique to the C_{60} molecule.

I INTRODUCTION

The discovery of efficient methods which allow the production of carbon molecular aggregates containing 60 and 70 (and higher) atoms in useful quantities [1] has enabled, in the last ten years, a great number of experimental and theoretical studies on the electronic properties of these molecules [2]. From the point of view of its possible behavior with respect to electron and positron scattering in the low-energy regime, the electron has been considered in some experimental studies, as we shall further report here, while positron collisions have not yet been studied experimentally although below we will make an attempt at predicting features of its cross sections from a computational model of positron scattering.

It is now fairly well known that the C_{60} molecule readily attaches electrons in solution and in the solid phase [3], where the intercalation of C_{60} with alkali metals can result in the formation of materials that range in electronic properties from superconductors to insulators depending on the occupation of the electronic levels

CP500, *The Physics of Electronic and Atomic Collisions*, edited by Y. Itikawa, et al.
© 2000 American Institute of Physics 1-56396-777-4/00/$17.00

of the carbon molecular cage. It has also been shown that C_{60}^- is stable in the gas phase [4] and that the fullerenes possess relatively large electron affinities: for C_{60} the value is 2.67 eV [5]. Several experiments have shown that there is a large cross section for electron attachment from just above threshold up to $\sim$10 eV with some resonant structure apparent on top of a broad background [6,7]. The structure in the cross section has been attributed to the excitation of a variety of excited states of the negative ion [6]. Further measurements which used high-n Rydberg atoms for attaching low-energy electrons [8,9] reached the conclusion that an attachment peak is present at nearly zero energy and that it is dominated by s-wave scattering in contrast to the dominance of p-wave scattering at low energy inferred from the electron scattering experiments. The threshold behavior of the scattered electrons from the highly symmetrical C_{60} molecule thus constitutes an intriguing problem, both experimentally and theoretically, which has spurred several attempts at both measuring such data and at interpreting the data as realistically as possible by using theoretical models of various levels of sophistication. The large size of this carbon cluster, the complexity of its nuclear motion, whether or not the latter is strongly coupled to the motion of the scattered electron, and the difficulty of describing the multicenter many-body nature of its interaction with the impinging electrons has, however, made very difficult the task of describing the forces and the dynamics of the collision from first principles. The present progress report gives a brief outline of what has been done in our laboratory on this specific problem.

II   THE COMPUTATIONAL MODEL

A   The Single-Center Expansion Approach

Resonant and non-resonant low-energy scattering of electrons from polyatomic targets can be studied theoretically (and computationally) at various levels of sophistication [10]. Within an *ab initio* approach one could start with the target nuclei being kept fixed at their equilibrium geometry. This simplifying scheme goes under the familiar name of the fixed nuclei (FN) approximation [11] and greatly reduces the dimensionality of the coupled equations for the dynamics. Furthermore, the target N-electrons bound in a specific molecular electronic state (which, for the present purpose, is taken as unchanged during the scattering) can be described within the self-consistent-field (SCF) approximation by using the single-determinant description of the $N/2$ doubly occupied molecular orbitals (MOs). In our implementation of the scattering equations, the occupied MOs of the target are expanded on a set of symmetry-adapted angular functions with their corresponding radial functions represented on a numerical grid [12]. In this approach, any arbitrary three-dimensional function describing a given electron, either one of the N bound electrons or the scattering electron, is expanded using a single-center expansion (SCE) with the origin of the SCE located at the center of the molecule under study

$$F^{p\mu}(r, \theta, \phi | \mathbf{R}) = \sum_{lh} r^{-1} f_{lh}^{p\mu}(r|\mathbf{R}) X_{lh}^{p\mu}(\theta, \phi). \tag{1}$$

In Eq. (1) the $p\mu$ refers to the μth element of the pth irreducible representation (IR) of the point group of the molecule at the nuclear geometry $\mathbf{R}$. The angular functions $X_{lh}^{p\mu}(\theta, \phi)$ are generalized harmonics given by proper combinations of spherical harmonics $Y_{lm}(\theta, \phi)$

$$X_{lh}^{p\mu}(\theta, \phi) = \sum_{m} b_{lhm}^{p\mu} Y_{lm}(\theta, \phi), \tag{2}$$

where the details about the computations of the $b_{lhm}^{p\mu}$ have been given by us before and will not be repeated here [12].

The quantum scattering equation for the radial part of the $(N+1)$th continuum electron is

$$\left\{ \frac{1}{2} \frac{d^2}{dr^2} - \frac{l(l+1)}{2r^2} + E \right\} f_{lh}^{p\mu}(r|\mathbf{R}) = \sum_{l',h'} \int_0^\infty V_{lh,l'h'}^{p\mu}(r, r'|\mathbf{R}) f_{l'h'}^{p\mu}(r'|\mathbf{R}) dr', \tag{3}$$

where E is the collision energy. The coupled partial integro-differential equations given in Eq. (3) contain the kernel of the integral operator V which is a sum of diagonal and nondiagonal terms that in principle can fully describe the electron-molecule interactions during the collision process. When the target is represented by a single determinant SCF wave function the potential V is the static-exchange (SE) representation of the electron-molecule interaction for the chosen electronic target state (usually the ground state) at the nuclear geometry $\mathbf{R}$. For a target which has a closed-shell electronic structure, as in the case for C_{60}, with n_{occ} doubly occupied orbitals ϕ_i, the potential can be written as

$$V_{SE}(\mathbf{r}) = \sum_{\alpha=1}^{M} Z_\alpha (\mathbf{r} - \mathbf{R}_\alpha)^{-1} + \sum_{i=1}^{n_{occ}} (2\hat{J}_i - \hat{K}_i), \tag{4}$$

where $\hat{J}_i$ and $\hat{K}_i$ are the usual local static potential and the non-local exchange potential operators, respectively. The index α labels one of the M nuclei located at the coordinate $\mathbf{R}_\alpha$ in the center-of-mass molecular frame of reference. A major shortcoming of such a model for studying electron-attachment processes is the lack of target response, i.e. the effect of static and dynamic electron correlation in the scattering process. At higher collision energies this is reflected in the fact that no electronically inelastic processes can be treated at the SE level of interaction. At the lower energy, of more direct interest in the present study, the lack of inclusion of target response leads to the neglect of important polarization effects which can significantly alter the energy location and the width of those resonances which occur below 10 eV. This difficulty could be avoided by the inclusion of additional target states in the expansion of the wave function although such an approach can significantly increase the computational complexity of solving the scattering equations [10].

In the present model, we have included the effects of dynamical correlation and polarization through the addition of a local, energy-independent model potential, $V_{CP}(\mathbf{r})$ already discussed before in some detail [13,14]. Briefly, the V_{CP} model potential contains a short-range correlation contribution, V_{corr}, which is smoothly connected to a long-range dipole polarization contribution, V_{pol}. The short-range term is obtained from the functional derivative of a Kohn and Sham correlation energy with respect to the electron density of the molecular target. The long-range part of V_{CP} is obtained by constructing a model polarization potential, V_{pol}, which asymptotically agrees with the potential obtained from the static polarizability of the target in its ground electronic state. One can construct such a model either by choosing a single "polarization" center or by partitioning the total static polarizability between different centers: in the present case we have divided the polarizability equally between the 60 carbon atoms. In the general case the long-range contribution, V_{pol}, does not exactly match the short-range correlation, V_{corr}, at any given value of r. We have therefore developed a procedure to connect smoothly V_{corr} and V_{pol} [13,14]. Using the sum of V_{SE} and V_{CP} then corresponds to carrying out the scattering calculations at the static-exchange-correlation-polarization (SECP) level of treatment for such a system.

B   The Model Adiabatic Potentials

In order to study in some detail the resonant features which may lead to low-energy electron attachment to the C_{60} molecule, even the above *ab initio* procedure, albeit somewhat simplified, still requires a substantial computational effort in order to compute the cross section behavior on a fine energy grid and to extract the resonant features at least for single-particle resonant attachment processes. Hence, in order to carry out such a detailed study, we solve Eq. (3) using the V_{SECP} potential, and then solve the scattering equations using a simpler, local approximation for the exchange part of the interaction, whereby the nonlocal exchange is replaced by a local model exchange (ME) potential denoted by V_{ME}. For this purpose we have used the Hara free-electron-gas-exchange (HFEGE) potential that has been discussed and employed many times before [13]. We found that the energy dependence of the HFEGE potential was fairly weak so that scattering results over a given range of energies (where their mean energy was used in the HFEGE potential) were very similar to the results where the actual scattering energy was used to compute the HFEGE potential.

The standard, symmetry-adapted angular momentum eigenstates, $X_{lh}^{p\mu}$, do not form the most compact angular basis set for the electron-molecule scattering problem. An alternative expansion basis set is the angular eigenfunctions obtained from diagonalizing the angular Hamiltonian at each radius r [13]. The angular functions obtained in this fashion are referred to as the adiabatic angular functions, $Z_k^{p\mu}$, which are linear combinations of the symmetry adapted harmonics defined in Eq. (2), *i. e.*

$$Z_k^{p\mu}(\theta, \phi, r) = \sum_{lh} X_{lh}^{p\mu}(\theta, \phi)C_{lh,k}(r), \tag{5}$$

where the expansion coefficients are solutions to the matrix eigenvalue equation

$$\sum_{lh} \left(V_{l'h',lh}^{p\mu}(r) + \delta_{l,l'}\delta_{h,h'}\frac{l(l+1)}{2r^2} \right) C_{lh,k}(r) = W_k^{p\mu}(r)C_{l'h',k}(r). \tag{6}$$

The eigenvalues $W_k^{p\mu}(r)$ then form an adiabatic radial potential for each index value k for scattering radial functions of $(p\mu)$ symmetry, which we will refer collectively to as the adiabatic static model exchange correlation polarization (ASMECP) potential V_{ASMECP}.

III   THE COMPUTATIONAL RESULTS

There are several approximations which have been introduced in developing the model outlined above. The calculations did not include the dynamical couplings between the electron and the target nuclei and the wave function employed includes only one target state and a rather limited basis set expansion for the bound molecular orbitals. Furthermore, the single-center expansion of the interaction with the impinging projectile was truncated at $l_{\mathrm{max}} = 40$, hence the SECP interaction included up to $l = 2l_{\mathrm{max}}$, and the correlation-polarization potential was also modeled through a local density functional scheme. The coupled equations which we solved for the scattered electrons in each IR of the icosahedral, I_h, symmetry point group varied from 9 in the a_u IR up to 77 for the h_g IR. The extent to which the above approximations may affect the results that we have obtained for this system are somewhat difficult to estimate since the C_{60} molecule, with 360 bound electrons, is so much larger than any of the molecular targets which we have studied previously with the present methods.

However, a possible qualitative gauge of the utility of the methods used can be provided by comparisons with existing experimental data. For example, our calculation can describe all the anion states which can be formed and characterized by an extra electron occupying any number of one-particle orbitals outside the closed-shell of the remaining 360 electron. When we computed the bound and resonant states of the ASMECP potential that included the radial couplings between the adiabatic channels, we found that the ground state of C_{60}^- was a state of ^2T_u symmetry with an energy of -2.12 eV. There were also excited states of $^2T_{1g}$ symmetry at -1.28 eV and of ^2H_g symmetry at -0.043 eV. At the same time we found a resonant scattering state of $^2T_{2u}$ symmetry located at 0.145 eV with a width of 8.28×10^{-6}eV [15]. The experimental electron affinity of C_{60} has been determined to be 2.67eV [5] and therefore our computational prediction is about 0.5 eV too low. Thus we might claim only qualitative accuracy for our evaluations of this quantity. Furthermore, when we computed the elastic, total differential cross sections at various angles, and as a function of collision energies [16], we found that the agreement with the

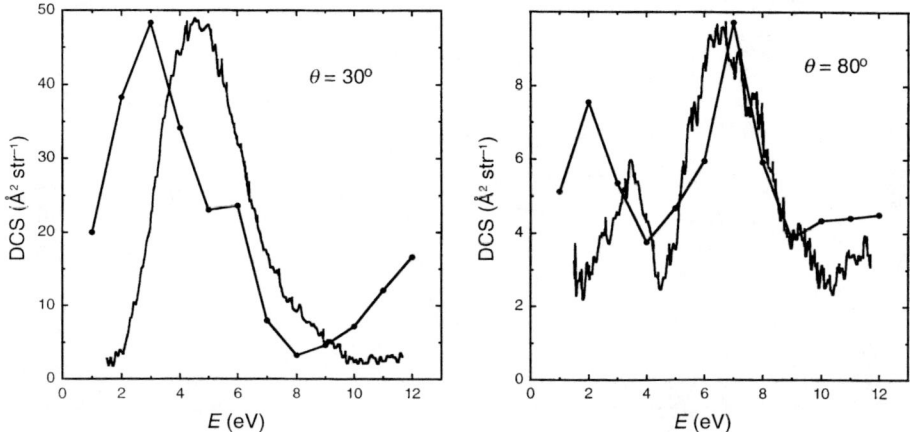

FIGURE 1. Taken from Ref. [16], computed and measured (Ref. [17]) differential cross sections at two fixed angles and over a range of collision energies from 1 eV up to 12 eV. The experiments are given by the noisy curves while the calculations are given by the straight line segments.

experimental data [17] was quite satisfactory considering the size of the target, and yielded reasonable agreement on an absolute scale as seen in the data given in Fig 1.

Another experimental puzzle, recently discussed by us [15] involves the assignment of the possible virtual state existing for electron scattering at threshold [9]. One useful computational tool for examining this problem is given by the shape of the adiabatic potential discussed in the previous section. We report, as examples, such potentials for the a_g IR (left panel) and the t_{1u} IR (right panel) in Fig. 2. Each diagram shows the lowest two partial waves and we see that the long-range attractive potential due to the large dipole polarizability of the target (558 au^3 [15]) leads to the long tail for the $l = 0$ a_g component and to a low centrifugal barrier for the $l = 1$ t_{1u} contribution.

If we now turn to the cross sections shown in Fig. 3, we see that the presence of a high threshold value of the cross section in the a_g symmetry followed by a Ramsauer-Townsend (RT) minimum at 0.08 eV as the eigenphase sum goes through zero (right panel of the figure) are features clearly due to the s-wave channel. In particular, the rising eigenphase sum is indicative of the presence of a virtual state for a metastable anion as surmised by the experiments [9]. Furthermore, the low-energy behavior of the elastic cross section associated with the t_{1u} contribution (also seen in Fig. 3) shows instead a rising behaviour at low energy followed by a marked decrease with energy and by the appearance of a RT-type structure away from the zero-energy threshold, *i. e.* at 0.4 eV, as surmised by another set of experiments [18]. A feature of the weakly bound states at low energy is that their behavior depends almost exclusively on the shape and features of the long-range interaction [15].

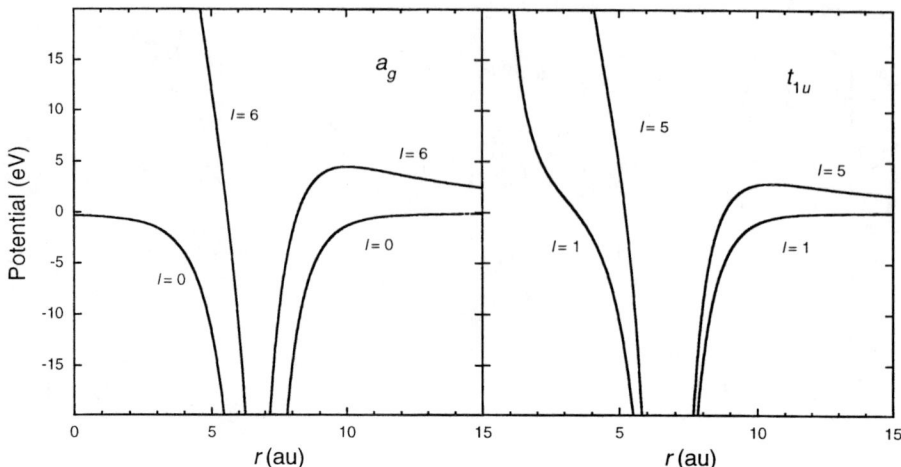

FIGURE 2. Computed adiabatic model potentials for the a_g (left) and t_{1u} (right) IR of the scattered electron. Each curve is labeled by its angular momentum value. The C nuclei are located at 6.806 au. Reprinted from Chem. Phys. Lett. 305, 413-418, Copyright 1999, with permission from Elsevier Science.

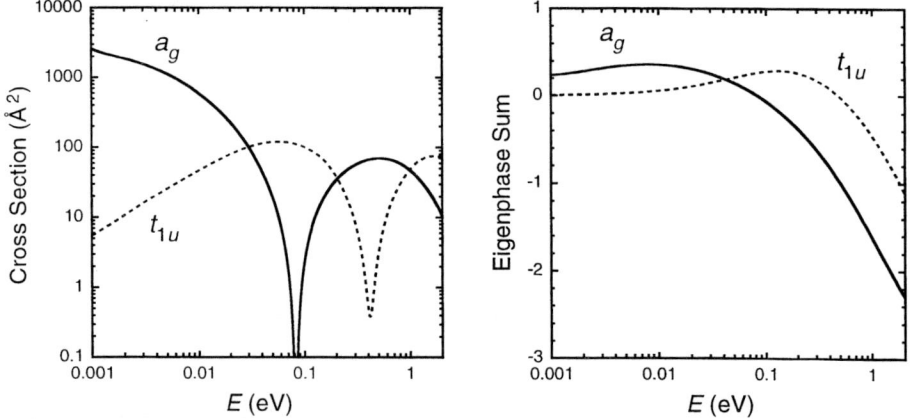

FIGURE 3. Computed cross sections (left panel) and eigenphase sums (right panel) at low collision energies for the two symmetries shown in Fig. 2. Reprinted from Chem. Phys. Lett. 305, 413-418, Copyright 1999, with permission from Elsevier Science.

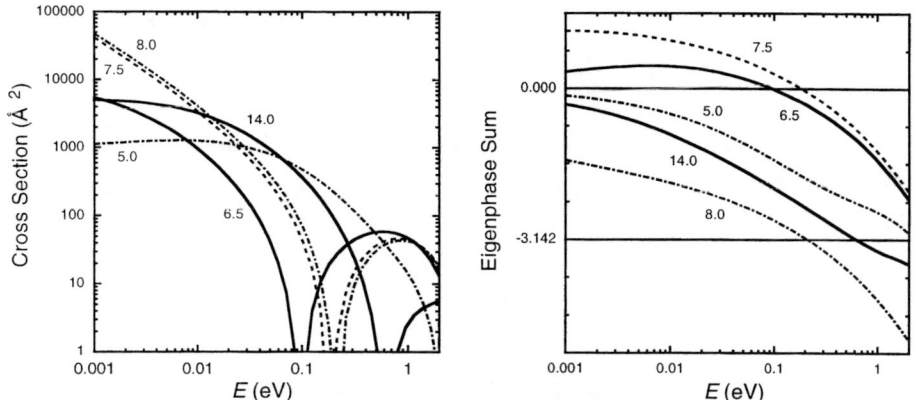

FIGURE 4. Model potential calculations of s-wave scattering in the a_g symmetry at very low energies. The left panel shows the changes in the RT structure of the cross section and the threshold cross section values as the potential strength is varied. The right panel shows the variations induced by the different potentials strengths on the eigenphase sum. The numbers next to each line indicate the value of C_G used as defined in Eq. (7).

This point can be furthered considered by examining a simple spherical model potential for the $e^- $-$C_{60}$ interaction, which consists of a long range polarization term added to an attractive gaussian centered at the same r distance as the carbon cage in C_{60} of the form

$$V(r) = -\frac{1}{2}\frac{\alpha_0}{r^4 + r_p^4} - C_G \exp\left\{-\gamma\left(r - r_G\right)^2\right\}. \tag{7}$$

In Eq. (7), the values of the constants in atomic units were taken to be $\alpha_0 = 558$ as was used in our SECP and ASMECP calculations [15], $r_p = 6.0$, $\gamma = 3.0$, $r_G = 6.0$, and C_G was varied from 5.0 to 14.0. The sensitivity of the cross section and eigenphase sum for s-wave scattering to changes in the short range part of the model potential can be seen in Fig. 4, where the potential is made more or less attractive by changing the coefficient which multiplies the added gaussian function. The value of $C_G = 6.5$ corresponds to the best modeling of the results we had already obtained with the full SECP potential. When the strength of the short range part of the potential is increased, one sees that at low energy the eigenphase sum continues to rise as $E \to 0$ until it reaches an integer multiple of π and shows the appearance of a bound state. On the other hand, the presence of only a virtual state is signalled by the eigenphase sum turning over and going back to zero at low energy. The corresponding cross sections are seen to be very large at threshold both before and after the virtual state becomes a bound state which occurs at a value of C_G somewhere between 7.5 and 8.0.

We have also recently carried out a computational experiment and analysis of $e^+ - C_{60}$ scattering [19] and found that the unusual shape of the interaction potential acting on the impinging positron can give rise to low-energy resonances for s-wave scattering. The resonant states are found to be trapped inside the C_{60} cage by the repulsive electrostatic barrier which exists at the position of the carbon atoms of the C_{60} cage. Experimental data for this system would be very useful in judging the model interaction potentials used for the positron-molecule interactions, and in particular for investigating the best method for including correlation effects in positron-molecule scattering.

IV   CONCLUSIONS

The foregoing discussion has tried to show that the study of electron scattering from large, polyatomic systems like C_{60} is today amenable to a computational analysis which starts from first principles and does not employ adjustable empirical parameters to describe the interaction of the electron and the molecule. The quantum scattering process needs to be treated in a somewhat simplified form because of the enormous number of electronic and nuclear degrees of freedom that would need to be considered in a more rigorous calculation. Our present results use the FN approximation and model short-range correlation effects with density functional theories which together greatly reduce the computational effort while still being able to provide a realistic representation of the scattering process [14,16,15]. Because of the large size of the cage and the special nature of its hollow interior, the C_{60} molecule is certainly seen as a very unique target for scattering electrons and positrons.

V   ACKNOWLEDGMENTS

We are grateful to the NATO Scientific Division for the award of Collaborative Research Grant no CRG950552, 922/94/JARC501. We also acknowledge the financial support of the Italian National Research Council (CNR), of the Italian Ministry for Universities and Research (MURST). One of us (R. R. L.) also wishes to thank the Welch Foundation (Houston) for its financial support under grant no A-1020 and to acknowledge the support of the Texas A&M University Supercomputing Facility.

REFERENCES

1. Kratschmer, W., Lamb, L. D., Fostiropoulos, K., Huffman, D. R., *Nature* **347**, 354-358 (1990).
2. Andreoni, W., *The Chemical Physics of Fullerenes 10 (and 5) years later: The Far-Reaching Impact of the Discovery of C_{60}*, Dordrecht: Kluwer Academic, 1996.
3. Haddon, R. C., *Phil. Trans. Roy. Soc. Lond. A* **343**, 53-62 (1993).

4. Yang, S. H., Pettiette, C. L., Conceicao, J., Cheshnovsky, O., and Smalley, R. E., *Chem. Phys. Lett.* **139**, 233-238 (1987).

5. Brink, C., Andersen, L. H., Hvelplund, P., Mathur, D., and Voldstad, J. D., *Chem. Phys. Lett.* **233**, 52-56 (1995).

6. Lezius, M., Scheier, P., and Märk, T. D., *Chem. Phys. Lett.* **203**, 232-236 (1993).

7. Huang, J., Carman, H. S., Jr., and Compton, R. N., *J. Phys. Chem.* **99**, 1719-1726 (1995).

8. Finch, C. D., Popple, R. A., Nordlander, P., and Dunning, F. B., *Chem. Phys. Lett.* **244**, 345-349 (1995).

9. Weber, J. M., Ruf, M.-W., and Hotop, H.,*Z. Phys. D* **37**, 351-357 (1996).

10. For recent reviews see: Huo, W. M., and Gianturco, F. A., *Computational Methods for Electron-Molecule Collisions*, New York: Plenum Press, 1995.

11. For example see: Gianturco, F. A., and Jain, A., *Phys. Rep.* **143**, 347-425 (1986).

12. Gianturco, F. A., Lucchese, R. R., Sanna, N., and Talamo, A., in *Electron Collisions with Molecules, Clusters and Surfaces*, edited by Ehrhardt, H., and Morgan, L. A., New York: Plenum Press, 1994, pp. 71-86.

13. Lucchese, R. R., and Gianturco, F. A., *Int. Rev. Phys. Chem.* **15**, 429-466 (1996).

14. Gianturco, F. A., Lucchese, R. R., *J. Chem. Phys.* accepted for publication (1999).

15. Lucchese, R. R., Gianturco, F. A., and Sanna, N., *Chem. Phys. Lett.* **305**, 413-418 (1999).

16. Gianturco, F. A., Lucchese, R. R., and Sanna, N., *J. Phys. B: At. Mol. Opt. Phys.* **32**, 2181-2193 (1999).

17. Tanaka, H., Boesten, L., Onda, K., and Ohashi, O., *J. Phys. Soc. Japan* **63**, 485-492 (1994).

18. Matejcik, S., Märk, T. D., Spanel, P., Smith, D., Jaffke, T., and Illenberger, E., *J. Chem. Phys.* **102**, 2516-2521 (1995).

19. Gianturco, F. A., and Lucchese, R. R., *Phys. Rev. A* submitted for publication (1999).

Electron capture by metallic clusters: theory and experiment

J.-P. Connerade

Laser Optics and spectroscopy Group,
Physics Department, Imperial College, London S.W.7 2BZ UK[1]

Abstract. Experiment and theory both reveal the existence of resonances in the attachment cross section of electrons on metallic clusters, the example we give being that of the alkali cluster K_8. The energies of these resonances correspond to the excitation energies of the giant resonances, due to the collective motion of all the delocalised electrons. We describe work done on this problem so far, and point towards future directions.

INTRODUCTION

This is an interim report on theoretical and experimental investigations of the attachment of electrons to metallic clusters. The theoretical investigations have been performed in collaboration with the following researchers:

Dr. Andrey Solov'yov (Ioffe Institute, St Petersburg, Russia)

Dr. Andrey Ipatov (St Petersburg Technical University, Russia)

Dr. Leonid Gerchikov (St Petersburg Technical University, Russia)

Prof. Walter Greiner (University of Frankfurt, Germany)

The experimental group involves:

Mr. Sukru Senturk (Ph.D. student, Imperial College, London)

Prof. D.D. Burgess (Imperial College, London)

Dr. N.J. Mason (Department of Physics and Astronomy, University College, London)

The structure of this Progress Report is as follows. First, some background to the problem of electron attachment to metallic clusters is described. Then, a summary of our theoretical investigations to date is presented. Finally, the experiments performed at Imperial College are described, including reference to previous studies in other laboratories as well as a description of our own work.

---

[1] Research supported by the E.P.S.R.C. (UK) and the Royal Society

CP500, *The Physics of Electronic and Atomic Collisions*, edited by Y. Itikawa, et al.
© 2000 American Institute of Physics 1-56396-777-4/00/$17.00

GENERAL BACKGROUND

Metallic clusters are characterised by the property that their valence electrons are fully delocalised. When the clusters are small (three or four atoms) their spectra are very complex and of molecular character. However, when they are of larger size (seven or eight atoms), a shell structure emerges [1,2] as a result of which their photoabsorption spectra are actually observed to become simpler: for a closed shell (eight atoms), the complex molecular structure is replaced by a single collective resonance, which is the earliest precursor of the plasmon in the metallic conduction band of a solid [3]. The transition from the cluster to the solid is an important issue. A number of authors have studied the way in which the giant resonance of the metallic cluster is related to the surface plasmon of the corresponding solid, and references to their work can be found in general texts on the subject [4]. One must bear in mind, however (a) that the geometry of a spherical cluster is not the same as that of an extended solid and (b) that as many different transitions exist from cluster to solid state behaviour as one can find independent physical observables to track the transition.

The large electronic shell of delocalised electrons extending over the whole cluster introduces a metallic force of quantum origin, which is is responsible for the stability of metallic clusters, as evidenced experimentally by the magic numbers just mentioned. Thus, a delocalised electronic shell is the defining property of a metallic cluster. Since it is large on an atomic scale, the delocalised shell is highly deformable. The collective resonance is actually a manifestation of the dynamic polarisability of this shell. In a collision with an electron, the dynamical polarisability also comes into play, and is believed to enhance the probability of capture. Resonantly enhanced electron attachment (REEA) to metallic clusters has been predicted [5] but not so far observed. It is expected, at the resonance energy, to be several orders of magnitude larger than the probability of capture via non-resonant bremsstrahlung. However, the experimental detection of REEA is rendered difficult for metallic clusters by the need for mass-selection, which dramatically reduces the number density in the interaction volume. This is why most work on metallic clusters involves laser excitation rather than electron capture.

EARLIER EXPERIMENTS

Metallic clusters are only formed in small sizes from metals with very low ionisation potentials. For example, alkali metals readily form very small metallic clusters. By way of contrast, mercury clusters in small sizes are of van der Waals type[2], and only become metallic for fairly large sizes [7], which corresponds to the fact that

---

[2] A van der Waals cluster is one for which the valence electrons remain attached to the atoms within the cluster. Thus, van der Waals and metallic behaviour represent two extremes in cluster physics

the ionisation potential of Hg is comparatively high. Consequently, alkali clusters are the most appropriate for the present investigation, in which the metallic delocalisation plays a crucial role.

A cluster is a fairly fragile object, and so rather gentle experimental techniques for producing negative ions are appropriate, to minimise the effects of fragmentation. The transfer of electrons by collisions with atoms in high Rydberg states has been extensively applied, both to molecular [8] and to alkali clusters in which Jahn–Teller deformations of the cluster shape have been shown to result in an even–odd alternation in the distribution of cross section versus cluster size [9].

The disadvantage of using Rydberg electrons is that the energy range which can be probed is restricted, and lies very close to the attachment threshold. In the present instance, this range would not encompass the interesting region in which resonances are predicted to occur.

For attachment to C_{60} (which in some respects has analogies with a metallic cluster since it contains 240 delocalised π electrons), collisions with low energy electron beams have also been used [10]. In such experiments, it is necessary to consider how the system relaxes, both during and after the attachment process, which can involve both electronic and core rearrangement. Under such circumstances, the distinction between van der Waals and metallic clusters is important. In a van der Waals cluster, the electron is first accommodated in an extended orbit, which then relaxes towards its ground state by localisation of the captured electron and an associated nuclear rearrangement [8]. Capture and relaxation are treated as separate processes, occurring on different timescales. Metallic clusters are expected to behave in a somewhat different way, because of the high mobility and characteristic correlation properties of delocalised electrons.

Another experimental approach has been applied to study attachment to metallic clusters by Kresin *et al* [11]. It consists in observing the *depletion* of a cluster beam crossed with an electron beam, as compared to the signal in absence of the electron beam. This method does not actually distinguish between capture and any other processes (fragmentation and ionisation) which would also deplete the beam. It has the merit of experimental simplicity, but the disadvantage that it is rather insensitive. One is looking for a very small attenuation in a rather large signal, and the statistical errors involved in this procedure are rather high. To date, high cross sections have been found and attributed to the formation of negative ions, but resonances as a function of electron energy were not visible, probably because of the large scatter of the data points. It is obviously more desirable to detect negative ions directly, since the experiment is then done on a 'dark' background.

Experiments in which negative ions have been detected directly were performed on C_{60} by Compton *et al* [10], and high probabilities of capture were found. The fullerenes are, as already noted, similar to metallic clusters, as they possess a large delocalised shell of π electrons (240 of them for C_{60}) which exhibits a giant resonance. However, the stability of C_{60} arises from molecular rather than metallic forces. This has the immediate consequence that a molecular beam of C_{60} is much more readily produced at comparatively high densities, with no need for mass selection. Consequently, attachment experiments with C_{60} are more readily performed than with mass selected alkali clusters. Also, the kinetic energy involved in collisions near the energy of the giant resonance is much higher for C_{60} than for metallic clusters.

EXPERIMENTS AT IMPERIAL COLLEGE

The above considerations all point to the need for an experiment in which an electron beam of adjustable energy is crossed with a cluster beam, negative ions of the cluster are detected and mass selection is also performed. This is the combination we have used. Our experimental method is based on the standard techniques of time-of-flight spectroscopy, applied in what we believe to be a novel way.

We perform the experiment in two steps. Step 1 determines the mass distribution of the beam without attachment (by laser ionisation), and step 2 is the attachment experiment (by crossing with an electron beam, with the laser turned off).

The cluster source is a two-temperature oven, the expansion chamber being maintained at a temperature about 50° higher than the primary oven in order to inhibit blockages in the escape nozzle, the pressure being about 10-20 bar inside the chamber. The beam escapes through a conical nozzle of about 10 μm diameter. The shape of this nozzle is extremely critical. It consists of a 10 μm hole with conical walls. Both a sonic nozzle (with the apex towards the outside) and a conical expansion nozzle (with the apex towards the inside) are used, giving different mass ranges. Essentially, the larger the hole and the smaller the angle of the cone, the higher the number of collisions, and the larger the clusters that can be produced. For the present work (cluster sizes less than fifty atoms) we used a sonic nozzle. The emerging cluster beam is directed downwards at an angle of 6°, so that neutral clusters cannot traverse the apparatus to the detector. Only cluster ions are detected: they are drawn into the flight tube by the accelerating voltages.

Step 1 allows us to set the cluster beam parameters. Under appropriate conditions, the source produces a beam of clusters with a distribution of sizes, which can be adjusted so that the heaviest readily detectable clusters are the ones we wish to study. In the present case, this can be eight atoms, and Fig. 1 shows a mass spectrum for this case. Note the very low peak for clusters containing nine or more atoms. The mass distribution of the cluster beam is determined by ionising the clusters with a nitrogen laser then accelerating them down a time-of-flight (TOF) tube. Our experiments show that the composition of the cluster beam changes according to experimental conditions on a timescale of a few hours, so that it is important to repeat this determination at regular intervals. Step 1 also yields the experimental flight time for cluster ions of a given size and beam temperature.

The TOF apparatus is designed so that a change can be made from the acceleration of positive to negative cluster ions in a matter of minutes. Furthermore, to accelerate negative ions, the accelerating voltage is pulsed with a rise-time of nanoseconds, so that the origin of time can be determined from the start of the acceleration pulse. In step 2, the laser is turned off, the electron beam is turned on, and the TOF voltages are inverted and pulsed. The detection time window is adjusted to the correct delay for the detection of negative ions of a given mass and the same beam temperature as in step 1. Since the laser and the onset of pulsed acceleration do not cross the cluster beam at exactly the same point, there is a small difference in flight time which must be corrected for.

It is important to note that, because of the distribution of cluster sizes, we do not detect fragmentation of larger masses if we set the detection window of step 2 to the appropriate cluster size (in our example, eight atoms). Because of the appreciable flight time of the clusters before detection (on a microsecond timescale), whatever attachment process is involved leads to energy loss by radiation, i.e. de-excitation occurs before detection.

The detector we use is a conventional pair of multichannel plates (MCP) in a chevron arrangement. It is important that the quality of the vacuum improves from the source end $(10^{-6}$ torr) down the flight tube to the detector end $(10^{-8}$ torr), since the MCP only functions well in a fairly clean environment. The lifetime of the detectors when used for cluster ions is many hours, but they must occasionally be baked clean when their performance drops. A more serious problem is the progressive decrease in the current from the electron gun, which is enclosed in a μ-metal shield. It was found necessary to replace the filaments fairly regularly, apparently because of progressive contamination by potassium from the cluster source. The electron current was therefore monitored, and the attachment signal was normalised by dividing it by the electron current, to remove the variation.

RESULTS AND DISCUSSION

The very simplest picture of attachment is the one described by Massey (1972): we suppose that there exists a Langevin attraction potential or polarisation force of the form

$$V(r) = -\frac{1}{2}\frac{\alpha e^2}{r^4} \tag{1}$$

outside a radius $r = a$, which is the radius of the cluster, and that the electron which attaches to the cluster cannot penetrate inside $r = a$. The quantity α is called the polarisability (units: length cubed) and is treated in the zeroth approximation as a constant (however, see below). One can then show that there is an orbiting cross-section

$$\sigma_{orb} = \pi \left\{ \frac{2\alpha e^2}{E} \right\}^{1/2} \tag{2}$$

which is the upper limit of the attachment cross section. E is the kinetic energy of relative motion. This simple treatment has the merit that it explains the general form of the cross section away from threshold, and allows estimates of the absolute cross section to be made.

The greatest weakness of this model is the treatment of α as an approximate constant. In fact, it has a complicated energy dependence, which is due to the dynamical polarisability of the metallic cluster. Thus, simple attempts to account for attachment by using measurements of the static polarizability α are not in accordance with observation [11].

Experiments on REEA for metallic clusters are of basic interest to test the range of validity of the jellium model. Within the jellium picture, which has been extensively used to describe the behaviour of metallic clusters [12], an electronic wavefunction is written for the whole cluster, treating it rather like a giant atom and making use of the observed spherical symmetry for closed shells. This model has the great merit of being simple, which allows electron-electron correlations to be computed, and a theoretical description of the giant resonances to be given [13]. On the other hand, it does *not* include effects of core excitation i.e. vibrations of molecular character to be included. The question of the range of validity of of the jellium picture is an open one. It involves such issues as the internal temperature of the cluster beam [14]. Direct calculations of the structure of metallic clusters by the methods of quantum chemistry [15] do not yield spherical shapes for closed shells but, typically, several isomers with different symmetries. One can imagine that the clusters are 'floppy' and (depending on temperature) that the clusters are constantly changing from one isomeric form into another, so that the average shape is actually spherical. For this to lead to a quantum mechanical shell structure, the fluctuations would need to be extremely short, on a timescale consistent with an energy difference of <0.1 eV between the isomers reported in [15]. The average response would then be described by the jellium model, but the possibility would then exist of some kind of phase transition, between a 'liquid' or jellium state of the cluster and a solid or 'frozen' state [14]. Attempts have indeed been made [16] to cool metallic cluster beams down and observe such a phase transition. Another, perhaps more direct explanation could be that, once delocalisation has occurred, the properties of the corresponding electronic shell are simply not very sensitive to the shape of the ionic core.

To date, however, photoabsorption experiments on metallic clusters have, as noted in the introduction, mainly revealed the disappearance of molecular complexity from spectra and its replacement by collective resonances.

To a first approximation, our observations are in principle complementary to the optical absorption spectrum, to which they are connected via Kramers-Kronig relations. However, other processes can also occur in electron capture which do not have analogues in photoabsorption, and differences between the structure in photoabsorption and the capture cross sections are therefore to be expected. In particular, electron capture can occur into excited states of the negative ion, leading to more complex spectral structure than observed by photoabsorption [17].

In the pure jellium picture, no notice is taken either of the possibility of different isomeric forms or of the detailed ionic structure of the cluster. One starts out simply by presupposing a spherical basis for closed shell clusters, and by constructing a model quite similar to the central field approximation for atoms although, in the case of metallic clusters, it is a less obvious approximation to make, since there is no massive charge centre. Resonant attachment of electrons results from an induced dipole, which in turn depends on the dynamical polarisability of the spherical cluster, and is therefore dominated by the giant resonance in the collective response.

Resonantly enhanced electron attachment to neutral metallic clusters was predicted in [5,6]. A number of recent papers [17–19,21] deal with various aspects of the process by both semi-empirical and *ab initio* methods. All these calculations are performed within the framework of the jellium model, which implies the existence of a shell structure and the excitation of collective resonances.

In line with the picture above, we may expect the jellium model to work well at the relatively high temperature at which a cluster beam is formed. Within this model, one can find a resonant de-excitation mechanism leading to attachment: as the electron approaches the cluster, it sets up a polarisation dipole, which rotates during the collision. A rotating dipole radiates (this mechanism is well-known for producing polarisational radiation [20]), and the attendant energy loss is one mechanism which enhances capture. Notice that it is again the dynamical polarisability which is involved, i.e. the same phenomenon which is responsible for the giant resonances in photoabsorption. In Fig. 2(a), a schematic representation of the effect is given, while Fig. 2(b) shows the competing, non-resonant effect known as bremsstrahlung, responsible for a much weaker background.

Experiment confirms the general predictions just described: our measurements so far have shown that resonances occur in the low-energy electron attachment spectrum of K_8. The experimental data are shown in Fig. 3. Two of these resonances dominate the spectrum. The lower one corresponds to the predicted energy for the collective resonance, which is extremely encouraging.

The higher energy resonance lies just above the ionisation threshold of K_8. A tentative explanation is that it corresponds to a shape resonance in the continuum, due to centrifugal barrier effects. High angular momentum states are predicted by theory [21], and very recent calculations [22] do indeed suggest that the cross section for attachment due to high angular momentum states is extremely high (especially the contribution from the $f \rightarrow g$ partial waves).

In addition to these two strong resonances, the experimental spectrum appears to contain weaker structure, but further work is required to confirm it, because the statistical errors in the data are comparable to the signal. The main problem in performing such measurements is of course the low density of the cluster targets and the limitations imposed by the restricted current available from the electron gun. Preliminary data on other cluster sizes (in particular: K_6) reveal similar structure, but with the peaks shifted somewhat in energy.

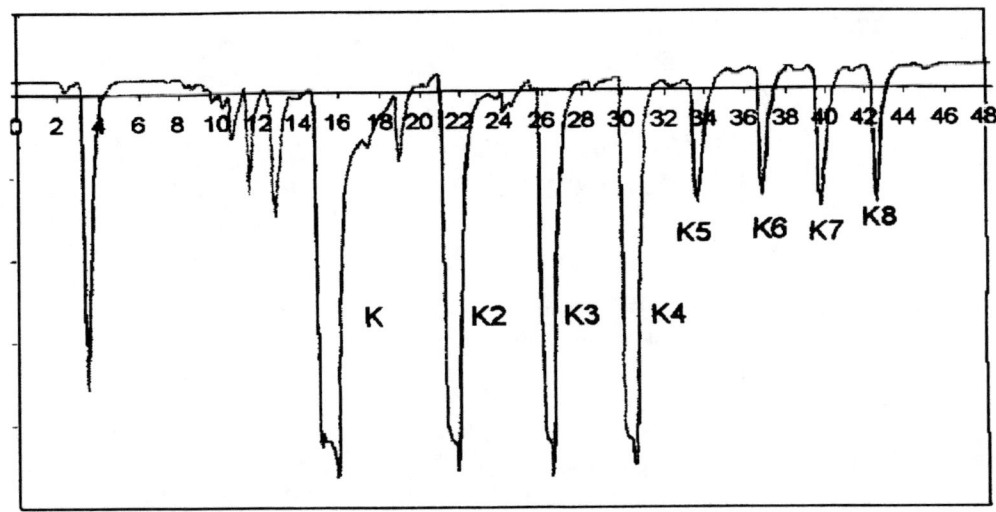

FIGURE 1. Composition of the potassium cluster beam as determined by laser ionisation and TOF spectroscopy.

CONCLUSION

Collaborative theoretical studies and preliminary experiments performed at Imperial College both indicate that low-energy resonances occur in the electron attachment spectrum of closed shell alkali clusters, the case studied being K_8 Experiments are still being pursued and, in particular, the dependence of the spectra on cluster size is being explored, within the limits imposed by the properties of our cluster source. It would be of interest to study the structure within the giant resonance for clusters with open shells. This is on the edge of feasibility at present, because sacrifices must be made in the resolution of the electron gun in order to achieve sufficient signal strength for detection. Another interesting measurement would be to explore the role of fragmentation in the attachment process, which is not described by the jellium model. To do this properly, however, would require a double time of flight experiment, with mass selection of the target clusters, their neutralisation, crossing with an electron beam, and mass selection of the negative ions (step 1 and step 2 performed simultaneously in the same experiment). At present, it does not seem possible to perform such an ambitious experiment. To be feasible, it would require a vast improvement in the performance of both electron guns and cluster sources.

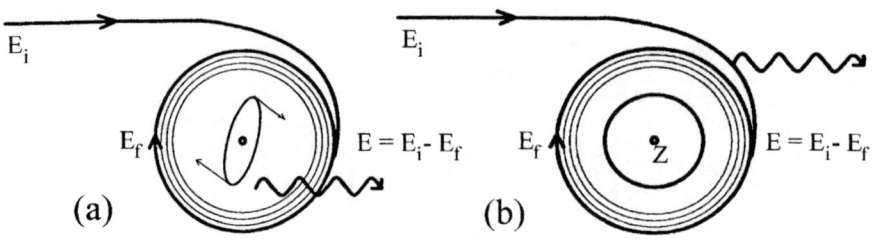

FIGURE 2. Schematic representations of (a) polarisational radiation and (b) bremsstrahlung for collision of an electron with a metallic cluster.

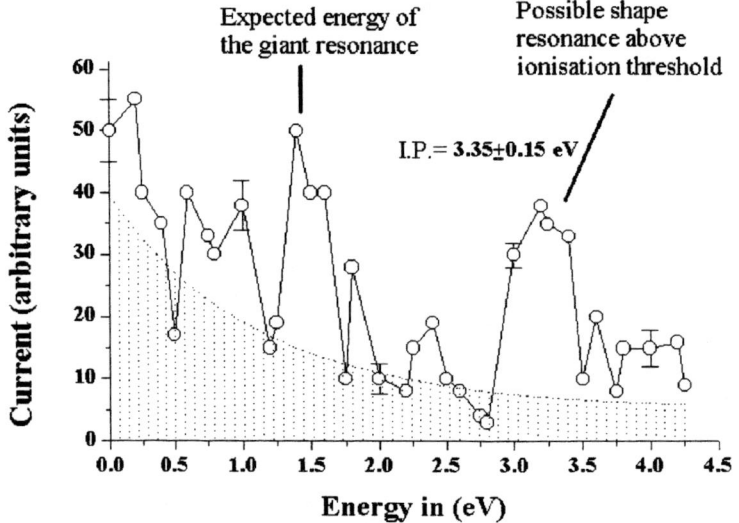

FIGURE 3. Experimental data for electron attachment to K_8 clusters.

REFERENCES

1. Knight W.D., Clemenger K., de Heer W.A., Saunders W.A., Chou M.Y., and Cohen M.L. *Phys. Rev. Lett.* **52**, 2141 (1984)
2. Clemenger K. *Phys. Rev.* **32**, 1359 (1985)
3. Bréchignac C., and Connerade J.-P. *J. Phys. B.* **27**, 3795 (1994)
4. Connerade J.-P. *Highly Excited Atoms*, Cambridge: Cambridge University Press, 1998, ch. 12
5. Connerade J.-P., and Solov'yov A.V. *J. Phys. B.* **29**, 365 (1996)
6. Connerade J.-P., and Solov'yov A.V. *J. Phys. B.* **29**, 3529 (1996)
7. C. Bréchignac C., Broyer M., Cahuzac Ph., Delacretaz G., Labastie P., Wolf J.P., and Wöste L. *Phys. Rev. Lett.* **60**, 275 (1988)
8. Kondow T. *J. Phys. Chem.* **91**, 1307 (1987)
9. Nagamine M., Someda K., and Kondow T. *Chem. Phys. Lett.* **229**, 8 (1994)
10. Huang J., Carman H.S., and Compton R.N. *J. Phys. Chem.* **99**, 1719 (1995)
11. Kresin V., Scheidermann A., and Knight W.A. *Electron Collisions with Molecules, Clusters and Surfaces* (Eds.) H. Eberhardt and L.A. Morgan, New York: Plenum Press, 1994, p. 183
12. Ekardt W. *Phys. Rev.* **B29**, 1558 (1984)
13. *Correlations in Clusters and Related Systems*, J.-P. Connerade (Ed.), Singapore: World Scientific Press 1996
14. Akulin V.M., Bréchignac C., and Sarfati A. *Phys. Rev. Lett.* **26**, 220 (1995)
15. Poteau R., and Spiegelmann F. *J. Chem. Phys.* **98**, 6549 (1993)
16. Haberland H., von Issendorf B., Yufeng J., and Kolar T. *Phys. Rev. Lett.* **69**, 784 (1992)
17. Connerade J.-P., Gerchikov L.G., Ipatov A.N., and Solov'yov A.V. *J. Phys. B* **32**, 877 (1999)
18. Ipatov A., Connerade J.-P., Gerchikov L.G., and Solov'yov A.V. *J. Phys. B* **31**, L27 (1998)
19. Gerchikov L.G., Solov'yov A.V., Connerade J.-P., and Greiner W. J. Phys. B. **30**, 4133 (1997)
20. Amusia M. Ya, and Korol A.V. *Phys. Lett.* **1186A**, 230 (1994)
21. Connerade J.-P., and Ipatov A. *J. Phys. B* **31**, 2429 (1998)
22. Ipatov A., and Gerchikov L.G. (1999, *private communication*)

Electron Stimulated Desorption of Charged Particles from Acetate Covered Rutile TiO$_2$(110)

Q. Guo[1] and E. M. Williams[2]

1. Nanoscale Physics Research Laboratory, School of Physics and Astronomy,
University of Birmingham, Birmingham B15 2TT UK
2. Surface Science Research Centre, University of Liverpool, Liverpool L69 3BX UK

Abstract. The interaction of low energy electrons with acetate adsorbed on the (110) surfaces of rutile TiO$_2$ has been studied using electron stimulated desorption ion angular distribution, low energy electron diffraction and scanning tunnelling microscopy. The adsorption of acetic acid at room temperature is dissociative, and produces a layer of carboxylate species and surface hydroxyl species in an ordered array. Irradiation of the adsorbate covered TiO$_2$(110) surface causes the dissociation of the surface species, leading to the desorption of a range of charged particles. The dominant desorption products are the H$^+$ ions. The H$^+$ ESD ion angular distribution can be resolved into two contributions: those ions desorbed from hydrogen atoms bonded at the oxide substrate, and those desorbed via the rupture of the C-H bonds of the acetate species. The geometry of the ESDIAD pattern led us to propose that the carboxylate species is bound to the Ti^{4+} ions with the molecular axis perpendicular to the surface.

INTRODUCTION

Metal oxides are widely used as catalysts supports in heterogeneous catalysis and thus the understanding of the catalytic properties of metal oxides is an important goal in surface science (1). The adsorption of organic molecules on metal oxide surfaces is also an area of growing interests due to the development of novel light emitting devices based on organic thin films (2). The properties of the interface between organic molecules and metal-oxide substrates are strongly dependent on the bonding and orientation of the molecular species. In this paper we reports our findings of the interaction of acetic acid with the TiO$_2$(110) surface studied using electron stimulated desorption ion angular distribution (ESDIAD). Apart from the relevance of carboxylic acids in catalysis, there is a wide range of organic molecules using the carboxyl functional group as the major rout for bond formation with solid substrates. The combination of ESDIAD with scanning tunnelling microscopy (STM) provides an opportunity of probing the detailed structure of the organic molecule-solid interfaces.

CP500, *The Physics of Electronic and Atomic Collisions*, edited by Y. Itikawa, et al.
© 2000 American Institute of Physics 1-56396-777-4/00/$17.00

EXPERIMENTAL

The experiments were performed in a UHV system equipped for STM, LEED, ESDIAD, Auger electron spectroscopy (AES) and time-of-flight, surface ion mass spectrometry, at a base pressure of $2x10^{-10}$ mbar. The LEED/ESDIAD equipment consists of two hemispherical grids ahead of a planar grid linked with two 40 mm diameter channel plates and a phosphorous screen. The electron beam for both LEED and ESDIAD is incident at 45 degrees from the surface normal, with electron current densities 100 nA/cm^2 and 10 μA/cm^2 for LEED and ESDIAD, respectively. The use of a higher current density in ESDIAD arose purely for the purpose of generating a satisfactory signal-to-noise ratio during image capture. For ESDIAD experiments, the energy of the exciting electrons was 400 eV with the sample at a positive potential of between 50–90 V relative to the system ground. The mass distribution of desorbed ions was analysed using a quadrupole mass spectrometer (Hiden model IP) with its ioniser turned off, and the kinetic energies of the desorbing ions could be determined by a time-of-flight technique with a pulsed electron beam. STM images were obtained at room temperature using an Omicron STM. A 10 mm x 10 mm x 1 mm single crystal of TiO$_2$(110) was clip-mounted on a tantalum plate which, in turn, was attached to the supporting electrical feedthroughs via 0.5 mm tantalum wires spot welded to the back of the plate. The sample could be resistively heated to 1100 K by passing a current through the tantalum wires, or cooled down to 170 K using liquid nitrogen. A Chromel-Alumel thermocouple pressed between the back of the sample and the tantalum plate was used for temperature measurements. Acetic acid (99.8% pure from Aldrich) was treated to a few freeze-pump-thaw cycles and admitted to the surface through a metal doser placed 5 mm away from the sample. The purity of the acid during dosing was checked with the mass spectrometer operating in RGA mode.

RESULTS

The clean surface of TiO$_2$(110) has been studied for some time with a variety of techniques and its structure is very close to the bulk truncated form (3-5). The structure of the carboxylic acid dosed surface became known to us only recently. Figure 1a shows an STM image of a TiO$_2$(110) surface, which has been exposed, to acetic acid at room temperature to saturation. Such a surface is covered by 0.5 monolayer (ML) of acetate and 0.5 ML of surface hydroxyl species (6). Each bright spot in the image is identified as one acetate molecule. The OH species are not visible in this image. The acetate species are separated by twice the lattice spacing along the [001] direction, while along the orthogonal direction, the spacing is identical to the lattice spacing of TiO$_2$. Thus, the overall adsorbate geometry is (2x1), in good agreement with the (2x1) diffraction pattern observed in LEED (6). The STM image does not provide any information about the orientation of the carboxylate species. The carboxylate overlayer is not stable under electron beam irradiation; when the current density in LEED is increased from 100 nA/cm^2 to 15 μA/cm^2, the (2x1) LEED pattern reverts to a (1x1) pattern within 30 seconds, as a result of electron

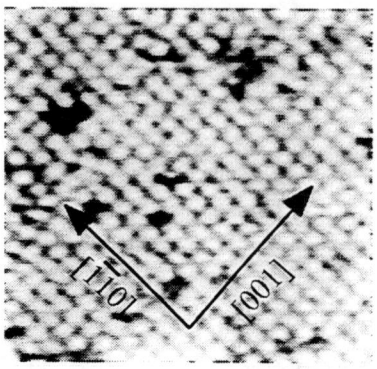

FIGURE 1. STM image of TiO$_2$(110) covered with 0.5 ML of acetate. The size of the image is 13 nm x 13 nm. Dark areas are not covered by acetate due to surface defects or contamination.

induced dissociation and decomposition.

The major ionic species observed to desorb from a clean TiO$_2$(110) surface when exposed to electrons was O$^+$ ions, with a sharp angular distribution peaked around the surface normal (7, 8). The kinetic energy distribution (KED) of O$^+$ ions was predominantly centred around a single peak at 4 eV, reflecting the dominance of one site of ion emission from the clean surface. When the TiO$_2$ (110) was exposed to CH$_3$COOH, the ESD signal of O$^+$ ions was greatly attenuated, with, in parallel, an increase of H$^+$ ion yield (6). Measurements using the mass spectrometer showed that, upon the formation of the (2x1) overlayer, 99% of the surface ion signal arises from H$^+$ ions, with the remaining 1% due to O$^+$ ions. Other desorption products have negligibly small intensities. Figure 2 shows the angular distribution of the H$^+$ ions, and figure 3 shows an intensity profile across the image shown in figure 2. It can be seen from these figures that the angular distribution has a normal emission component and an off-normal component. The intensity profile shown in figure 3 has been fitted with three peaks of Gaussian distribution, as shown in the figure. The central peak represents the desorption intensity along the surface normal, and the two

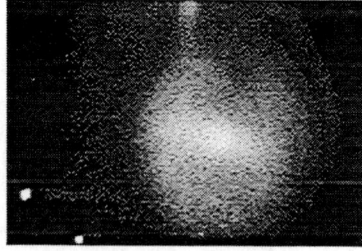

FIGURE 2. The angular distribution of H$^+$ ions desorbed from TiO$_2$(110) following a saturation exposure to acetic acid at room temperature.

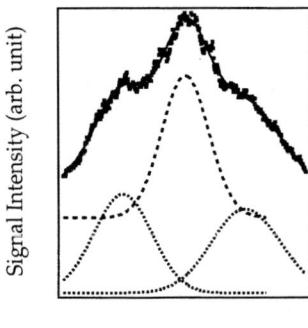

Signal Intensity (arb. unit)

Azimuth angle (degree)

FIGURE 3. Intensity profile across the image shown in figure 2. The central peak is due to H^+ ions emitted along the surface normal while the off-normal component arises from H^+ ions emitted with a polar angle of 35 degrees.

peaks at each side of the central peak are related to the off-normal emission. Since off-normal emission has a circular symmetry, the integrated image shown in figure 2 can be viewed as a superposition of a central lobe and a doughnut-shaped halo. Simulation of the ion trajectories using the computer programme SIMION leads to the identification of the off-normal feature with a field-free polar angle of 35 degrees (relative to the surface normal).

The ESDIAD pattern was found to change rather quickly under normal imaging conditions using 100 nA current (current density $10\mu A /cm^2$). The changes involve a significant increase of the intensity of the central lobe with the breakdown of the circular symmetry of the halo, and the transformation of the LEED pattern from a (2x1) to a (1x1). As a part of this study, a brief experiment with formic acid was also conducted. The ESDIAD pattern of H^+ ions from the formic acid dosed surface is composed of mainly a central lobe, without the off-normal features observed with acetic acid. Fig. 4 shows the kinetic energy distribution (KED) of the H^+ ions from a (2x1) overlayer of acetate, as determined by time-of-flight experiments. Two maxima can be seen at 3.5 eV and 5.5 eV. The surface species giving the 3.5 eV peak is more sensitive to electron beam exposure, and is completely removed after exposure to 300 eV electrons (current density $\sim 100\mu A/cm^2$) for 10 minutes. By rotating the sample, it was found that the 3.5 eV to 5.5 eV peak ratio was higher at greater emission angles from the surface normal, suggesting that 3.5 eV ions make a greater contribution to the off-normal emission features shown in Fig. 2 and 3.

In an attempt to test the association of the 3.5 eV and the 5.5 eV peaks with specific adsorbed species at the surface, experiments with isotopically labelled CH_3COOD were performed, with the KED of both H^+ and D^+ ions being monitored. During dosing, however, a significant amount of CH_3COOH was seen to be present, associated with the reaction of CH_3COOD with residual water in the gas line, which

395

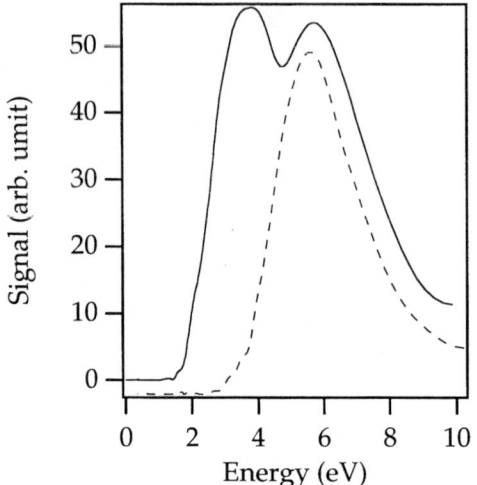

FIGURE 4. Kinetic energy distributions for H^+ ions (solid line) and D^+ ions (dashed line) emitted from a surface that has been dosed with CH_3COOD.

persisted even with long-term conditioning. Moreover, mixed minority isotopic species of the forms CH_2DCOOD, CHD_2COOD were known to be present at a level ~2% in the original liquid sample. Fig. 4 shows the KED for both H^+ ions and D^+ ions from the dosed surface. The overall yield of D^+ is two orders of magnitude lower than H^+, as would be expected from the isotope effect in ESD (9). As can be seen from the figure, the distribution for D^+ ions favours the higher energy peak at 5.5 eV. The H^+ ion KED is similar to that from a surface dosed with CH_3COOH.

In order to gain insight into the disappearance of the (2x1) LEED pattern and the disruption of the ESDIAD pattern under electron beam exposure, parallel studies were performed of the gaseous desorption products from the surface accompanying exposure to electrons. The procedure was first to expose the $TiO_2(110)$ surface to saturation coverage with acetic acid at room temperature, then introduce a 300 eV electron beam, current of 1 μA (current density 100μA/cm²). The desorption products were followed using the mass spectrometer in RGA multi-ion detection mode, filament switched on, with its ion source positioned close to the sample so as to detect both neutral and ionic species from the surface. Fig. 5 shows that the signals at mass number 1, 15, 16, and 42 all respond immediately to the electron beam. When repeating this measurement with the mass spectrometer ioniser turned off, mass number 15 and 42 remained at their background level, thus associating these species uniquely with neutral desorption. Both CH_3 radicals and cracking from CH_4 will contribute to the signal of mass number 15. Mass number 42 is possibly from ketene (CH_2CO) which is a major thermal decomposition product from adsorbed acetate on a number of oxide surfaces (10,11).

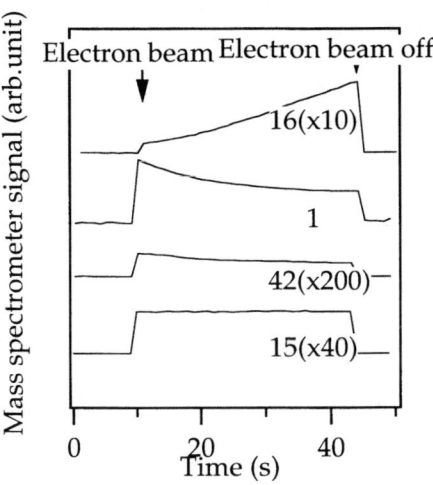

FIGURE 5. Desorption product distribution under electron beam irradiation. The H^+ signal decreases with time, accompanied by an increase of the O^+ signal, as the acetate species are damaged by electrons.

DISCUSSION

The adsorption of carboxylic acids on faceted $TiO_2(001)$ surfaces has been characterised using XPS and TPD by Kim and Barteau (12). Both molecular and dissociative adsorption of acetic acid are found at 200 K, but only dissociative adsorption is found at 300 K. Thermal desorption experiments show that 15% of the dissociatively adsorbed acetate desorbs as acetic acid at 390K, via recombination with surface hydroxyls, and the remaining acetates decompose at 600 K. At room temperature, all the adsorbed acetic acid can be regarded as having gone through dissociation to form chemisorbed acetate and hydroxyls. According to a recent review by Rajadurai (13), dissociative adsorption is favoured for CH_3COOH on many metal oxides at room temperature. The (2x1) overlayer formed at room temperature in our study can be regarded as ordered acetate and hydroxyl species. In keeping with the (2x1) LEED pattern, the coverage of acetate is taken as 0.5 ML (relative to the density of the five-fold co-ordinated Ti^{4+} ions at the surface), with the distance between acetates along the [001] direction being twice the unit cell length of the clean surface. The situation would appear to be similar to that with formate adsorption which has been shown by STM to attain half-monolayer coverage (14). The symmetrical features of the angular distribution of H^+ ions from the adsorbate can best be described as due to adsorbed acetate in an upright adsorption geometry. Since the C–H bond axis from acetate is in the region of 70° away from the C-C axis (15), electron stimulated desorption leading to the rupture of the C–H bonds is expected to produce off-normal directed H^+ ions. The actual measured polar angle of ion emission at large values will be significantly affected by the neutralisation of ions due to the proximity of the surface (16). Thus whilst a change in the C-H bond axis may arise for the surface acetate, its identification will be masked by the strong influence of neutralisation on ions desorbing with larger polar angles (≥ 50 degrees). In these terms, ESDIAD can be taken only as a qualitative indication of a significant degree of off normal bonding.

The halo of the off-normal emission would result from the free rotation of the methyl group around the C-C axis. At sufficiently low temperature, below the lower limit of 150 K in the present study, the halo would be expected to transform into three azimuthal zones of polar emission once the rotation of the methyl group is frozen. The central lobe can be interpreted as due to desorption of H^+ ions from surface hydroxyls. The preferred hydroxyl site would be at the position of the bridging oxygen sites at the $TiO_2(110)$ surface. The ESDIAD result from formate can also be explained by assuming that the adsorbed formate has the same configuration as acetate on the surface. Therefore, the C–H bond axis in formate would be perpendicular to the surface, and the breaking of C–H bonds by electron stimulation would only produce H^+ ions ejected along the surface normal.

The association of the off-normal H^+ ion signal with acetate and the normally emitted H^+ ions with surface OH species receives further support from the kinetic energy distribution (KED) of the H^+ ions. According to the result shown in Fig. 4, the KED for D^+ gives prominence to the higher energy peak at 5.5 eV. Since the deuterium containing species on the surface with the dissociative adsorption of CH_3COOD is OD, the 5.5 eV peak is assigned to D^+ ion desorption from OD. The higher sensitivity of the 3.5 eV peak to electron beam irradiation is, moreover, consistent with a great efficiency for electron beam induced decomposition of the adsorbed acetate, leading to a rapid decrease in coverage.

The intensity of desorbed H^+ ions is greatest along the surface normal, arising as discussed earlier from the hydroxyl species formed at the bridging oxygen rows with the O–H bond axis perpendicular to the (110) surface. It is of interest to note that if the hydrogen atoms were bound with the in-plane oxygen, their desorption signal would be rather attenuated by acetate due to the proximity of the hydroxyls and the acetates.

The change of LEED and ESDIAD patterns from adsorbed acetate under electron beam exposure confirms that the acetate has undergone electron-induced reaction. This points to the difficulties in obtaining good LEED patterns with this adsorbate. Onishi et al (17) have observed a faint (2x1) LEED pattern for formate, which transforms to a (1x1) pattern after annealing to 350 K. In our experiment the (2x1) pattern from both formate and acetate was clear and stable up to the higher temperature of 450 K. Since a very low electron current density was used in the present experiment, the lower temperature of transformation observed by Onishi et al for formate may possibly be related to a higher rate of electron induced damage of the overlayer. Based on the mass resolved desorption products shown in Fig. 5, we propose that the following reactions take place when an acetate overlayer is exposed to electrons:

The first process involves the elimination of hydrogen from the methyl group, causing the formation of either atomic hydrogen or a proton. The desorption of the latter forms the off-normal features in ESDIAD images shown in Fig. 2.

The remaining fragments decompose to form ketene and leave an oxygen atom on the surface. The second process involves the breaking of the carbon-carbon bond and generating a methyl radical, plus a CO_2 molecule which then desorbs. The methyl radical can either leave the surface or recombine with a surface hydrogen atom to form a methane molecule which then leaves the surface.

CH_2CO is a major thermal decomposition product from acetate on oxide surfaces when annealed in vacuum to 600 K (10, 12). The same reaction is seen to take place here at room temperature under electron excitation. The common rate-limiting step in ketene formation is likely to be the elimination of a hydrogen atom from the methyl group. The loss of a hydrogen atom would destroy the sp3 hybridisation around the methyl carbon with the tendency to form double carbon-carbon bonds. However, there is a difference between electron beam induced reaction and thermal activated reaction. Electron beam excitation produces a larger amount of CH_3/ CH_4. This may be related to the high efficiency with which the electrons can break the C–C bond.

CONCLUSIONS

The dissociative adsorption of acetic acid on $TiO_2(110)$ forms an ordered (2x1) overlayer of acetate at room temperature with a coverage of 0.5 ML. The acetate is proposed to be bonded to the Ti^{4+} ions in an upright form with its major molecular axis perpendicular to the surface. The acid hydrogen from the dissociation of acetic acid is believed to bind with the bridging oxygen atoms forming 0.5 ML of hydroxyls at saturation coverage. Efficient decomposition of acetate takes place under electron beam exposure to form H^+ ions, ketene, methyl radicals and methane.

REFERENCES

1. Henrich, V. E., and Cox, P. A., *The Surface Science of Metal Oxides*, Cambridge University Press, Cambridge, 1994. and referenced therein.
2. Carter, F. L., Siatowski, R. E., and Wholtjen, H., (eds.) *Molecular Electronic Devices*, North Holland, Amsterdam, 1988.
3. Sander, M., and Engel, T., Surf. Sci. **302**, L263 (1994).
4. Murray, P. W., Condon, N. G., and Thornton, G., Phys. Rev. B **51**, 10989 (1995).
5. Pan, J. M., Maschhoff, B. L., Diebold, U., and Madey, T., Surf. Sci. **291**, 381 (1993).
6. Guo, Q., Cocks, I., and Williams, E. M., J. Chem. Phys. **106**, 2924 (1997).
7. Guo, Q., Cocks, I., and Williams, E. M., Phys. Rev. Lett. **77**, 3851 (1996).
8. Kurtz, R., Surf. Sci. **177**, 526 (1986).
9. Madey, T. E., Yates, Jr. J. T., King, D. A., and Uhlaner, C. J., J. Chem. Phys. **52** 5215 (1970).
10. Bowker, M., Houghton, H., and Waugh, K. C., J. Catal. **79**, 431 (1983).
11. Parrott, S. L., Rogers Jr, J. W., and White, J. M., Appl. Surf. Sci. **1**, 443 (1978).
12. Kim, K. S., and Barteau, M. A., J. Catal. **125**, 353 (1990).
13. Rajadurai, S., Catal. Rev. Sci. Eng. **36**, 385 (1994).
14. Onishi, H., and Iwasawa, Y., Chem. Phys. Lett. **226**, 111 (1994).
15. Handbook of Chemistry and Physics, CRC Press Inc. 1974/75.
16. Miskovic, Z., Vukanic, J., and Madey, T. E., Surf. Sci. **141**, 285 (1994).
17. Onishi, H., Aruga, Egawa, T. C., and Iwasawa, Y., Surf. Sci. **193**, 33 (1988).

Field Effects on the Recombination of Atomic Ions Studied at a Storage Ring Electron Cooler

S. Schippers*, G. Gwinner[†], T. Bartsch*, A. Hoffknecht*,
A. Müller*, D. Schwalm[†], and A. Wolf[†]

*Institut für Kernphysik, Justus-Liebig-Universität, 35392 Giessen, Germany
[†]Max-Planck-Institut für Kernphysik, 69177 Heidelberg, Germany

Abstract. Recent results on the recombination of atomic ions with free electrons in the presence of external electromagnetic fields are reported. While dielectronic recombination via high Rydberg states is known to be influenced by external electric fields for already more than two decades its sensitivity to additional crossed magnetic fields has only been demonstrated recently. This novel effect is reported here as well as the discovery of a magnetic field dependence of the yet unexplained recombination rate enhancement at low energies.

INTRODUCTION

In the past decade electron coolers at heavy ion storage rings have developed into the most successful experimental tool for electron-ion recombination studies [1,2]. Currently, research programmes involving highly charged atomic ions are carried out at the heavy ion storage rings ESR of GSI in Darmstadt [3], TSR of the Max-Planck-Institute of Nuclear Physics in Heidelberg [4,5] and at CRYRING of the Manne-Siegbahn-Laboratory in Stockholm [6,7].

The unique possibility of merging a cooled ion beam with electrons in storage ring electron coolers has facilitated the exploration of many novel phenomena in atomic collision physics such as the spectroscopy of doubly excited states in highest Z ions [8,9] where the atomic structure is determined by relativistic and QED effects, recombination at lowest energies where an unexplained enhancement of the experimental recombination rate beyond theoretical expectations is observed [10,11], interference effects between dielectronic recombination (DR) and radiative recombination (RR) [12,13], and the influence of external electromagnetic fields on DR rates [14,15]. The ability of the merged beams approach to provide absolute recombination rate coefficients with low statistical and systematical uncertainties has been exploited for providing accurate recombination rates for plasma applica-

CP500, *The Physics of Electronic and Atomic Collisions*, edited by Y. Itikawa, et al.

tions [16–18]. The current status of all these activities has been reviewed recently [19,20].

We here report experimental results on the dependence of recombination rates on external electric and magnetic fields which have been obtained during the past two years at the heavy ion storage ring TSR of the Max-Planck-Institut für Kernphysik in Heidelberg. Two novel phenomena are presented. Firstly, the dependence of DR rates on magnetic fields in $\vec{E} \times \vec{B}$ configurations and, secondly, the influence of magnetic fields on the unexpected recombination rate enhancement at very low energies.

DR IN THE PRESENCE OF EXTERNAL FIELDS

DR is a two step process

$$e^- + A^{q+} \rightarrow [A^{(q-1)+}]^{**} \rightarrow A^{(q-1)+} + h\nu \tag{1}$$

where a multiply-excited intermediate state is created by a resonant dielectronic capture (inverse Auger) process in a first step and, subsequently, that intermediate state decays by photon emission. The enhancement of DR rates by external electric fields has already been predicted by Jacobs et al. [21] in the 1970s. Briefly, the effect arises from the Stark mixing of ℓ states and the resulting influence on the autoionization rates which, by detailed balancing, determine the capture of the free electron. Autoionization rates strongly decrease with increasing ℓ and, therefore, only low ℓ states significantly contribute to DR. Electric fields mix low and high ℓ states and thereby increase the autoionization rates of the high ℓ states and consequently the contribution of high Rydberg states to DR. For an illustrative example of the effect the reader is referred to Fig. 1 which shows experimental DR spectra for Li-like Cl^{14+} ions [15] taken in the presence of different external electric fields. Clearly the strength of high n (where ℓ-mixing is most effective) DR resonances due to $Cl^{13+}(1s^2 2p_j n\ell)$ intermediate states increases with increasing field strength.

The control of external fields in experiments using intense ion and electron beams is not easy. In fact, results from early recombination experiments [22,23] could only be brought in agreement with theory under the assumption that external electric fields had been present in the interaction region. The first experiment where external fields were applied under well controlled conditions was performed by Müller and coworkers [24] who investigated DR in the presence of external fields (DRF) with singly charged Mg^+ ions. They observed an increase of the measured DR cross section by a factor of about 1.5 increasing the motional $\vec{v} \times \vec{B}$ electric field from 7.2 to 23.5 V/cm.

The first DRF experiment using highly charged ions at a storage ring was carried out with Si^{11+} ions by Bartsch et al. [14]. It produced results with an unprecedented accuracy enabling a detailed comparison with theory. Whereas the overall agreement between experiment and theory as for the magnitude of the effect (up to

401

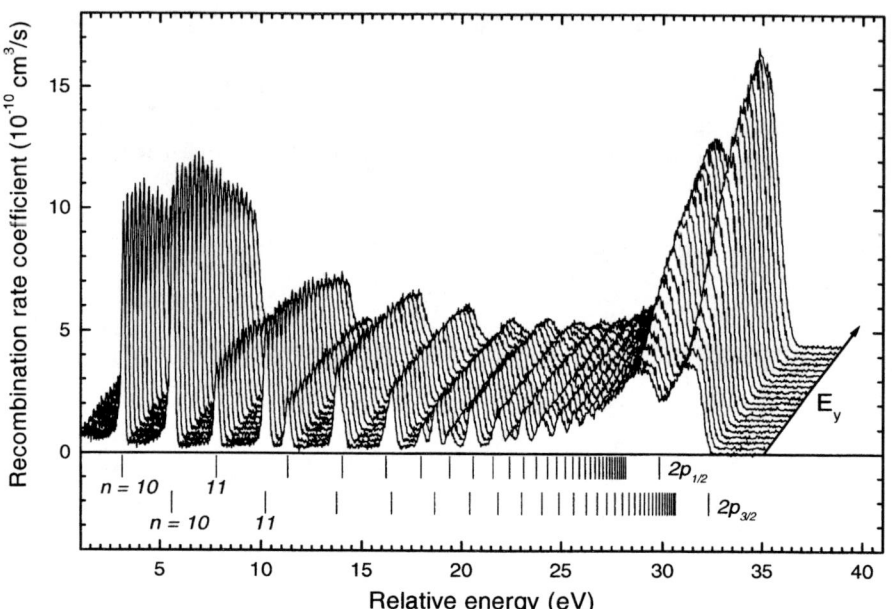

FIGURE 1. Absolute recombination rate coefficients measured for 250 MeV Cl^{14+} ions at applied motional electric fields E_y increasing nearly linearly from 0 to 380 V/cm; longitudinal magnetic field $B_z = 69\,\text{mT}$, electron density $0.5 \times 10^7\,\text{cm}^{-3}$. Energetic positions of the $2p_{1/2}\,nl$ and $2p_{3/2}\,nl$ resonances according to the Rydberg formula are indicated. The figure is taken from Ref. [15]

a factor of 3 when increasing the field from 0 V/cm to 183 V/cm) was fair, discrepancies remained in the functional dependence of the rate enhancement on the electric field strength. This finding stimulated theoretical investigations into the role of the additional magnetic field which in storage ring DR experiments always is present, since it guides and confines the electron beam within the electron cooler. In a model calculation Robicheaux and Pindzola [25] found that in a configuration of crossed $\vec{E}$ and $\vec{B}$ fields indeed the magnetic field through the mixing of m levels influences the rate enhancement generated by the electric field via ℓ level mixing. More detailed calculations [26,27] confirmed these results.

Inspired by these predictions we performed DRF experiments using Li-like Cl^{14+} [15], Ti^{19+} [28], and Ni^{25+} [29] ions where in an $\vec{E} \times \vec{B}$ configuration additionally to the external electric field $\vec{E}$ also the magnetic guiding field strength $\vec{B}$ directed along the axis (z-direction) of the electron cooler has been varied. Cl^{14+} recombination spectra for a fixed value of $B_z = 69$ mT and electric fields $0 \leq E_y \leq 380$ V/cm (pointing into the y-direction perpendicular to the z-axis) are shown in Fig. 1.

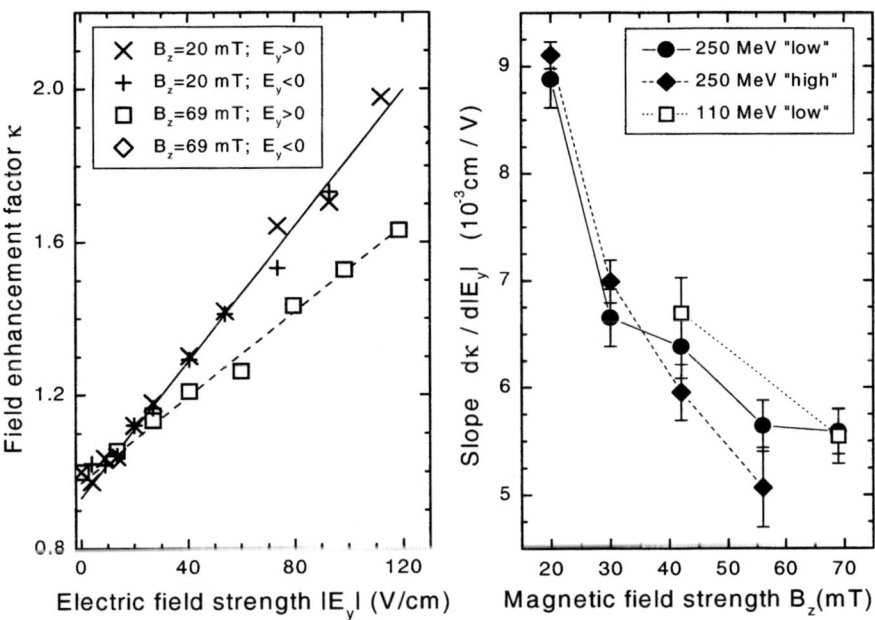

FIGURE 2. Left panel: Measured Cl^{14+} field enhancement factor κ as a function of the applied motional electric field strength $|E_y|$. Results are shown for the largest and the smallest longitudinal magnetic field B_z as indicated. Right panel: Dependence of the slope $d\kappa/d|E_y|$ on the longitudinal magnetic field strength B_z. The different curves represent measurements under different experimental conditions, i. e. ion energies of 110 and 250 MeV and "high" $(0.8 - 1.2 \times 10^7 \text{ cm}^{-3})$ and "low" $(0.46 - -0.48 \times 10^7 \text{ cm}^{-3})$ electron density. Both plots are taken from Ref. [15].

We quantify the E_y-field induced increase of resonance strength over the energy interval 24.6–35 eV by considering the integrated recombination rate normalized to the $E_y = 0$ V/cm integral. The resulting field enhancement factor κ as a function of electrical field strength is plotted in the left panel of Fig. 2 for two values of the magnetic field strength B_z. For a given B_z the field enhancement factor increases linearly over the experimental range of electric field strengths independently from the sign of E_y.

Clearly there is an influence of B_z on the field enhancement factor's slope $d\kappa/d|E_y|$ which is lower for the higher magnetic fields. The right panel of Fig. 2 summarizes our experimental findings about the influence of the magnetic field on the slope of the field enhancement factor which has been repeatedly measured under different experimental conditions. We also observe a decrease of the slope $d\kappa/d|E_y|$ with increasing magnetic field strength for Li-like Ti^{19+} ions [28] over the range 30 mT $\leq B_z \leq$ 80 mT as shown in the left panel of Fig. 3. A decrease

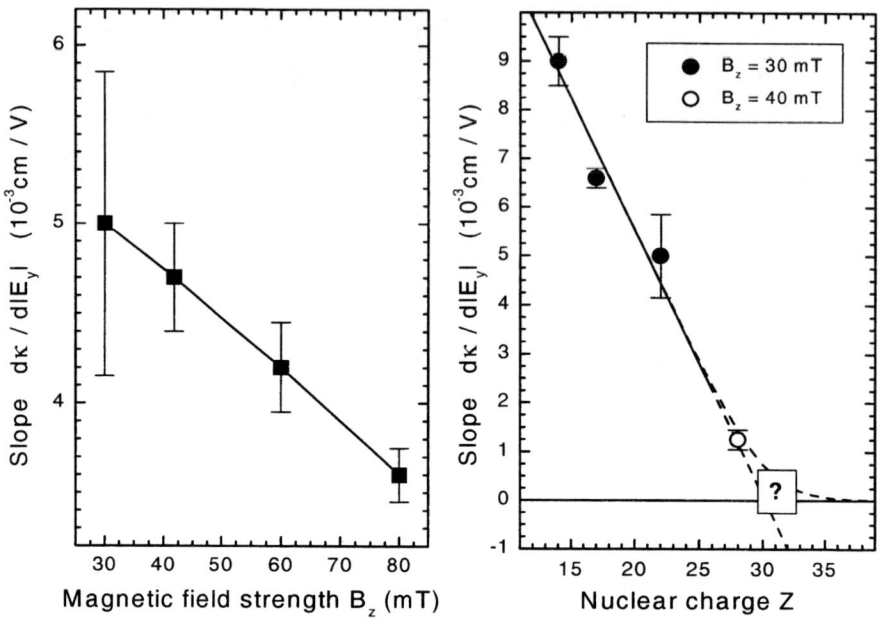

FIGURE 3. Dependence of the slope $d\kappa/d|E_y|$ on B_z for Ti^{19+} ions (left panel) and on the nuclear charge Z for iso-electronic Li-like Si^{11+}, Cl^{14+}, Ti^{19+}, ($B_z = 30$ mT) and Ni^{25+} ($B_z = 40$ mT) ions (right panel). The extrapolation to high Z is ambiguous (see text).

of the electric field enhancement by a crossed magnetic field is also predicted by the model calculation of Robicheaux and Pindzola [25] for magnetic fields larger than approximately 20 mT. At lower fields the calculations yield an increase of the enhancement with increasing field. A corresponding experimental observation has been made recently by Klimenko and coworkers [30] who studied recombination of Ba$^+$ ions from a continuum of finite bandwidth which they previously had prepared by laser excitation of neutral Ba atoms. For a given electric field strength of 0.5 V/cm the recombination rate is increasingly enhanced by crossed magnetic fields up to about 20 mT. However, there is no effect of the magnetic field when it is directed parallel to the electric field vector.

Qualitatively this behaviour can be explained by an increasing number of states taking part in DR through m-mixing by a weak magnetic field on top of the ℓ-mixing induced by the electric field. At higher magnetic field strengths the magnetic interaction energy becomes larger than the electric one. This leads to a reduced ℓ-mixing and correspondingly a lower number of states participating in DR. For the m-mixing to occur the *crossed* $\vec{E}$ and $\vec{B}$ arrangement is essential. In the case of *parallel* $\vec{B}$ and $\vec{E}$ fields m remains a good quantum number and no influence of the magnetic field is expected.

By choosing isoelectronic ions for our DRF studies we also aim at obtaining information about the dependence of magnitude of the field induced enhancement of the DR rates on the nuclear charge Z. Because of the Z^4 scaling of radiative rates it is expected that with higher Z less ℓ states of a given Rydberg n level take part in DR and therefore the sensitivity to ℓ-mixing decreases [24]. Results of Griffin and Pindzola [31] who calculated decreasing DR rate enhancements for increasing charge states of iron ions point into the same direction. The right panel of Fig. 3 shows our experimentally determined slopes $d\kappa/d|E_y|$ for different Li-like ions. Indeed the field induced enhancement becomes weaker with increasing Z. The question remains to what extent at even higher Z the field enhancement effect prevails. In any case, a reduction of DR strength by external fields, i. e. a negative slope, as suggested by a linear extrapolation of our experimental results is not predicted by any of the existing theories.

RECOMBINATION AT VERY LOW ENERGIES

Unexpectedly high recombination rates at 0 eV relative energy between the electron and the ion beam exceeding the theoretical predictions by factors up to 365(!) for Au^{25+} [32] have been observed in a number of storage ring and single pass merged beams experiments [33]. To date the effect is not fully understood. Partly it has been explained with the fortuitous presence of DR resonances close to threshold [10,12,32,34]. With bare Ar^{18+} ions, where DR is not possible, still enhancement factors up to 10 have been found [35].

In order to find out the origin of the observed recombination rate enhancement a number of studies of this effect has been performed recently where experimental parameters such as the the ion charge, the electron density, and the magnetic guiding field have been varied systematically. Gao et al. [36] used a range of bare ions (D^+, He^{2+}, N^{7+}, Ne^{10+}, and Si^{14+}) under nearly constant experimental conditions. They discovered a $Z^{2.8}$ scaling of the excess rate $\Delta\alpha$ at 0 eV relative energy. Although a similar scaling is expected from three body recombination (TBR) processes these cannot be held responsible for the recombination rate enhancement for two reasons: i) If the measured excess rate was due to TBR it should exhibit a strong dependence on the electron density. In experiments where the electron density has been varied by more than an order of magnitude such a dependence has not been found [37,38]. ii) TBR effectively only populates high Rydberg states [39] which, however, within the usual experimental arrangement instead of being detected are field ionized in the charge analyzing dipole magnet. It should be noted that from density of states arguments an approximate Z^3 scaling can be expected for any recombination process which populates Rydberg states within a fixed, Z-independent range of binding energies.

Another experimental parameter which has been varied is the magnetic guiding field strength B_z within the electron cooler which somewhat surprisingly exhibited

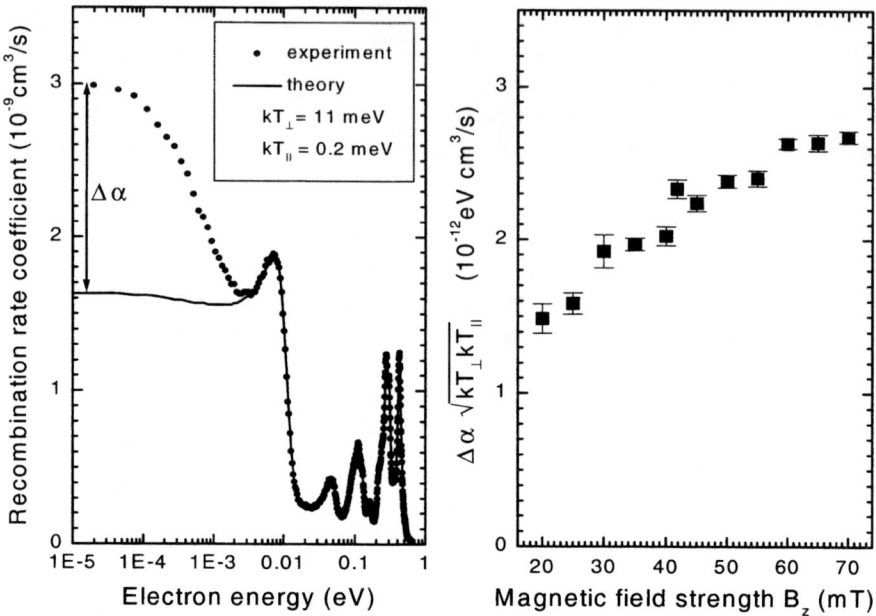

FIGURE 4. Recombination of Li-like F^{6+} ions at low energies (left panel). The DR resonances are due to $F^{5+}(1s^22p6\ell)$ intermediate states. The theoretical spectrum has been obtained by convoluting the theoretical F^{6+} recombination cross section [43] with an electron velocity distribution characterized by the temperatures $T_\perp = 11$ meV/k and $T_{\parallel} = 0.2$ meV/k (k is the Boltzmann constant). The unexplained excess rate $\Delta\alpha$ at 0 eV increases with increasing magnetic guiding field strength in the electron cooler (right panel).

an influence on the recombination rate at 0 eV. For Au^{25+} ions Hoffknecht et al. [32] found a linear increase of the recombination rate by a factor of 2.4 when varying B_z by only 27% from 0.24 to 0.33 T. However, the interpretation of this result remained ambiguous because the change of the magnetic field could have altered the electron beam temperature which also has a direct influence on the recombination at 0 eV [40]. Thus, without monitoring the electron beam temperature it cannot be decided whether the observed effect is due to the magnetic field change or due to an accompanying change of the electron velocity distribution.

In an electron cooler the electron velocity distribution can be described by an anisotropic Maxwellian with two temperatures $T_{\parallel}$ and $T_\perp$ parallel and perpendicular to the electron beam direction, respectively [2]. In a recombination experiment both temperatures can be inferred from the shape of dielectronic recombination resonances close to threshold [20]. At higher energies the shape of DR resonances is determined by $T_{\parallel}$ only and no information on $T_\perp$ can be extracted. An ion exhibiting an isolated DR resonance at only 10 meV which proved to be well suited for temperature diagnostics is Li-like F^{6+} [37,41]. The procedure for the temperature determination consists of essentially fitting a convoluted theoretical recombination cross section to the experimental data and thereby treating $T_{||}$ and $T_{\perp}$ as fit parameters.

Using F^{6+} ions we performed a systematic study of the recombination at low energies [42]. In this investigation the magnetic guiding field strength has been varied from 20 to 70 mT and at the same time the electron beam temperatures have been monitored by applying the procedure described above. A theoretical description which is in very good agreement with our experimental recombination spectra (left panel Fig. 4) and which we have used for our temperature fits has been devised by Lindroth [43] within the framework of relativistic many-body perturbation theory. The temperature variations due to different magnetic field settings in the electron cooler turned out to be less than 50%. In the right panel of Fig. 4 we have plotted the scaled excess rate $\Delta\alpha\sqrt{kT_{\perp}kT_{||}}$ as a function of B_z. In this representation temperature effects are removed [42] and the influence of the magnetic field strength on the excess rate can be seen directly. The scaled excess rate increases by a factor of 1.8 when increasing B_z from 20 to 70 mT. This establishes a distinct effect of the magnetic guiding field strength on the yet unexplained recombination rate enhancement at very low energies.

CONCLUSIONS

During our experimental investigations of electron-ion recombination in external fields we have found hitherto unmeasured magnetic field dependencies. The $\vec{B}$ field dependence of DR via high Rydberg states in $\vec{E} \times \vec{B}$ field configurations bears implications on the diagnostics of astrophysical and fusion plasmas where electromagnetic fields are ubiquitous. We hope that the novel finding of a magnetic field dependence of the recombination rate enhancement at low energies helps to pave the road towards a theoretical explanation of this effect.

ACKNOWLEDGEMENTS

We would like to thank M. Beutelpacher, S. Böhm, C. Brandau, H. Danared, G. H. Dunn, N. Eklöw, P. Glans, M. Grieser, R. A. Phaneuf, G. Saathoff, A. A. Saghiri, D. W. Savin, R. Schuch, and G. Wissler for actively having taken part in the experiments and for valuable discussions. We also thank E. Lindroth for making available her theoretical results prior to publication. This work has been supported by BMBF, Bonn, through contracts no. 06 GI 848 and no. 06 HD 854.

REFERENCES

1. Schuch, R., *Reviews of Fundamental Processes and Applications of Atoms and Ions*, Singapore: World Scientific, 1993.
2. Müller, A. and Wolf, A., *Accelerator Based Atomic Physics Techniques and Applications*, Woodbury, New York: AIP Press, 1997, p. 147.
3. Brandau, C. *et al*, *GSI Scientific Report* 1997, (available on request), p. 104.
4. Müller, A. and Wolf, A., *Hyperfine Interact.* **109**, 233 (1997).
5. Müller, A., Bartsch, T., Brandau, C., Gwinner, G., Hoffknecht A., Kozhuharov, C., Saghiri, A. A., Schippers, S., Schmitt, M., and Wolf, A., *Atomic Processes in Plasmas, AIP Conference Proceedings Vol. 443*, New York: American Institute of Physics, 1998, p. 241.
6. Schuch, R., DeWitt, D. R., Gao, H., Mannervik, S., Zong, W., and Badnell, N. R., *Phys. Scr.* **T73**, 114 (1997).
7. Schuch R., Zong, W., Lindroth, E., DeWitt, D. R., Gao, H., Spies, W., and Danared, H., *Photonic, Electronic and Atomic Collisions*, Singapore: World Scientific, 1998, p. 323.
8. Spies, W., Müller, A., Linkemann, J., Frank, A., Wagner, M., Kozhuharov, C., Franzke, B., Beckert, K., Bosch, F., Eickhoff, H., Jung, M., Klepper, O., König, W., Mokler, P. H., Moshammer, R., Nolden, F., Schaaf, U., Spädtke, P., Steck, M., Zimmerer, P., Grün, N., Scheid, W., Pindzola, M. S., and Badnell, N. R., *Phys. Rev. Lett.* **69**, 2768 (1992).
9. Brandau, C., Bosch, F., Franzke, B., Groening, L., Hoffknecht, A., Knopp, H., Kozhuharov, C., Müller, A., Stachura, Z., Steck, M., Stöhlker, T., and Winkler, T., *Hyperfine Interact.* **114**, 45 (1998).
10. Gao, H., DeWitt, D. R., Schuch, R., Zong, W., Asp., S., and Pajek, M., *Phys. Rev. Lett.* **75**, 4381 (1995).
11. Uwira, O., Müller, A., Spies, W., Linkemann, J., Frank, A., Cramer, T., Empacher, L., Becker, R., Kleinod, M., Mokler, P. H., Kenntner, J., Wolf, A., Schramm, U., Schüssler, T., Schwalm, D., and Habs, D., *Hyperfine Interact.* **99**, 295 (1996).
12. Schippers, S., Bartsch, T., Brandau, C., Linkemann, J., Gwinner, G., Müller, A., Saghiri, A. A., and Wolf, A., *J. Phys. B* **31**, 4873 (1998).
13. Schippers, S., Bartsch, T., Brandau, C., Linkemann, J., Müller, A., Saghiri, A. A., and Wolf, A., *Phys. Rev. A* **59**, 3092 (1999).
14. Bartsch, T., Müller, A., Spies, W., Linkemann, J., Danared, H., DeWitt, D. R., Gao, H., Zong, W., Schuch, R., Wolf, A., Dunn, G. H., Pindzola, M. S., and Griffin, D. C., *Phys. Rev. Lett.* **79**, 2233 (1997).
15. Bartsch T., Schippers, S., Müller, A., Brandau, C., Gwinner, G., Saghiri, A. A., Beutelspacher, M., Grieser, M., Schwalm D., Wolf, A., Danared, H., and Dunn, G. H., *Phys. Rev. Lett.* **82**, 3779 (1999).
16. Müller, A., *Atomic and Plasma-Material Interaction Data for Fusion, Suppl. to Nucl. Fusion, Vol. 6*, Vienna: IAEA, 1995), pp. 59–100.
17. Savin, D. W., Bartsch, T., Chen, M. H., Kahn, S. M., Liedahl, D. A., Linkemann, J., Müller, A., Schippers, S., Schmitt, M., Schwalm, D., and Wolf, A., *Astrophys. J. (Lett.)* **489**, L115 (1997).

18. Savin, D. W., Kahn, S. M., Linkemann, J., Saghiri, A. A., Schmitt, M., Grieser, M., Repnow, R., Schwalm, D., and Wolf, A., Bartsch, T., Brandau, C., Hoffknecht, A., Müller, A., Schippers, S., Chen, M. H., and Badnell N. R. *Astrophys. J. (Suppl.)* , (1999).
19. Schippers, S., *Phys. Scr.* **T80** 158 (1999).
20. Müller, A., *Phil. Trans. R. Soc. Lond. A* **357**, 1279 (1999).
21. Jacobs, V. L., Davies,J., and Kepple, P. C., *Phys. Rev. Lett.* **37**, 1390 (1976).
22. Dittner, P. F. and Datz, S., *Recombination of Atomic Ions, NATO ASI Series, Vol. 296*, New York: Plenum Press, 1992, p. 133.
23. Andersen, L. H., *Recombination of Atomic Ions, NATO ASI Series, Vol. 296*, New York: Plenum Press, 1992, p. 143.
24. Müller, A., Belić, D. S., DePaola, B. D., Djurić, N., Dunn, G. H., Mueller, D. W., and Timmer, C., Phys. Rev. Lett. **56** 127 (1986); *Phys. Rev. A* **36**, 599 (1987).
25. Robicheaux, F. and Pindzola, M. S., *Phys. Rev. Lett.* **79**, 2237 (1997).
26. Griffin, D. C., Robicheaux, F., and Pindzola, M. S., *Phys. Rev. A* **57**, 2798 (1998).
27. Robicheaux, F., Pindzola, M. S., and Griffin, D. C., *Phys. Rev. Lett.* **80**, 1402 (1998).
28. Bartsch T., et al., to be published.
29. Schippers S., Bartsch, T., Brandau, C., Müller, A., Gwinner, G., Wissler, G., Beutelspacher, M., Grieser, M., Wolf, A., and Phaneuf, R. A., to be published.
30. Klimenko, V., Ko, L., and Gallagher, T. F., to be published.
31. Griffin, D. C., and Pindzola, M. S., *Phys. Rev. A* **35**, 2821 (1987).
32. Hoffknecht, A., Uwira, O., Schennach, S., Frank, A., Haselbauer, J., Spies, W., Angert, N., Mokler, P. H., Becker, R., Kleinod, M., Schippers S., and Müller, A., *J. Phys. B* **31**, 2415 (1998).
33. Müller, A., and Wolf, A., *Hyperf. Interact.* **109**, 233 (1997) and references therein.
34. Mitnik, D. M., Pindzola, M. S., Robicheaux, F., Badnell, N. R., Uwira, O., Müller, A., Frank, A., Linkemann, J., Spies, W., Angert, N., Mokler, P. H., Becker, R., Kleinod, M., Ricz, S., and Empacher, L., *Phys. Rev. A* **57**, 4365 (1998).
35. Uwira, O., Müller, A., Spies, W., Frank, A., Linkemann, J., Brandau, C., Cramer, T., Kozhuharov, C., Klabunde, J., Angert, N., Mokler, P. H., Becker, R., Kleinod, M., and Badnell, N. R., *Hyperfine Interact.* **108**, 167 (1997).
36. Gao, H., Schuch, R., Zomg, W., Justiniano, E., DeWitt, D. R., Lebius, H., and Spies, W., *J. Phys. B* **30**, L499 (1997).
37. Hoffknecht, A., Bartsch, T., Schippers, S., Müller, A., Eklöw, N., Glans, P., Beutelspacher, M., Grieser, M., Gwinner, G., Saghiri, A. A., and Wolf, A., *Phys. Scr.* **T80**, 298 (1999).
38. Hoffknecht, A., Bartsch, T., Schippers, S., Müller, A., Gwinner, G., Saghiri, A. A., Linkemann, J., and Wolf, A., to be published.
39. Pajek, M., and Schuch, R., *Hyperf. Interact.* **108**, 185 (1997).
40. Pajek, M., and Schuch, R., *Phys. Rev. A* **45**, 7894 (1992).
41. Glans, P., Lindroth, E., Eklöw, N., Zong, W., Gwinner, G., Saghiri, A. A., Pajek, M., Danared, H., and Schuch, R., *Nucl. Instrum. Meth. Phys. Res. B* **154**, 97 (1999).
42. Gwinner, G., et al., to be published.
43. Lindroth, E., et al., to be published.

Positronium formation in positron-atom collisions

J.W. Humberston

Department of Physics and Astronomy
University College London, Gower Street, London WC1E 6BT, U.K

Abstract. Positronium formation has been the subject of extensive theoretical investigations, especially since it became possible to compare experimental results with theoretical predictions. This article describes some recent theoretical studies, mainly in the vicinity of the positronium formation threshold where accurate results for positron scattering by atomic hydrogen and helium reveal interesting structure in the elastic scattering and positronium formation cross sections.

INTRODUCTION

Of all the possible processes in positron-atom collisions, positronium formation is of particular interest because it has no counterpart in electron-atom collisions. The threshold energy for this rearrangement process, in which an electron in the target system becomes attached to the positron projectile, is $E_{Ps} = E_i$ - 6.8 eV, where E_i is the ionization energy of the target atom. Thus, for atoms with ionization energies < 6.8 eV, such as the alkali atoms, the positronium formation channel is open even at zero incident positron energy, when the positronium formation cross section is infinite. For most target systems, however, the threshold is at a positive energy which, for atoms but not for molecules, is lower than that for any other inelastic process. Accordingly, there is an energy range between the positronium formation threshold and the the lowest positron impact excitation threshold of the atom, known as the Ore gap, in which elastic scattering and positronium formation are the only two possible processes (apart from positron-electron annihilation with its very much smaller cross section). It is in the Ore gap that the most detailed and accurate theoretical investigations of positronium formation have been made.

For most atomic targets the positronium formation cross section rises quite steeply with increasing incident positron energy throughout the Ore gap; indeed, as we shall consider in more detail later, this cross section has an infinite slope at the threshold itself. The cross section continues to rise as the energy is increased further, but with a progressively smaller slope, up to a maximum at an energy which is approximately three times that of the positronium formation threshold.

CP500, *The Physics of Electronic and Atomic Collisions*, edited by Y. Itikawa, et al.
© 2000 American Institute of Physics 1-56396-777-4/00/$17.00

Thereafter, the cross section falls away moderately quickly so that it has already become a small component of the total scattering cross section when the incident positron energy is a few times the ionization energy of the target.

In this progress report we concentrate on theoretical studies of positronium formation in low energy positron scattering by atomic hydrogen and helium, for which the Ore gaps are (6.80 - 10.2) eV and (17.78 - 20.6) eV respectively. These systems are sufficiently simple that very accurate results can be obtained, enabling detailed comparisons to be made with increasingly accurate experimental results. Such theoretical results can also test the range of validity of general theoretical concepts and reveal structures which, although not yet amenable to experimental investigation, may become so in the not too distant future.

One of the most widely used methods of approximation for determining positronium formation cross sections is the coupled-state method in which the wave function of the system is expanded in terms of several states, and possibly pseudostates, of the target and of positronium. In principle this expansion is doubly complete, but convergence of the results with respect to increasing the number of terms is slow unless several states of both the target and positronium are included. The method has been used throughout the energy range in which positronium formation contributes significantly to the total scattering cross section, most notably by Hewitt *et al.* [1], Higgins and Burke [2], Kernoghan *et al.* [3], Campbell *et al.* [4], Mitroy and Ratnavelu [5] ,Kuang and Gien [6], and Chaudhuri and Adhikari [7]. The results obtained from the best of these calculations are in very satisfactory agreement with the most accurate experimental measurements.

An alternative method of approximation, which has been used extensively by the author and his co-workers to produce very accurate results at low energies, is the Kohn variational method, in which a trial wave function is constructed consisting of the sum of many algebraic functions multiplied by associated linear variational parameters to represent the various interparticle correlations of the system. This function is then substituted into the stationary Kohn functional for the K-matrix elements,

$$K_{pq}^v = K_{pq}^t - \langle \Psi_p | L | \Psi_q \rangle \quad (p, q = 1, 2), \tag{1}$$

where $L = 2(H - E)$. The requirement that the functional be stationary with respect to variations in all the linear parameters provides a set of linear simultaneous equations, the solution to which is the set of optimum values of the linear variational parameters. Substituting these values back into the Kohn functional yields the variational values of the K-matrix elements, the errors in which are of second order in the errors in the wave function. Although not providing a rigorous bound on the scattering parameters, the values of the diagonal K-matrix elements obtained using the Kohn variational method usually become more positive as the number of variational parameters in the trial wave function is increased, and therefore the results obtained are usually lower bounds on the exact values.

POSITRON-HYDROGEN SCATTERING

Within the Ore gap the two-component wave function used in the variational calculations of Humberston *et al.* [8] is of the form, for s-wave scattering,

$$\Psi_1 = Y_{0,0}(\hat{\boldsymbol{r}}_1)\phi_H\sqrt{k}\left\{j_0(kr_1) - K_{11}^t n_0(kr_1)\left[1 - \exp(-\lambda r_1)\right]\right\}$$

$$- Y_{0,0}(\hat{\boldsymbol{\rho}})\phi_{Ps}\sqrt{2\kappa}K_{21}^t\{n_0(\kappa\rho) + e^{-\mu\rho}(1 + a\rho + b\rho^2)/\kappa\rho\} + \sum_{i=1}^{N} c_i\phi_i, \tag{2}$$

$$\Psi_2 = Y_{0,0}(\hat{\boldsymbol{\rho}})\phi_{Ps}\sqrt{2\kappa}\left\{j_0(\kappa\rho) - K_{22}^t\left[n_0(\kappa\rho) + \exp(-\mu\rho)(1 + a\rho + b\rho^2)/\kappa\rho\right]\right\}$$

$$- Y_{0,0}(\hat{\boldsymbol{r}}_1)\phi_H\sqrt{k}K_{12}^t n_0(kr_1)\left[1 - \exp(-\lambda r_1)\right] + \sum_{j=1}^{N} c_j\phi_j, \tag{3}$$

where $\phi_i = \exp\left[-(\alpha r_1 + \beta r_2 + \gamma r_{12})\right] r_1^{k_i} r_2^{l_i} r_{12}^{m_i}$, c_i are variational parameters, $\boldsymbol{r}_1$ and $\boldsymbol{r}_2$ are the position vectors of the positron and the electron respectively, and $r_{12} = |\boldsymbol{r}_1 - \boldsymbol{r}_2|$. All terms with $k_i + l_i + m_i \leq \omega$, where k_i, l_i, m_i and ω are non-negative integers, are included in the summation, so that increasing the value of ω provides a systematic means of improving the wave function. The convergence of the results with respect to ω can then be investigated, and very accurate estimates of the exact results can sometimes be obtained. For example, the pattern of convergence for K_{11}^v is found empirically to be well represented by

$$K_{11}^v(\omega) = K_{11}^v(\infty) + \frac{p}{\omega^n}. \tag{4}$$

Extrapolation to infinite ω then yields what is presumably a very accurate approximation to the exact result.

In terms of the $\boldsymbol{K}$-matrix elements, the cross sections for the various processes are

$$\sigma_{pq} = \frac{4\pi(2l+1)}{k_p^2}\left|\left(\frac{\boldsymbol{K}}{1 - i\boldsymbol{K}}\right)_{pq}\right|^2 \qquad (p, q = 1, 2), \tag{5}$$

where σ_{11} and σ_{12} are the elastic scattering and positronium formation cross sections, σ_{el} and σ_{Ps} respectively; k_1 and k_2 are the wave numbers of the positron and the positronium, k and κ respectively, which are related by energy conservation;

$$E = \frac{k^2}{2} - \frac{1}{2} = \frac{\kappa^2}{4} - \frac{1}{4}. \tag{6}$$

The s-wave positronium formation cross section obtained in this way is plotted in figure 1. Very similar results have also been obtained using several states of the target and of positronium in the coupled-state method [5,6].

Expressed as a function of the incident positron energy, the s-wave positronium formation cross section exhibits an infinite slope at the positronium formation threshold. This is in accord with the prediction of Wigner's [9] threshold law;

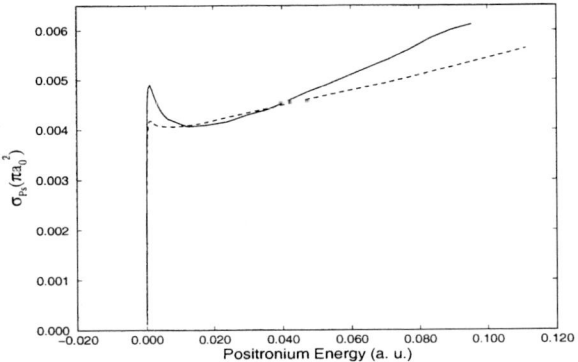

FIGURE 1. The variation of $\sigma_{Ps}(l = 0)$ with positron energy for positron collisions with hydrogen and helium. Full line; helium: dashed line; hydrogen.

$$K_{12}(l) \propto \kappa^{(l+1/2)}; \qquad \sigma_{Ps}(l) \propto E_1^{(l+1/2)}, \qquad (7)$$

where E_1 is the kinetic energy of the positronium. However, there is no prediction about the energy range over which the law is valid, but we see in figure 2, where $K_{12}(l = 0)$ is plotted as a function of $\kappa^{1/2}$ that the range of validity is very narrow. According to Ward *et al.* [10], the smallness of $\sigma_{Ps}(l = 0)$ relative to $\sigma_{el}(l = 0)$, and also to $\sigma_{Ps}(l > 0)$, can be explained using hidden crossing theory as arising from the almost complete destructive interference of two terms in the s-wave amplitude for positronium formation.

The opening of the positronium formation channel produces structure in the elastic cross section, both above and below the threshold energy. From the unitarity

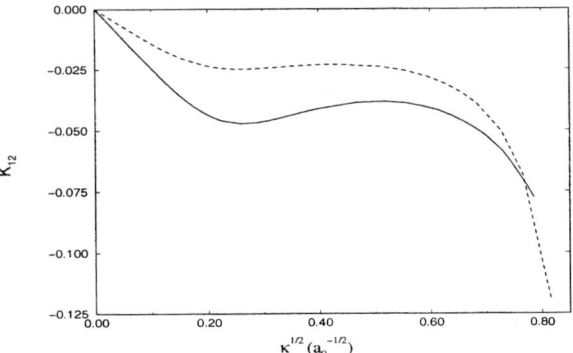

FIGURE 2. The variation of the $K_{12}(l = 0)$ matrix element with $\kappa^{1/2}$. Full line; helium: dashed line; hydrogen.

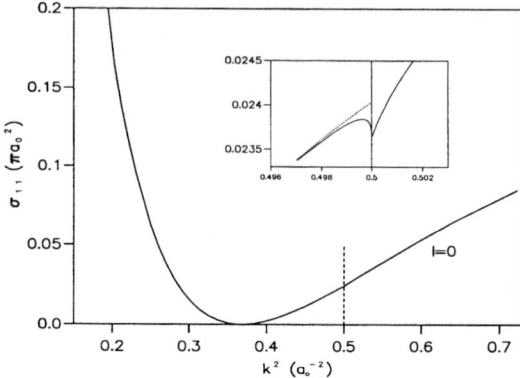

FIGURE 3. The variation of $\sigma_{el}(l=0)$ with positron energy for positron-hydrogen scattering

of the **S**-matrix it follows that the elastic scattering cross section in the vicinity of the threshold can be expressed as

$$\sigma_{el}(l=0) = \frac{4\pi}{k^2} \sin^2 \eta_0 - \sigma_{Ps}(l=0) \begin{cases} 2\sin^2 \eta_0 & E \geq E_{Ps} \\ \sin 2\eta_0 & E < E_{Ps}. \end{cases} \tag{8}$$

Thus the infinite slope in $\sigma_{Ps}(l=0)$ at the threshold creates a cusp or rounded step (depending on the sign of the elastic scattering phase shift, η_0) in $\sigma_{el}(l=0)$. In this particular case the phase shift is negative and therefore $\sigma_{el}(l=0)$ exhibits

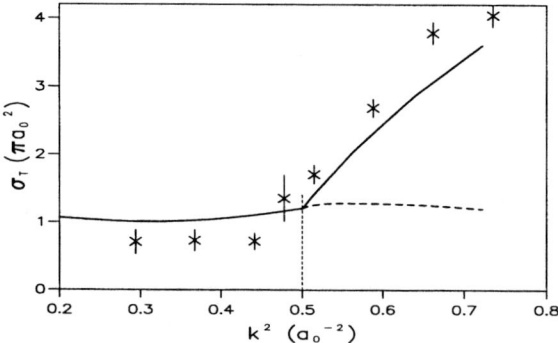

FIGURE 4. Comparison of the present theoretical total cross sections (the solid line) with the total cross section measurements of Zhou *et al* [11] (✳). The broken line is the theoretical elastic scattering cross section above the threshold. The dotted line is at the positronium formation threshold.

a rounded step, although the fractional size of the feature is very small because $\sigma_{Ps}(l = 0)$ is so small, as can be seen in figure 3. Higher partial wave contributions to σ_{el} continue smoothly through the threshold with no discontinuity in slope [8]. The total elastic and positronium formation cross sections are displayed in figure 4, together with the most recent experimental results of Zhou *et al.* [11]. The theoretical results have almost certainly converged to within a few per cent of the exact results, whereas the experimental measurements suffer from substantial statistical and systematic errors, arising mainly from the low density of atomic hydrogen in the target gas cell.

POSITRON-HELIUM SCATTERING

The only very detailed theoretical studies of positronium formation in positron-helium scattering in the Ore gap are those of Van Reeth and Humberston [12–14]. These authors used a similar technique to that described above for hydrogen, with trial wave functions containing many Hylleraas-type correlation functions involving all six interparticle coordinates, and a very accurate correlated Hylleraas wave function for the target helium atom. Although this system is much more complicated than the positron-hydrogen system, and the computational effort required is much greater, a study of the convergence with respect to systematic improvements in the trial wave functions, similar to that carried out for hydrogen, leads us to believe that the results presented here are accurate to within approximately 10% and that any structural features in the cross sections are genuine.

A remarkable feature of the results for the two systems is the very close similarity between the s-wave contributions to σ_{Ps}, when expressed as functions of the energy of the positronium, as shown in figure 1. Each displays a very steep rise from the threshold, followed by an abrupt change of slope to a gently rising plateau. Van Reeth and Humberston [13] have speculated that the similarities arise because, although the two atoms are very different, the residual ions, the proton and the compact He^+ ion, appear to the relatively diffuse positronium in the final state as being similar. If this hypothesis is valid we might expect all atoms with tightly bound singly-charged positive ions, and particularly the noble gases, to have rather similar values of $\sigma_{Ps}(l = 0)$ in the Ore gap. There are no such similarities in the higher partial wave contributions to σ_{Ps}, which are much larger for hydrogen than for helium.

Another interesting feature of the results for both systems is that an extrapolation of $\sigma_{el}(l = 0)$ from some way below the threshold, where the influence of the threshold is slight, into the Ore gap is very similar to $\sigma_{total}(l = 0)$ above the threshold, i.e. $\sigma_{el}(l = 0) + \sigma_{Ps}(l = 0)$, suggesting that the magnitude of $\sigma_{Ps}(l = 0)$ is related to the amount by which $\sigma_{el}(l = 0)$ is reduced as a consequence of the threshold. This feature is displayed in figure 5 for the case of helium.

Individual partial wave contributions to σ_{Ps}, and their sum, for positron-helium scattering in the Ore gap are shown in figure 6. The partial wave contributions

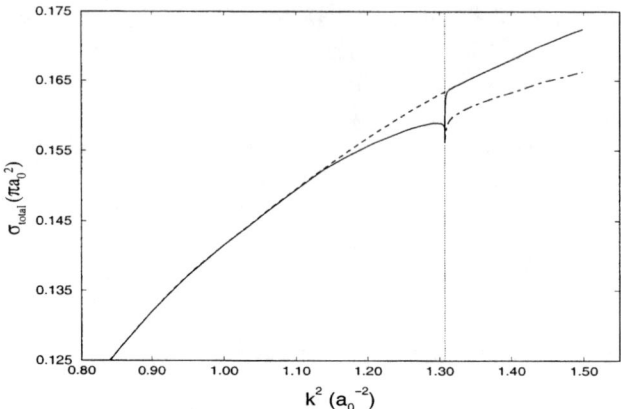

FIGURE 5. The total cross section for s-wave positron- helium scattering. The dashed line is an extrapolation of $\sigma_{el}(l = 0)$ from some way below the threshold, and the dashed-dotted line is the variational result for $\sigma_{el}(l = 0)$ above the threshold.

for $l \leq 3$ have been calculated as described above, and the contributions from all higher partial waves have been obtained using the Born approximation. A log-log plot of the dependence of the total positronium formation cross section on the energy of the positronium yields a reasonably good straight line whose slope is approximately 1.5. When plotted in a similar manner, the experimental data of Moxom *et al.* [16] for the positronium formation cross section in helium also lie close to a straight line with slope ≈ 1.5, a fact which Moxom *et al.* [16] interpreted, using equation 7, as evidence that the dominant contribution to σ_{Ps} comes from the p-wave. However, this conclusion is incorrect, as can be seen from the theoretical results of figure 6, where the d-wave contribution to σ_{Ps} is seen to be the largest

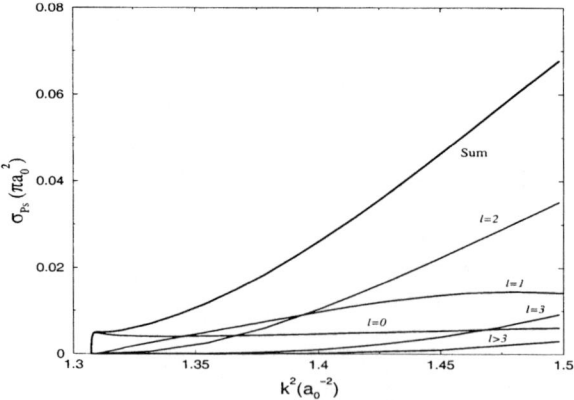

FIGURE 6. The variation of σ_{Ps} with positron energy for positron-helium scattering. The $l = 0, 1, 2$ and 3 contributions were obtained using the Kohn variational method. For $l > 3$ see text.

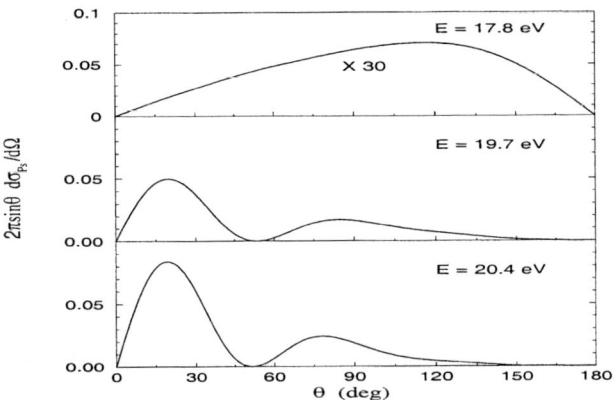

FIGURE 7. The angular distributions for positronium formation in positron-helium scattering at three energies in the Ore gap.

throughout most of the Ore gap, although not overwhelmingly so. Nevertheless, the sum of all the contributions has an energy dependence which approximates quite well to $E_1^{1.5}$ [14]. The discrepancy between the theoretical and experimental positronium formation cross sections for helium amounts to approximately 25%, with the experimental results being larger. However, the theoretical results may only be accurate to within 10%, and the pattern of convergence suggests that they are likely to be too low rather than too high. Also, the experimental results are believed to be accurate to within only 20%. The measure of agreement between the theoretical and experimental results is therefore not unreasonable.

The theoretical differential positronium formation cross sections at three energies in the Ore gap, expressed as the angular distribution $2\pi (d\sigma_{Ps}/d\Omega) \sin \theta$, are plotted in figure 7. Just above the threshold, where the s-wave contribution is dominant, the differential cross section is approximately isotropic, but the angular distribution rapidly becomes quite strongly peaked around the forward direction as the energy is increased, a feature which has been exploited in the production of experimental positronium beams (Garner *et al.* [17]).

It is easier to make accurate measurements of the total scattering cross section than of individual partial cross sections, and much better overall agreement is obtained between the theoretical and experimental values of the total positron-helium scattering cross sections, as can be seen in figure 8. The total cross section is, of course, just the elastic scattering cross section below the positronium formation threshold and the sum of the elastic scattering and positronium formation cross sections above the threshold. The step-like rise in the positronium formation cross section at the threshold which, as we have seen, arises from the s-wave contribution, manifests itself as an abrupt 2% increase in the total cross section. Such a feature is too small and too narrow to be resolved experimentally using existing conventional positron beams because their energy spreads are too wide, typically (0.2 - 0.8)

417

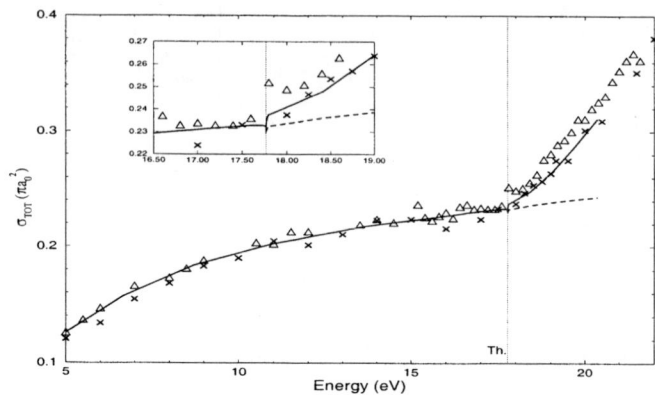

FIGURE 8. Comparison of theoretical and experimental values of the total positron-helium scattering cross section. The experimental data are those of Mizogawa *et al.* (1985) ($\triangle$) and Stein *et al* (1978) ($\times$). The dashed line is the theoretical elastic scattering cross section above the positronium formation threshold.

eV. Convoluting the theoretical total cross section with an energy spread function for the beam of this magnitude smears out the step-like feature to such an extent that it becomes invisible [14]. However, recent developments using cooled trapped positrons have enabled positron beams to be produced with energy spreads as narrow as 0.02 eV [18], and it might be possible to observe the threshold feature with such a beam. If so, the very sharp rise in the total cross section within an energy interval of less than 0.01 eV would provide a precise energy calibration point for the positron beam at 17.78 eV.

Further investigations of these and other systems are in progress, and it is to be hoped that improvements in the energy resolution of positron beams will enable detailed comparisons to be made between theory and experiment.

ACKNOWLEDGEMENTS

I am particularly grateful to Dr P. Van Reeth, with whom I have collaborated closely in these investigations. I also wish to thank Dr M. Charlton, Dr G. Laricchia and Prof. C. Surko for several useful discussions. This work was supported by the United Kingdom EPSRC on grant number GR/L3843.

REFERENCES

1. Hewitt R.N., Noble C.J., and Bransden B.H., J. Phys. B **25**, 557 (1992)
2. Higgins K., and Burke, P.G., J. Phys. B **26**, 4269 (1993)
3. Kernoghan, A.A., Robinson, D.J.R., McAlinden, M.T., and Walters, R.J., J. Phys. B **29**, 2089 (1996)

4. Campbell, C.P., McAlinden, M.T., Kernoghan, A.A., and Walters, R.J., Nucl. Instrum. Methods B **143**, 41 (1998)

5. Mitroy, J., and Ratnavelu, K., J. Phys. B **28**, 287 (1995)

6. Kuang, Y.R., and Gien, T.T., Phys. Rev. A **55**, 256 (1997)

7. Chaudhuri, P., and Adhikari, S. K., J. Phys. B **31**, 3057 (1998)

8. Humberston, J.W., Van Reeth, P., Watts, M.S.T., and Meyerhof, W.E., J. Phys. B **30**, 2477 (1997)

9. Wigner, E.P., Phys. Rev. **73**, 1002 (1948)

10. Ward, S.J., Macek, J.H., and S.Yu. Ovchinnikov, Phys. Rev. **59**, 4418 (1998)

11. Zhou, S., Li, H., Kauppila, W.E., Kwan, C.K., and Stein, T.S., Phys. Rev. A **55**, 361 (1997)

12. Van Reeth, P., and Humberston, J.W., J. Phys. B **30**, L95 (1997)

13. Van Reeth, P., and Humberston, J.W., J. Phys. B **31**, L621 (1998)

14. Van Reeth, P., and Humberston, J.W., J. Phys. B **32**, 3651 (1999)

15. Van Reeth, P., and Humberston, J.W., J. Phys. B **32**, L103 (1999)

16. Moxom, J., Laricchia, G., Charlton, M., Kover, A., and Meyerhof, W.E., Phys. Rev. A **50**, 3129 (1994)

17. Garner, A.J., Laricchia, G., and Ozen, A., J. Phys B **29**, 5961 (1996)

18. Kurz, C., Gilbert, S.J., Greaves, R.G., and Surko, C.M., Nucl. Instrum. Methods B **143**, 188 (1998)

Atomic Collisions Involving Pulsed Positrons

J. P. Merrison[1], H. Bluhme[1], D. Field[2], H. Knudsen[1], S. Lunt[2], K. A. Nielsen[1], S. Stahl[2] and E. Uggerhøj[1]

[1]Institute of Physics and Astronomy, University of Aarhus, Aarhus 8000C, Denmark
[2]Institute for Storage Ring Facilities, University of Aarhus, Aarhus 8000C, Denmark

Abstract. Conventional slow positron beams have been widely and profitably used to study atomic collisions and have been instrumental in understanding the dynamics of ionisation. The next generation of positron atomic collision studies are possible with the use of charged particle traps. Not only can large instantaneous intensities be achieved with in-beam accumulation, but more importantly many orders of magnitude improvement in energy and spatial resolution can be achieved using positron cooling. Atomic collisions can be studied on a new energy scale with unprecedented precion and control. The use of accelerators for producing intense positron pulses will be discussed in the context of atomic physics experiments.

INTRODUCTION

Atomic physics has benefited greatly from collision studies using positrons, they are light (like electrons) but positive (like protons), mass and charge effects can in this way be studied independently (1). As an antiparticle they may be used to probe symmetry and (especially in atomic systems) to test quantum mechanical predictions, specifically for example by spectroscopic studies of positronium and other positronic systems. Positrons are dynamically similar to electrons, but lack exchange. The study of exotic positronic systems, such as positron-atom bound systems, can provide a unique tool for investigating many body physics. Such compounds are currently a hot topic in the positron community (2). Experiments on such systems require large improvements in beam quality and thus more sophisticated positron technology. These improvements in technology will be discussed in the context of the new physics which can be studied.

In this report three stages of slow positron beam development will be charted. Beginning with the use of conventional slow positron beams, experimental details will be presented and results discussed. The benefits and limitations of such beams will also be outlined. The application of electron accelerators for slow positron production will be discribed and how this development has affected the scope of positron atomic physics. Specifically the Aarhus slow positron facility will be described. Finally positron trapping will be discussed and how this is shaping the future of positron atomic physics.

CONVENTIONAL RADIOACTIVE SOURCE BASED BEAMS

Conventional source based beams have been usefully utilised in studies of atomic collisions, specifically ionisation. A large body of data involving single ionisation of atoms and simple molecules has been obtained such that a consistent picture has emerged. The low intensity and poor resolution of such beams, however, prohibits the more ambitious fundamental studies which could be performed using slow positrons. The precision of modern positron atomic collision studies are also severely limited with this conventional technology.

A so called conventional slow positron beam is based on a radioactive β^+ emitter, these high energy positrons are then slowed down in a moderating material. Positrons entering a well prepared metal (usually tungsten) foil are stopped and diffuse thermally until a small fraction reach the surface. Due to the negative positron work function of this material they are emitted with a few eV kinetic energy. Under normal ultra high vacuum conditions (10^{-9} Torr) they are scattered at the metal surface and acquire at least 1eV in perpendicular energy spread. Typical moderating efficiencies (slow positron per β^+ produced) are of the order 0.01%, giving typical beam intensities of 100,000 slow positrons per second or less. More efficient slow positron moderators exist for example rare gas solids, though these are inconvenient (requiring cryogenics) and have reduced energy resolution. Figure 1 shows a typical conventional electrostatic slow positron beam, utilising electrostatic lenses to transport the beam and incorporating a cylindrical mirror analyser to remove fast particles and gamma rays. Other slow positron beams utilise axial magnetic fields for radial confinement during slow positron transport. One of the most sophisticated of conventional slow positron beams was contructed at Aarhus in order to perform a collision experiment between a proton beam and a positronium target.

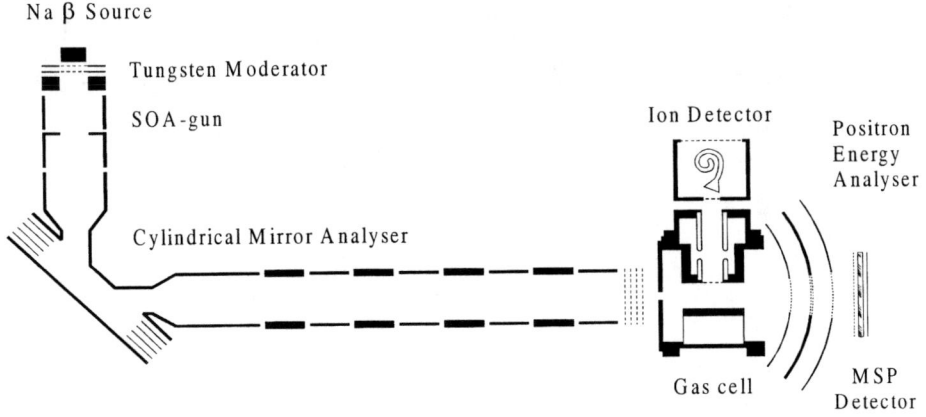

Figure 1 A conventional radioactive source based slow positron beam, utilising a metal moderator and electrostatic transport (6).

It utilised several new technologies at that time, rare gas solid moderation, beam bunching, magnetic velocity analysers and also used both magnetic and electrostatic transport. This experiment successfully observed ionisation of positronium by proton impact (3).

A typical atomic collision experiment involves passing the positron beam through a gas target and detecting the positrons and/or any ions produced. Early slow positron studies measured , for example, beam attenuation and obtained total elastic and inelastic cross sections (4). Of particular success for beam of this kind have been the large number of studies, by various groups, of ionisation of atoms and simple molecules following positron impact. The beam depicted in figure 1 detects only the ions produced in coincidence with pulsing of the slow positron beam, all ionisation channels are thus included in this detection scheme, specifically positronium formation. Other studies detecting the emitted projectile positron clearly only measure direct ionisation.

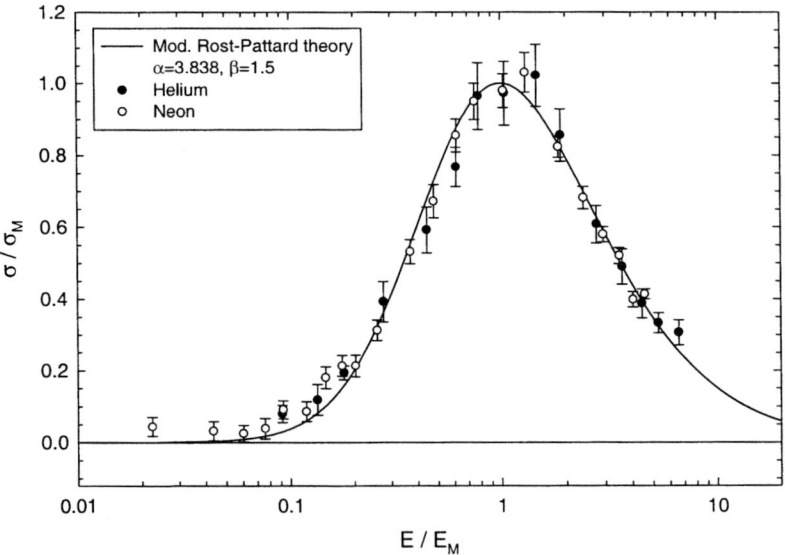

Figure 2 Positron induced double ionisation of He and Ne plotted on scaled axis and showing a fitted curve based on a modified Rost-Pattard parameterisation (5,6).

A consistent picture of single ionisation, also including positronium formation, has emerged. For direct single ionisation a parameterisation developed by Rost and Pattard has been found to give good agreement also for proton, antiproton and electron impact

(5). This model merely combines the expression for the threshold behavior with the well accepted high energy Born approximation. Since this high energy dependence is charge sign and mass independent, the differing dynamics involved in ionisation are governed by the threshold dependence. Extension of these studies has been made to double ionisation (6,7). A modification of this model has allowed positron impact double ionisation to be successfully incorporated into this scheme (see figure 2). Positron induced (total) single ionisation is seen to be dominated at low energy by positronium formation. Interestingly investigations of double ionisation have failed to observe the expected positronium channel. A suppression of positronium formation together with double ionisation is observed over all energies for the light noble gases (He and Ne). The heavier noble gases Ar, Kr and Xe share this supression in the near threshold region, but some transfer ionisation is observed at intermediate energies (around 100eV). These phenomena have yet to be explained. It is as a result of the suppression of transfer ionisation that our results may be used in comparison with a modified Rost-Pattard parameterisation, since this theory is only valid for direct ionisation (see figure 2).

Apart from ionisation, other positron induced atomic collision processes have been poorly studied experimentally. A few energy loss experiments have observed atomic excitation, though not to sufficient precision to challenge theory (8). Multiply differential cross sections (ionisation, excitation etc...) contain much more information and are much more critical tests of theory. Some pioneering work has been performed, specifically measuring ionistion by positron impact where angular information was obtained. These experiments proved to be extremely difficult and limited in resolution and accuracy (9). The conclusion seems to be that such conventional slow positron technology is insufficient in beam intensity and spatial resolution for such studies.

ACCELERATOR BASED BEAMS

An alternative to a conventional source based slow positron beam is the use of an electron accelerator to generate positrons by pair production. There are several advantages, most importantly is that the beam is pulsed and the instantaneous intensity is thus enhanced by many orders of magnitude. There are unfortunately few such facilities in operation and only two perform atomic physics, one of which is based in Aarhus. The high instantaneous beam intensity is ideal for studying short lived positronic systems.

Intense sources of positrons may also be produced using nuclear reactors (neutron sources) or ion accelerators. These produce intense positron emitting isotopes. Since they are not pulsed, the instantaneous intensity is still relativley low and such facilities are not used for atomic physics investigations.

The advent of intense pulsed positron sources has allowed several groups to perform precision studies of the positronium atom which are otherwise made extremely difficult due to its short self annihilation time in the ground state (142ns for the ortho state). Positronium is a simple atom containing two structureless particles and can be treated theoretically to high precision. It therfore makes a good candidate for tests of QED. Despite this fundamental interest only two precise (spectroscopic) series of studies have been performed using positronium; one measuring the 1S-2S transition and another the fine structure splitting (10,11). The production of dense positronium targets also allows collision studies involving this exotic atom. The interaction of positronium with intense photon fields is of particular interest. Being orders of magnitude lighter than other atoms makes it dynamically unique (12).

Intense pulsed positron beams are also well suited to studying other (short lived) exotic positronic systems such as positron-atom bound systems (eg Lie^+). The interest here is to shed light on the many body problem. The first of this type of compound was observed (in vacuum) at Aarhus, but no detailed study was possible in this energy loss type experiment (13). Such compounds are presently a hot topic in the positron atomic physics community not just for studying many body physics, but also to describe the observed high positron annihilation probability on some molecules. If extremely high positron densities can be produced, it may be possible (and is being considered) to produce exotic multi-positronic systems such as molecular positronium (Ps_2). The structure and dynamics of such an extremely exotic (light) molecule is of considerable interest.

It is now possible to perform kinematically complete collision experiments with the use of cold target atoms (atomic jets) and performing time of flight measurments and position sensitive detection of the products (projectile and ejected ion/electron). The momentum of all particles before and after the collision may thus be obtained using this Cold Ion Recoil Ion Mass Spectroscopy (COLTRIMS) technique. This constitutes arguably ideal multiply differential ionisation cross section measurments and provide a challenge to theory. Such studies require an intense, pulsed and preferably well resolved beam. It would thus be possible to extend these studies by using slow positrons as projectiles using an accelerator based slow positron beam.

Recently the construction of an intense pulsed slow positron beamline has been completed at Aarhus based on a Microtron electron accelerator (14), this is shown in the schematic (figure 3). As with other accelerator based beams (15) a dense target is used to stop high energy (100MeV) electrons, this generates high energy Bremsstrahlung gamma rays and pair production subsequently occurs also within the target. The high energy positrons produced can then be moderated to low energy using moderating material as before, in this case an array of tungsten mesh is used. In order to improve the beam quality a single crystal tungsten remoderator is used which reflects the accelerated positron beam into a subsequent magnetic transport system. The positrons are guided into an adjacent laboratory around 10m away in order to be free from radiation.

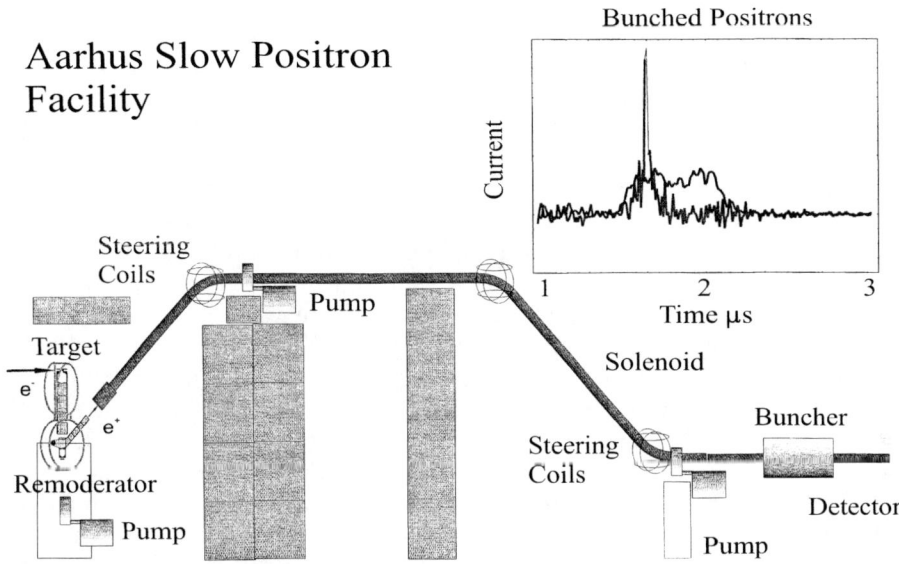

Aarhus Slow Positron Facility

Figure 3 The Aarhus accelerator based slow positron facility. The 1μs pulse length is compressed using a beam buncher to less than 100ns, as shown in the displayed detector signal (14). Part of the figure reprinted from (14), with permission from Elsevier Science.

The first pulsed positron beam was obtained at this facility in early 1998. Typical pulses contained around 20000 positrons at a frequency of up to 15Hz. The inherant pulse duration of around 1μs generated by the Microtron accelerator is too long for most experiments involving positrons (or positronium). Hence a beam buncher has been employed, this is a 1m long array of accelerating electrodes which compresses these positron pulses to around 20ns. The first collaborative experiment using this facility involved conversion of the positron pulses into positronium which could then be ionised by an intense pulsed IR laser. The process of interest involves multiple photon absorption and possible above threshold ionisation. Here excess photon absorption occurs resulting in energetic electron (and positron) emission. Such processes have been studied using several atomic systems. The dynamic of positronium is however rather different, with the positron ('nucleus') also taking part in coupling to the strong electromagnetic field. Theoretical interest has since been generated and this experimental program continues (12).

POSITRON TRAPPING

Charged Particle traps are proving to be useful tools in many fields of science and technology and no field is benefiting more than that of antiparticle research. The fact that particles can be stored almost indefinitely, cooled to cryogenic temperatures and accumulated in large quantities allows unprecedented precision and control. Penning traps can be used in combination with a conventional slow positron beam to accumulate large numbers (10^9) of positrons (16). Large instantaneous beam intensity can be achieved, superior even to accelerator based devices and without the large scale investment. Once positrons are trapped various cooling techniques may be applied. This allows the positron energy to be reduced to those corresponding to cryogenic temperatures (less than 1meV for a liquid helium cooled device). Once trapped positrons are trapped and cooled they may be centered onto the beam axis by exciting the so called magnetron motion (precession around the trap axis). In principle it is in this way possible to reduce the beam radius to around 20µm without loss of intensity.

All of these beam enhancement techniques may be acheived with the use of an in-beam Penning trap. This constitutes such a huge improvement in beam quality that a new generation of collision experiments becomes possible. Most obviously studies may now be performed at energies much below 1eV which complements studies being performed with electrons and planned experiments using equivelocity antiprotons and protons (17). At this high energy resolution the all important near threshold region in ionisation could be studied in great detail, this is of considerable theoretical interest (18).

For efficient trapping from a source based beam one requires rapid cooling of the axial energy. Conventional (ultra high vacuum) cooling mechanisms used in Penning traps, such as resistive cooling or sychrotron radiation, are typically too slow (10-100ms). Early trapping schemes were therefore inefficient (19). Efficient trapping in a dense buffer gas is successfully being applied as an in-beam cooler and accumulator. The buffer gas technique is inconvenient however, due to the need for high pressure gas and thus has poor vacuum compatability.

As part of a European collaboration (EUROTRAPS) a superconducting Penning trap has been installed at the Aarhus positron facility in order to trap and cool positron bunches. An ultra high resolution extraction beamline is also under construction. The purpose is to cool highly charged ions using cold positrons, such that more precise charge/mass measurements can be performed. As will be discussed, however, the potential benefits of positron trapping and cooling are tremendous and an extended experimental program is planned using this trap.

The bunched positron pulses are accelerated (to around 2keV) before being remoderated using a single crystal annealed tungsten foil (1000Å thick) held at +180V. The trapping is performed using a Penning Malmberg arrangment in a 5.8T magnetic field. With the remoderator at a magnetic field strength of only 0.4T some energy broadening occurs on injection into the trap due to conversion of axial to radial energy components. The axial energy spread becomes around 4eV FWHM. The trap region is of length around 20cm and is floated at around +163V. Trapping is performed by raising a confining endcap potential to +188V after passage of the positron pulse. Another endcap electrode (also at +188V) reflects the positrons at the other end of the trap. Energetically therefore the positrons are injected close to the top of the (25V deep) trap potential. The positrons are stored for various times and extracted by lowering the far encap electrode. Detection of the positrons was performed using a particle detector placed around 1m from the trap. Varying the depth to which the extraction electrode is lowered allows the axial positron energy distribution within the trap to be analysed. The spectrum obtained is commonly termed a retarding field analysis. In our case a fitting function was then applied which could then be differentiated to obtain the true energy distribution.

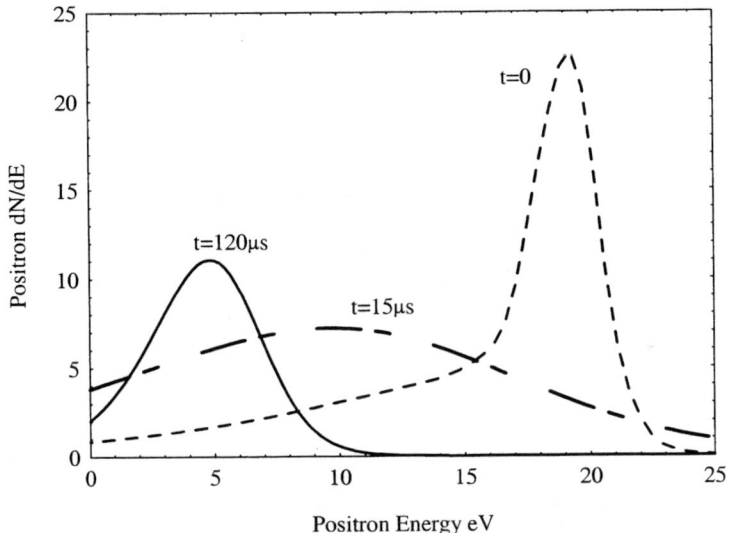

Figure 4 Positron axial energy distribution after different storage times (t) in the Penning trap.

Figure 4 shows three positron energy spectra taken after different storage times, as can be seen the axial energy rapidly (15μs) broadens and at a later time (120μs) becomes narrow again at the bottom of the trap potential. Figure 5 shows the variation in mean axial positron energy, the axial energy spread (FWHM) and the number of stored positrons all as a function of trapping time. One notes a rapid decrease in the mean energy and a rapid rise in the energy spread with a time constant of around 20μs. This can be understood in terms of collisional mixing of the axial and radial degrees of freedom, thus causing an increased energy spread (presumably also in the radial direction), but decreased mean axial energy. Following equilibriation, at longer times, a decrease of both mean positron energy and energy width is observed (around 100µs). This time constant also corresponds also to that of an exponential decay observed in the stored particle number. These observations are consistent with evaporative cooling (ie loss of the most energetic component) thus causing a reduction of the mean positron energy and also the positron energy spread or "temperature" (see figure 4). It should be noted at this point that after storing the particles for around 120µs an increase in the observed number density at low positron energies is observed, this can only occur as a result of a cooling mechanism as opposed to a loss mechanism for higher energy particles for example. This observed rapid energy exchange can be explained by the large radial compression of the beam on entering the high magnetic field of the Penning trap. A crude calculation based on in-beam coulomb scattering predicts an initial energy exchange of 10eV in the first 20µs, in reasonable agreement with our observations. It should be noted that the Debye length of the trapped positrons in our study was around 15cm, this is large compared to our beam size and we are therefore far from the plasma regime.

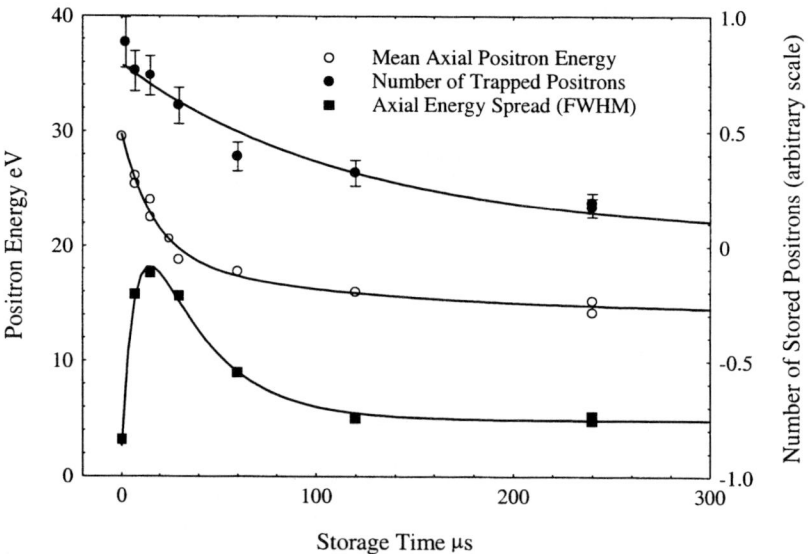

Figure 5 Variation of the number of stored positrons, the mean axial energy and the axial energy spread (FWHM) as a function of storage time in the Penning trap.

Evaporative cooling is known to be effective only down to around 1/5 of the trap depth since particle loss becomes insignificant, in this case this corresponds to around 5eV. This is in good agreement with the minimum value observed. Although inefficient in our experiment if performed correctly evaporative cooling can be an effective method for the all important initial cooling phase. Some particle loss is involved, though this can be minimal. At longer times a rather long lived component is present (up to 100ms were observed) though this has yet to be studied in detail. During these investigations the trap was generally held at around 200K and in a vacuum of around 10^{-9} Torr, collisional effects with rest gas are therefore insignificant on these timescales.

CONCLUSION

Although conventional slow positron beams have been widely and profitably used for atomic collision studies, especially of atomic and molecular ionisation, modern collision experiments and more ambitious fundamental studies require major improvement in beam quality (and intensity). At electron accelerator facilities one can produce intense pulsed positron beams. These are ideal for studies of positronium and other short lived (exotic) positronic systems which are of fundamental importance in testing QED and studying many body effects. Accelerator based beams are, however, inconvenient and limited in beam resolution.

The application of charged particle (Penning) traps as in-beam accumulators and coolers is having a profound effect on slow positron physics. It not only allows huge enhancement in instantaneous beam intensity, but most importantly many orders of magnitude improvement in energy and spatial resolution. Collision studies can be performed with meV resolution, this is a previously unexplored energy scale. The cold, dense positron plasmas which can be produced in a cryogenic trap environment gives the possibility for new precision studies of exotic atoms such as positronium. Here the key lies in producing the atom at low temperatures, at high density and preferably in excited states where the lifetime is long.

We present preliminary results which show that injecting slow positrons into a high field Penning trap can allow collisional energy exchange to be a rapid process, on a time scale of the order of 10µs. Evaporative cooling can then also be performed on a similar time scale. This constitutes a potentially new and simple technique for in-beam, ultra high vacuum, positron trapping which may be applied to source based beams.

ACKNOWLEDGMENTS

We would like to acknowledge the various contributions by M. Charlton, the collaboration of P. Balling and M. Raarup, the support of our collaborators in the positron trapping program, R. Ley and G. Werth. We would like to acknowledge the financial support from the EUROTRAPS European network contract number ERBFMRXCT97-0144, the financial support from the Danish Natural Science Research Council and the financial support by the Carlsberg Foundation of J.Merrison.

REFERENCES

1) Knudsen, H., and Reading, J.F., *Physics Reports* **212** 107 (1992)
2) Connerade, J.P., *Nature* **391**, 439 (1998)
3) Merrison, J.P., Charlton, M., Chevallier, J., Deutch, B., Hangst J.S., Hvelplund, P., Jørgensen, L.V., Knudsen, H., Laricchia, G., and Poulsen, M.R., *Phys. Rev. Lett.* **78**, 2728 (1997)
4) Charlton, M., *Rep. Prog. Phys.*, **48**, 737 (1985)
5) Rost, J.M., and Pattard, T., *Phys. Rev. A* **55** R5 (1997)
6) Bluhme, H., Knudsen, H., Merrison, J.P., and Poulsen, M.R., *Phys. Rev. Lett.* **81**, 73 (1998)
7) Bluhme, H., Knudsen, H., Merrison, J.P., and Nielsen, K.A., *submitted to J.Phys. B: At. Mol. Opt. Phys.* (1999)
8) Coleman, P.G., Hutton, J.T., Cook, D.R., and Chandler, C.A., *Can J. Phys.* **60** 584 (1982)
9) Schmitt, A., Cerny, U., Moeller, H., Raith, W., and Weber, M., *Phys. Rev. A* **49** R5 (1994)
10) Fee, M.S., Mills, A.P., Chu, S., Shaw E.D., Danzmann, K., Chichester, R.J., Zuckerman, D.M., *Phys. Rev. Lett.* **70** 1397 (1993)
11) Hagena, D., Ley, R., Weil, D., Werth, G., Arnold, W., and Schneider, H., *Phys. Rev. Lett.* **71** 2887 (1993)
12) Madsen., L.B., and Lambropoulos, P., *Phys. Rev. A* **59** 4774 (1999)
13) Scrader, D.M., Jacobsen, F.M., Frandsen, N.P., and Mikkelsen, U., *Phys. Rev. Lett.* **69** 57 (1992)
14) Merrison, J.P., Hertel, N., Knudsen, H., Stahl, S., Uggerhøj, E., *Appl. Surf. Science* **149** 11 (1999)
15) Dahm, J., Ley, R., Niebling, K.D., Schwarz, R., Werth, G., *Hyp. Int.* **44** 246 (1988)
16) Surko, C.M., Greaves, R.G., and Charlton, M., *Hyperfine Interactions* **109** 181 (1997)
17) Gilbert, S.J., Greaves, R.G., and Surko, C.M., *Phys. Rev. Lett.* **82**, 5032 (1999)
18) Van Reeth, P., and Humberstone, J.W., *J.Phys. B: At. Mol. Opt. Phys.* **32** 1 (1999)
19) Haarsma, L.H., Abdullah, K., and Gabrielse, G., *Phys. Rev. Lett.* **75** 806 (1995)

Absolute Fully Resolved (e,3e) Cross Sections for the Double Ionization of Helium

A. Lahmam-Bennani*, A. Duguet*, I. Taouil*, M.N. Gaboriaud*,
L. Avaldi[1], J. Berakdar[2] and A. Kheifets[3]

*Laboratoire des Collisions Atomiques et Moléculaires (UMR 8625), Bât. 351, Université Paris XI,
91405 Orsay cedex, FRANCE
[1] IMAI del CNR, Area della Ricerca, CP 10, 00016 Monterotondo Scalo, ITALY
[2] Max-Planck Institute für Mikrostrukturphysik, Weinberg 2, 06120 Halle, GERMANY
[3] Research School of Physical Sciences, The Australian National University, Canberra ACT 0200,
AUSTRALIA

We have measured for the first time full sets of coplanar (e,3e) angular distributions for double ionization of He, at an impact energy of 5 and ·1 keV. Ejected electrons are detected with equal energies, 10 and 4 eV. Comparison is made with analog (γ,2e) results, and deviations from the dipolar limit are pointed out. The data are also compared with two calculations, using either a four-body final-state wavefunction for the three electrons moving in the field of He^{2+}, or the convergent close coupling method.

WHY DOING THESE EXPERIMENTS?

Main objective is the investigation of the double ionization (DI) process under electron impact, in a kinematically complete experiment in which the energies and emission angles, and hence the momenta of all participating particles are determined in the final state, and all these particles are detected in a triple coincidence. For the purpose, the ideal target is He, the simplest two-electron system that yields a pure 4-body problem in the final state. This simplicity is essential since the N-body problem, with N=3 or larger, is one of the most fundamental problems not yet solved in atomic physics. Such experiments yield the most detailed insight into the fundamentals of the DI process.

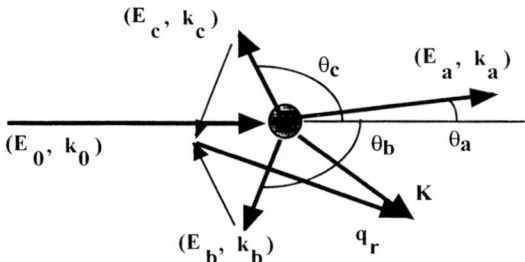

FIGURE 1. Schematic momentum diagram of a coplanar (e,3e) experiment.

We briefly report here results of the first kinematically completely determined (e,3e) experiment for He. More details may be found in Refs. (1,2). Figure 1 shows an (e,3e) momentum diagram : the incident electron, denoted 0, with energy E_0 and momentum $\mathbf{k_0}$ is scattered under the angle θ_a, with energy E_a and momentum $\mathbf{k_a}$, while 2 electrons denoted b and c are ejected from the target, respectively in the directions θ_b and θ_c, with E_b, $\mathbf{k_b}$ and E_c, $\mathbf{k_c}$. To fully determine the kinematics, one needs to measure all three energies and angles and detect the 3 final electrons in a triple coincidence. Here, $\mathbf{K} = \mathbf{k_0}\text{-}\mathbf{k_a}$ is the momentum transfer to the target, and $\mathbf{q_r}$ is the ion recoil momentum, given by the difference "$\mathbf{K}$ minus the sum momentum for the pair of ejected electrons". $\mathbf{q_r}$ is of course known from the measurement of all other quantities.

The (e,3e) spectrometer has been extensively described in (3). Briefly, a 1 to 10 keV electron beam crosses at right angle the gas jet. The fast a electron is analysed in a cylindrical analyser, and detected on a scintillator-photomultiplier arrangement. The electron gun rotates about the gas jet axis, which allows to vary the scattering angle θ_a. The slow ejected b and c electrons are analysed in a double toroidal analyser. This system includes multiangle detection of the ejected electrons, the key point being that the angular information contained in the collision plane is preserved upon arrival on the two position sensitive detectors.

Before presenting the results, it should be stressed that the measurements are obtained on an absolute scale. This is very important in order to be able to disentangle between different theoretical models which might yield similar results as to the shape of the angular distributions, but might differ by large factors as to the magnitude. Our absolute scale determination is based on the relationship between the measured triple coincidence count rate and the (e,3e) cross section, via a number of experimental parameters. These parameters are not determined directly, which would be a tedious and quite inaccurate procedure. Rather, we relie first on the measuremt of DDCS and TDCS obtained under exactly the same kinematical parameters as in the (e,3e) experiments, and second on their comparison with well established theoretical DDCS and TDCS. The overall final accuracy reached with this method is roughly 30%.

RESULTS

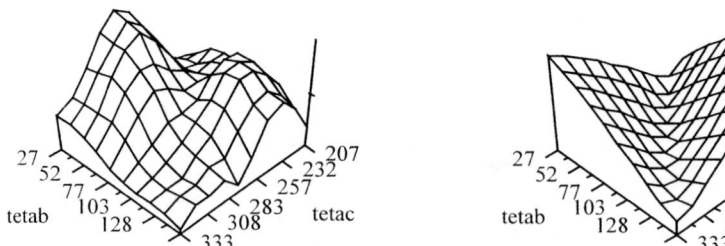

FIGURE 2. Left : 3-D plot of the measured (e,3e) cross section for He, versus the θ_b- and θ_c-angles. The incident direction is along the diagonal, from right to left. The scattered electron is detected at a fixed angle, 0.45° (K=0.24 au), with an energy of 5.5 keV. Right : Magnitude of the ion recoil momentum, versus θ_b and θ_c.

Figure 2 presents results for the DI of He in the so-called equal-energy sharing case, $E_b = E_c = 10$ eV. We can here make two important observations. First, there are mostly two structures in this surface, that is, the b and c electrons are preferentially emitted either *both simultaneously* forward, or *both simultaneously* backward with respect to the incident direction. Such forward or backward emission is not *a priori* an obvious expectation, as it corresponds to an ion recoil momentum which has to be rather large.

Indeed, one can see (Fig. 2) a striking similarity between the 3D surface for He and the one representing the magnitude of the ion recoil momentum, q_r : the cross section is minimum along the 'valley' where q_r is minimum, and the two intensity peaks strikingly correspond to the maximum ion momentum. The reason for this behaviour can be understood as follows : at our high incident energy and small momentum transfer, the optical limit is quite closely approached. However, the optical transition is forbidden for two free electrons, (which is the condition for the Bethe sphere), that is for photon absorption without participation of the nucleus. This is because a photon imparts to the system energy, but basically no momentum, therefore the electrons must recoil off the massive nucleus.

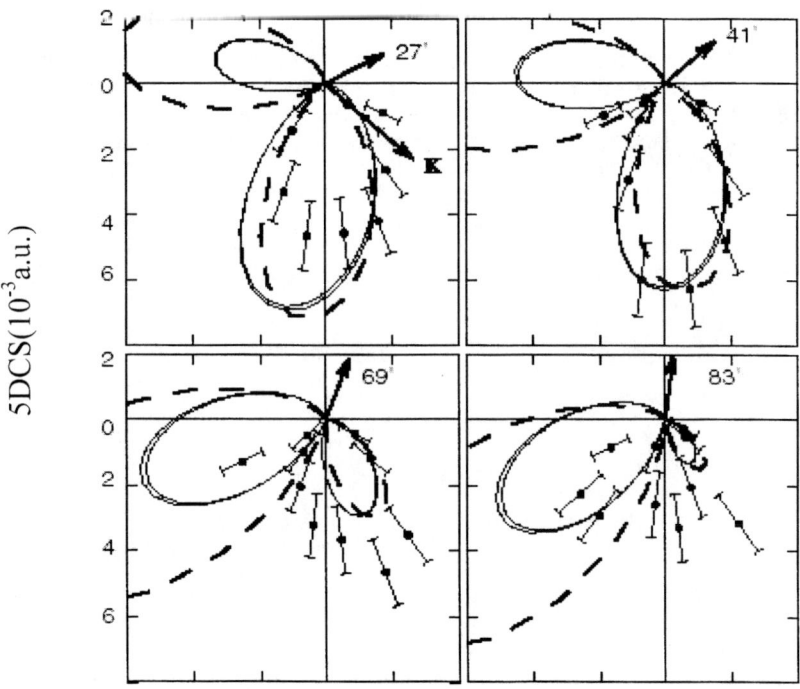

FIGURE 3. A selection of the measured angular distributions, with one ejection angle fixed (as shown by the arrow and the labeling), and the other one variable. The dots are the experimental data. The dashed curves are calculated results using the correlated 4-body final state (C4FS x 0.7) model, whereas the full curves are calculations using the Convergent Close Coupling (CCC x 3.2) method.

The second observation is the following : there are at present a few theoretical calculations dealing with these results. This short presentation does not allow to discuss them here. We will only briefly refer to Fig. 1 of Ref. (1) to illustrate the point that all of them reasonably agree with the experiments as far as a global picture like the one in Fig. 2 is concerned : we see the two peaks, forward and backward, and the valley in between. But noticeable differences do appear when one examines detailed cuts of these surfaces. A sample of such cuts is shown in Figure 3. We note here that first, as to the magnitude, one theory (CCC) is a factor of 3.2 too small with respect to the experiment, whereas the other theory (C4FS) is about a factor of 1.5 too large. Second, as to the shape, the agreement between theory and experiment is good for some fixed angles, whereas for some others it is less satisfactory, if not bad.

Another observation is the following : at our high incident energy and small momentum transfer, it is well known that the electron impact ionization is approaching the photoionization. Therefore, it is certainly of interest to compare the (e,3e) results with photo-double ionization (PDI) results. This is done in Figure 4. We first note that the two electrons do not fly out in the same direction with the same velocity : this is trivial, due to the Coulomb repulsion in the final state. Moreover, the back to back emission corresponding to a mutual emission angle of π is clearly not the most likely one.This can be easily understood, as we are very closely approaching the optical limit. In PDI of He, it has been shown (4) hat there is a node in intensity at a mutual angle of π, due to the $^1P^o$ symmetry for the pair of outgoing electrons. The minimum observed here in the (e,3e) data means that the collision is still dominated by dipolar contributions, but the fact that it is a non-zero minimum means that non dipole contributions are also present, and several electron final states are accessible. Note that the possibility that the experimental non-zero minimum might be due to finite angular and/or energy resolutions has been ruled out in Ref. (2).

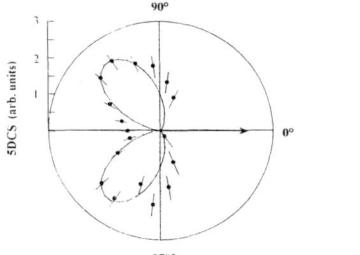 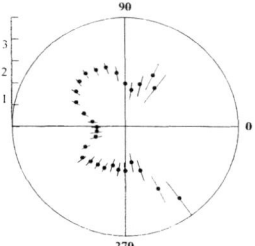

FIGURE 4. Mutual angle representation of the (e,3e) cross section. Left : $E_b=E_c=10eV$. Right : $E_b=E_c=4eV$. Dots: Present experiments. Full curve : parametrization of the PDI results according to (5).

However, while the situation for 10eV/10eV outgoing energies (Fig.4a) seems to be understood, for 4eV/4eV ejected electrons there seems to be 4 lobes in the mutual angle representation (Fig. 4b), with a minimum at about $\pm 90°$, quite different from the PDI observations. The origin of these additional structures is not clear. Of course, we have checked and counter-checked the experimental results, and we think this is not an experimental artefact. What else? May be a strong non-first Born contribution? The question remains open, and more theoretical as well as experimental investigations are needed to confirm or not this observation.

CONCLUSION

We have presented a sample of results from the first kinematically complete (e,3e) experiments on He, at ~5 keV impact energy, and 10+10 and 4+4 eV outgoing energies. Reasonable global agreement is obtained with available theories, but more work is still to be done on the detailed comparison. The (e,3e) data resemble the PDI ones, however significant differences are observed, showing that the optical limit is not fully reached.

The measurements are presently being extended to a wider range of kinematical variables (lower E_0, larger K, unequal E-sharing, larger E_{ej}, etc ...).

REFERENCES

1. Lahmam-Bennani, A., Taouil, I., Duguet, A., Lecas, M., Avaldi, L., and Berakdar, J., *Phys. Rev. A* **59**, 3548-55 (1999).
2. Kheifets A., Bray, I., Lahmam-Bennani, A., Duguet, A., and Taouil, I., *J. Phys. B* in press (1999).
3. Duguet, A., Lahmam-Bennani, A., Lecas, M., and El Marji, B., *Rev. Sci. Instrum.* **69**, 3524 36 (1998).
4. Huetz, A., Selles, P., Waymel, D. and Mazeau, J., *J. Phys. B* **24**, 1917 (1991).
5. Malegat, L., Selles, P., and Huetz, A., *J. Phys. B* **30**, 251 (1997).

Angular Momentum Transferred in Inelastically Scattered P to S State Electron-Atom Collisions

M. Shurgalin[1], A. J. Murray[2], W. R. MacGillivray[3], M. C. Standage[3], D. Madison[4], K. D. Winkler[4] & I. Bray[5]

[1] *Harvard-Smithsonian Centre for Astrophysics, Cambridge, MA 02138, USA*
[2] *Quantum Dynamics Group, University of Manchester, Manchester, M13 9PL, UK*
[3] *Laser Atomic Physics Laboratory, Griffith University, Brisbane, QLD 4111, Australia.*
[4] *Atomic, Molecular & Optical Research, University of Missouri-Rolla, Missouri, 65401, USA*
[5] *Electronics Structure of Materials Centre, Flinders University, SA, 5001, Australia.*

Abstract: Coherent laser preparation of atoms has been employed in superelastic and inelastic electron scattering from the excited $3^2i'_{3/2}$ state of sodium. This allows Atomic Collision Parameters to be deduced for the 3S to 3P and 3P to 4S transitions respectively. By employing micro-reversibility, a long standing proposal relating the sign of the transferred angular momentum to the sense of energy transfer has been tested, and is not supported in general by experiment.

INTRODUCTION

Detailing the processes which occur when an electron excites or ionises an atom is fundamental to understanding many processes which occur both in nature and in manufacturing. These processes play a dominant role in the dynamics of stars, plasmas, lightning, the ionosphere, fluorescent lighting and ion laser production to name a few. Indeed, since almost all energy used by mankind is transported by electricity, understanding the interaction of electrons with atoms and molecules is of great importance.

Experimental study of these interactions at a fundamental level has proceeded in a number of ways, the most detail being when coincidence or related techniques are used. These techniques maximise information by studying scattering between *individual* atoms and electrons, allowing the most exacting tests of theoretical models to be implemented.

Much of this effort has concentrated on the interaction with atoms in their ground state, and a number of models have been formulated to account for observation with varying degrees of success. The simplest first Born approximation is reasonably successful at very high impact energies, however at lower energies the interaction is far more complex than explainable by this model. At these energies, interactions between the incident electron, electrons in the atom and the ion core all play a role in incident and exit reaction channels, the dynamics being a complex problem which is difficult to solve.

New techniques allow data to be taken from *excited* targets, although these experiments are usually difficult. These experiments include electron interactions with metastable targets[1,2], and electron interactions with laser excited targets [3]. In the experiments

CP500, *The Physics of Electronic and Atomic Collisions*, edited by Y. Itikawa, et al.
© 2000 American Institute of Physics 1-56396-777-4/00/$17.00

detailed here, laser preparation of sodium atoms in the $3^2P_{3/2}$ excited state is used, the electrons being scattered both inelastically and superelastically.

These results have prompted a number of new models [4,5], and in particular have highlighted deficiencies in methods successful in explaining electron scattering from ground state atoms. In particular, the 2nd order Distorted Wave Born Approximation [DWB2] [5] is deficient in regions so far studied. By contrast, the Convergent Close Coupling [CCC] method [5] produces results in good agreement with experiment. This is especially highlighted for the angular momentum $L_\perp$ imparted to the atomic charge cloud. Contrasts between data and these two leading theories are discussed here.

In addition to producing new data from excited states, one of the aims of this work was to consider a long standing proposal regarding the sign of $L_\perp$ imparted during excitation and de-excitation for an S to P state transition. From the first Born approximation no angular momentum can be imparted during the collision, which has been refuted by many experiments. Using a higher order expansion of the Born Series by Madison and Winters [6], Andersen and Hertel [7] proposed that for positive scattering angles, $L_\perp$ should be *negative* for de-excitation from an S to P state. This proposal was based upon symmetries in the experiment compared to calculations involving positron scattering.

In this paper results are presented where excitation and de-excitation processes are measured in a "time-inverse" geometry. Excitation uses *superelastic* scattering from the laser prepared P-state to a lower S-state, whereas de-excitation uses *inelastic* scattering from the laser prepared P-state to a higher S-state. Adopting the principles of micro-reversibility, these processes can be equated once an appropriate geometry is chosen. To explain these results, the apparatus is detailed, together with the time inverse geometry used. Finally, the results are presented together with calculations using DWB2 and CCC theories, and conclusions regarding the hypothesis of Andersen and Hertel are drawn.

THE EXPERIMENTAL APPARATUS

The apparatus consists of a μ-metal shielded vacuum chamber pumped by a 500l/s pump. Located inside was an energy selected electron gun and a Faraday cup mounted on a rotary table. A hemispherical analyser fixed to the lower flange of the vacuum system was used for measurements. Scattered electrons were momentum selected and detected using a channeltron located at the exit aperture of the hemispherical selector. Output from the channeltron was amplified and counted using standard Ortec NIM electronics.

The sodium beam was created in a two stage crucible oven. Atoms effusing from the oven passed through a liquid nitrogen trapped collimation stage to reduce the Doppler profile to 100MHz. A cold trap opposite the oven was used for deposition of the sodium beam. Laser excitation used two Spectra Physics 380D dye lasers, the beams being combined and directed into the chamber through a window located on the top of the vacuum system. The laser beams were injected orthogonally to the scattering plane defined by the incident and scattered electrons.

Polarisation control was accomplished using a linear polariser followed by a $\lambda/4$ retardation plate. The lasers were tuned between the ground states and the closely spaced hyperfine states of the upper 3P-state, allowing efficient laser excitation of this state.

437

SUPERELASTIC SCATTERING

Figure 1 shows the superelastic scattering process. The two resonant laser beams interact with sodium, exciting the 3S ground state to the 3P state with a photon energy ~2.1eV. By controlling the laser polarisation, direction and power, the m_L substates of the P-state are coherently excited, creating a superposition state given by :

$$|j\rangle = \sum_{m_L} |L=1, m_L\rangle = a_{-1}|1,-1\rangle + a_0|1,0\rangle + a_{+1}|1,+1\rangle$$

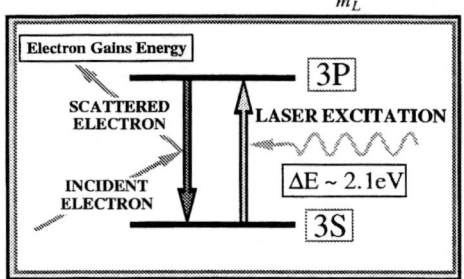

Of significance is laser pumping using circular polarisation. In this case, stimulated emission and absorption rapidly reduces Rabi cycling to a dynamic two-state system between the $m_L = \pm 1$ states and the ground state. A fully oriented P-state is produced with $|a_{\pm 1}|^2 = 1$, the sign of $L_\perp$ being derived from the handedness of the laser radiation.

FIG 1. Superelastic scattering from sodium.

Electrons produced in the electron gun which are scattered from a laser prepared P-state can *gain* energy from the atom. This process is known as *superelastic* scattering, the electron gaining 2.1eV from the collision. Preparation of the atom using circularly polarised radiation results in the electron 'seeing' a fully oriented atom, whereas after the collision the atom is in the ground state and has zero angular momentum. The scattered electron removes angular momentum from the atom, resulting in the electron being scattered with respect to the incident direction. The probability of scattering in a given direction depends upon details of the reaction. There is zero probability of scattering through 0° and 180° at any incident energy, since at these angles no angular momentum can be removed from the atom by the electron.

By controlling the amplitudes of the superposition state using the laser, this scattering can be varied. Assuming no electron spin flip occurs, the spin averaged scattering is described by three Atomic Collision Parameters (ACP's) given in the natural frame by :
- The alignment angle γ
- The linear shape or degree of anisotropy of the charge cloud P_{lin}
- The angular momentum transferred to the atom $L_\perp$.

STEPWISE INELASTIC SCATTERING

In addition to superelastic scattering, the incident electron may *lose* energy to the atom which is excited to a higher state as shown in figure 2. In this case, sodium is excited to the 4S state, gaining ~1.1eV energy. The incident electron loses energy and is scattered.

In an analogous way to superelastic scattering, the probability of scattering into a given direction depends on the reaction dynamics of the interaction process. Again, there can be no flux in either the forward or backward scattering directions for a fully oriented

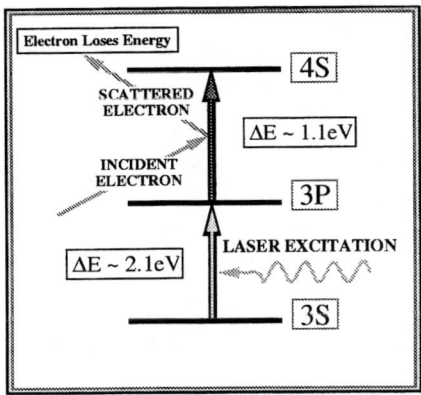

FIG 2. The Inelastic scattering process.

atom. Further, if the upper state is an S-state, the spin-averaged scattering can be described by the same ACP's used for the superelastic process.

TIME INVERSE GEOMETRY

One of the main aims of this study was to consider the proposition by Andersen and Hertel [7] regarding the sign of $L_\perp$ for excitation and de-excitation. The experiments discussed here allow this test to be made by using a 'time inverse' geometry. This is a standard technique used in superelastic scattering, and can be applied to the inelastic scattering studied here.

Firstly, consider the 'time direct' geometry as shown in figure 3. In this case, the sodium is initially in the ground 3S state, and an incident electron of energy E_{in} excites it to the 3P state, losing energy and scattering through an angle θ_{scat} where it is detected. The atom subsequently decays, fluorescence $I(\theta_{scat})$ being detected and analysed in coincidence with the detected electron. A Stokes analysis of the circular component of the radiation $P_3(\theta_{scat})$

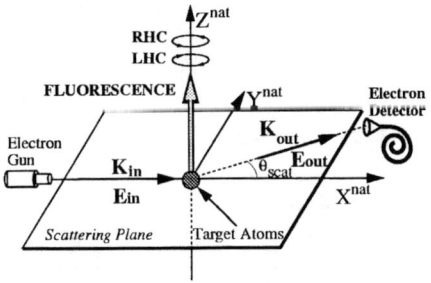

FIG 3. The 'time-direct' scattering process

then gives information as to the angular momentum $L_\perp$ imparted in the reaction [8] :

$$L_\perp(\theta_{scat}) = -P_3(\theta_{scat}) = \frac{I_{LHC}(\theta_{scat}) - I_{RHC}(\theta_{scat})}{I_{LHC}(\theta_{scat}) + I_{RHC}(\theta_{scat})}$$

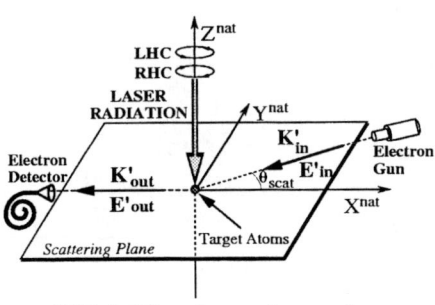

FIG 4. 'Time-inverse' scattering.

Figue 4 shows the reaction in a 'time-inverse' geometry, where the experiment effectively runs backwards in time. In this case, the atom is prepared in a well defined state using polarised laser radiation directed in the opposite direction to the detected photon shown in figure 3. The incident electron is now directed at the angle θ_{scat}, whereas the detector is placed in the same position as the electron gun in figure 3.

The detected electron *gains* energy from the reaction. Hence for equivalence, the incident energy E_{in} for superelastic scattering is 2.1eV *lower* than E_{in} for the coincidence experiment : $E_{in} = E_{in} - 2.1eV$. The superelastic experiment measures the properties of electron *excitation* from the S to the P-state as in the coincidence experiments.

A similar argument applies for inelastic scattering. In this case, E_{in} is 1.1eV *higher* than E_{in} in the coincidence experiment : $E_{in} = E_{in} + 1.1eV$. This is equivalent to *de-excitation* from an S to P-state. The Andersen and Hertel hypothesis can hence be tested.

One major difference between coincidence and laser experiments is the coherent laser field used to prepare the atom, compared to spontaneous emission for a coincidence experiment. This difference leads to an optical pumping parameter K, describing the laser interaction. $L_\perp$ is then determined using the pseudo Stokes parameter :

$$L_\perp(\theta_{scat}) = -\frac{1}{K} \cdot P_3^S(\theta_{scat}) = \frac{1}{K} \cdot \left(\frac{S_{LHC}(\theta_{scat}) - S_{RHC}(\theta_{scat})}{S_{LHC}(\theta_{scat}) + S_{RHC}(\theta_{scat})} \right)$$

$S_{LHC \atop RHC}(\theta_{scat})$ measures superelastic yield at a scattering angle θ_{scat} for left and right hand circular polarisation respectively. The laser pumping parameter is found to be $K = -0.99$ for the 3S - 3P hyperfine excitation under the experimental conditions used here.

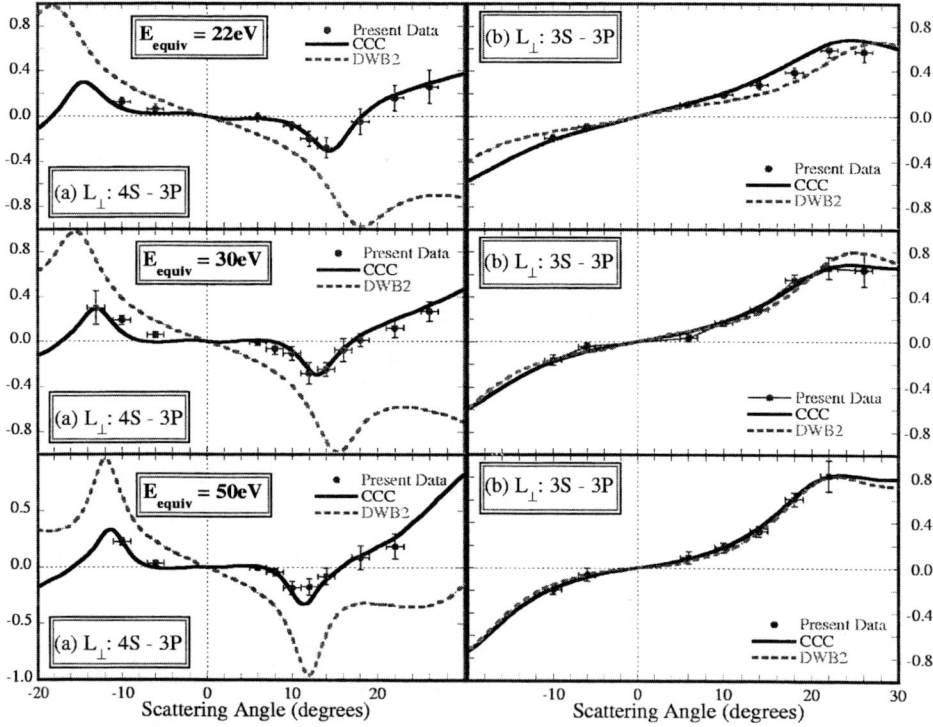

FIG 5. Experimental results compared to the CCC and DWB2 theoretical calculations

RESULTS & CONCLUSIONS

Measurements were conducted for 3S-3P *excitation*, and 4s-3P *de-excitation* to mini-mise experimental variations. The equivalent incident energies were 22eV, 30eV and 50eV, the incident energy being 2.1eV lower than this for superelastic measurements, and 1.1eV higher for inelastic experiments as previously discussed [9].

Figure 5 shows the results, the 4S - 3P de-excitation data labelled (a) and 3S - 3P excitation labelled (b). The CCC & DWB2 calculations are also shown for comparison. Considering first the superelastic results, it is seen that both models are in excellent agreement with experiment. For all angles from 0° to +30°, $L_\perp$ is positive, increasing towards almost full charge cloud orientation around 20°. By contrast, the 4S - 3P de-excitation results are very different. At low scattering angles up to 10° almost no angular momentum is imparted, in line with the first Born approximation. The results are slightly negative as predicted by Andersen and Hertel, but are not always negative as expected from this hypothesis. At an angle around 16° $L_\perp$ passes through zero and becomes positive. The 'dip' becomes sharper and moves to lower angles as energy increases. If de-excitation had been the same as excitation as proposed, the 4S - 3P results would have been negative at all scattering angles.

Striking about these data is the success of the CCC theory. The calculation passes through almost all data within measured uncertainty. By contrast, the DWB2 theory fails to predict the results, and is more in line with the hypothesis of Andersen and Hertel. This is perhaps not suprising since the hypothesis was based upon a Born expansion. It should be noted that for other ACP's, both theories are in agreement with experiment.

The disagreement between the DWB2 theory and the data indicates this model is incomplete, and that further work is needed to establish the cause of this discrepancy. Success for excitation from the ground state indicates that much of the physics has been included, however inadequacies still exist. The CCC theory is principally a low energy model and has difficulty converging for high impact energies and heavier atoms, whereas the DWB2 theory has been successful in these regions. Since it is important to gain an understanding of electron impact not only from ground state targets but also from excited targets, further experimental and theoretical work is clearly required.

The authors would like to thank the Australian Research Council and U.S. Natural Science Foundation for supporting this work.

REFERENCES

[1].J.B. Boffard et al, Phys Rev A **59**, 2749 (1999)
[2]. I.Y. Baranov, N.B. Kolokolov & N.P. Penkin Opt Spect. **58** 268 (1985)
[3]. W.R. MacGillivray & M.C. Standage Phys. Rep. **168** 1 (1988)
[4]. K. Bartschat & V. Zeman, Phys Rev A **59** R2552 (1999)
[5].M. Shurgalin et al, Phys. Rev. Letts. **81**, 4604 (1998).
[6].D. H. Madison and K.H. Winters, Phys. Rev. Letts. **47**, 1885 (1981).
[7].N. Andersen and I.V. Hertel, Comments At. Mol. Phys. **19**,1 (1986).
[8].N. Andersen, J.W. Gallagher and I.V. Hertel, Phys. Rep. **165**, 1 (1988).
[9].M. Shurgalin et al, J. Phys. B: At. Mol. Phys. **32**, 2439 (1999).

Low Energy High Resolution Dissociative Electron Attachment to Ozone

G Senn[*], H Drexel[*], N J Mason[†],
J D Skalny[‡], A Stamatovic[§], P Scheier[*] and T D Märk[*,‡]

[*] Institut für Ionenphysik, Leopold Franzens Universität, Technikerstrasse 25,
A-6020, Innsbruck, Austria
[†] Molecular Physics Laboratory, Department of Physics and Astronomy,
University College London, Gower Street, London WC1E 6BT, United Kingdom
[‡] Department of Plasma Physics, Comenius University, Mlynska dolina F2,
84215 Bratislava, Slovak Republic
[§] Faculty of Physics, Beograd, P.O.Box 368, 11001, Beograd, Yugoslavia

Abstract. The production of O^- and O_2^- by dissociative electron attachment to ozone has been studied between about 0 and 10 eV with a high resolution crossed beam apparatus recently developed in the Innsbruck laboratory. A previously unobserved sharp structure is observed in the formation of O^- ions at zero incident energy. This large additional cross section peak has important consequences for the role of ozone in anion formation processes in the terrestrial ionosphere.

INTRODUCTION

The observation of the 'ozone hole' above Antarctica in 1985 has led to a major international research programme to understand the chemical reactions of ozone responsible for dramatic ozone loss. The catalytic destruction of ozone by halogen free radicals is now largely understood and the major mechanisms by which ozone is lost in the stratosphere have been identified. The role of ozone in the D-region of the ionosphere is, however, less well established.

The D-layer is the lowest part of the Earth's ionosphere lying at altitudes between 60 km and 90 km. The D-layer is of great importance in radio communication since most of the radio wave absorption in the HF and MF bands occurs in this region. It is also the critical region in the interaction between the ionosphere and the lower stratosphere where terrestrial ozone chemistry is dominant. Uniquely in the Earth's atmosphere the D-region is dominated by the formation and chemical reactions of anions. To date there have been rather few detailed mass spectrometric studies of the negative ion concentrations within the region but the major anions appear to be CO_3^-, NO_3^-, HCO_3^-, Cl^- and the hydrates $NO_3^-(H_2O)_n$ and $CO_3^-(H_2O)_n$ (1). These ions dominate the local chemistry and may be a source of nucleation for aerosol growth. Indeed the migration of such ions into the lower stratosphere may have important consequences for global ozone depletion since gaseous ions may allow ion-catalysed reactions to occur. Large

CP500, *The Physics of Electronic and Atomic Collisions*, edited by Y. Itikawa, et al.

cluster ions react with gas-phase species in a manner similar to surface catalysed reactions, the reactant molecule being absorbed on the surface of the cluster ion. Ions promote nucleation via several processes, of which the growth of large clusters and the formation and growth of stable ions by ion-ion recombination are the most important.

The mechanisms for formation of these ions are complex and remain unclear. Nevertheless several models have been developed to try to reproduce the measured concentrations (2, 3, 4). In such models it is essential to understand the primary ion production channels. Current models assume that the major negative ion formation process in this region arises from the exothermic non-dissociative three body electron attachment to molecular oxygen (5) i.e.,

$$e + O_2 + O_2 \rightarrow O_2^- + O_2 \qquad [1]$$

The product oxygen anions may then be subsequently lost either by the associative detachment reaction:

$$O_2^- + O \rightarrow O_3 + e \qquad [2]$$

or by charge transfer reactions with ozone. The O_3^- anions produced in these charge transfer reactions may then undergo further reactions with H, CO_2 and NO_2 to form CO_3^-, NO_3^-, HCO_3^-.

Dissociative electron attachment to molecular oxygen (6) i.e

$$e + O_2 \rightarrow O^- + O \qquad [3]$$

is not considered in these models since the cross section is only significant at electron energies well above those thermal energies available in the D-region. However electron impact dissociative attachment to ozone may also form significant concentrations of molecular oxygen anions at thermal energies. Two possible dissociative attachment channels exist (7):

$$e + O_3 \rightarrow O_2^- + O \qquad [4]$$

$$e + O_3 \rightarrow O^- + O_2 \qquad [5]$$

Current ionospheric models (2-4) have used the rate coefficients estimated for reaction [1] from early experiments of Curran (7) and Phelps and co-workers (8) to compare the anion formation rate from electron attachment to ozone via the reactions [4] and [5] with the three body electron attachment to molecular oxygen via reaction [1] and concluded that the electron attachment process to O_3 is only a minor contribution to the total oxygen anion production. However recent experiments (9-13) have shown that DEA to ozone is considerably more complex than these early experiments suggest with many different anion channels being accessible at low electron impact energies. It is therefore important to study low energy electron attachment to ozone and to quantify the magnitudes of the cross section for the formation of both O⁻ and O_2^- anions. As part of a continuing investigation of electron scattering from aeronomic molecules (14-16) we have studied low energy dissociative electron attachment to ozone. In particular we have studied those processes leading to the dissociation of ozone at energies comparable to those in the D-region of the ionosphere (less than 2eV).

EXPERIMENTAL APPARATUS

The apparatus (Figure 1) has been described in several earlier publications (17) so only the most salient features will be discussed here. In the present study a high resolution trochoidal electron monochromator produces an electron beam with a resolution of approximately 30 meV FWHM incident upon the ozone sample within a collision region. The anions formed in the collision are extracted by a weak electric field (<1 V/cm) prior to analysis in a quadrupole mass spectrometer. The mass-selected ions are detected in a single ion pulse counting mode using a channeltron. Data acquisition during the experiment was controlled by a PC. The electron energy scale and resolution were calibrated and monitored throughout the experiment using the well know DEA from CCl_4 and SF_6.

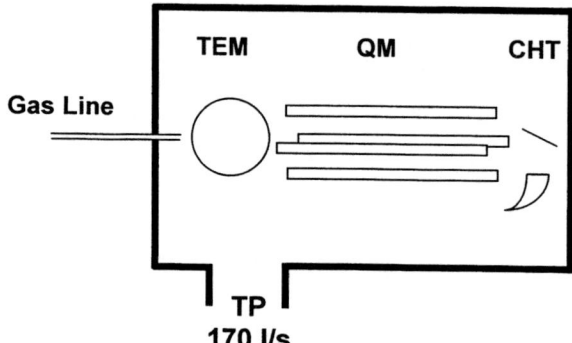

FIGURE 1a: Schematic view of the experimental apparatus consisting of a crossed electron/molecular beam source, a trochoidal monochromator (TEM) a quadrupole mass spectrometer (QM) and a channeltron (CHT)

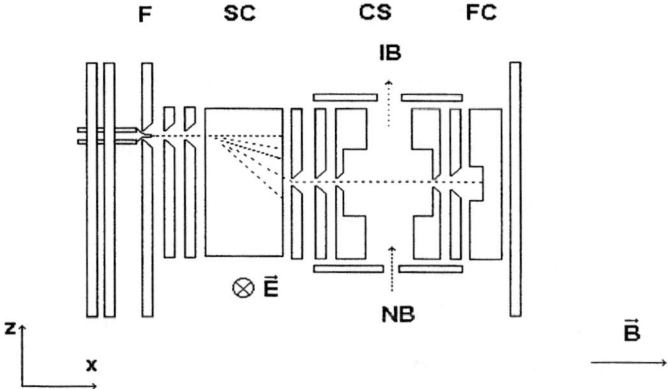

FIGURE 1b: The trochoidal electron monochromator (TEM). F is the filament, IB the ion beam, SC the scattering chamber, CS the collision region, FC the Faraday cup and NB the nozzle gas beam. Electrons are collimated and focused by electric (E) and magnetic (B) fields

Due to the high reactivity of ozone on surfaces and in order to quantify any charge transfer processes or ion-molecular reactions two ozone samples were used of greatly different purities. High purity sources of ozone (> 80%) were prepared using the UCL mobile ozone generator (18) while low ozone concentrations (1%) were produced with a negative corona discharge ozoniser (11). The measured cross sections were found to be independent of the ozone concentration and its production method, and the geometry or construction materials of the gas inlet system.

RESULTS

Figure 2(a) shows the measured O^- anion cross section as a function of incident electron energy in the low energy region up to 2 eV. A broad peak centred at 1.3eV is observed and found to be in excellent agreement with earlier experimental data (9-13). Fine vibrational structure as reported by Allan et al (9) is observed superimposed on the peak and may be attributed (9) to the opening of dissociation channels with a vibrationally excited molecular oxygen fragment. However evidence for dissociative excited states of O_3^- in this energy range has also been reported in laser photo-dissociation studies and thus this structure could equally arise from complex curve crossings between O_3^- anion and the ground state of the neutral ozone molecule. Figure 2(b) shows the measured O_2^- anion cross section as a function of incident electron energy in the low energy region up to 2 eV. A single broad smooth peak is observed with a maximum cross section at 1 eV. These results are once again in excellent agreement with earlier experiments (9-13).

However the major observation in the present study is the observation, for the first time (12), of a rather large and sharp peak at or close to zero energy in the measured O^- anion cross section. The height and width of this sharp peak is dependent solely upon the incident electron beam resolution, indeed for resolutions greater than 100 meV the feature is broadened to such a degree that it can no longer be distinguished from the background.

The incident electron beam resolution was monitored by measuring the well known dissociative electron attachment cross sections for formation of Cl^- from CCl_4 and SF_6^- from SF_6. Thus the effects of electron energy beam resolution may be de-convoluted from the measured O^- anion cross section as a function of incident electron energy in the low energy region up to 2 eV. These cross sections may then be converted to a rate constant as a function of electron temperature and are shown in Figure 3.

The present O_2^- anion rate constant is in fairly good agreement with those currently used in ionospheric models, at least in shape. The higher resolution of the present experiments and well defined incident electron scale however leads to higher values at all temperatures (at the peak the present value 3×10^{-10} cm^3 s^{-1} is compared with 6×10^{-11} $cm^3 s^{-1}$ reported by Skalny et al (11)).

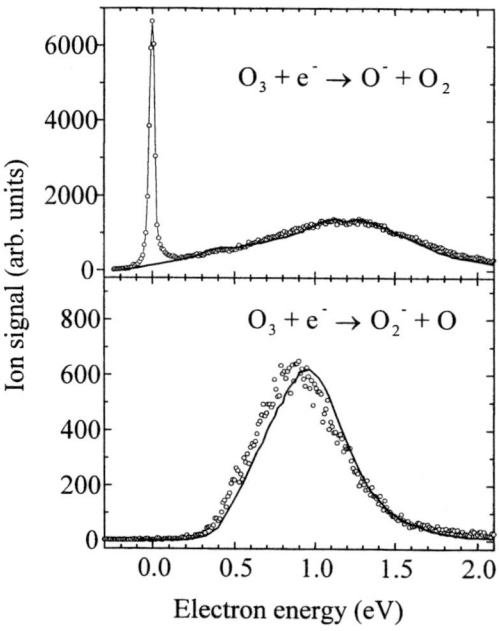

Figure 2. (a) O^- cross section and (b) O_2^- cross section for dissociative electron attachment to ozone. The solid line is data of Allan et al (9).

However due to the observation of the near zero energy peak the present derived O^- anion rate constant is considerably different from that derived from other experiments. The early beam data of Curran (7) or the swarm derived data of Phelps et al (8) and Ferguson et al (19, 20) suggest that no O^- anions are produced at thermal energies

(Figure 3) while the present experiment shows that the O⁻ anion rate constant at low electron temperatures is now not only finite but reaches its highest values (3 x 10^{-9} cm^3s^{-1}) at energy electron temperatures below 10K.

DISCUSSION

Such results may have important consequences for the role of direct electron impact dissociative excitation of ozone within the ionosphere. In current models the exothermic non-dissociative three body electron attachment to molecular oxygen reaction [1] above, is believed to be the dominant formation process for anions in the D-layer with a rate constant of 4 x 10^{-30} exp (-193/T) cm^3/s. Electron impact dissociative attachment to ozone, reaction [5] above, was assumed to be a minor anion process with a rate constant given by 9.1 x 10^{-12} $(T/300)^{1.46}$ cm^3/s. Since the concentration of molecular oxygen is high compared to ozone at low (thermal) electron energies the negative ion chemistry was therefore dominated by the initial formation of O_2^-. In contrast our results would now suggest that the local chemistry in the D-layer may be driven by the initial formation of O⁻ anions via reaction [5] and be strongly dependent upon the local concentrations of ozone. Thus at some altitudes the O⁻ anion may be the dominant anion rather than O_2^- in turn leading to different ratios of the terminal anions CO_3^-, NO_3^-, HCO_3^-, Cl⁻ and the hydrates $NO_3^-(H_2O)_n$ and $CO_3^-(H_2O)_n$.

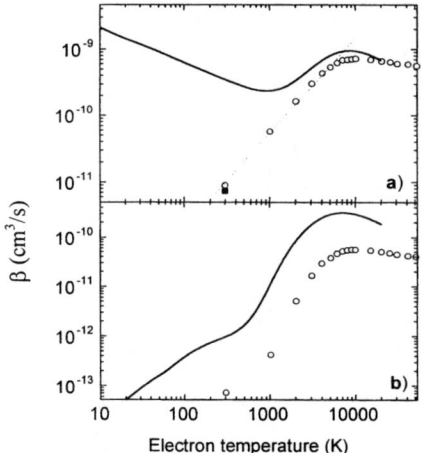

FIGURE 3. Absolute rate coefficients as a function of electron temperature for the production of (a) O⁻ and (b) O_2^- anions from ozone, calculated from the present cross sections. ■Measured rate constant of Fehsenfeld et al(19,20); ○ data of Skalny et al (11).

It is therefore timely to adopt the present values for anion formation and incorporate them into the ionospheric models (21). These models should then be compared with observational data and the time dependence of product terminal anions reviewed in relation to the changing ozone concentrations.

ACKNOWLEDGEMENTS

This work has been supported in part by FWF and BMWV, Wien, Austria and the Aktion Österreich-Slowakei science co-operation (contract 21S17). NJM thanks the Royal Society and the EPSRC and Universität Innsbruck for the opportunity to perform these joint studies and PS the Austrian Academy of science for the award of an APART grant. We would also like to acknowledge the assistance of Dr Peter Cicman in taking some of the data.

REFERENCES

1. Wayne, R. P., *Chemistry of atmospheres*, Oxford University Press (1991)
2. Brasseur, G. and Chatel, A, *Ann. Geophys* **1**, 173 (1983)
3. Thomas, L. and Bowman, M. R., *J. Atmos. Terr. Phys.* **47**, 547 (1985)
4. del Pozo, C.F., Hargreaves, J., and Aylward, A.D., *J Atmos and Solar Phen.* **59** 1919
5. Ferguson, E.E., Proc of the NATO Advanced study Institute on Atmospheric ozone, Portugal 517
6. Curtis, M.G., and Walker, I.C., *J Chem Soc Farad Trans.,* **88,** 2805 (1992)
7. Curran, R. K., *J Chem Phys* **35**, 1849, (1961)
8. Stelman, D., Moruzzi, J. L., and Phelps, A. V., *J. Chem Phys.,* **56,** 1783 (1972)
9. Allan, M., Asmis, K.R., Popovic, D. B., Stepanovic, M., Mason, N. J., and Davies, J. A. *J Phys B*, **29,** 4742 (1996)
10. Walker, I.C., Gingell, J. M., Mason, N. J., and Marston, G., *J Phys B*, **29,** 4749, (1996)
11. Skalny, J. D., Matejcik, S., Kiendler, A., Stamatovic, A., and Märk, T. D., *J Chem Phys Lett*, **25,** 112 (1996)
12. Senn, G., Mason, N. J., Skalny, J. D., Stamatovic, A., Scheier, P., and Märk, T. D. *Phys Rev Lett,* **85,** 5308- (1999)
13. Rangwala, S., Kumar, S. V., Krishnakumar, E., and Mason, N. J., *J Phys B in press* (1999)
14. Mason, N. J., Gingell, J. M., Davies, J. A., Zhao, H., Walker, I. C. and Siggel, M. R. F., *J.Phys B,* **29,** 3075, (1996)
15. Mason, N. J., Pathak, S. K., *Contemporary Physics,* **38,** 289 (1997)
16. Marston, G., Walker, I. C., Mason, N. J., Gingell, J. M., Zhao, H., Brown, K. L., Motte-Tollet, F., Delwiche, J., and Siggel, M. R. F., *J. Phys. B* **31,** 3387, (1998)
17. Matejcik, S., Senn, G., Scheier, P., Kiendler, A., Stamatovic, A., and Märk, T. D. *J Chem Phys* **107,** 8955, (1997) and Matejcik, S., Kiendler, A., Stampfli, P., Stamatovic, A., and Märk, T. D., *Phys. Rev. Lett.,* **77,** 3771 (1996)
18. Newson, K. A., Luc, S. M., Price, S. D., and Mason, N. J., *Int J Mass Spect Ion Proc* **148,** 203, (1995)
19. Fehsenfeld, F. C., Schmeltekopf, A. L., Schiff, H. I., and Ferguson, E. E., *Space Sci* **15,** 373, (1967)
20. Fehsenfeld, F. C., and Ferguson, E. E., *Planet Space Sci.* **15,** 701, (1968)
21. Schunk, R.W., and Soika, J. J., Ionospheric models, Modern Ionospheric Science, eds. H Kohl, R Rüster, and K Schlegel, European Geophysical Society, Kathenburg-Lindau, Germany, 181-215, (1996)

Ionization Cross Sections of Be^{2+}, B^{3+} and C^{4+} by Electron Impact

Seishirou Nasu[†], Shinobu Nakazaki[†] and Keith A. Berrington[‡]

[†] *Department of Applied Physics,Faculty of Engineering, Miyazaki University, Miyazaki 889-2192, Japan*
[‡] *School of Science and Mathematics, Sheffield Hallam University, Sheffield S1 1WB, UK*

Abstract. A unified theoretical approach is applied to calculate electron impact ionization cross sections for He-like ions(Be^{2+}, B^{3+} and C^{4+}). The 25 target states including 10 pseudostates are used in the R-matrix calculation. The cross sections are obtained as the sum of the cross sections into all states lying above the ionization threshold. The present results are in good agreement with those of the experiment.

INTRODUCTION

The electron impact ionization cross section of light He-like ions typically shows a background which rises to a maximum with impact energy. This background is due to the ionization of one of the 1s-shell electrons. At sufficiently high energies, both electrons can be ionized, giving a rise in the background: the inner-shell edge. Quantum mechanical interference between the one- and two-electron processes gives rise to resonance structures embedded on the background cross section both below and above the inner-shell edge. Such effects are small, however, and are difficult to observe experimentally, though there is certainly some experimental evidence [1]. The aim of this paper is to quantify such innershell effects in ionization of Be^{2+}, B^{3+} and C^{4+} from *ab initio* calculations.

THE UNIFIED TREATMENT

A unified theoretical treatment of different ionization processes in electron and photon impact on the atoms and ions was examined within the same wavefunction [2,3].

In the unified approach for the electron + C^{4+} and photon + C^{3+}, we involve the following processes at a total energy around the threshold for K-shell excitation. Here, nl indicates the principal and angular momentum quantum numbers of an

CP500, *The Physics of Electronic and Atomic Collisions*, edited by Y. Itikawa, et al.

excited outer electron:

$$
\left.\begin{array}{l}
e^- + C^{4+}(1s^2) \\
e^- + C^{4+}(1s2s)^{3,1}S \\
\gamma + C^{3+}(1s^2 2s)
\end{array}\right\} \rightarrow
\left\{\begin{array}{ll}
C^{4+}(1snl) + e^- & \\
C^{5+}(1s) + 2e^- & \\
C^{3+}(2s^2 nl) & \rightarrow C^{5+}(1s) + 2e^- \quad \text{READI} \\
C^{4+}(2s2l) + e^- & \rightarrow C^{5+}(1s) + 2e^- \quad \text{EA} \\
C^{3+}(2s2pnl) & \rightarrow C^{5+}(1s) + 2e^- \quad \text{REDA}.
\end{array}\right.
$$

The final states are essentially of two kinds: the first two are the direct processes of excitation and ionization for electron impact, corresponding to single and double photoionization; the others are indirect processes, involving the short-lived intermediate ($C^{4+} + e^-$) states. These latter processes are conventionally categorized as resonant excitation auto-double-ionization (READI), excitation autoionization (EA) and resonant excitation double autoionization (REDA). The last four processes can each lead to electron-impact ionization or double photoionization. We treat the unified approach of all these processes with the same wavefunction description.

TARGET WAVEFUNCTIONS AND CALCULATIONS

We include the following 25 target states in the present ionization calculations:

(1) bound states: $1s^2$ ^1S, $1s2s$ ^3S, $1s2s$ ^1S, $1s2p$ ^3P, $1s2p$ ^1P, $1s3s$ ^3S, $1s3s$ ^1S, $1s3d$ ^3D, $1s3d$ ^1D, $1s\overline{2}p$ ^3P, $1s\overline{2}p$ ^1P;

(2) continuum states: $1s\overline{1}s$ ^3S, $1s\overline{1}s$ ^1S, $1s\overline{1}d$ ^3D, $1s\overline{1}d$ ^1D, $1s\overline{1}f$ ^3F, $1s\overline{1}f$ ^1F, $1s\overline{1}p$ ^3P, $1s\overline{1}p$ ^1P;

(3) continuum state(hollow states): $2s^2$ ^1S, $2s2p$ ^3P, $2s2p$ ^1P, $2p^2$ ^1S, $2p^2$ ^3P, $2p^2$ ^1D.

Each state is represented by a configuration interaction wavefunction with the twelve orbitals (1s, 2s, 2p, 3s, 3d, $\overline{1}$s, $\overline{1}$p, $\overline{1}$d, $\overline{1}$f, s, p and $\overline{2}$p). To allow for inner-shell vacancies and polarization we introduce a set of short-range correlation orbitals $\overline{1}l$ and long-range polarized orbital $\overline{2}$p.

This strategy has been shown in [2,3] to give reliable cross section near the inner-shell thresholds, but it is not expected to be valid over an extended energy range. Table 1 shows osscilator strengths and the the ground state($1s^2$ ^1S) dipole polarizabilities of Be^{2+}, B^{3+} and C^{4+} to illustrate the accuracy of our wavefunctions.

The R-matrix theory of electron-ion collisions has been described in [4]. The total wavefunction representing the electron-ion collision system is expanded in the inner region as follows:

$$
\Psi_k = \mathcal{A} \sum_{ij} c_{ijk} \Phi_i(1,2,\hat{r}_3,\sigma_3) u_{ij}(r_3) + \sum_j d_{jk} \phi_j(1,2,3), \tag{1}
$$

where $\mathcal{A}$ is the anti-symmetrization operator, Φ_i the channel function representing the target state coupled with the spin and angular functions for the scattering

TABLE 1. Oscillator strengths and the ground state dipole polarizabilities of Be^{2+}, B^{3+} and C^{4+}.

Property	Be²⁺	Ions B³⁺	C⁴⁺
Oscillator strength 1s² ¹S→1s2p¹P			
Present	0.5488	0.6068	0.6455
Schiff et al. [9]	0.5516	0.6089	0.6471
The 1s² ¹S state dipole polarizability			
Present	5.256-2	1.971-2	8.984-3
Thornbury and Hibbert [10]	5.227-2	1.964-2	8.962-3

electron, u_{ij} the continuum basis orbitals for the scattered electron, and ϕ_j three-electron wavefunctions formed from twelve orbitals. The coefficients c_{ijk} and d_{jk} are determined by diagonalizing the total Hamiltonian of the whole system with the basis set expansion defined by Eq.(1).

We use the computer code [5] to calculate the R-matrix on the boundary of the sphere, whose radius (r_a) is taken to be 18.8, 15.4, 11.4a.u. for Be^{2+}, B^{3+} and C^{4+}, respectively. We include 38 continuum orbitals for each ion. In the outer region of the sphere, a set of close-coupling equation is solved for the partial waves $L = 0 - 20$, using the asymptotic code of Seaton(The Opacity Project team 1996).

CROSS SECTIONS

We calculate total cross section $Q(E)$ at energy E for electron impact ionization as the sum of the cross sections into all states lying above the ionization threshold of Be^{2+}, B^{3+} and C^{4+},

$$Q(E) = \sum_l Q_{1s\bar{1}l}(E) + \sum_{l,l'} Q_{2l2l'}(E). \tag{2}$$

Figures 1-2 show the resulting ionization cross sections of Be^{2+}, B^{3+} and C^{4+} from the ground state for electron impact processes, comparing with experiments. Present ionization features give rise to two resonances, $2s^2 2p$ $^2P^o$ and $2s2p^2$ $^2D^e$, which are the same as those of the experiments of Müller et al. [1] in the e^- + Li^+ ionization. As can be seen from these figures, present results are in good agreement with experiments those of Crandall et al. [6], and the recommended data of [7,8]. Their data are based on the simple formula

$$Q(E) = (IE)^{-1} \left(A \, ln \left(\frac{E}{I} \right) + \sum_{i=1}^{2} B_i \left(1 - \frac{I}{E} \right)^i \right). \tag{3}$$

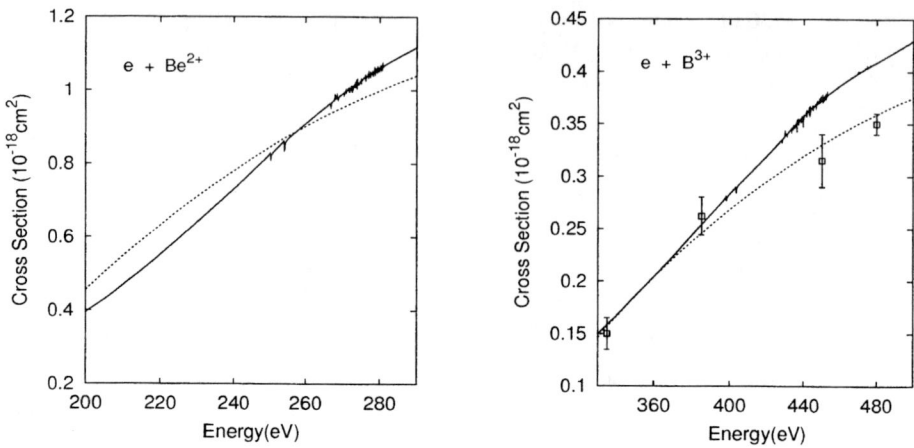

FIGURE 1. Ionization cross sections of Be^{2+} and B^{3+} for the $1s^2$ state by electron impact. Present calculation(———). Experimental results [6](□). Approximate analytical formula [7] (·····).

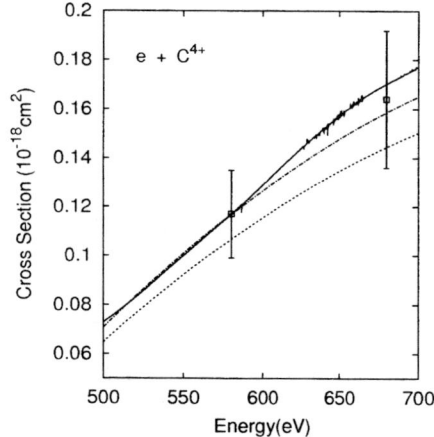

FIGURE 2. Ionization cross sections of C^{4+} for the $1s^2$ state by electron impact. Present calculation(———). Experimental results [6](□). Approximate analytical formula: [7](·····), [8](— · —).

where I is the ionization potential, the parameter A is a Bethe coefficient and the coefficients B_i are determined by a least square fitting procedure.

We also calculate ionization cross sections of Be^{2+}, B^{3+} and C^{4+} for the metastable 1s2s states. We identified prominent resonance structures below the

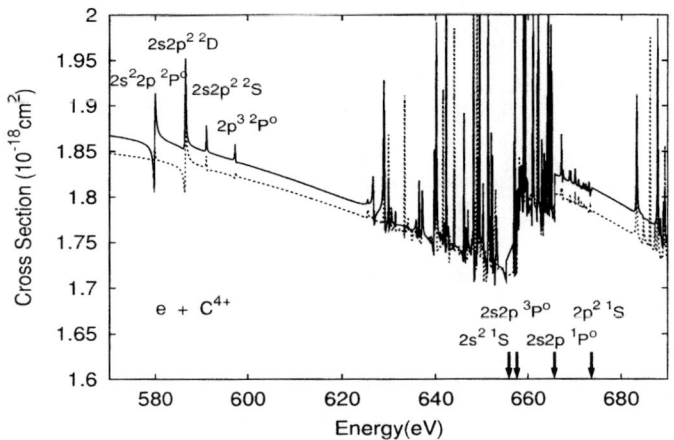

FIGURE 3. Ionization cross sections of C^{4+} for the 1s2s ^{1}S and 1s2s ^{3}S states by electron impact. Present calculations: 1s2s ^{1}S(———), 1s2s ^{3}S (· · · · ·).

double excited state as $2s^2 2p\ ^2P^o$, $2s2p^2\ ^2D$, $2s2p^2\ ^2S$ and $2p^3\ ^2P^o$. Figure 3 shows the $C^{4+}(1s2s)$ ^{1}S and ^{3}S ionizations. We note the differences in resonance behavior for ionizing 1s2s ^{3}S and ^{1}S: the lowest doublet state ($2s^2 2p\ ^2P$) is much more significant for ^{1}S than for ^{3}S, since the latter(^{3}S) requires a spin flip in the 1s2s $\rightarrow$ $2s^2$ core transition.

REFERENCES

1. Müller, A., Hofmann, G., Weissbecker, B., Stenke, M., Tinschert, K., Wagner, M. and Salzborn, E., *Phys. Rev. Letters* **63**, 758 (1989).
2. Berrington, K. A., Pelan, J. and Quigley, L., *J. Phys. B* **30**, 4973 (1997).
3. Berrington, K. A. and Nakazaki, S., *J. Phys. B* **31**, 313 (1998).
4. Burke, P. G., and Robb, W. D., *Adv. At. Mol. Phys.* **11**, 143(1975).
5. Berrington, K. A., Eissner, W. B. and Norrington, P. N., *Comp. Phys. Commun.* **92**, 290 (1995).
6. Crandall, D. H., Phaneuf, R. A. and Gregory, D. C., *Oak Ridge National Laboratory Report*, ORNL-7020/TM (1979).
7. Bell, K., Gilbody, H. B., Hughes, J. G., Kingston, A. E. and Smith, F. J., *Culham Laboratory Report*, CLM-R216 (1982).
8. Phaneuf, R. A., Janev, R. K. and Pindzola, M. S., *Oak Ridge National Laboratory Report*, ORNL-6090/V5 (1987).
9. Schiff, B., Pekeris, C. L. and Accad, Y., *Phys. Rev. A* **4**, 885 (1971).
10. Thornbury, J. F. and Hibbert, A., *J. Phys. B* **20**, 6447 (1987).

Positronium - Atom Scattering

Jennifer E.Blackwood*, C.P.Campbell*, Mary T.McAlinden+ and H.R.J.Walters*

*Department of Applied Mathematics and Theoretical Physics, The Queen's University of Belfast, Belfast BT7 1NN, UK
+School of Computing and Mathematical Sciences, Oxford Brookes University, Headington, Oxford OX3 0BP, UK

Abstract. Results of recent large coupled - pseudostate calculations of Ps scattering by H and He are presented. The parlous state of both theory and experiment for He is highlighted.

With the advent of monoenergetic energy tunable positronium (Ps) beams [1] a whole new area of atomic collision physics has been opened up. Positronium is special in that it is a **light** neutral projectile. The states of Ps fall into two classes, ortho (o) and para (p), depending upon whether the electron and positron are in a spin triplet or spin singlet combination respectively. The significance of this classification lies in the different lifetimes of these spin states against annihilation of the electron and positron into photons. Thus, in its electronic ground level, Ps(1s), o-Ps has a lifetime of 142ns and decays predominantly into three photons, while p-Ps has a lifetime of 0.125ns and decays predominantly into two photons. Like a hydrogen atom (H), positronium can be created in any electronic state Ps(nlm). It is therefore necessary that the electronic condition of the beam be defined. In the present state of the art, Ps beams consist of essentially o-Ps in the ground 1s state [1], p-Ps(1s) is too short - lived to be transportable as a beam. Experimental capability is presently at an early stage and, except for a very limited amount of rough data on differential scattering [1], is confined to total cross section measurements. So far, such measurements have been made for Ps scattering by He, Ar, H_2 and O_2 [1]. In addition to the beam measurements there are also some cross section data at very low energies deduced from observations of the annihilation rate of o-Ps(1s) in various gases [2-5].

From a theoretical viewpoint, Ps scattering is a very difficult problem. Unlike electron and positron scattering, the projectile, Ps, now has internal degrees of freedom which must be taken into account as well as those of the target, this is a significant complication [6]. Because the centre of charge of the Ps coincides with its centre of mass, the direct Coulomb interaction between the Ps and the atomic or molecular target is very much weakened compared with that arising from electron exchange. However, the exchange process is very difficult to calculate since it involves electron swapping between two different centres, the target and the Ps. Because of these difficulties, progress in the theoretical treatment of Ps - atom collisions has been slow. The archetypical system is Ps + H, which is also a classic example of the Coulomb four - body problem. Pioneering theoretical work on this system was done by Massey and Mohr, Fraser, and Drachman and Houston. Early theoretical studies of Ps + He scattering were also made by Fraser, Drachman and Houston, and Barker and Bransden. In more recent times McAlinden/Campbell et al [6,7], Ghosh and co - workers, Biswas and Adhikari, and Peach have made noteworthy contributions to the field. A complete set of references on these works may be found in a forthcoming article by Blackwood et al [8].

Here we report recent results for Ps + H and Ps + He scattering obtained using a large coupled - pseudostate expansion [7,8]. To be specific, let us consider Ps + H scattering. In the coupled pseudostate approximation the collisional wave function, Ψ, is expanded as [9]

$$\Psi = \sum_{a,b} \left[G_{ab}(\mathbf{R}_1)\phi_a(\mathbf{t}_1)\psi_b(\mathbf{r}_2) + (-1)^S G_{ab}(\mathbf{R}_2)\phi_a(\mathbf{t}_2)\psi_b(\mathbf{r}_1) \right] \tag{1}$$

where $\mathbf{R}_i \equiv (\mathbf{r}_p + \mathbf{r}_i)/2$ is the position vector of the Ps centre of mass, $\mathbf{t}_i \equiv \mathbf{r}_p - \mathbf{r}_i$ is the Ps internal coordinate, and $\mathbf{r}_p$ ($\mathbf{r}_i$) is the position vector of the positron (ith electron), all position vectors being referred to the atomic nucleus as origin. In (1) S specifies the total electronic spin which, assuming a non - relativistic purely Coulombic Hamiltonian, is conserved in the collision. The functions ϕ_a (ψ_b) are Ps (H) states. The sets of states ϕ_a and ψ_b consist of both pseudostates and eigenstates satisfying

$$\langle \phi_a | H_{Ps} | \phi_{a'} \rangle = E_a \delta_{aa'} \qquad \langle \psi_b | H_A | \psi_{b'} \rangle = \epsilon_b \delta_{bb'} \tag{2}$$

where H_{Ps} (H_A) is the Ps (H) Hamiltonian. The expansion (1) leads to coupled equations of the form [7,8]

$$\left(\nabla_R^2 + p_{ab}^2 \right) G_{ab}(\mathbf{R}) = 4 \sum_{a'b'} [V_{ab,a'b'}(\mathbf{R})G_{a'b'}(\mathbf{R})$$

$$+ (-1)^S \int L_{ab,a',b'}(\mathbf{R}, \mathbf{R}')G_{a'b'}(\mathbf{R}')d\mathbf{R}' \Big] \tag{3}$$

where p_{ab} is the momentum of the Ps in the "ab" channel. The local potentials $V_{ab,a'b'}(\mathbf{R})$ give the direct Coulombic interaction between the Ps and the atom. The non - local couplings $L_{ab,a'b'}(\mathbf{R}, \mathbf{R}')$ arise from electron exchange between the Ps and the atom. It is not difficult to show that $V_{ab,a'b'} = 0$ if the Ps states ϕ_a and $\phi_{a'}$ have the same parity. In particular this implies that $V_{ab,ab} = 0$ so that the direct Coulomb interaction is non - diagonal. This property has the consequence that the direct term is weakened relative to the exchange interaction which has non - zero diagonal elements $L_{ab,ab}(\mathbf{R}, \mathbf{R}')$, hence the pronounced importance of exchange in Ps - atom scattering.

From (1) it is clear that the scale of the calculation grows as the product of the number of Ps states ϕ_a times the number of atom states ψ_b. To contain the siize of the calculation it has therefore been assumed that the atom remains in its initial (ground) state, ie, only one atom state, ψ_b, is used in the expansion (1), that being the ground state. However, for the dynamics of the Ps considerable flexibility has been retained. In the results shown here 22 Ps states, consisting of 3 eigenstates 1s, 2s, 2p and 19 pseudostates, $\overline{3s}$ to $\overline{7s}$, $\overline{3p}$ to $\overline{7p}$, $\overline{3d}$ to $\overline{7d}$, $\overline{4f}$ to $\overline{7f}$, have been used [7,8]; we refer to this as a 22ST approximation.

Some results for Ps(1s) scattering by H(1s) are shown in figures 1 and 2. Figure 1 reveals an interesting sequence of resonances which appear in the electronic spin singlet partial waves up to H - wave. These resonances may be understood as unstable bound states of the positron orbiting H^- [10]. Of interest is the conversion of o-Ps into p-Ps, figure 2 shows such a cross section for o-Ps(1s) $\longrightarrow$ p-Ps(1s). This cross section is interestingly structured, the structure near 5eV coming from the resonances of figure 1. A complete picture of the 22ST total cross section over the energy range 0 to 40eV has been given in reference [7], it is similar in pattern to that shown in figure 4 below for a He target.

More interesting, because of the availability of experimental data, is the Ps + He system. Figure 3 shows the 22ST total cross section for o-Ps(1s) + He(1^1S) scattering. Here it is compared with the experimental data and with two other theoretical approximations, those of Biswas and Adhikari [11] and Peach [1,12]. From figure 3 we see that there is considerable disagreement between the low energy measurements, with the most recent of these, by Skalsey et al [5], being by far the smallest. The largest cross section, that of Nagashima et al [4], is a factor of 5 bigger than the Skalsey et al datum point, at $13 \pm 4 \, \pi a_0^2$ it is in good agreement

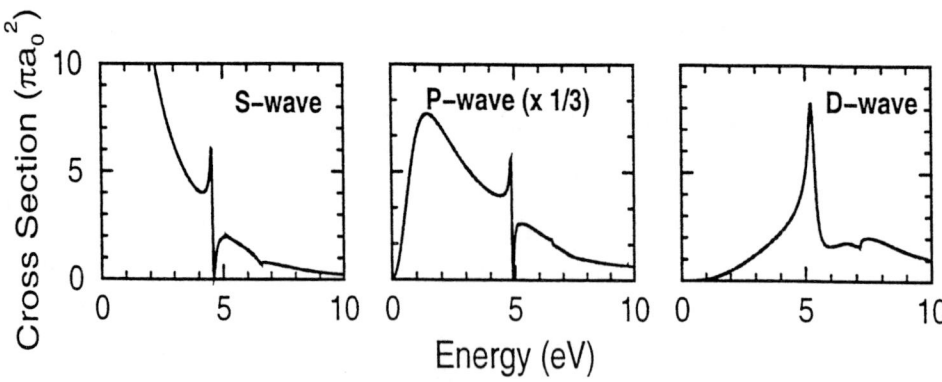

FIGURE 1. Electronic spin singlet partial wave cross sections for Ps(1s) + H(1s) elastic scattering. The P - wave cross section has been reduced by a factor of 3.

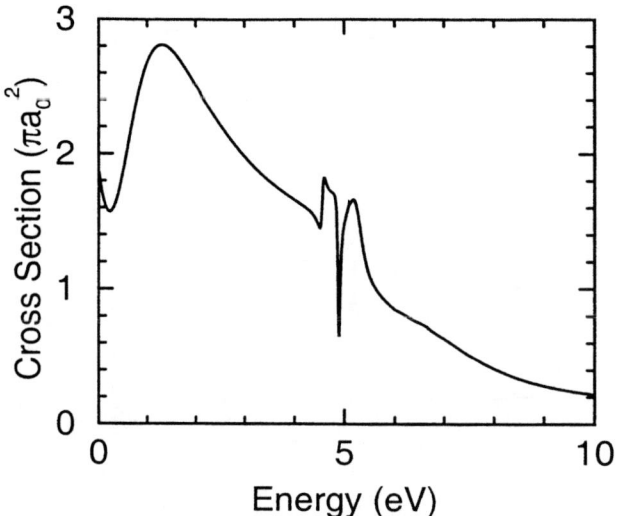

FIGURE 2. Cross section for o-Ps(1s) + H(1s) $\longrightarrow$ p-Ps(1s) + H(1s)

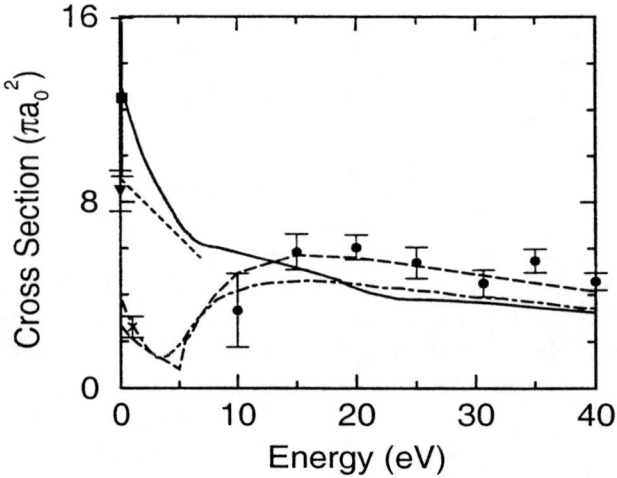

FIGURE 3. Total cross section for o-Ps(1s) + He(1^1S) scattering. Theoretical results : solid curve, 22ST; long - dashed curve, Biswas and Adhikari [11]; dash - dot curve, Peach [1]. Experimental data : solid circles, Garner et al [1]; solid square, Nagashima et al [4]; solid triangle down, Canter et al [2]; cross, Skalsey et al [5]; short - dashed curve, Coleman et al [3] (no error estimates are given by these authors).

457

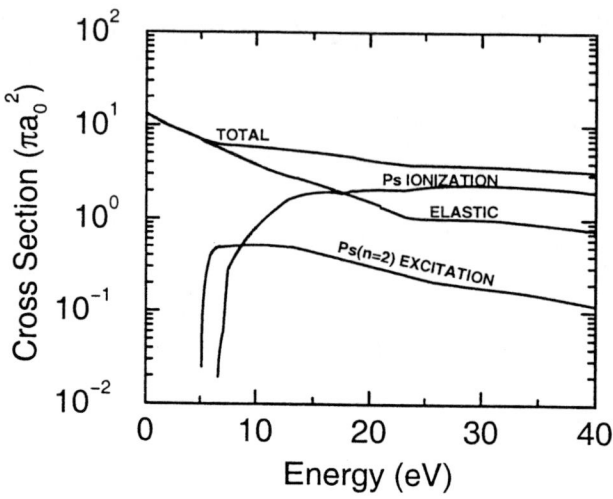

FIGURE 4. The 22ST total cross section and its components for o-Ps(1s) + He(1^1S) scattering.

with the 22ST calculation. By contast, the theories of Biswas and Adhikari and of Peach go through the measurement of Skalsey et al and even predict the down - turn in the data of Garner et al [1] at 10eV. A critique of the three approximations shown in figure 3 has been presented in reference [8]. There it is pointed out that the much smaller low energy cross section of Biswas and Adhikari results from a modification of the three - state Ps(1s, 2s, 2p) + He(1^1S) close - coupling approximation that they use, a modification which we consider to be unfounded. The approximation of Peach is a more serious contender. Unlike the 22ST approximation, the work of Peach allows for distortion of the He atom in elastic scattering (the only process possible below 5.1eV). However, there are two weaknesses in her treatment of elastic collisions. Firstly, local model potentials are used to represent the interaction between the Ps and the atom, the true interaction, as equation (3) shows, is non - local; secondly, Peach adopts an adiabatic approximation to the scattering, whether this is reasonable for a light projectile like Ps is an interesting question; by contrast the 22ST approximation is fully dynamic. The main criticism of the 22ST approximation lies in its restriction to the He ground state. Yet, we find it hard to believe that the full inclusion of He states in the expansion for Ψ would lead to so substantial a change in the 22ST cross section as to give agreement with the measurement of Skalsey et al. Clearly there are some very serious problems with our understanding of Ps + He collisions, both experimentally and theoretically !

Finally, in figure 4 we show the total cross section from the 22ST approximation dissected into its principal components. Below the first inelastic threshold at 5.1eV, corresponding to Ps(n=2) excitation, elastic scattering is the only possibility; between 5.1eV and about 17eV it is the main process, but at higher energies Ps ionization is dominant. Ps(n=2) excitation is relatively unimportant at any energy. Figure 4 highlights a very significant strength of the coupled - pseudostate approach, namely, that it gives a complete picture of the fate of the Ps, it also emphasizes the importance of including ionization channels in the approximation, this is the main function of the pseudostates in our calculations.

REFERENCES

1. Garner A.J., Özen A., and Laricchia G., *Nucl. Instrum. Methods B* **143**, 155 (1998).
2. Canter K.F., McNutt J.D., and Roellig L.O., *Phys. Rev. A* **12**, 375 (1975).
3. Coleman P.G., Rayner S., Jacobsen F.M., Charlton M., and West R.N., *J. Phys. B* **27**, 981 (1994).
4. Nagashima Y., Hyodo T., Fujiwara F., and Ichimura I., *J. Phys. B* **31**, 329 (1998).
5. Skalsey M., Engbrecht J.J., Bithell R.K., Vallery R.S., and Gidley D.W., *Phys. Rev. Lett.* **80**, 3727 (1998).
6. McAlinden M.T., MacDonald F.G.R.S., and Walters H.R.J., *Can. J. Phys.* **74**, 434 (1996).
7. Campbell C.P., McAlinden M.T., MacDonald F.G.R.S., and Walters H.R.J., *Phys. Rev. Lett.* **80**, 5097 (1998).
8. Blackwood J.E., Campbell C.P., McAlinden M.T., and Walters H.R.J., *Phys. Rev. A*, to be published.
9. We shall use atomic units (au) in which $\hbar = m_e = e = 1$. The symbol a_0 is used to denote the Bohr radius.
10. Drachman R.J., *Phys. Rev. A* **19**, 1900 (1979).
11. Biswas P.K., and Adhikari S.K., *Phys. Rev. A* **59**, 363 (1999).
12. Peach G., private communication.

COLLISIONS INVOLVING
HEAVY PROJECTILES

Relativistic Effects in Collisions of High-Z Ions

Paul H. Mokler

GSI - Darmstadt and University of Giessen
GSI, Planckstr.1, 64291 Darmstadt, Germany
e-mail: P.Mokler@gsi.de

Abstract. Highly-charged, high-Z ions provide unique relativistic systems for atomic structure and collision studies. Typical examples for relativistic effects observed in those systems by photonic, electronic and atomic collisions will be discussed. These examples are radiative electron capture (REC), resonant transfer and excitation (RTE) and single excitation (EXC). In all these processes the complete interaction - including retardation and full multipole expansion - has to be taken into account. In contrast to non-relativistic systems the magnetic interaction plays an essential role. A review on experiments confirming these effects at moderate relativistic collision velocities is given.

INTRODUCTION

Few-electron ions with $Z_p \cdot \alpha \rightarrow 1$ are unique laboratories in the relativistic domain (where Z_p is the atomic number of a highly-charged projectile ion and $\alpha \approx 1/137$ the fine-structure constant). For instance, in a H-like U^{91+} ion with $Z_p = 92$ the K-shell electron has a classical orbital velocity in the relativistic regime of $Z_p \cdot \alpha \approx 0.67$ ($\text{ß} = v/c \approx Z_p \cdot \alpha$) and it probes a mean electric field of $F \approx 2 \cdot 10^{+16}$ V/cm, cf. Ref. (1). This field is close to the Schwinger limit, where by moving the electron over its own extension characterized by the Compton wavelength the electron-positron pair energy is involved. It is evident that relativistic effects dominate the atomic structure of and the dynamics in these systems. The interaction mechanisms for such an electron with its surroundings are strongly influenced by the strong central fields of the high-Z ion, i.e. by relativistic effects. On the other hand, due to the strong confinement of the electron and its high binding energy (for U^{91+} we have about $r_K \approx 500$ fm and $E_K \approx 130$ keV) high momenta have to be transferred in all interactions, which means collisions in the relativistic region. Hence, we have to consider two points (*i*) relativistic effects caused by the strong fields at high $Z_p \cdot \alpha$ values and (*ii*) relativistic effects by the interaction dynamics. For a detailed overview on the whole field of relativistic collisions the reader is referred to the book Eichler and Meyerhof (2).

CP500, *The Physics of Electronic and Atomic Collisions*, edited by Y. Itikawa, et al.
© 2000 American Institute of Physics 1-56396-777-4/00/$17.00

In this presentation, we will focus on selected interactions in photonic, electronic and atomic collisions for high-Z_p projectiles at relativistic collision velocities. In most cases we concentrate on the intermediate relativistic domain where the collision velocity is roughly in the order of the orbital velocity for the innermost electron in the high-Z projectile ion. Using as a guideline the classical adiabaticity parameter with η = E_{kin}/E_{bin} (where E_{kin} corresponds to the collision velocity of an equivalent fast electron and E_{bin} to the binding energy of the innermost projectile electron), we will mainly consider collisions roughly in the region with $\eta \geq 1$; this means for U ions about a velocity $\beta \geq 0.6$, a relativistic factor $\gamma \geq 1.25$, and a specific projectile energy in the region of 200 MeV/u. Moreover, we treat only processes related to high-Z projectiles where the targets consists of comparable low-Z_t atoms.

THE INTERACTIONS AND THEIR SIGNATURES

The interactions depicted here for collisions at high central fields ($Z_p \cdot \alpha \rightarrow 1$) are
(*i*)   for photonic collisions, ***radiative electron capture*** (REC)
        – the interaction of an electron with the photon field,
(*ii*)   for electronic collisions, ***resonant transfer and excitation*** (RTE)
        – the electron-electron interaction and
(*iii*)   for atomic collisions, projectile ***excitation*** (EXC) or also ***ionization*** (ION)
        – the interaction of a target atom (Z_t) with the projectile electron.
All these processes (except pure ION) lead in the end (directly or by de-excitation) to specific and – due to the high Z_p – to prompt x-ray emission (Fig. 1).

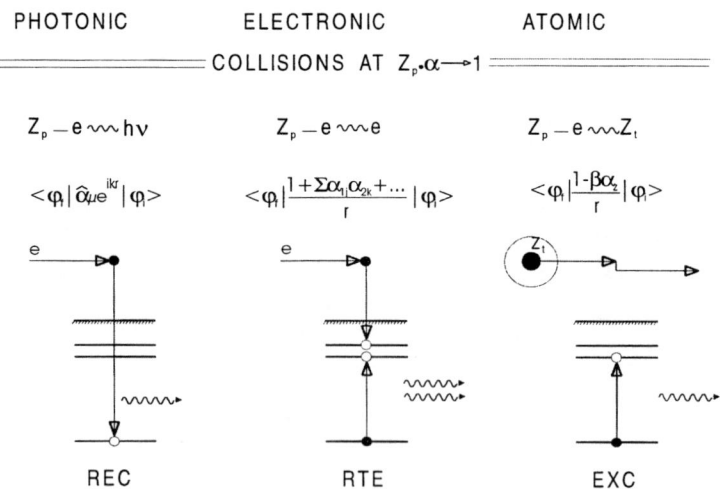

FIGURE 1. Interaction processes discussed for $Z_p \cdot \alpha \rightarrow 1$, for details see text.

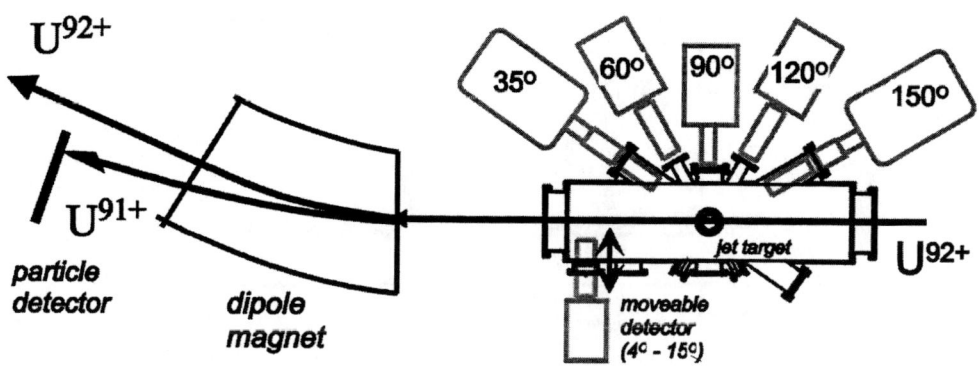

FIGURE 2. Typical arrangement for x-ray – particle coincidence measurements.

Within the impulse approximation REC and RTE are equivalent to radiative recombination (RR) and to dielectronic recombination (DR), respectively (3, 4), and can be treated as the time reversed processes to photo-ionization (PI) and to Auger ionization (AI). The kinetic energy of the quasi-free target electron with respect to the projectile ion (E_{kin}) and the binding energy gained is transferred in REC to the photon and in RTE to another bound projectile electron. Whereas REC has no limitations for the projectile energy, and hence for E_{kin}, RTE is a resonance process active only at specific E_{kin}. Due to the confinement of the quasi-free electron in the target atom its Compton profile has to be taken into account for the energy balance, leading to a broadening of the transitions. First observations on REC and RTE were made in the early seventeenth and early eighteenth (5 – 8); a survey on the fields can be found in the books of Eichler and Meyerhof (2) and of Graham et al. (9), respectively. An overview on recent developments in the fields of REC and DR is given in the invited papers of Eichler (10) and Schippers (11) in this proceedings. For EXC and ION the reader is also referred to Eichler and Meyerhof (2). Here, already a quasi-relativistic treatment in first order perturbation theory seems to describe ION reasonably well, see below (12, 13).

For collisions at high fields, i.e. at $Z_p \cdot \alpha \to 1$, both fully relativistic wave functions and the full, relativistic correct interactions have always to be taken into account. That means, in particular instead of the Coulomb interaction the full Liénard-Wiechert potential and the full multipole expansion have to be considered (2). In Fig. 1 the essentials of the corresponding matrix elements for the cases considered are exemplified. In the photonic case the full multipole expansion determines the transition: $\alpha \cdot \mu \cdot \exp(i \cdot kr)$, with α the Dirac matrices and μ the polarization vector. For the electronic case current and spin interactions between both the electrons – the Breit term (14) – is essential in addition to the 1/r Coulomb interaction: $1/r \cdot (1 - \Sigma \, \alpha_{1j} \cdot \alpha_{2k} + \ldots)$, for the full term see e.g. Refs. (15 – 17). Also in the atomic case the inclusion of the full interaction, i.e. the Liénard-Wiechert Potential $1/r \cdot (1 - \beta \cdot \alpha_z)$ is crucial. In all the cases retardation has to be included into the considerations.

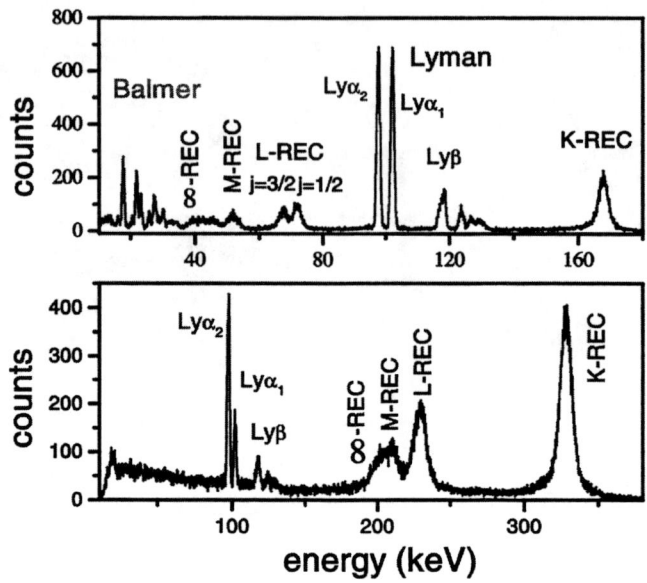

FIGURE 3. X-ray spectra for bare U^{92+} ions capturing one electron from a N_2 gas-jet target at 68 and 358 MeV/u (top and bottom, respectively). The x-ray emission is observed under 132° and the x-ray energies are transferred to the frame of the fast ion.

For the interaction detected the x-ray emission in connection with the charge of the projectile after the collision is a unique signature. For REC and RTE a quasi-free target electron is captured and the x-ray emission is measured in coincidence with the capture event, i.e. with the projectile after a magnetic charge state separation. For EXC both, coincidence and anti-coincidence techniques can be used in order to determine the x-ray emission associated with no change in ion charge. (For pure ION the change in ion charge due to the electron loss is detected.) In Fig. 2 a typical arrangement for those collision experiments is displayed (18): For the case of bare U^{92+} ions (coming from the right) colliding with atoms - here of a gas-jet target (19) - the x-ray emission associated with one-electron capture is detected under a bunch of different observation angles.

Fig. 3 depicts typical x-ray spectra measured in coincidence with one-electron capture for bare U^{92+} – N collisions at 68 and 385 MeV/u (20). The spectra were taken by a Ge(i)–x-ray detector under 132° observation at the gas-jet target (19) of the heavy ion storage ring ESR (21). Two different energy scales displaying the x-ray energy with respect to the projectile emitter frame are used for convenience. Two kinds of x-ray lines can be observed: (*i*) Characteristic lines from x-ray transitions within the projectile ions – in the shown case within H-like U^{91+}; these narrow lines are primarily the Lyman transitions to the ground state (at around 100 keV the Ly-α_1 line, $2p_{3/2}$ – $1s_{1/2}$ transitions and the Ly-α_2 line, $2p_{1/2}$, $2s_{1/2}$ – $1s_{1/2}$ transitions) and to a smaller extent also the Balmer lines (around 20 keV). (*ii*) Broad lines from REC transitions primarily to the projectile ground state, K-REC, and to some extent also to excited projectile shells, L-REC and M-....REC. The REC line centroides shift with the projectile energy accordingly ($E_{h\nu} = E_{bin} + \gamma \cdot (mc^2 - 1) - \gamma \cdot E_t$; with F_{bin} and F_t projectile and target binding energies).

Two cases are shown in Fig. 3: In the high-energy case (358 MeV/u at the bottom) all REC lines are in energy far beyond the characteristic lines; at the lower collision energy (68 MeV/u at the top) the Lyman lines are imbedded in between the K- and L-REC lines. At an intermediate collision energy in the region around 110 – 140 MeV/u the L-REC lines will coincide with the Ly-α lines. This is exactly the resonance region for RTE. At this energy the L-REC is still resolved into two components with j = 1/2 and 3/2 ($2s_{1/2}$, $2p_{1/2}$ and $2p_{3/2}$ levels – the REC width increases with collision velocity). Each of these L-REC components can be at resonance with the Ly-α_2 or Ly-α_1 transition energies. This is the resonance case for KLL- RTE. Correspondingly we have three resonances named $KL_{1/2}L_{1/2}$- , $KL_{1/2}L_{3/2}$- and $KL_{3/2}L_{3/2}$- RTE (using the Auger notation). At these resonance energies L-REC and KLL-RTE are competing processes and may interfere with each other, cf. Ref. (22).

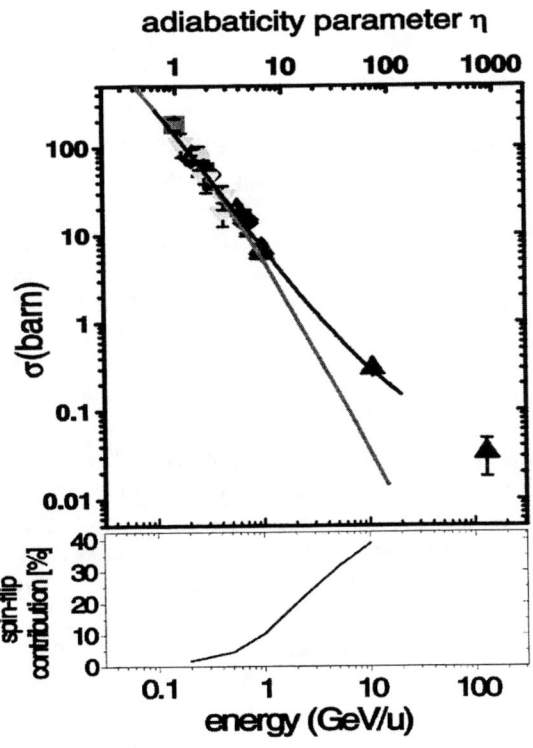

FIGURE 4. Total K-REC cross sections for bare ions and per quasi-free electron as function of the adiabaticity parameter η (top graph and top scale) in comparison with theory (dipole approximation and full relativistic calculation - lower and higher curves, respectively). The bottom graph gives the spin-flip contributions to the cross section for Au^{79+} ions with the corresponding energy scale.

PHOTONIC COLLISIONS – REC

As mentioned above, REC is the time reversed process to photo-ionization, see also the article of Eichler (10) in this volume and references cited there. In particular, REC gives access to PI for single electron systems – both in the ground or in excited states - at extreme central fields where higher multipoles contribute to the transitions. The total cross sections for REC (and PI) can be deduced from the corresponding x-ray emission taking into account the correct angular distribution, cf. the spectra in Fig. 3.

In Fig. 4 (top graph) the total K-REC cross sections measured for bare projectiles are summarized as a function of the adiabaticity parameter η, top energy scale (23). At the bottom a direct energy scale is given for Au^{79+} projectiles for comparison. All data are normalized to the number of quasi-free target electrons and coalesce within the unified η-plot onto one common curve. In the moderate relativistic collision regime, i.e. for $\eta < 10$, abundant data are available, cf. (24, 25), which can be well described already by the non-relativistic dipole approximation developed originally for PI by Stobbe (3). In the high relativistic regime ($\eta > 10$, heavy Z) experimental cross sections deviate considerably to higher values. In the whole energy region the experimental results agree perfectly with the fully relativistic calculations (26, 27). The theory is so perfect that K-REC emission can be used ideally for absolute calibration of other collision processes.

In the high relativistic regime only two data points are available; they refer to measurements at AGS, Brookhaven with 11 GeV/u Au ions and at SPS, CERN with 160 GeV/u Pb ions, i.e. γ about 13 and 168, respectively (28, 29). The highest energy data point shown here has to be taken with care. It was extracted from the measured capture cross section for light Z_t by subtracting the pair production value (30), which is the dominant electron capture mechanism (ECPP) at this ultra relativistc energy. (Originally, the authors of Ref. (29) put REC theory into the data evaluation for ECPP; here however, using ECPP theories as reliable the given REC cross section was deduced. Hence, this value is questionable and has a large uncertainty. The ECPP mechanism is not treated in the present paper; the reader is referred e.g. to Refs. (2, 28, 29, 31) and the literature cited there.)

The rigorous relativistic treatment of REC includes the full interaction of the electron with the photon field (26, 27). In particular, all multipoles are taken into account, including magnetic ones leading to spin-flip transitions. In the bottom part of Fig. 4 the spin-flip contributions for K-REC to bare Au^{79+} are displayed as function of ion energy, cf. (23). Beyond 1 GeV/u these spin-flip contributions increase considerably.

Spin-flip contributions will in particular modify the emission pattern for the REC radiation - and this already at lower projectile energies. The emission is governed by the selection rules for radiative transitions. In the non-relativistic case we expect a dipole distribution for the K-REC emission pattern. For higher projectile velocities we have to consider on the one hand retardation, i.e. the inclusion of higher multipoles, which shifts the distribution to backwards angles in the projectile frame (which is the

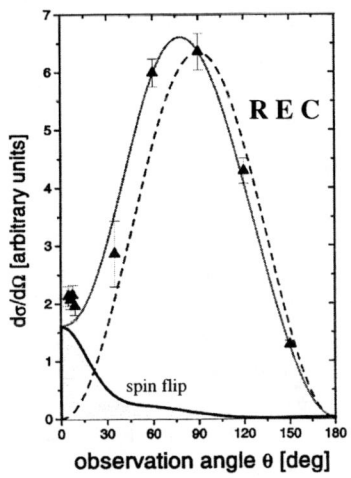

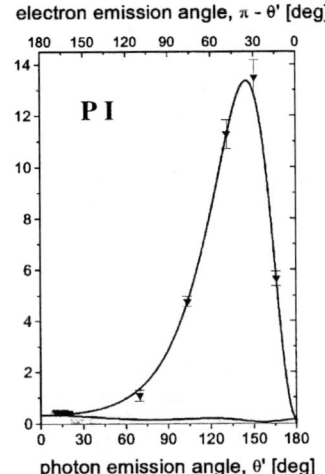

FIGURE 5. K-REC angular distribution (left graph) for 310 MeV/u U^{92+} – N_2 collisions compared to relativistic calculations (the spin-flip part is shown, the dotted line corresponds to a dipole distribution). The right graph gives the transformation for PI (; for the electron emission angles see the top scale).

direction of the electron hitting the projectile). On the other hand, the Lorentz transformation shifts this distribution back again, so that in the lab frame almost the original dipole distribution is approached, see e.g. (2, 32)

In Fig. 5, left side, an example for an emission pattern of K-REC radiation is given for the case of 310 MeV/u bare U^{92+} ions (33). The spin-flip contributions caused by magnetic multipole transitions manifest themselves uniquely by the emission near $0°$. The rigorous relativistic calculation agrees beautifully with the experimental findings (23, 27, 33). It is emphasized that for K-REC, the $0°$ emission – which is caused by the magnetic spin-flip transitions - is a pure relativistic effect for the electron-photon interaction mediated finally by the presence of the strong central field. In photon-ionization this basic effect is difficult to observe, as – due to the Lorentz transformation – the $0°$ enhancement is smeared out in that case over a wide angular range in backward directions, see the right graph in Fig.5. Moreover, scattering and background effects may dominate there measurements and, finally, electron correlations can not be excluded in PI experiments for neutral atoms.

REC investigations enable to penetrate into a region of the fundamental electron - photon interaction especially at extreme central fields. This is true not only for K-REC: First measurements are available for L- and M-REC showing also clearly the higher multipole contributions (32, 34). REC into excited states with successive radiative cascades to the ground state penetrate - by applying the principle of time reversal - into the field of resonance ionization at strong central fields and at photon energies around 100 keV (35). Moreover, the threshold behavior of PI can be studied now by REC measurements using decelerated ions at the storage ring ESR (36). REC studies down to 25 MeV/u Pb^{82+} ions, i.e. at $\eta \approx 0.12$ (!) colliding with a hydrogen gas-jet target have just been performed (37).

ELECTRONIC COLLISIONS – RTE

For electron-electron interaction in the presence of a strong central field we consider here essentially the time reversed Auger effect for $Z \cdot \alpha \to 1$. As pointed out above, this resonance process is strongly influenced by relativistic effects leading to enhanced transition strengths (15 – 17). First RTE measurements in the high-Z range were performed with He- and Li-like U projectiles studying the total charge exchange (38, 39). RTE leads to specific intermediate (doubly-excited) states which decay radiatively; this radiation gives detailed information on the formed intermediate resonance states. In Fig. 6 x-ray spectra are shown for the three $KL_jL_{j'}$ resonances (j, j' = 1/2 and 3/2) in the case of He-like U^{92+} – C collisions (40). According to the levels involved the K-α_2 or the K-α_1 lines show up selectively at the different resonance energies: the K-α_2 line for the $KL_{1/2}L_{1/2}$ resonances at 116 MeV/u, the K-α_1 line for $KL_{3/2}L_{3/2}$ the resonances at 132 MeV/u and both the lines for the $KL_{1/2}L_{3/2}$ resonances at 124 MeV/u. At these projectile energies the L-REC centroides coincides with the corresponding K-α lines; the REC contributes here as a broad background (41). Subtracting the REC contributions from the K x-ray emission the strength for the resonances can be extracted.

In Fig. 7, left side, the differential K emission cross sections are given for U^{90+} – C collisions under $30°$ observation (40). On the right side a comparison of the summed radiation with calculations is shown (42). The calculations including the full interaction, i.e. Coulomb and Breit terms, agree fairly well both in total and relative resonance strengths. This is also true for both the K-α components separately. Omitting the Breit interaction, i.e. mainly the magnetic part of the interaction, the first resonance strength is reduced by more than 30% proofing the importance of this contribution. The same is true for the total emission integrated over all angles.

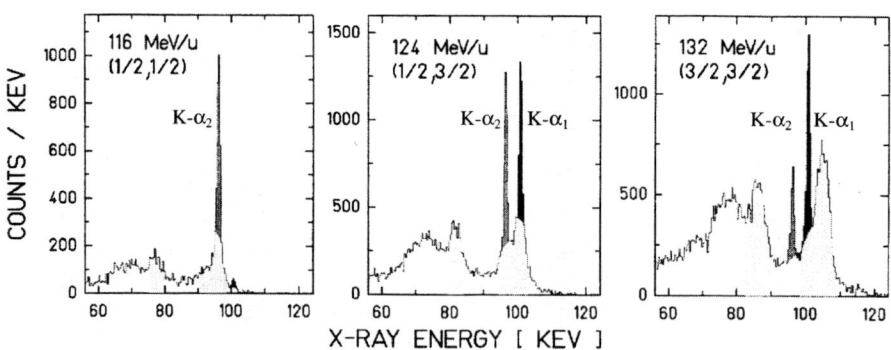

FIGURE 6. X-ray spectra observed at $30°$ for U^{90+} – C collisions at the three $KL_jL_{j'}$ - RTE resonance energies. (The X-ray energies refer to the projectile frame; the regions for L- and M- REC radiation are indicated by the gray areas.)

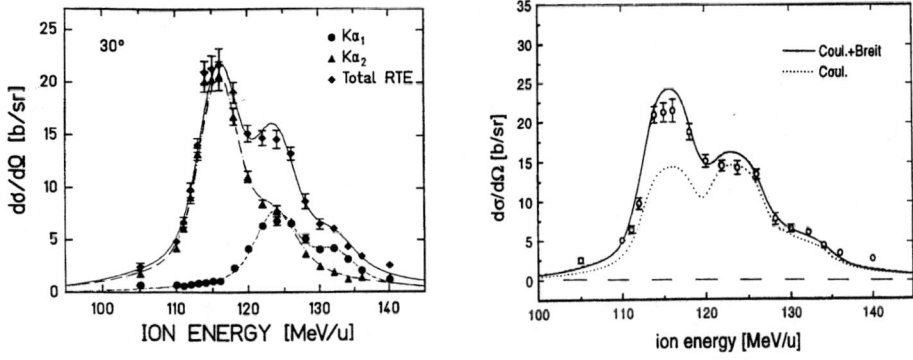

Figure 7. Differential x-ray emission cross sections as function of projectile energy for 30° observation of U^{90+} − C collisions (left side: experimental results). On the right side, the summed K emission is compared to theory with full interaction (full line) and without Breit interaction (dotted line).

The emission pattern for the K radiation induced by RTE is also well described by theory (42). In Fig. 8 the data points are compared to theory for the selected cases of K-α_2 emission at $KL_{1/2}L_{1/2}$ − RTE and K-α_1 emission at $KL_{1/2}L_{3/2}$ − RTE. In the $KL_{1/2}L_{1/2}$ case we have only transitions between $j = 1/2$ states leading necessarily to an isotropic emission in the projectile frame. The curve shown gives an isotropic emission pattern transferred from the ion frame into the Lab system. In the second case we have $3/2 − 1/2$ x-ray transitions which are peaked towards $45°$ in the lab system. In the projectile frame this corresponds to a strong dipole pattern preferring $90°$ emission with an anisotropy parameter $\beta_A = -0.5 \pm 0.04$ (43). That means that the formed intermediate states with $j = 3/2$ are strongly aligned perpendicular to the collision direction, i.e. $m_j = \pm 1/2$. As pointed out in Ref. (43), for uni-directional electron impact what can be assumed for RTE no angular momentum is transferred in collision direction. This was also confirmed by calculations for dielectronic satellites in initially H-like U^{91+} ions (44). Here, the anisotropy for the single intermediate DR states varies strongly with Z, i.e. with the strength of the central field. However, single resonances can only be resolved by direct DR measurements. First DR measurements with highly charged U ions were performed in the super-EBIT with mixed charge states (22).

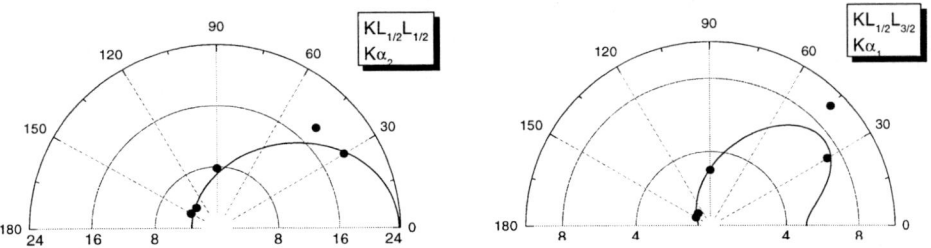

FIGURE 8. Emission characteristic for the K-α_2 line at the $KL_{1/2}L_{1/2}$ resonance and for the K-α_1 line at the $KL_{1/2}L_{3/2}$ resonance (right and left side, respectively) compared to theory.

ATOMIC COLLISIONS – ION; EXC

ION and EXC of a heavy and highly charged ion by a light-to-medium heavy target atom can usually be well described by first order perturbation theories (2). As long as the cross sections are small we have to consider close collisions and the screening of the target nucleus can normally be neglected. At relativistic velocities the Coulomb potential has to be replaced by the Liénard-Wiechert potential which particularly increases with γ perpendicular to the collision direction. This causes a corresponding cross sections enhancement, especially of the so-called transversal part, roughly with $\ln\gamma$ (2, 12, 13). Additionally, spin-flip contributions are induced by the motional magnetic field also increasing with increasing ion energy. Within this quasi-relativistic first order perturbation theory and using the direction of momentum transfer as reference axis the transversal and "more standard" longitudinal part of the interaction are orthogonal and, hence, can be added incoherently (12, 13).

In Fig. 9 data for electron loss in H-like Au^{78+} projectiles colliding with C atoms are compared with a quasi-relativistic first order perturbation theory for ION where additionally the longitudinal part is given separately (45, 46). At low energies, additionally recent results for Pb^{81+} ions are shown (47); for the high-relativistic region the AGS-Brookhaven point for Au^{78+} (28) and the SPS-CERN point for Pb^{81+} (29) are included.

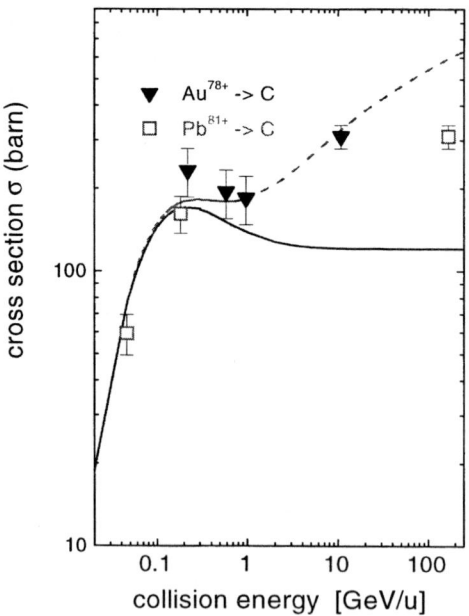

FIGURE 9. Ionization cross section for H-like Au^{78+} projectiles colliding with C target atoms in comparison to the quasi-relativistic predictions (dashed line); the longitudinal part is shown separately (full line). Data for Pb^{81+} ions are also included.

There are abundant data for ION of heavy H- and He-like projectiles available for adiabaticity parameters η roughly below around 3, all agreeing quite well with first order perturbation theory and fitting to a general (non-relativistic) η scaling (45, 48). Only the one ultra-relativistic data point falls clearly far below the quasi-relativistic prediction. In how far this deviation is caused by a breakdown of the quasi-relativistic first order approximation or by restrictions in screening distance for Z_t by the target electrons is under discussion. Taking into account in theory a cutoff impact parameter - also determined by screening - yields reasonable agreement with the high-relativistic data (49). An independent, very recent fully relativistic perturbation calculation in first order including the screening of the neutral target atom yields also good agreement for the ultra-relativistic data (50).

Rigorous relativistic calculations for ION are difficult to perform since they include a summation over all relativistically correct continuum states, cf. (2, 49 – 53). For EXC only a few bound state wave functions (initial and final states) have to be incorporated and, hence, full relativistic calculations are easier accessible. On the other hand, data for projectile excitation are rare. Only recently those data were published for H- and He-like Bi ions colliding with C targets at 82 and 119 MeV/u (54, 55); and new data have been taken only very recently for H-like Au^{78+} projectiles excited in 256 MeV/u collisions with Ar target atoms at the gas-jet target of the heavy ion storage ring ESR (56, 57). As shown in Fig.10 clean spectra for projectile excitation are obtained after subtracting the x-ray emission induced by capture from the total emission (56).

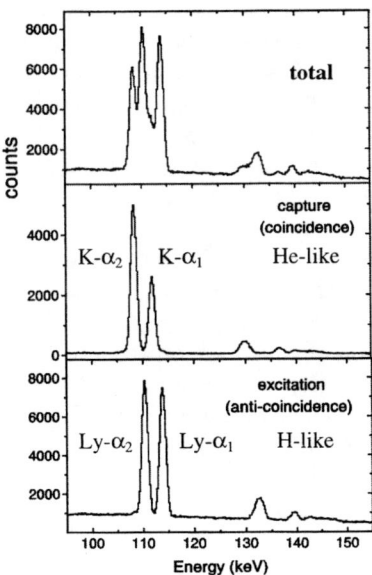

FIGURE 10. K x-ray spectra for H-like Au^{78+} ions colliding at 256 MeV/u with Ar target atoms; top: total emission, middle: emission associated with capture (He-like spectrum with K lines), bottom: emission associated with excitation (H-like spectrum with Lyman lines).

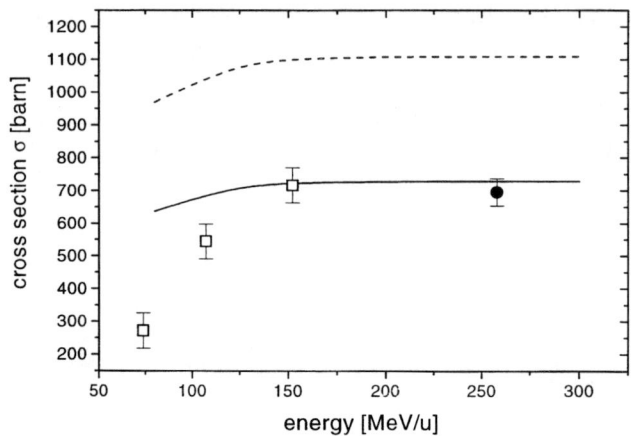

FIGURE 11. Cross sections for $1s_{1/2} - 2p_{3/2}$ excitation in H-like Au ions. Bi ions (open symbols) scaled in a proper way are also included (full line: full interaction; dashed line: without magnetic part).

The x-ray emission from projectile excitation can be accurately calibrated by the REC radiation and than compared to theory. Fully relativistic first order perturbation calculations using relativistic correct wave functions and the full interaction were performed by Ionescu (58) and Fritsche (59) leading to almost perfect agreement in EXC cross sections (54 – 57). On the other side, excluding the magnetic term in the interaction results in an "*increase*" in cross section by about 30 %. This means, that parts of the electric and the magnetic amplitudes of the interaction interfere which each other (54, 55). In Fig. 11 this fact is elucidated for the case of $1s_{1/2} - 2p_{3/2}$ excitation (Lyman-α_1 emission) in Au^{78+} ions colliding with Ar target atoms (60). Also included are the data for EXC in the neighboring Bi^{82+} ions scaled to the Au-Ar collision system. The theoretical curves are based on calculations for only a few energy points and are still preliminary. Nevertheless, the full calculations (full line) give good agreement with experiments, whereas omitting the magnetic part of the Liénard-Wiechert potential (dashed line) leads to an appreciable cross section enhancement (obey the offset of 200 barn in the figure). This so-called "interference effect" is a newly discovered and very unique feature of target-electron interaction at high projectile Z and moderate relativistic velocities (54, 55). As is evident from the calculations spin-flip transitions are essential for EXC.

Similar findings were made for the excitation into the $j = \frac{1}{2}$ levels (Ly-α_2 emission) as well as for He-like projectiles (54, 55). In order to get access to the impact parameter of concern first measurements on two-electron processes were started for He-like ions (223 MeV/u U^{90+}) investigating simultaneous ionization-excitation (61). This process is within the independent particle model confined to small impact parameter by the product of both probabilities. Also here a good agreement with a full relativistic first order perturbation theory (58) is claimed, establishing the importance of the found effect.

SUMMARY

Relativistic effects in collisions of high-Z ions have been discussed for selected cases. In all the cases it was essential to apply rigorous relativistic calculations to describe all the interaction processes correctly. In photonic collisions – represented by REC – spin-flip transitions are an essential part of the photon-electron interaction at $Z \cdot \alpha \rightarrow 1$, uniquely demonstrated by the emission pattern showing 0° degree emission. Further work especially near threshold as well as involving exited states will lead to a still deeper understanding of this fundamental interaction. Highly aligned excited states are produced during uni-directional REC collisions.

In electronic collisions – represented by RTE – the electron-electron interaction was studied. For $Z \cdot \alpha \rightarrow 1$ the Breit interaction, i.e. the interactions between currents and spins, leads to an appreciable cross section enhancement in particular involving $j = \frac{1}{2}$ levels, an effect which is difficult to observe so clearly in the corresponding fundamental Auger ionization. Perfectly aligned states are produced by the electron-electron interaction in these uni-directional collisions. Measurements resolving single resonances in the different $KL_jL_{j'}$ groups are a valuable goal for future experiments. In those measurements the interesting question for an interference between RTE and REC, i.e. between electron-photon and electron-electron interaction, can be explored.

The interaction of a target atom with the electron of a high-Z projectile – ION and EXC – is in the relativistic collision regime determined by "transversal" interaction, where also the magnetic part has to be included. ION is already well described by the quasi-relativistic treatment. Whereas EXC shows at moderate collision velocities quite clearly the influence of the magnetic part of the interaction yielding in total a reduction (!) in cross sections. More detailed experimental and theoretical work both for EXC and ION will give a deeper insight into this topic. Differential measurements, as simultaneous "ION $*$ EXC" will give additional information on the importance of the magnetic interaction at small impact parameters.

The present review concentrated on processes related to highly-charged high-Z projectiles since only in one- and few-electron systems the interactions can be studied most clearly. Special emphasis was given here to moderate relativistic collision velocities. In the ultra-relativistic region other interactions as pair production mechanisms will dominate. Moreover, heavy target species were not the subject of this presentation despite the fact that at moderate velocities higher field strengths can transiently be present in a collisionally formed "super-heavy" quasi-molecule, see e.g. Ref. (62). Those measurements involving only a few electrons are still to come.

ACKNOWLEDGEMENTS

The author highly appreciates the support from colleagues of the various collaborations working in this field – inside and outside GSI – for preparing this manuscript. The offer to include also most recent, partially unpublished results is especially acknowledged.

REFERENCES

1. Mokler, P.H., *Hyperfine Interactions* **114**, 21 – 43 (1998)
2. Eichler, J., and Meyerhof, W.E., *Relativistic Atomic Collisions*, San Diego: Academic Press, 1995
3. Stobbe, M., *Ann. Phys.* (Leipzig) **7**, 661 (1930)
4. Burgess, A., *Astrophys. J.* **139**, 776 (1964)
5. Raisbeck, G., and Yiou, F., *Phys. Rev. Lett.* **4**, 1858 (1971)
6. Schnopper, H.W. et al., *Phys. Rev. Lett.* **29**, 898 (1972)
7. Tanis, J.A., et al., *Phys. Rev. Lett.* **49**, 1325 (1982)
8. Tanis, J.A., et al., *Phys. Rev.* **A31**, 4040 (1985)
9. Graham, W.G., et al. (eds.), *Recombination of Atomic Ions*, New York: Plenum Press (1992)
10. Eichler, J., "Radiative electron capture and the photoelectric effect in hydrogen-like high Z systems", in *Proceedings of this Conference, XXI ICPEAC,* 1999, pp.
11. Schippers, S., "Recent storage ring results on electron-ion recombination", in *Proceedings of this Conference, XXI ICPEAC,* 1999, pp.
12. Anholt, B., *Phys. Rev.* **A19**, 1004 (1979)
13. Anholt, B., and Becker, U., *Phys. Rev.* **A36**, 4628 (1987)
14. Breit, G., *Phys. Rev.* **34**, 553 (1929)
15. Zimmerer, P., Grün, N., and Scheid W., *Phys. Lett.* **A148**, 457 (1990)
16. Chen, M.H., *Phys. Rev.* **A41**, 4102 (1990)
17. Pindzola, M.S., and Badnell, N.R., *Phys. Rev.* **A42**, 6526 (1990)
18. Stöhlker, Th., et al., *GSI Scientific Report 1997*, Darmstadt: GSI report **98-1**, 106 (1998)
19. Reich, H., et al., *Nucl. Phys. A626*, 473c (1997)
20. Stöhlker, Th., et al., *Hyperfine Interactions* **108**, 29 (1997)
21. Franzke, B., *Physica Scripta* **T22**, 41 (1988)
22. Knapp, D.A., et al., *Phys. Rev. Lett.* **74**, 54 (1995)
23. Eichler, J., and Stöhlker, Th., *Atomic physics with heavy ions*, Heidelberg: Springer, 1999, pp. 249
24. Mokler, P.H., and Stöhlker, Th., *Adv. At. Mol. Phys.* **37**, 297 (1996)
25. Stöhlker, Th., et al., *Phys. Rev.* **A51**, 2098 (1995)
26. Ichihara, A., Shirai, T., and Eichler, J., *Phys. Rev.* **A49**, 1875 (1994)
27. Eichler, J., Ichihara, A., and Shirai, T., *Phys. Rev.* **A51**, 3027 (1995)
28. Claytor, N., et al., *Phys. Rev.* **A55**, R842 (1997)
29. Krause, H.F., *Phys. Rev. Lett.* **80**, 1190 (1998)
30. Baltz, A.J., Rhoades-Brown, M.J., and Weneser, J., *Phys. Rev E54*, 4233 (1996)
31. Gould, H., "Electron capture and loss experiments at relativistic energies", Proceedings of the XX ICPEAC, *Photonic, Electronic and Atomic Collisions*, Singapore: World Scientific, 1997, pp. 465
32. Stöhlker, Th., et al., *Comments At. Mol. Phys.* **33**, 271 (1997)
33. Stöhlker, Th., et al., *Phys. Rev. Lett.* **82**, 3232 (1999)
34. Kandler, T., et al., *Z. Phys.* **D35**, 15 (1995)
35. Stöhlker, Th., et al., *Phys. Rev. Lett.* **79**, 3270 (1997)
36. Stöhlker, Th., et al., *Phys. Rev.* **A58**, 2043 (1998)
37. Krämer, A., et al., to be published (1999)
38. Graham, W.G.,et al., *Phys. Rev. Lett.* **65**, 2773 (1990)
39. Tanis, J.A., et al., *Nucl. Instr. Meth.* **B53**, 442 (1991)
40. Kandler, T., et al., *Phys. Lett.* **A204**, 274 (1995)
41. Kandler, T., et al., *Nucl. Instr. Meth.* **B98**, 320 (1995)
42. Gail, M., Grün, N., and Scheid, W., *J.Phys.* **B31**, 4645 (1998)
43. Mokler, P.H., et al., Physica Scripta T73, 247 (1997)
44. Chen, M.H., and Scofield, J.H., Phys. Rev. A52, 2057 (1995)
45. Stöhlker, Th., et al., *Nucl. Instr. Meth.* **B124**, 160 (1997)
46. Stöhlker, Th., private communication (1999)

47. Bräuning, H., et al. (1999) to be published
48. Scheidenberger, C., et al., *Nucl. Instr. Meth.* **B142**, 441 (1998)
49. Sorensen, A.H., *Phys. Rev.* **A58**, 2895 (1998).
50. Voitkiv, A.B., Grün, N., and Scheid, W., *Phys. Lett.* **A**, (1999) submitted
51. Mehler, G., et al., *Z. Phys.* **D13**, 193 (1989)
52. Momberger, K., et al., *J. Phys.* **B23**, 2293 (1990)
53. Momberger, K., Belkacem, A., and Sorensen, H.A., *Phys. Rev.* **A53**, 1605 (1996)
54. Stöhlker, Th., et al., *Phys. Lett.* **A238**, 43 (1998)
55. Stöhlker, Th., et al., *Phys. Rev.* **A57**, 845 (1998)
56. Krämer, A., et al., Physicy Scripta T80B, ppp (1999)
57. Krämer, A., et al., in *GSI scientific report 1998, GSI* **99-1**, 95 (1999)
58. Ionescu, D.C., private communications (1998)
59. Fritsche, S., private communications (1998)
60. Stöhlker, Th., private communication (1999)
61. Ludziejewski, T., et al., *Phys. Rev.* **A**, (1999) submitted
62. Mokler, P.H., and Liesen, D., *Progress in atomic spectroscopy,* New York: Plenum, 1984, part C, ch. 8, pp. 321

Charge–Changing Ion-Ion Collisions

Frank Melchert* and Erhard Salzborn+

*Physikalisch-Technische Bundesanstalt, 38116 Braunschweig, Germany
+Institut für Kernphysik, Justus-Liebig-Universität, 35392 Giessen, Germany

Abstract. Collisions between ions belong to the elementary processes occurring in all types of plasmas in astrophysical objects as well as in laboratory discharges. In recent years, the impetus for accurate cross section data for such collisions has become greater due to research in thermonuclear fusion using either magnetic or inertial confinement. Data are reviewed for collisions between light ions, in which only one or two electrons are involved.

INTRODUCTION

The investigation of charge-changing ion-ion collisions at keV energies is of considerable interest for several reasons. First, thermonuclear fusion research requires comprehensive data for those processes. In magnetically-confined fusion plasmas, ion-ion reactions may provide effective tools for plasma diagnostics. Furthermore, the fusion plasma needs auxiliary heating by injection of powerful neutral hydrogen beams. Negative hydrogen ions H^- accelerated to energies of about $10^5 eV$ can be neutralized most effectively in so-called plasma-neutralizers [1] where the interaction between H^- and multiply-charged plasma ions X^{q+} dominates the neutralization process. If inertial confinement fusion is driven by heavy ion impact, large storage rings are projected to accumulate intense ion beams of about 10 GeV energy [2]. Intra-beam ion-ion collisions at keV energies cause intensity losses [3] which can be minimized by a proper choice of the driver ion species.

Second, more fundamental reasons motivate the study of ion-ion collisions which are different from those between ions and atoms. When ions and atoms collide at low velocities $v < $ 1a.u., the charge clouds of the target atom and projectile ion merge and a transient molecule is formed. Electron capture and ionization processes can be understood in terms of transitions between states of these quasi-molecules. When two ions collide, the process is effected by long-range Coulomb forces absent in ion-atom collisions. At very low velocities it is obvious that the positive ions' charge clouds can not merge due to the Coulomb repulsion and then no quasi-molecule can be formed. Therefore trajectory and polarisation effects of the two ions may be important.

CP500, *The Physics of Electronic and Atomic Collisions*, edited by Y. Itikawa, et al.

While ion-atom, electron-atom and, more recently, electron-ion collisions have been widely investigated, the study of ion-ion collisions is still in its infancy (see e.g. [4]). Collisions between ions occur in all kind of plasmas at high rates, but the preparation of a single collision event requires isolation of ions in a given charge state and with given kinetic energies.

EXPERIMENTAL METHODS AND TECHNIQUES

The straightforward idea to investigate ion-ion collisions is the so-called *crossed-beams* technique where two ion beams are made to intersect. Although this technique, in principle, seems to be rather simple, inherent difficulties arise from the tenuous particle densities provided by typical ion beams.

Mutual Coulomb repulsion between the ions within their beams causes them to expand along their beam path. To ensure experimental resolution, the beam diameter and divergence have to be limited which, as a consequence, limits the ion beam currents. For example, a beam of singly-charged Ar^+ ions (energy 5 keV, diameter 4 mm, max. divergence 1°) can not transport more than a few μA of ion beam current. This beam contains less than 10^7 ions per cm^3 which is comparable to the particle density within ultrahigh vacuum (UHV). While two ion beams penetrate each other, ion-ion collision rates can be expected which are equal to the rates observed when an ion beam travels through UHV, and that is almost nothing. Depending on the particular cross section, only 10^{-10}-10^{-14} of the incoming flux undergoes an ion-ion collision. Furthermore, the ion beams travel through UHV along their complete trajectory, while they overlap only in the interaction region which has typically mm dimensions for a *crossed-beams* arrangement. That is why the ion-ion reaction signal is masked by 10^2-10^4–times more intense background contributions which arise from ionic collisions with the residual gas, even at prevailing UHV-conditions of 10^{-11} mbar.

The experimental challenge of *crossed-beams* experiments is to single out the ion-ion events, which occur at absolute rates of 10^{-2}-10^1 per second, from background rates of ions in the same charge state in the order of 10^2-10^4 per second, and from the incoming flux in the order of 10^{10}-10^{13} per second. For this purpose, coincidence [5] and beam modulation [6] techniques have to be applied. In spite of these difficulties and the need for a complex apparatus, the *crossed-beams* technique is very attractive. The reactants and the type of reaction can be clearly defined and the range of accessible energies spans seven or more orders of magnitude. If two ion beams cross at an interaction angle α, the collision energy E_{cm} in the center-of-mass frame (cm) is given by

$$E_{cm} = \mu \left(\frac{E_1}{m_1} + \frac{E_2}{m_2} - 2\sqrt{\frac{E_1 E_2}{m_1 m_2}} \cos \alpha \right), \qquad (1)$$

where E_i and m_i are lab frame energies and masses of beams 1,2 and μ is the reduced mass $\mu = (m_1 \cdot m_2)/(m_1 + m_2)$. The interaction angle α determines the cm-energy range that is accessible for given lab frame energies.

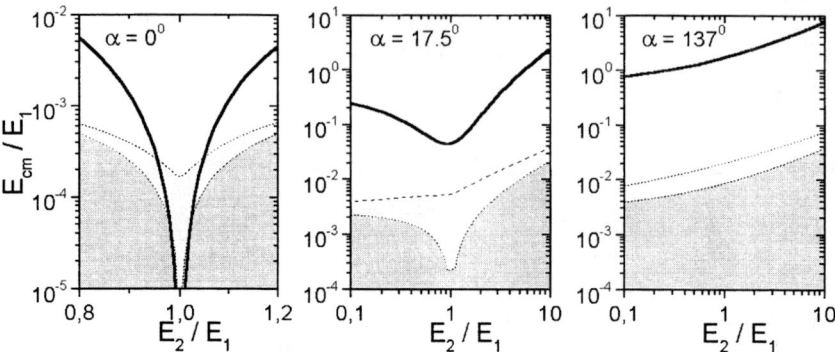

FIGURE 1. Center-of-mass energy E_{cm} (solid line) as a function of the lab frame kinetic energy E_2 of ion beam 2, measured in units of the lab frame kinetic energy E_1, for 3 different interaction angles α between the two ion beams: $\alpha = 0^o$ (left), $\alpha = 17.5^o$ (center) and $\alpha = 137^o$ (right). The shaded areas represent contributions to the energy resolution ΔE originating from the lab frame energy spreads $\Delta E_i, i = 1, 2$ (dark gray) and from the collision angle uncertainty $\Delta \alpha$ (light gray). $\Delta E_i = 0.005 \cdot E_i$, $\Delta \alpha = 1^o$ and $m_1 = m_2$ were assumed.

When both ion beams merge ($\alpha = 0^o$), very low cm-energies are attainable and special kinematic effects occur [7]. In Fig.1 energies E_{cm} are plotted for a *merged-beams* setup ($\alpha = 0^o$) and for *crossed-beams* setups using acute ($\alpha = 17.5^o$) or obtuse ($\alpha = 137^o$) angles. All energies are normalized by the kinetic energy E_1 in beam line 1. The energy resolution ΔE_{cm} stems from the lab frame energy spreads ΔE_1 and ΔE_2 in both ion beams. Furthermore, the collision angle uncertainty $\Delta \alpha$ which is a consequence of individual ion trajectories spread around the beam symmetry axis contributes to ΔE_{cm}. To access low cm-energies in *merged-beams* experiments, a careful beam collimation is required to reduce the angular spread $\Delta \alpha$ within the ion beams.

If two ion beams cross at an angle α, the cross section σ can be calculated from the equation

$$\sigma = \frac{v_1 v_2 \sin \alpha}{v_{rel}} \cdot \frac{R \, q_1 q_2}{I_1 I_2} \cdot F, \qquad (2)$$

where v_i, q_i and I_i are the velocities, charges and ion currents of both ion beams $i = 1, 2$. The so-called formfactor F describes the spatial overlap of the ion beams and is determined by sliding a slotted shutter along the z-axis, which is perpendicular to the interaction plane, through the ion beams. v_{rel} denotes the relative velocity of the ions.

Among all quantities in eq. (2), the event rate R of ion-ion reactions is the most difficult to determine. Far more particles originating from interactions other than ion-ion collisions reach the detectors. If both collision partners alter their charge state, e.g. in an electron capture reaction (6), both reaction products can be detected in coincidence. Since they are generated simultaneously and since their flight times from the beam intersection to the detectors are fixed, the corresponding output pulses of single particle detectors show a fixed time delay in the case of true ion-ion signals, whereas there is no time correlation between background events. Hence, time-to-amplitude-converter (TAC) spectra display a sharp peak produced by ion-ion collisions on top of a flat background due to random coincidences. A typical TAC spectrum is shown in the upper right part of Fig.2; details of this technique are discussed in Ref. [5].

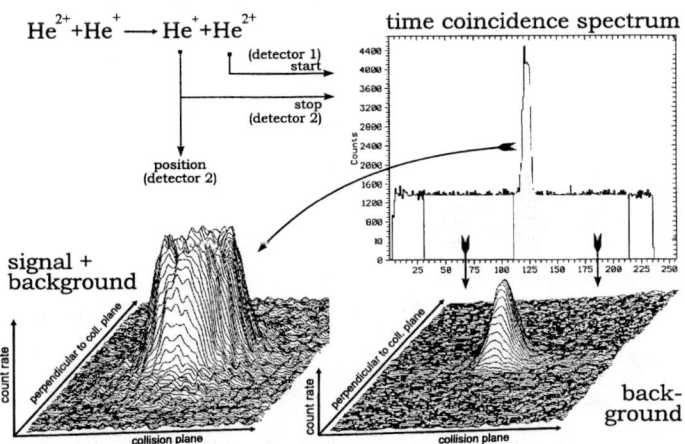

FIGURE 2. Position-sensitive detection of reaction products, illustrated for the electron capture reaction $He^{2+} + He^+ \longrightarrow He^+ + He^{2+}$ (6). He^{2+} product ions, recorded by single particle detector 1, start a time-to-amplitude-converter (TAC) which is stopped by He^+ product ions recorded by single particle detector 2. Simultaneously, detector 2 records the event position. All events in the TAC coincidence window (dark gray) are accumulated in one position matrix (signal + background), the other events (light gray) in a second position matrix (background). The spatial distribution of ion-ion events is obtained by suitable matrix subtraction. The matrix area is $40mm \cdot 40mm$.

If only one collision partner alters its charge state, e.g. in an ionization reaction (4), a coincidence technique does not apply. The ion-ion signal can be separated from background events by means of a beam pulsing technique. Both beams are chopped, e.g. by electrostatic deflectors upstream the interaction region, and the time dependence of the detector count rate is recorded. When both beams are switched on, ion-ion reactions are recorded on top of background contributions produced by either of the beams, and by electronic noise. In additional time periods, only beam 1, beam 2 or neither of the two is switched on in order to record the different background contributions. The true ion-ion signal is obtained by suitable subtraction. This method is described in detail in Ref. [6].

Recently, the first angular differential cross sections were measured for ion-ion collisions [8]. These measurements require position-sensitive product-ion detection. The idea is outlined in Fig.2 for the resonant electron capture reaction (6). As in the coincidence technique described above, a TAC spectrum is measured which is started by He^{2+} products, which stem from their parent He^+ beam, and stopped by He^+ reaction products, separated from their parent He^{2+} ion beam. Detector 2 consists of a channelplates-array and a position-sensitive resistive anode which allows to record the event time and position simultaneously. The TAC coincidence time window (dark gray) contains signal and background events which were accumulated in the left position matrix 1, while non-coincident windows (light gray) contain background events only, accumulated in the right position matrix 2. After suitable normalisation procedures, matrix 2 can be subtracted from matrix 1 in order to obtain the spatial distribution of He^+ reaction products.

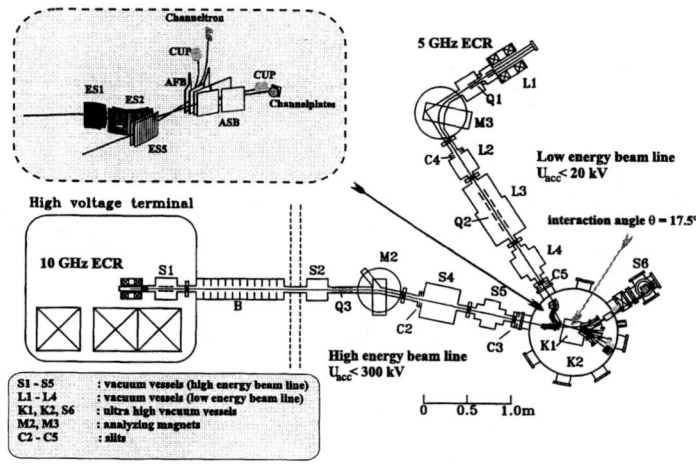

FIGURE 3. Schematic set-up of the Giessen ion-ion *crossed-beams* facility.

Note that a spatial distribution matrix is accumulated at rates ranging from 10^{-2} to 10^1 events per second. For a single distribution, data acquisition times in the order of a few days are usually required.

The extraction of differential cross sections $d\sigma/d\Omega$ from a measured spatial distribution matrix requires, on one hand, the transformation between lab- and cm-system. On the other hand, the primary ion beam is not a mathematical line which is infinitely narrow, but has a finite beam profile $\mathcal{P}$ (beam radius less than 1 mm, divergence half angle $\vartheta_d \approx 0.1°$) which broadens the observed scattering distribution and smears out structures. The experimental data can be deconvoluted with respect to $\mathcal{P}$, if sufficient angular scattering occurs, i.e. if the main part of the distribution is scattered into solid angles $d\Omega(\vartheta)$ with ϑ larger or at least comparable to ϑ_d. Bayesian deconvolution [9] or Fourier methods were introduced by Kruedener [10] for ion-ion angular differential cross section measurements.

A typical example of a modern *crossed-beams* experiment is the Giessen ion-ion collision facility (Fig.3), which has been described in full detail by Meuser et al. [11]. Ions are produced in ECR sources and accelerated to energies up to $q_1 \cdot 300kV$ and $q_2 \cdot 20kV$, respectively. Both beams are momentum-analyzed by magnets and pass through differentially pumped vacuum tanks providing increasingly lower residual gas pressure before they intersect in an UHV vacuum chamber which is operated at pressures in the low 10^{-11} $mbar$ region. The interaction angle α can be adjusted at 17.5^o or 137° which gives access to the cm-energy ranges plotted in Fig.1. The ion optical system of the apparatus must allow to distinguish a true ion-ion event from the incoming flux which can be up to 10^{14}-times more intense. The complete system is ion-optically optimized in extensive computer simulations and test measurements. Detailed information can be found in Ref. [11].

COLLISIONS BETWEEN LIGHT IONS

One-Electron Systems

The investigation of collisions between protons and singly-charged helium ions is of considerable interest in plasma physics, especially in thermonuclear fusion research. Not only do these encounters occur inevitably in fusion plasmas but they also represent the simplest ion-ion collision system involving only one electron. Description of electron capture σ_C

$$H^+ + He^+ \longrightarrow H^0 + He^{2+} \qquad (3)$$

and ionization σ_I

$$H^+ + He^+ \longrightarrow H^+ + He^{2+} + e^- \qquad (4)$$

is therefore an ideal testing ground for different theoretical models. Since the wavefunctions are known exactly, the accuracy of the calculated cross sections directly reflects the quality of the theoretical approach. Of special interest is the intermediate energy range where the relative velocity of the colliding particles is comparable with that of the orbiting electron.

Three experimental groups [12,13,5] have studied these collisions independently. While capture cross sections σ_C were measured directly by coincident detection of H^0 atoms and He^{2+} ions, ionization cross sections σ_I were measured indirectly by subtraction $\sigma_I = \sigma_{2+} - \sigma_C$, where σ_{2+} is the cross section for the total production of He^{2+} ions

$$H^+ + He^+ \longrightarrow He^{2+} + ... \qquad (5)$$

which is obviously the sum of reactions (3) and (4). It can be measured by means of a beam pulsing technique [6]. For low collision energies the cross sections σ_{2+} and σ_C are almost equal; therefore, subtraction yields considerably larger error bars for σ_I.

483

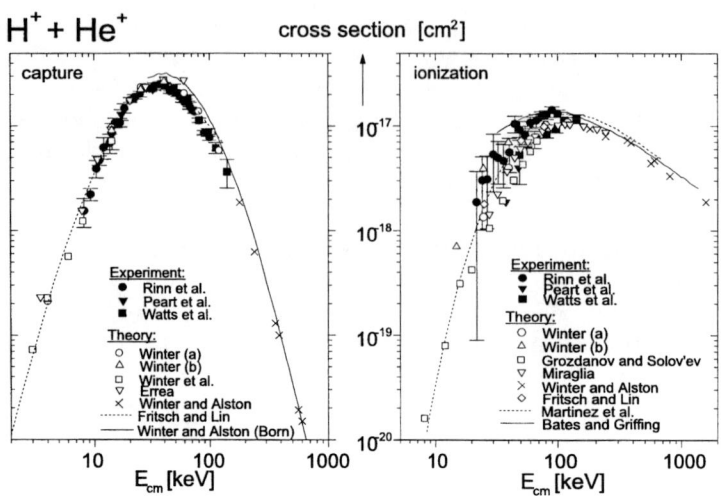

FIGURE 4. Cross sections for electron capture (3) and ionization (4) in $H^+ + He^+$ collisions, as a function of the cm-collision energy E_{cm}. Experimental *crossed-beams* results by Rinn et al. [5,6], Peart et al. [12] and Watts et al. [13]. Most theoretical results are obtained by close-coupling (CC) calculations: Winter (a) [14]: Sturmian basis and pseudostates; Winter (b) [15]: 3-center atomic orbital (AO) basis; Winter et al. [16]: MO basis; Errea [17]: molecular orbital (MO) basis; Winter and Alston [23]: high energy CC; Fritsch and Lin [19,20]: AO^+ basis; Grozdanov and Solov'ev [21]: MO basis; Miraglia [18]: CDW; Martinez et al. [22]: CDW-EIS; Winter and Alston [23] and Bates and Griffing [24]: Born approximation.

The experimental cross sections are compared with theory in Fig.4. In the high energy limit, the Born approximation is valid while for lower energies the continuum distorted wave calculation CDW-EIS with eikonal initial states by Martinez et al. [22] agrees as perfect with the ionization data as the atomic orbital AO^+ calculation by Fritsch and Lin [19] with the capture results. Since various authors confirm these findings, total cross sections for the reactions (3) and (4) are well described by theory.

For about two decades, theory was tested by comparing calculated and measured total cross sections. The first angular differential cross section has been measured by Kruedener et al. [8] for the electron capture of α-particles from He^+ ions:

$$He^{2+} + He^+ \longrightarrow He^+ + He^{2+} \tag{6}$$

This one-electron, charge-symmetric collision is resonant for $1s \rightarrow 1s$ electron capture and provides a unique testing ground for the study of the long-range Coulomb interaction in the quantum three-body problem. Kruedener et al. found an oscillatory structure in the deconvoluted scattering distribution of He^+ ions produced in reaction (6) which is displayed in Fig.5. The minimum in forward direction is surrounded by symmetric structures with rapidly decreasing amplitudes

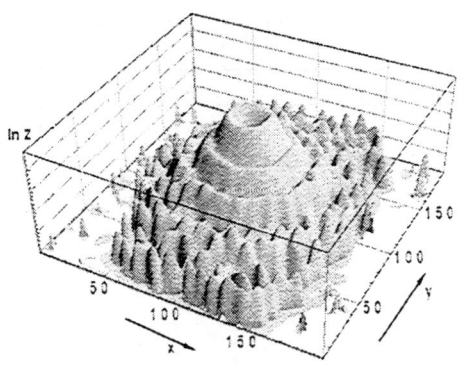

FIGURE 5. Three-dimensional view of the scattering distribution of He^+ ions produced in $He^{2+} + He^+$ collisions at $E_{cm} = 2.5 keV$, drawn on a natural logarithmic z scale. Channel numbers on x and y axes equal 0.156 mm/channel, i.e. 0.00815^o/channel. Data are deconvoluted with respect to the primary He^{2+} ion beam profile. Figure taken from Ref. [8]

(note the ln z scale). For larger scattering angles numerical fluctuations due to very low signal rates are introduced by the deconvolution algorithm. Integration over the azimuthal angle φ averages these fluctuations more or less and yields the differential cross section $d\sigma/d\Omega$, plotted in Fig.6 as a function of the cm-scattering angle ϑ_{cm}.

In addition to the minimum in forward direction, another four minima of $d\sigma/d\Omega$ are resolved experimentally. The oscillation period increases with scattering angles. Forster et al. [25] used a semiclassical eikonal calculation which describes well the experimental data for low scattering angles ϑ_{cm}, but "runs out of phase" for larger ϑ_{cm}. The observed oscillatory structures in the differential electron capture cross section can be interpreted as interference between *gerade* and *ungerade* electronic states of the ionic molecule. Uskov and Presnyakov [8] presented a quantum approach based on a partial wave analysis using a molecular basis. They also included the nonadiabatic rotational $2p\sigma_u - 2p\pi_u$ coupling describing the capture into the excited $2p$ state and related scattering matrix element corrections and finally delivered a very good description of the experimental data. We note that theory and experiment are compared on an absolute scale. It is mainly counting statistics and related deconvolution effects which limit the experimental resolution.

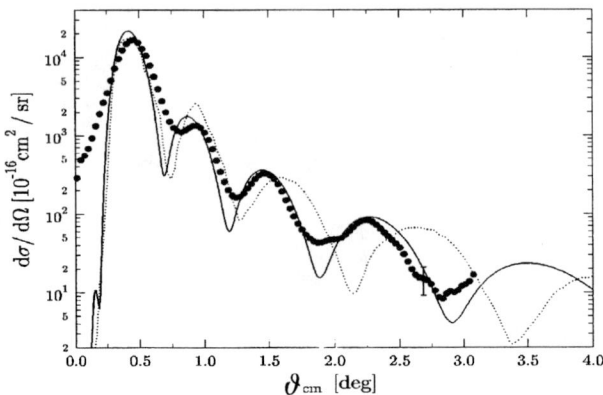

FIGURE 6. Angular differential cross sections for the electron capture (6) in $He^{2+} + He^+$ collisions at $E_{cm} = 2.5keV$. •: experiment, Kruedener et al. [8]; dotted line: semiclassical calculation, Forster et al. [25]; solid line: quantal calculation, Uskov and Presnyakov [8]. Figure taken from Ref. [8]

Two-Electron Systems

Electron-electron interaction is introduced by the second electron present in the collision, which enhances considerably the difficulties of the theoretical approaches. Subject of fundamental studies is either the neutral (H − H) complex consisting of two electrons and two protons, or the charged $(He - He)^{2+}$ complex where two α-particles attract two electrons. We will first concentrate on the charged complex.

The $(He - He)^{2+}$ complex can be investigated either by colliding α-particles with atomic helium or, more sophisticated, by ion-ion collisions. The principle of detailed balance suggests that the cross section for electron capture σ_C

$$He^+ + He^+ \longrightarrow He^{2+} + He^0 \tag{7}$$

in $He^+ + He^+$ collisions equals the cross section for the reverse reaction

$$He^{2+} + He^0 \longrightarrow He^+ + He^+ \tag{8}$$

in $He^{2+} + He^0$ collisions. This argument holds for state-specified transitions only. If two He^+ ions collide, the electrons can orientate their spins parallel or antiparallel to each other, forming spin singlet or triplet states in the exit channels with statistical weights 1:3. Only 25% of the incoming flux will therefore populate singlet states. As helium gas is usually completely in the singlet ground state, the ion-atom cross section (8) is expected to be four times larger than the ion-ion cross section (7).

Furthermore, it is clear that at very low collision energies mutual Coulomb repulsion between the ions suppresses electron transfer (7) whereas the Coulomb wall is absent in ion-atom collisions (8).

Modern close-coupling calculations are able to predict electron capture in the intermediate energy range, displayed in Fig.7. Gramlich et al. [33] used Gaussian type orbitals, whereas Kimura [34] employed molecular, and Fritsch and Lin [35] atomic orbitals. All calculations correspond within the experimental uncertainty with *crossed-beams* results by Peart et al. [26], Melchert et al. [27,28] and Murphy et al. [29].

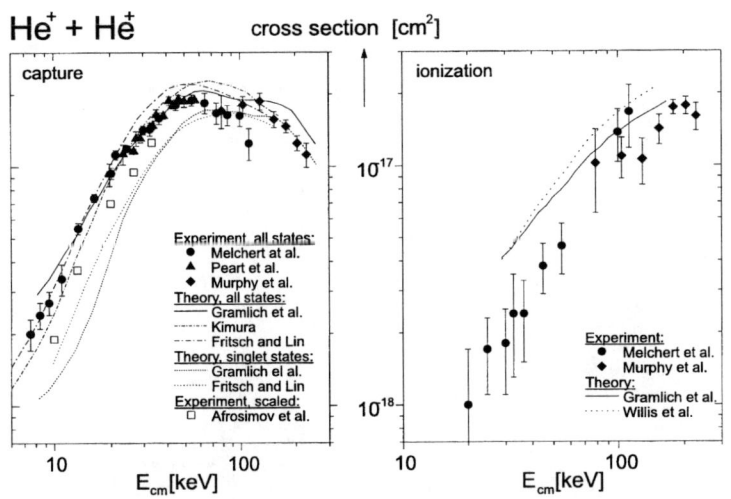

FIGURE 7. Cross sections for electron capture (7) and ionization (9) in $He^+ + He^+$ collisions, as a function of the cm-collision energy E_{cm}. Experimental *crossed-beams* data by Peart et al. [26], Melchert et al. [27,28] and Murphy et al. [29]. Theoretical results by Fritsch and Lin [35] (atomic orbitals), Kimura [34] (molecular orbitals), Gramlich et al. [33] (Gaussian orbitals) and Willis et al. [32]. Partial cross sections denote electron capture into singlet states only. Cross sections (8) for $He^{2+} + He^0$ collisions, measured by Afrosimov et al. [30], are plotted divided by a factor of four.

Gramlich et al. and Fritsch and Lin calculated also partial cross sections for electron capture into singlet states. They showed that at low energies almost 50% of the incoming flux populates singlet states, and this fraction increases with collision energy. Afrosimov's measured cross sections for electron capture (8) slightly overestimate the singlet results when divided by the statistical weight of four.

Electron capture (7) into the singlet ground state $He^0(1s^2)$ has a Q-value of -30 eV while capture into triplet states $He^0(1s, n = 2)$ requires $Q \approx -50$ eV. Translational energy spectroscopy would be an extremely sophisticated attempt in ion-ion collisions and has not been done yet, but the peculiar features of *crossed-beams* kinematics allow an estimate of Q. Pfeiffer et al. [31] observed a scattering

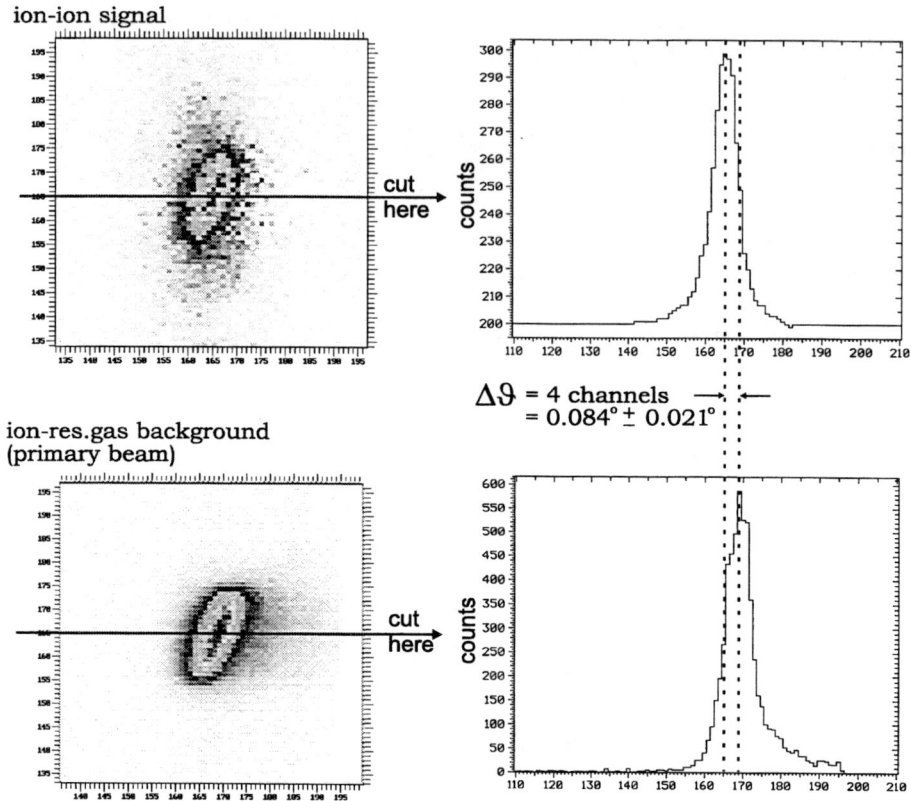

FIGURE 8. Scattering distribution $\mathcal{D}$ of He^0 atoms produced in He$^+$ + He$^+$ collisions (7) (top left) and distribution $\mathcal{P}$, produced in collisions of He$^+$ with the residual gas (bottom left). Both distributions were measured simultaneously by Pfeiffer et al. [31]. Horizontal cuts through the same vertical channel are displayed on the right side. The intense parts of both distributions are shifted by 4 channels (0.62 mm), which corresponds to $\Delta\vartheta = (0.084 \pm 0.021)^\circ$ and $Q = (-36.5 \pm 9.1)~eV$. All scales are linear.

distribution $\mathcal{D}$ (Fig.8) of He^0 atoms produced in reaction (7). $\mathcal{D}$ is shifted with respect to the primary He$^+$ ion beam axis, defined experimentally by He^0 atoms produced in collisions with the residual gas which form the distribution $\mathcal{P}$ (Fig.8). Cuts through $\mathcal{D}$ and $\mathcal{P}$ exhibit a shift between them from which $Q = -(36 \pm 9)~eV$ can be calculated. As a consequence, the intense inner part of $\mathcal{D}$ can be associated with charge transfer into the singlet ground state. The structure reflects the primary beam profile $\mathcal{P}$ – deflection by the small scattering angles can be neglected in comparison with the primary beam dimensions.

$\mathcal{D}$ shows a second feature: an elliptical structure perpendicular to the horizontal

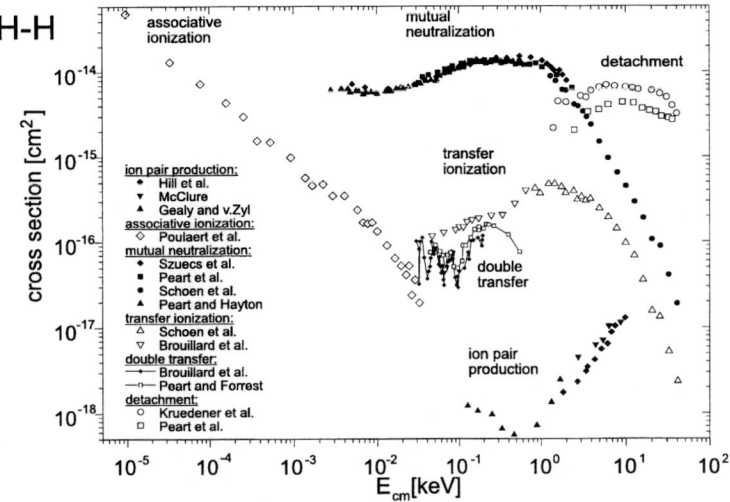

FIGURE 9. Measured cross sections for various processes within the H-H collision complex as a function of the cm-collision energy E_{cm}. Ion pair production (10): Hill et al. [47], McClure [38], Gealy and v.Zyl [39]; associative ionization (11): Poulaert et al. [36]; mutual neutralization (12): Szuecs et al. [40], Peart et al. [37], Schoen et al. [41], Peart and Hayton [42]; transfer ionization (13): Schoen et al. [48], Brouillard et al. [48]; double transfer (14): Brouillard et al. [43], Peart and Forrest [44]; detachment (15): Kruedener et al. [45], Peart et al. [46]

collision plane is observed (light gray in Fig.8). *Crossed-beams* kinematics determines its symmetry since the relevant scattering angles are large in comparison with the beam profile for which $\mathcal{P}$ is a measure. This feature probably originates from electron capture into the triplet states, because these states contribute to the total cross section as discussed above. The proof would be a shift of 7 channels ($Q \approx -50\ eV$) for this feature, but poor counting statistics of $\mathcal{D}$ hinders the exact verification.

Ionization in $He^+ + He^+$ collisions

$$He^+ + He^+ \longrightarrow He^{2+} + He^+ + e^- \tag{9}$$

is treated by Gramlich et al. [33] within the same framework as electron capture (7). They generate two-electron eigenstates with positive energies which they interpret as pseudostates describing the ionized continuum of He^+. Fig.7 shows a fair agreement of their results and of calculations by Willis et al. [32] with the experimental work of Melchert et al. [28] and Murphy et al. [29]. Note, however, that ionization in $He^+ + He^+$ collisions is far less well understood than electron capture.

The second basic two-electron collision complex $(H-H)$ has been widely studied: intense experimental and theoretical effort has been concentrated on this system. A detailed discussion would be beyond the scope of this article; therefore, in Fig.9, we show only experimental data for ion pair production

$$H^0 + H^0 \longrightarrow H^+ + H^- \tag{10}$$

in $H^0 + H^0$ collisions and for various processes in reverse $H^+ + H^-$ collisions:

$$\text{associative ionization:} \quad H_a^+ + H_b^- \longrightarrow H_2^+ + e^- \tag{11}$$

$$\text{mutual neutralization:} \quad H_a^+ + H_b^- \longrightarrow H_a^0 + H_b^0 \tag{12}$$

$$\text{transfer ionization:} \quad H_a^+ + H_b^- \longrightarrow H_a^0 + H_b^+ + e^- \tag{13}$$

$$\text{double transfer:} \quad H_a^+ + H_b^- \longrightarrow H_a^- + H_b^+ \tag{14}$$

$$\text{detachment:} \quad H_a^+ + H_b^- \longrightarrow H_a^+ + H_b^0 + e^- \tag{15}$$

Indices a and b indicate the fact that in *crossed-beams* experiments an ion of beam b can be distinguished by its momentum vector from an ion of beam a.

At low collision energies, associative ionization (11) occurs with cross sections as large as $5 \cdot 10^{-14}$ cm^2, but it decreases rapidly with E_{cm}. In the medium energy range, mutual neutralization (12) is the dominating process with cross sections in the order of 10^{-14} cm^2. Data obtained in *crossed-beams* and *merged-beams* experiments by different experimental groups correspond almost perfectly. At collision energies above 2.5 keV, the H^- electron is not primarily captured by the proton, but ionized instead; electron detachment (15) surpasses mutual neutralization (12). Measurements by Kruedener et al. [45] show deviations from the pioneering data by Peart et al. [46] for the detachment (15). An actual calculation by Belkic [49] seems to confirm Kruedener's results.

A great variety of calculations has been performed for the (H − H) collision complex; for clarity, they have been omitted in Fig.9.

Collisions between Multiply-Charged Ions

Ion-ion collision data required for various applications are sparse; for collisions between multiply-charged ions they are practically absent. It is due to the recent development of powerful Electron Cyclotron Resonance (ECR) ion sources that *crossed-beams* experiments can be carried out with intense beams of multiply-charged ions. Up to now, however, only a few experiments, in which at least one ion carries a charge higher than one, have been performed.

Here, we limit the discussion to electron capture in the quasi-one-electron collision system

$$^3He^{2+} + {}^{13}C^{3+} \longrightarrow {}^3He^+ + {}^{13}C^{4+} \tag{16}$$

which was experimentally studied by Melchert at al. [50] and calculated by Sidky and Lin [50]. The ^{3}He and ^{13}C isotopes were chosen to avoid contamination of the primary ion beams by molecular $^1H_2^+$ and $^4He^+$ ions, which have virtually the same charge-to-mass ratio as $^4He^{2+}$ and $^{12}C^{3+}$ ions, respectively. In their close-coupling calculation Sidky and Lin expand the active electron wave function as a sum of

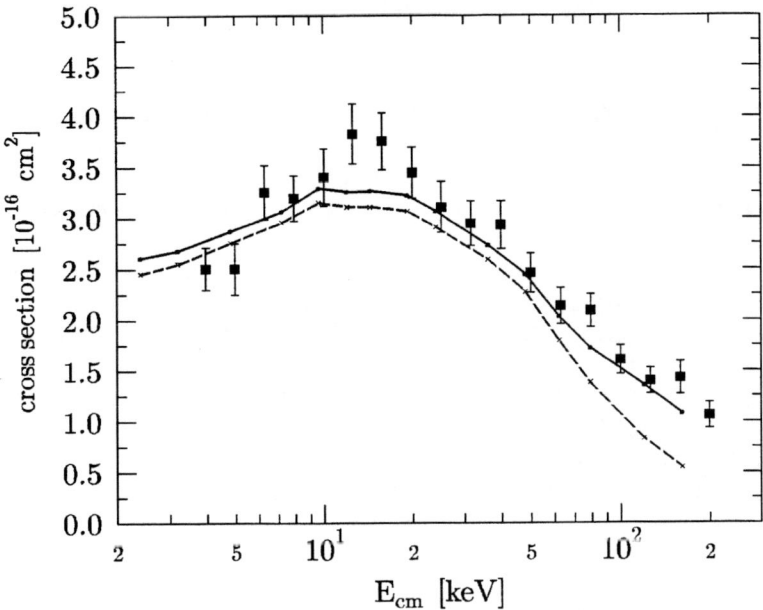

FIGURE 10. Cross sections for electron capture (16) in $^3\text{He}^{2+} + ^{13}\text{C}^{3+}$ collisions. Experimental data points by Melchert et al. [50] include 90% of statistical error. Close coupling calculation by Sidky and Lin [50] for capture to all He^+ states (solid curve) and to the $\text{He}^+(1s)$ state only (dashed curve).

atomic states on both charge centers. The two inner electrons of the lithium-like C^{3+} ion are inert and enter as a model screening potential for the outer active electron. In Fig.10 measured and calculated cross sections are compared. Up to a collision energy of 50 keV the capture goes primarily to the $\text{He}^+(1s)$ channel, since the $\text{He}^+(1s)$ level is closest to the initial C^{3+} binding energy. As soon as one reaches the matching velocity, where the incoming He^{2+} ion is traveling at about the average speed of the C^{3+} valence electron, the capture spreads to other states of He^+, and the $1s$ state represents only half of the total capture probability (at $E_{cm} = 150\ keV$).

He^+ ions produced in the capture reaction (16) at $E_{cm} = 5.15\ keV$ were position-sensitively detected by Melchert et al.. Taking into account the *crossed-beams kinematics*, the Q-value of (16) can be estimated to be less than 36 eV. This experimental finding confirms that electron capture to the $1s$ ground state of the He^+ ion ($|Q| = 10\ eV$) contributes mainly to the total cross section while capture to $n = 2$ ($|Q| = 51\ eV$) or even higher states play a minor role.

CONCLUSION

Ion-ion collisions in one- and two-electron systems are well described by modern theoretical approaches, at least as far as total cross sections are concerned. The measurement of differential ion-ion cross sections is still in its infancy.

Ion-ion collisions differ fundamentally from ion-atom collisions. Mutual Coulomb repulsion bends the particle trajectories, changes the potential energy of the system, and therefore influences the locations of potential energy curve crossings as well as the interaction energies. In general, cross sections for ion-ion collisions can not be derived from results obtained in ion-atom collisions by the principle of detailed balance since the electronic levels populated in ion-ion collisions usually differ from those populated in ion-atom collisions.

In the future, modern ECR ion sources will allow the systematic investigation of collisions between multiply-charged ions. Position-sensitive single-particle detection will enhance the information about the collision process itself: angular differential and state selective cross sections can be measured for appropriate collisions.

ACKNOWLEDGEMENT

The authors appreciate the fruitful collaboration with numerous colleagues and co-workers. Support by the Bundesministerium für Erziehung, Wissenschaft, Forschung und Technologie (BMBF), Bonn, and by the Deutsche Forschungsgemeinschaft (DFG), Bonn, is gratefully acknowledged.

REFERENCES

1. A.S. Schlachter, K.N. Leung, J.W. Stearns, R.E. Olson: In *Production and neutralization of Negative Hydrogen Ions and Beams*, ed. by J.G. Alesi, AIP Conference Proceedings No. 158 (Brookhaven 1986) 631
2. *HIBALL II: An Improved Conceptual Heavy-Ion-Driven Fusion Reactor Study*, Kernforschungszentrum Karlsruhe, Report KfK 3840 (July 1985)
3. D. Budicin, I. Hofmann, M. Conte, R. Schulze, F. Melchert, E. Salzborn: Il Nuovo Cimento Vol. **106 A (11)**, 1621 (1993)
4. E. Salzborn: In *The Physics of Electronic and Atomic Collisions*, AIP Conference Proceedings No. 205 (1990) 290
5. K. Rinn, F. Melchert, E. Salzborn: J. Phys. B **18**, 3783 (1985)
6. K. Rinn, F. Melchert, K. Rink, E. Salzborn: J. Phys. B **19**, 3717 (1986)
7. F. Brouillard, W. Claeys: In *Physics of Ion–Ion and Electron–Ion Collisions*, ed. by J.W. McGowan, Nato Advanced Study Institutes Series, Series B, Vol. 83 (Baddeck, Nova Scotia, Canada 1981) pp 415-459
8. S. Kruedener, F. Melchert, K. v.Diemar, A. Pfeiffer, K. Huber, E. Salzborn, D.B. Uskov, L.P. Presnyakov: Phys. Rev. Lett **79**, 1002 (1997)

9. W.H. Richardson: J. Opt. Soc. Am. **62**, 55 (1972)
10. S. Kruedener: In *Proceedings of the XIXth International Conference on the Physics of Electronic and Atomic Collisions*, ed. by L.J. Dube, J.B.A. Mitchell, J.W. Mc-Conkey, C.E. Brion, AIP Conf. Proc. 360 (Whistler, Canada, 1995) p. 677
11. S. Meuser, F. Melchert, S. Kruedener, A. Pfeiffer, K. v.Diemar, E. Salzborn: Rev.Sci.Instrum. **67**, 2752 (1996)
12. B. Peart, K. Rinn, K. Dolder: J. Phys. B **16**, 1461 (1983)
13. M.F. Watts, K.F. Dunn, H.B. Gilbody: J. Phys. B **19**, L355 (1986)
14. T.G. Winter: Phys. Rev. A **35**, 3799 (1987)
15. T.G. Winter: Phys. Rev. A **37**, 4656 (1988)
16. T.G. Winter, G.J. Hatton, N.F. Lane: Phys. Rev. A **22**, 930 (1980)
17. L.F. Errea, J.M. Gomez-Llorente, L. Mendez, A. Riera: J. Phys. B **20**, 6089 (1987)
18. J.E. Miraglia: Phys. B **16**, 1029 (1983)
19. W. Fritsch, C.D. Lin: J. Phys. B **15**, 1255 (1982)
20. W. Fritsch, C.D. Lin: In *Thirteenth International Conference on the Physics of Electronic and Atomic Collisions* (Berlin 1983), Abstract of Contributed Papers, p 502
21. T.P. Grozdanov, E.A. Solov'ev: Phys. Rev. A **38**, 4333 (1988)
22. A.E. Martinez, R. Deco, D. Rivarola, P.D. Fainstein: Nucl. Instr. Meth. in Phys. Res. B **43**, 24 (1989)
23. T.G. Winter, S.G. Alston: Phys. Rev. A **45**, 1562 (1992)
24. D.R. Bates, G.W. Griffing: Proc. Phys. Soc. London A **66**, 961 (1953)
25. C. Forster, R. Shingal, D.R. Flower, B.H. Bransden, A.S. Dickinson: Phys. B **21**, 3941 (1988)
26. B. Peart, K. Rinn, K. Dolder: J. Phys. B **16**, L361 (1983)
27. F. Melchert, K. Rink, K. Rinn, E. Salzborn, N. Gruen: J. Phys. B **20**, L223 (1987)
28. F. Melchert, K. Rink, K. Rinn, E. Salzborn: J. Phys. B **20**, L797 (1987)
29. J.G. Murphy, K.F. Dunn, H.B. Gilbody: J. Phys. B **27**, 3687 (1994)
30. V.V. Afrosimov, A.A. Basalaeu, G.A. Leiko, M.N. Panov: Sov. Phys. JETP 47, 837 (1978)
31. A. Pfeiffer, F. Melchert, E. Salzborn: unpublished (1994)
32. S.L. Willis, M.R.C. McDowell and J. Banerji: J. Phys. B **18**, 3939 (1985)
33. K. Gramlich, N. Gruen, W. Scheid: J. Phys. B **22**, 2567 (1989)
34. M. Kimura: J. Phys. B **21**, L19 (1988)
35. W. Fritsch, C.D. Lin: Phys. Lett. A **123**, 128 (1987)
36. G. Poulaert, F. Brouillard, W. Claeys, J.W. McGowan, G. van Wassenhove: J. Phys. B **11**, L671 (1978)
37. B. Peart, M.A. Bennett, K. Dolder: J. Phys. B **18**, L439 (1985)
38. G.W. McClure: Phys. Rev. **166**, 22 (1968)
39. M. Gealy, B. van Zyl: Phys. Rev. A **36**, 3100 (1987)
40. S. Szuecs, M. Karemera, M. Terao, F. Brouillard: J. Phys. B **17**, 1613 (1984)
41. W. Schoen, S. Kruedener, F. Melchert, K. Rinn, M. Wagner, E. Salzborn: J. Phys. B **20**, L759 (1987)
42. B. Peart, D.A. Hayton: J. Phys. B **25**, 5109 (1992)
43. F. Brouillard, W. Claeys, G. Poulaert, G. Rahmat, G. van Wassenhove: J. Phys. B

12, 1253 (1979)

44. B. Peart, R.A. Forrest: J. Phys. B **12**, L23 (1979)

45. S. Kruedener, F. Melchert, K. Huber and E. Salzborn: J. Phys. B At. Mol. Opt. Phys., **32** (1999) L139

46. B. Peart, R. Grey, K. Dolder: J. Phys. B **9**, 3047 (1976)

47. J. Hill, J. Geddes, H.B. Gilbody: J. Phys. B **12**, 3341 (1979)

48. W. Schoen, S. Kruedener, F. Melchert, K. Rinn, M. Wagner, E. Salzborn, M. Karemera, S. Szuecs, M. Terao, D. Fussen, R.K. Janev, X. Urbain, F. Brouillard: Phys. Rev. Lett. **49**, 1565 (1987)

49. D. Belkic: J. Phys. B **30**, 1731 (1997)

50. F. Melchert, S. Meuser, S. Kruedener, A. Pfeiffer, K. v.Diemar, E. Salzborn, E.Y. Sidky, C.D. Lin : J.Phys.B: At.Mol.Opt.Phys.**30**, L697 (1997)

Complete Solutions of the Landau-Zener-Stueckelberg Curve Crossing Problems, and Their Generalizations and Applications

Hiroki Nakamura

Department of Theoretical Studies,
Institute for Molecular Science,
and
Department of Functional Molecular Science,
The Graduate University for Advanced Studies,
Myodaiji, Okazaki 444-8585, Japan

Abstract. The compact analytical complete solutions recently obtained for the two-state Landau-Zener-Stueckelberg problems are reviewed and explained. The theory covers both Landau-Zener (LZ) type in which the two diabatic potential curves cross with the same sign of slopes and the nonadiabatic tunneling (NT) type in which the potentials cross with different signs of slopes. The theory is applicable virtually in the whole range of energy and coupling strength and is convenient for practical use. The new theory for time-dependent nonadiabatic transition can be formulated from the time-independent theory of the LZ-type. The utilizability of the theory to various multi-channel problems and also to multi-dimensional problems is demonstrated and explained. The intriguing phenomenon of complete reflection which appears in the NT-case is explicitly utilized to propose a new type of molecular switching and to control molecular processes such as molecular photodissociation. Also proposed is a new way of controlling molecular processes by using time-dependent external fields. Finally, a trial to formulate a unified analytical theory to cover both Landau-Zener-Stueckelberg and Rosen-Zener-Demkov types of nonadiabatic transitions is briefly touched upon.

INTRODUCTION

Needless to say, nonadiabatic transitions due to potential curve crossings play very important roles in atomic and molecular collisions and spectroscopic processes. They are actually playing crucial roles in many other fields of physics, chemistry, and biology. [1, 2] Without nonadiabatic transitions this world would have been dead. The recent remarkable progress in laser technology has made it possible to figure out various ways of controlling molecular processes and has actually opened up additional significance of nonadiabatic transitions due to curve crossing, al-

CP500, *The Physics of Electronic and Atomic Collisions*, edited by Y. Itikawa, et al.
© 2000 American Institute of Physics 1-56396-777-4/00/$17.00

though it requires a time-dependent version of the theory. This is because dynamic processes occur only effectively through curve crossings, which are created by external fields. Because of the importance mentioned above, the theory of nonadiabatic transition naturally has a long history. The pioneering works were done, as is well known, by Landau, Zener and Stueckelberg independently in the same year of 1932. [3–5] Since then in order to overcome the various defects of the Landau-Zener-Stueckelberg (LZS) theory, tons of papers have been published by many authors, and many books and review articles are available. [1, 2, 6–13] However, there have been many important aspects of the LZS curve crossing probelms left unsolved. In the last several years, we have been succcessful in solving all these problems and providing complete solutions for both LZ type in which two diabatic potentials cross with the same sign of slopes and NT type in which the diabatic potentilas cross with different signs of slopes. [2, 14] The latter case gives a transmission under the influence of nonadiabatic coupling, which is quite different from the ordinary single barrier penetration. Quantum mechanically exact solutions have been obtained for the linear potential model, and practically useful compact solutions have been derived semiclassically. [15, 16] These are applicable to general curved potentials whatever the energy and diabatic coupling strength are. Furthermore, the theory for the LZ case can be easily transferred to a new time-dependent theory applicable to various control problems. [17]

COMPLETE SOLUTIONS OF THE LZS PROBLEMS

The most basic model of curve crossing is the linear diabatic potentials coupled by a constant diabatic coupling. By using Fourier transform, the corresponding two coupled time-independent differential Schrödinger equations can be transformed into two coupled first order differential equations, which are further reduced to a second order single differential equation with a quartic polynomial as its coefficient, i.e. to a four-transition-point(zero-point) problem. This single differential equation was solved with use of the complex WKB solutions, which provide exact solutions at infinity, by analyzing the corresponding Stokes phenomenon associated with the asymptotic solutions of the differential equation. [15, 18] The details of the mathematical procedure are omitted here, but the scattering matrices for both LZ and NT cases are turned out to be expressed in terms of one Stokes constant U_1 which is actually a function of two basic parameters a^2 and b^2 defined by

$$a^2 = \frac{\hbar^2}{2\mu}\frac{F(F_1 - F_2)}{8V_X^3} \quad \text{and} \quad b^2 = (E - E_X)\frac{F_1 - F_2}{2FV_X}, \tag{1}$$

where μ is the mass, $F_j(j = 1, 2)$ is the slope of the $j-$th diabatic potential, V_X is the diabatic coupling and E_X is the energy at the crossing point. Without loss of generality $F_1 - F_2$ is assumed to be positive. The semiclassical analysis which replaces the four-transition-point problem by two two-transition-point problems enables us to derive compact analytical expressions of the Stokes constant U_1 and thus to obtain compact analytical expressions for all the necessary probabilities and phases. [16, 19]

NEW FORMULAS AND NICE CHARACTER OF THE NEW THEORY

The final recommended formulas for the two-channel problems are summarized below. As can be understood from the summary below, our theory has the follwoing nice features and can be easily utilized for various applications. (1) All the probabilities are expressed in simple analytical forms. (2) All the necessary phases are provided in compact analytical forms. (3) All the basic parameters can be directly estimated only from adiabatic potentials on the real axis. This means that (1) no non-unique diabatization is needed, (2) no complex calculus is necessary, and (3) no nonadiabtic coupling information is required. Furthermore, the theory works for whole range of energy and coupling strength for general curved potentials. [19]

Landau-Zener Type (see Fig.1)

The transition matrix I_N at the avoided crossing point is defined by

$$\begin{pmatrix} C \\ D \end{pmatrix} = I_N \begin{pmatrix} A \\ B \end{pmatrix}, \tag{2}$$

where A and B (C and D) are the coeffcients of the wave functions at $R = R_0 + 0$ ($R = R_0 - 0$). The nonadiabtic transition matrix I_N is given by

$$I_N = \begin{pmatrix} \sqrt{1-p}\,e^{-i(\sigma-\psi)} & -\sqrt{p}\,e^{i\sigma_0} \\ \sqrt{p}\,e^{-i\sigma_0} & \sqrt{1-p}\,e^{i(\sigma-\psi)} \end{pmatrix}, \tag{3}$$

where p represents the nonadiabatic transition probability for one passage of avoided crosssing. In the final formulas given below some empirical corrections are introduced in order to make them cover better the whole range of coupling strength.

The basic parameters originally defined by Eq.(1) can now be estimated directly from adiabatic potentials as follows (see Fig.1):

$$a^2 = \sqrt{d^2 - 1}\,\frac{\hbar^2}{\mu(T_2^{(0)} - T_1^{(0)})^2(E_2(R_0) - E_1(R_0))} \tag{4}$$

and

$$b^2 = \sqrt{d^2 - 1}\,\frac{E - (E_2(R_0) + E_1(R_0))/2}{(E_2(R_0) - E_1(R_0))/2}, \tag{5}$$

where

$$d^2 = \frac{[E_2(T_1^{(0)}) - E_1(T_1'^{(0)})][E_2(T_2^{(0)}) - E_1(T_2'^{(0)})]}{[E_2(R_0) - E_1(R_0)]^2}. \tag{6}$$

It should be noted that the Landau-Zener formula is given by

$$p_{LZ} = \exp[-\frac{\pi}{4a|b|}]. \tag{7}$$

$E \geq E_X$ (crossing energy) $(b^2 \geq 0)$

The various quantities in Eq.(3) are given by

$$p = \exp\left[-\frac{\pi}{4a|b|}\left(\frac{2}{1 + \sqrt{1 + b^{-4}(0.4a^2 + 0.7)}}\right)^{1/2}\right] , \tag{8}$$

$$\sigma_0 = \frac{\sqrt{2}\pi}{4\sqrt{a^2}} \frac{F_-^C}{F_+^2 + F_-^2}, \tag{9}$$

$$\psi = \sigma - \phi_S, \tag{10}$$

$$\sigma = \int_{T_1}^{R_0} K_1(R)dR - \int_{T_2}^{R_0} K_2(R)dR + \sigma_0, \tag{11}$$

$$K_j(R) = \sqrt{\frac{2\mu}{\hbar^2}(E - E_j(R))}, \tag{12}$$

$$\phi_S = \frac{\delta}{\pi} - \frac{\delta}{\pi}\ln(\frac{\delta}{\pi}) + \arg\Gamma(i\frac{\delta}{\pi}) + \frac{\pi}{4}, \tag{13}$$

$$\delta = \delta_0 = \frac{\sqrt{2}\pi}{4\sqrt{a^2}} \frac{F_+^C}{F_+^2 + F_-^2}, \tag{14}$$

$$F_\pm = \sqrt{\sqrt{(b^2 + \gamma_1)^2 + \gamma_2} \pm (b^2 + \gamma_1)} + \sqrt{\sqrt{(b^2 - \gamma_1)^2 + \gamma_2} \pm (b^2 - \gamma_1)}, \tag{15}$$

$$F_+^c = F_+(b^2 \longrightarrow [b^2 - \frac{0.16b_x}{\sqrt{b^4 + 1}}]), \quad F_-^c = F_-(\gamma_2 \longrightarrow \frac{0.45\sqrt{d^2}}{1 + 1.5e^{2.2b_x}|b_x|^{0.57}}), \tag{16}$$

$$b_x = b^2 - 0.9553, \gamma_1 = 0.9\sqrt{d^2 - 1} , \quad \text{and} \quad \gamma_2 = 7\sqrt{d^2}/16. \tag{17}$$

$$E \leq E_X \ (b^2 \leq 0)$$

The nonadiabtic transition probability for one passage is given by

$$p = [1 + B(\sigma/\pi)\exp(2\delta) - g\sin^2(\sigma)]^{-1}, \tag{18}$$

where

$$B(x) = \frac{2\pi x^{2x} e^{-2x}}{x\Gamma^2(x)}, \tag{19}$$

$$\delta = \int_{T_1}^{R_0} |K_1(R)|dR - \int_{T_2}^{R_0} |K_2(R)|dR + \delta_0, \quad \sigma = \sigma_0 \tag{20}$$

and

$$g = \frac{3\sigma}{\pi\delta} l_n(1.2 + a^2) - \frac{1}{a^2}. \tag{21}$$

The phase ψ in Eq.(3) is given by

$$\psi = \arg(U), \tag{22}$$

where

$$\mathrm{Re}U = \cos(\sigma)\left\{\sqrt{B(\sigma/\pi)}e^\delta - he^{-\delta}\sin^2(\sigma)/\sqrt{B(\sigma/\pi)}\right\}, \tag{23}$$

$$\mathrm{Im}U = \sin(\sigma)\left\{B(\sigma/\pi)e^{2\delta} - h^2\sin^2(\sigma)\cos^2(\sigma)e^{-2\delta}/B(\sigma/\pi) + 2h\cos^2(\sigma) - g\right\}^{1/2}, \tag{24}$$

and

$$h = 1.8(a^2)^{0.23}e^{-\delta}. \tag{25}$$

The quantities σ_0 and δ_0 are given by Eqs. (9) and (14).

Total Scattering Matrix

The scattering matrix in the adiabatic representation is explicitly given by

$$S_{nm} = S_{nm}^R \exp[i\eta_n + i\eta_m], \tag{26}$$

even if the energy E is lower than E_X. Here η_n is the elastic scattering phase by the n−th adiabatic potential and S^R is the reduced scattering matrix. The total S−matrix is explicitly given by

$$S = P_{\infty X} I_N^t P_{XTX} I_N P_{X\infty} = QI^t IQ = QS^R Q, \tag{27}$$

where I_N^t (I^t) is the transpose of I_N (I),

$$(P_{\infty X})_{nm} = (P_{X\infty})_{nm} = \delta_{nm} \exp[i \int_X^{\infty} (K_n(R) - K_n(\infty))dR - iK_n(\infty)X], \tag{28}$$

$$(P_{XTX}) = \delta_{nm} \exp[2i \int_{T_n}^{X} K_n(R)dR + i\frac{\pi}{2}], \tag{29}$$

$$Q_{nm} = \delta_{nm} \exp[i\eta_n], \tag{30}$$

$$\eta_{1,2} = \lim_{R\to\infty} [\int_{T_{1,2}}^{R} K_{1,2}(R)dR - K_{1,2}(R)R + \frac{\pi}{4}], \tag{31}$$

$$I = \begin{pmatrix} \sqrt{1-p}e^{-i(\sigma-\psi)} & -\sqrt{p}e^{i\sigma} \\ \sqrt{p}e^{-i\sigma} & \sqrt{1-p}e^{i(\sigma-\psi)} \end{pmatrix}, \tag{32}$$

and

$$S^R = I^t I. \tag{33}$$

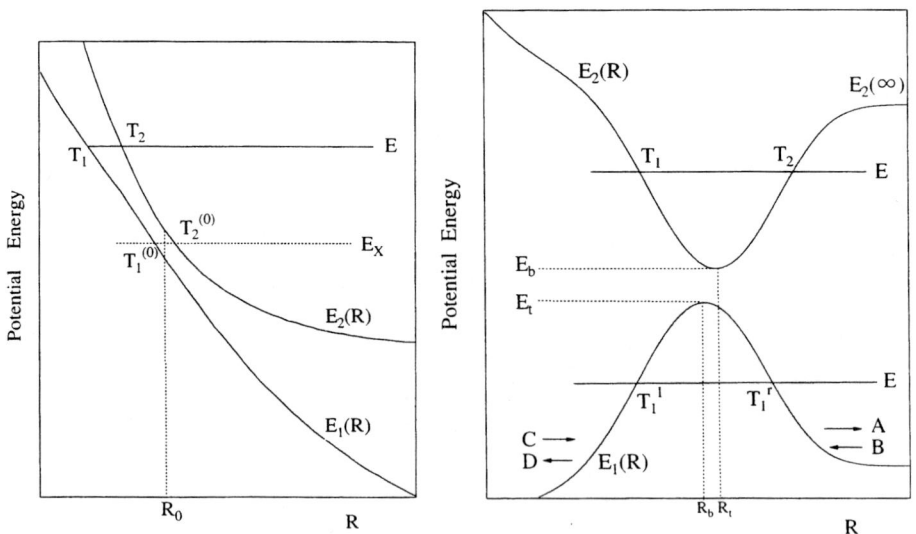

FIGURE 1. Landau-Zener Type   FIGURE 2. Nonadiabatic Tunneling Type

Nonadiabatic Tunneling Type (see Fig.2)

The I_N matrix for $E \geq E_b$ ($b^2 \geq 1$) is given by

$$I_N = \begin{pmatrix} \sqrt{1-p}\,e^{-i\phi_S} & \sqrt{p}\,e^{i\sigma_0} \\ -\sqrt{p}\,e^{-i\sigma_0} & \sqrt{1-p}\,e^{i\phi_S} \end{pmatrix}, \tag{34}$$

where

$$p = \exp\left[-\frac{\pi}{4ab}\left(\frac{2}{1+\sqrt{1-b^{-4}(0.72-0.62a^{1.43})}}\right)^{1/2}\right], \tag{35}$$

$$\sigma_0 = \frac{R_b - R_t}{|R_b - R_t|}\,\frac{2\sqrt{1-\gamma^2}}{3a}\,\frac{[2b^2+\sqrt{b^4-1}]}{\sqrt{b^2+1}+\sqrt{b^2-1}}, \tag{36}$$

and ϕ_S is the same as Eq.(13). At $E > E_2(\infty)$, this I_N can be used in the same way as I_N in the LZ case.
The total (S) and the reduced (S^R) scattering matrices are defined as follows:

$$S_{mn} = S^R_{mn}\exp[i(\eta_m + \eta_n)], \tag{37}$$

$$\eta_1 = \lim_{R\to\infty}\left[\int_{T_1^r}^R K_1(R)dR - K_1(R)R + \frac{\pi}{4}\right], \tag{38}$$

$$\eta_2 = \lim_{R\to-\infty}\left[-\int_{T_1^l}^R K_1(R)dR + K_1(R)R + \frac{\pi}{4}\right], \tag{39}$$

$$S^R = \frac{1}{1+U_1U_2}\begin{pmatrix} e^{i\Delta_{11}} & U_2 e^{i\Delta_{12}} \\ U_2 e^{i\Delta_{12}} & e^{i\Delta_{22}} \end{pmatrix} \tag{40}$$

and

$$U_2 = 2i\mathrm{Im}(U_1)/(|U_1|^2 - 1). \tag{41}$$

Note that T_1^l and T_1^r should be replaced by R_t for $E \geq E_b$ and the state 1(2) corresponds to the right (left) side of the barrier. The parameters b^2 and a^2 can also be estimated directly from adiabatic potentials by

$$b^2 = \frac{E - (E_b + E_t)/2}{(E_b - E_t)/2} \quad \text{and} \quad a^2 = \frac{(1-\gamma^2)\hbar^2}{\mu(R_b - R_t)^2(E_b - E_t)}, \tag{42}$$

where

$$\gamma = \frac{E_b - E_t}{E_2\left(\frac{R_b + R_t}{2}\right) - E_1\left(\frac{R_b + R_t}{2}\right)}. \tag{43}$$

When $R_b = R_t$, $\gamma = 1$ and

$$a^2 = \frac{\hbar^2}{4\mu(E_b - E_t)}\left[\frac{\partial^2 E_2(R)}{\partial R^2}\Big|_{R=R_b} - \frac{\partial^2 E_1(R)}{\partial R^2}\Big|_{R=R_t}\right]. \tag{44}$$

The transfer matrix N for $E < E_2(\infty)$ is defined by

$$\begin{pmatrix} C \\ D \end{pmatrix} = N \begin{pmatrix} A \\ B \end{pmatrix}, \tag{45}$$

where $A(B)$ is the coefficient of the wave running to the right (left) along $E_1(X)$ on the right side of the barrier, and C (D) is the coefficient of the wave running to the right(left) along $E_1(X)$ on the left side of the barrier. The N matrix is related to the S−matrix as

$$N_{11} = 1/S_{12}^R, \quad N_{12} = S_{22}^R/S_{12}^R, \quad N_{21} = N_{12}^*, \quad N_{22} = N_{11}^*. \tag{46}$$

$$E \geq E_b$$

The various quantities in Eq.(40) are given as follows:

$$\Delta_{12} = \sigma, \quad \Delta_{11} = 2\int_{T_1}^{R_b} K_2(R)dR - 2\bar{\sigma}_0 \quad \text{and} \quad \Delta_{22} = 2\int_{R_b}^{T_2} K_2(R)dR + 2\bar{\sigma}_0, \tag{47}$$

$$\bar{\sigma}_0 = \left(\frac{R_b - R_t}{2}\right)\left\{K_1(R_t) + K_2(R_b) + \frac{1}{3}\frac{[K_1(R_t) - K_2(R_b)]^2}{K_1(R_t) + K_2(R_b)}\right\}, \tag{48}$$

$$U_1 = i\sqrt{1 - p}\exp\left(i\sigma + i\phi_S\right), \tag{49}$$

$$\sigma = \int_{T_1}^{T_2} K_2(R)dR, \quad \text{and} \quad \delta = \frac{\pi}{8ab}\frac{1}{2}\frac{\sqrt{6 + 10\sqrt{1 - \frac{1}{b^4}}}}{1 + \sqrt{1 - \frac{1}{b^4}}}. \tag{50}$$

The nonadiabatic transition probability p is given by Eq.(35). The overall transmission probability is given by

$$P_{12} = \frac{4\cos^2(\sigma + \phi_S)}{4\cos^2(\sigma + \phi_S) + p^2/(1 - p)}. \tag{51}$$

It should be noted that when $\sigma + \phi_S = (n + 1/2)\pi$ $(n = 0, 1, 2, ...)$, P_{12} becomes zero. Namely, complete reflection occurs.

$$E_b \geq E \geq E_t$$

The necessary quantities to define S^R are given as follows:

$$\Delta_{12} = sigma \,, \quad \Delta_{11} = \sigma - 2\bar{\sigma}_0 \quad \text{and} \quad \Delta_{22} = \sigma + 2\bar{\sigma}_0, \tag{52}$$

$$\bar{\sigma}_0 = -\frac{1}{3}(R_t - R_b)K_1(R_t)(1 + b^2), \tag{53}$$

$$U_1 = i[\sqrt{1 + W^2}e^{i\phi} - 1]/W, \tag{54}$$

$$\phi = \sigma + \arg\Gamma(\frac{1}{2} + i\frac{\delta}{\pi}) - \frac{\delta}{\pi}\ell n(\frac{\delta}{\pi}) + \frac{\delta}{\pi} - g_4, \tag{55}$$

$$g_4 = 0.34 \frac{a^{0.7}(a^{0.7} + 0.35)}{a^{2.1} + 0.73}(0.42 + b^2)\left(2 + \frac{100b^2}{100 + a^2}\right)^{0.25}, \tag{56}$$

$$\sigma = -\frac{1}{\sqrt{a^2}}[0.057(1 + b^2)^{0.25} + \frac{1}{3}](1 - b^2)\sqrt{5 + 3b^2}, \tag{57}$$

$$\delta = \frac{1}{\sqrt{a^2}}[0.057(1 - b^2)^{0.25} + \frac{1}{3}](1 + b^2)\sqrt{5 - 3b^2}, \tag{58}$$

$$W = \frac{g_3}{a^{2/3}}\int_0^\infty \cos\left[\frac{t^3}{3} - \frac{b^2}{a^{2/3}}t - \frac{g_1}{2a^{2/3}}\frac{t}{g_2 + a^{1/3}t}\right]dt, \tag{59}$$

$$g_3 = 1 + \frac{0.38}{a^2}(1 + b^2)^{1.2-0.4b^2}, \quad g_2 = 0.61\sqrt{2 + b^2}, \text{and} \quad g_1 = \frac{\sqrt{a^2} - 3b^2}{\sqrt{a^2} + 3}\sqrt{1.23 + b^2}. \tag{60}$$

The overall transmission probability is given by

$$P_{12} = \frac{W^2}{1 + W^2}. \tag{61}$$

$$E \leq E_t$$

The required quantities are given by

$$\Delta_{12} = \Delta_{11} = \Delta_{22} = -2\sigma, \tag{62}$$

$$\text{Re}U_1 = \sin(2\sigma_c) \left\{ \frac{0.5\sqrt{a^2}}{1 + \sqrt{a^2}} \sqrt{B(\sigma_c/\pi)}e^{-\delta} + \frac{e^{\delta}}{\sqrt{B(\sigma_c/\pi)}} \right\}, \tag{63}$$

$$\text{Im}U_1 = \cos(2\sigma_c) \sqrt{\frac{(\text{Re}\,U_1)^2}{\sin^2(2\sigma_c)} + \frac{1}{\cos^2(2\sigma_c)} - \frac{1}{2\sin(\sigma_c)} \left| \frac{\text{Re}\,U_1}{\cos(\sigma_c)} \right|}, \tag{64}$$

$$\sigma_c = \sigma(1 - 0.32 \times 10^{-2/a^2} e^{-\delta}), \tag{65}$$

$$\delta = \int_{T_1^l}^{T_1^r} |K_1(R)| dR, \quad \text{and} \quad \sigma = \frac{\pi}{8a|b|} \frac{1}{2} \frac{\sqrt{6 + 10\sqrt{1 - \frac{1}{b^4}}}}{1 + \sqrt{1 - \frac{1}{b^4}}}, \tag{66}$$

where the function $B(x)$ is defined by Eq.(19). The overall transmission probability is equal to

$$P_{12} = \frac{B(\sigma_c/\pi)e^{-2\delta}}{\left[1 + \frac{0.5\sqrt{a^2}}{\sqrt{a^2}+1} B(\sigma_c/\pi)e^{-2\delta} \right]^2 + B(\sigma_c/\pi)e^{-2\delta}}. \tag{67}$$

It should be noted that when $a^2 \to 0$, we have

$$P_{12} = \frac{e^{-2\delta}}{1 + e^{-2\delta}}, \tag{68}$$

which agrees with the ordinary single potential barrier penetration probability.

TIME-DEPENDENT VERSION OF THE THEORY

The above solutions derived in the time-independent framework can be easily transferred to the solutions in the corresponding time-dependent scheme. [17] The exact solutions in the time-independent linear potential model, for instance, provide the exact solutions for the time-dependent quadratic potential model, covering not only the two-crossing case, but also tangentially touching and diabatically avoided crossing cases. The following replacements of the parameters are good enough for this transfer.

$$a^2 \Leftrightarrow \alpha = \frac{\sqrt{d^2 - 1}h^2}{2V_0^2(t_t^2 - t_b^2)}, \tag{69}$$

$$b^2 \Leftrightarrow \beta = -\sqrt{d^2 - 1}\frac{t_b^2 + t_t^2}{t_t^2 - t_b^2}, \tag{70}$$

and

$$\sigma + i\delta = \frac{1}{\hbar}\left[\int_0^{t_t} E_-(t)dt - \int_0^{t_b} E_+(t)dt + \Delta_1\right] \tag{71}$$

with

$$\Delta_1 = \frac{t_x - (t_b + t_t)/2}{\sqrt{\alpha}(\beta^2 + i)(t_b - t_t)}\sqrt{\frac{d^2}{d^2 - 1}}, \tag{72}$$

$$V_0 = \frac{1}{2}\Big(F_+(t_x) - F_-(t_x)\Big), \tag{73}$$

and

$$d^2 = \frac{\Big[E_+(t_b) - E_-(t_b)\Big]\Big[E_+(t_t) - E_-(t_t)\Big]}{\Big[E_+(t_x) - E_-(t_x)\Big]^2}, \tag{74}$$

where t_x is the position at which $E_+(t) - E_-(t)$ becomes minimum, and $t_b(t_t)$ is the bottom (top) of the potential $E_+(t)(E_-(t))$.

It should be noted that the NT case in the time-independent scheme does not show up in the time-dependent framework. This is simply because the time is unidirectional. The above replacement should be thus applied only to the LZ case.

GENERALIZATIONS AND POSSIBLE APPLICATIONS

Multi-Channel Problems

Our basic theory is only for one-dimensional two-channel problems; but thanks to the localizability of the nonadiabatic transitions and the accuracy of the new theory, the theory enables us to treat general multi-channel problems. Applicability of the theory was demonstrated in the case of multi-channel scattering with heavy overlapping resonances, which were actually very nicely reproduced. [20, 21] Very sharp Feshbach type resonances in the two-channel collinear O+HO reaction, the widths of which are as small as $10^{-5}cm^{-1}$, are also nicely reproduced by the theory. [22]

Multi-Dimensional Problems

There are two ways of treating multi-dimensional problems; One is a reduction to one-dimensional multi-channel system and the other is to use classical trajectories somehow. As an example of the first treatment, we have applied the theory to analyze the three dimensional electronically adiabatic chemical reactions $O(^3P) + HCl$ and Br+HCl. [23] The parameters a^2 proves to be a nice criterion for picking up dynamically important avoided crossings for reactive transitions and the cumulative reaction probabilities turned out to be well reproduced by the theory. For the second kind of treatment, our theory can be incorporated into the frameworks of the trajectory surface hopping (TSH) method [24], the semiclassical propagation based on the initial value representation (IVR), [25] and the cellular frozen Gaussian propagation method. [26] In the case of TSH, only the nonadiabatic transition probability is required, but for the other two methods not only the probability but also all kinds of necessary phases are required and actually all of them can be furnished by our theory. These would provide useful methodologies to attack large systems.

Utilization of the Unusual Phenomenon of "Complete Reflection"

In the case of NT there occurs an unusual phenomenon of complete reflection at certain discrete energies due to phase interference. This intriguing phenomenon may be utilized somehow. One possibility is the molecular switching in a finite periodic NT-type potential units. Complete transmission which exists whenever we have a pure periodic potential system can be switched off and on by creating reversibly an impurity potential unit, the complete reflection position of which is designed to agree with the original complete transmission position. [27,28] Another possible utilization is a control of photodissociation process. Dressing up the ground state potential energy surface by applying a strong laser field and creating the complete reflection condition by appropriately adjusting the laser frequency, we can stop the dissociation into that direction and let the molecule dissociate into the other channel. This makes it possible to dissociate a molecule into any desirable channel, and actually even into the channel which cannot be reached by an ordinary photodissociation. Preliminary theoretical studies are being carried out for the model systems mimicking CH_3SH and HOD.

Control of Nonadiabatic Processes, or Molecular Processes, by Time-Dependent External Fields

The recent progress of laser technology is remarkable and various ways of controlling molecular processes have been proposed. [29] We believe, however, that avoided crossings created by the external fields play crucial roles in these processes, whenever new processes induced by lasers occur effectively. Thus we think that it is most fundamental and efficient to control nonadiabatic transitions at avoided crossings among the dressed states created by the external field. This is actually possible by sweeping the field periodically at the relevant avoided crossoing. [30,31] By appropriately adjusting the sweeping velocity, the sweeping amplitude and the number of sweeping, we can make any desirable transition from any specified initial state with hundred percent efficiency. The control parameters can be found analytically with the help of the analytical theory of time-dependent nonadiabatic transition described in section IV. Numerical examples carried out so far are (1) quantum spin tunneling in a magnetic field [30] and (2) one-dimensional model of vibrational and tunneling transitions of a molecule by lasers. [31] When avoided crossings among dressed states are appropriate (not too sharp and not too wide), we can use the laser frequency sweeping and the Landau-Zener type of transitions. If it is possible to keep the laser intensity constant during the transition period, the required intensity is rather low. When the avoided crossing is too sharp, it is recommended to sweep the laser intensity with the laser frequency fixed at an appropriate value. We do not recommend, however, to use the resonant frequency at the sharp avoided crossing, i.e.the $\pi-$pulse, because small fluctuation in frequency creates large deviation in the final transition probability and thus the method becomes unstable. It is much better to use a certain off-resonant frequency and more than one intensity pulses, although the time interval between the pulses should be controlled. If the shaping of the frequency and intensity pulses is possible, the exponential potential model is recommended, in which both diabatic potentials and coupling are exponential functions of time. In general, an appropriate path may be found in the two-dimensional (ω, I) space, where ω is the laser frequency and I is the intensity.

Unified Theory of the LZS and the RZD Problems(?)

There is another basic type of nonadiabatic transition called Rosen-Zener-Demkov (RZD) type, in which the two diabatic potentials are pararell and the diabtic coupling is an exponential function. It would be very useful, if we can formulate a unified theory which can cover both LZS and RZD type of transitions. The exponential potential model which was first proposed by Nikitin in the semiclassical time-dependent scheme can actually provide the Landau-Zener and Rozen-Zener transition probabilities in certain limits. [7] We are now trying to generalize this model to try to formulate a unified theory. [32,33]

FUTURE PERSPECTIVES

Various applications of the complete solutions of the LZS problems should be possible in many fields of physics, chemistry and biology. Nonadiabatic transitions due to curve crossing are one of the most important mechanisms of state and phase changes in a variety of fields. Multi-channel as well as multi-dimensional applications should be carried out in near future and various new physical and chemical phenomena could be discovered. An effort to formulate a unified theory should also be continued. Considering the recent technological innovation of manipulating various time-dependent fields, control of molecular dynamic processes would be an important and interesting subject in the future and the theory of nonadiabatic transitions should play a crucial role there. This will be a challenging subject for the next century.

ACKNOWLEDGMENTS

The works reported here have been carried out in collaborations with Dr.C.Zhu, Dr.Y.Teranishi, Mr.K.Nagaya, Mr.L.Pichl, Prof.V.I.Osherov, Dr.V.G.Ushakov, Dr.O.I.Tolstikhin, Dr.K.Nobusada, and Dr.G.V.Mil'nikov. The research was partially supported by a Grant-in-Aid for Scienctific Research on Priority Area "olecular Physical Chemistry" and a Research Grant 10440179 from The Ministry of Education, Science, Culture and Sports of Japan.

REFERENCES

1. H.Nakamura, Int.Rev.Phys.Chem. **10**, 123 (1991).
2. H.Nakamura, "Nonadiabatic Transitions: Beyond Born-Oppenheimer", in *Dynamics of Molecules and Chemical Reactions*, eds. R.E.Wyatt and J.Z.H.Zhang (Marcel Dekker Inc., 1995).
3. L.D.Landau, Phys.Zts.Sowjet, **2**, 46 (1932).
4. C.Zener, Proc.Roy.Soc. **A137**, 696 (1932).
5. E.C.G.Stueckelberg, Helv.Phys.Acta **5**, 369 (1932).
6. M.S.Child, *Molecular Collision Theory* (Academic Press, 1974); *Semiclassical Mechanics with Molecular Applications* (Oxford Univ.Press, 1991).
7. E.E.Nikitin and S.Ya.Umanskii, *Theory of Slow Atomic Collisions* (Springer, 1984).
8. E.S.Medvedev and V.I.Osherov, *Radiationless Transitions in Polyatomic Molecules* (Springer, 1994).
9. Yu.N.Demkov and V.N.Ostrovsky, *Zero-Range Potentials and Their Applications in Atomic Physics* (Plenum, 1988).
10. D.S.F.Crothers, Adv.Phys. **20**, 405 (1971); Adv.Atom.Molec.Phys. **17**, 55(1981).
11. K.S.Delos,Adv.Chem.Phys. **65**, 161 (1986).
12. K.S.Lam and T.F.George, in *Semiclassical Methods in Molecular Scattering and Spectroscopy* ed. M.S.Child (Reidel, 1979).
13. H.Nakamura, Adv.Chem.Phys. **82**, 243 (1992).
14. H.Nakamura and C.Zhu, Comm.Atomic and Molec.Phys. **32**, 249 (1996).
15. C.Zhu, H.Nakamura, N.Re and V.Aquilanti, J.Chem.Phys. **97**, 1892 (1992).

16. C.Zhu and H.Nakamura, J.Chem.Phys. **97**, 8497 (1992); ibid **98**, 6208 (1993); ibid **101**, 4855 (1994); Erratum, J.Chem.Phys. **108**, 7501 (1998).

17. Y.Teranishi and H.Nakamura, J.Chem.Phys. **107**, 1904 (1997).

18. C.Zhu and H.Nakamura, J.Math.Phys. **33**, 2697 (1992).

19. C.Zhu and H.Nakamura, J.Chem.Phys. **101**, 10630 (1994); Erratum, J.Chem.Phys. **108**, 7501 (1998); J.Chem.Phys. **102**, 7448 (1995); ibid **109**, 4689 (1998).

20. C.Zhu and H.Nakamura, Chem.Phys.Lett. **258**, 342 (1996); ibid **274**, 205 (1997).

21. C.Zhu and H.Nakamura, J.Chem.Phys. **106**, 2599 (1997); ibid **107**, 7839 (1997); Erratum, J.Chem.Phys. **108**, 7501 (1998).

22. G.V.Mil'nikov, C.Zhu, H.Nakamura, and V.I.Osherov, Chem.Phys.Lett. **293**, 448 (1998).

23. C.Zhu, H.Nakamura and K.Nobusada, "Chemical reactions analyzed by the semiclassical theory of nonadiabatic transitions:Heavy-Light-Heavy systems." XXI-ICPEAC (Sendai,1999) Post-post deadline abstracts TU154.

24. see for instance, S.Chapman, Adv.Chem.Phys. **82**, 423 (1992).

25. see for instance, W.H.Miller, J.Chem.Phys. **95**, 9428 (1991).

26. see for instance, E.J.Heller, J.Chem.Phys. **94**, 2723 (1991).

27. H.Nakamura, J.Chem.Phys. **97**, 256 (1992); ibid **110**, 10253 (1999).

28. S.Nanbu, H.Nakamura and F.O.Goodman, J.Chem.Phys. **107**, 5445 (1997).

29. see for instance, R.J.Gordon and S.A.Rice, Annu.Rev.Phys.Chem. **48**, 601 (1997).

30. Y.Teranishi and H.Nakamura, Phys.Rev.Lett. **81**, 2032 (1998).

31. Y.Teranishi and H.Nakamura, J.Chem.Phys. **111**, 1415 (1999).

32. V.I.Osherov and H.Nakamura, J.Chem.Phys. **105**, 2770 (1996); Phys.Rev. **A59**, 2486 (1999).

33. V.I.Osherov, V.G.Ushakov and H.Nakamura, Phys.Rev. **A57** 2672 (1998).

Cusp formation and threshold effects in break-up collisions

R. O. Barrachina[*1], J. Fiol[*], V. D. Rodríguez[†1] and P. Macri[‡]

* Centro Atómico Bariloche and Instituto Balseiro[2],
8400 S. C. de Bariloche, Río Negro, Argentina.
† Departamento de Física, FCEyN, Universidad de Buenos Aires,
1428 Buenos Aires, Argentina.
‡ Instituto de Astronomía y Física del Espacio[3],
C.C. 67, Suc. 28, 1428 Buenos Aires, Argentina.

Abstract. We show how the single-particle double differential cross sections (DDCS) in breakup collisions are intertwined by dynamical constraints. In particular, we study the corresponding relations among threshold and cusp structures and identify some of their properties by means of a final-state interaction theory. We provide general expressions for the cusp and threshold structures that any theoretical description of the collision process has to fulfil. Finally, we show how these structures change with the relative mass ratios of the three particles in the final state.

INTRODUCTION

A large number of scattering processes in nuclear, atomic and molecular physics leads to final states with three fragments. This is the case -for instance- of a proton-deuteron collision, the ionization of atoms by the impact of photons, electrons or ions, or a fragmentation process in molecular scattering. Besides its practical interest, the three-particle breakup scattering has a further attribute which contributes to its appeal. It is the simplest scattering process beyond the elastic, excitation or rearrangement collisions, but it is also complicated enough to yield no exact classical or quantum mechanical solution.

A kinematically complete description of a three-body continuum final-state would not require the knowledge of all its nine momentum components, since many of them can be determined from conservation and symmetry relations. However, instead of a so-called kinematically complete measurement, the great majority of the experimental techniques analyses only one of the particles in the final state.

---

[1]) Also member of the Consejo Nacional de Investigaciones Científicas y Técnicas, Argentina.
[2]) Comisión Nacional de Energía Atómica and Universidad Nacional de Cuyo, Argentina.
[3]) Consejo Nacional de Investigaciones Científicas y Técnicas, Argentina.

Here we show how the corresponding double differential cross sections (DDCS) are interrelated by dynamical constraints.

A straightforward example of this relation among different DDCSs is provided in ion-atom collisions by a recent measurement of the target ionization cross section, doubly differential in the projectile energy loss and scattering angle [1]. Both the experimental data of Schulz *et al.* [1] and the CDW-EIS calculation of Rodríguez and Barrachina [2] show a distinctive shoulder at an energy that corresponds to electrons moving at the same velocity than the projectile. Thus, this DDCS associated to the projectile shows a clear fingerprint of the conspicuous "electron capture to the continuum" (ECC) cusp-shaped structure observed in electron spectroscopy experiments by Crooks and Rudd three decades ago [3]. Furthermore, the presence of a threshold structure in the longitudinal recoil - ion momentum associated to the ECC cusp in the electron momentum distribution was theoretically predicted some years ago [4,5].

These examples from atomic collisions only represent particular cases of a deeper and more general structure, where the momentum distribution of one particle is necessarily related, in a sense that is described here, to the momentum distribution of any other of the three particles in the final state. In particular, we are interested in studying the relation among threshold and cusp structures in different DDCSs. The energy conservation law imposes a limitation to the values that the different momenta can attain. Thus, one momentum would reach its maximum value whenever another complementary momentum is zero; and vice versa. Thus, the cross sections for both momenta are connected in this limit.

In order to analyse the relation among threshold structures in the three possible single-particle DDCSs, we need some additional information on the momentum dependence at the threshold itself. To this end we employ the final-state interaction theory of cusp formation [6]. Being only related to the analytic properties at zero energy of the regular two-body scattering wave-functions, this approach will help us to study the problem under very general grounds.

THREE-BODY KINEMATICS

In the center-of-mass reference system, the three-body problem can be described by any of three possible sets of Jacobi coordinates [7], $(\mathbf{r}_T, \mathbf{R}_T)$, $(\mathbf{r}_P, \mathbf{R}_P)$ and $(\mathbf{r}_N, \mathbf{R}_N)$ as shown in figure 1. In our notation, $\mathbf{r}_T$, $\mathbf{r}_P$ and $\mathbf{R}_N$ are the position vectors of the particle of mass m relative to the recoiling fragment T, the particle P and the centre-of-mass of T + P, respectively. $\mathbf{R}_P$ is the position vector of the centre-of-mass of m + P relative to T; and $\mathbf{r}_N$ and $\mathbf{R}_T$ are the coordinates of P relative to T and the centre-of-mass of m + T, respectively. In momentum space, the system is described by the associated pairs $(\mathbf{k}_T, \mathbf{K}_T)$, $(\mathbf{k}_P, \mathbf{K}_P)$ and $(\mathbf{k}_N, \mathbf{K}_N)$. The kinetic energy of the three-body system in the centre-of-mass reference frame can be written in terms of any of these Jacobi pairs $E = k_j^2/2m_j + K_j^2/2\mu_j$, $(j = T,$ P or N). Here we have defined the reduced masses m_j and μ_j $(j = N, T$ or $P)$ of each two- and three-body system, respectively.

Switching back to the Laboratory reference frame, $\mathbf{k}$, $\mathbf{K}_R$ and $\mathbf{K}$ are the final momenta of the particle of mass m, the recoiling (target) fragment of mass M_T and the particle (projectile) of mass M_P. All these momenta in the Laboratory reference frame can be written in terms of the Jacobi impulses $\mathbf{K}_j$ as

$$\mathbf{k} = m\mathbf{v}_{CM} + \mathbf{K}_N \qquad \mathbf{K} = M_P\mathbf{v}_{CM} + \mathbf{K}_T \qquad \mathbf{K}_R = M_T\mathbf{v}_{CM} - \mathbf{K}_P \,,$$

where $\mathbf{v}_{CM}$ is the velocity of the centre-of-mass. The differential cross section for the fragmentation collision can be written in terms of any pair of Jacobi impulses $(j = T, P \text{ or } N)$,

$$\frac{d\sigma}{d\mathbf{k}\,d\mathbf{K}\,d\mathbf{K}_R} = \delta\left(\mathbf{k} + \mathbf{K} + \mathbf{K}_R - M\mathbf{v}_{CM}\right)\frac{d\sigma}{d\mathbf{k}_j\,d\mathbf{K}_j}\,,$$

where $M = m + M_T + M_P$ and

$$\frac{d\sigma}{d\mathbf{k}_j\,d\mathbf{K}_j} \propto |t_{if}|^2\,\delta\left(\frac{1}{2m_j}k_j^2 + \frac{1}{2\mu_j}K_j^2 - E\right)\,.$$

FINAL-STATE INTERACTION THEORY

By energy conservation, K_j reaches its maximum value $K_j^{max} = \sqrt{2\mu_j E}$ whenever $k_j = 0$; and vice versa, k_j is maximum (i.e. $k_j^{max} = \sqrt{2m_j E}$) for $K_j = 0$. Here we want to analyse the behaviour of the cross sections $d\sigma/d\mathbf{k}_j$ and $d\sigma/d\mathbf{K}_j$ in the first of these limits, i.e. when the relative momentum of any two particles in the final

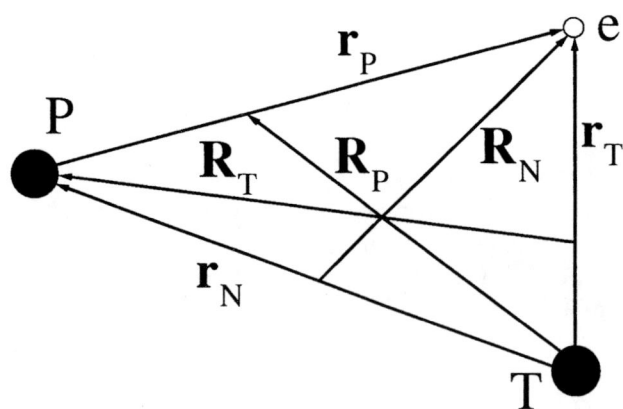

FIGURE 1. Jacobi coordinates for the three-body problem.

512

state is very small. In this case, the *final-state interaction* theory indicates that, under suitable but very general conditions, the transition matrix element for the fragmentation collision can be written in the following way [6]

$$t_{if} = \frac{1}{f_0(k_j)} \tilde{t}_{if} \ .$$

Here $f_0(k_j)$ is the s-wave Jost function for the corresponding two-body elastic scattering and $\tilde{t}_{if}$ can be approximated by the T matrix for the fragmentation process in the absence of that two-body interaction in the final state. The Jost function $f_\ell(k_j)$ is defined by the $r_j \to 0$ limit of the normalized radial wavefunction for the two-body system

$$\psi_{\ell,k_j}(r_j) \underset{r_j \to 0}{\approx} \frac{(k_j r_j/\hbar)^{\ell+1}}{(2\ell+1)!!} \frac{1}{f_\ell(k_j)} \ .$$

The main advantage of this decomposition is that it decouples in a very simple way the *final-state interaction* of any two particles in the continuum from the actual process of fragmentation. Actually, with this separation of the transition matrix, it can be shown that, in the close vicinity of the j-channel threshold (i.e. for $K_j \to K_j^{max}$ or $k_j \to 0$), the double differential cross section (DDCS) in the Jacobi momentum $\mathbf{K}_j$ reads

$$\frac{d\sigma}{d\mathbf{K}_j} = \int \frac{d\sigma}{d\mathbf{k}_j \, d\mathbf{K}_j} \, d\mathbf{k}_j \approx \frac{m_j k_j}{\mu_j K_j^{max}} \frac{B_j(\hat{\mathbf{K}}_j)}{|f_0(k_j)|^2} \Theta(K_j^{max} - K_j) \ ,$$

where $\Theta(x)$ is Heaviside's step function. Similarly, the DDCS in the Jacobi momentum $\mathbf{k}_j$ reads

$$\frac{d\sigma}{d\mathbf{k}_j} = \int \frac{d\sigma}{d\mathbf{k}_j \, d\mathbf{K}_j} \, d\mathbf{K}_j \approx \frac{A_j(\hat{\mathbf{k}}_j)}{|f_0(k_j)|^2} \ ,$$

where both functions $A_j(\hat{\mathbf{k}}_j)$ and $B_j(\hat{\mathbf{K}}_j)$ are trivially related by angular integration

$$\int A_j(\hat{\mathbf{k}}_j) d\hat{\mathbf{k}}_j = \int B_j(\hat{\mathbf{K}}_j) d\hat{\mathbf{K}}_j \ .$$

These limiting behaviours decouple the dependence on the modulus of $\mathbf{k}_j$, as given by a factor proportional to the distortion function $1/|f_0(k_j)|^2$, from the angular dependence introduced by the functions A_j and B_j, respectively. One further attribute of this decomposition is that, while these latter functions depend on the actual scattering process through the reduced transition matrix element $\tilde{t}_{if}$, the distortion factor $F(k_j) = 1/|f_0(k_j)|^2$ only depends on the s-wave Jost function for the final-state interaction among the two particles of reduced mass m_j and relative momentum $\mathbf{k}_j$. Note that we are not concerned with how this particular final-state configuration is reached. For the analysis of the behaviour of the DDCS at lower order in k_j, we only need to inquire into the analytic properties at zero energy of the regular two-body scattering wave-functions, through its s-wave Jost function.

PROPERTIES OF THE DISTORTION FACTOR

Let us identify some general properties of the distortion factor $F(k) = 1/|f_o(k)|^2$ in the limit $k \to 0$, for a given two-body final-state interaction $V(r)$. Firstly, it can be shown that $F(k) \to 1$, for $k \to \infty$ or if the potential vanishes. Thus, the distortion produced on the cross section by the factor $F(k)$ is dominant for small values of the relative impulse k. Secondly, $F(k) \geq 1$ for any generally attractive interaction, and $F(k) \leq 1$ for repulsive potentials. In this latter case, the distortion factor F displays an exponential or linear fall-off as $k \to 0$. But far more interesting is the case of a slowly decreasing potential with a long-range attractive power-law "tail" of the form $V(r) \approx - Z/r^\nu$ with $\nu < 2$, which leads to a distortion factor that diverges in the $k \to 0$ limit as $F(k) \propto 1/k$ [6].

Note that we are not making any statement about the overall structure of the potential, but only about its asymptotic behaviour. This is so, since the divergence of the distortion factor for an attractive power-law "tail" is related to the fact that the corresponding discrete spectrum accumulates at zero energy [8]. Developing this continuation argument a little further, it can be shown that the corresponding cross section for excitation or capture to a highly excited s-wave state with energy ε_n (i.e., principal quantum number n) has to obey the following scaling rule

$$\sigma_n \propto \frac{d\varepsilon_n}{dn} .$$

In particular, for a final-state interaction with an attractive Coulomb tail, we obtain the well-known Jackson - Schiff scaling rule, $\sigma_n \propto 1/n^3$ [9]. However, in its generalized form, it is valid not only for a Coulomb potential, but for any final state interaction whose bound states accumulate at zero energy. Note that this result does not depend on the dynamical details of the collision, but only on the final-state. Thus the previous scaling rule applies to any scattering process leading to a highly excited bound state of any two of the three fragments in the final state.

These results are valid for attractive potentials vanishing slower than $1/r^2$ at infinite. On the other hand, for a short range potential $V(r)$, there is a completely different behaviour. Under very general conditions, the small-k behaviour of the distortion factor shows a characteristic Lorentzian shape [6,10]

$$F(k) \approx \frac{a_o^2}{1 + a_o^2 k^2} .$$

The "scattering length" a_o is associated to the presence of a low-lying bound ($a_o > 0$) or virtual ($a_o < 0$) state of the potential $V(r)$. Whenever the energy of this s-wave bound or virtual state is very close to zero, a_o is large, and the distortion factor would present a sharp structure. In particular, in the case of a "zero-energy resonance", a_o is infinite and the distortion factor diverges like $1/k^2$, producing a sharp cusp at threshold. These results are strictly valid for potentials that fall off faster than any power of $1/r$. In other cases, the Jost function might have a branch-point singularity at the origin, and some distortion of the previous Lorentzian shape might occur [11]

With this final analysis, we have exhausted all the possible small-k dependence for the distortion factor $F(k)$. They can be roughly classified in three groups: i) a decay to zero for repulsive interactions, ii) a $1/k$ divergence for the case of potentials with long-range power-law tails $- Z/r^\nu$ with $\nu < 2$, and iii) a lorentzian $a_o^2/(1 + a_o^2 k^2)$ shape for other attractive short-range interactions, with some kind of distortion for power-law tails.

SOME EXAMPLES FROM ATOMIC PHYSICS

Typical examples of the first two groups in our classification are provided by the Coulomb interaction of two charged particles of equal and opposite signs, respectively. The third group can be exemplified by the interaction of electrons with neutral polarisable systems. Except for some logarithmic distortion due to its long

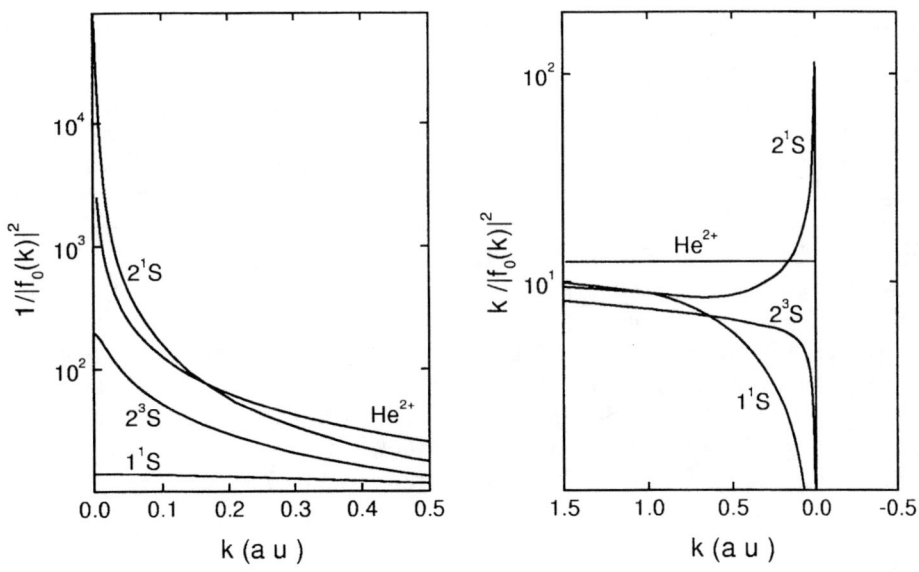

FIGURE 2. Distortion factors $1/|f_o(k)|^2$ and $k/|f_o(k)|^2$ at threshold for an e + He system with the helium atom in the ground 1^1S state, the metastable 2^1S and 2^3S states, and in an ionic He^{2+} state.

$-\alpha/2r^4$ tail, the corresponding distortion factor $F(k)$ would present a characteristic Lorentzian shape for small values of k. Let us consider, for instance, an $e^- +$ He system. The corresponding s-wave Jost function was recently evaluated on an absolute scale for the ground 1^1S and metastable 2^1S and 2^3S states [11]. Figure 2 shows the distortion factor $F(k)$ for these three cases. Note the giant increase of $F(k)$ at small-energies for the $e^- +$ He(2^1S) interaction. This sharp effect is due to the presence of a low-lying virtual state with a scattering length of about a_o = - 330 a.u. [10,11]. The distortion factor for the $e^- +$ He(2^3S) system shows a much more moderate enhancement at low velocities, which can be ascribed to the proximity of a ^2S resonance to the 2^3S excitation threshold. In spite of being much smaller and broader than for the 2^1S state, it still produces a sizable effect.

In 1989 Sarkadi and co-workers [12] observed a very sharp cusp in coincidence with neutral outgoing helium projectiles in He + He and He + Ar collisions. Taking into account the common understanding of a finite peak for any screened interaction, this result came as a very big surprise, even though the possibility of a sharp cusp for such interaction had already been predicted in 1983 [13]. But now, with the final state interaction model, we can understand this effect as due to a fraction of metastable He(2^1S) states either already present in the incoming beam or produced during the collision. This hypothesis was later tested by measuring the dependence of the peak's yield with respect to the incoming metastable fraction [14]. Recent results [15], in which the cusp electron production for the 2^1S and 2^3S metastable states were distinguished, have provided a conclusive confirmation for the theoretical model.

The $e^- +$ H interaction is an example of a limiting case between the short and long- range potentials. Owing to its degeneracy in ℓ, an excited state of an hydrogen atom can acquire a dipole momentum, producing a final-state interaction with a dipolar "tail". The corresponding discrete s-wave spectrum might accumulate at zero energy, leading to a distortion factor $F(k)$ with the same $1/k$ diverging behaviour than for a long-range attractive interaction, but influenced by a Gailitis-Damburg oscillatory term in $\log k$ [16]. The existence of this $1/k$ cusp for the $e^- +$ H(n=2) system was experimentally proved by Penent et $al.$ [17] in their study of the collisional $electron$ $detachment$ process from the H$^-$ ion, by detecting the emitted electrons in coincidence with the Lyman-α photons stemming from the decay of H($2p$) outgoing projectiles (See also Ref. [18]). In previous non-coincident measurements of H$^-$ detachment collisions, the contribution to the cusp from the final-state Coulomb interaction in $double$ $electron$ $loss$ processes had not been separated.

THRESHOLD EFFECTS AND CUSP FORMATION

In this section we discuss in general the appearance of threshold structures in breakup collisions. Recalling the relation between the momenta of the three outgoing particles in the laboratory frame and the corresponding Jacobi impulses we write the momentum distribution of each particle as follows:

$$\frac{d\sigma}{dK_j} = \frac{(2\pi)^4}{v} m_j k_j \int d\hat{k}_j |t_{if}|^2 \ ,$$

where k_j is given by energy conservation, $k_j = \sqrt{m_j(K_j^{max2} - K_j^2)/\mu_j}$. Now let us analyse the behaviour of this double differential cross section when the system is close to a given ℓ-channel threshold $\mathbf{K}_\ell \to \mathbf{K}_\ell^{max}$ (i.e. $k_\ell \to 0$), with ℓ not necessarily equal to j. It can be shown that any fingerprint on $d\sigma/dK_j$ of the $k_\ell = 0$ threshold would have to be concentrated near the surface of a sphere in momentum space given by

$$K_j^2 = 2\left(\mu_j - (1 - \delta_{j\ell}) m_\ell\right) E \ .$$

For $j = \ell$, this equation gives the outermost boundary of the single-particle momentum distribution $d\sigma/dK_j$, while for $j \neq \ell$ it defines the *locus* of any possible structure arising from the $k_\ell = 0$ threshold. For each cross section $d\sigma/dK_j$, these three spheres are concentric and centred at $\mathbf{v}_{CM}$ in velocity space. Thus, we conclude that any cusp structure in $d\sigma/dk_\ell$ at $k_\ell = 0$ would produce a structure in the actual single-particle cross section $d\sigma/dK_j$, as measured in the laboratory frame, which in principle would be spread along the surface of a sphere in momentum (or velocity) space, rather than be located at one point. This is clearly seen in figure 3 for the ionization of Helium by the impact of 435 eV positrons. Besides the kinematical threshold and a "soft collision" cusp-shaped peak at $\mathbf{v}_{CM} \approx 0$, the

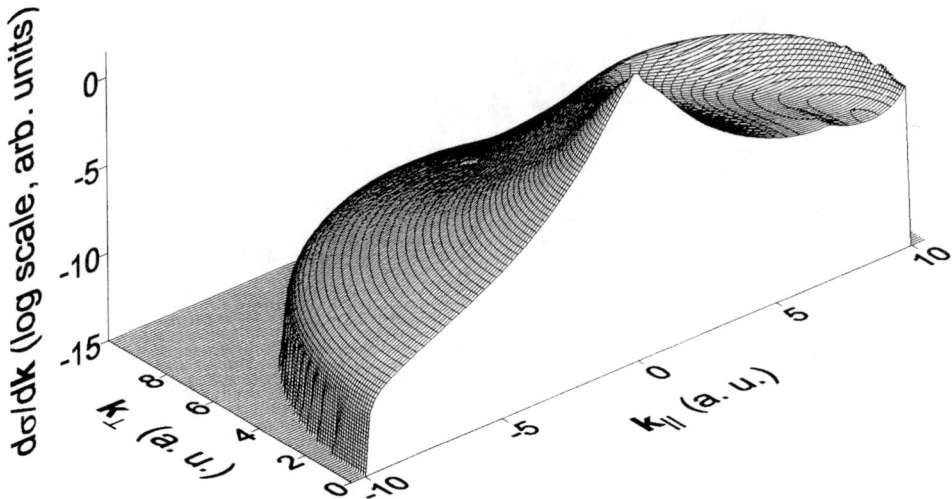

FIGURE 3. Momentum distribution of the electrons emitted in the ionization of Helium by the impact of positrons (velocity = 8 a.u.). $\mathbf{k}_\parallel$ and $\mathbf{k}_\perp$ are the electron momentum components parallel and perpendicular to the initial impulse of the projectile, respectively.

517

electron's momentum distribution clearly shows a rather mild ECC structure distributed along a circle of radius $[2(\mu_N - m_P)E/m^2]^{1/2}$. Actually, a cusp structure would be located at one point in momentum space, as it is the case for the cusp structure in figure 3, only for certain particular relations between the masses of the three particles in the final state. For, instance, this might occur whenever the radius of the circle vanishes, i.e. for $\mu_j - m_\ell \approx 0$.

It is possible to demonstrate that a sharp cusp (or anticusp) structure, associated to a given attractive (repulsive) two-body interaction in the final state, can be observed in a single particle DDCS whenever its mass is much smaller than that of one of its collision partner. In the collision, the particle of larger mass practically remains undisturbed while the other gets all the movement. In figure 4 we show the momentum distribution of the electrons emitted in the ionization of Helium by the impact of protons (velocity = 5 a.u.). The so-called *soft-collision* peak structure at zero velocity, and the "electron capture to the continuum" (ECC) cusp at the projectile's initial velocity **v** are typical examples of this situation. Whenever both particles have comparable masses, there is no clear direction where equivelocity ejected particles may be looked for, and the peak is spread along the surface of the $K_j^2 = 2(\mu_j - m_\ell)E$ sphere.

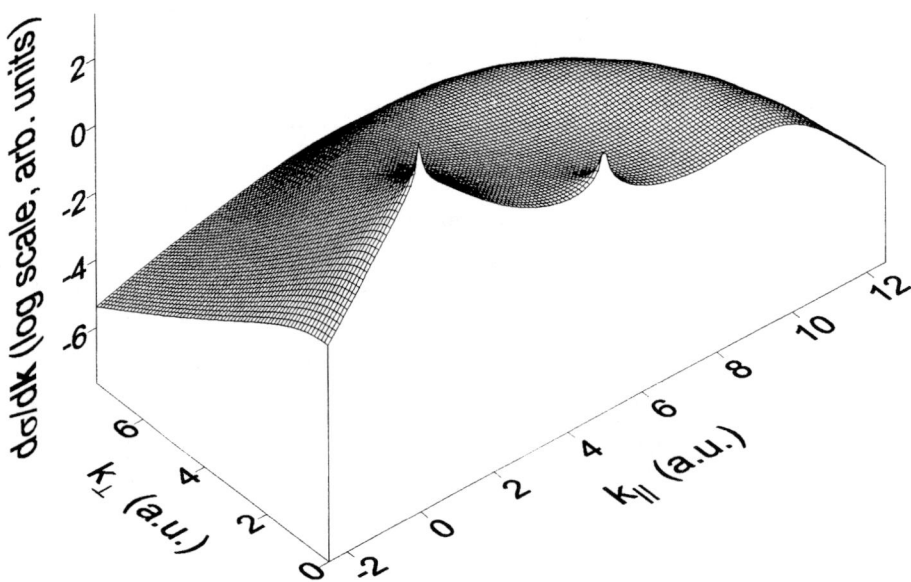

FIGURE 4. Momentum distribution of the electrons emitted in the ionization of Helium by the impact of protons (velocity = 5 a.u.). $\mathbf{k}_\parallel$ and $\mathbf{k}_\perp$ are the electron momentum components parallel and perpendicular to the initial impulse of the projectile, respectively.

CONCLUSIONS

In this communication we have demonstrated that it is possible to identify some general properties of the threshold and cusp structures in break-up collisions. We remark that this analysis was based only on the nature of the interaction between a pair of fragments in the final state and on some weak assumptions about the general dependence of $\tilde{t}_{if}$ on the transferred impulse. For this reason it is independent of the mechanisms that lead to the breakup process and the pair of particles involved. Thus, the present approach provides general expressions for the momentum dependence at threshold that any theoretical description of a collision process has to fulfil.

ACKNOWLEDGEMENTS

This work was partially supported by the Agencia Nacional de Promoción Científica y Tecnológica (Contrato de Préstamo BID 802 OC-AR, Grants PICT No. 03-00986 and 03-04021), Argentina.

REFERENCES

1. Schulz M., Vajnai T., Gaus A. D., Htwe W., Madison D. H. and Olson R. E., *Phys. Rev.* A **54**, 2951 (1996).
2. Rodríguez V. D. and Barrachina R. O., *Phys. Rev.* A **57**, 215 (1998).
3. Crooks G. B. and Rudd M. E., *Phys. Rev. Lett.* **25**, 1599 (1970).
4. Rodríguez V. D., Wang Y. D. and Lin C. D., *Phys. Rev.* A **52** R9 (1995).
5. Rodríguez V. D. and Barrachina R. O., *Phys. Rev.* A **53**, 3335 (1996).
6. Barrachina R. O., *Nucl. Inst. Meth.* B **124**, 198 (1997).
7. Macek J. H. and Shakeshaft R., *Phys. Rev.* A **22**, 1441 (1980).
8. Rudd M. E. and Macek J., *Case Studies in Atomic Physics* **3**, 48 (1972).
9. Jackson J. D. and Schiff H. *Phys. Rev.* **89**, 359 (1953).
10. Barrachina R. O., *J. Phys.* B **23**, 2321 (1990).
11. Macri P. A. and Barrachina R. O., *J. Phys.* B **31**, 1303 (1998).
12. Sarkadi L., Pálinkás J., Köver A., Berényi D. and Vajnai T., *Phys. Rev. Lett.* **62**, 527 (1989).
13. Garibotti C. R. and Barrachina R. O., *Phys. Rev.* A **28**, 2792 (1983).
14. Kuzel M., Sarkadi L., Pálinkás J., Závodszky P. A., Maier R., Berényi D. and Groeneveld K. O., *Phys. Rev.* A **43**, 1745 (1993).
15. Báder A., Sarkadi L., Víkor., Kuzel M., Závodszky P. A., Jalowy T., Groeneveld K. O., Macri P. A. and Barrachina R. O., *Phys. Rev.* A **55**, 14 (1997).
16. Gailitis M. and Damburg R., *Proc. Phys. Soc.* **82**, 192 (1963).
17. Penent F., Grouard J. P., Montmagnon J. L. and Hall R. I., *J. Phys.* B **25**, 2831 (1992).
18. Menendez M. G. and Duncan M. M., *Phys. Rev.* A **36**, 1653 (1987).

Radiative Electron Capture and the Photoelectric Effect in Hydrogen-like High-Z Systems

Jörg Eichler

Bereich Theoretische Physik, Hahn-Meitner-Institut Berlin, 14109 Berlin, Germany[1]
and
Fachbereich Physik, Freie Universität Berlin, 14195 Berlin, Germany

Abstract. Radiative electron capture (REC) from a low-Z target atom into a bare high-Z projectile ion has recently been studied in great detail. This process plays an important role in plasmas and in the spectroscopy of high-Z one- or few-electron ions. It also provides a means to investigate the inverse reaction, the photoelectric effect, in a regime that is otherwise inaccessible. The theoretical development and some recent results are presented.

INTRODUCTION

The interpretation of the photoelectric effect by Einstein in 1905 provided the first simple and fundamental confirmation of Planck's quantum hypothesis. Since that time, the photoeffect has continued to be an object of research. Besides a basic test of quantum theory, an important goal of early theoretical treatments has been to assess the contribution of the photoelectric effect to the absorption of radiation in matter. In these applications and early experiments, the photon ionizes a neutral target atom, so that, in general, one has to deal with many-electron effects, which render calculations difficult and – in most cases – approximate.

In recent years, it has been possible to approach the pure situation of a one-electron problem from another side, namely by studying the inverse process of radiative recombination (RR), in which a bare nucleus captures a free electron with the simultaneous emission of a photon. Bare nuclei, as for example Au^{79+}, Pb^{82+}, or U^{92+}, can be produced in high-energy atomic collisons by stripping off all electrons. This is possible for projectile velocities comparable to the speed of light, that is with kinetic energies of 100 MeV/u or more. In these cases, the electronic motion requires a relativistic quantum description by the Dirac equation. On the

---

[1] E-Mail: eichler@hmi.de

other hand, the nuclear motion can, to a very good approximation, be described by classical relativistic kinematics.

In practice, no free-electron target of sufficient density for studying radiative recombination is available, so that one uses loosely bound electrons in low-Z target atoms. These electrons can be considered as quasi-free, which is a very good approximation for high-energy high-Z projectiles. Experimentally, radiative electron capture (REC) can be identified by detecting the down-charged projectile in coincidence with an x-ray photon.

The most recent technical development in the field of relativistic heavy-ion physics consists in relativistic heavy-ion storage rings, for example the ESR at the GSI in Darmstadt. This electron cooler ring provides the basis for atomic collision experiments dealing with beams of unprecendented quality [1]. Here, the interaction of intense beams of cooled high-Z ions with low-density gaseous matter can be studied without any beam collimation, guaranteeing almost completely background-free experimental conditions, see, e.g. [2].

THEORETICAL TREATMENT OF RADIATIVE ELECTRON CAPTURE

For free electrons, nonradiative capture is excluded by energy and momentum conservation. If, however, the electron transfer is accompanied by the emission of electromagnetic radiation, the emitted photon acts as a third body carrying away energy and momentum released by the formation of the final bound state.

In the following, we assume that (i) the projectile charge is large compared to the target charge, $Z_P \gg Z_T$, and (ii) the projectile velocity is large compared to the electron velocity in the target atom, $v \gg v_i$. Under these conditions, the target electron can be considered as quasi-free and its momentum spread within the target can be disregarded. Radiative electron capture (REC) is then equivalent to radiative recombination (RR) in the moving projectile frame. This process, in turn, is the inverse of the photoeffect.

Approximate approach to the photoelectric effect and to radiative recombination

We start with the nonrelativistic Born approximation for the photoeffect. For hydrogenlike ions with charge Z, this assumption implies the conditions

$$\hbar\omega \ll m_e c^2$$
$$\alpha Z \ll 1$$
$$\varphi_f = e^{i\mathbf{k}_f \cdot \mathbf{r}}. \tag{1}$$

Here $\hbar\omega$ is the energy of the incident photon, m_e is the electron mass, $\alpha \approx 1/137$ the fine-structure constant, φ_f represents the wave function of the emitted electron with the wave vector $\mathbf{k}_f$. If $\mathbf{p}_{op} = \frac{\hbar}{i}\nabla$ is the momentum operator, $\hat{\mathbf{u}}$ the polarization vector of the photon and $\mathbf{k}$ its wave vector, the matrix element for the transition from the initial bound-state wave function φ_i to the final state is given by

$$M_{fi} = \int \varphi_f^*(\mathbf{r})(\mathbf{p}_{op} \cdot \hat{\mathbf{u}}) e^{i\mathbf{k}\cdot\mathbf{r}} \varphi_i(\mathbf{r}) d^3r$$
$$= (\mathbf{k}_f \cdot \hat{\mathbf{u}}) \int e^{i\mathbf{q}\cdot\mathbf{r}} \varphi_i(\mathbf{r}) d^3r, \qquad (2)$$

where $\mathbf{q} = \mathbf{k} - \mathbf{k}_f$. This is the Fourier transform of the initial state. Now, if $\mathbf{k} \parallel \hat{\mathbf{e}}_z$ and $\hat{\mathbf{u}} = \hat{\mathbf{e}}_x$, then $\mathbf{k}_f \cdot \hat{\mathbf{u}} = k_f \sin\theta \cos\varphi$ with θ being the angle between the electron and the photon direction and φ the azimuthal angle. Inserting the Fourier transform of the initial 1s state [3] and averaging over the photon polarization, we obtain for the differential cross section

$$\frac{d\sigma_{ph}}{d\Omega} \propto |M_{fi}|^2 \propto \frac{\sin^2\theta}{(1 - \beta\cos\theta)^4}, \qquad (3)$$

where $\beta = v/c$. If retardation is neglected, $\exp(i\mathbf{k}\cdot\mathbf{r}) \to 1$, we have the dipole approximation, $\mathbf{q} \to \mathbf{k}_f$, so that the denominator is a constant and

$$\frac{d\sigma_{ph}^{dip}}{d\Omega} \propto \sin^2\theta. \qquad (4)$$

Radiative recombination of an electron at rest with a moving projectile is the inverse reaction to the photoeffect in the projectile system. We then have to replace $\theta \to \pi - \theta$ and apply the Lorentz transformation to the laboratory frame, i.e.

$$\cos\theta = \frac{\cos\theta_{lab} - \beta}{1 - \beta\cos\theta_{lab}}$$
$$\frac{d\Omega}{d\Omega_{lab}} = \frac{1}{\gamma^2(1 - \beta\cos\theta_{lab})^2} \qquad (5)$$

with $\gamma = 1/(1 - \beta^2)^{1/2}$. Inserting into the differential cross section (3) *including retardation*, i.e. retaining the denominator, one obtains the result

$$\frac{d\sigma_{RR}}{d\Omega_{lab}} \propto \sin^2\theta. \qquad (6)$$

This peculiar effect of cancellation between the effects of retardation and Lorentz transformation in the nonrelativistic limit has been first observed by Spindler [4]. It is a remarkable fact that this behavior is still approximately valid for rather high projectile energies and high charge numbers.

Exact treatment of the photoelectric effect and of radiative recombination

In a rigorous relativistic description [5], one has to adopt the Dirac theory for the electron subject to a Coulomb potential in the initial bound and final continuum state. Assuming that the electron spin is not detected and the incoming photons (whose momenta define the quantization axis) are unpolarized, one obtains the differential cross section for a single electron [6,7]

$$\frac{d\sigma_{ph}}{d\Omega} = \frac{\alpha\, m_e c^2}{8\, \hbar\omega} \frac{\lambda_c^2}{2j_i + 1} \sum_{\mu_i} \sum_{m_s} \sum_{\lambda} |M_{\mathbf{p},i}(m_s, \lambda, \mu_i)|^2 , \tag{7}$$

where we have averaged over the $(2j_i + 1)$ angular momentum projections μ_i in the initial bound state and have summed over the spin components $m_s = \pm\frac{1}{2}$ of the emitted electron. Furthermore, we have averaged over the circular polarizations $\lambda = \pm 1$ of the incoming photon.

Adopting relativistic units $\hbar = c = m_0 = 1$, the transition matrix element can be written as

$$M_{\mathbf{p},i}(m_s, \lambda, \mu_i) = \int \psi^\dagger_{\mathbf{p},m_s}(\mathbf{r})\, (\boldsymbol{\alpha} \cdot \hat{\mathbf{u}}_\lambda)\, e^{i\mathbf{k}\cdot\mathbf{r}}\, \psi_{j_i,\mu_i}(\mathbf{r})\, d^3 r, \tag{8}$$

where $\psi_{j_i,\mu_i}(\mathbf{r})$ and $\psi_{\mathbf{p},m_s}(\mathbf{r})$ are exact bound and continuum Coulomb-Dirac wave functions, the latter being given by a partial-wave expansion [5]. Eq. (7) represents the rigorous relativistic cross section for photoionization of a hydrogenlike atom to first order in the photon interaction. The effect of retardation, that is, contributions from all multipole orders are included. The formulas have to be evaluated numerically [7–9].

In the following denoting coordinates with respect to the projectile frame with a prime, the cross section $\sigma_{RR}(E', \theta')$ for radiative recombination is related by the principle of detailed balance to the cross section $\sigma_{ph}(\omega', \theta')$ of its inverse reaction, namely the photoelectric effect. By multiplying $\sigma_{RR}(E', \theta')$ with the phase space ratio k'^2/p'^2, we replace the phase space factor of the outgoing electron by that of the emitted photon. For the transition of an electron into a specific substate of the final shell, we hence obtain

$$\frac{d^2\sigma_{RR}(E', \theta')}{dE'\, d\Omega'_{ph}} = \frac{(\gamma - 1 + |\epsilon_i|/m_e c^2)^2}{\gamma^2 - 1} \frac{d^2\sigma_{ph}(E', \theta')}{dE'\, d\Omega'_{el}}. \tag{9}$$

In applications to relativistic ion-atom collisions, the z-direction is usually defined as the direction of the projectile (or photon) momentum. This is opposite to the direction of the electron (or photon) momentum as seen from the projectile. Hence for REC, the angle θ' of the photoelectric effect or of the radiative recombination has to replaced by $\pi - \theta'$, or $\cos\theta'$ is replaced by $-\cos\theta'$. As a final step, one has to transform all relevant quantities from the projectile into the laboratory system using Eq. (5).

Specifically, within the nonrelativistic dipole approximation of Stobbe [10], an explicit expression for the RR cross section can be given. In the nonrelativistic limit, the factor in Eq. (9) is $(\hbar\omega'/m_ec^2)^2 (1/\beta^2\gamma^2) = (\hbar\omega/2m_ec^2)(1 + \nu^2)$, where $\nu = Ze^2/\hbar v$ is the Sommerfeld parameter. Then the Stobbe cross section for radiative electron capture into an empty K-shell ($j_i = \frac{1}{2}$) can be written as

$$\sigma_{\mathrm{RR}}^{\mathrm{Stobbe}} = \frac{2^8\pi^2\alpha}{3}\, \lambda_c^2 \left(\frac{\nu^3}{1+\nu^2}\right)^2 \frac{e^{-4\nu\arctan(1/\nu)}}{1 - e^{-2\pi\nu}}. \tag{10}$$

The constants in front of the ν-dependent terms make up a factor of 9164.7 barn.

Radiative electron capture

A loosely bound target electron may be considered as approximately free in a high-energy collision. In this limit, REC is identical with radiative recombination (RR), in which an electron, initially moving with the velocity $-\mathbf{v}$ towards the projectile, is captured into a bound state with the simultaneous emission of a photon of energy $\hbar\omega'$ and wave number $\mathbf{k}'$.

If the cross section σ_{RR} for radiative recombination is known, it is natural to refer all momenta to the projectile frame. If the target electron has the momentum $\mathbf{q}$ with respect to the target nucleus, its momentum in the projectile frame $\mathbf{q}'$ is obtained by a Lorentz transformation. Assuming that the momentum $\gamma m_e v$ of an electron traveling with the relative speed of the target towards the projectile is large compared to the electron momentum q in a low-Z target atom, one may use the impulse approximation [11] to write the double-differential REC cross section in the projectile frame (primed coordinates) as

$$\frac{d^2\sigma_{\mathrm{REC}}}{d\Omega'\,d(\hbar\omega')} = \int d^3q\, \frac{d\sigma_{\mathrm{RR}}(q')}{d\Omega'}\, |\tilde{\varphi}_i(\mathbf{q})|^2\, \delta(\hbar\omega' - |\epsilon_f'| + \gamma|\epsilon_i| - T_e + \gamma v q_z), \tag{11}$$

where $\tilde{\varphi}_i(\mathbf{q})$ is the Fourier transform of the initial electronic target wave function. The delta function expresses the energy conservation between the final electronic energy $E_f' = m_ec^2 - |\epsilon_f'|$ and the photon energy $\hbar\omega'$ in the projectile frame on the one hand and the initial electronic energy $E_i' = \gamma(E_i - vq_z)$ (also in the projectile frame) on the other hand. Here $|\epsilon_i|$ and $|\epsilon_f'|$ are the binding energies of initial target and final projectile states. Furthermore, $T_e = m_ec^2(\gamma - 1)$ is the kinetic energy of an electron traveling with the same speed as the projectile.

The initial momentum distribution of the target electron centered around $\mathbf{q} \approx 0$ gives rise to a peak in the double-differential cross section (11) around

$$\hbar\omega_0' = \gamma E_i - E_f' = |\epsilon_f'| - \gamma|\epsilon_i| + T_e. \tag{12}$$

The Doppler broadening described by the integral over the electronic momentum $\mathbf{q}$ is approximately related to the Compton profile of the initial state. The required momentum wave function can be obtained from a hydrogenic model or from a Hartree-Fock calculation [7,12].

In order to demonstrate the effect of electron binding more clearly, we consider differential cross sections at 90° for 2.0 GeV/u Au^{79+} ions impinging on an Au target. Certainly, distortion effects by the target charge can not be ignored in this case. Nevertheless, it is instructive to examine the deviations from radiative recombination for various shells. For the purpose of demonstration, we confine ourselves to ns subshells. While for the $1s_{1/2}$ shell the ratio $\sigma_{\mathrm{REC}}/\sigma_{\mathrm{RR}}$ is still 0.59 showing a strong deviation of REC from RR, the deviation decreases to 1% for the $4s_{1/2}$ shell and is considerably less than 1% for the $5s_{1/2}$ and $6s_{1/2}$ shells [13]. We may conclude that for high projectile energies of a few hundred MeV/u and for $Z_{\mathrm{P}} \gg Z_{\mathrm{T}}$, it is a very good approximation to substitute RR for REC in differential and total cross sections.

DIFFERENTIAL CROSS SECTIONS

Angular distributions are particularly sensitive to the REC reaction mechanism and to details of atomic wave functions. We should remark here that for high projectile energies, high-Z projectiles and low-Z target atoms, the momentum distribution within the target is negligible and amounts to a correction of the order of a percent [7,9]. It is therefore indeed justified to identify radiative electron capture with radiative recombination.

The photoeffect versus radiative recombination

A particularly clear-cut example for angular distributions is provided by REC into the K shell of a relativistic high-Z ion. The nonrelativistic Born approximation, see Eq. (3), as well as the rigorous treatment shows that for high velocities, $\beta \rightarrow 1$, the peak of the distribution for the photoeffect is increasingly shifted towards forward angles owing to the effect of retardation. This means that the detailed structure is compressed into a narrow forward cone. On the other hand, the Lorentz transformation (5) to the laboratory system associated with the manifestation in the inverse process, namely RR, decompresses the peak so that details become more visible. This is illustrated in Fig. 1, which contains two pairs of identical cases, i.e. with the same photon energy, viewed as the photoelectric effect and, alternatively, as radiative recombination [9]. This demonstrates that REC (or RR) is the most practical way to study the photoelectric effect at high energies [14]. Similarly, REC into the L shells [15] and the strong alignment caused by two-step photoionization have been studied in the inverse process of REC into excited states [16,17].

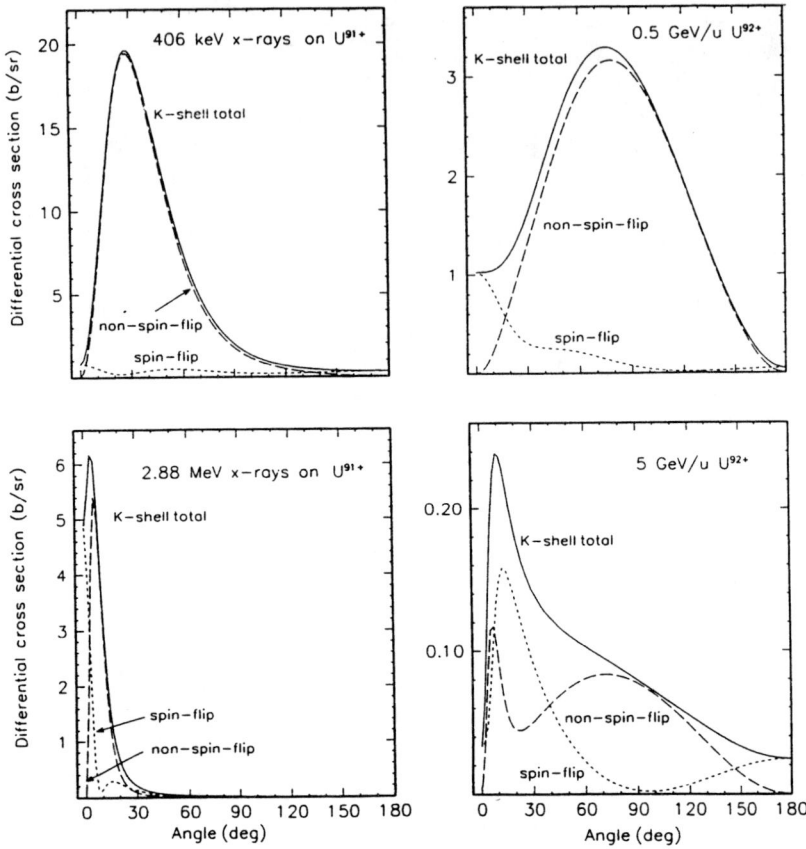

Fig. 1: Differential cross sections as a function of the laboratory angle for photoionization (left side) of hydrogen-like uranium. For comparison, the right side shows the corresponding cross sections for radiative recombination at projectile energies of 0.5 and 5.0 GeV/u. Adapted from [9].

Effects of magnetic transitions

A systematic examination of radiative recombination with the K shell [8] reveals that at moderately relativistic energies of a few hundred MeV/u still *roughly* behaves like a nonrelativistic $\sin^2\theta$ distribution (6) including retardation, i.e., all multipoles. However, at forward angles the cross section has been predicted [7] to be finite owing to magnetic spin-flip transitions. Indeed, conservation of angular momentum rigorously forbids the emission of electrons at angles of 0° or 180° without spin-flip. This can be seen in Fig. 1, which also shows that at 5 GeV/u projectile energy, the spin-flip transitions at forward angles are dominant. Recently, measurements performed near 0° at the GSI Darmstadt [18] have confirmed the predicted angular distribution [7], see Fig. 2. In this way, the electron spin and

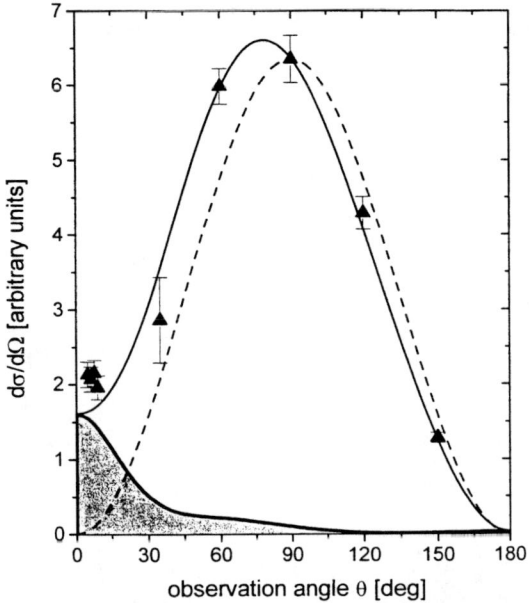

Fig. 2: Angular distribution for K-REC in $U^{92+} \rightarrow N_2$ at 309.7 MeV/u projectile energy as a function of the laboratory angle. The triangles refer to the experiment, the solid line to complete relativistic calculations and the shaded area to spin-flip contributions [7,8]. The $\sin^2 \theta$ shape of the nonrelativistic theory is given by the dashed line. The data and the nonrelativistic results are normalized to the relativistic curve at 90° [18].

magnetic transitions manifest themselves in a particularly clear-cut manner.

Radiative corrections

With the increasing accuracy in measuring angular distributions and the rigorous relativistic treatment of radiative recombination developed recently [7], it may become possible one day to identify radiative corrections to the cross sections. The effects will increase with increasing collision energy and increasing projectile charge. The theory of QED corrections to the photoeffect is very involved. While early calculations [19] were carried out to the lowest order in $Z\alpha$, a current approach uses the bound-state lepton propagator, which includes the interaction with the nucleus in all orders and hence is applicable also to high-Z systems. As a first result [20], Fig. 3 displays the relative contribution of one part of the full QED correction, due to the Uehling potential. Although this is not a final result, it may be indicative of the order of magnitude to be expected. The self-energy correction is more difficult to calculate and is presently being studied [20,21].

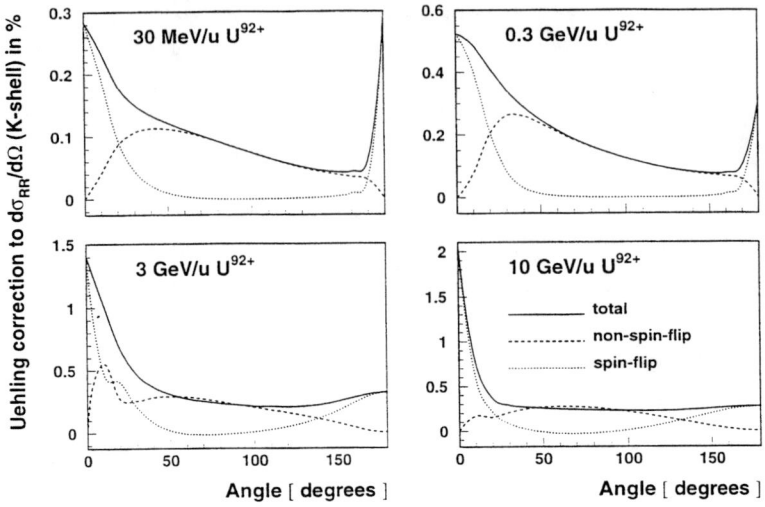

Fig. 3: The contribution of the Uehling potential to the differential cross section is given in percent for various projectile energies. Reprinted from [20], with permission from Elsevier Science.

SUMMARY AND OUTLOOK

The possibility to accelerate high-Z ions to relativistic velocities and to strip them of all or almost all of their electrons, has opened up a new field of atomic physics. One has gained access to the spectroscopy of heavy one-electron and few-electron ions up to uranium and hence to new tests of quantum electrodynamics in strong fields.

Certainly the best understood process is radiative electron capture, which is closely connected with radiative recombination and the photoelectric effect. For this reaction, calculated differential and total cross sections are in very good agreement with experimental results for K, L, and M shell capture. It has been shown that at high photon energies, radiative recombination is the most practical way to study the dynamics of the photoeffect in a regime that is not accessible otherwise. At small forward angles, it is possible to identify magnetic spin-flip transitions in a unique fashion by comparing experimental results with the theoretically predicted behavior of the angular distribution. It is furthermore expected that concrete predictions for the effects of QED corrections to the photoeffect and to radiative recombination will be available in the near future.

Acknowledgements

This work would not have been possible without the intense and delightful collaboration with A. Ichihara and T. Shirai on the theoretical side, with Th. Stöhlker on the experimental side and with Th. Beier, V.M. Shabaev, V.A. Yerokhin and G. Soff within the field of radiative corrections. The author is also indebted to the the Japan Atomic Energy Research Institute for continuing support and hospitality and to the St. Petersburg State University for the hospitality during a fruitful visit.

REFERENCES

1. F. Bosch, *Nucl. Instr. and Meth. B* **23**, 190, (1987); *Nucl. Instr. and Meth. A* **314**, 1992 (1992).
2. Th. Stöhlker, *Phys. Scripta* **T73**, 29 (1997).
3. H.A. Bethe and E.E. Salpeter, *Quantum Mechanics of One- and Two-Electron Atoms* (Academic, New York, 1957).
4. E. Spindler, H.-D. Betz, and F. Bell, *Phys, Rev. Lett.* **42**, 832 (1979).
5. For a monograph on the subject, see, e.g., J. Eichler and W.E. Meyerhof, *Relativistic Atomic Collisions*, (Academic Press, San Diego, 1995).
6. R.H. Pratt, A. Ron, and H.K. Tseng, *Revs. Mod. Phys.* **45**, 273 (1973).
7. A. Ichihara, T. Shirai, and J. Eichler, *Phys. Rev. A* **49**, 1875 (1994).
8. J. Eichler, A. Ichihara, and T. Shirai, *Phys. Rev.A* **51**, 3027 (1995).
9. A. Ichihara, T. Shirai and J. Eichler, *Phys. Rev. A* **54**, 4954 (1996).
10. M. Stobbe, *Ann. Phys. (Leipzig)* **7**, 661 (1930).
11. M. Kleber and D.H. Jakubassa, *Nucl. Phys. A* **252**, 152 (1975).
12. E. Clementi and C. Roetti, *At. Data Nucl. Data Tables* **14**, 177 (1974).
13. A. Ichihara and J. Eichler, submitted for publication.
14. Th. Stöhlker, P.H. Mokler, C. Kozhuharov, and A. Warczak, *Comments At. Mol. Phys.* **33**, 271 (1997).
15. Th. Stöhlker, H. Geissel, H. Irnich, T. Kandler, C. Kozhuarov, P.H. Mokler, G. Münzenberg, F. Nickel, C. Scheidenberger, T. Suzuki, M. Kucharski, A. Warczak, P. Rymuza, Z. Stachura, A. Kriessbach, D. Dauvergne, B. Dunford, J. Eichler, A. Ichihara, T. Shirai, *Phys. Rev. Lett.* **73**, 3520 (1994).
16. Th. Stöhlker, F. Bosch, A. Gallus, C. Kozhuharov, G. Menzel, P.H. Mokler, H.T. Prinz, J. Eichler, A. Ichihara, T. Shirai, R.W. Dunford, T. Ludziejewski, P. Rymuza, Z. Stachura, P. Swiat and A. Warczak, *Phys. Rev. Lett.* **79**, 3270 (1997).
17. J. Eichler, A. Ichihara and T. Shirai, *Phys. Rev. A* **58**, 2128 (1998).
18. Th. Stöhlker, T. Ludziejewski, F. Bosch, R.W. Dunford, C. Kozhuharov, P.H. Mokler, H.F. Beyer, O. Brinzanescu, F. Franzke, J. Eichler,A. Griegal, S. Hagmann, A. Ichihara, A. Krämer, D. Liesen, H. Reich, P. Rymuza, Z. Stachura, M. Steck, P. Swiat and A. Warczak, *Phys. Rev. Lett.* **82**, 3232 (1999).
19. J. McEnnan and M. Gavrila, *Phys. Rev. A***15**,1537 (1977); D.J. Botto and M. Gavrila, *Phys. Rev. A* **26**, 237 (1982).
20. Th. Beier, A.N. Artemyev, J. Eichler, V.M. Shabaev, and V.A. Yerokhin, Physica Scripta, in press. T. Beier, A.N. Artemyev, J. Eichler, V.M. Shabaev, G. Soff, and V.A. Yerokhin, *Nucl. Inst. Meth. B* **154**, 102-112 (1999).
21. V.M. Shabaev, T. Beier, J. Eichler, and V.A. Yerokhin, to be published.

Hidden crossings theory beyond single electron systems

Predrag Krstić

Physics Division, Oak Ridge National Laboratory, Oak Ridge,TN
37831, USA

Abstract. What are the hidden crossings, in what atomic systems do they emerge?
What does one gain in styding and developing the underlying theory? Where and
when the hidden crossings theory is applied? Meaning of the hidden crossings and
techniques for their detection and application are discussed for multielectron systems.
Particular attention is paid to the slowly colliding H+H and H^++He systems.

INTRODUCTION

A unique feature of the Hidden Crossings (HC) theory [1] is that it provides
a relatively easy description of the physical processes where other theories have
difficulties. Thus, ionization in slow ion-atom collisions, at least at the level of cross
sections is simply and accurately described and calculated with the HC theory. On
the other hand, Molecular and Atomic Orbital Close Coupling (MOCC and AOCC,
respectively) methods, based on the expansion in discrete basis functions, require
large bases and quasi-discretization of the continuum to reach the results often
with questionable accuracy. Even the methods based on a direct solution [2] of the
time-dependent Schrödinger equation on numerical lattice (LTDSE) are ineffective
at low energies due to a large number of time steps needed. In these cases the
HC-based calculations are hundreds to thousands times more effective concerning
the CPU time, even if one counts the time needed to sfind the topology of the
hidden crossings for a considered collision system. Besides, the method provides
the treatment of all inelastic processes "on the same footings" and with the same
accuracy [3]. To illustrate, Fig. 1 shows comparisons of the HC results with the
"benchmarking", LTDSE calculations for excitation and ionization for the low-
energy heavy particle collisions.

The HC approach is limited to the phenomena which are slowly varying on the
time scale of the considered quantum system, like are atomic systems in slowly
varying external fields or slow heavy-particle collisions. It is an asymptotic approach, exact in the adiabatic limit. For a collision at relative velocity v_{coll} the
range of validity is, thus, defined by the adiabatic condition

$$v_{coll} < v_{at} , \tag{1}$$

where v_{at} is the characteristiv velocity of the active electron. A significant part of the phase of an ionizing electron wave function is built through the Rydberg states where adiabaticity is lost and, thus, the electron angular spectra cannot be accurately calculated with the standard HC approach. This drawback might be corrected to a certain extent and probably removed by expansion of the (Z_1,e,Z_2) wavefunction in terms of the complex-energy two-center Sturmians [4].

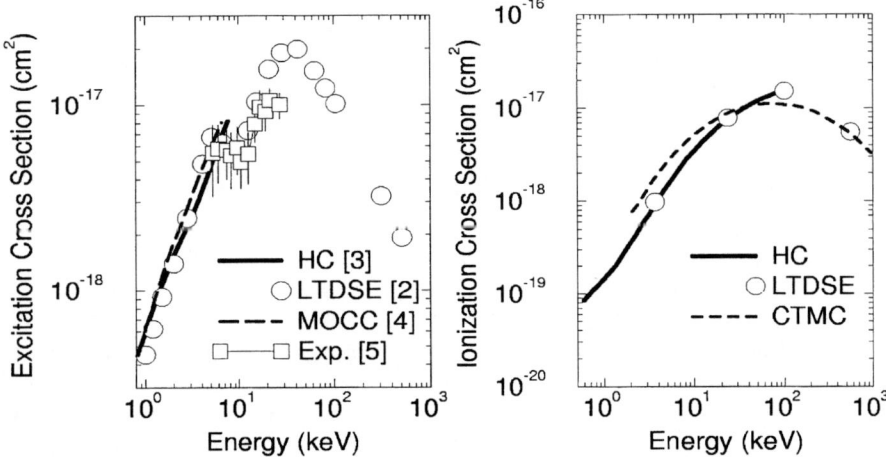

FIGURE 1. Comparison of the HC and LTDSE results for a) excitation H(1s)→H(2s) by proton impact and b) ionization of He$^+$(1s) by antiproton impact [7,8].

The HC theory strongly relies on existence of well separated electronic part, H_{el}, of the total Hamiltonian H of the system. The conditions for separation of fast "internal" (electronic) and slow "external" (nuclear) motion (Born-Oppenheimer approximation) rely on the large difference in masses between an electron and a nucleus, i.e. on the smallness of the characteristic electron momentum in comparison to the nuclear one. This defines the lower bound of v_{coll} for the applicability of the HC theory.

The hidden crossings are crossings of the complex eigenenergy surfaces of the same symmetry at complex values of R. These emerge form the electronic adiabatic problem

$$H_{el}(\{\vec{r}\}, \vec{R})\Phi_{el}(\{\vec{r}\}, \vec{R}) = E_{el}(R)\Phi_{el}(\{\vec{r}\}, \vec{R}) \tag{2}$$

when solved for a fixed complex "parameter" R. Since the adiabatic quasi-molecular terms of the same symmetry do not cross, the HC's can appear only for complex

R, when the H_{el} is not Hermitian, and the Neuman-Wigner non-crossing rule is not applicable. $\{\vec{r}\}$ is here a set of electronic coordinates, while the complex parameter R characterises the (slowly varying) perturbation of the problem, which causes transitions: Internuclear distance (collisions), external field (electric or magnetic), envelope of a laser pulse (in Floquet space), ... The physical conditions for appearance of a hidden crossing for a particular R are discussed in Section II, on the example of a three-body (Z_1, e, Z_2) collision system. Multielectron Hidden Crossings theory (MEHC), discussed in Section III, is conceptually straightforward multi-body generalization of the three-body HC theory, but technically this is a formidable problem which reduces to a multielectron molecular eigenvalue problem in the plane of complex internuclear distance R.

I SINGLE-ELECTRON HC THEORY

Collisionaly induced inelastic transition cannot occur in the adiabatic limit, $v_{coll} \to 0$ (with exception for the exothermic ones). As v_{coll} increases from zero, a tunneling may take place at the internuclear distances R_{LZ} of the so called accidental (dynamic) quasi-resonances, known also as Landau-Zener or avoided crossings of the adiabatic quasi-molecular terms. This is shown in Fig. 2a), for the example of N^{7+}+H. At $R \sim R_{LZ} \sim 30$ a.u. the small coupling between the states $|7i\sigma>$ and $|6h\sigma>$, H_{12}, is proportional to the small energy gap ΔE_{12} between the terms $E_{7i\sigma}$ and $E_{6h\sigma}$, causing a slow tunneling through the radial potential barrier between the two centers, (Fig. 2b)), with an exponentially small, velocity dependent probability $p \sim \exp(-a/v_{coll})$. The tunneling lasts only as long as the quasiresonance, which implies a transition localized for particular values of R. The terms which experience LZ avoided crossings cross for a complex internuclear distance R_c^{LZ}. These have very small imaginary part $\text{Im}\{R_c^{LZ}\} \sim \Delta E_{12}$, and are not generally considered as hidden crossings.   Besides the avoided crossings, which for single-electron-two-center systems occur only if $Z_1 - Z_2 > 1$, there is a physically different type of localized transitions, which occur for internuclear distances where the top of potential barrier touches the populated energy term [9]. A prominent example is the case of the radial barrier between a two Coulomb centers (Fig. 4). For $Z_1 = Z_2 = Z$ and the coordinate origin at $\vec{r} = 0$, along the internuclear z-axis the electronic potential energy is

$$V_{el}(r, R) = -\frac{Z}{|z - R/2|} - \frac{Z}{|z + R/2|} \tag{3}$$

which maximizes to $V_{el}^{max} = -4\frac{Z}{R}$ at z=0. As long as the barrier separates the two centers at larger internuclear distances, the electronic wave function has atomic character, being localized at one of the centers. V_{el}^{max} decreases as the colliding partners approach to each other. When the top of barrier taches the populated electronic term of the state $|0>$, the wave function is free to "diffuses" to the other center and is now shared between the two centers, becoming the molecular

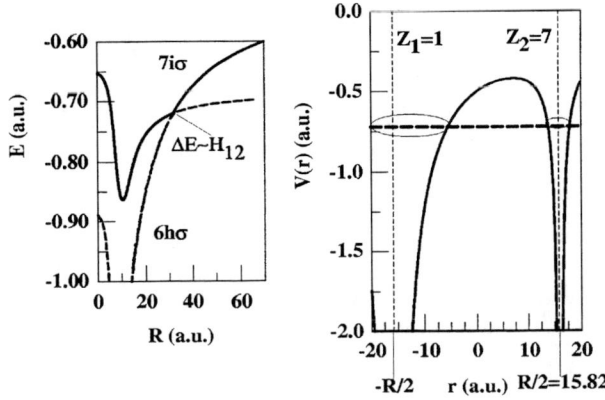

FIGURE 2. a) Avoided (LZ) crossings of the molecular terms results in b) slow tunneling through the radial potential barrier.

in character. The other eigenfunctions of the problem adapt to the new situation. This sudden change of the wave function with R results in its large derivative $\partial/\partial R$, and thus in a peak in the nonadiabatic radial matrix element $H_{i0} \sim <i|\partial/\partial R|0>$, $i = 1, 2, ...$, where i and 0 belong to the same symmetry group of radially coupled adiabatic eigen-levels. A peak in the matrix element H_{i0} for $Re\{R_{c_i}\}$ is associated to its complex-conjugated poles [10] at R_{c_i} and $R_{c_i}^*$ (Fig.5), to a root singularity in the normalization coefficients of the relevant eigenfunctions and to a square root branch point. The relevant eigenenergies coincide (cross) at complex R_{c_i}, i.e. $\Delta E_{i0} \backsim \sqrt{R - R_{c_i}}$, $\Delta E^* \backsim \sqrt{R - R_{c_i}^*}$. These branch points, shared between the complex terms i and 0 form the so called Q-series of closely localized branch points. The largest peak of all H_{i0} and thus, the smallest imaginary part of $R_{c_i} = R_{c_0}$ is for adjacent terms. Collection of all R_{c_0} obtained when the barrier touches various levels of the system in course of changing R is called a Q-superseries. This promotes terms to higher excited states and continuum in the receding phase of the collision. In calculations of the system dynamics it is in most cases enough to consider only contributions from superseries, neglecting the higher order transitions.

For the same example-system, dynamics of the electron in the united-atom limit (UA) is described by atomic wavefunctions, generated by the potential $V_{ell}^{UA}(r, R \to 0) \sim -2Z/r + \ell(\ell+1)/r^2$. The emerging centrifugal barrier while moving toward the $R \to 0$ pulls the consecutive terms upward in a localized region around $R \backsim 0$. During that process the molecular character of the electronic wave functions is changed to the atomic one while passing a narrow range of R, thus causing highly localized peaks in matrix elements of $\partial/\partial R$ between the states of a fixed ℓ. This is further associated with the poles in the radial matrix elements for complex $R_c^{(\ell)}$ as well as to the branch points between the relevant eigenenergies. These branch

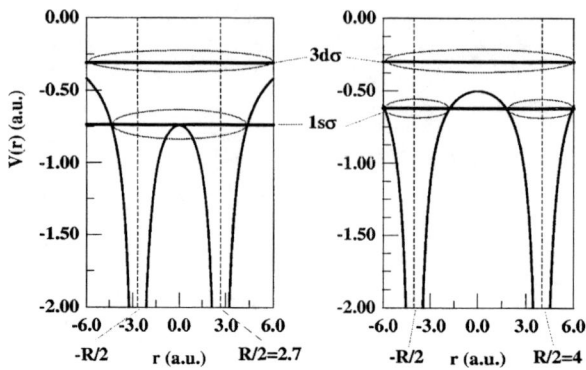

FIGURE 3. Hidden crossings occur when top of the potential barrier touches the populated energy term, causing sudden change of the wave function. Shown is an example of the radial barrier (Q-type HC).

points, localized for small R promote in consecution and pairwise all terms of a fixed ℓ in the incoming phase of the collision and constitute a Coulomb-like $S^{(\ell)}$- superseries, which is in many cases the most effective channel for ionization at low energies.

While the S-series do not exist for the terms with zero angular momentum there is an important exception from that rule: Collision of an antiparticle of charge $Z_1 < 0$ with a single electron ion of charge Z_2 [15]. For small $R \ll r$ electronic potential contains an electric dipole in addition to the unbalanced Coulomb charge,

$$V_{ell}^a \sim \frac{(Z_1 - Z_2)\vec{R} \cdot \hat{r}}{r^2} - \frac{Z_1 + Z_2}{r} + \frac{\ell(\ell+1)}{r^2} \qquad (4)$$

Since the dipole term has the same structure as the centrifugal term, the S-series emerges even for the $\ell = 0$ state. This is the main mechanism for ionization of the low-lying states of ions in collision of antiprotons, in a broad collision-energy range, since there is no radial potential barrier to support the Q-superseries [8].

Knowledge on topology of the hidden crossings in the plane of complex R is equivalent to knowing where the strong, localized couplings for a particular system are. In addition, the LZ avoided crossings constitute a physically different category of localized transitions which are treated separately. All these define the low-energy dynamics of a collision system and an important issue is how to extract from them the most detailed and accurate information on the system evolution. Thus one can use the topology of the hidden crossings to estimate the nonadiabatic matrix elements, which in the adiabatic limit takes the form [11,12,10]

$$< i|\partial/\partial R|j > \approx \frac{1}{2} \frac{\text{Im}\{R_c\}}{(R - \text{Re}\{R_c\})^2 + (\text{Im}\{R_c\})^2} \tag{5}$$

for each particular hidden crossing R_c. These matrix elements do not suffer from the defects of the usual radial matrix elements between the adiabatic states (like are incorrect $R \rightarrow \infty$ limit, dependence on electron origin) and could be used in the MOCC-like system of coupled equations to obtain the transition amplitudes among the low laying states, with either classical ($R = R(t)$) or quantum internuclear motion. This approach leads to the correct transition amplitudes (including their phases) as long as the adiabatic condition (1) is preserved within the coupled states. A more appropriate treatment for obtaining ionization cross sections is the "quasi-elastic" approach [1,10], which relies on features of the adiabatic Schrödinger equation when analytically extended in the plane of complex R. The relevant Hamiltonian loses Hermiticity and has only one, though multvalued and multiple connected complex eigenenergy surface $E(R)$, and the branch points (hidden crossings) are the only singularities. By deforming the evolution path of the system through the complex R plane one evolves quasi-elastically. Various Riemann (branch) surfaces $E_i(R)$ coincide with the eigenenergy terms for $\text{Re}\{R\}$, which enables proper definition of the initial and boundary conditions of the problem. The elastic evolution operator, in the adiabatic limit and for a classical $R = R(t)$ is $\alpha_i(t) = \exp\{i \int_C E_i[R(t)]dt\}$, where a path C connects initial and final states through the complex time (i.e. complex R). Starting from E_i at $R \rightarrow \infty$, the path is promoted to a different Riemann surface whenever C encircle a branch point . At any R the path can make exit to the real axis, ending at a state $E_j(\text{Re}\{R\})$. Thus, to encircle a branch point is equivalent to an inelastic transition. The transition amplitude between states $|i>$ and $|j>$ is

$$\alpha_{ij}(t) = \exp\{i \int_{C_{ij}} E_i[R(t)]dt\} = \exp\{i \int_{C_{ij}} E(R)dR/v_R\} \tag{6}$$

where $v_R = dR/dt$ is the radial velocity. This yields the transition probability in form of $P_{ij} = \exp(-2\Delta_{ij}/v)$, where Δ_{ij} is the Massey parameter, $\Delta_{ij} = \text{Im}\{\int_{C_{ij}} E(R)dR/[v_R/v]$. In a good approximation $\Delta_{ij} \approx \Delta E_{ij} \text{Im}\{R_{c_{ij}}\}$. Shape of the path-curve C is arbitrary as long as it keeps a track of causality for real times, and can be, for example, deformed to ionize a ground state level through an S-superseries by either a set of pairwise and consecutive excitations or by a path encircling the whole superseries. Similar quasi-elastic evolution of the system through the complex R-plane can be realized for semiclassical (WKB) motion of the nuclei [8].

II MULTI-ELECTRON HIDDEN CROSSINGS THEORY

Occurrence of the hidden crossings is likely in any system for which the electronic motion is influenced by a parametrically dependent potential barrier. The system dynamics induced by slowly changing parameter can be described similarly as described in the previous Section. Still, nonseparability of the problem Hamiltonina, as is a multielectron case, may cause a lot of difficulties in searching for the hidden crossings, even if one restricts to only low lying states.

The adiabatic problem of a two-center-two-electron system like is H+H is described by the electronic Hamiltonian of the form

$$H_{el}(\{\vec{r}\}, \vec{R}) = h(\vec{r}_1, \vec{R}) + h(\vec{r}_2, \vec{R}) + h_{12}(\vec{r}_1, \vec{r}_2), \tag{7}$$

$$h(\vec{r}_i, \vec{R}) = -\frac{1}{2\mu}\nabla_{\vec{r}_i}^2 - \frac{1}{\left|\vec{r}_i - \frac{\vec{R}}{2}\right|} - \frac{1}{\left|\vec{r}_i + \frac{\vec{R}}{2}\right|}, i = 1, 2, \tag{8}$$

$$h_{12}(\vec{r}_1, \vec{r}_2) = \frac{1}{|\vec{r}_1 - \vec{r}_2|} \tag{9}$$

where $\vec{r}_i$ are the electronic coordinates, $\{\vec{r}\} = \{\vec{r}_1, \vec{r}_2\}$, $\vec{R}$ the internuclear vector, μ is the electron reduced mass. If the electron correlation is neglected, the two electrons evolve separately producing the identical single-electron hidden crossings topologies. The electron correlation term, h_{12}, introduces novelty in the hidden crossings, either by modifying the single-electron ones that emerge from h_i, or by inducing new series. If the electrons are not equivalent in the initial configuration, one can expect separate promotive superseries for each of them. New types of series associated to the double-excited quasi-molecular terms could be expected, too. If the electron correlation is weak any significant couplings, associated to the hidden crossings is more likely to describe single-electron excitation, while keeping the second electron inactive.

An important feature of all matrix elements $\Omega_{ij}(R)$ of an operator Θ analytically extended into the complex plane of parameter R is their symmetry $\Omega_{ij}(R) = \Omega_{ji}(R)$, a consequence of the analyticity requirement to both Hamiltonian and the wave function [13,14]. This can be fulfilled only with the definition

$$\Omega_{ij}(R) = \int \Phi_i^*(\vec{r}, R^*)\Theta\Phi_j(\vec{r}, R)d\vec{r} \tag{10}$$

It has been shown that the variational principle is valid [13], for real and imaginary parts of the eigenenergies, separately ("bi-variational principle"), for a complex, nonhermitian but symmetric hamiltonian. Thus, the adiabatic eigenvalue problem for a nonseparable systems like H+H or H$^+$+He can be solved for complex R by expanding the wave function in a convenient truncated bases, and with application of the bi-variational principle. This is enough to support generalized Hartree-Fock and Configuration Interaction (CI) procedures [15–17].

An important step in construction of a complex solver is scaling of the electronic radial coordinates r_i by complex R, thus introducing hypergeometric coordinates $q_i = r_i/R$, and then rotating $q'_i s$ back to the real axis. This scaling enables good definitions and convergence of all analytical matrix elements for complex R. The scaled Gaussian primitives $\exp(\alpha R^2(\vec{q} - \hat{R}/2)^2)$ become nonintegrable functions in q if the complex R is in the upper half of the first quadrant. This was handled [15] to a certain extent by redefining the exponent $\alpha \rightarrow \alpha_R$ as a complex, R-dependent quantity, and $\alpha_R = \alpha$ when R is real. Validity of the redefinition must be checked for each system and for various ranges of R. Unrestricted Hartree-Fock (UHF), combined with the single-excited CI codes were developed for the complex R and applied to the H+H and H$^+$+He systems [15,17], using appropriate uncontracted bases of gaussian primitives, optimized for Re$\{R\}$. The branch points were detected by "triple coincidence" of crossings of the complex eigenenergies, of singularities in the corresponding $\partial E/\partial R$ and singularities in the relevant radial matrix elements of $\partial/\partial R$.

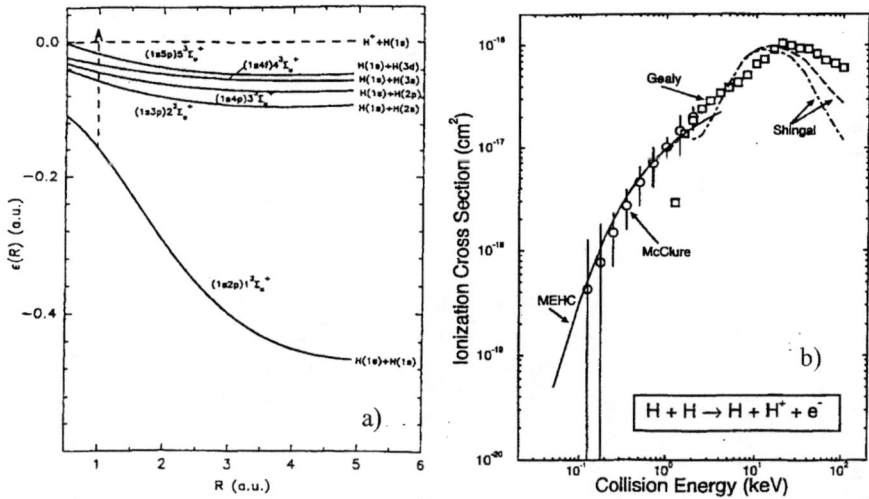

FIGURE 4. Single-electron ionization of H+H. a) S-superseries promotion (dashed arrow) to the single-electron continuum edge (H$^+$+H(1s)) through the single-excited electronic triplet terms (solid lines). b) MEHC (solid line); experiments: McClure et al [19], Gealy et al [18]; MOCC: Shingal et al [20].

Fig. 4a) shows the term diagram of the single-excited triplet ungerade states of the H+H system, n$^3\Sigma^+$. The ground level of H$^+$+H system (single-electron ionization edge) is chosen for zero of energy. A single excited superseries of hidden crossings of the S-type was found about $R = 1$ a.u. (dashed-line arrow in Fig. 4a))and used to calculate the single-electron ionization cross section (Fig. 4b). Excellent agreement with the experiment in the range of 50 ev - 4 keV is obtained. This imply that the principal mechanism for the single-electron ionization in this case is the interaction of an electron with the nucleus of the incoming atom. The ionization is realized when the two clouds penetrate into each other.

Somewhat different is the mechanism for single ionization of the He ground state

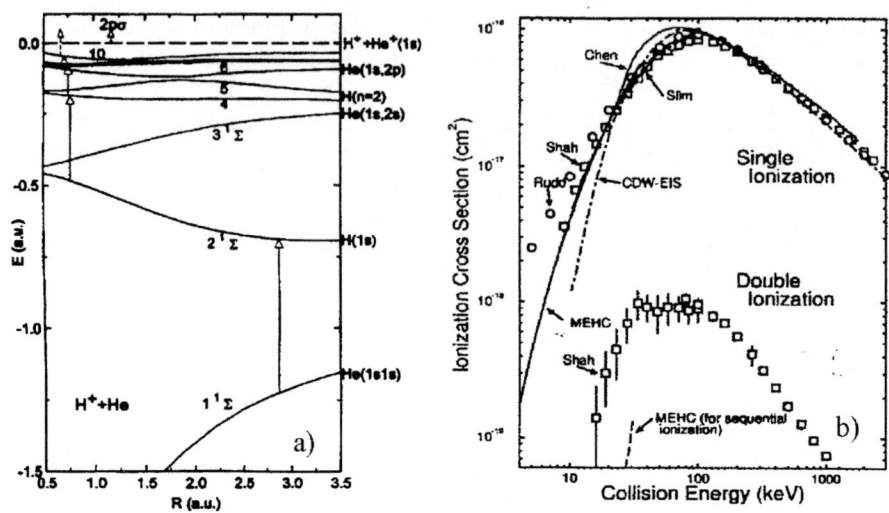

FIGURE 5. Ionization of He by proton impact. a) Single excited singlet electronic terms; localizations of the hidden crossings transitions are indicated by arrows. b) The MEHC results in comparison to various experimental and theoretical data.

by proton impact. [16,17]. An electron first makes the transition from the ground to the first excited singlet state $(2^1\Sigma^+)$ by the Q-transition at the branch point $\mathrm{Re}\{R_{c_1}\}=2.87$ a.u.. Because of the large energy splitting of these two states (Fig. 5a)) the transition is weak at low collision energies and this is responsible for suppression of both single and double ionization cross sections (Fig. 5b)). The $2^1\Sigma^+$ is the charge exchange state, the active electron is localized at proton at large internuclear distances. During a further approach of the nuclei, the electron is promoted to the continuum through the S-type series which starts at $\mathrm{Re}\{R_{c_2}\}=0.76$ a.u. and runs through only these excited states which are asymptotically localized at hydrogen. This yields the single ionization cross section in good agreement with the measurements of Shah et al [21], as well as with the coupled channel calculations of Slim et al [22] and Chen et al [23]. Double ionization cannot be explained with the sequential mechanism (Fig. 5b)). It requires a ladder type mechanism through the double excited states.

III CONCLUSIONS

Hidden crossings are associated with the top(s) of potential barrier(s) and their occurrence and positions in an arbitrary atomic system can be predicted by analysis of the electronic adiabatic potentials. Application of hidden crossings to study the system dynamics in response to a slowly changing perturbation is limited by

538

the near-adiabatic requirements. Multielectron hidden crossings theory has been successfully applied to two-electron (H+H, H$^+$+He, $\bar{p}$+He) as well as to six-electron (C$^+$+H) collision systems, to the processes which involve excitations of only one electron (for example, single electron ionization). This was accomplished by the performing Hartree-Fock and Single-Excited- Configuration Interaction calculations in the plane of complex internuclear distance. Study of highly correlated processes, with inclusion of the multiple electron excitations is underway.

ACKNOWLEDGEMENTS

I gratefully acknowledge support from the U.S. DoE, Office of Fusion Energy Sciences through Oak Ridge National Laboratory, managed by Lockheed Martin Energy Research Corp., under contract No. DE-AC05-96OR22464.

REFERENCES

1. Solov'ev, E. A., *Zh. Eksp. Teor. Fiz.* **81**, 1681 (1981). [*Sov. Phys. JETP* **54**, 893 (1981)].
2. Schultz, D. R., Strayer, M. R., and Wells, J. C., *Phys. Rev. Lett.* **82**, 3976 (1999).
3. Krstic, P. S. and Janev R. K., *Phys. Rev. A* **47**, 3894 (1993).
4. Macek, J. H., and Ovchinnikov, S. Yu., *Phys. Rev. Lett.* **80**, 2298 (1998).
5. Kimura, M., and Thorson, W. R., *Phys. rev. A* **54**, 1780 (1981).
6. Morgan, T. J., Geddes, J., and Gilbody, H. B., *J. Phys. B* **6**, 2118 (1973).
7. Schultz, D. R., Krstic, P. S., Reinhold, C. O., and Wells, J. C., Phys. Rev. Lett **76**, 2882 (1996).
8. Krstic, P. S., Schultz, D. R. and Janev, R. K., *J. Phys. B* **29**, 1941 (1996).
9. Bugdörfer, J., private communication (1999).
10. Solov'ev, E. A., *Usp. Fiz. Nauk* **157**, 437 (1989). [*Sov. Phys. Usp.* **32**, 228 (1989)].
11. Krstic, P. S., Reinhold, C. O., and Schultz, D. R., *J. Phys. B* **31,** L155 (1998).
12. Pieksma, M., and Ovchinnikov, S. Yu., *Comment. At. Mol. Phys.* **31**, 21 (1995).
13. Lowdin, P. O., *Advances in Quantum Chemistry* **19**, 87, Academic, New York,(1988).
14. Jaffe, R. L., George, T. F. and Morokuma, K., C., *J. Chem. Phys.* **61**, 4717 (1974) and refrences therein.
15. Krstic, P. S. , Schultz, D. R. and Bent, G. D., *Phys. Rev. Lett.* **77**, 2428 (1996).
16. Krstic, P. S., Schultz, D. R. and Bent, G., *J. Phys. B* **30**, 1 (1997).
17. Bent, G., Krstic, P. S., and Schultz, D. R., *J. Chem. Phys.* **108**, 1461 (1998).
18. Gealy, M. W., and Van Zyl, B., *Phys. Rev. A* **36**, 3100 (1987).
19. McClure, G. W., *Phys. Rev.* **22**, 166 (1968).
20. Shingal, R., Bransden, B. H. and Flower, D. R., *J. Phys. B* **22**, 855 (1989).
21. Shah, M. B., and Gilbody, H. B., *J. Phys. B* **18**, 899 (1985).
22. Slim, H. A., Bransden, B. H., and Flower, D. R., *J. Phys. B* **28**, 1623 (1995).
23. Chen, Z., and Msezane, A. Z., *Phys. Rev. A* **49**, 1752 (1994).

Antiproton Capture by Simple Atoms and Molecules

James S. Cohen

Theoretical Division, Los Alamos National Laboratory, Los Alamos, New Mexico 87545

Abstract. Antiprotonic atoms form in low-energy collisions between an antiproton and a normal atom or molecule. The fermion molecular dynamics (FMD) method enables us to go beyond the atomic hydrogen target and study multielectron effects with helium and neon targets and molecular effects with hydrogen and deuterium targets. Multiple ionization and ro-vibrational excitation are found to greatly increase the maximum energy of capture, with the latter leading to a significant dependence on the projectile mass and target isotope. The kinetic energies of the ionized electrons are found to increase somewhat with the target nuclear charge.

INTRODUCTION

Capture of negative exotic particles, principally the negative muon μ^- (of mass $206.77m_e$), the negative pion π^- (of mass $273.14m_e$), and the antiproton $\bar{p}$ (of mass $1836.15m_e$),[1] have been studied experimentally and theoretically since the discovery of these particles (μ^- in 1937, π^- in 1947, and $\bar{p}$ in 1955). However, most of the theory has been done for the hydrogen *atom*, while capture by this simplest target is yet to be examined experimentally.[2] Several theoretical treatments have shown that, for the hydrogen atom, the capture energy primarily goes to the target ionization potential with the ionized electron carrying off little kinetic energy.

Beyond the one-electron atomic target there are two key questions: (i) are multielectron effects important and (ii) do vibrations and rotations in *molecules* affect the capture. The few previous calculations on multielectron atoms have treated ionization as sequential and uncorrelated. Alternatively, the capture cross sections have sometimes simply been assumed proportional to the stopping power. For molecules, the capture cross sections have often been represented by the sum of the cross sections of the constituent atoms, e.g., the cross section for the hydrogen

---

[1] The ultimate fate of the leptonic muon is often quite different from that of the hadronic pion and antiproton, but for purposes of the *atomic physics* they are the same except for mass and thus provide a succinct test of mass effects.

[2] The ASACUSA collaboration plans an experiment on $\bar{p}$ capture by the hydrogen atom using the new Antiproton Decelerator (AD) at CERN.

molecule taken as twice that of the hydrogen atom, *a la* Bragg's rule for stopping power. Actually it is now understood that there is no direct relationship between the stopping power and capture since the stopping occurs principally at much higher energies. As understanding of the capture mechanism developed, it then became expected that capture depended mainly on the first ionization potential, so would be about the same for H and H_2. Before the calculations described in the present paper, it was not anticipated that molecular structure of the target would have much effect on capture.

What is really needed to test these hypotheses is a theoretical method that can account for the dynamics of electron correlation as well as electronic, vibrational, and rotational molecular structure during a collision. This is a daunting task, exceeding the capability of any existing fully quantum-mechanical method, but treatable by a recently developed quasiclassical method known as fermion molecular dynamics (FMD). The FMD method uses pseudopotentials to approximate quantum-mechanical constraints and formulate the problem within Hamilton's equations of motion. It accounts for the three-dimensional, correlated motion of all electrons as well as the molecular degrees of freedom.

The targets thus calculated include the H atom (for comparison with other calculations), the D atom (for atomic isotope effect), the He and Ne atoms (for multielectron and mixture effects), and the H_2, D_2, and HD molecules (for two-center, rotational, vibrational, and dissociative effects). Most of the results discussed in this paper are for $\bar{p}$ capture, but a third key question, which will be discussed a bit, is how capture depends on the projectile mass.

FERMION MOLECULAR DYNAMICS

The classical-trajectory Monte Carlo (CTMC) method, which treats the electron quasiclassically, has been very useful in atomic physics, esp. for ion-atom collisions. Quasiclassical approaches offer a number of practical advantages for capture of heavy $(m \gg m_e)$ negative particles:

- The classical equations of motion are easier to solve than the Schrödinger equation.

- Full dynamics is done for all particles. Thus correlation is taken into account, all rearrangement channels are treated on equal footing, and electron translation factors are not needed.

- The electronic continuum presents no difficulty.

Capture of heavy negative particles is well suited for the quasiclassical approach:

- The collisions involve a quasicontinuum of electronic orbitals.

- Capture is into high n, l orbitals, so the correspondence principle applies.

- Slowing down and capture are treated consistently.

541

But the CTMC method is essentially limited to one-electron atoms because multielectron atoms are classically unstable (autoionize). A more general quasiclassical method is needed to treat multielectron atoms without using explicit wave mechanics. One solution is molecular dynamics carried out with the Kirschbaum-Wilets (KW) model [1] of multielectron atoms.

The simple *ansatz* of Kirschbaum and Wilets to "cure" the classical instability problem places constraints on the electrons (designated i and j): (i) $p_i r_i \geq \xi_H \hbar$, *a la* the Heisenberg uncertainty principle, which with equality is just the deBroglie-Bohr condition for the hydrogen atom and (ii) $p_{ij} r_{ij} \geq \delta_{s_i s_j} \xi_P \hbar$, *a la* the Pauli exclusion principle, implying fermion-like behavior. To this effect, KW added effective potentials, v_H and v_P, to the usual Hamiltonian:

$$H_{\text{KW}} = T + V_{\text{Coul}} + \sum [v_H(r_i, p_i) + \delta_{s_i s_j} v_P(r_{ij}, p_{ij})], \tag{1}$$

where T is the total kinetic energy and V_{Coul} is the total Coulomb energy. The pseudopotentials are taken of the form

$$v_c(r, p) \propto r^{-2} \exp[-\alpha_c(rp/\xi_c \hbar)^4] \text{ for } c = H \text{ or } P \tag{2}$$

where ξ_H and ξ_P, of order unity, can be chosen to match the atomic energies and α_c is a hardness parameter. The associated dynamical calculations using Hamilton's equations of motion are called fermion molecular dynamics [2].

The advantages offered by a quasiclassical approach to capture by the H_2 *molecule* are even greater than for the H *atom* because of the several rearrangement channels and a large number of ro-vibrational states. The KW *ansatz* suffices to stabilize simple molecules (like H_2^+ and H_2) but does not give a sufficiently accurate description. We have found that a small modification termed KWC dramatically improves the molecular treatment [3].

KWC is like KW (same as KW in the case of atoms) except additional one- and two-electron potentials with respect to *molecular* symmetry positions are introduced (the atomic symmetry point is the nucleus; the geometric center of the molecule is designated by o). The physical motivation is similar to that for the Heisenberg potential — preclude unphysical localization, which is forbidden by quantum mechanics. The effective Hamiltonian becomes

$$H_{\text{KWC}} = H_{\text{KW}} + v_{m1}(r_{o1}, p_{o1}) + v_{m1}(r_{o2}, p_{o2}) + v_{m2}(r_{12}, p_{12}) \tag{3}$$

with v_{m1} and v_{m2} of form similar to v_H and v_P except with r_{12} instead of r in the prefactor. The associated clamped-nuclei (Born-Oppenheimer) potential curves (minimum of H_{KWC} for fixed $r_{12} = R$) for H_2 and H_2^+ are in excellent agreement with the exact potential curves in the range $0.5 < R < 6a_0$ important in the present application [3].

CAPTURE BY THE HYDROGEN ATOM

Many more calculations have been made for negative-muon capture than for antiproton capture. The physics of the two is similar, so the muon calculations serve for comparing theoretical methods. Results of various calculations are shown in Fig. 1. The first method applied was the adiabatic-ionization (AI) model by Wightman [4], which was based on the observation by Fermi and Teller [5] that there exists a critical dipole strength for binding an electron. With the additional allowance for trajectory curvature [6,7], the AI model well illuminates the essential mechanism for negative-particle capture by the hydrogen atom, and an idea of this essence using diabatic states (DS) [8] yields quantitatively correct results. Unfortunately there then followed a diversion by several Born- and Coulomb-Born-approximation calculations [9, for example, shown in Fig. 1]. These perturbative results can barely be viewed in the frame of Fig. 1. The lesson learned here is that the target is really greatly distorted and the transition to the electronic continuum must be treated carefully.

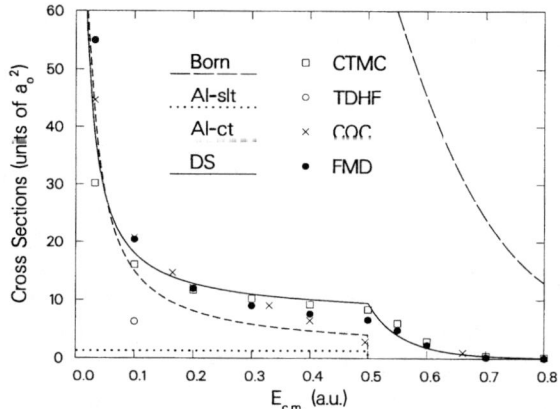

FIGURE 1. Comparison of calculations of μ^- capture by the hydrogen atom.

The time-dependent Hartree-Fock (TDHF) calculation [10], with wave packet for the projectile muon, is a completely quantum-mechanical approach. Its failure teaches the next lesson that correlation of the muon and electron motions, neglected by TDHF, is essential. The classical-quantal coupling (CDC) formulation [11] overcame this problem by coupling a Schrödinger equation for the electron with a Newton equation for the muon. The CQC result may be considered the current benchmark.

Now we come to the quasiclassical methods. The CTMC [12] result is in good agreement with the DS and CQC calculations. The FMD method [13] (with $\xi_H = 0.943$, $\alpha_H = 4.0$), which has the advantage of also being applicable to more complex targets, gives a result similar to CTMC though it is rather closer to CQC at very low collision energies and at energies around the ionization threshold.

All of the nonperturbative methods agree that the energy loss is mainly to the electron, which escapes with quite low kinetic energy, as can be deduced in Fig. 1 from the rapid falloff of the capture cross section at energies above the ionization potential of 0.5 a.u.; the average ejected electron kinetic energy is ~ 0.09 a.u. For

$\bar{p}$ capture by the H atom (see Fig. 5), the behavior is even more adiabatic, the average electron kinetic energy being ~ 0.05 a.u.

CAPTURE BY HELIUM AND NEON ATOMS

The only existing exotic-atom formation experiments that have previously been done with *atomic* targets are for noble-gas atoms capturing muons and pions. Those results allow us to gauge the accuracy of our results for antiproton capture by helium and neon. Capture by multielectron atomic targets raises a number of questions: (i) are more than one electron ionized, (ii) is the target left in its ground electronic state immediately after capture, and (iii) are the electrons ionized with very low kinetic energy? The analog to the adiabatic-ionization model would suggest the answer to (i) is the minimum number consistent with a positive answer to (ii) and (iii).[3] However, our findings indicate otherwise.

The capture and total ionization cross sections for antiproton capture by helium and neon ($\xi_H = 0.934$ for He; $\xi_H = 1.131$, $\xi_P = 1.511$ for Ne; $\alpha_H = 2.0$, $\alpha_P = 1.0$) are shown in Fig. 2. As required for conservation of energy, the capture and total ionization cross sections coincide at collision energies below the first ionization potentials of the respective targets. In the case of helium, it is still generally true that $\bar{p}$ capture occurs with ionization of a single electron, but that electron is found to carry off about twice as much kinetic energy (average ~ 0.1 a.u.) as compared with the hydrogen atom (this is consistent with findings using the diabatic-states model [6]). The second electron usually promptly follows in ionization via an Auger process, with energy provided by the deexciting $\bar{p}$, before the first electron is far

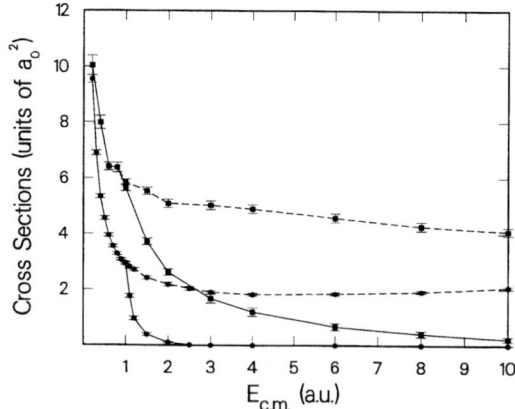

FIGURE 2. Capture (solid curves) and total ionization (dashed curves) for capture of antiprotons by helium (circles) and neon (squares). The error bars reflect Monte Carlo statistics.

---

[3] *Strictly* adiabatic ionization cannot occur with He or Ne targets since the negative ions H^- and F^-, attained in the united-atom limit, are bound.

removed. In fact, it was found more practical to characterize the $\alpha\bar{p}$ state, which determines the subsequent cascade, with hydrogen-like quantum numbers after both electrons were removed. The results for n and l are shown in Fig. 3. However, at the longest time integrated (~ 0.3 ps), there are still some targets retaining an electron, as can be inferred from the population at $n \gtrsim 40$. These are generally high angular-momentum states and candidates for the so-called $\alpha\bar{p}e$ "atomcules" observed experimentally in 1991 [14], though most of these complexes will still decay before the ~ 3 μs lifetime of the $\sim 3\%$ that form truly metastable states.

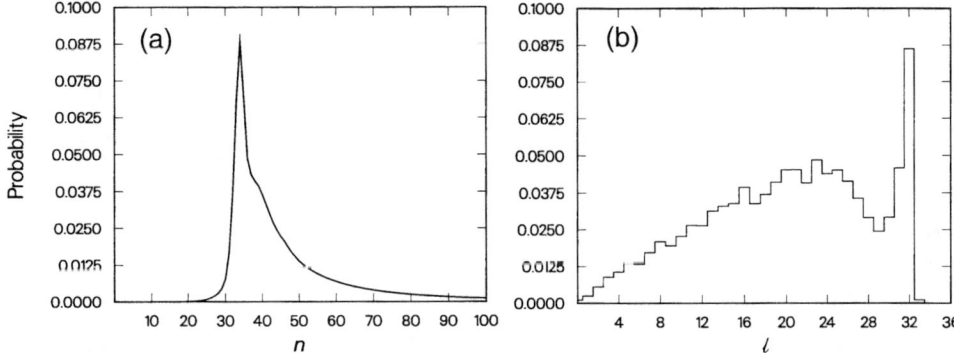

FIGURE 3. Capture of antiprotons stopped in helium, integrated over capture energies. (a) Distribution of quantum number n (at $t \sim 0.3$ ps). For $n \lesssim 40$, both target electrons have been ionized; for large n one electron is still bound. (b) Distribution of quantum number l for $n = 33$.

For neon targets the situation is more complex. The capture cross section reaches considerably higher collision energies, still significant at ~ 10 a.u. in Fig. 2. The average kinetic energy of the first-ejected electron is ~ 0.2 a.u., higher than for helium, but not high enough to enable such capture without ionization of additional electrons. Capture at the higher energies thus indicates ionization of more than four electrons, though the target is often left in a shakeup state and some of the ionization dynamics may reflect electron-electron Auger processes. The second and later electrons are ejected with somewhat higher kinetic energies on the average.

The larger, more extensive capture cross section for the Ne target implies that the *per-atom* capture probability of Ne is greater than that of He in a He-Ne mixture. This enhancement is found to range from a factor of 2.7 for mixtures that are predominantly Ne to 4.3 for mixtures that are predominately He. The concentration dependence arises from the relative energy dependence of the total ionization cross sections in the energy range of capture.

An important quantity for modeling the x-ray cascade is the distribution of angular-momentum states, which largely determine the allowed radiative steps as well as nuclear capture. The l distribution for the most likely n value after both electrons of He are ejected is shown in Fig. 3(b). It is fairly statistical ($\propto 2l + 1$) though there is a dip just below and a bump at the largest possible l.

The calculations on $\bar{p}$+Ne were done with a different strategy than those on

$\bar{p}$+He. As already noted, many of these captures require ionization of more than one electron, though we certainly want to characterize the state before all ten electrons are ionized. The "instant" of capture is not a well-defined quantity, nor is the initial n quantum number, which has been identified with and without screening effects.[4] We follow the trajectory until some more-or-less arbitrary n value is reached; a value of 45 was chosen. The resulting l distribution is shown in Fig. 4. In either interpretation, the distribution is flat-to-statistical at low-to-moderate l with a significant pileup at large l. The fact that the result with screening seems to cut off at $l < n - 1$ is presumably an indication of the inadequacy of the hydrogenic approximation.

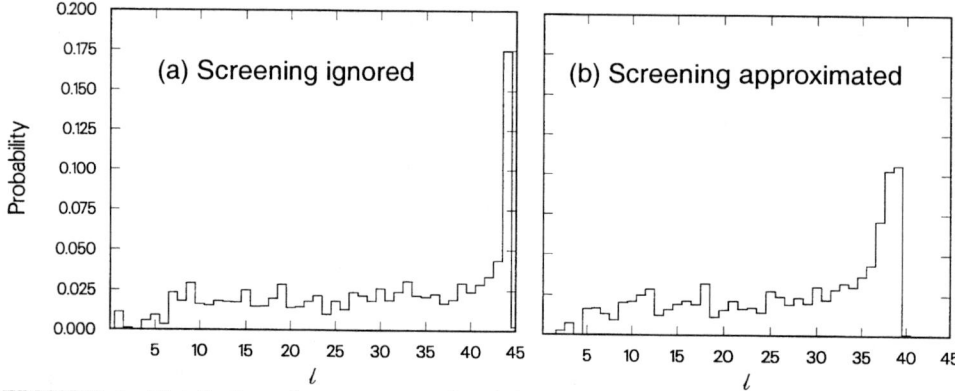

FIGURE 4. Distribution of quantum number l for antiprotons captured in level $n = 45$ of neon, integrated over capture energies. Results are obtained (a) ignoring electron screening and (b) taking screening into account but still treating the antiproton orbital as hydrogenic (Note that this is just a matter of *interpretation*; the dynamical calculations are the same).

CAPTURE BY MOLECULAR HYDROGEN

The antiproton-capture cross section for the molecular H_2 target is shown in Fig. 5 (same ξ_H and α_H as for the H atom; $\xi_{m1} = 0.90$, $\xi_{m2} = 1.73$, $\alpha_{m1} = \alpha_{m2} = \alpha_H$). It is larger than the analogous cross section for the H atom, also shown, and extends to much higher energies. The energies of the ionized electrons, usually only one per capture, are similar for the H and H_2 targets, so that cannot be the explanation. In most cases, the ionization of the first electron is prompt, followed after several vibrational periods by dissociation with final products $\bar{p}p$ and H. More rarely ($\lesssim 1\%$ of the time), the accompanying products are H^- or H^+.

---

[4] This ambiguity is not due to the quasiclassical treatment, but rather to the hydrogenic approximation to the orbital where there are actually several electrons as well as the antiproton present. The angular momentum l is definite, but the "effective" one-particle n value is only qualitatively defined. The approximate treatment of screening yields $n_{eff} = (-\mu Z_{eff}^2 / 2E_{eff})^{\frac{1}{2}}$, where E_{eff} is the effective binding energy of the $\bar{p}$ with reduced mass μ and Z_{eff} is the effective nuclear charge.

There are three possible sources of this difference: (i) the two-center, two-electron structure of H_2 (even though the first ionization potentials of H and H_2 are similar), (ii) molecular vibrations and dissociation, and (iii) molecular rotations. In order to separate these effects we have done two treatments intermediate between the atom and fully dynamical molecule: (i) a calculation with the H_2 constrained as a rigid rotor with $R = R_0 = 1.4a_0$ but otherwise unhindered and (ii) a calculation with the H_2 constrained as a rigid "nonrotor" which inhibits rotation as well as vibration (but still allows recoil and conserves energy and angular momentum).

The results of these two approximations are also shown in Fig. 5. The rigid-rotor result lies in between the results for the H atom and vibrating H_2 molecule at all energies. It is close to the H_2 result at collision energies below ~ 0.8 a.u. but falls to zero much more rapidly than does the cross section for the unconstrained target. Only rotational excitation is probable at low impact energies while vibrational excitation becomes probable at the higher energies. The dividing energy is close to the energy ($E_{c.m.} \approx 1.0$ a.u.) where the cross sections peak for vibrational excitation of H_2 by H^+ impact [15]. At energies above 2.0 a.u. the $\bar{p}$ capture cross section essentially vanishes when vibration is disabled. With the vibrational degree of freedom enabled, it extends to much higher energies and falls off in a manner similar to the $H^+ + H_2$ vibrational excitation cross section.

The rigid-nonrotor result falls off even more rapidly than the rigid-rotor result. As in the atomic case, the rigid-nonrotor capture cross section rapidly decreases at energies above the ionization potential, which is 0.605 a.u. for R fixed at 1.4 a_0. However, the rigid-nonrotor cross section is still substantially larger than the atomic cross section, showing that the two-center, two-electron effect is also important.

These observations suggest that the molecular enhancement of the capture cross section is due to the efficient energy transfer from the antiproton projectile to the

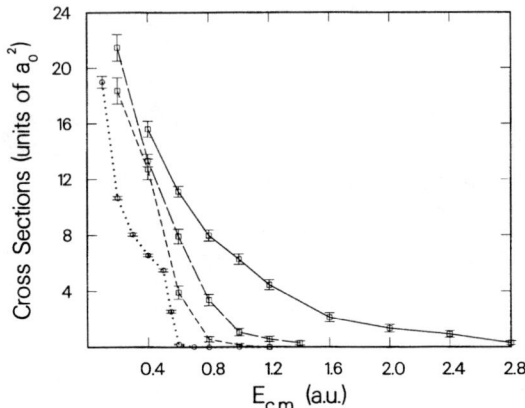

FIGURE 5. Cross sections for capture of antiprotons by the hydrogen atom (dotted curve) and hydrogen molecule (solid curve). The results for H_2 treated as a rigid rotor (long-dashed curve) and rigid nonrotor (short-dashed curve) are also shown.

equal-mass proton nucleus in the H_2 target, which can lead to vibrational and/or rotational excitation. Thus it may be expected that capture will be optimum where the mass of the negative projectile matches that of an atom in the target molecule and that the molecular enhancement will diminish as the mass match deteriorates. As shown in Fig. 6, calculations performed for capture of $\bar{p}$ by D_2 and for capture of μ^- by H_2 and D_2 confirm this hypothesis [16]. The magnitudes of the cross sections decrease in the same order as the projectile-to-target-nucleus mass ratios: 1.00 for $\bar{p}+H_2$, 0.50 for $\bar{p}+D_2$, 0.113 for μ^-+H_2, and 0.056 for μ^-+D_2. The vibrational degree of freedom is most important in distinguishing the behavior of the molecule from the atom and in distinguishing different molecular isotopes. However, the effects of rotation, two-center charge distribution, and mass-dependent nonadiabaticity (which tends to favor μ^- over $\bar{p}$ capture) are also significant in determining the cross sections.

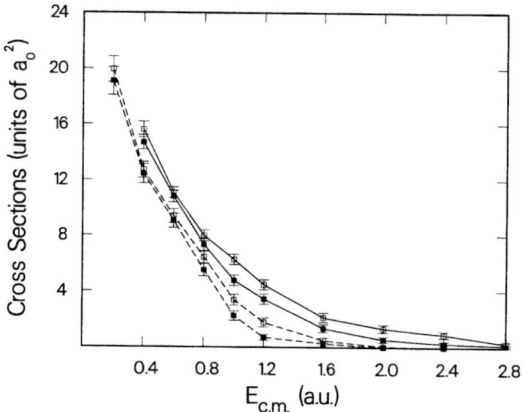

FIGURE 6. Capture cross section for $\bar{p}+H_2$ (solid curve with open data points), $\bar{p}+D_2$ (solid curve with filled data points), μ^-+H_2 (dashed curve with open data points), and μ^-+D_2 (dashed curve with filled data points).

The n and l distributions of the exotic atoms formed with molecular targets, shown in Fig. 7, are found to be quite different from those for atomic targets. In the case of the molecular targets, the maxima of both the n and l distributions are shifted to lower values and the very large n values are suppressed. The difference between the n distributions for atomic and molecular targets is largely due to the breakup dynamics of the intermediate complex formed in molecular capture.

The difference between the l distributions for atomic and molecular capture is more interesting and potentially important for modeling the ensuing cascade. For the atomic targets the distributions are fairly statistical up to some peak, even with a pileup at large l. The situation for the molecular targets is quite different due to their two-center structure. The preponderance of electron density, lying between the nuclei, tends to emphasize impact parameters equal to or less than half the internuclear distance. This leads to a broader peak at lower l than with

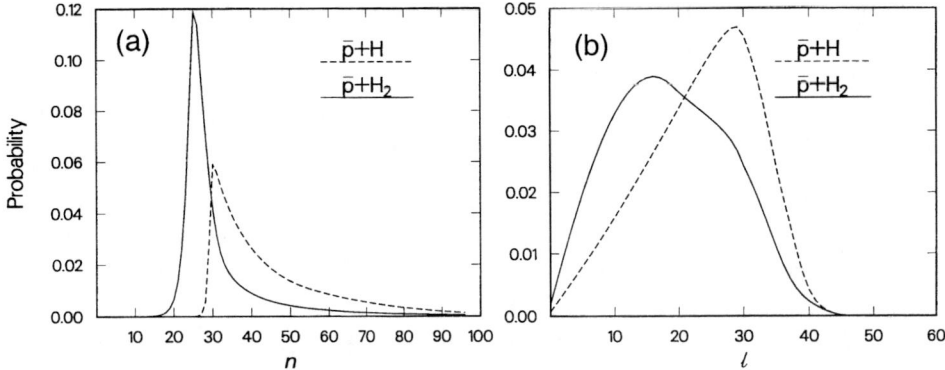

FIGURE 7. Antiproton capture by the hydrogen atom (dashed curves) and hydrogen molecule (solid curves), integrated over capture energies. (a) Distribution of n quantum numbers. (b) Distribution of l quantum numbers *summed over all n*.

an atomic target. Also, rotation of the molecule tends to broaden the distribution of angular-momentum states.

The larger capture cross section for H_2 than for D_2 implies that the per-atom capture probability in a H_2-D_2 mixture will be larger for H_2 than for D_2. The enhancement is found to be a factor of ~ 1.6, with only a relatively weak dependence on the concentration ratio. Another interesting situation is presented by the heteronuclear HD target, where the relative probabilities of forming $\bar{p}p$ or $\bar{p}d$ are determined by the dissociation dynamics. This calculation has not been done, but the calculation for π^- capture by HD, where experimental data exist, has been done [17]. It showed that the dissociation dynamics favors the heavier isotope, just opposite the initial capture, and thereby explained some previously puzzling experimental results [18].

CONCLUSIONS

The detailed dynamics, taken into account by the FMD method, has shown clear trends in the Coulomb capture of antiprotons and other exotic particles. For capture by atoms:

1. The adiabatic-ionization model is qualitatively correct for the hydrogen atom, but not for other atoms, for which no one-electron model can be expected to be adequate.

2. The kinetic energy of the ionized electrons increases somewhat with Z, but the increase in the capture cross section is mainly due to multielectron ionization and excitation.

3. The isotope effect is quite small for atomic targets.

4. The angular-momentum distribution is approximately statistical ($\propto 2l + 1$) at low-to-moderate l, but tends to pile up at the end point.

For capture by H_2, D_2, or HD molecules:

1. Capture cross sections are larger and extend to higher collision energies for the molecular targets than for the corresponding atomic targets. The effect is greater for $\bar{p}$ than for μ^- or π^-.

2. The isotope effect is large for H(or D)-containing molecular targets.

   - Initial capture favors the lighter isotope.
   - Subsequent dissociation favors the heavier isotope.

3. The n and l distributions are narrower and shifted to lower values relative to the distributions for the corresponding atom.

ACKNOWLEDGMENTS

I am grateful to D. Horváth and Y. Yamazaki for helpful comments.

REFERENCES

1. Kirschbaum C. L. and Wilets L., *Phys. Rev. A* **21**, 834 (1980).
2. For a review, see Wilets L. and Cohen J. S., *Contemp. Phys.* **39**, 163 (1998).
3. Cohen J. S., *Phys. Rev. A* **56**, 3583 (1997).
4. Wightman A. S., *Phys. Rev.* **77**, 521 (1950).
5. Fermi E. and Teller E., *Phys. Rev.* **72**, 399 (1947).
6. Cohen J. S., Martin R. L., and Wadt W. R., *Phys. Rev. A* **27**, 1821 (1983).
7. Cohen J. S., in *Electromagnetic Cascade and Chemistry of Exotic Atoms*, edited by L. M. Simons, D. Horváth, and G. Torelli, New York: Plenum, 1990, pp. 1-22.
8. Cohen J. S., Martin R. L., and Wadt W. R., *Phys. Rev. A* **24**, 33 (1981).
9. Korenman G. Ya. and Rogovaya S. I., *J. Phys. B* **13**, 641 (1980).
10. Garcia J. D., Kwong N. H., and Cohen J. S., *Phys. Rev. A* **35**, 4068 (1987).
11. Kwong N. H., Garcia J. D., and Cohen J. S., *J. Phys. B* **22**, L633 (1989).
12. Cohen J. S., *Phys. Rev. A* **27**, 167 (1983).
13. Cohen J. S., *J. Phys. B* **31**, L833 (1998).
14. Iwasaki M. *et al.*, *Phys. Rev. Lett.* **67**, 1246 (1991).
15. Phelps A. V., *J. Phys. Chem. Ref. Data* **19**, 653 (1990).
16. Cohen J. S., *Phys. Rev. A* **59**, 1160 (1999).
17. Cohen J. S., *Phys. Rev. A* **59**, 4300 (1999).
18. Aniol K.A. *et al.*, *Phys. Rev. A* **28**, 2684 (1983).

The Control of Dynamical Systems
– Recovering Order from Chaos –

Louis J. Dubé and Philippe Després

Département de Physique, Université Laval
Québec, Canada G1K 7P4

Abstract. Following a brief historical introduction of the notions of chaos in dynamical systems, we will present recent developments that attempt to profit from the rich structure and complexity of the chaotic dynamics. In particular, we will demonstrate the ability to **control chaos** in realistic complex environments. Several applications will serve to illustrate the theory and to highlight its advantages and weaknesses. The presentation will end with a survey of possible generalizations and extensions of the basic formalism as well as a discussion of applications outside the field of the physical sciences. Future research avenues in this rapidly growing field will also be addressed.

INTRODUCTION

In his 1985 Gifford Lectures, Freeman Dyson expressed his opinion on the matter of chaos. In his subsequently published words [1], the chapter entitled "Engineers' Dreams" contains the following statement:

> A **chaotic motion** is generally neither predictable nor controllable. It is **unpredictable** because a small disturbance will produce exponentially growing perturbation of the motion. It is **uncontrollable** because small disturbances lead only to other chaotic motions and not to any stable and predictable alternative. Von Neumann's mistake was to imagine that every unstable motion could be nudged into a stable motion by small pushes and pulls applied at the right places.

As one can see, the assertion was also meant as an answer to comments made by von Neumann in the early 1950s and who obviously held a less pessimistic point of view. Dyson's position represents well the traditional wisdom until 1990 ...

In order to appreciate why the juxtaposition of the two words chaos and control is so counter-intuitive and why Dyson's statement was so representative and sensible, an operational definition of chaos will be helpful. In our exposition, the word chaos has a technical and precise meaning to be distinguished from its greek origin where it designated "the primeval emptiness of the universe before things came into being of the abyss of Tartarus, the underworld" . A universal definition is not available, but most researchers would agree that *deterministic chaos* could

CP500, *The Physics of Electronic and Atomic Collisions*, edited by Y. Itikawa, et al.

be described as follows:

Chaos is a long-term aperiodic behaviour of a dynamical system that possesses the property of sensitivity to initial conditions.

– long-term aperiodic behaviour means that the time evolution of the system does not tend towards a stationary or periodic state, i.e. regularity of the motion is absent.

– dynamical system describes a process whose future behaviour is strictly determined by its past state, i.e. determinism is present and the source of the irregularity is inherent and not to be found in a stochastic component.

– sensitivity to initial conditions implies that a very small deviation in the initial conditions is sufficient to create large deviations in the future states (the so-called "butterfly effect"), i.e. despite the presence of determinism, practical long-term predictability is lost.

This is the type of motion that Dyson had in mind. It is not new of course and it is clear that Maxwell and Boltzmann, the founders of statistical physics, were acutely aware of the property of sensitivity to initial conditions and its consequences. Not before Poincaré [2] however, could one ascertain the existence of this property in a system with *few* degrees of freedom, namely the reduced 3-body problem. In the continuing history of nonlinear dynamical systems, the first evidence of physical chaos is associated with the name of Edward Lorenz [3], whose 1963 discovery of the first *strange attractor* in a simplified meteorological model containing only 3 state variables has led to a remarkable explosion in the sudsy of chaos and its properties. It was not until 1990 however that Ott, Grebogi and Yorke (OGY) [4] addressed the question of control of chaos and described, very much in the spirit of von Neumann, the theoretical steps necessary to achieve this goal. This work was rapidly followed by experimental verification [5]: von Neumann's dream had become reality.

This Progress Report will describe some practical implementations for the recovery of order from chaos. The theoretical foundations of the methods will be first explained for 1D and 2D systems and then demonstrated with some recent calculations taken from mathematics and physics. Our conclusions and a glimpse at future applications and extensions make the last part of the presentation. Lack of space precludes completeness, and the interested reader may wish to consult the special 1997 December issue of *Chaos* for further information[1].

THE CONTROL STRATEGY

All stable processes, we shall predict.
All unstable processes, we shall control.
JOHN VON NEUMANN, circa 1950

---

[1] As an indication of the rapid growth of the literature, approximately 1500 articles have been published on the subject of Control and Synchronization between 1990-1999.

552

In this Section, we show how the richness, the complexity and the sensitivity of chaotic dynamics can be used to select and stabilize at will, with small programmed perturbations, an otherwise unstable state of the natural dynamics. The goal is to achieve this feat *without* altering appreciably the original system. It is precisely the properties that differentiate a chaotic motion from an irregular or unstable behaviour that are the solution to the control task. The important ingredients are:

– unstable periodic orbits (UPO) are typically dense in the chaotic attractor of dissipative systems or in the stochastic web of conservative systems, i.e. there are practically an infinity of unstable states to choose from.

– chaotic motion is ergodic, meaning that a chaotic trajectory will revisit infinitely often the neighbourhood of any point within the available phase space.

– chaotic dynamics is sensitive to initial conditions, implying that *small* perturbations will naturally induce large effects.

To go beyond qualitative description, we establish some working conditions:

1. we suppose that the dynamics can be represented by a d-dimensional nonlinear map (either given explicitly or reconstructed from the observations)

$$\mathbf{x}_{n+1} = \mathbf{F}(\mathbf{x}_n, p) \tag{1}$$

where p is an accessible system parameter, the *control parameter*.

2. there exists one or more specific UPOs for a given *nominal* value p_0 of the parameter, defined by

$$\{\mathbf{x}(i, p_0) : \mathbf{x}(i, p_0) = \mathbf{F}^{(m)}(\mathbf{x}(i, p_0), p_0) , \quad \forall\, i = 1, m\} \tag{2}$$

for an orbit of period m, around which one wishes to stabilize the dynamics.

3. control is activated only when points of the trajectory $\{\mathbf{x}_n\}$ fall in a small neighbourhood of the UPOs, usually taken to be a ball $\mathcal{B}_\delta$ of radius δ around $\{\mathbf{x}(i, p_0)\}$,

$$|\mathbf{x}_n - \mathbf{x}(i, p_0)| \leq \delta \qquad \text{for some} \quad i = 1, m \quad , \tag{3}$$

hereafter referred to as the *control* or $\delta-neighbourhood$.

4. we restrict the parameter variations δp, necessary to achieve control, to a maximum small perturbation

$$|\delta p| \leq |\delta p_{\max}| \ll |p| \tag{4}$$

defining the *control range*.

5. since the position of a periodic orbit is a function of p, and we assume that the local dynamics does not vary much within $|\delta p|$, a linear representation of the dynamics is possible.

Obviously, the control range and neighbourhood are not independent and experience shows that a judicious choice is to take them of the same order of magnitude.

1D Control

In a 1D system where the nonlinear map is given explicitly by

$$x_{n+1} = F(x_n, p) \quad ,$$
(5)

and where a target UPO of period m, i.e. $\{x(i, p_0)\}$, exists at nominal value p_0 of the control parameter, the perturbations necessary to stabilize the orbit can be calculated directly. Indeed, assume that at the n-th iteration, x_n comes in $\mathcal{B}_\delta$ of the i-th component of the target UPO, i.e. $|x_n - x(i, p_0)| \leq \delta$, then equation(5) can easily be linearized around $x(i, p_0)$ and $p = p_0$ such that

$$x_{n+1} - x(i+1, p_0) \sim U_i \left[x_n - x(i, p_0) \right] + V_i \left[p_n - p_0 \right] \equiv U_i \, \delta x_{n,i} + V_i \, \delta p_n$$
(6)

where U is the Jacobian of the map and V expresses the parametric variation of the map

$$U = D_x F(x, p) = \frac{\partial}{\partial x} F(x, p) \qquad V = D_p F(x, p) = \frac{\partial}{\partial p} F(x, p) \quad .$$
(7)

The notation U_i and V_i indicates that the partial derivatives are evaluated at $[x = x(i, p_0), p = p_0]$. To obtain an expression for δp_n, one imposes the *control criterion* (not unique) that equation(6), taken as a strict equality, should be equal to zero, namely that

$$x_{n+1} - x(i+1, p_0) = 0 \quad .$$
(8)

(control criterion 1D)

Solving for δp_n, equation(6) leads immediately to

$$\delta p_n = -\frac{U_i}{V_i} \, \delta x_{n,i} \quad .$$
(9)

In order to complete the procedure, one should make sure that $|\delta p_n|$ just obtained is $\leq \delta p_{\max}$ (we will always take $\delta p_{\max} > 0$). If it is so, $p_n = p_0 + \delta p_n$ for the next iteration; if not, one could set $p_n = p_0$ and wait until the trajectory reenters $\mathcal{B}_\delta$. We have found that a more robust choice is to apply the corrections during the control stage according to the prescription

$$\delta p_n \longrightarrow \delta p_{\max} \tanh(\delta p_n / \delta p_{\max}) \qquad \text{for} \qquad |\delta p_n| > \delta p_{\max} \quad .$$
(10)

This is a minor point however since typically δp_n decreases rapidly after the first few iterations as we are getting ever closer to the target orbit ($\delta p_n \propto \delta x_n$).

2D Control

Stabilization in higher dimensions is qualitatively different from the 1D case since phase space is endowed with a much richer structure. For simplicity, we will confine our discussion to two dimensions. In 2D, the generic local neighbourhood of a UPO is equipped with a stable and unstable manifold. A chaotic trajectory entering the neighbourhood will move toward the UPO along the stable direction and escape along the unstable one. This is the "saddle dynamics" illustrated in Figure (1).

OGY [4] realized that a possible solution of controlling chaos could be obtained by locally displacing the manifolds such that the component of the motion along the unstable direction could be eliminated (at least to first order) such that subsequent evolution will naturally lead the orbit to the unstable point along the stable direction. This idea is geometrically presented in Figure (1) for the situation of an unstable fixed point denoted $\mathbf{x}_F(p_0)$, much like the task of bringing a ball bearing to rest on a saddle. The remaining part of this subsection is devoted to the mathematical translation of the applications of "small pushes and pulls applied at the right places".

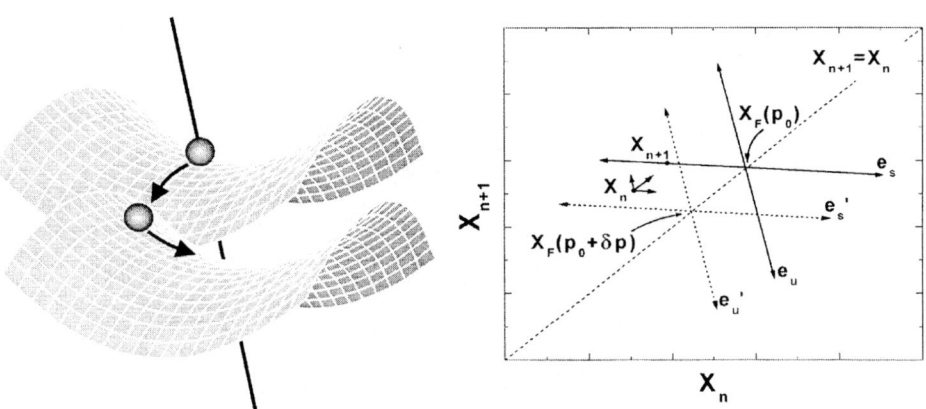

FIGURE 1. Local Geometry of Control: (*left*) 2D "saddle dynamics" and (*right*) linearization of the stable and unstable manifolds.

Starting from the 2D nonlinear map

$$\mathbf{x}_{n+1} = \mathbf{F}(\mathbf{x}_n, p) \tag{11}$$

with $\mathbf{x}_n \in \mathcal{R}^2$ and $p \in \mathcal{R}$, one traces the steps to achieve control through the following algorithm.

1. locate a UPO of period m for the nominal value of the parameter p_0

$$\mathbf{x}(1, p_0) \rightarrow \mathbf{x}(2, p_0) \rightarrow \cdots \mathbf{x}(m, p_0) \rightarrow \mathbf{x}(m+1, p_0) = \mathbf{x}(1, p_0) \tag{12}$$

2. linearize the dynamics the δ-neighbourhood of $\mathbf{x}(i, p_0)$:

$$\mathbf{x}_{n+1} - \mathbf{x}(i+1, p_0) \sim \mathbf{U}_i \; [\mathbf{x}_n - \mathbf{x}(i, p_0)] + \mathbf{V}_i \; \delta p_n \tag{13}$$

where $\mathbf{U}$ is the 2×2 Jacobian matrix and $\mathbf{V}$ is a 2×1 parametric variation vector

$$\mathbf{U} \equiv D_{\mathbf{x}} \; \mathbf{F}(\mathbf{x}, p) \qquad \text{and} \qquad \mathbf{V} \equiv D_p \; \mathbf{F}(\mathbf{x}, p) \tag{14}$$

with the partial derivatives evaluated at $[\mathbf{x} = \mathbf{x}(i, p_0), \; p = p_0]$.

3. characterize the local dynamics by the *stable* $\mathbf{e}_{s,i}$ and the *unstable* $\mathbf{e}_{u,i}$ directions.

4. construct the contravariant vectors defined by

$$\mathbf{f}_{u,i} \cdot \mathbf{e}_{u,i} = \mathbf{f}_{s,i} \cdot \mathbf{e}_{s,i} = 1 \qquad \mathbf{f}_{u,i} \cdot \mathbf{e}_{s,i} = \mathbf{f}_{s,i} \cdot \mathbf{e}_{u,i} = 0 \quad . \tag{15}$$

5. stabilize the orbit by demanding that it falls, on the next iteration, on the stable direction, i.e.

$$\mathbf{f}_{u,i+1} \cdot [\mathbf{x}_{n+1} - \mathbf{x}(i+1, p_0)] = 0 \quad . \tag{16}$$

(control criterion OGY)

Therefore, according to step 2., we obtain the relation

$$\mathbf{f}_{u,i+1} \cdot \{\mathbf{U}_i \; [\mathbf{x}_n - \mathbf{x}(i, p_0)] + \mathbf{V}_i \; \delta p_n\} = 0 \quad . \tag{17}$$

6. calculate the perturbation necessary to satisfy equation(17)

$$\delta p_n = -\frac{\mathbf{f}_{u,i+1} \cdot \{\mathbf{U}_i \; [\mathbf{x}_n - \mathbf{x}(i, p_0)]\}}{\mathbf{f}_{u,i+1} \cdot \mathbf{V}_i} \tag{18}$$

and apply only if $|\delta p_n| < \delta p_{\max}$ otherwise set e.g. $\delta p_n = 0$ or use equation (10).

In summary, the stabilization procedure can be divided in three separate stages: the *learning stage*, where one identifies the desired UPOs, extracts the Jacobian matrices, calculates the corresponding stable and unstable directions $\mathbf{e}_{s,i}$, $\mathbf{e}_{u,i}$ to construct the contravariant vectors $\mathbf{f}_{u,i}$; the *transient stage*, where, after randomly choosing an initial condition, the system is let to evolve freely at the nominal parameter value p_0 until, at the *control stage*, once the chaotic trajectory has entered the prescribed δ-neighbourhood, the control is attempted by means of small parameter perturbations.

Alternative Control Algorithms

The algorithm just described is *dynamically optimal* in that it uses a control criterion and perturbations that involve the complete local structure of the system's dynamics. From a practical (experimental) point of view, it is also most demanding considering that the underlying dynamical law is often not even known *a priori*. It requires the determination of the Jacobian matrices and the parametric variation of the map along the UPOs and the corresponding stable and unstable directions. Several modifications of the original method have been proposed and we now present some of them with an emphasis towards algorithms that are simpler and in some instances more practicable.

If instead of equation(13), one writes the linearization as

$$\mathbf{x}_{n+1} - \mathbf{x}(i+1, p_n) \sim \mathbf{U}_i \; [\mathbf{x}_n - \mathbf{x}(i, p_n)] \tag{19}$$

where the dependence of $\mathbf{U}_i$ on p has been ignored, and one introduces the parametric variation of the periodic points as

$$\mathbf{g}_i \equiv \left. \frac{d}{dp} \mathbf{x}(i, p) \right|_{p=p_0} \sim \frac{\mathbf{x}(i, p_0 + \delta p) - \mathbf{x}(i, p_0)}{\delta p} \tag{20}$$

or equivalently,

$$\mathbf{x}(i, p_0 + \delta p) \sim \mathbf{x}(i, p_0) + \mathbf{g}_i \, \delta p \quad , \tag{21}$$

one arrives, under the criterion (16), to the perturbations

$$\delta p_n = -\frac{\mathbf{f}_{u,i+1} \cdot \{\mathbf{U}_i \; [\mathbf{x}_n - \mathbf{x}(i, p_0)]\}}{\mathbf{f}_{u,i+1} \cdot (\mathbf{g}_{i+1} - \mathbf{U}_i \, \mathbf{g}_i)} \quad . \tag{22}$$

The expression (22) has the advantage that the variables $\mathbf{g}_i$ can easily be obtained from observations of the shift of the periodic points under small parameter change.

One further simplification arises if it is sufficient to intervene on the dynamics only once per period. The modifications to the formula are straightforward. Equation (19) becomes

$$\mathbf{x}_{n+m} - \mathbf{x}_{F,i}(p_n) \sim \mathbf{U}_i^{(m)} \; [\mathbf{x}_n - \mathbf{x}_{F,i}(p_n)] \tag{23}$$

where the notation is chosen to emphasize that $\mathbf{x}_{F,i}(p) = \mathbf{x}(i, p)$ is a fixed point of $\mathbf{F}^{(m)}$ (the m times application of the map $\mathbf{F}$). Furthermore, the Jacobian matrix $\mathbf{U}_i^{(m)} = D_{\mathbf{x}}\mathbf{F}^{(m)}(\mathbf{x}, p)$, evaluated at $[\mathbf{x}_{F,i}(p_0), p_0]$, can be expanded in terms of its eigenvectors (the stable and unstable manifolds) and eigenvalues as $\lambda_{u,i} \, \mathbf{e}_{u,i}\mathbf{f}_{u,i} + \lambda_{s,i} \, \mathbf{e}_{s,i}\mathbf{f}_{s,i}$ to modify (22) to

$$\delta p_n = -\frac{\lambda_{u,i}}{(1 - \lambda_{u,i})} \frac{\mathbf{f}_{u,i} \cdot [\mathbf{x}_n - \mathbf{x}_{F,i}(p_0)]}{\mathbf{f}_{u,i} \cdot \mathbf{g}_i} \quad . \tag{24}$$

Until now the control criterion has not been modified, but in situations where the stable and unstable manifolds may be difficult to obtain (e.g. in high dimensions) or for the sake of simplicity, one might choose to minimize the deviation from the target orbit instead of projecting onto the stable manifold, i.e. we demand that

$$||\mathbf{x}_{n+m} - \mathbf{x}_{F,i}|| = \text{minimum} \quad , \tag{25}$$

(control criterion MED)

where an estimate of $\mathbf{x}_{n+m}$ is given by equations (23) and (21), namely

$$\mathbf{x}_{n+m} \sim \mathbf{x}_{F,i}(p_0) + \mathbf{U}_i^{(m)} \left[\mathbf{x}_n - \mathbf{x}_{F,i}(p_0)\right] + (\mathbf{g}_i - \mathbf{U}_i^{(m)} \mathbf{g}_i) \, \delta p_n \quad . \tag{26}$$

The solution of the minimization (25) is then

$$\delta p_n = -\frac{(\mathbf{g}_i - \mathbf{U}_i^{(m)} \mathbf{g}_i) \cdot \left\{\mathbf{U}_i^{(m)} \left[\mathbf{x}_n - \mathbf{x}_{F,i}(p_0)\right]\right\}}{(\mathbf{g}_i - \mathbf{U}_i^{(m)} \mathbf{g}_i) \cdot (\mathbf{g}_i - \mathbf{U}_i^{(m)} \mathbf{g}_i)} \quad . \tag{27}$$

This technique was first introduced in [6] and goes by the name of minimal expected deviation (MED) method[2].

The perturbations δp_n transform the original autonomous systems to nonautonomous ones. One could therefore consider a formulation extending phase space by one dimension with p_n as the new dynamical variable. Alternatively, as was first realized in [7], one could account explicitly for the p_n dependence by introducing in the mapping itself the past history of the perturbations, namely

$$\mathbf{x}_{n+1} = \mathbf{F}_c(\mathbf{x}_n, p_n, p_{n-1}) \quad . \tag{28}$$

We restrict ourselves to the last two perturbations. The sub-index on $\mathbf{F}_c$ is to remind us that the mapping differs from the original one and is only identical to it when $p_n = p_{n+1} = p_0$. In analogy to equation (26), one can write

$$\mathbf{x}_{n+m} \sim \mathbf{x}_{F,i}(p_0, p_0) + \mathbf{U}_i^{(m)} \left[\mathbf{x}_n - \mathbf{x}_{F,i}(p_0, p_0)\right]$$
$$+ (\mathbf{g}_{u,i} - \mathbf{U}_i^{(m)} \mathbf{g}_{u,i}) \, \delta p_n + (\mathbf{g}_{b,i} - \mathbf{U}_i^{(m)} \mathbf{g}_{b,i}) \, \delta p_{n-1} \quad . \tag{29}$$

with

$$\mathbf{g}_u = \frac{d}{dp} \, \mathbf{x}_F(p, p') \bigg|_{p=p'=p_0} \qquad \mathbf{g}_b = \frac{d}{dp'} \, \mathbf{x}_F(p, p') \bigg|_{p=p'=p_0} \quad . \tag{30}$$

To complete the modification, a control condition must be imposed and a reasonable choice is

(control criterion RPF)

$$\left[\mathbf{x}_{n+2m} - \mathbf{x}_{F,i}(p_0, p_0)\right] = 0 \qquad \text{and} \qquad \delta p_{n+1} = 0 \quad . \tag{31}$$

---

[2] For conciseness, we only give the FIRST reference for each technique, although the methods are continuously being improved and modified: this applies to the entire report.

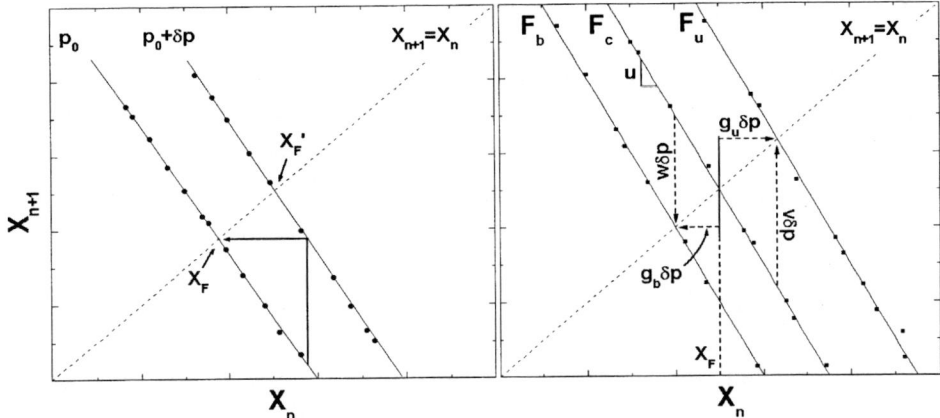

FIGURE 2. Alternative Local Geometries of Control: (*left*) the Occasional Proportional Feedback (OPF) method and (*right*) the Recursive Proportional Feedback (RPF) method.

The conditions (31) are sufficient to solve the linearized equation (29) for δp_n as

$$\delta p_n = -\frac{\mathbf{X}_i^{(m)} \cdot \{\mathbf{U}_i^{(m)}\ \mathbf{U}_i^{(m)}\ [\mathbf{x}_n - \mathbf{x}_{F,i}(p_0, p_0)]\}}{||\mathbf{X}_i^{(m)}||^2} - \frac{\mathbf{X}_i^{(m)} \cdot (\mathbf{U}_i^{(m)}\ \mathbf{W}_i^{(m)})}{||\mathbf{X}_i^{(m)}||^2}\ \delta p_{n-1} \quad (32)$$

with $\mathbf{X}_i^{(m)} = \mathbf{U}_i^{(m)}\ \mathbf{V}_i^{(m)} + \mathbf{W}_i^{(m)}$, $\mathbf{V}_i^{(m)} = \mathbf{g}_{u,i} - \mathbf{U}_i^{(m)}\ \mathbf{g}_{u,i}$ and $\mathbf{W}_i^{(m)} = \mathbf{g}_{b,i} - \mathbf{U}_i^{(m)}\ \mathbf{g}_{b,i}$. Our derivation is somewhat different from [8] and , in view of the numerous parameters to determine, it should be taken as a serious alternative only for 1D (or quasi 1D) systems. In the latter case, the method has a simple geometrical interpretation as illustrated in the right panel of Figure (2) for an unstable fixed point $x_F(p_0)$: U is the local slope of the original map, $g_u \delta p$ and $g_b \delta p$ correspond to the shifts of the fixed point to positions $x_{F,u}(p_0 + \delta p, p_0) \sim x_F(p_0) + g_u \delta p$ and $x_{F,b}(p_0, p_0 + \delta p) \sim x_F(p_0) + g_b \delta p$ respectively, whereas $V \delta p$ denotes the map displacement from $F_c(x, p_0, p_0) \equiv F(x, p_0) \rightarrow F_u(x, p_0 + \delta p, p_0)$ and $W \delta p$ from $F(x, p_0) \rightarrow F_b(x, p_0, p_0 + \delta p)$. These parameters are readily obtained from the observations as was first demonstrated by [8] who gave the algorithm the name of recursive proportional feedback (RPF).

Our last modification, which we only quote for 1D system where it is most likely to be used, consists of ignoring in the RPF formula the dependence on p_{n-1} which amounts to setting $W = 0$ in the previous equations to obtain

$$\delta p_n = -\frac{U}{(g - U\ g)}\ [x_n - x_{F,i}(p_0)] \quad . \quad (33)$$

One should not be deceived by the apparent simplicity of equation (33). In its regime of applicability, it can be enormously successful as we will see shortly and as Hunt [9] first demonstrated. Its geometrical interpretation is shown in the left panel of Figure (2) where it amounts to choose a perturbation such that upon the next iteration, x_n is mapped onto the target fixed point. This occasional proportional feedback (OPF), as it was called originally, is eminently suited for experimental control since it requires feedback signals proportional to the deviations to the target orbit and two parameters U and g readily available. Even more, if one is bold enough, one might just as well adjusted the proportionality constant until control is established.

MATHEMATICAL AND PHYSICAL APPLICATIONS

We have selected five examples of increasing complexity and novelty. The first three correspond to dissipative systems confined to a chaotic attractor, whereas the last two are conservative Hamiltonian systems where the attractor is replaced by bounded regions of phase space where the dynamics is chaotic.

We will apply the perturbations only once per period for all cases except the last. Save for the two discrete maps, ALL the relevant control informations are obtained numerically in an effort to simulate more closely an experimental setting. Reliable methods (see e.g. [10]) exist to locate the positions of the UPOs and we will assume thereafter that the locations are known prior to the control session. The numerical construction of the Jacobian matrices from time series is often a subtle task and is beyond the scope of this article. The reader is referred to [11] for technical details.

1D Logistic Map

Our first example is the *logistic map*

$$x_{n+1} = r\, x_n(1 - x_n) \tag{34}$$

which is known to have a broad range of parameters values $r \in [\sim 3.57, 4.0]$ where chaotic motion can be observed. The variables entering the control signals are simply

$$U = r(1 - 2x) \qquad \text{and} \qquad V = x(1 - x) \tag{35}$$

and for a given UPO, $\{x(i)\}$, the perturbations are

$$\delta p_n = r_0 \frac{2x(i) - 1}{x(i)(1 - x(i))} [x_n - x(i)] \tag{36}$$

for $|x_n - x(i)| \le \delta \ll 1$.

The left panel of Figure (3) shows the asymptotic behaviour of the orbits for different values of r. We have chosen the nominal value, $r_0 = 3.8$, and control settings of $\delta = 10^{-4}$, $\delta r_{max} = 10^{-3}\, r_0$, and requested from our controller to successively

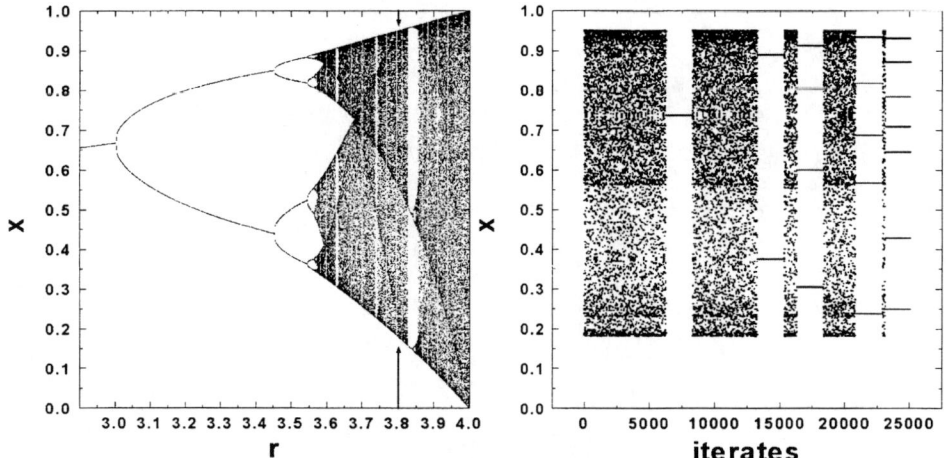

FIGURE 3. Logistic Map: The left panel shows the bifurcation diagram as a function of the parameter r. The right panel illustrates the successive control of UPOs of period 1, 2, 4, 5, 7 for $r_0 = 3.8$ (indicated by the arrow in the left panel). The control is held for 2 000 iterates with $\delta = 10^{-4}$ and $\delta r_{max} = 10^{-3} r_0$.

stabilize periods $m = 1, 2, 4, 5, 7$ and hold control for 2 000 iterates each. The right panel of Figure (3) shows the flexibility of the procedure letting ergodicity bring the trajectory near the next target orbits after a successful control sequence.

A few remarks are in order. First, note that the control process does not create the UPOs, they exist already in the natural (free) dynamics and build, so to speak, the lattice upon which the chaotic trajectory wanders. The control mechanism simply picks them out of the background. Second, the transients in between controlled periods are of varying lengths and could be considerably reduced by optimizing the control variables and/or by steering the chaotic orbit to the UPOs by a technique called *targeting* [13] .

2D Dissipative Hénon Map

The Hénon map [14] has been a paradigmic example in nonlinear dynamics ever since its inception. It has the form

$$
\begin{aligned}
x_{n+1} &= 1 - a\,x_n^2 + y_n \\
y_{n+1} &= b\,x_n
\end{aligned}
\tag{37}
$$

and a Jacobian matrix given by

$$
\mathbf{U} = \begin{pmatrix} -2ax & 1 \\ b & 0 \end{pmatrix}
\tag{38}
$$

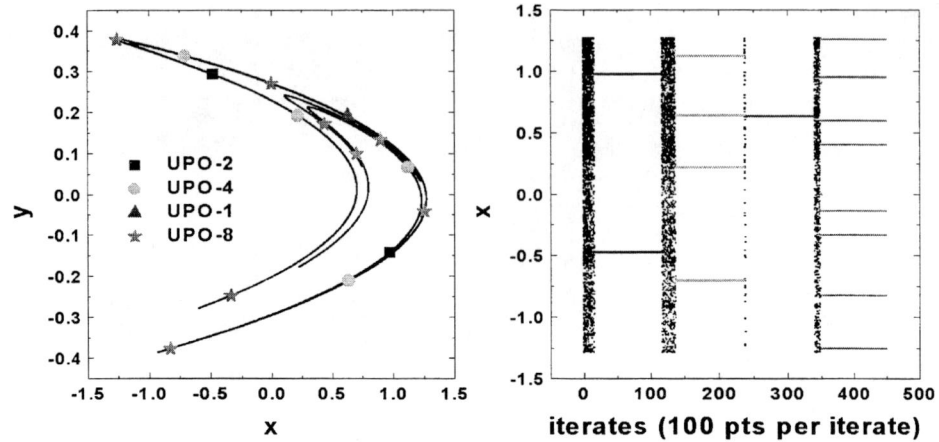

FIGURE 4. Hénon Map: The left panel shows the attractor with a number of embedded UPOs. The right panel shows the successive OGY control of UPOs of period 2, 4, 1, 8 for $a_0 = 1.4, b_0 = 0.3$. The control is held for 10 000 iterates with $\delta = 5 \times 10^{-3}$ and $\delta a_{\max} = 10^{-2} \, a_0$.

with determinant (the Jacobian) equal to b. It is dissipative for $|b| < 1$ and possesses a non trivial strange attractor for various combinations of the parameters $a, |b| < 1$. For our purpose, it serves as a benchmark, since the complete OGY strategy (eqns (18) or (22) or (24)) can be performed (semi-) analytically and compared with other implementations. For example, the Jacobian matrices $\mathbf{U}_i^{(m)}$ are the product of m individual Jacobian matrices along the periodic orbit, i.e. $\mathbf{U}_{i+m-1} \, \mathbf{U}_{i+m-2} \, \ldots \, \mathbf{U}_{i+1} \, \mathbf{U}_i$, whose eigenvectors can be written down analytically.

Our experiment (Fig. 4) shows the attractor for $a_0 = 1.4, b_0 = 0.3$ with the locations of embedded UPOs of periods 1, 2, 4, 8. The OGY control algorithm was used and the perturbations applied to the parameter a_0. The right panel of Figure (4) shows the values of the x variable for a scenario involving the successive stabilizations of period 2, 4, 1, and 8. The target orbit is changed once a UPO has been controlled for 10 000 iterates. The control region was set to $\delta = 5 \times 10^{-3}$ and the maximum perturbation allowed was $\delta a_{\max} = 10^{-2} \, a_0$. Under the same operational conditions, the MED minimization gave equally satisfactory results.

3D Rössler Flow

Until now, our examples have described discrete dynamics and our formalism is also derived for maps. We now show how to appropriately discretize a flow to achieve control with the methods derived thus far. The Rössler system [15] consists of 3 coupled differential equations

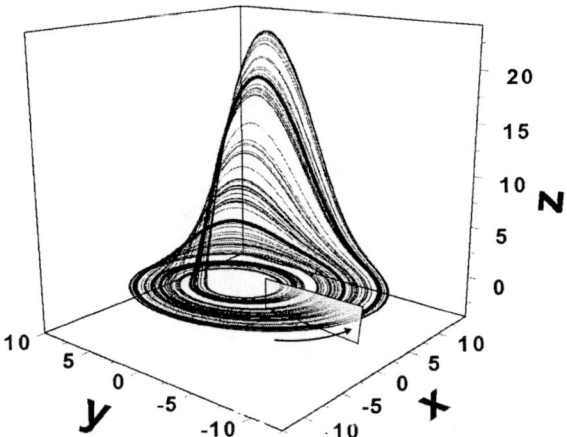

FIGURE 5. Rössler Attractor with an embedded UPO of period 3 in darker line. The flow parameters are $a_0 = b_0 = 0.2$ and $c_0 = 5.7$. Also indicated is the plane $x = 2$ where dynamical informations are gathered.

$$\dot{x} = -y - z$$
$$\dot{y} = x + a\,y \qquad\qquad (39)$$
$$\dot{z} = x\,z - c\,z + b$$

and a chaotic orbit is bounded to a funnel-like attractor (Fig. 5: $a_0 = b_0 = 0.2, c_0 = 5.7$) where the motion is mostly in the $x - y$ plane with rapid excursions in the z direction.

The last observation leads us to one possible discretization: one can register intersections of the flow with a plane $x =$ cte (the Poincaré section) and accumulate the pairs (y_n, z_n) from which one could subsequently infer a map (the Poincaré map) $(y_{n+1}, z_{n+1}) = \mathbf{F}_P((y_n, z_n))$. For uniqueness, one must choose *directed* intersections by monitoring $\dot{x}$ on the Poincaré section (see Fig. 5).

The intersections with the Poincaré section $(x = 2, \dot{x} > 0)$ is represented in the lower left panel of Figure (6). One notices little variation in the z component and our quasi 1D control methods should be appropriate. In the same panel, one has indicated the positions of 3 UPOs to be stabilized. The OPF control of the y component is shown in the lower right panel of Figure (6) resulting in 3 continuous trajectories (upper panel) embedded in the attractor (compare with Fig. 5). We believe that this success indicates just how robust this type of linear feedback is. Remember that we intervene in the dynamics *only* on the Poincaré section with a small perturbation (here $< 1\%$ of c_0) at every m intersections.

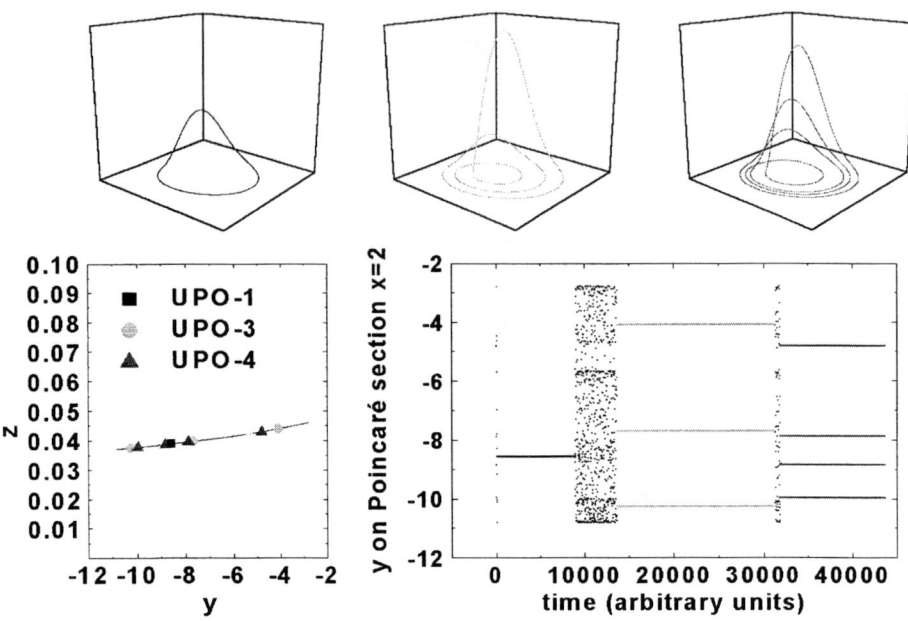

FIGURE 6. Rössler Flow: (*upper panel*) 3 stabilized continuous trajectories embedded in the attractor of Fig. 5; (*lower left*) locations of the UPOs on the Poincaré section $x = 2, \dot{x} > 0$; (*lower right*) OPF control on the Poincaré section of the y components of the UPOs. Control is held during 1 500, 1 000, and 500 cycles for period 1, 3, and 4 respectively (1 cycle = 1 complete orbit) with $\delta = 10^{-2}$ and $\delta c_{\max} = 10^{-2} \, c_0$.

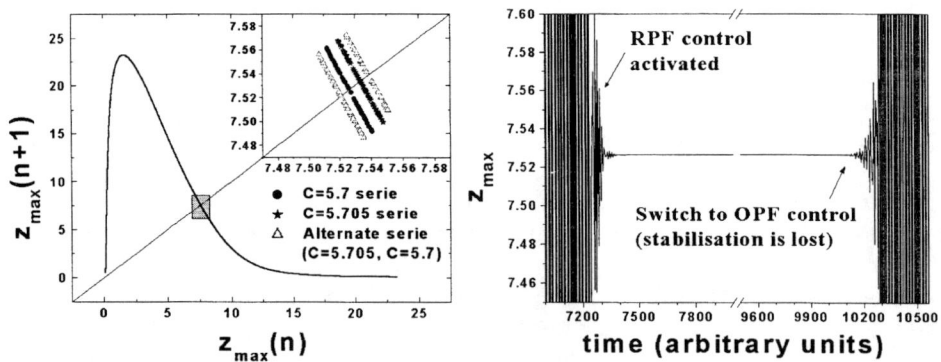

FIGURE 7. Rössler Flow: (*left*) first return map of successive maxima of the z variable (Lorenz map) for the attractor of Fig. 5; (*right*) success of the RPF control and failure of the OPF method applied to the $z_{\max}$ series with $\delta = 10^{-1}$ and $\delta c_{\max} = 10^{-2} \, c_0$.

We have performed another experiment on the Rössler flow. It consists of collecting subsequent maxima $z_{n,max}$ of the z variable along a chaotic trajectory to construct a discretization of the flow. By plotting $z_{n+1,max}$ versus $z_{n,max}$, one obtains a (first) return map which is usually called a Lorenz map after the man who first defined the procedure. Our map is shown on the left of Figure (7) and it has all the characteristics of a chaotic 1D map. Again this indicates that our quasi 1D control (RPF or OPF) methods should be applicable. However, according to Figure (5), this should be rather delicate since the trajectory spends most of its time in the $x - y$ plane. In other words, the Lorenz map may not be adequate to gather enough dynamical information for control.

We have therefore used the RPF algorithm where g_u and g_b are first estimated by alternating the value of c_0 between 5.700 and 5.705. As expected, this creates two new applications F_u and F_b as clearly seen in the enlarged section of Figure (7). The right side of Figure (7) shows the RPF stabilization of a period 1 UPO which is maintained for 5 000 iterates before switching to the OPF algorithm where control is rapidly lost. The refinement incorporated in the RPF method has served us well and this experiment reveals nicely the limitation of the simplest strategy.

Billiard Dynamics

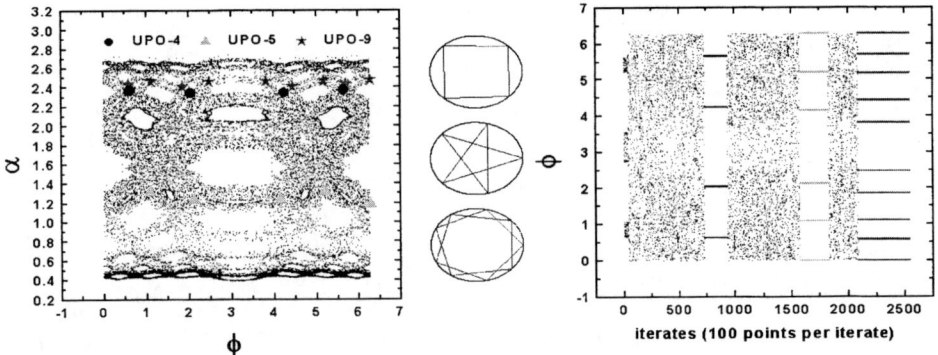

FIGURE 8. Cosine Billiard: (*left*) mixed chaotic (filled) and regular (open islands) phase space with embedded UPOs for $\epsilon_0 = 0.3$; (*middle*) MED controlled UPOs of period 4, 5 , 9; (*right*) stabilized ϕ variable of the corresponding UPOs held for 5 000 cycles each with $\delta = 10^{-2}$ (1 cycle = 1 complete orbit).

The study of the frictionless motion of a particle bounded by a closed surface where it is specularly reflected is known as *billiard dynamics* and dates back to Birkhoff [16]. It serves to illustrate the transition from strict regularity (integrability) to chaos (ergodicity) in Hamiltonian systems [17] and bears important connections to *quantum chaos* as well [18]. We have chosen to study the 2D *cosine billiard* where the surface is parameterized in polar coordinates by the relation

$$r(\phi) = 1 + \epsilon \cos \phi \quad . \tag{40}$$

The parameter ϵ is a measure of nonintegrability since for $\epsilon = 0$, the curve is a circle and represents the integrable limit. For all $\epsilon \neq 0$, there are finite regions of phase space that contain chaotic trajectories. Figure (8) shows on the left the mixed and complex structure of phase space for $\epsilon_0 = 0.3$: the state variables are the incident angles on the surface, $\{\alpha_n\}$, and the polar angles of the point of impact, $\{\phi_n\}$. Since motion is free in between collisions with the surface, our example belongs to the class of 2D area-preserving mappings, where attractors are absent and replaced by stochastic bands mixed with regular regions. Within these bands, the motion is ergodic: the blackened region is produced by *one* single chaotic orbit. Embedded in this stochastic web, one observes a number of UPOs whose physical trajectories inside the boundary are shown in the middle portion of Figure (8). By pulsating the deformation parameter ϵ about its nominal value, we have achieved control of 3 UPOs of period 4, 5 and 9. We used the MED control algorithm with a neighbourhood of $\delta = 10^{-2}$. A numerical OGY method gives identical performance.

We mention that the successful control of billiard dynamics may offer a solution to the degradation of finesse in resonant optical microcavities [19]. It has been inferred that the loss of lasing activities might be associated with ray chaos (geometrical limit) in the optical resonators where the photons are transported (via chaotic diffusion) to regions of phase space where refractive escape (Snell's law) becomes possible. The dielectric droplets making up the resonators behave very much like 2D billiards and we propose that programmed variations of the asymmetry may help reduce photon leakage. The viability of the proposal is currently being investigated.

Diamagnetic-Kepler Hamiltonian

Our final example is a continuous, 2 degrees of freedom (4D phase space) Hamiltonian system. It represents the motion of an electron under the combined influence of a Coulomb and a magnetic field. It goes under the name, *diamagnetic Kepler problem* (DKP), and occupies central stage in classical and quantum chaos research [20]. We use scaled semi-parabolic coordinates and write the resulting scaled (pseudo-) Hamiltonian as (for angular momentum $L = 0$)

$$\hat{h}_{DK} = \frac{1}{2}(p_\nu^2 + p_\mu^2) - \epsilon (\nu^2 + \mu^2) + \frac{1}{8}\nu^2\mu^2(\nu^2 + \mu^2) \equiv 2 \quad . \tag{41}$$

The scaled energy ϵ is related to the physical energy E by $\epsilon = \gamma_0^{-2/3} E$ where the parameter $\gamma_0 = B/B_0$ denotes the strength of the magnetic field relative to the unit $B_0 \simeq 2.35 \ 10^5 \ T$. The classical flow covers a wide range of Hamiltonian dynamics reaching from bound, nearly integrable behaviour to completely chaotic and unbound motion as the scaled energy is varied [21].

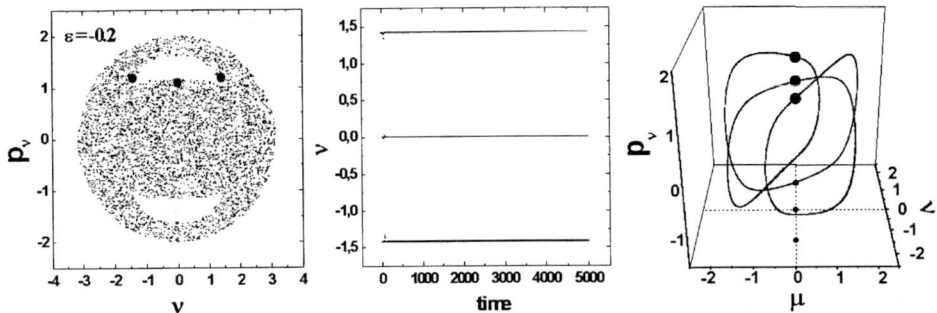

FIGURE 9. Diamagnetic Kepler Problem for scaled energy $\epsilon_0 = -0.2$: (*left*) Poincaré section $\mu = 0, \dot{\mu} > 0$ showing one chaotic trajectory (filled space) and an OGY controlled period 3 orbit (black dots); (*middle*) the stabilized ν variable on the Poincaré section; (*right*) corresponding 3D trajectory. The control settings are $\delta = 10^{-3}$ and $\delta\epsilon_{\max} = 7 \times 10^{-2}$.

The dimension reduction (from 4D to 2D) and discretization is performed by observing the dynamics on the Poincaré section defined by $\mu = 0, \dot{\mu} > 0$. The energy shell is then mapped to an area bounded by the condition $p_\nu^2 - 2\epsilon \, \nu^2 = 4$ which represents an ellipse in the (ν, p_ν) plane. The left panel of Figure (9) shows the collection of points $\{\nu_n, p_{\nu,n}\}$ obtained by numerical integration of the equations of motion for $\epsilon_0 = -0.2$. One notices, for this energy, that phase space has few regular structures: apart from two lobes of regularity, the rest of the ellipse is filled by the successive piercings of *one* chaotic trajectory. The three dots indicate the positions of a UPO of period 3. We have succeeded in stabilizing a number of UPOs for the system, one of them is displayed with its 3D trajectory in Figure (9). We have employed a complete numerical implementation of the OGY strategy.

In attempting to bring order to the DKP dynamics, we had to overcome a number of difficulties not encountered in our previous examples. First, a typical trajectory spends a lot of time away from the Poincaré section and because of the sensitivity of the dynamics we had to device an efficient variable step *symplectic* integrator thereby preserving the geometrical structure of the Hamiltonian. Second, we had to obtain numerical Jacobians for all members of the UPOs since it was found to be necessary to intervene at *every* crossing of the Poincaré section. Third, the eigenvalues of area-preserving Jacobians are often complex and the stable and unstable manifolds are no longer along the directions of their eigenvectors. A new method had to be implemented. Details of the ingenious solutions to these problems can be found in [22].

We should comment that this is the first control of a realistic Hamiltonian system. It still remains an open question however if manipulations of the magnetic field to induce stabilization of a classical unstable orbit can be extended to the control, for example, of Rydberg wave packet dynamics.

Properties of the Control Procedure

The lessons learned through the previous examples and many more not reported here allow us to draw a list of the properties and advantages of the adopted control 'philosophy' and to point to remaining difficulties.

– no model dynamics is required *a priori* and only *local information* is needed;
– computations at each step is minimal;
– *gentle touch*: the required changes in p_0 can be quite small ($< 1\%$);
– *multi-purpose flexibility:* different periodic orbits can be stabilized for the *same* system in the *same* parameter range;
– control can be achieved even with imprecise measurements of eigenvalues and eigenvectors: the methods are robust;
– the methods can also be applied to *synchronization* of several chaotic systems.

At least three complications come to mind when one considers the implementation of chaos control strategies to the laboratory. The presence of noise, ignored so far, may induce occasional loss of control or hinder it altogether. The average waiting time to fall in the δ-neighbourhood may be very (too) long, especially for Hamiltonian systems and a targeting strategy [13] should complement the control method. Furthermore, the system's parameters may drift with time and this non-stationarity should be accounted by updating the control informations. *Tracking* [23] is the name given to this procedure.

CONCLUSIONS AND FUTURE PERSPECTIVES

CHAOS often breeds life,
When ORDER breeds habit.
HENRY BROOKS ADAMS

We have presented the basic techniques for controlling chaos and we have reported on some of our efforts to recover order from chaos. Applications of the control of chaos have been reported in such diverse areas as aerodynamics, chemical engineering, communications, electronics, fluid mechanics, laser physics, as well as, biology, finance (not confirmed!), medicine, physiology, epidemiology and the list is constantly growing. It is perhaps instructive at this point to quote some of the earliest *experimental* successes of the methods to gain an idea of the breadth and diversity of the systems considered: in *solid state devices* and *condensed matter*, magneto-elastic ribbon [5], spin-wave system [24], electric diode [9]; in *fluid mechanics*, regularization of chaotic convection [25]; in *chemistry*, mechanisms of control of autocatalytic reactions [26]; in *laser systems*, stabilization of coupled ensemble of lasers [27]; in *physiology*, cardiac arrhythmia [28], (anti)-control of epileptic seizures [29].

The last decade has seen much accomplishments and challenges for the future are numerous. The following items seem to provide a glimpse of things to come:

generalization to spatio-temporal chaos, adaptive control for non-stationary dynamics, effective control in the presence of noise (dynamical and/or observational), adaptive synchronization of chaos.

However, the greatest challenge will remain for some times the application to complex biological systems and in particular to brain dynamics [30]. Despite early efforts, euphoria has been replaced by a healthy skepticism. Indeed, complex natural systems are noisy, contain a strong stochastic component and not endowed with a behaviour called chaos (at least not in its mathematical rigourous sense). Yet, one would like to believe that "the controlled chaos of the brain is more than an accidental by-product of the brain complexity" [31]. The perspective of unifying the techniques of deterministic chaos control with a statistical stochastic description as a possible therapeutic strategy against dynamical diseases is surely something to consider.

Acknowledgments. $\mathcal{LJD}$ thanks the members of his research group for their contributions and stimulating discussions on the subject. He is also grateful for the hospitality of G. Deco and B. Schürmann (München) where this work took its flight and to J. Honerkamp and J. Timmer (Freiburg) where this report was completed.

REFERENCES

1. Dyson, F., *Infinite in All Directions*, New York: Harper and Row Publishers, 1988, chap. 10, pp. 182-184.
2. Poincaré, H., *Acta Mathematica* **13**, 1 (1890).
3. Lorenz, E.N., *J. of Atmos. Sci.* **20**, 130 (1963).
4. Ott, E., Grebogi, C., and Yorke, J.A., *Phys. Rev. Lett.* **64**, 1196 (1990).
5. Ditto, W.L., Rauseo, S.N., and Spano, M.L., *Phys. Rev. Lett.* **65** , 3211 (1990).
6. Reyl, C., Flepp, L., Baddi, R., and Brun, E., *Phys. Rev. E* **47**, 267 (1993).
7. Dressler, U., and Nitsche, G., *Phys. Rev. Lett.* **68**, 1 (1992).
8. Rollins, R.W., Parmananda, P., and Sherard, P., *Phys. Rev. E* **47**, R780 (1993).
9. Hunt, E.R., *Phys. Rev. Lett.* **67**, 1953 (1991).
10. Ott, E., Sauer, T., and Yorke, J.A., eds., *Coping with Chaos* , New York: John Wiley & Sons Inc., 1994.
11. Eckman, J.P., Kamphorst, S.O., Ruelle, D., and Ciliberto, S., *Phys. Rev. A* **34** , 4972 (1986); Sanon, M., and Sawada, Y., *Phys. Rev. Lett.* **55** , 1082 (1985).
12. Takens, F., *Lectures Notes in Math.* **898** (1981); Sauer, T., Yorke, J.A., and Casdagli, M., *J. Stat. Phys.* 65, 579 (1991).
13. Shinbrot, T., Ott, E., Grebogi, C., and Yorke, J.A., *Phys. Rev. Lett.* **65** , 3215 (1990); Kostelich, E.J., Grebogi, C., Ott, E., and Yorke, J.A., *Phys. Rev. E* **47** , 305 (1993).
14. Hénon, M., *Comm. Math. Phys.* **50**, 69 (1976).
15. Rössler, O.E., *Phys. Lett. A* **71** , 155 (1979).
16. Birkhoff, G.D., *Acta Math.* **50**, 359 (1927).

17. Berry, M.V., *Eur. J. Phys.* **2**, 91 (1981); Robnik, M., *J. Phys. A* **16** , 3971 (1983).

18. Stöckmann, H.J., and Stein, J., *Phys. Rev. Lett.* **64** , 2215 (1990); Stein, J., and Stöckmann, H.J., *Phys. Rev. Lett.* **68** , 2867 (1992).

19. Nöckel, J.U., and Stone D., *Nature* **385**, 45 (1997).

20. Blümel, R., and Reinhardt, W.P., *Chaos in Atomic Physics*, Cambridge: Cambridge Uni. Press, 1997.

21. Delos, J.B., Knudson, S.K., and Noid, D.W., *Phys. Rev. A* **30** , 1209 (1984); Friedrich, H., and Wintgen, D., *Phys. Rep.* **183**, 37 (1989).

22. Pourbohloul, B., *Control and Tracking of Chaos in Hamiltonian Systems*, Ph.D. Thesis (Université Laval, May 1999); Pourbohloul, B., and Dubé, L.J., *Control and Tracking in the Diamagnetic Kepler Problem* (submitted to Phys. Rev. Lett.)

23. Schwartz, I.B., Carr T.W., and Triandaff, I., *Chaos* **7** , 64 (1997).

24. Azevedo, A., and Rezende, S.M., *Phys. Rev. Lett.* **66** , 1342 (1991).

25. Singer, J., Wang, Y., and Bau, H., *Phys. Rev. Lett.* **66** , 1123 (1991).

26. Peng, B., Petrov, V., and Showalter, K., *J. of Phys. Chem.* **95**, 4957 (1991).

27. Roy, R., Murphy Jr., T.W., Maier T.D., Gills, Z., and Hunt, E.R., *Phys. Rev. Lett.* **68** , 1259 (1992).

28. Garfinkel, A., Spano, M.L., Ditto, W.L., and Weiss, J.N., *Science* **257**, 1230 (1992).

29. Schiff, S.J., Jerger, K., Duong, D.H., Chang, T., Spano, M.L., and Ditto, W.L., *Nature* **370**, 615 (1994).

30. Goldberger, A.L., in *Applied Chaos*, J. H. Kim and J. Stringer (Eds.), New York: John Wiley & Sons Inc., 1992, pp. 321-331.

31. Freeman, W.J., *Scientific American* **Feb.**, 78 (1991).

Kinematically Complete Ion-Atom Collision Experiments:
Ionization of Atoms in Strong Fields

R. Moshammer[1], M. Schulz[3,1], W. Schmitt[1,2], H. Kollmus[1], R. Mann[2],
B. Bapat[1], S. Hagmann[4], P.D. Fainstein[5], R.E. Olson[3], and J. Ullrich[1]

[1] Universität Freiburg, Hermann-Herder-Str. 3, 79104 Freiburg, Germany
[2] Gesellschaft für Schwerionenforschung, 64220 Darmstadt, Germany
[3] Department of Physics, University of Missouri-Rolla, Rolla, Missouri 65409
[4] Department of Physics, Kansas State University, Manhattan, Kansas 66506-2601
[5] Centro Atomico Bariloche, 8400 Bariloche, Argentina

Abstract. When fast highly charged ions pass a target atom or molecule at distances of a few atomic units electric fields of up to 10^{11} V/cm are acting on the target electrons. Depending on the field strength one or several electrons are released from the target atom by these super-strong electric pulses within the collision time of typically 10^{-18} s. In a series of kinematically complete experiments the dynamics of the ionization process has been studied for single up to triple ionization of atoms. In these experiments the complete many-particle final state in momentum space has been mapped with high resolution.

Introduction

Single and multiple ionization of a target atom is the most simple and therefore most fundamental reaction channel in fast charged particle atom collisions. Many questions are still open concerning the theoretical treatment of this many particle breakup of atoms in the extremely strong and time dependent electric field generated by a fast passing ion. These questions are closely related to the basic and unsolved problem in physics: what is the evolution in time of mutually interacting particles under the action of a time dependent force. Neither the classical equations of motion nor the Schrödinger equation are solvable in closed form for more than two interacting particles. Appropriate numerical methods are not at hand to solve this problem even approximately for many particles interacting via the long-range Coulomb force. As a result, even the most simplest atomic collisions, involving only 3 or 4 particles, are not fully understood. Only for single ionization successful theories have been developed (1,2,3), but their extension to the case of double (multiple) ionization turned out to be extremely complicated. Double ionization is treatable theoretically only for very specific and restricted collision systems (4,5), where first order theories are applicable.

A decisive experimental breakthrough in atomic collision physics has been established during the last few years by the development of advanced recoil ion

CP500, *The Physics of Electronic and Atomic Collisions*, edited by Y. Itikawa, et al.
© 2000 American Institute of Physics 1-56396-777-4/00/$17.00

momentum spectrometers in combination with completely new and extremely efficient electron detection techniques (6). These so called "reaction-microscopes" allow to identify simultaneously the momenta of all reaction products emerging from a single collision with high resolution. Up to 10 electrons in coincidence with the recoiling target ion can be detected. This makes reaction-microscopes the most advanced devices to study atomic collisions and enables kinematically complete experiments on single and multiple ionization reactions. Those experimental data sets provide an ideal test ground for theories and allow for the first time the separation of different mechanisms involved in the ionization process.

In this paper recent experimental data on single, double and triple ionization of Ne and double ionization of He are presented for 3.6 MeV/u Au^{53+} projectiles, corresponding to a very high perturbation of $q/v = 4.4$ (projectile charge to projectile velocity ratio; $v = 12$ a.u.). The dynamics of the many particle continuum after ionization induced by fast highly-charged ions will be discussed. Whereas the validity of first order approaches and the close relationship to single and double photoionization facilitate the theoretical treatment at low perturbations ($q/v < 1$) our understanding of the collision dynamics in the regime of large perturbation is still rudimentary.

Experiment

In an ionizing ion-atom collision the final state is composed of at least 3 particles in the continuum (the electron(s), the projectile, and the recoiling target ion). Hence, a kinematically complete experiment for single ionization requires to measure the momentum vectors of 2 particles. The third momentum vector can then be deduced via momentum conservation. In the experiments described below, the momentum vectors of up to 3 ionized electrons were measured in coincidence with the recoil ion. The experiments were performed at the UNILAC Accelerator of GSI providing a collimated (0.5 mm diameter) and charge state selected beam of 3.6 MeV/u Au^{53+} ions. The projectiles are crossed with an atomic beam from a supersonic gas jet (2.8 mm diameter). The outgoing projectiles are charge state analyzed (no charge change) after the collision chamber and detected by a fast scintillator in coincidence with the target fragments. Electrons and recoil ions created in the collision zone are momentum-analyzed in our combined multi-electron recoil ion spectrometer shown in Fig. 1 (for details see (7)). It consists of two resistive plates 22 cm in length used to generate a weak electric field (typically a few V/cm) extracting electrons and recoil ions along the ion-beam into opposite directions. After traversing a 22 cm long field free drift tube following the extraction region, the recoil ions and electrons are detected by two two-dimensional channel plate detectors with diameters of 50 and 80 mm, respectively. The electric extraction field is not strong enough to guide a sufficiently large fraction of the electrons onto the detector. An additional solenoidal magnetic field generated by Helmholtz-coils (1.5m diameter) confines the electrons transverse motion. This way all electrons with transverse energies below some 100 eV

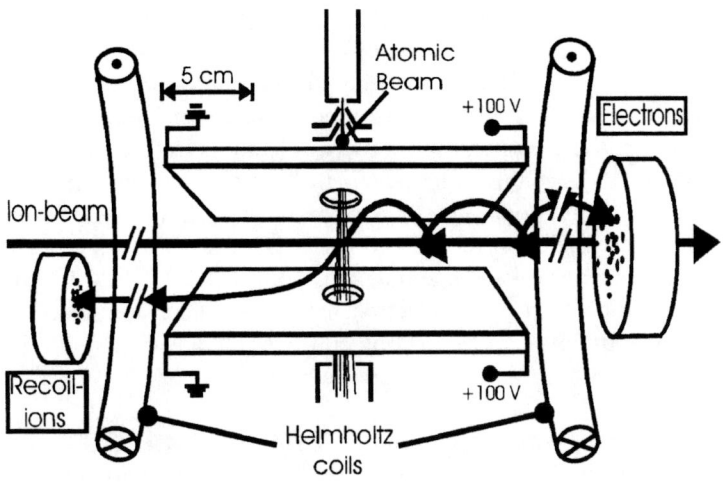

FIGURE 1. Schematic drawing of the combined recoil-ion electron momentum spectrometer.

(depending on the magnetic field strength) and all recoil ions are projected onto position sensitive multi-hit detectors. The multi-hit resolution, i.e. the minimum time delay between two electrons required to identify them as separate particles, is 10 ns. As a result of this deadtime, all events are recorded in which the longitudinal momentum difference between any two electrons is larger than $\Delta p_{12} \approx 0.1$ a.u. The recoil-ion charge state and the full momentum vector of both, recoil-ion and electron, can be calculated from their measured absolute flight times and their positions on the detectors. In the high resolution mode (low electric and magnetic fields) an optimum electron momentum resolution of $\Delta v_{\parallel} = 1\cdot10^{-2}$ a.u. in the longitudinal direction and $\Delta v_{\perp} = 1.4\cdot10^{-2}$ a.u. in the transverse direction has been achieved. This corresponds to an electron energy resolution of $\Delta E_e = 2.5$ meV for electron energies $E_e = 0$ eV. Since for each event the momentum vectors of up to three electrons were recorded the experimental data are kinematically complete for single, double and triple ionization.

Results

Low Energy Electron Emission in pure Single Ionization

Mainly because of large uncertainties of conventional electron spectrometers at energies below a few eV the low-energy part of the electron spectrum is basically unexplored even though it strongly contributes to the total ionization cross section. Hence, little is known about the many-particle dynamics when the final electron energy is small or even zero: Do these electrons recombine with their parent ion? What is the momentum balance between all collision partners? Theories predict the cross section differential in electron velocity to diverge like $d\sigma/dv_e \propto 1/v_e$. This so called electron Cusp peak arises because of phase space arguments with properties depending on the velocity of the projectile and the multi-electron initial state of the target atom. Results of continuum-distorted-wave-eikonal-initial-state (CDW-EIS) calculations predict the emission of low energy electrons in fast ion induced ionization to depend on the details of the target potential (2). These calculations have never been verified experimentally for targets heavier than Helium. Moreover, a reliable experimental test of single ionization theories in the non-perturbative regime (for $q/v > 1$) was missing because it was experimentally impossible to unambiguously separate pure single ionization from multi-electron processes. Using our reaction-microscope we were able to study low-energy electron emission for several targets and for the first time for well defined degrees of ionization with a unprecedented resolution of $\Delta E_e = 2.5\,meV$ at $E_e = 0\,eV$.

In Fig. 2 (a) the Doubly Differential Cross Sections (DDCS) for pure single ionization of Neon by 3.6 MeV/u Au^{53+} are shown in comparison with Continuum Distorted Wave (CDW-EIS) calculations. Good agreement between theory and experiment is observed in shape as well as in absolute magnitude. The emerging

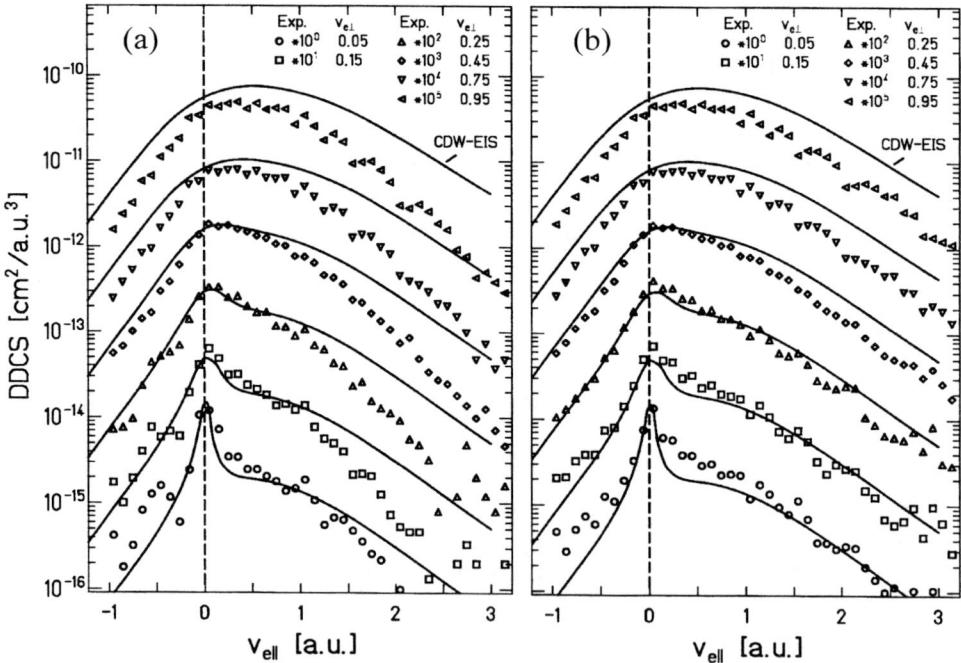

FIGURE 2. Double differential cross section DDCS = $d^2\sigma/(dv_\parallel dv_\perp 2\pi v_\perp)$ as function of the longitudinal electron velocity (parallel to ion beam) for certain transverse velocity cuts in 3.6 MeV/u Au^{53+} on Neon collisions. (a): DDCS for pure single ionization. (b): DDCS with no condition on the recoil-ion charge state (exp. data are divided by 1.5). DDCS at different $v_\perp$ are multiplied by factors of ten, respectively. Lines: CDW-EIS results.

highly-charged projectile pulls the electrons into the forward direction resulting in a distinct forward-backward asymmetry (post collision interaction PCI (3)). Furthermore, a peak, the already mentioned "target-cusp", appears at ultra-low electron energies (8). Systematic discrepancies between experiment and theory are observed at higher electron energies. This is not surprising because CDW-EIS is a "single active electron" theory where the whole impact parameter range only and exclusively contributes to single ionization. But, in particular at small impact parameters, single ionization is competing with double and multiple ionization. These multi-electron processes are important because they contribute considerably to the total ionization cross section. Experimentally, these close collision contributions can approximately be included by taking the sum of the experimental DDCS over several recoil-ion charge states, i.e. by plotting the DDCS irrespective of the degree of ionization (Fig. 2 (b)). This quantity corresponds to the one measured by conventional electron spectroscopy. The considerably improved agreement in shape between experiment and theory over the whole range of electron energies supports our interpretation. It further demonstrates that reliable ionization data are obtainable only when the charge state of the remaining target ion is known in the experiment. Moreover, a comparison of DDCS data from different targets reveals significant and target specific structures attributable to the nodal structure of the initial bound state in momentum space (9). Thus, low energy electrons emitted in fast ion-atom collisions are extremely sensitive to both, the many particle collision dynamics and the multi electron initial state.

Multiple Ionization Dynamics

In recent kinematically complete experiments on ionization of atoms by impact of fast electrons (11), 100MeV/u C^{6+} (4), 1000 MeV/u U^{92+} (10) and ultra-short laser pulses (12) the relationship between charged particle and photon induced ionization has been studied in detail. These experiments uncovered, depending on the projectile properties, in a very clean manner the similarities and the differences between photon and fast ion induced ionization. Under certain conditions the fast charged projectile has been interpreted as an extreme strong and ultra short electromagnetic pulse interacting with the target atom thus revealing the equivalence with intense laser fields.

In some cases, however, it is more descriptive to compare not with photoionization but to consider the interaction with the projectile as it is, namely an almost purely electric interaction. This interpretation is appropriate particularly in the regime of very high perturbations (i.e. at large q/v ratios) or equivalently, at large electric field strengths. Then, impact parameters large compared to the mean radius of the target atom dominantly contribute. This is true not only for single ionization but also for the production of higher charge states. Ionization can be interpreted as a pulling apart of target nucleus and electrons in the strong electric field generated by the passing ion. In particular in fast collisions, where the interaction time is short (smaller than the bound electron revolution time) it can easily be shown that the force is acting dominantly in a direction perpendicular to the initial projectile velocity vector. Since for distant collisions the electric field generated by the projectile is almost the same for the target electron and the target nucleus both particles get the same amount of momentum but into opposite directions because of different sign of charge. Hence, the total momentum transferred to the target fragments is negligibly small. Electrons are pulled towards the side where the projectile is passing and the recoil ion is pushed away. Once one or several electrons are in the continuum the receding projectile still perturbs and influences the created many-particle continuum in a sense that electrons are dragged to the forward and the recoil-ion is pushed to the backward direction (PCI). This is in contrast to photoionization or charged particle induced ionization in the regime of low perturbation where the projectile is either simply not present in the final state or where the influence of the receding projectile on the continuum electrons can be neglected. But, in any case the momentum transferred to the target is very small and the final state momenta of the target fragments bare an additional contribution arising from the initial state momentum distribution.

The data obtained in single and triple ionization of Ne with 3.6 MeV/u Au^{53+} support the above given interpretation. These projectiles represent almost the highest perturbation available at any accelerator. In the collision the target atom is exposed to

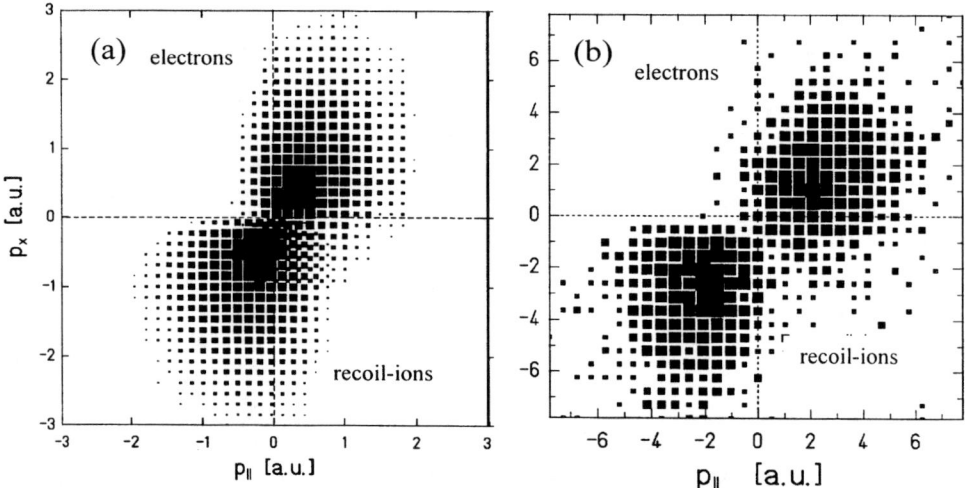

FIGURE 3. Distributions of the electron and the target ion momenta projected onto the collision plane (see text) for 3.6 MeV/u Au^{53+} on Ne single (a) and triple (b) ionization. For triple ionization (b) the electron sum-momentum is shown. Attention should be paid to the different scales.

electric fields of typically 10^{11} V/cm exceeding the inter-atomic field by almost an order of magnitude. In Fig. 3 a collision plane is defined by the initial projectile beam direction (the $p_{\parallel}$-axis) and the outgoing recoil-ion momentum vector ($\mathbf{p}_r = (-p_x, p_{\parallel})$). The recoil-ion and the electron momentum vector (or the electron sum-momentum vector, respectively) are projected onto this collision plane for single and triple ionization of Ne (Fig. 3). All the above discussed features are clearly visible in these spectra: first, the momentum distributions of recoil ions and electrons are very similar in shape and in magnitude. Second, their momentum vectors are pointing into opposite directions indicating a back-to-back emission. In fact they balance each other almost perfectly demonstrating that the momentum transferred by the projectile is very small. Third, there is again a distinct forward-backward asymmetry in the longitudinal direction with the sum momentum vector of the electrons pointing in the forward direction and the recoil momentum vector pointing in the backward direction.

These fairly global features of the multiple ionization dynamics are in excellent agreement with results of purely classical CTMC calculations based on the independent electron model, i.e. neglecting the electron-electron interaction in the initial as well as in the final state (13). But, exactly this contribution has a significant influence on the details of the many-electron continuum. Choosing an appropriate representation the influence of the electron-electron interaction and of the outgoing projectile on the two emitted electrons in double ionization is clearly seen in the experimental data. In fig. 4 the differential cross section as a function of the electrons relative momenta and the angle between both momentum vectors is plotted for double ionization of He by 3.6 MeV/u Au^{53+}. The dominance of small angles is a result of the final state interaction with the outgoing projectile (PCI). On the other hand, the counteracting electron-electron interaction prohibits the occurrence of equal energy electrons emitted into the same direction (Coulomb repulsion).

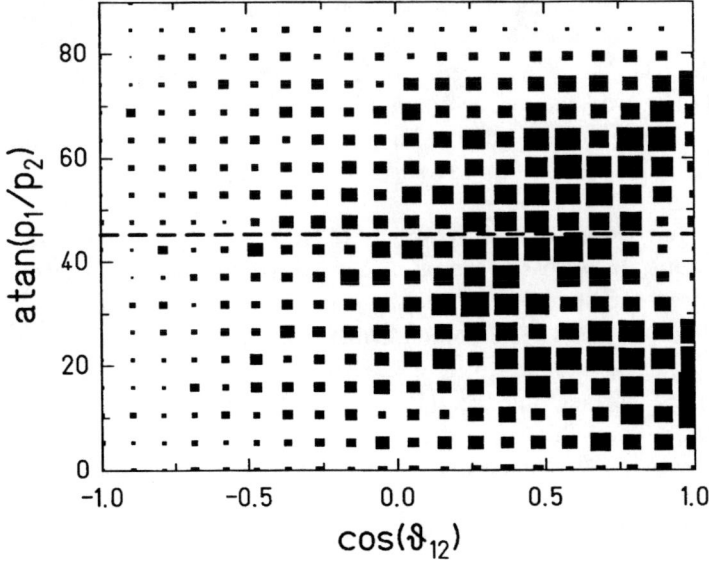

FIGURE 4. Differential cross section for double ionization of He by 3.6MeV/u Au^{53+} plotted as function of the electrons momentum ratio and the total angle between both momentum vectors (hyperspherical coordinates). The dashed line represents the case where both electrons have the same energy.

Electron Correlation

The momentum difference between two ejected electrons is apparently a very sensitive measure of the inter-electronic repulsion. But, as discussed in the preceding section, the many-electron continuum is influenced by both, the electron-electron interaction and the PCI with the outgoing highly-charged ion. As a consequence, the separation and characterization of electron correlation in ion-atom collisions is persistently difficult.

To circumvent these difficulties we introduced a new method of analyzing the two-electron final state momentum distributions after double ionization (14). The method is based on the so called Intensity interferometry, which originally has been invented in astro-physics to determine the sizes of distant stars (15). It is also widely used in nuclear collision physics to analyze effects due to the symmetry of the many-particle wavefunction (16). Intensity interferometry reveals electron correlations very sensitively, but at the same time it is remarkably insensitive to the kinematics and dynamics of the ion-atom collision. It thus enables to analyze electron correlations in a very clean manner independent on the projectile perturbation.

A correlation function is defined based on the electron momentum difference obtained in double ionization normalized to the momentum difference of non-coincident electron pairs. First, the momentum difference $\Delta p = |\mathbf{p_1}\text{-}\mathbf{p_2}|$ is taken from two electrons emitted in the same collision and measured in coincidence. This yields a distribution I_{cor}. Second, exactly the same set of events is used, however, now Δp is determined for two completely independent electrons which were emitted in two independent collisions (event mixing) and which were not measured in coincidence with each other. This leads to a second distribution I_{unc} as function of Δp for unrelated electrons. If the two electrons emitted in double ionization would be completely independent, then I_{cor} and I_{unc} should be identical for all Δp. We therefore define the correlation function $R = I_{cor}/I_{unc} -1$, so that $R = 0$ refers to completely independent electrons.

In Fig. 5 the above defined correlation function is shown for three different collision systems. In all cases a pronounced and similar structure is observed. The negative values near $\Delta p = 0$ indicates the suppression of two electron emission with equal momentum vectors. The common maximum around $\Delta p \approx 2$ a.u. means that these intermediate momentum differences are favored. In any case the data reveal clear signatures of electron correlation. The common shape for very different projectiles shows that this type of representation is surprisingly insensitive to the collision dynamics and to the specific reaction mechanisms contributing to double ionization at different perturbations. Thus, the correlation function represents a quantity ideally suited to study electron correlation in dynamical, time-dependent processes essentially free of complications due to kinematic and dynamic effects like for example the PCI. This is an important aspect because of theoretical difficulties in the simultaneous treatment of both, the many-particle dynamics and electron correlations. For the correlation function, however, it might be sufficient to describe

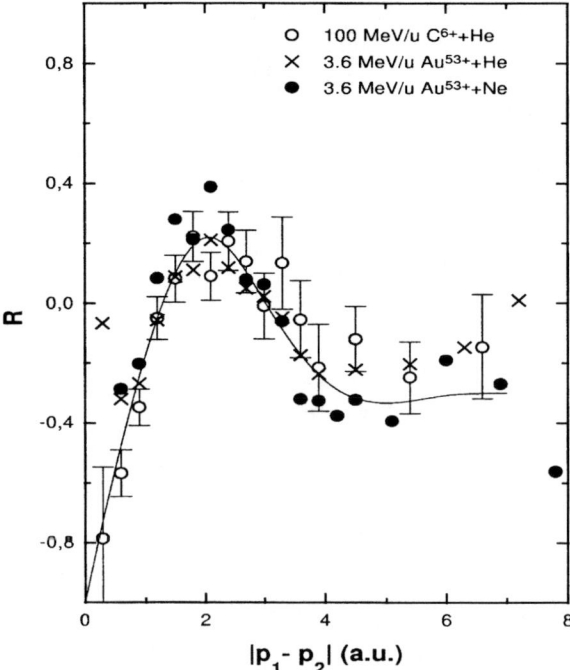

FIGURE 5. Correlation function $R = I_{cor}/I_{unc}-1$ (see text) extracted from double ionization data of different collision systems. The solid line is just to guide the eyes. Error bars are shown only for the C^{6+} data. For the other two collision systems the error bars are generally smaller.

the dynamics with a relatively simple model putting the main efforts in the investigation of the electron-electron interaction in the initially bound and the final continuum state.

Conclusions

Just a few aspects of the manifold information obtained in kinematically complete collision experiments are presented in this paper with special emphasis on multiple ionization. This process is ideally suited to study under very clean and well defined conditions the dynamical behavior of electrons in the two-center potential of target and projectile. Its investigation is fundamental for the understanding of static and dynamic electron correlation in multi electron transitions. Ion-atom collision experiments in connection with upcoming complete experiments of ionization in ultra-short intense laser fields (12) will allow to study in very detail the response of many electron systems on ultra-strong electric pulses covering the time scale from attoseconds to femtoseconds. This will provide a new insight into the dynamical mechanisms of atomic and molecular fragmentation as function of magnitude, shape and duration of intense electric and electromagnetic pulses.

Acknowledgement

We are grateful to GSI for providing excellent beams. We acknowledge financial support from GSI and the Deutsche Forschungsgemeinschaft within the Leibniz-Program and SFB-276.

References

1. S.F.C. O´Rourke et al., J. Phys. B **30**, 2443 (1997).
2. P.D. Fainstein, L. Gulyás, F. Martín, and A. Salin, Phys. Rev. A **53**, 3243 (1996).
3. N. Stolterfoht, R.D. DuBois and R.D. Rivarola, *Electron Emission in Heavy-Ion Atom Collisions*, edited by J. Toennies (Springer, Berlin 1997) Vol. 20.
4. B. Bapat et al. J. Phys. B **32**, 1859(1999).
5. S. Keller, H.J. Lüdde, and R.M. Dreizler, Phys. Rev. A **55**, 4215 (1997).
6. J. Ullrich et al., J. Phys. B **30**, 2917 (1997).
7. R. Moshammer et al., Nucl. Instrum. Methods B **108**, 425 (1996).
8. W. Schmitt et al., Phys. Rev. Lett. **81**, 4337 (1998).
9. R. Moshammer et al., submitted to Phys. Rev. Lett.
10. R. Moshammer et al., Phys. Rev. Lett. **79**, 3621 (1997).
11. A. Dorn et al., Phys. Rev. Lett. **82**, 2496 (1999).
12. R. Dörner et al., submitted to Phys. Rev. Lett.
    R. Moshammer et al., submitted to Phys. Rev. Lett.
13. M. Schulz et al. submitted to J. Phys. B.
    R.E. Olson et al., phys. Rev. A **58**, 270 (1998).
14. M. Schulz et al. submitted to Phys. Rev. Lett.
15. R. Hanbury-Brown and R.Q. Twiss, Nature **178**, 1046 (1956).
16. D.H. Boal, C.K. Gelbke, and B.K. Jennings, Rev. Mod. Phys. **62**, 553 (1990).

Momentum spectroscopy in ion-atom collision

Lokesh C Tribedi[1]

Tata Institute of Fundamental Research, Homi Bhabha Road, Colaba, Mumbai 400 005, India.

Abstract. In this article, we review the recent progress in the field of momuntum spectroscopy in fast-ion atom ionization based on the measurements of electron double differential cross section (DDCS). It is shown that the energy and angular distributions of the low energy electron emission not only provide informations on two center effects but also can be used to study the final-state longitudinal momentum distributions of the electrons, recoil-ions and projectiles. To demonstrate this we have used the recently measured data of low energy electron DDCS for three collision systems: C^{6+} | He/H_2 ($v = 10$) and p+H ($v = 2.14$). The complementary nature of the electron spectroscopy and the recoil-ion momentum spectroscopy (RIMS) have been investigated using a new formulation based on three-body kinematics to explore the ionization dynamics in detail. The separation of the soft and hard collision branches of recoil-ion distributions is a novel feature of the present technique. Atomic units are used unless specified.

INTRODUCTION

The energy and angular distributions of the low energy electrons measured by standard electron spectroscopic techniques have enriched our understanding regarding the ion-atom ionization mechanisms. The distinct features characterizing the ionization process are the soft collision (SE), the electron capture in continuum (ECC) cusp and the binary encounter (BE). The low energy electrons produced in the soft collisions largely dominate the double differential ionization cross section (DDCS) spectrum. The cusp electrons observed at zero degree are produced by electron capture to the continuum and are identified as a cusp-like structure at an electron velocity which matches with the projectile velocity (v) [1]. The third is the broad peak of binary encounter electrons which are elastically scattered target electrons from the projectile nucleus and centered around an electron velocity which is twice the projectile velocity. The high resolution Auger electron spectroscopy has been used to study various phenomena in ion-atom ionization such as, resonant transfer and excitation, inelastic resonant excitation, and electron-electron interactions [2–4]. The energy and angular distributions of the low energy electrons

---

[1] lokesh@tifrc3.tifr.res.in

CP500, *The Physics of Electronic and Atomic Collisions*, edited by Y. Itikawa, et al.

581

in ion-atom collisions have provided important inputs in understanding two-center mechanism of ionization [5–10]. Recently these measurements are also extended for pure three-body collision systems involving fully stripped heavy [9] and light ions [11,12] colliding on atomic hydrogen target. The recent developments in this field has been reviewed by Stolterfoht et al. [13].

Recent studies on ionization using EES and RIMS

In case of ionization by highly charged heavy ions experimental data are scarce and the theoretical methods often find difficulties in describing the final electronic state where the emitted electrons travel in presence of two Coulomb potentials centered at the target and the moving ion. In B1 approximation it is assumed that the projectile potential acts as a perturbation which causes the transition from initially bound to final continuum state. This is also termed as single center approximation. At intermediate to high energies the theoretical method commonly employed, to incorporate the two-center effect, is CDW-EIS (continuum distorted wave-eikonal initial state) approximation [14–16]. In Fig. 1(a) and (b) we present the electron DDCS [9,10] measured for $\theta_e = 15°$, $45°$ for $C^{6+}+H$ ($v = 10$) as a function of electron energy (ε_e). For extreme forward angle a large deviation from the B1 calculation (dotted line in Fig 1(a))) is obvious and CDW-EIS calculation (solid line) although gives a better agreement some deviation still exists. In case of $\theta_e = 60°$ the binary encounter peak is also visible (Fig. 1(b)) which is separated from the soft collision peak. The CDW-EIS provides a beautiful agreement with the data, throughout the energy range. The post collision interaction between the ionized electron and the passing projectile has a large influence for the electrons moving in the extreme forward direction. The deviation from B1 calculations and the two center effect can be visualized by plotting the DDCS ratio R(data divided by B1 prediction) as a function of electron energies as demonstrated by Stolterfoht et al. [6]. In Fig. 1(c) we show R for emission angles 45°, 90° and 135° in ionization of He by bare C ions ($v = 10$) [10,17]. The ratio is larger than one at forward angles and less than one for backward angles, therefore indicating enhancement of electron emission in the forward direction and depletion in the backward direction a signature of two center effect, in heavy ion collision. An example of the e-DDCS for light ions on atomic hydrogen (114 keV p+H) and comparison with CDW-EIS calculation as shown in Fig. 1(d) for $\theta_e=15°$ [11].

While electron spectroscopy has been used extensively to study the ionization mechanism, recoil-ion momentum spectroscopy (RIMS) is relatively new [18-27] (also see the article by R. Moshammer in this volume). The high resolution RIMS technique has enriched our understanding regarding the different inelastic processes like ionization and capture. The separation of the electron-electron interaction from the nuclear-electron interaction in ion-atom ionization has been investigated using this technique [25,26]. Kinematically complete experiments on single ionization [20] double ionization [22] and transfer ionization [21] have been carried out recently

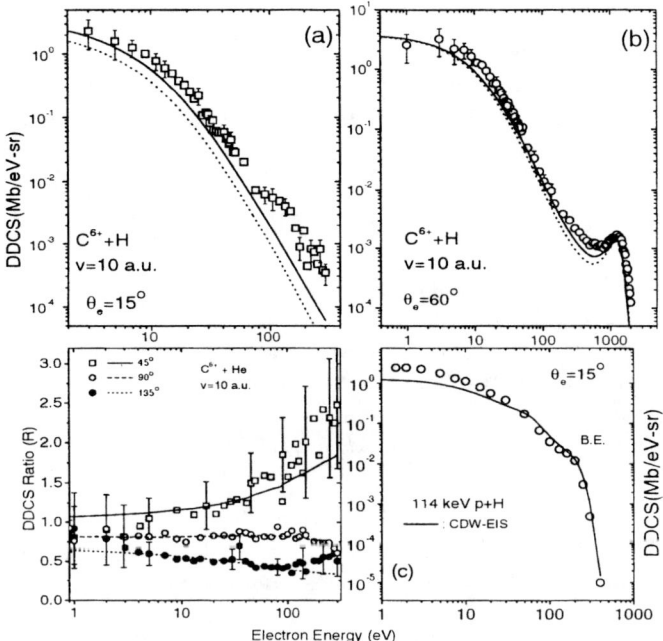

FIGURE 1. The electron DDCS ($d^2\sigma/d\varepsilon_e d\Omega_e$) for $C^{6+} + H$ (v=10) measured at (a) $\theta_e = 15°$ and (b) $\theta_e = 60°$. (c) The ratios of measured electron DDCS to the B1 calculations for three different θ_e for $C^{6+} + He$ (v=10). The lines represent the ratios of CDW-EIS to B1 calculations [10,17]. (d) e-DDCS for 114 keV p+H measured at $\theta_e = 15°$ [11].

using RIMS along with a cold jet target(COLTRIM) [23]. The single differential cross sections ($d\sigma/dp_{R\|}$ and $d\sigma/dp_{e\|}$) are measured which involves the detection of recoil-ions in coincidence with soft electrons (and projectiles) having energy below 50 eV ($p_{R\|}$ and $p_{e\|}$) being the longitudinal momenta of recoil-ion and electron, respectively). The large shifts in the electron and recoil-ion longitudinal momentum distributions observed for highly charged heavy-ion projectiles [20,28] have been interpreted as due to strong post collision interaction (PCI). The PCI strongly depends on the perturbation strength q/v (q being the charge state of the projectile) of the collision and a systematic analysis in terms of CTMC (classical trajectory Monte Carlo) calculations can be found in [28]. Recently, Kravis *et al.* [24] have studied the ejected electron momentum distributions in ionization of He by low velocity proton and C^{6+} ions in order to probe the saddle point mechanism of ionization. The RIMS technique has also been used [29–31] for state selective electron capture studies.

Both the EES (ejected electron spectroscopy) and RIMS have been used to investigate different phenomena of ion-atom collisions. The relationship and complementary nature of these two techniques have been addressed only recently [32–34].

It is shown that some of the aspects which are studied using the RIMS also can be addressed from EES, although the later is not a kinematically complete experiment. The standard electron spectroscopy experiment has an added advantage in that it does not need a cold jet target and electrons with high momenta can be detected easily.

RESULTS AND DISCUSSIONS

Zero degree emission angle

The final-state electron longitudinal and transverse momentum distributions in terms of electron DDCS can be expressed as,

$$\frac{d^2\sigma}{dp_{e\parallel}d\Omega_e} = \frac{|p_{e\parallel}|}{\cos^2\theta_e}\frac{d^2\sigma}{d\varepsilon_e d\Omega_e}. \tag{1}$$

In figure 2(a) we display the DDCS spectrum (in terms of longitudinal momentum) for electrons observed in zero degree in single ionization of He by 1.5 MeV/u F^{9+} [35,32]. Three ionization mechanisms such as SE, ECC, BE can be easily identified.

The key question is to obtain the momentum distributions of recoil-ions from the measured ejected electron spectra. For a kinematically complete experiment, we need to have simultaneous measurement of five out of nine variables that are required to determine the three particles in the final continuum states. The energy-momentum conservation relations can be used to determine the other four. Recently Rodríguez et al. [36] investigated the role of three-body kinematics in understanding the longitudinal recoil-ion momentum distribution. For single ionization in fast-ion atom collision where the residual ion is in ground state, the longitudinal momentum conservation is related to the to the Q value of the reaction through the following equation:

$$-p_{P\parallel} = p_{e\parallel} + p_{R\parallel} = Q/v = (\varepsilon_e - \varepsilon_i)/v \tag{2}$$

where $p_{P\parallel}$ is the longitudinal momentum transfer of the projectile and $|\varepsilon_i|$ is the binding energy of the target atom in the initial state. This relation is valid if the projectile energy loss is very small compered to the initial energy. Equation (2) provides the necessary link to derive the doubly differential recoil-ion longitudinal momentum distributions from electron DDCS. Having obtained the longitudinal recoil-ion momentum, we can calculate $\frac{d^2\sigma}{dp_{R\parallel}d\Omega_e}$ from the following equation,

$$\frac{d^2\sigma}{dp_{R\parallel}d\Omega_e} = |\frac{vp_e}{p_e - v\cos\theta_e}|\frac{d^2\sigma}{d\varepsilon_e d\Omega_e}. \tag{3}$$

The derived doubly differential recoil-ion longitudinal momentum distribution is shown in Fig. 2(b) for $\theta_e = 0°$ in ionization of He by 1.5 MeV/u bare F ions. It

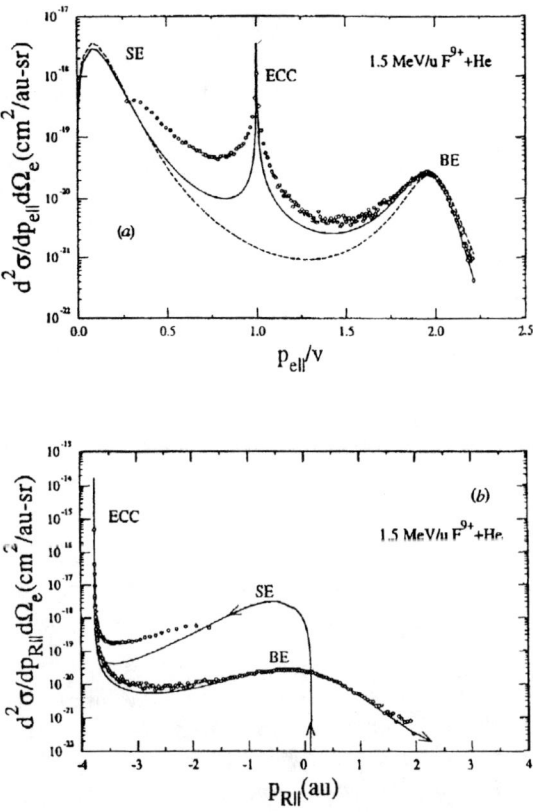

FIGURE 2. (a) Zero-degree doubly differential electron longitudinal momentum distribution for F^{9+}+He.(b) The doubly differential longitudinal recoil-ion momentum distribution derived from the e-DDCS spectrum in (a). (figure taken from [32,35]). The solid and dotted lines are the CDW-EIS and B1 calculations.

may be seen that the spectrum has two branches. To understand the different phenomena associated with these branches we have to understand the solution of Eq. (2) which is displayed in Fig 3 for C^{6+}+He for $v = 10$. The longitudinal recoil-ion momentum ($p_{R\parallel}$) can be determined from Eq.(2), for a given ε_e and θ_e. It may be seen from Fig 3(a) and (b) that, for a given $p_{R\parallel}$, there are two branches of electron energy for θ_e less than 90°. As an example, for $p_{R\parallel} = 0$ and for $\theta_e = 0°$ Eq.(2) gives $p_e = v \pm \sqrt{(v^2 - 2|\varepsilon_i|)}$ which, for $\varepsilon_i \approx 0$, can be further simplified to give $p_e \approx 2v$ and 0 (as shown in Fig. 3(a)). The first solution implies the binary encounter process while the second solution corresponds to very low energy electron emission in a large impact parameter collision resulting in very small recoil-ion momentum in a three-body ionization. The two energy branches corresponding to the electron

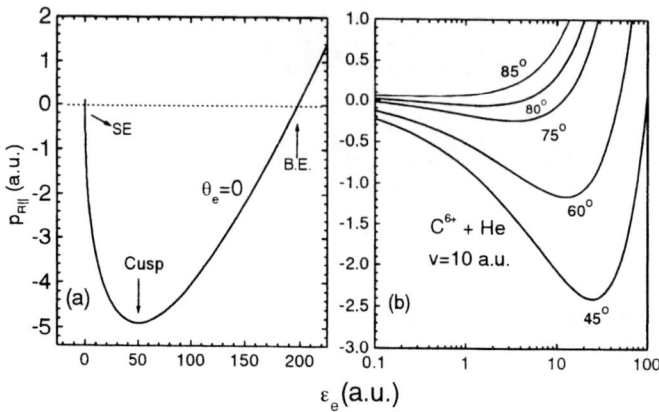

FIGURE 3. The relation between longitudinal momentum of the recoil-ions and the electron energy, for (a) $\theta_e = 0°$; and (b) for other forward angles as obtained from Eq. (2).

energies ε_e^+ and ε_e^- join together at electron capture into the projectile continuum (ECC), $\varepsilon_e^+ = \varepsilon_e^- = \frac{1}{2}v^2$, since $\theta_e = 0$ and $p_e = v$ [36]. The longitudinal recoil-ion momentum acquires an absolute minima at the ECC, $p_{R||}^{min} = -\frac{v}{2} + \frac{|\varepsilon_i|}{v}$, marking the threshold of longitudinal recoil-ion momentum distribution.

We show in this figure the relations for different angles since we will discuss the momentum distributions for non-zero degree electron emission, in next section. It is clearly seen that for $\theta_e < 90°$ two different values of ε_e can have the same longitudinal momentum of the recoil-ions. For backward angles the $p_{R||}$ becomes a single-valued function of ε_e. The $p_{R||}$ is expected to be zero at the BE since it is a two-body collision between the electron and the projectile. Since the BE peak energy ($\varepsilon_{BE} = 4t\cos^2\theta_e$, t being the cusp energy $v^2/2 = 50$ since $v = 10$) becomes smaller for higher emission angles, the turnover point gradually shifts (Fig 2(b)) towards low electron energies for higher emission angles. The curve $p_{R||}(\varepsilon_e, \theta_e)$ becomes a monotonically increasing single-valued function of ε_e for $\theta_e \geq 90°$ (not shown) [34], and no binary encounter peak is observed for back angles.

Now it can be seen from Fig. 2(b) that the recoil ion spectrum $d^2\sigma/dp_{R||}d\Omega_e$ is divided into two branches. The upper branch is for low energy electrons i.e. between $0 \leq \varepsilon_e \leq \frac{1}{2}v^2$ or recoil-ions with longitudinal momenta between $p_{R||}^s$ ($\approx +0.12$) and $p_{R||}^{min}$ ($\approx +0.12$). The two branches join together at $p_e = v\cos\theta_e$ where the distribution has a divergence which arises because of the denominator in Eq.(2). This divergence may not have any physical meaning except for $\theta_e = 0$ for cusp electrons. The recoil-ions are generally having negative longitudinal momentum i.e. moving against the electrons and projectiles except in the region close to $p_{R||}^s$. Around $p_{R||}^s$ the electrons are emitted with nearly zero energy and the recoil ions move in the forward direction. At this point the forward moving recoil ions compensate the momentum transfer (loss) from the projectile necessary to overcome

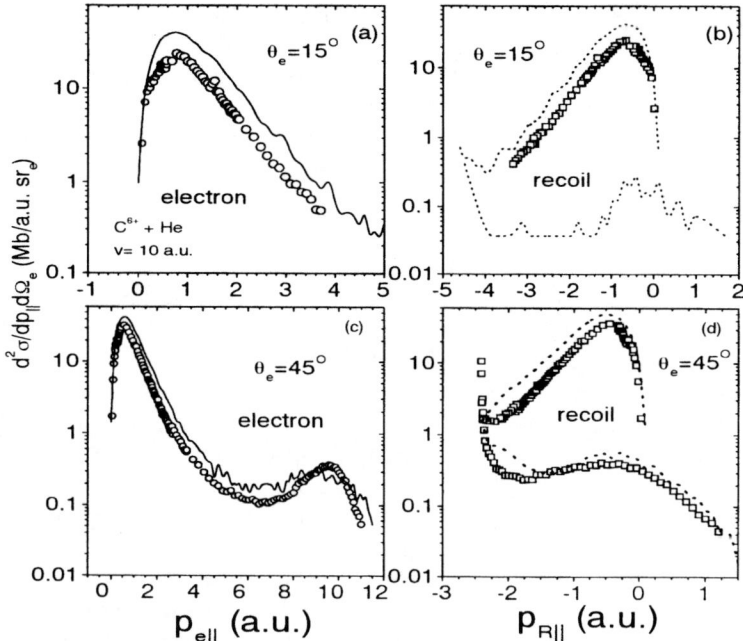

FIGURE 4. The double differential longitudinal momentum distributions of electrons (a), and recoil-ions (b), in collision of 30 MeV C^{6+}+He ($v = 10$) for $\theta_e = 15°$. The similar distributions for $\theta_e = 45°$ are shown in (c) and (d). The lines represent the CTMC calculations.

the target binding. The lower branch is mostly dominated by the peak around $p_{R||} \approx 0$ which correspond to the binary encounter peak in electron spectra.

Results for non-zero degree emission angles

Till now we have discussed only the results based on zero degree electron spectroscopy. Recently there have been several experimental studies on the angular distributions of low energy electron DDCS. In these measurements the electrons with energies between 0.1 eV to about 1 keV have been detected at several angles between 15° and 160°. We give a few examples of the results on the longitudinal recoil-ion momentum distributions in collisions of C^{6+}+H_2/He and p+H. For forward angles ($\theta_e = 15°$ and 45°), the electron longitudinal momentum distributions peak at some positive $p_{e||}$ as expected (Figs 4(a) and 4(c)). For $\theta_e = 45°$ the measured electron distribution shows both peaks corresponding to the soft electrons (around 0.5) as well as to the binary encounter peak (around 10). Accordingly, the recoil-ion distributions for the forward (electron emission) angles peak at negative $p_{R||}$ values ((Figs 4(b) and 4(d)), as can be predicted from Eq. (1). The higher branch for recoil-ions has a peak around $p_{R||} \approx -0.7$ for $\theta_e = 15°$ which has a corresponding electron counterpart at $p_{e||} \approx +0.8$. The recoil-ion distributions clearly have two branches. The higher branch corresponds to the low energy electrons produced in three-body soft collisions (soft electrons). These peaks originate from the lowest energy part (below ~ 10 eV) of the electron DDCS spectrum. Hence the detection of these low energy electrons is important in order to have the complete peak in the recoil-ion momentum distribution. The lower branch has a broad peak around $p_{R||}=0.0$ which corresponds to the binary encounter peak in the electron spectrum. Since this peak essentially is produced in a two-body collision between the electrons and the projectiles, the recoil ions are not expected to have any momentum (i.e. $p_{R||} \approx 0.0$) at the peak of the distributions. The width of the distribution arises due to the Compton profile of the target nucleus, since in the center-of-mass frame of the target system the nucleus has the same Compton profile as the electron. As in the electron DDCS spectrum, the soft electrons, i.e., the upper branch contributes the most to the total ionization cross section. In fact the cross sections for the recoil ions produced in the binary encounter, i.e., in the lower branch, is two to three orders of magnitude smaller compared to the upper branch. Therefore, the present method allow one to distinguish clearly, the separate contributions of the three-body collision and the binary encounter mechanisms of the recoil-ion production. It may be emphasized here that the present method allows one to determine the cross sections of electrons and projectiles with very high momenta that are not easily achieved in RIMS. Similar examples are also shown in Fig. 5 for backward angles ($\theta_e = 135°$ and $160°$) in collision of bare ions with He and H_2. The electron distributions peak at negative value of $p_{e||}$ and accordingly the recoil distributions have peaks at positive momentum.

In order to understand these distributions we have performed CTMC calculations for the present collision system. The electron double differential cross sections have been already compared in detail with the CDW-EIS and B1 calculations (not shown here; see Ref. [10]). However, these calculations use the same transformations to derive the recoil-ion distributions, as used above. The CTMC, on the other hand, does not require any such transformation and hence can provide an independent check on the method we have used. In Fig. 4 we show such comparisons with the data for $\theta_e = 15°$ and $45°$ for C^{6+}+He. It may be noted from Figs. 4(b) and 4(d) that the two branches in the recoil-ion distributions are reproduced in these calculations and in general the qualitative agreement is very good. The two branches in recoil-distributions could be reproduced by the CTMC calculations only after identifying the collisions producing low and high energy ($\varepsilon_e \geq \frac{1}{2}\cos^2\theta_e$) electrons. This was achieved by introducing coincidence conditions on the recoil-ions and electrons in the calculations. The observation of two branches in the recoil-ion longitudinal momentum distribution, and its independent check by the CTMC calculations, is a novel feature of the present technique. For small angles like 15° the calculations overestimate the observed data for electrons and recoils. But for higher angles ($15° \leq \theta_e \leq 90°$) the agreement is much better. Fig. 5(a) and 5(b) show the distributions for backward angles in case of C^{6+} ions on H_2 and He.

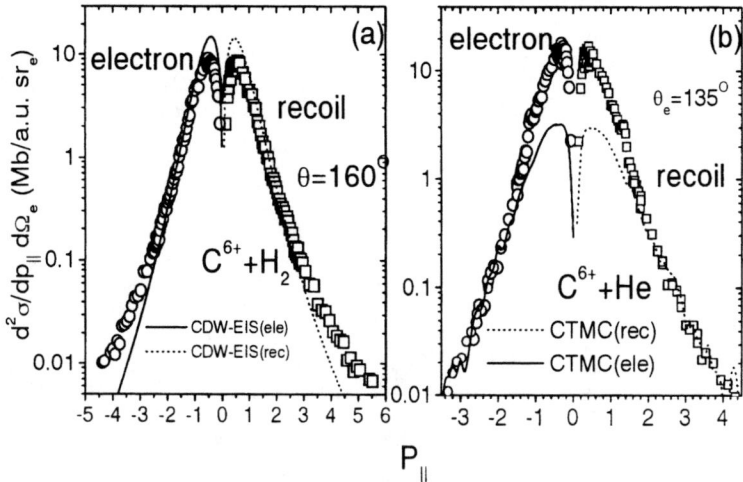

FIGURE 5. (a) The double differential longitudinal momentum distributions of electrons and recoil-ions for backward forward angles. (a) For $C^{6+}+H_2$ ($v = 10$) and $\theta_e = 160°$; lines are the CDW-EIS calculations. (b) For $C^{6+}+He$ ($v = 10$) and $\theta_e = 135°$; lines are the CTMC calculations.

The CDW-EIS reproduces the H_2 data except for high momenta whereas CTMC underestimates the He data except at large momenta.

Contribution due to double ionization was also included in the CTMC calculations. It was found that the total electron spectra contain 5% contribution from the double ionization at small angles, rising to about 15% for the largest backward angle studied (amounting to 30% electron yields). However the discrepancy between the theory and experiment is too large to be explained by the double ionization contribution. Such doubly differential measurements can be carried out by detecting the recoil-ions in coincidence with the ejected electrons emitted in a given direction, using a cold jet target. No such measurements have been reported.

We also show, in Fig. 6(a) and 6(b) the electron and recoil-ion longitudinal momentum distributions for low energy (114 keV) proton colliding on atomic hydrogen target which are derived from the electron DDCS spectrum as shown in Fig.1(d). It may be noticed that CTMC calculations reproduce the cross sections and two branches quite well. The recoil ion distributions are also shown (Fig 6(c)) for several other forward angles i.e. for $\theta_e = 15°$, 30°, 50°, 70° and 90°. It may be noticed that the point of sharp divergence shifts towards positive values of $p_{R\parallel}$ and vanishes for $\theta_e = 90°$. In case of $\theta_e = 70°$ the CDW-EIS calculation is also shown.

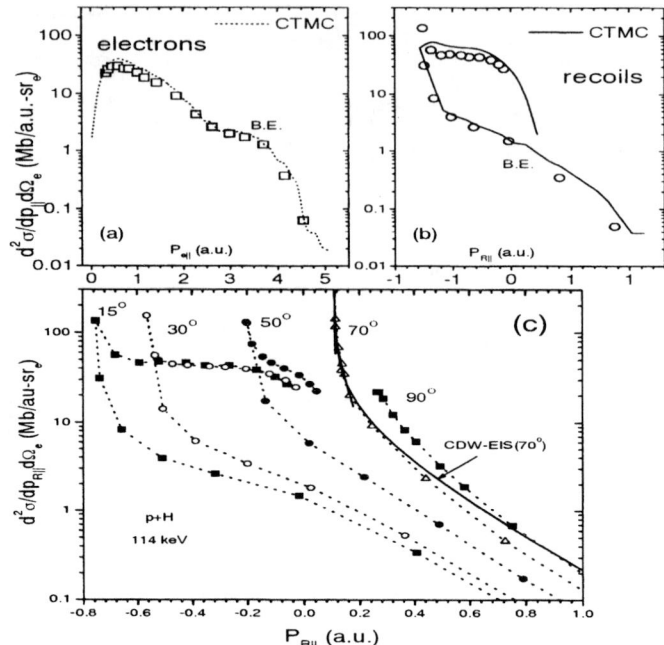

FIGURE 6. The similar distributions of electrons (a), and recoil-ions (b), in collision of 114 keV p+H ($v = 2.14$) for $\theta_e = 15°$ along with CTMC calculations. (c) The same distributions for recoil-ions for different values of θ_e i.e. for 15°, 30°, 50°, 70° and 90° The CDW-EIS calculation is shown for $\theta_e = 70°$ (thick solid line).

Single differential distributions

Single differential distributions, such as $d\sigma/dp_{R\|}$, $d\sigma/dp_{e\|}$, have been derived by performing numerical integration on the corresponding double differential distributions at different angles [34,37]. We show the longitudinal momentum distributions of the electrons and recoil-ions for C^{6+}+He ($v = 10$) (Fig. 7), C^{6+}+H_2 ($v = 10$) and p+H ($v = 2.14$) (Fig. 8). It may be seen from Fig. 7 that the electron distribution is slightly shifted towards positive momentum by ~ 0.1 (which is approximately the same as Q/v (=.0903 for He and .057 for H_2) of the reaction. The recoil-ion longitudinal momentum distribution peaks around 0. Since the forward shift of the electron distribution can be accounted for almost fully by the projectile energy loss for the emission of lowest energy electrons, it may be concluded that the PCI causes a much smaller shift with respect to that observed in the case of Ni ions [20,28]. This is consistent with the fact that we have in the present case a much smaller value of Z/v(= 0.6) and the PCI is expected to increase with increasing Z/v. The improved version of CTMC calculations [28] which, include the model potential based on HFS calculations, show a good agreement with both the distributions

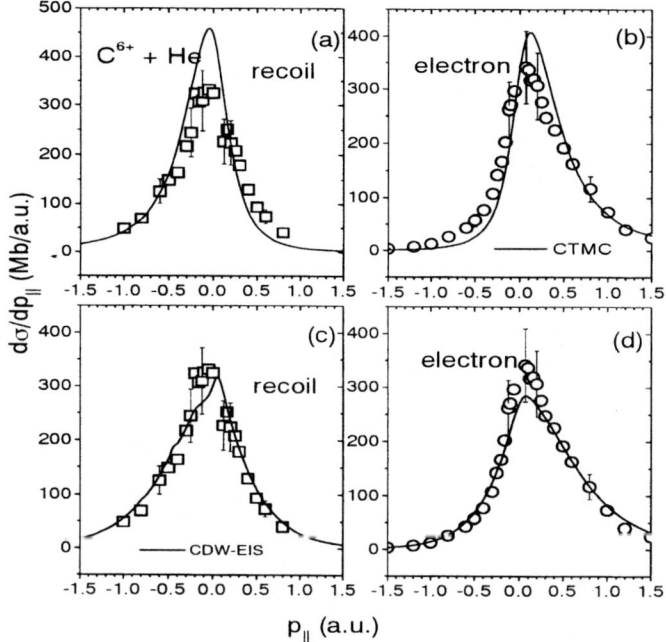

FIGURE 7. The single differential longitudinal momentum distributions of recoil-ions (a) and electrons (b) for . The lines in (a) and (b) represent the CTMC calculations. The same distributions are shown in (c) and (d) along with CDW-EIS calculations [37], with kind permission from The Royal Swedish Academy of Sciences.

(solid lines in Fig. 7(a) and 7(b)). The widths are reproduced well whereas the magnitude of the cross sections are overestimated slightly. However, the CDW-EIS calculations, shown in Figs. 7(c) and 7(d) also provide a very good agreement with the distributions in collision with He as well as for H_2 (Figs. 8(a) and 8(b)). In case of p+H the electron distribution has a large shift (Fig. 8(c)) which is entirely due to the large value of $Q/v(=0.24$ a. u.) for the collision system and the recoil distribution peaks around $p_{R||}=0.0$ (Fig. 8(d)).

The transverse momentum of the recoil-ions is a result of a complicated interplay among the three particles in the final state. Using the present method it is not possible to derive the transverse momentum distributions of the recoil-ions and projectiles since no simple relation among the transverse momenta of the ionization products exists similar to Eq. (1). Extensive measurements on the recoil ion transverse momentum distribution have been carried out by Dörner et al. [27]. However, the single and double differential electron transverse momentum distributions can be derived using present technique, along with the projectile longitudinal momentum distributions and a detail analysis can be found in Ref. [34].

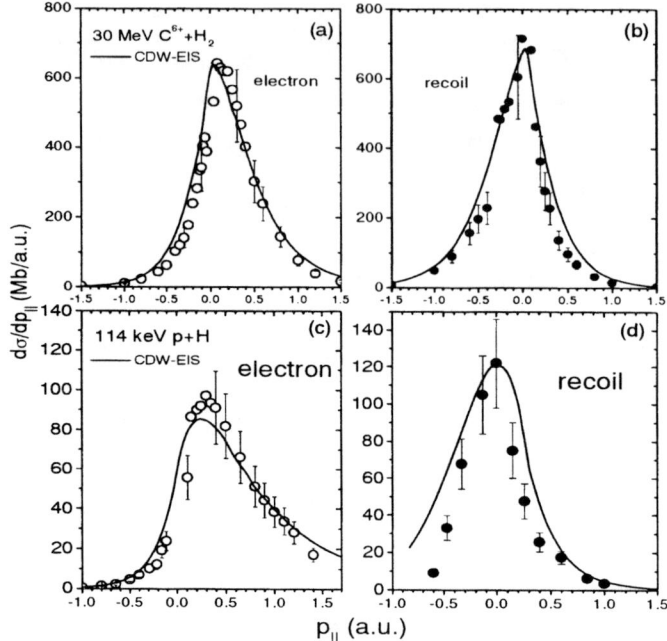

FIGURE 8. Same as in Fig. 7 except for (a) and (b) C^{6+}+He ($v = 10$); (c) and (d) p+H ($v = 2.14$) along with CDW-EIS calculations (Ref. [33].

CONCLUSION

In conclusion, we have shown that the doubly differential longitudinal momentum distributions of the recoil-ions and electrons can be derived from the measured energy and angular distributions of electron DDCS in ion-atom ionization. The kinematical relations are worked out to show that the recoil ion longitudinal momentum distributions have two branches and the separation of these two branches is a novel feature of this method. The examples are given using the electron spectroscopic measurements of ionization of atomic, molecular hydrogen and He in collision with fast heavy ion like bare carbon and light ion such as proton. The complementary nature of electron spectroscopy and the RIMS has been explored in details to show that some of the features of RIMS can be addressed using EES technique. The present method does not require a cold jet target and hence can be used for other collision systems. Many of the existing electron angular distribution data can be reanalyzed to get much more informations regarding the recoil-ion distributions and post collision interactions. The improved CTMC and the CDW-EIS calculations reproduce the distributions extremely well.

ACKNOWLEDGMENTS

I have a great pleasure to thank P. Richard, R.E. Olson, C.D. Lin and L. Gulyás for a fruitful collaboration in this work. Various discussions with C.L. Cocke and M.E. Rudd are gratefully acknowledged. I am thankful to R.E. Olson for providing theoretical support whenever it was necessary. I wish to express sincere thanks to my colleagues at T.I.F.R. for their encouragement and supports.

REFERENCES

1. M.E. Rudd and T. Jorgensen Jr., Phys Rev **131**, 666 (1963); M.E.Rudd, C.A. Sautter and C.L. Bailey, Phys. Rev. **151** 20 (1966).
2. N. Stolterfoht, Phys. Reports **146**, 315 (1987).
3. N. Stolterfoht, J. El. Spectr. **67**, 309 (1994).
4. T.J.M. Zouros, D.H. Lee, P. Richard, Phys. Rev. Lett. **62**, 2261 (1989).
5. S. Suárez, C. Garibotti, W. Meckbach, and G. Bernardi, Phys. Rev. Lett. **70**, 418 (1993).
6. N. Stolterfoht, H. Platten, G. Schiwietz, D. Schneider, L. Gulyás, P.D. Fainstein and A. Salin, Phys. Rev. A**52**, 3796 (1995); Stolterfoht et al., Europhys. Lett. **4**, 899 (1987).
7. J.O.P. Pedersen, P. Hvelplund, A. Petersen and P. Fainstein, J. Phys. B**24**, 4001 (1991).
8. Lokesh C. Tribedi, P. Richard, D. Ling, Y.D. Wang, C.D. Lin, R. Moshammer, G.W. Kerby III, M.W. Gealy and M.E. Rudd, Phys. Rev. A**54**, 2154 (1996).
9. Lokesh C. Tribedi, P.Richard, W. DeHaven, L. Gulyás, M.W. Gealy, and M.E. Rudd, J. Phys. B. (Letters) **31**, L369 (1998).
10. Lokesh C. Tribedi, P.Richard, Y.D. Wang, C.D. Lin, L. Gulyás and M.E. Rudd, Phys. Rev A.**31**, 3619 (1998).
11. M.W. Gealy, G.W. Kerby III, Y.-Y. Hsu, and M.E. Rudd, Phys. Rev. A**51**, 2247 (1995).
12. Y.-Y. Hsu, M.W. Gealy, G.W. Kerby III, and M.E. Rudd, Phys. Rev. A **53**, 297 (1996);
13. N. Stolterfoht, R.D. DuBois and R.D. Riverola, 'Electron emission in heavy-ion atom collision', Springer Series on Atoms and Plasmas, 1997.
14. D.S.F. Crothers and J.F. McCann, J. Phys. B**16**, 3229 (1983).
15. P.D. Fainstein, V.H. Ponce and R.D. Rivarola, J. Phys. B**21**, 287 (1988); P.D. Fainstein, V.H. Ponce and R.D. Rivarola, J. Phys. B**24**, 3091 (1991).
16. L. Gulyás, P.D. Fainstein and A. Salin, J. Phys. B**28**, 245 (1995).
17. Lokesh C Tribedi, to appear in 'Trends in Atomic and Molecular Physics', Plennum Press, New York (1999).
18. C.L. Cocke and R.E. Olson, Phys. Rep. **205**, 153 (1991).
19. R. Ali, V. Frohne, C.L. Cocke, M. Stöckli and M. Raphaelian, Phys. Rev. Lett. **69**, 2491 (1992).
20. R. Moshammer et al., Phys. Rev. Lett. **73**, 3371 (1994).

21. V. Mergel et al., Phys. Rev. Lett. **79**, 387 (1997).
22. R. Moshammer et al., Phys. Rev. Lett. **77**, 1242 (1996)
23. J. Ullrich et al., Comments At. Mol. Phys. **30**, 285 (1994).
24. S.D. Kravis et al., Phys. Rev. A **54**, 1394 (1996).
25. W. Wu et al., Phys. Rev. Lett. **72**, 3170 (1994).
26. R. Dörner et al., Phys. Rev. Lett. **72**, 3166 (1994).
27. R. Dörner, V. Mergel, L. Zhaoyuan, J. Ulrich, L. Spielberger, R. E. Olson and H. Schmidt-Böcking, J. Phys. B **28**, 435 (1995).
28. R.E. Olson et al., Phys. Rev. A **58** 270, (1998)
29. V. Mergel et al., Phys. Rev. Lett. **74**, 2200 (1995).
30. A. Cassimi et al., Phys. Rev. Lett. **76**, 3679 (1996).
31. T. Kambara et al., J. Phys B**30**, 1251 (1997).
32. Y.D. Wang, Lokesh C. Tribedi, P. Richard, C.L. Cocke, V.D. Rodríguez, and C.D. Lin, J. Phys. B **29**, L203 (1996).
33. Lokesh C. Tribedi, P. Richard, Y. D. Wang, C. D. Lin and R. E. Olson, Phys. Rev. Lett. **77** 3767 (1996).
34. Lokesh C. Tribedi, P. Richard, Y. D. Wang, C. D. Lin and R. E. Olson and L. Gulyás, Phys. Rev. A **58** 3626 (1998).
35. D.H. Lee, P. Richard, T.J.M. Zouros, J.M. Sanders, J.L. Shinpaugh and H. Hidmi, Phys. Rev. A **41**, 4816 (1990).
36. V.D. Rodríguez, Y.D. Wang and C.D. Lin, Phys. Rev. A **52**, R11 (1995).
37. L.C. Tribedi, P. Richard, C.D. Lin, R.E. Olson and L. Gulyás, Physica Scripta **T80**, 333 (1999).

Impact-Parameter Dependence of the Electronic Energy Loss

P.L. Grande[*] and G. Schiwietz[†]

[*] *Instituto de Física da Universidade Federal do Rio Grande do Sul,*
Avenida Bento Goncalves 9500, 91501-970, Porto Alegre, Brazil
[†] *Bereich F, Hahn-Meitner-Institut Berlin,*
Glienicker Str. 100, D-14109 Berlin, Germany

Abstract.
    In this work, we discuss the application of the coupled-channel method for the calculation of the electronic energy loss of ions in matter. Special emphasis will be given to the formulation of simple models that account for the basic energy-loss processes without being a large-scale calculation. The results are compared to experimental channeling data as well as to other current energy-loss models.

INTRODUCTION

    The electronic energy loss, which is related to the first moment of the single differential cross section $d\sigma/d\epsilon$, has been studied for many years because of its direct application in problems concerning material damage and ion beam analysis. The theoretical treatment of the energy loss in atomic collisions has been greatly improved over the last decades and relies on an accurate treatment of target-continuum states up to high emitted electron energies. The numerical solutions require much more continuum partial-waves as are, e.g., necessary for the computation of total ionization cross sections. Calculations of the electronic energy loss have been performed by using traditional methods known from atomic physics investigations such as the plane wave Born approximation (PWBA) [1,2], the high-energy solution by Bethe [3] and the semi-classical approximation (SCA) [4]. More advanced models are the continuum distorted wave (CDW-EIS) [5], the classical trajectory Monte Carlo (CTMC) [6,7] and finally the atomic orbital coupled-channel method (AO) [8–10] that yields reliable values of the impact-parameter dependent electronic energy loss. The methods based on atomic physics calculations offer reliable ways to obtain detailed information on the energy-loss processes in gases as well as for the inner-shell electrons in solids. Of course, other approaches have to be adopted for conduction-band electrons of solid-state targets [11–13] in order to obtain an accurate description of the energy loss due to the valence electrons.

CP500, *The Physics of Electronic and Atomic Collisions*, edited by Y. Itikawa, et al.
© 2000 American Institute of Physics 1-56396-777-4/00/$17.00

During the last years we have investigated the electronic energy loss of bare and screened ions for light targets using the coupled-channel method. This first principle calculation [8–10], based on an expansion of the time dependent electronic wave function in terms of atomic orbitals, has been applied to evaluate the impact-parameter and angular dependence of the electronic energy loss and the total stopping cross-section of ions (antiprotons, H and He) colliding with H and He atoms at energies of 1 to 500 keV/amu. It has also been applied successfully to calculate the entrance angle dependence of the stopping power for He ions channeling along the Si main crystal directions [14].

These benchmark calculations are also used to check simplified models that account for the basic energy loss processes without the need of large scale calculations [15,16]. In particular, simple models for the impact parameter dependent energy loss are needed to be included in computer simulation codes as well as for channeling data analysis. Here we describe a model for fast ions that yields accurate results for the electronic energy loss from small to large impact parameters. The physical inputs are the electron density and oscillators strengths of the atoms. If not indicated otherwise, atomic units ($e = m = \hbar = 1$) will be used throughout the paper.

THE COUPLED-CHANNEL METHOD

Here we will focus the attention on atomic treatments of the energy transfer process. Thus we will not consider solid-state effects such as intra-band transitions, collective excitations (bulk and surface plasmons) and the corresponding dynamic projectile screening. In a full quantum mechanical description, the ion-atom collision process is described by the many-body Schrödinger equation. For incident energies above a few eV/u the motion of the heavy particles may be described by classical nuclear trajectories $\vec{R}(t)$. Then the electronic system obeys the time dependent Schroedinger equation

$$\mathcal{H}_e(\{\vec{R}(t)\})\Phi_e(\{\vec{r}\}, t) = i\frac{\partial}{\partial t}\Phi_e(\{\vec{r}\}, t) \tag{1}$$

where $\{\vec{r}\}$ represents all electronic coordinates and $\{\vec{R}(t)\}$ is the set of projectile and target nuclear coordinates. The nuclear trajectories may be obtained dynamically for each impact parameter b and each time t through the classical Hamilton equations for an averaged heavy-particle Hamiltonian [8] or, as in most current theories, they are simply replaced by straight lines.

The electronic many-body Hamiltonian in Eq.(1) is treated in the framework of the independent-electron frozen-core model. This means that there is only one active electron, whereas the other electrons are passive (no dynamic correlation is accounted for and no relaxation occurs). In this model the electron-electron interaction is replaced by an initial-state Hartree-Fock- Slater potential [17]. Moreover, the independent-electron approximation allows to distinguish target electrons and projectile-centered electrons which screen the projectile nuclear charge and in this way, target and projectile ionization/excitations can be calculated easily.

The time-dependent Schrödinger equation may be solved by expanding $\Phi_e(\{\vec{r}\}, t)$ in terms of eigenfunctions ϕ_i of the unperturbed target and projectile systems [8]. Thus, Eq.(1) is replaced by a set of coupled first-order differential equations, the so-called coupled-channel equations, containing matrix elements of the projectile and target potentials as well as the overlap between projectile and target centered states. For bare incident ions, the active-electron projectile interaction is just the Coulomb potential. However, in the case of projectiles carrying electrons, we use a screened potential made up of the Coulomb part due to the projectile-nuclear charge and the static potential produced by the target electrons that screen the projectile-nuclear charge [10]. Dielectronic processes [18] involving bound projectile and target electrons are not included in the present calculations. The coupled-channel equations are then solved numerically in order to compute all important transition probabilities for the active electron. The average electronic energy loss $Q(b)$ at a given impact parameter b is evaluated from the calculated probabilities times the corresponding energy transfer.

The results of the coupled-channel method results agree with the predictions of first order perturbation theory (SCA) in the case of a small perturbation (fast projectiles, large impact parameters or small projectile charges). Thus, the advantages of coupled channel calculations compared to first order theories should show up especially at intermediate incident energies and for small impact parameters. In contrast to other coupled-channel calculations we do not use pseudo states to represent the electron continuum wave functions. Instead we use a large number of continuum wave-packets that are composed of a superposition of exact continuum eigenstates (up to 500 gerade target-centered states with partial waves up to l=10 and 10 projectile-centered states), since the computation of the stopping power demands high accuracy of the emitted electron energy spectrum. The details of the present calculations such as the numerical treatment of continuum states, adopted basis set, projectile screening may be found elsewhere [8–10,19].

Fig.1 shows the comparison of the present coupled-channel results (solid curve) with other calculations for the total ionization cross-section as well as for the electronic stoping power S_e of antiprotons on H. The dashed curve represents also coupled-channel calculations [20] using a large number of pseudo states. Both coupled-channel calculations provide similar results and are in rather good agreement with recent measurements [21](symbols). Also displayed are results of first-order Born (PWBA) and the continuum distorted wave (CDW-EIS) model. As it can be see from this figure, higher-order effects become very important at low projectile energies. The PWBA calculations yield too large values of the electronic energy loss, since for antiprotons the polarization effect leads to a reduced electronic density along the ion path. In the Continuum-Distorted Wave Eikonal-Initial-State (CDW-EIS) model [5] (solid line) the initial and final states partially include the effect of the projectile potential (approximate two-center initial and final states) and as a consequence the results at intermediate to high energies are significantly improved. However, a breakdown of this model is observed for energies below 70 keV. This is attributed to the incomplete treatment of the two-center effects and to the neglect of higher-order residual projectile-target interactions in the CDW-EIS model. Furthermore, at low energies it is not able to describe the Fermi-Teller effect responsible for the slow decrease of the stopping power as a function of the projectile energy. The curve denoted by AI provides a simple model for this adiabatic ionization [22]. In this model the adiabatic potential curves for the electronic states in the field of the quasidipole formed by p and $\bar{p}$ are taken into account and a very good agreement with AO results is observed for low energies.

The higher-order effects are visualized in Fig. 2 in a contour plot of the time-dependent electron density of a hydrogen atom disturbed by a positively (displayed on the left) and negatively (displayed on the right) charged particle at 10keV for an impact parameter of 1 a.u. These electronic densities correspond to a cut in the collision plane and were obtained directly from the solution of the coupled-

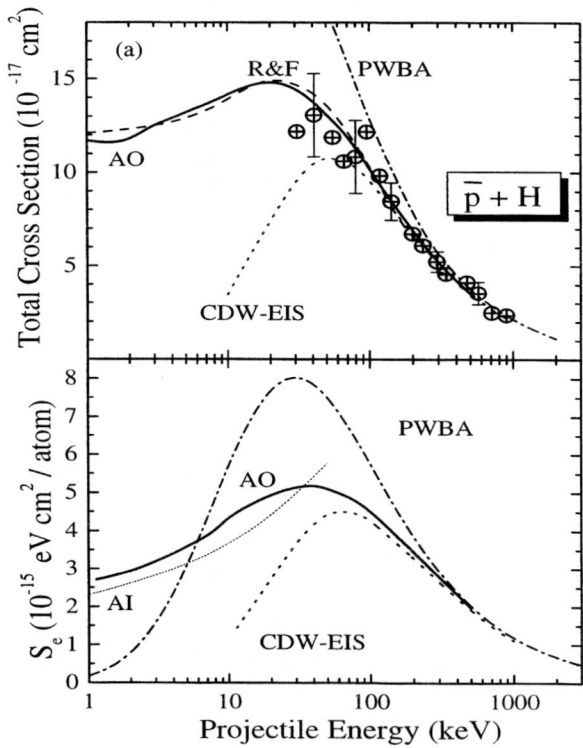

FIGURE 1. Total ionization cross-section and stopping power for antiprotons on hydrogen atoms.

channel equations using about 200 gerade target-centered states. An inspection of this figure shows several interesting features. First, the positively charged particle (proton) attracts the electron on the incoming path; the so-called polarization process. One may see that the electron density moves towards the projectile. The opposite effect takes place for the antiproton. Second, for protons at the distance of closest approach, the maximum of the electron-density points to the backward direction at an angle of about 120 degrees with respect to the beam axis. It is clearly visible that the electron density lies behind the projectile, although the proton is attracting the electron. The reason for this delayed response of the electron cloud is the inertia due to the electron mass. Third, the proton enables electron-capture in the outgoing path of the collision and a large fraction of the electron density is finally bound to and moving with the projectile. Since an antiproton repels the target-electron, the electron density near the projectile on the outgoing path of the collision is almost zero. Futhermore, for collisions of antiprotons with atomic hydrogen, the dipolar antiproton-proton system does not support bound states for inter-particle distances below 0.64 a.u. [22]. For finite velocities and larger impact parameters b (in the figure, b = 1) there is still a significant ionization contribution. As can be observed in the figure at the distance of closest approach there is a high

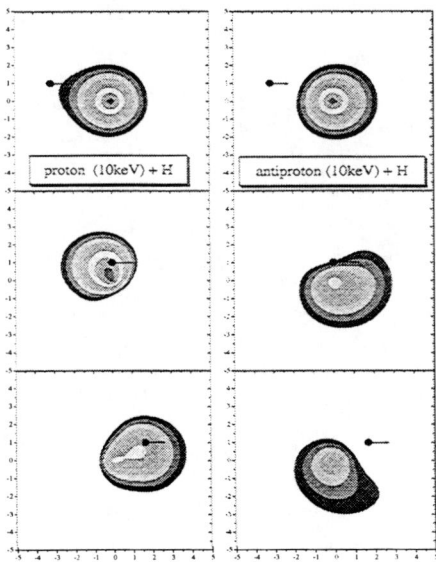

FIGURE 2. Contour plot of the time-dependent electronic density of a hydrogen atom disturbed by a 10 keV proton (on the left) and antiproton (on the right) at b =1.

transition probability (blowing up of the density).

COMPARISON WITH MEASUREMENTS

The first coupled-channel calculations for total and differential energy losses were performed for very simple systems such as H on H, He [9,10,23]. Later theses calculations have been extended for more complex systems such as He channeling along the main directions of Si. Good agreement with experimental data has been found and the remaining discrepancies have been attributed to multielectron processes. Here, we shortly describe the comparison with experimental channeling energy loss of He ions in Si [14,24].

Coupled-channel calculations were performed for He^+ and He^{2+} on the Si inner shells. The energy-loss term associated with the Si-valence electrons was obtained from the experimental stopping cross-section of ref. [25] by subtracting the calculated contributions involving Si inner-shell and He electrons. The He charge-state distribution was determined from experimental results under channeling conditions from ref. [26] (see insert in Fig 3) and the sum of the energy loss for each Si atom located across the channel was averaged according to the ion flux distribution [14]. Further details of the present energy-loss model may be found in ref [14].

Fig. 3 shows the stopping power of He ions moving through the Si crystal in the < 100 > channeling direction. The symbols correspond to recent experimental

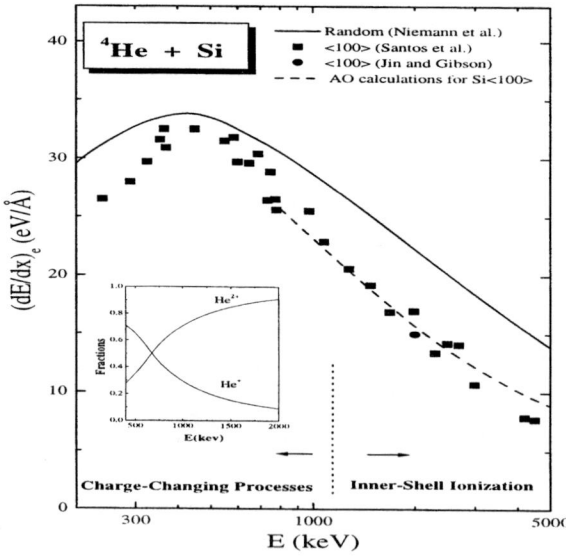

FIGURE 3. Electronic stopping power as a function of the ^{4}He energy for the < 100 > Si direction.

data [14,27] for the channeled energy-loss and the solid line represents accurate experimental stopping values for a random direction [25]. Experiments at 800 keV with He^+ and He^{2+} ions show that charge equilibrium is reached at a depth of about 50 Å. We expect this distance to increase by an order of magnitude for 5 MeV He ions. Since the mean charge state of fast ions is close to 2 and the measurements above 1000 keV were performed with He^{2+} ions there should be no significant deviation from the assumed equilibrium charge-distribution.

The results of the AO calculations (dashed-line) for the projectile-energy dependence of the electronic stopping power under channeling conditions agree with the data to within the experimental uncertainty. For ion energies above 1.2 MeV (see insert in Fig. 3 for the He charge-state fractions) , the He^{2+} fraction is dominant and the main physical process responsible for the reduction of the energy loss under channeling conditions is the suppressed inner-shell ionization (L-shell) of Si atoms. The energy region from 1.2 MeV up to 5 MeV is close to the maximum of the stopping cross section due to Si L-shell electrons and only non-perturbative calculations (including many higher-order terms) are reliable in this energy region. By comparing the AO results with first-order ones at 2 MeV we obtain a difference of about 40% for $b = 1.3$Å (middle of $< 100 >$ channel). For energies below 1.2 MeV, the influence of charge-changing processes begins to be significant. The present energy-loss results as a function of the projectile energy are most sensitive to the computation of the inner-shell contribution at random directions, since under channeling conditions they are determined by the contribution of the valence excitations. The inner-shell contribution under channeling condition is suppressed by 75% at 5MeV). Thus, a comparison with the angle dependent energy-loss data provides more information about the on the impact-parameter dependence of the energy loss [14].

SIMPLE MODELS FOR THE ENERGY LOSS

The coupled-channel calculations are used to check simple models of the impact parameter dependence of the electronic energy loss. A detailed description of such models (convolution approximation) may be found elsewhere [15,16]. Here we present only a short outline of the method. The electronic energy loss involves a sum over all final target states for each impact parameter. Usually this demands an computational effort that precludes its direct calculation in a computer simulation code. Therefore, we search for an approximate solution without the necessity of performing a large-scale calculation.

In recent works [15,28] we have proposed a simple formula for $Q(b)$ (called PCA) that virtually reproduces SCA results for all impact parameters for bare and screened projectiles.

The following simple formula

$$Q(b) = \int d^2r_T \, \mathcal{K}(\vec{b} - \vec{r}_T) \int dz \, \rho(\vec{r}_T, z) \qquad (2)$$

with

$$\mathcal{K}(b) = \frac{2Z^2}{v^2b^2} \times h(2vb/\eta) \times \sum_i f_i \, g\left(\frac{\omega_i b}{v}\right) \qquad (3)$$

joins smoothly all regions of impact parameters b for which two-body ion-electron (small b) and dipole (large b) approximations can be used. The function $h(2vb)$ (see ref. [15]) approaches zero for $b \ll 1/v$ (relative impact parameter smaller than the electron de Broglie wavelength in the projectile frame) and it reaches 1 for large values of b. The first two terms in Eq.(3) resemble the classical energy transfer to a statistical distribution of electrons at rest and describe violent binary collisions. The last term, involving the g function (see ref. [15]) and the oscillator strengths f_i, accounts for the long ranged dipole transitions. The first integral $\int d^2r_T \dots$ in Eq. (2) describes a convolution with the initial electron density also outside the projectile path and yields nonlocal contributions to the energy loss. It

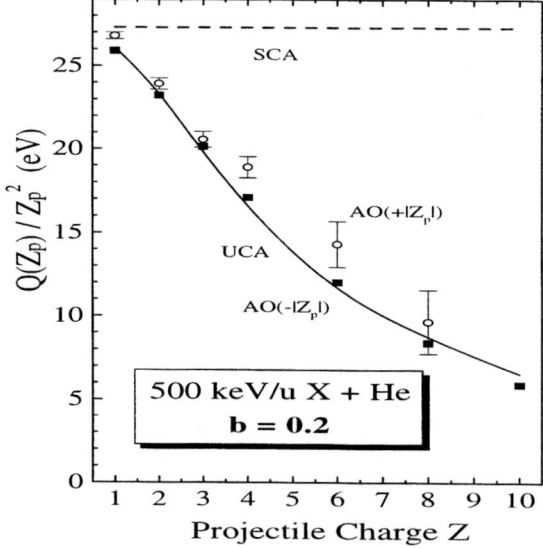

FIGURE 4. Non-perturbative results for the energy loss at a small impact parameter in 500 keV/u X^{Z_p+} + He collisions, compared to the values from first-order perturbation theory (SCA, dashed line). Atomic orbital (AO) coupled-channel results for positively charged particles (open circles) and for anti-nuclei (closed squares). Present results using the UCA model: solid curve.

602

is noted that these nonlocal contributions are neglected in most previous simple energy loss models. With the parameter η equal to one, this formula mimics the first-order Born approximation very well [15] and it is denoted *PCA* (perturbative convolution approximation). For increasing projectile-charge first order calculations (on which PCA is based) break down. They do not take in account, for instance, that each electronic transition gives rise to an increased final-state population and a corresponding reduction of the initial state population. It is clear that the ionization probability cannot increase indefinitely with the strength of the perturbation (the so-called saturation effect). Since these ionization processes come mostly from small impact parameters, we have introduced in ref. [16], a scaling parameter η in the function h that enforces unitarity in accordance with the Bloch model [29].

Fig. 4 displays calculated scaled energy losses (Q/Z_p^2) as a function of the projectile charge Z_p for a small impact parameter ($b = 0.2$ a.u.) compared to the He 1s-shell radius ($r_o = 0.6$ a.u.). The SCA results show up as a horizontal dashed line, since they scale with Z_p^2. The AO results for positively (open circles) and negatively charged projectiles (solid squares) are shown separately in this plot. The error bars of the AO results for positive bare ions are estimated from the numerical convergency and integration properties and are mainly related to the accuracy of the capture matrix-elements. The uncertainties for the antiparticle energy losses are only 3%, since a large basis set of target-centered states is sufficient for accurate AO calculations. AO calculations for positive ions at 500 keV/u were performed with an explicit consideration of 10 bound projectile states, for an improved treatment of electron capture, in addition to 210 target states. It is clearly evident from this figure that the deviations between results for heavy particles and antiparticles is much smaller than the deviation from the SCA. Thus, the even orders of an Z_p expansion, as included in the UCA, dominate the non-perturbative effects. The present UCA results are plotted as a solid curve. This curve lies close to the average of the AO results for particles and antiparticles. Hence, although the present UCA does not include sign-of-charge effects it perfectly describes the majority of the energy transfer processes (dominated by ionization) of fast heavy particles at small impact parameters.

CONCLUSIONS

We have described calculations of the electronic energy loss using the coupled-channel (AO) method with atomic orbitals and continuum wave packets. This non-perturbative method provides full information of each basic single-electron mechanism leading to an energy loss of bare and screened particles as a function of the impact parameter.

Through the coupled-channel method AO we have calculated the electronic energy loss under channeling conditions for He slowing down into Si, a system of interest for ion beam analysis. A full analysis of the channeled stopping power as a function of the projectile energy was performed by considering the impact-parameter-dependent energy loss, projectile-charge-state and ion-flux distribution. The independent particle model was used to obtain the total energy loss. Near the stopping power maximum, the stopping power under channeling conditions is reduced basically due to projectile electron loss and capture processes from Si inner shell electrons. The contribution of the Si inner-shell ionization to the total stopping power becomes larger for increasing projectile energies at a random direction. Since the L-shell ionization is almost suppressed under channeling conditions, the ratio (channel to random) decreases at large energies. Very good agreement with the existing experimental data has been achieved without using any free parameters or further approximations.

Hence, for atoms, insulators or inner shells of conductors accurate stopping cross sections may be computed (including excitation, ionization and electron capture) using the atomic-orbital (AO) coupled-channel method. This is a time-consuming task, since it has to be done for each subshell and each projectile charge-state separately. For slow heavy particles, however, electron-exchange and dynamic mean-field effects will be important and have to be incorporated in the treatment.

We have also devised a simple perturbative convolution approximation that is in accordance with the exact solution within first-order perturbation theory at intermediate to high projectile velocities. This model can be extended to take into account non-perturbative effects using Bloch's stopping power derivation [16]. The main improvement over perturbation theory is the conservation of unitarity, which restricts the sum over all reaction probabilities to a maximum of 100%. The resulting unitary convolution approximation (UCA) includes the main features of heavy-ion stopping, as is shown by comparison with large-scale quantum-mechanical AO coupled-channel results for the impact-parameter dependent electronic energy transfer. Finally, the computation of the energy loss within the UCA is much simpler and by many orders of magnitude faster than the full numerical solution of the time-dependent Schrödinger equation.

ACKNOWLEDGMENTS

This work was partially supported by CAPES (Coordenação de Aperfeioamento de Pessoal de Nível Superior) and by the Alexander-von-Humboldt foundation.

REFERENCES

1. D.R.Bates and G.Griffing, Proc.Phys.Soc. **A66**, 961 (1953).

2. E.J. McGuire, Phys. Rev. **A3**, 267 (1971); ibid **A28**, 2096 (1983);ibid **A 57**, 2758 (1998).

3. H.Bethe, Ann. Phys. **5**, 325 (1930).

4. N.M.Kabachnik,V.N.Kondratev and O.V.Chumanova, Phys. Stat. Sol(b) **145**, 103 (1988).

5. P.D. Fainstein, V.H.Ponce and A.E.Martinez, Phys.Rev. **A47**, 3055 (1993).

6. R.E. Olson, Rad. Eff. and Deff. in Solids **110**, 1 (1989).

7. P.L. Grande and G. Schiwietz, J. Phys. B: At. Mol. Opt. Phys. **28**, 425 (1995).

8. G. Schiwietz, Phys. Rev. **A42**, 296 (1990).

9. P.L. Grande and G. Schiwietz, Phys. Rev. **A44**, 2984 (1991).

10. G. Schiwietz and P.L. Grande, Nucl. Instr. and Meth. **B69**, 10 (1992); P.L.Grande and G. Schiwietz, Phys. Rev. **A47**,1119 (1993).

11. I. Campillo, J. M. Pitarke, and A. G. Eguiluz, Phys. Rev. **B 58**, 10307 (1998).

12. J.J. Dorado and F. Flores, Phys. Rev. **A47**, 3092 (1993).

13. P.M. Echenique, R.M. Nieminem and R.H. Ritchie, Solid State Commun. **37**, 779 (1981); P.M. Echenique, R.M. Nieminem, J.C. Ashley and R.H. Ritchie, Phys. Rev. **A33**, 897 (1986).

14. J.H.R. dos Santos, P.L. Grande, M. Behar, H. Boudinov, and G. Schiwietz, Phys. Rev. **B 55**, 4332 (1997).

15. P. L. Grande and G. Schiwietz, Phys. Rev. **A 58**, 3796 (1998).

16. G. Schiwietz and P. L. Grande, Nucl. Instr. and Meth. **B 153**, 1 (1999).

17. F. Herman and S. Skillmann, in Atomic Structure Calculations, (Prentice-Hall, Inc. Englewood Cliffs, New Jersey,1963).

18. E. C. Montenegro, W. E. Meyerhof and J. H. McGuire, Adv. At. Mol. Opt. Phys. **34**, 249 (1994).

19. P.L. Grande and G. Schiwietz, Nucl. Instr. and Meth. **B 132**, 264 (1997).

20. J.F. Reading, T. Bronk, A.L. Ford, L.A. Wehrman and K.A. Hall, J. Phys. B: At. Mol. Opt. Phys. **30**, L189 (1997).

21. H. Knudsen, U. Mikkelsen, K. Kirsebom, S.P. Moeller, E. Uggerhoef, J. Slevin, M. Charlton and E. Morenzoni, Phys. Rev. Lett. **74**, 4627 (1995).

22. G. Schiwietz,U. Wille, R. Diez Muiño, P.D. Fainstein and P.L. Grande, Journal of Physics B : At. Mol. Opt. Phys. **29**, 307 (1996).

23. G. Schiwietz, P.L. Grande, C.Auth, H. Winter and A. Salin, Phys. Rev. Lett. **14**, 2159 (1994).

24. P.L. Grande and G. Schiwietz, Nucl. Intr. and Meth. **B 136-138**, 125 (1998).

25. D. Niemann, P. Oberschachtsiek, S. Kalbitzer, H. P. Zeindl, Nucl. Instrum. and Meth. **B80/81**, 37 (1993).

26. R.J. Petty and G. Dearnaley, Phys. Lett. **50A**, 273 (1974).

27. H. S. Jin, W. M. Gibson, Nucl. Instrum. and Meth. **B13**, 76 (1986).

28. G.M. Azevedo, P.L. Grande and G. Schiwietz, to be published in Nucl. Instrum. and Methods (2000).

29. F. Bloch, Ann. Physik **16**, 285 (1933).

Production of doubly excited He atoms in collisions of He²⁺ ions with alkaline-earth atoms

Yasuyuki Kanai, Kazuaki Iemura[†], and Xiao-Min Tong[*]

RIKEN, Wako, Saitama 351-0198, Japan
[†]*Inst. Laser Science, Univ. Electro-Communications, Chofu, Tokyo 182-8585, Japan*
[*]*ICORP, JST, Axis Chofu 3F, 1-40-2 Fuda, Chofu, Tokyo 182-0024, Japan*

Abstract. We have observed Auger electrons from doubly excited He**($nln'l'$), $n = 2$, 3, and 4, produced by the double-electron transfer processes in the collisions between He²⁺ and alkaline earth atoms. From these results, we see that highly excited states such as ($4lnl'$) states can be produced by the double-electron transfer collisions between He²⁺ and alkaline-earth atoms. Relative population of such states increased as the target atomic number increased. The mechanism of the two electron transfer processes is discussed by using potential curves.

INTRODUCTION

We have studied double-electron transfer processes in low-energy highly charged ion collisions with atoms by measuring the Auger electrons (1-10). Our main purpose is to understand the multi-electron transfer processes in general (11). Double-electron transfer is the simplest case of the multi-electron transfer processes and contains the basic feature of multi-electron transfer processes. Furthermore, in the case of double-electron transfer, a comparison between experiment and theory should be relatively easy. In this paper, we present the results of our recent studies of 20-40 keV He²⁺ collisions with alkaline-earth atoms (9). These collision systems have the following features:

1. Alkaline-earth atoms have a fully filled core and two electrons in the outermost electronic shell and can be treated as quasi-two-electron atoms. By changing the target, we can change the binding energy of the two outermost electrons I_{1+2} (= first ionization potential + second ionization potential) as I_{1+2} = 22.7, 18.0, 16.7, and 15.2 eV for Mg, Ca, Sr, and Ba, respectively.

2. By double-electron transfer from an alkaline-earth atom to a He²⁺ ion, neutral doubly excited He**($nln'l'$) atoms are produced. He**($nln'l'$) atoms are the simplest hollow atoms. One of the characteristic features of this process is that both the neutral ground-state alkaline-earth atoms and He**($nln'l$) have large polarizabilities. The strong polarization potential in

CP500, *The Physics of Electronic and Atomic Collisions*, edited by Y. Itikawa, et al.
© 2000 American Institute of Physics 1-56396-777-4/00/$17.00

both the initial and the final channels may affect the double-electron transfer from the alkaline-earth atoms to He^{2+} ions (12).

Based on these considerations, highly excited states such as $He^{**}(nln'l')$, $n' \geq n \geq 3$, are expected to be produced by the double-electron transfer. Indeed we observed relatively strong Auger electron signals from $He^{**}(nln'l')$, $n = 2$, 3, and 4, in the collisions of He^{2+} with alkaline-earth atoms. We will discuss the production mechanism of $He^{**}(nln'l')$ by using the potential curves with the large polarizabilities of $He^{**}(nln'l')$.

EXPERIMENT

The experiment was carried out by using a 14-GHz ECR ion source (13) at the Center for Nuclear Study, University of Tokyo. A beam of He^{2+} was extracted from the source with a voltage of 5 to 20 kV applied on the plasma chamber. The resultant He^{2+} ions, with kinetic energies of 10 to 40 keV, were selected by a 90° charge state analyzer. The beam was further transported through a 45° switching magnet and an Einzel lens to pass through a beam defining aperture of an oven for alkaline-earth atoms. The vacuum in the charge state analyzer and the beam transport system was less than 10^{-7} Torr. The He^{2+} beam current was typically 200-400 μA, as measured just after the charge state analyzer.

The oven for alkaline-earth atoms has an entrance aperture with a diameter of 1.5 mm and an exit hole with a diameter of 2 mm. The length of the oven is 50 mm. A small piece of alkaline-earth atom was in this oven and heated to a temperature at which the vapor pressure is about 10^{-4} Torr. The resulting target thickness was sufficiently thin to guarantee single collision conditions. The oven was electrically insulated from the vacuum chamber and a variable negative voltage was applied on the oven where the Auger electrons were produced.

Auger electrons emitted from doubly excited $He^{**}(nln'l')$, which was produced by double-electron capture from targets, were measured by the zero-degree Auger electron spectroscopy technique (14). To analyze the electron energy, we used an electron energy analyzer which consists of a 45° parallel-plate electrostatic pre-analyzer and a high energy-resolution electrostatic analyzer of the simulated hemispherical type with a mean radius of 104 mm. The pre-analyzer was used as a deflector to steer the electrons out of the ion beam as well as to suppress background electrons produced along the beam within the analyzer. To improve the electron-energy resolution the deflected electrons were decelerated in the region between the two analyzers to about 6 eV by a retarding electric field. The overall resolution of the electron energy was about 0.1 eV (FWHM). During the measurement, the ion-beam current was monitored by a Faraday cup after the pre-analyzer. Typical ion-beam current was 50-200 nA at the Faraday cup.

EXPERIMENTAL RESULTS

We measured the ejected electron spectra in the following collisions:

$$20\text{-}40\,\text{keV}\quad He^{2+} + A \longrightarrow He^{**}(nln'l') + A^{2+}$$
$$\longrightarrow He^{+}(n''l'') + e. \qquad (1)$$

A: Mg, Ca, Sr, or Ba

Typical experimental results are shown in Figs. 1 and 2. The intensity of each spectrum is normalized by the target thickness and the beam currents. If one may postulate that the angular distribution of the Auger electrons is the same for different targets and different collision energies, these normalized intensities are proportional to the double-electron transfer cross sections (15).

Figure 1 shows the ejected electron spectra in the collisions of 20-40 keV He^{2+} on Ca. There are clear structures between 33 and 41 eV and below 7.5 eV. Based on the theoretical calculations for the energy levels of $He^{**}(nln'l')$ states (16-19), we tentatively identify the peaks between 33 and 41 eV and the structures below 7.5 eV as due to the Auger electrons from $He^{**}(2ln'l')$ and $He^{**}(nln'l')$, $n \geq 3$, respectively. As the collision energy increased from 20 to 40 keV, the ratio of $I(nln'l';\ n \geq 3)/I(2lnl')$ became larger, where $I(2lnl')$ is the intensity of Auger electrons from $He^{**}(2lnl')$ and $I(nln'l';\ n \geq 3)$ the intensity of Auger electrons from $He^{**}(nln'l')$, $n \geq 3$. This tendency was observed in all collision systems in a similar impact energy region.

When we changed the targets, this ratio $I(nln'l';\ n \geq 3)/I(2lnl')$ also changed. The target dependence of the ejected electron spectra is shown in Fig. 2. In collisions of He^{2+} on Mg, the intensity of Auger electrons from $He^{**}(2lnl')$ states is stronger than that from $He^{**}(nln'l')$, $n \geq 3$. On the other hand, in the collisions of He^{2+} on Ba, the intensity of Auger electrons from $He^{**}(2lnl')$ states is weaker than that from $He^{**}(nln'l')$, $n \geq 3$. For comparison, the ejected electron spectra in the collisions of 40 keV He^{2+} on Xe are also shown in Fig. 2. For Xe target, highly excited states such as $He^{**}(nlnl')$, $n \geq 3$, were not produced by the double-electron transfer. This must be due to the difference in I_{1+2} between Xe and alkaline-earth atoms, the I_{1+2} for Xe being 33.3 eV and much larger than the values for alkaline-earth atoms. This tendency may be explained by using the potential curves. We will discuss this point later.

Figures 3 and 4 show detailed Auger spectra from $He^{**}(2lnl')$ and $He^{**}(nln'l')$, $n \geq 3$, respectively, produced in the collisions between He^{2+} and Ba. In these spectra, series limits are observed clearly. Especially, the limit of $(3lnl')$ series is very clear in Fig. 4. This indicates that the population of the doubly excited $He^{**}(3lnl')$ with high n is relatively large. Some theoretical values for Auger electron energies (16-19) are indicated in Figs. 3 and 4. The theoretical values for $He^{**}(2l2l')$ states are in good agreement with our measurements. Identification of the clear peaks for Auger electrons from the $He^{**}(2l3l')$ and $He^{**}(3lnl')$ states is in progress.

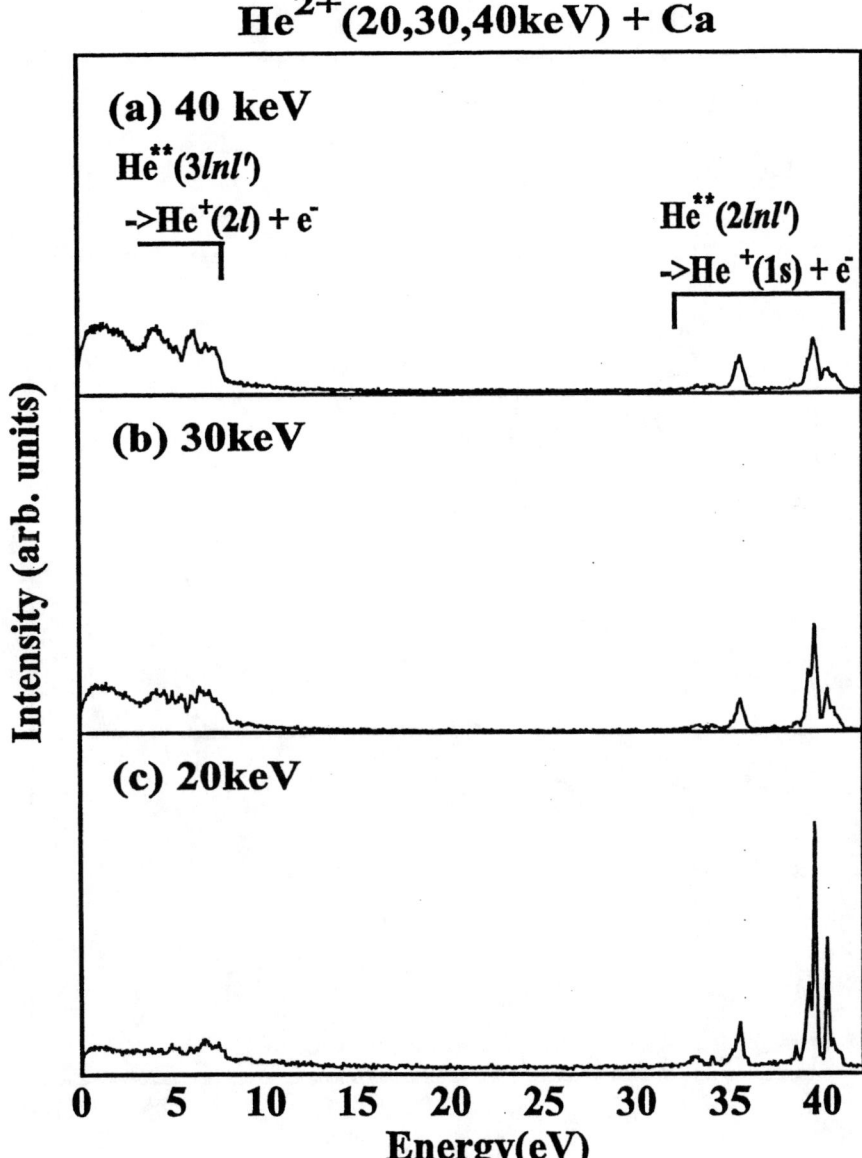

FIGURE 1. Ejected electron spectra from the collisions of 20-40 keV He²⁺ on Ca.

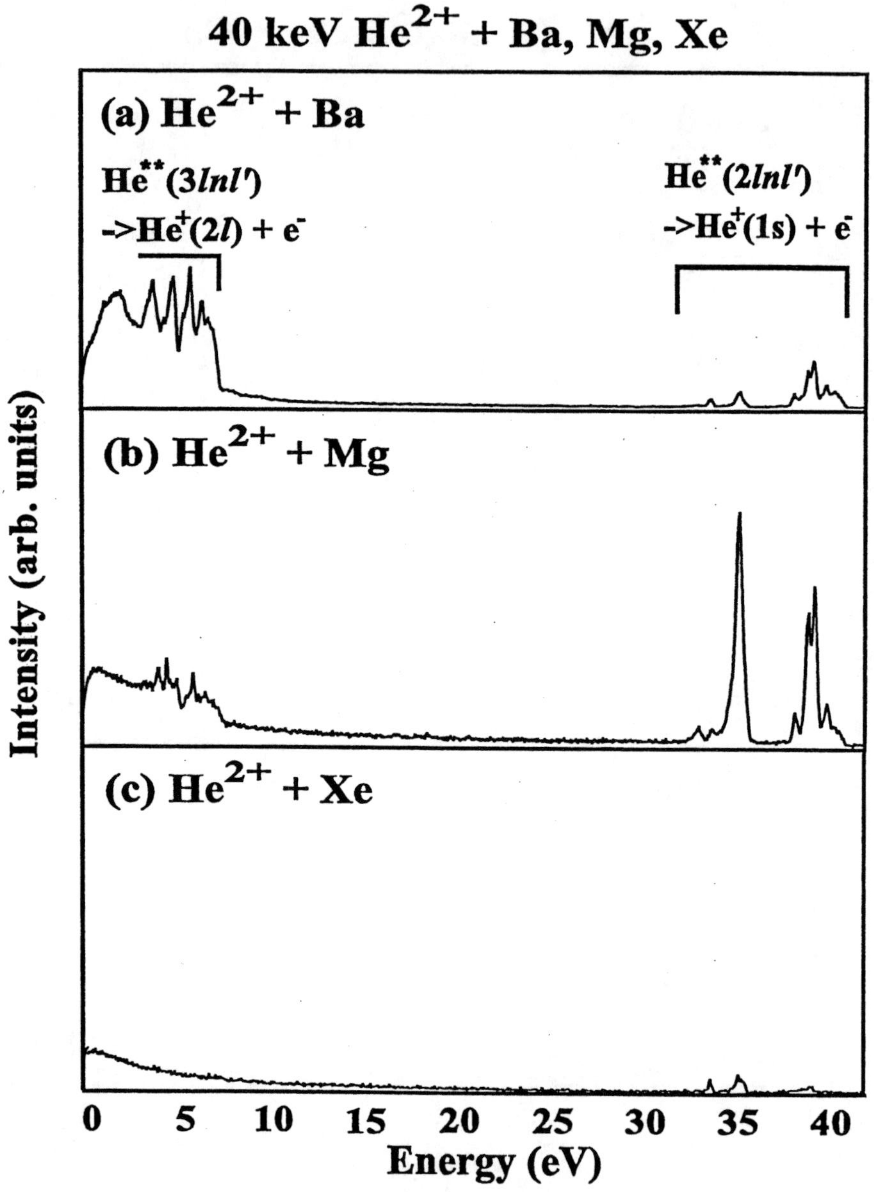

FIGURE 2. Ejected electron spectra from the collisions of 40 keV He^{2+} on Mg, Ba, and Xe.

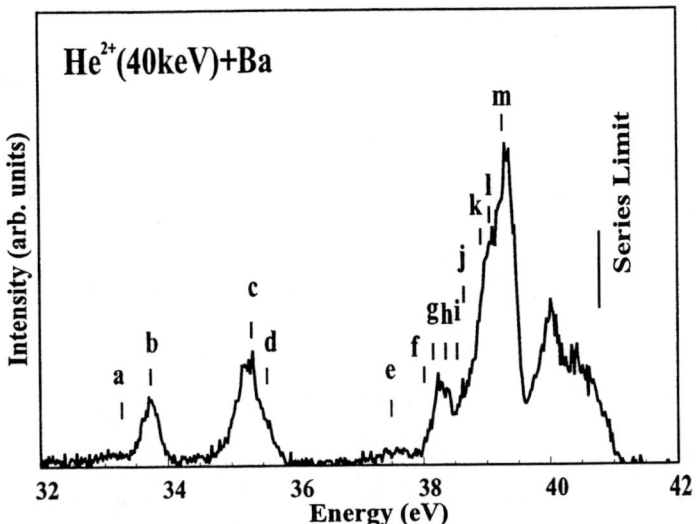

FIGURE 3. Typical Auger electron spectra from He**($2lnl'$) in the collisions between 40 keV He^{2+} and Ba. The alphabetical notations represent peaks which are a: 2s^2 ^{1}S^e, b: 2s2p ^{3}P^o, c: 2p^2 ^{1}D^e, d: 2s2p ^{1}P^o, e: 2p2 ^{1}S^e, f: 2p3p ^{3}S^e, g: sp23- ^{1}P^o, h: 2s3s ^{1}S^e, i: 2p3p ^{3}D^e, j: 2p3p ^{3}D^o, k: 2p3p ^{1}D^e, l: sp23+ ^{1}P^o, and m: 2s3d ^{1}D^e. The vertical lines represent the theoretical energies (16,17).

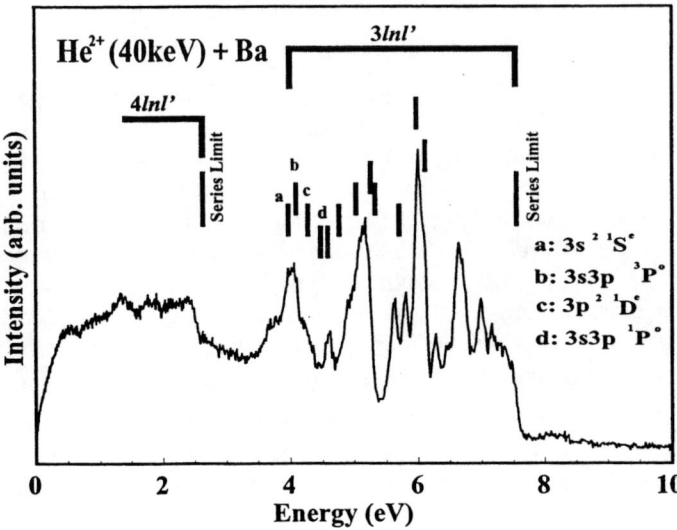

FIGURE 4. Typical Auger electron spectra from He**($nlnl'$), n $\geq$3 in the collisions between 40 keV He^{2+} and Ba. Some theoretical Auger energies are indicated in the figure (18,19).

General features of our experimental results are summarized as follows:

1. Auger electrons from He**(2lnl') and He**(3lnl') were observed for Mg, Ca, Sr, and Ba targets.
2. Auger electrons from the He**(4lnl') states were observed for Ca, Sr, and Ba targets.
3. Clear series limits of (2lnl'), (3lnl'), and (4lnl') states were observed in Auger electron spectra.
4. $I(3lnl')/I(2lnl')$ became larger as the target atomic number increased.
5. $I(3lnl')/I(2lnl')$ became larger as the collision energy increased.

DISCUSSION

As we mentioned in the introduction, He**($nln'l'$) has a large polarizability α ($nln'l'$). To estimate the potential curves for the collisions of He^{2+} on alkaline-earth atoms, we calculated the polarizabilities of He**($nln'l'$) states by the linear density response method with density functional theory (20, 21). Some calculated polarizabilities are listed in Table 1. For highly excited states, we used scaled values of the polarizability $\alpha (nlnl') = (n/2)^3 \times \alpha$ (2s2s) and $\alpha (nln'l') = (n/2)^3 \times \alpha$ (2s3s) for $n > n'$. Potential curves for He^{2+} on Mg and He^{2+} on Ba estimated by using these polarizabilities are shown in Fig. 5. Due to the large polarizability of He**($nln'l'$), potential curves of He**(3lnl') and He**(4lnl') cross with the incident-channel potential curve. This means that He**(3lnl') and He**(4lnl') states can be produced by the "simultaneous" double-electron transfer at the crossing points.

Here, we compare the potential curves for two different collision systems He^{2+} - Mg and He^{2+} - Ba shown in Fig. 5. We evaluate approximately the double-electron transfer cross sections by using the angle between the initial- and the final-channel potential curves at the crossing points. For example, the incident channel and the curves of He**(3lnl') states for the Mg target cross with each other nearly at right angle at about 11 a.u. On the other hand, for the Ba target, the crossing occurs at a smaller angle of 45° at about 12 a.u. in the present plot. This indicates that the double-electron transfer cross section for the Ba target must be larger than that for the Mg target according to the following consideration. The electron transfer probability p at a crossing point depends on the angle Θ between the initial- and final-channel potential curves. The probability p is almost zero when the angle Θ is 90° and almost one for 0°. Since the electron transfer cross section is proportional to $p \times (p-1)$, it must be larger for $\Theta \simeq 45°$ than for $\Theta \simeq 0°$ or 90° (11). Similar argument can also be applied to the He^{2+} - Xe system. For the Xe target, the binding energy for outermost two electrons I_{1+2} is much larger than that for Mg; the I_{1+2} for Xe is 33.3 eV and the I_{1+2} for Mg is 22.7 eV. Because of this, the incident potential curve crosses with all potential curves for He**($nln'l'$) at ~ 90°. Therefore, the double-electron transfer cross section in the collisions of He^{2+} on Xe must be small as shown in Fig. 2. To create highly excited He**($nln'l'$) in He^{2+} collisions, we have to choose targets with a large polarizability and with a low binding energy. The validity of this approximate consideration needs to be confirmed by a more precise calculation of the double-electron transfer cross sections (22) for these collision systems. Such a calculation is in progress and the results will be compared qualitatively with the experimental results. A detailed report including all the experimental data and theoretical discussion will be given elsewhere (23).

Table 1. Calculated polarizabilities of He.

States	Polarizability (Å³)
1s2s (singlet)	61.25
1s2s (triplet)	35.07
2s2s (singlet)	17.84
2s2p (singlet)	17.79
2p2p (singlet)	22.76
2s3s (singlet)	246.95

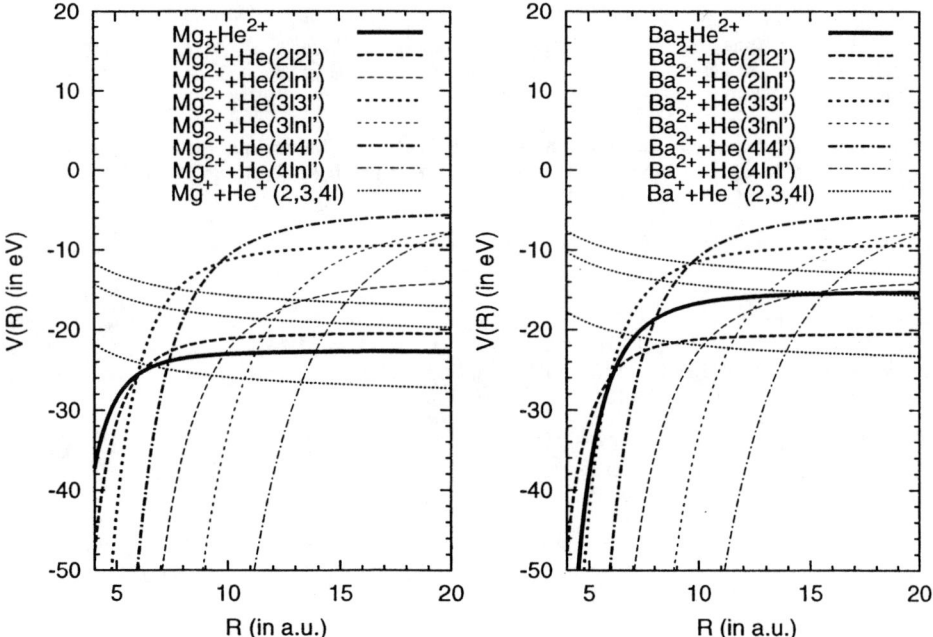

FIGURE 5. Potential energy curves for He^{2+} + Mg and He^{2+} + Ba.

613

SUMMARY

The Auger electrons from doubly excited He**($nln'l'$), $n = 2$, 3, and 4, produced by double-electron transfer from alkaline earth atoms, were observed. The intensities of the Auger electrons from He**($3lnl'$) and He**($4lnl'$) increase as the target atomic number increases. By taking into account the large polarizabilities of the doubly excited He**($nln'l'$), this tendency can be explained qualitatively by examining the potential curve crossings.

ACKNOWLEDGMENTS

We would like to thank Prof. Hirosi Suzuki, Prof. Tsutomu Watanabe, Prof. Syunsuke Ohtani, Prof. Kazuyoshi Wakiya, Dr. Toshinobu Takayanagi and Dr. Daiji Kato for their helpful discussions, and Prof. Masayuki Sekiguchi and the staff of Center for Nuclear Study, University of Tokyo, for the operation of the ECR ion sources.

REFERENCES

1. Sakaue, H. A., Kanai, Y., Ohta, K., Kushima, M., Inaba, T., Ohtani, S., Wakiya, K., Suzuki, H., Takayanagi, T., Kambara, T., Danjo, A., Yoshino, M., and Awaya, Y., *J. Phys. B: At. Mol. Opt. Phys.* **23**, L401-L405 (1990).

2. Sakaue, H. A., Awaya, Y., Danjo, A., Kambara, T., Kanai, Y., Nabeshima, T., Nakamura, N., Ohtani, S., Suzuki, H., Takayanagi, T., Wakiya, K., Yamada, I., and Yoshino, M., *J. Phys. B: At. Mol. Opt. Phys.* **24**, 3787-3795 (1991).

3. Kanai, Y., Sakaue, H. A., Ohtani, S., Wakiya, K., N., Suzuki, H., Takayanagi, K., Kambara, T., Danjo, A., Yoshino, M., and Awaya, Y., *Z. Phys D* **21**, S225-S226 (1991).

4. Kanai, Y., Sakaue, H. A., Kambara, T., Awaya, Y., Nakamura, N., Kitazawa, S., Koide, M., Ohtani, S., Suzuki, H., Nabeshima, T., Negishi, T., Takayanagi, T., and Wakiya, K., in *Physics of Highly-Charged Ions* eds. P. Richard, M. Stockli, C. L. Cocke, and C. D. Lin (AIP Conference Proceedings 274, AIP Press), 1992, pp. 63-70

5. Nakamura, N., Nabeshima, T., Currell, F. J., Kanai, Y., Kitazawa, S., Koide, M., Sakaue, H. A., Ida, H., Matsui, Y., Wakiya, K., Takayanagi, T., Kambara, T., Awaya, Y., Suzuki, H., Ohtani, S., and Safronova, U. I., *J. Phys. B: At. Mol. Opt. Phys.* **27**, L785-L793 (1994).

6. Kanai, Y., Kambara, T., Awaya, Y., Ida, H., Matsui, Y., Takayanagi, T., Wakiya, K., Nakamura, N., Koide, M., Kitazawa, S., and Ohtani, S., *Nucl. Instrum. Methods* **B98**, 81-84 (1995).

7. Nakamura, N., Ida, H., Wakiya, K., Takayanagi, T., Koide, M., Currell, F. J., Kitazawa, S., Suzuki, H., Ohtani, S., Safranova, U. I., and Sekiguchi, M., *J. Phys. B: At. Mol. Opt. Phys.* **28**, 4743-4758 (1995).

8. Nakamura, N., Awaya, Y., Currell, F. J., Kambara, T., Kanai, Y., Kitazawa, S., Koide, M., Ohtani, S., Safronova, U. I., Suzuki, H., Takayanagi, T., and Wakiya, K., *J. Phys. B: At. Mol. Opt. Phys.* **29**, 1995-2006 (1996).

9. Iemura, K., Currell, F. J., Ohtani, S., Kitazawa, S., Koide, M., Suzuki, H., Sekiguchi, M., Machida, S., Takayanagi, T., and Wakiya, K., *Physica Scripta* **T73**, 205-206 (1997).

10. Kitazawa, S., Tanabe, K., Machida, S., Matsui, Y., Ida, H., Takayanagi, T., Wakiya, K., Iemura, K., Currell, F., Ohtani, S., Suzuki, H., Sekiguchi, M., and Safranova, U. I., *J. Phys. B: At. Mol. Opt. Phys.* **31**, 3233-3243 (1998).

11. Barat, M. and Roncin, P., *J. Phys. B: At. Mol. Opt. Phys.* **25**, 2205-2243 (1992).

12. Watanabe, T., Tong, X. M., Ohtani, S., Iemura, K., Kanai, Y., and Suzuki, H., Abstracts of Contributed Papers of 21st ICPEAC, 1999, P.549

13. Sekiguchi, M., Ohshiro, Y., Fujita., M., Yamazaki, T., Yamashita, Y., Isoya, Y., Yamada, T., and Kitagawa, A., *Proc. 12th Int. Conf. on Cyclotron and Their Applications* (World Scientific, Singapore),1992, pp. 326-329.

14. Stolterfoht, N., *Physics Reports* **146**, 315-424 (1987).

15. Prior, M. H., Holt, R. A., Schneidor, D., Randall, K. L., and Hutton, R., *Phys. Rev.* **A48**, 1964-1974 (1993).

16. Tang, J. Z, Watanabe, S., and Matsuzawa, M., *Phys. Rev.* **A46**, 2437-2444 (1992).

17. Morishita, T., Hino, K., Watanabe, S., and Matsuzawa., M., *Phys. Rev.* **A53**, 2345-2358 (1996).

18. Ho, Y. K. and Callaway, J., *J. Phys. B: At. Mol. Opt. Phys.* **18**, 3481-3486 (1985).

19. Lipsky, L., Anania, R., and Conneely, M. J., *Atomic Data and Nuclear Data tables* **20**, 127-141 (1977).

20. Tong, X. M. and Chu, S. I., *Phys. Rev.* **A55**, 3406-3416 (1997).

21. Mahan, G. D. and Subbaswamy, K. R., *Local Density Theory of Polarizability* (Plenum Press, New York, 1990).

22. Tong, X. M. and Chu. S. I., *Phys. Rev.* **A57**, 452-461(1998).

23. Iemura, K. and Tong, X. M., unpublished work.

Slow ions scattering on C_{60}: The electronic response of the C_{60} molecule and its ions

H. Cederquist, A. Fardi, K. Haghighat, A. Langereis, H.T. Schmidt, and S.H. Schwartz

Atomic Physics, Stockholm University, Frescativ. 24, S-104 05 Stockholm, Sweden

J. Levin and I.A. Sellin

Department of Physics and Astronomy, University of Tennessee, Knoxville, TN37996-1200, USA

H. Lebius and B. Huber

Département de Recherche Fondamentale sur la Matiere Condensée/ S12A CEA-Grenoble, 17 rue des Martyrs, F- 30854, Grenoble Cedex 9, France

M.O. Larsson and P. Hvelplund

Institute of Physics, University of Aarhus, DK-8000 Aarhus C, Denmark

Abstract. In this progress report we discuss the electronic response of the C_{60} molecule and its positive ions by means of comparisons between experimental results and two different models for the charge mobility in C_{60}. We have measured projectile angular differential cross sections, $d\sigma/d\theta$, and mean projectile energy gain/loss, ΔE, as functions of the number, s, of electrons stabilized on the projectile in 16 and 26.4 keV $Ar^{8+}+C_{60} \rightarrow Ar^{(8-s)+}+C_{60}^{r+}+(r-s)e^-$ collisions. In the infinitely conducting sphere (ICS) model the high mobility averages out all effects of localization of individual charge carriers. In the movable-holes (MHo) model 'positive holes' are localized as point charges in their equilibrium positions on the 'molecular surface' within the times (down to 10^{-16} s) between sequential over-the-barrier electron transfers. The two sets of predictions for θ are close for $r \leq 8$, and for $r \leq 5$ they are also in agreement with experimental results indicating ultrafast electronic response of ionized C_{60}. For $r > 5$, both models underestimate θ strongly which is due to close interactions with individual Carbon atoms.

CP500, *The Physics of Electronic and Atomic Collisions*, edited by Y. Itikawa, et al.

I INTRODUCTION

The electrical properties of fullerene materials, such as individual fullerene molecules, nanowires or crystals of fullerenes may be manipulated by introducing foreign atoms. For example, fullerite is formed by means of van der Waals forces which weakly bind individual C_{60} molecules in an insulating crystal. This crystal may become a good electric conductor when it is doped with alkali atoms in appropriate amounts [1]. The reason is that the alkali atoms donate electrons to the Lowest Unoccupied Molecular Orbital of neighboring C_{60} molecules, which then form the conduction band of the fullerite crystal. In pure fullerite, the Highest Molecular Orbitals of C_{60} are fully occupied and thus the valence band of the crystal is completely filled and the conduction band is empty.

Here, instead of adding electrons to the lowest unoccupied orbital, we will investigate the electrical response of single C_{60} molecules after transfer of one or several electrons to Ar^{8+} ions in

$$Ar^{8+} + C_{60} \rightarrow Ar^{(8-s)+} + C_{60}^{r+} + (r-s)e^- \tag{1}$$

collisions. The number of electrons stabilized on the projectile ranges from $s=1$ to $s=8$. The collision time, during which r electrons are removed from C_{60}, is about 10 femtoseconds at the present collision energies of 16 and 26.4 keV. The main experimental observables are; the projectile angular differential cross sections $d\sigma/d\theta$ and the mean projectile energy gain/loss ΔE as functions of s. These results will be discussed within two different classical over-the-barrier models in which the electrons are transferred sequentially with time-separations of 0.1-1 fs. In the infinitely conducting sphere (ICS) model the electronic response time is infinitely short, i.e. the electronic motion averages out all effects of localization of individual charge carriers. In the model with r positive and localized holes moving on 'the surface' of C_{60}^{r+}, the charge mobility can be controlled. This is done by adjusting the time required to reach new equilibrium charge configurations after each (sequential) electron transfer. A response time of 10^{-16} s in the Movable Holes (MHo) model is found to yield results in agreement with experiments and the ICS model for small and moderate r ($r \le 5$).

The main concepts of the ICS and the MHo models have been used before and in both cases rather good agreement with various earlier experimental results have been obtained. The metal (infinitely conducting) sphere concept was first used in 1994 by Walch et al to account for their pioneering experimental results on charge transfer in Ar^{8+}-C_{60} collisions [2]. Somewhat later, Thumm et al [3] used the same model to calculate translational energy gain partly in agreement with measurements by Selberg et al [4]. The metal sphere concept has also been used to successfully account for critical electron transfer distances [5], the polarizability of C_{60} [6], surface plasmon excitation in collisions with electrons [7], the sequence of ionization potentials [8] and radiative cooling of hot C_{60}^- [9]. Further, Bárány

The Infinitely Conducting Sphere model (ICS)

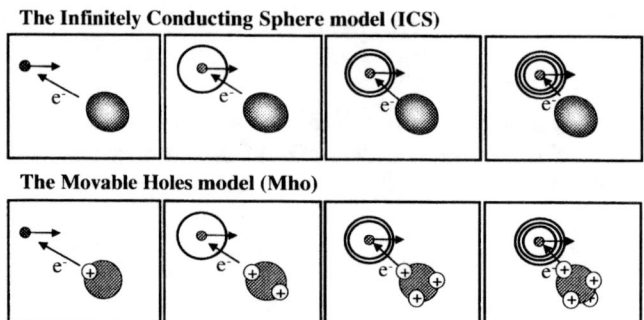

The Movable Holes model (Mho)

FIGURE 1. Schematic pictures of the electron-transfer sequences according to the Infinitely Conducting Sphere (ICS) and the Movable Holes (MHo) models (cf. text).

and Setterlind [10] developed a general model in which C_{60} is treated as a dielectric sphere.

In 1995, Shen *et al* [11] measured large cross sections for *non-fragmenting* single and double-electron capture in fullerene-fullerene collisions (C_{60}^{q+}-C_{60}). Electrons may thus obviously be transferred at distances large enough to avoid strong direct interactions between the molecular cages. Shen *et al* [11] invoked the model with movable and localized charges to rationalize their results. The hole model has also been used to account for charge insensitive electron-capture cross sections in near thermal collisions between multiply charged fullerenes and various atoms [12,13].

In the following we will give brief descriptions of the main features of electron transfer using the Infinitely Conducting Sphere and the Movable Holes models. This is followed by short descriptions of the techniques yielding $d\sigma/d\theta$ and ΔE as functions of the numbers of electrons stabilized. Finally, we compare experimental and model results and discuss the region of validity for the models in view of the influence from close collisions with individual Carbon atoms in the C_{60} molecule [14]. There are several more intriguing features of collisions between highly charged ions and C_{60} (or other fullerenes or clusters) which will not be discussed here. Such issues concern the stability limit (against fragmentation) for highly ionized C_{60} [15,16] and the detailed relations between collision induced energy transfer and the different subsequent fragmentation modes [17,18].

II MODELS OF HCI-C_{60} INTERACTIONS

In the Infinitely Conducting Sphere model we assume that the conduction electrons are free to move without resistance inside the model sphere and thus that an inhomogeneous surface-charge distribution develops during the approach of the projectile. This distribution is such that the electric field inside the sphere is zero at all times (it is a perfect metal), while the field *outside* the sphere is due

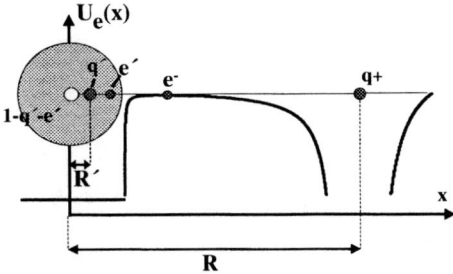

FIGURE 2. A schematic of the potential between the model (ICS) sphere and the projectile when the first electron is transferred. The potential is given by the ion of charge q at a distance R from the sphere center, its image charge q' at R', the electrons self-image charge e and the central charge of the sphere $1-q'-e'$.

to the projectile ion *and* the inhomogenous charge distribution. For simplicity we calculate the dynamically changing electric potential at the position of the ion by means of the method of electrostatic images. For an ion of charge q at a distance R from the center of the model sphere, the image charge must be placed at a distance $R' = a^2/R$ (from the sphere center) and have the magnitude $q' = -aq/R$. We have taken the model sphere radius to be $a = \alpha^{1/3}$, where $\alpha = 543\ a_0^3$ is a consensus value for the dipole polarizability of C_{60} [6]. All points on the surface of a conducting sphere must have the same electrical potential and in order to preserve its total charge (zero before electron transfer) we place a charge of opposite sign, $-q' = aq/R$, in the center. We then assume that electron transfer proceeds in a step-wise manner controlled by a sequence of classical over-the-barrier conditions. The first condition is fulfilled at the critical distance, R_1, where the binding energy of the outermost target electron equals the maximum of the potential barrier seen by this electron in transit towards the projectile. The former quantity becomes $I_1^* = I_1 + q/R_1$, in the field of the ion, where I_1 is the ionization potential of C_{60}. The potential barrier is obtained from the potential of an electron moving between the ion and the Infinitely Conducting Sphere:

$$U_e^1(x) = -\frac{q}{R-x} - \frac{q'}{x-R'} + \frac{q'-1}{x} + \frac{1}{2}\left(\frac{a}{x^2} - \frac{a}{x^2-a^2}\right),$$

where $x > a$ is the distance between the center of the sphere and the active electron and R' is the distance to the image charge of the ion q' (cf. figure 2 and ref. [14]). We thus assume a quasi-continuum of projectile capture states. In order to calculate the angular differential cross section, $d\sigma/d\theta$, and the translational energy gain (or loss) as functions of r we go through the following steps. (i) Solve the over-the-barrier conditions numerically for a head-on trajectory, i.e. a trajectory with zero impact parameter with respect to the center of the sphere ($b_{C60} = 0$). (ii) Select b_{C60} randomly and calculate the trajectory numerically under the consideration of

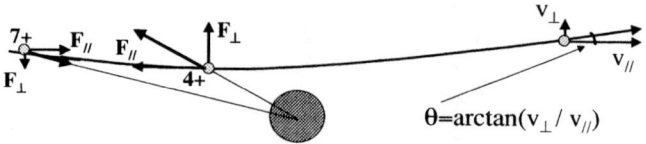

$\theta = \arctan(v_\perp / v_{//})$

FIGURE 3. A schematic projectile trajectory according to the ICS model. The positions (times) for electron transfers number one and four are indicated together with the time dependent forces used to calculate the final projectile velocity components. The force is subject to the complete, over-the-barrier controlled, flow of electrons towards the projectile.

the step-wise changes of the target and projectile charges at the critical distances. The final projectile velocity is obtained by integrating the force along the trajectory and we thus arrive at θ and the change in kinetic energy as indicated in figure 3. By running many such randomly selected trajectories in a Monte Carlo program we build the differential cross sections [14]. In the calculation we assume electron transfer only on the incoming part of the trajectory and, further, we neglect the possibility of autoionization during the collision.

In the Movable Holes model we again use classical over-the-barrier conditions to identify critical distances for sequential electron transfer. As electrons are removed from the sphere they leave behind positive holes moving on the surface in a localized fashion. The motion for one of the holes is given by the field from the projectile ion, the other holes and a friction force representing the electrical resistance due to scattering of the hole against the molecular lattice and electrons [14]. The field strength from the projectile is highest at the closest point on the sphere surface and therefore new positive holes are always introduced in this position. We have chosen the damping friction force in such a way that the holes reach their new equilibrium positions (minimum potential energy for the ion-holes system) between sequential electron transfers. For 26.4 keV Ar^{8+}-C_{60} collisions this gives damping times down to 10^{-16} s as indicated in figure 4.

The calculations of critical distances, projectile scattering angles and energy gain basically follow the same procedure as for the ICS-model. First we calculate the sequence of critical distances for a head-on trajectory ($b_{C60}=0$) under consideration of the hole dynamics during the approach of the projectile. Second, we select b_{C60} randomly in a Monte Carlo program and calculate the trajectories, final velocity vectors and the charge for the scattered ion. Here we assume that the outgoing projectile charge is given by the incident charge and the number of electrons removed from the sphere. That is, we assume that autoionization life times are considerably longer than the collision time (about 10 fs) and thus that downstream relaxation only has minor influences on θ and ΔE. We again use the final velocity vectors to build $d\sigma/d\theta$ and ΔE as functions of the number of active electrons, r. We have used the same sphere radius ($a=8.2\ a_0$) as in the ICS model.

We compare results from the ICS and MHo models and the critical dis-

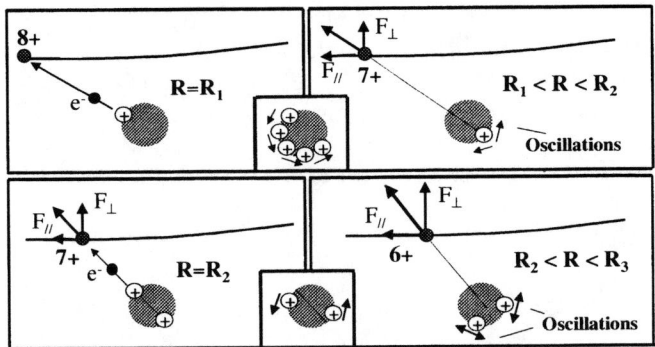

FIGURE 4. A schematic trajectory for the Movable Holes model. The trajectory is indicated at four different times; at the transfer of the first electron, before and at the transfer of the second electron and after transfer of the second electron. The insets show the hole motion between electron transfers. The forces used to calculate the final velocity components are indicated.

tances for electron transfer, R_r, are shown in figure 5. In both models, we have assumed full screening of the projectile charge by earlier transferred electrons, such that the effective charge seen by the target decreases in a step-wise manner. For ion-*atom* collisions the same screening assumption is known to indicate [19] that the projectile is not more than half-way neutralized. With the present C_{60}-model targets, however, we obtain *full neutralization* of the incoming Ar^{8+} projectile at distances several atomic units above the C_{60}-cage. This is due to the large polarizability of C_{60} which is included to some extent in both models, but in different ways. This result can be understood within the ICS model since the (negative)

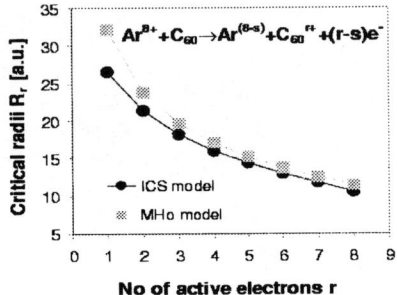

FIGURE 5. The critical radii for electron transfer, R_r, according to the infinitely conducting-sphere (ICS: black circles) and the movable-hole (MHo: grey squares) models. The lines between the data points are to guide the eye.

image charge helps to push the active electron towards the projectile even when the net target charge is larger than the effective projectile charge. The two sets of R_r-predictions get closer as the number of active electrons increases, since a larger number of positive holes more easily may give a similar external field as an Infinitely Conducting Sphere. In figure 6 we compare the ICS and MHo results for the most

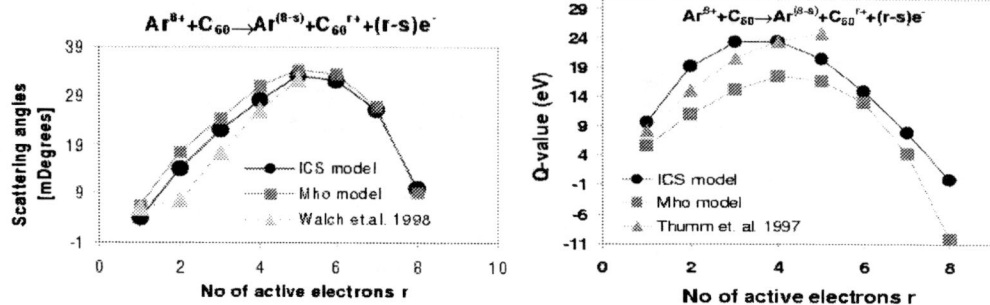

FIGURE 6. Left figure: The most likely scattering angles θ as functions of the number of active electrons (r) according to the ICS (black circles) and the MHo models (grey squares). Energy scaled experimental values by Walch *et al* [20] measured in coincidence with intact (i.e. non-fragmenting) C_{60}^{r+} molecules are shown as grey triangles for $r=1-5$. Right figure: Q-values from the ICS (black circles) and the MHo (grey squares) models. The model results by Thumm *et al* [3] are shown as grey triangles. The lines are to guide the eye.

likely projectile scattering angle and the energy gain. The two sets of predictions for $\theta(r)$ are very close in particular for large values of r. The energy gain, ΔE, is predicted to follow similar bell-shaped curves as functions of r by both models. The Movable Holes model predict energy loss for $r=8$.

III   EXPERIMENTAL TECHNIQUE

The experimental set-ups are shown in figure 7. The Ar^{8+} beam is provided by an Electron Beam Ion Source and separated at an energy of 26.4 keV by an analyzing magnet. This beam is collimated and its angular resolution was measured to be $\pm 0.02°$ before entering a resistively heated C_{60}-cell of length 30 mm. The C_{60} pressure was low enough to make the corrections for multiple collisions small but not insignificant. The projectile charge state after the collision was measured by means of a cylindrical analyzer or a pair of straight deflector plates. The angular distribution for a given s is obtained from the spatial distribution of events on the position sensitive detector following the respective charge-state analyzer. The position events are sorted according to their distance from the center of the image taking the focusing of the analyzing device into account. The two sets of experimental results for $d\sigma/d\theta$ were fully consistent with each other. The ΔE-distributions

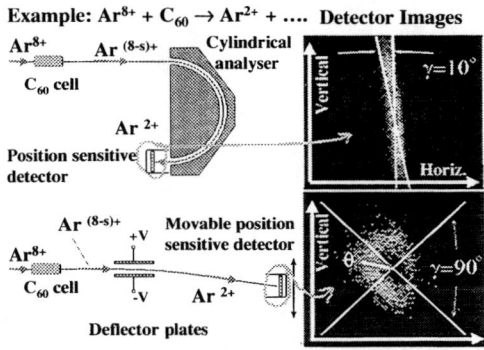

FIGURE 7. The two techniques used to measure $d\sigma/d\theta$ as functions of s in Ar^{8+}-$C_{60} \rightarrow Ar^{(8-s)+}$-... collisions at 26.4 keV. The images on the detector for the $s=6$ process are shown to the far right, where the bow-tie apertures, with opening angles γ, indicate the areas used for extraction of $d\sigma/d\theta$. The distances from the center of the bow tie is used to calculate θ for each event taking the focusing of the respective analysing system into account.

were measured by means of a hemispherical electrostatic analyzer.

IV   RESULTS AND DISCUSSIONS

In order to compare the present experimental results for projectile scattering angles and energy gain with the model results (ICS and MHo), we must remember that the latter are functions of the number of active electrons (r) while the former depend on s. For this purpose, we connect the average number of active electrons $\langle r \rangle$ to a given number of stabilized electrons s. In the specific case of Ar^{8+}-C_{60} collisions this relation is $\langle r \rangle = s+2$ for $s > 1$ [21]. For $s=1$, the average number of active electrons is $\langle r \rangle = 1.4$ [14]. In figure 8 we compare experimental results for θ and ΔE now as functions of $\langle r \rangle$ with the corresponding model results as functions of r. The agreement for θ is excellent when $r \leq 5$, while the model and experimental results deviate strongly for larger values of $\langle r \rangle$ and r. The results by Walch et al [20], who measured coincidences between *intact* C_{60}^{r+} molecular ions and scattered projectiles extending up to $r=5$, are also in excellent agreement with the models for $r \leq 5$. Thus, the ICS or the MHo model may be used to describe the angular scattering for large impact parameter collisions for which the number of active electrons is low or moderate. The discrepancy for large values of r, is mainly due to the strong influence from close collisions with individual Carbon atoms. It is not surprising that the ICS and MHo models, which both assume that the C_{60} is a smooth sphere, much better predict θ for distant than for close collisions. From *this* perspective *it is* surprising that both models give ΔE-values close to the measured ones for the whole range of $\langle r \rangle / r$. Although there is no perfect agreement in terms

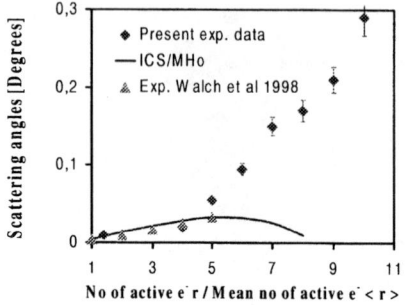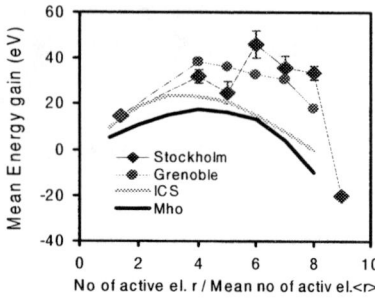

FIGURE 8. Left figure: The present most likely experimental scattering angles θ as functions of the mean numbers of active electrons $\langle r \rangle$ in 26.4 keV Ar^{8+}-C_{60} collisions. We have used the relation $\langle r \rangle = s+2$ [21] for $r \geq 2$. For $s=1$, we use $\langle r \rangle = 1.4$ (cf. text). The results from the ICS and MHo models, as functions of r, are included by a common curve. The results from [20] (triangles) are shown as functions of r. Right figure: The present mean experimental energy gain as a functions of $\langle r \rangle$ (Stockholm 26.4 keV; black diamonds, Grenoble 16 keV; grey circles). The lines between the data points are to guide the eye. The lower curves yield results from the ICS (grey) and MHo (black) models.

of absolute numbers the characteristic bell-shaped model ΔE-curves are reproduced by two independent measurements (Stockholm and Grenoble). Before the present experiment it was already well known that the outgoing projectile charge-state distribution due to Ar^{8+}-C_{60} collisions has a secondary maximum for six and seven stabilized electrons [2]. This observation has later been confirmed by the present collaboration [14] and by Martin *et al* [21] and it is connected to the approach of charge-state equilibration for passage of a single C_{60}-molecule. We thus know that a large part of the cross sections associated with many active electrons are due to trajectories through or very close to the molecule. The present angular distributions for large s have indeed been reproduced in a separate Monte Carlo calculations of scattering on sixty Carbon atoms [14]. At present, we do not understand why the smooth-sphere models work better for predictions of ΔE than for θ at large r.

V   CONCLUSIONS

In this report we have discussed the electronic response of C_{60} in collisions with slow Ar^{8+} ions. We have used two models of C_{60} - an Infinitely Conducting Sphere model and a model with Movable and localized positive Holes on a sphere surface. Both models yield mean energy gain values in good agreement with the measurements even for collisions with many active electrons. The model results for the angular scattering give good predictions up to a moderate number of active electrons (five). In the Movable Holes model we assumed that the electronic response time was ultrafast $(10^{-16}$ s) and it is important to note that the agreement with the ICS model *and the experimental results* would have been lifted with a longer response time. We are thus able to conclude, indirectly, that the electronic response time of real ionized C_{60} is 10^{-16} seconds or shorter. Both smooth-sphere models contain ingredients that appear sound and they have been used, in slightly different forms, in successful comparisons with earlier experimental results. It is clear that the ICS model is the more realistic one for the early part of the collision due to the inclusion of the polarization force from the neutral target. However, the idea of positive holes created closest to the projectile and then transported in a localized fashion is also appealing. Partly similar concepts have been used successfully to describe charge transport in insulators charged by electron transfer. A possible new development could thus be to merge the two smooth sphere models and to include effects of close collisions.

REFERENCES

1. M.S. Dresselhaus, G. Dresselhaus, and P.C. Eklund, *Science of Fullerenes and Carbon Nanotubes*, (Academic Press, San Diego, 1995).
2. B. Walch, C.L. Cocke, R. Voelpel, and E. Salzborn, Phys. Rev. Lett. **72**, 1439 (1994).
3. U. Thumm *et al*, Phys. Rev. A, **56**, 4799 (1997).
4. N. Selberg *et al*, Phys. Rev. A **53**, 874 (1996).
5. U. Thumm, J. Phys. B: At. Mol. Opt. Phys. **27**, 3515 (1994) and U. Thumm, J. Phys. B: At. Mol. Opt. Phys. **28**, 91 (1995).
6. A.A. Scheidemann, V.V. Kresin, and W.D. Knight, Phys. Rev. A **49**, R4293 (1994).
7. Sohmen; Fink and Krätschmer, Z. Phys. B **86**, 87 (1992).
8. Matt *et al*, Z. Phys. D **40**, 389 (1997).
9. J.U. Andersen *et al*, Phys. Rev. Lett. **77**, 3991 (1996).
10. A. Bárány and C.J. Setterlind, Nucl. Instrum. Methods Phys. Res. B **98**, 184 (1995).
11. H. Shen *et al*, Phys. Rev. A **52**, 3847 (1995).
12. S. Petrie, J. Wang, and D.K. Bohme, Chem. Phys. Lett. **204**, 473 (1993).
13. D.B. Cameron and J.H. Parks, Chem Phys. Lett. **272**, 18 (1997).
14. H. Cederquist *et al*, Phys. Rev. A, *accepted for publication 1999*.
15. G. Seifert, R. Gutierrez, and R. Schmidt, Physics Letters A **211**, 357 (1996).
16. T. Bastug *et al*, Phys. Rev. B **55**, 5015 (1997).
17. E.E.B. Campbell, T. Raz and R.D. Levine, Chem. Phys. Lett. **253**, 261 (1996).
18. T. Schlathölter, O. Hadjar, R. Hoekstra and R. Morgenstern, Phys. Rev. Lett. **82**, 73 (1999).
19. A. Bárány *et al*, Nucl. Instrum. Methods, Phys. Res. B **9**, 397 (1985).
20. B. Walch *et al*, Phys. Rev. A, **58**, 1261 (1998).
21. S. Martin *et al*, The European Physics Journal D4, 1 (1998).

X-ray Signatures of Charge Transfer Reactions Involving Cold, Very Highly Charged Ions

P. Beiersdorfer[a1], R. E. Olson[b], L. Schweikhard[c], P. Liebisch[c]
G. V. Brown[a], J. Crespo López-Urrutia[a], C. L. Harris[d]
P. A. Neill[a], S. B. Utter[a], K. Widmann[a]

[a] Department of Physics, Lawrence Livermore National Laboratory, Livermore, CA 94550, USA
[b] Department of Physics, University of Missouri-Rolla, Rolla, MO 65401, USA
[c] Institut für Physik, Johannes Gutenberg-Universität, D-55099 Mainz, Germany
[d] Department of Physics, University of Nevada Reno, Reno NV 89557, USA

Abstract.
Charge transfer reactions involving highly charged ions up to U^{91+} are being studied using x-ray spectroscopic techniques. The measurements are performed using ions produced *in situ* in a cylindrical trap with very low kinetic energy (≤ 3 eV/amu). The observed K-shell and L-shell x-ray emission show prominent features from high-n levels decaying to the ground level. Such emission can serve as a diagnostic marker for charge transfer processes in plasmas, for example in cometary comae or in ion traps. The emission is enabled by electron capture into states with low orbital angular momentum, mainly $\ell = 1$. The observed spectra, thus, provide a handle on determining the fraction of electron capture into low-ℓ states that can be used for testing theoretical models in the limit of low collision energy. The present paper summarizes the progress made to date.

INTRODUCTION

As an ion approaches a neutral atom or molecule, one or more electrons may transfer from the neutral to the ion, thereby lowering the ion's charge. This process, dubbed charge exchange or charge transfer, is an important process in situations when ions come in contact with neutrals. In laboratory plasmas this invariably happens near the plasma edge region, for example in the divertor region of magnetic fusion devices. It also happens when energetic neutral beams are used for plasma heating, or when the plasma is fueled by the injection of solid pellets. Highly charged ions circulating in a storage ring undergo charge exchange with background neutrals. Similarly, the ion storage time in a Penning trap is limited by charge

---

[1] Email address: beiersdorfer@llnl.gov

CP500, *The Physics of Electronic and Atomic Collisions*, edited by Y. Itikawa, et al.
© 2000 American Institute of Physics 1-56396-777-4/00/$17.00

transfer reactions. An accurate description of charge transfer processes is thus not only important for understanding basic atomic processses but has important implications for describing and controlling laboratory plasmas and ion trapping. Low-energy collisions below about $100 \, \text{eV/amu}$, in particular, have now also become of great interest in describing non-terrestrial phenomena, such as the x-ray emission from comets. Cometary x-ray emission is presumed to emanate from the interaction between highly charged heavy ions in the solar wind and gases in the coma. The collision energy is thought to be on the order of about $50 \, \text{eV/amu}$.

Measurements of charge transfer have mostly focused on determining the total cross section of the interaction, as this is the fundamental quantity needed to describe ionization balance in plasmas or ion storage times in storage rings and traps. Such measurements involve detailed accounting of the number and type of particles involved in the process before and after a charge changing collision. Measurements of the charge exchange cross sections have been obtained this way for a variety of collision partners (e.g. [1–3]).

Another way to study charge transfer is by looking at the photons emitted in the reaction. Photon emission results from the fact that the electron is captured by the ion into an orbital with high principal quantum number n. As illustrated in Fig. 1, this electron will radiatively deexcite by sequentially giving off one or more photons:

$$A^{q+} + B^0 \rightarrow A^{(q-1)+*} + B^+ \tag{1}$$

$$A^{(q-1)+*} \rightarrow A^{(q-1)+**} + h\nu_1 \rightarrow A^{(q-1)+***} + h\nu_2 \text{ etc.} \tag{2}$$

Observations of photon emission have concentrated on visible light. This is because of the interest in magnetic fusion to spectroscopically observe charge transfer between energetic neutral beams and plasma ions [4–6]. In such high-energy collisions, the electron is captured into high angular momentum states, and deexcitation proceeds along the yrast chain where each deexcitation step reduces the angular momentum and principal quantum number by unity, as indicated in Fig. 1(a).

Charge transfer processes also produce x-ray emission. Even if the initial decay is from one high-n state to a state of n-1 by giving off a photon in the visible, this process eventually will produce photons of higher and higher energy as n gets smaller. Ions with an open K-shell will give off a K-shell x ray when the electron deexcites from the $n=2$ shell to $n=1$. X rays with even higher energy are produced if selection rules allow the electron to make the jump from higher n shells to ground. This is the case, if the angular momentum of the captured electron plus that of other bound electrons of the ion, if any, differs by that of the ground level by 0 or ± 1 (if the ground level has an angular momentum of 0, then the excited state cannot differ by 0). Studies of the x-ray emission pattern produced in charge exchange thus provide information on the angular momentum values associated with the electron capture.

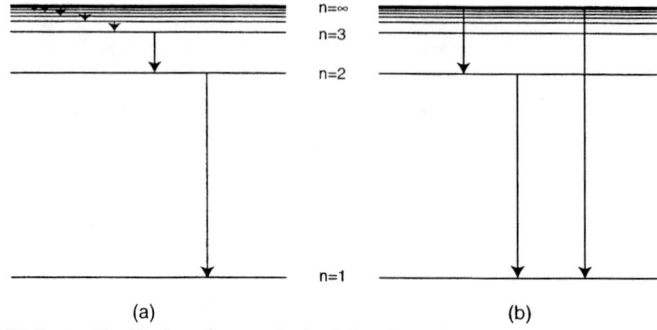

FIGURE 1. Radiative decay chain following electron capture into a high n level. (a) transitions along the yrast chain ($\Delta \ell = 1$) after high-angular momentum capture in a high-energy collision; (b) transitions after s, p, and d angular momentum capture in a low-energy collision.

Studies of the x-ray emission from charge exchange have been made in the case of very energetic ion-atom collisions. For example, the x-ray emission of 50 – 360 MeV/amu U^{92+} ions colliding with various atomic and molecular species in a gas-jet target was measured by Stölker *et al.* [7]. Very recently, work has begun to look at x-rays emitted from ions colliding with neutrals at 0.3 – 7.5 keV energies produced in ECR ion sources [8]. By contrast, x-ray measurements of very low-energy charge transfer collisions in the eV energy range have been made for several years now using the Livermore electron beam ion trap (EBIT) [9,10].

The first x-ray measurements using the Livermore EBIT were performed to demonstrate the utility of the device as an ion trap in the absence on an electron beam [9]. The electron beam was simply used to produce highly charged ions, but it is then turned off. In the absence of the electron beam, the ions are stored in a Penning trap defined by the 3-T axial magnetic field and a 10 to 100 V potential on the end cap electrodes. The net effect is that the trap is filled *in situ* and thus does not suffer any transport losses. The trap can thus easily contain 10^5 or 10^6 ions of interest. This is in stark contrast to various schemes in which traps are filled from external ion sources where only 10^0 to 10^2 ions of interest are trapped at any given time [11]. This mode of operation of the Livermore EBIT was dubbed the magnetic trapping mode [12]. Depending on the base pressure and trapping potential, ion storage times as high as several seconds were found in these measurements [9,12,13]. The energy of the ions stored in the trap is limited by the depth of the potential well to values between about 0.5 – 50 eV/amu [14–16].

The very low collision energies that can be studied using the magnetic trapping mode enable the observation of phenomena that cannot be observed in high-energy collisions 250 eV/amu or higher. For example, the angular momentum values of the captured electron are no longer distributed statistically favoring high ℓ values but concentrate near the smaller values [17]. The result is that the x-ray emission changes character and on average moves toward higher energies. The change-over

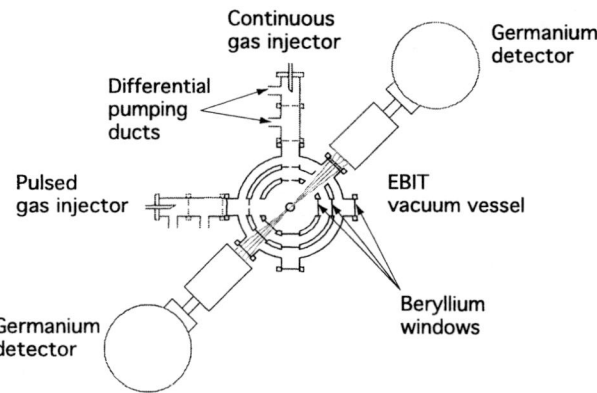

FIGURE 2. Cross-sectional view of the radial ports viewing the EBIT trap. The ports are used for gas injection and monitoring of the x-ray emission from the trap.

between statistical and non-statistical population of angular momentum values is predicted to occur at collision energies on the order of a few hundred eV/amu. This means that ion accelerator or even ECR-source based measurements cannot access this low-energy region, and sources like the Livermore EBIT need to be employed.

In the following we present an overview of the spectra obtained at low collision energy. The spectra are in part explained by classical trajectory Monte Carlo calculations of charge exchange reactions in ion-atom collisions. The possibility to study charge exchange reactions in a heretofore inaccessible regime continue to make this area of investigation novel and exciting.

EXPERIMENTAL ARRANGEMENT

The x-ray emission produced in charge exchange reactions were recorded with a high-purity Ge detector, which looked directly into the trap, as illustrated in Fig. 2. A typical K-shell spectrum of krypton obtained during the magnetic mode is shown in Fig. 3. Only charge exchange into bare and hydrogenic ions contributes to the observed emission, as only these ions have the necessary K-shell vacancy to allow the emission of a K-shell x ray. The spectrum not only contains features from the $2 \rightarrow 1$ transitions but also emission from the $n \rightarrow 1$ prompt decay indicating that a substantial fraction of electrons are captured into $\ell = 1$ angular momentum states.

An L-shell emission spectrum of very highly charged uranium produced via charge exchange reactions is shown in Fig. 4. L-shell emission can only be produced if the capturing ion has one or more vacancies in the L-shell. A total of ten charge states (bare through fluorinelike) can thus contribute to the L-shell emission spectrum. In the present case, the charge balance was peaked near helium-, lithium-, and berylliumlike uranium. No bare or hydrogenic uranium was produced. The ob-

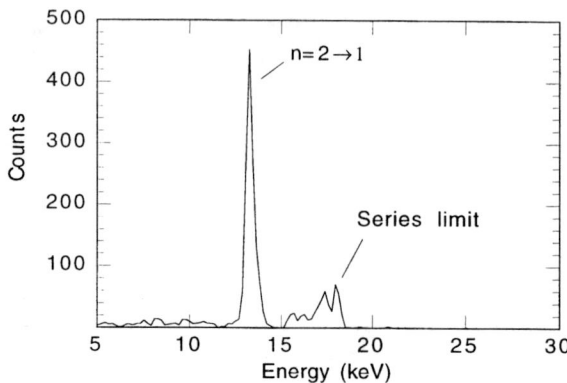

FIGURE 3. X-ray spectrum of K-shell krypton ions produced by charge exchange reactions of bare Kr^{36+} and hydrogenic Kr^{35+} ions with neutral krypton atoms.

served L-shell emission spectrum has one dominant peak. This peak is assigned to transitions of the type $3s_{1/2} \to 2p_{3/2}$. X-ray features at higher energy are seen as well, extending all the way to the series limit of $n \to 2s$ prompt decay in lithiumlike uranium at 32.5 keV.

The charge-exchange-induced spectra can be compared to spectra produced by electron-impact excitation. In Fig. 4(b) we show the L-shell emission spectrum of uranium produced by direct electron impact recorded before the beam was turned off. Instead of one prominent line, three prominent lines are seen. These are assigned to transitions of the type $3s_{1/2} \to 2p_{3/2}$, $3d_{3/2} \to 2p_{5/2}$, and $3p_{1/2} \to 2s_{1/2}$. The intensity of x-ray emission from the series limit $n \to 2$ is comparatively much weaker in Fig. 4(b) than Fig. 4(a).

Because the background neutral pressure in the EBIT trap is low ($\leq 10^{-9}$ torr) charge transfer processes involving background atoms happen on a rather long time scale. This was borne out by measuring the x-ray signal produced as a function of time in the magnetic mode. As illustrated in Fig. 5, the x-ray signal decays on a time scale of hundreds of milliseconds in EBIT. Measurements on the high-energy SuperEBIT facility, which has an even lower base pressure ($\leq 10^{-10}$ torr), have shown that charge transfer processes happen on time scales of multiple seconds [9,12,13,18].

The rate of decay can be influenced by injecting gases at varying pressures. The x-ray emission and ion storage time as a function of injection pressure was studied in Ref. [9]. Gas injection takes place by using a ballistic gas injector, as indicated in Fig. 2. We operate the injector either in a continuous mode [9] or in a pulsed mode [18].

In order to concentrate the charge exchange-induced x-ray emission into a a very short time interval, it is best to use the pulsed injection system. This is illustrated in Fig. 6, which shows the x-ray emission from the uranium L-shell as a function

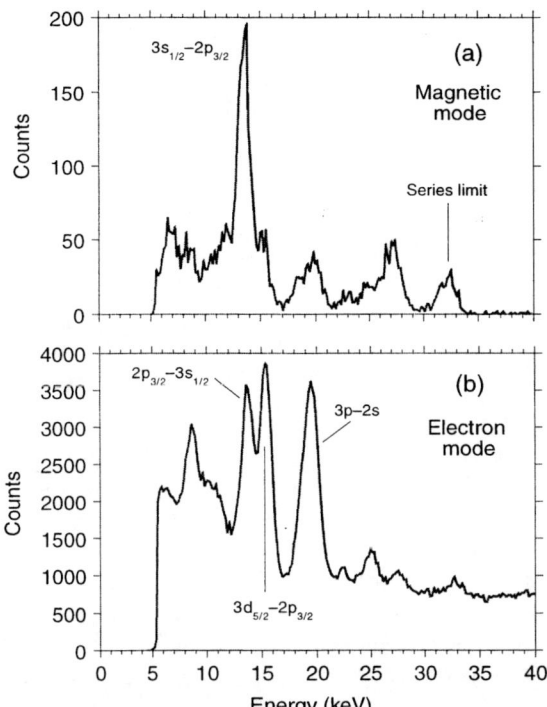

FIGURE 4. X-ray emission of L-shell uranium ions U^{q+}, $q = 83+, ...90+$. (a) spectrum produced by charge exchange reactions with neon atoms; (b) spectrum produced by electron-impact excitation with a 120-keV electron beam.

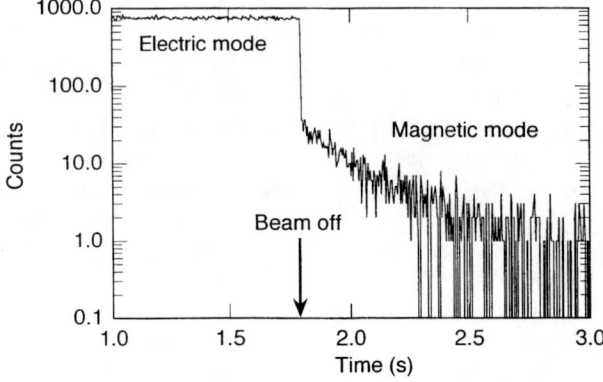

FIGURE 5. Time dependence of the K-shell x-ray emission produced by charge transfer reaction of bare Fe^{26+} and hydrogenic Fe^{25+} ions with background gases.

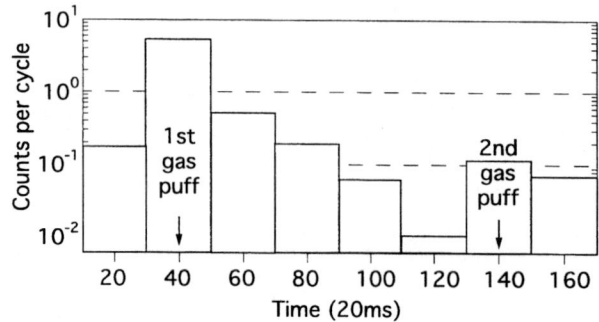

FIGURE 6. L-shell x-ray emission produced by charge transfer reactions of highly charged uranium ions U^{q+}, $q = 83+, ...90+$ with neon introduced into the trap with a pulsed gas injector.

of time. A puff of neon gas is introduced into the trap via a fast acting ($\approx 20\mu s$) gas valve. The puff strongly increases the number of charge exchange reactions and concentrates the measured x-ray flux into a single time bin. A second puff, delayed by by 100 msec, is used to check whether all or most ions have recombined. It produces an x-ray flux that is fifty times less than that produced by the first puft. This demonstrates that the first puff indeed is sufficient to force charge exchange recombination of most uranium ions with an open L-shell in the trap.

RESULTS AND DISCUSSION

The observed spectral emission can be understood by noting that the impact parameter is very small for cold collisions. The spectra are, therefore, very different from those observed in high-energy collsions where the impact parameter is large. In high energy collisions, the values of angular momentum of the captured electron are statistically populated. This is not the case in low-energy collisions. A smaller impact parameter reduces the angular momentum value of the captured electron and limits it to a small subset of values [17]. A dramatic illustration of this is given in Fig. 7, where we plot the predictions of classical trajectory Monte-Carlo calculations [19] for the capture cross section into a particular ℓ value in the reaction $Xe^{54+} + H \rightarrow (Xe^{53+})^* + H^+$. The calculations were carried out for three different collision energies: 250 eV/amu, 2.5 eV/amu, and 0.625 eV/amu. The angular momentum values for 250 eV/amu are clearly statistical. This is not the case for the two lower energies. In the case of 2.5 eV/amu the distribution peaks at $\ell = 5$. The peak drops to $\ell = 3$ for the 0.625 eV/amu collision energy. The calculations clearly show that the x-ray emission patterns will differ from the that produced by statistical population of the ℓ values only if the collision energy is below 250 eV/amu. Prompt x-ray emission from transitions that proceed from the capturing level to ground is extremely weak in high-energy collisions. The prompt emission

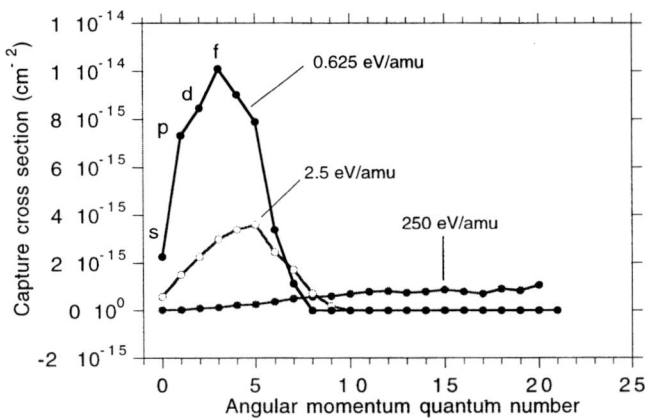

FIGURE 7. Predicted distribution of angular momentum values of electrons captured in the reaction $Xe^{54+} + H \rightarrow Xe^{53+*} + H^+$. Results of classical trajectory Monte Carlo calculations are shown for the different collision energies indicated.

that we observe can only be detected in the low collision energy regime inaccessible to measurements performed with ions from heavy-ion accelerators or ECR sources.

As the collision energy drops, the fraction of electrons captured into an $\ell = 1$, i.e. p, state increases. p electrons can decay directly to the s ground level in hydrogenlike ions. Electrons captured with angular momentum $\ell = 1$ produce prompt x-ray emission by decaying directly from the n level into which they were captured to the $1s$ grond level. Electrons with other angular momentum values will deexcited via intermediate steps, typically by ultimately producing a $n = 2 \rightarrow n = 1$ x ray. The ratio of the intensity of the prompt x rays from high n levels to the intensity of the x rays from the $n = 2$ level, therefore, gives a measure of the fraction of electrons captured into $\ell = 1$ angular momentum states. In Fig. 8 we plot the Monte Carlo prediction for this ratio. The figure clearly shows the increase in the amount of prompt x rays as with decreasing collision energy.

Emission from high-n levels is not only seen in K-shell spectra but also in the L-shell spectra. This was already illustrated in Fig. 4. The intensity of this emission is larger than it is in a comparable spectrum produced by electron-impact excitation. A quantitative measure of this enhancement is given in Table I in the case of B-like Fe^{21+}. This ion has a partially open L-shell; the $2s$ subshell is, however, filled. The comparison between the emission produced by electron-impact excitation and that produced by charge exchange shows that the emission from the $n=7$ level, which is most likely to capture a charge exchange electron, is enhanced by about a factor of five. Like in the case of K-shell ions, the enhancement can serve as a diagnostic indicator that charge-transfer processes contribute to or even dominate the x-ray production.

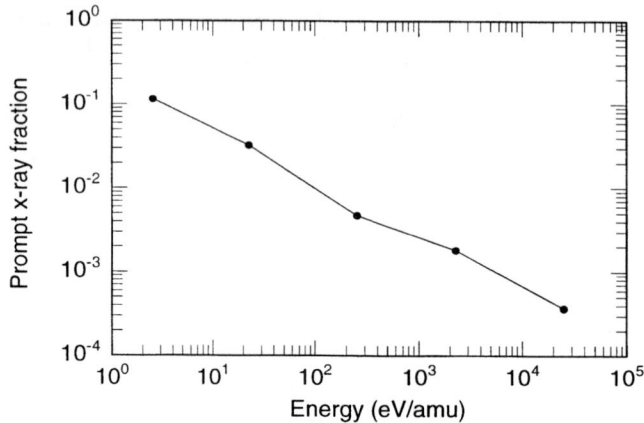

FIGURE 8. Fraction of emitted x rays from prompt radiative decay $np \rightarrow 1s$ as a function of collision energy. The results shown are for the process Xe^{54+} + H $\rightarrow$ Xe^{53+*} + H^{+}.

CONCLUSION

Using the unique capabilities afforded by operating the Livermore electron beam ion trap in the magnetic trapping mode, it was possible to study charge transfer reactions in a regime that was heretofore inaccessible to experimental scrutiny. Direct line-of-sight access to the trap allowed the recording of the x-ray emission produced in the reactions. X-ray emission patterns were observed that were different from the high-energy limit observed using ions from ECR sources or heavy-ion accelerators. We showed that "high" energy refers to energies ≥ 250 eV/amu. The emitted x-ray pattern above this limit is virtually independent of collision energy. Below this energy limit, the angular momentum of the captured electron changes with energy resulting in a change of the x-ray emission pattern. By studying this emission pattern, measurements allow us to determine the angular momentum values of the captured electron states. In particular, we can determine the fraction of capture into a p state directly from the prompt x-ray emission in K-shell spectra. Studies are now under way to exploit this fact for detailed testing of atomic calculations and for developing spetral plasma diagnostics.

TABLE 1. Results from L-shell measurements of boronlike Fe^{21+}. The results are normalized to the intensity of the $3 \rightarrow 2$ x rays.

Transition	Intensity Electron-Impact Excitation	Intensity Charge-Transfer Excitation
$3 \rightarrow 2$	100	100
$7 \rightarrow 2$	12.2	60
Series limit	6.2	26

ACKNOWLEDGMENTS

This work was supported in part by the Office of Basic Energy Sciences of the U. S. Department of Energy and performed under the auspices of the Department of Energy by Lawrence Livermore National Laboratory under contract W-7405-ENG-48. We are also grateful for travel support received from the Deutscher Akademischer Austausch Dienst.

REFERENCES

1. R. A. Phaneuf, R. K. Janev, and H. T. Hunter, Nuclear Fusion Special Supplement **7**, 7 (1985).
2. M. Pieksma and C. C. Havener, Phys. Rev. A **57**, 1892 (1998).
3. H. Cederquist and A. Bárány, Phys. Scripta **T37**, 94 (1991).
4. M. G. von Hellermann, W. Mandl, H. P. Summers, H. Weisen, A. Boileau, P. D. Morgan, H. Morsi, R. Koenig, M. F. Stamp, and R. Wolf, Rev. Sci. Instrum. **61**, 3479 (1990).
5. R. C. Isler, Plas. Phys. Contr. Fusion **36**, 171 (1994).
6. D. G. Whyte, R. C. Isler, M. R. Wade, D. R. Schultz, P. S. Krstic, C. C. Hung, and W. P. West, Phys. Plasma **5**, 3694 (1998).
7. Th. Stöhlker *et al.*, Phys. Rev. A **58**, 2043 (1998).
8. A. Chutjian, private communication (1999).
9. P. Beiersdorfer, L. Schweikhard, J. Crespo López-Urrutia, and K. Widmann, Rev. Sci. Instrum. **67**, 3818 (1996).
10. P. Beiersdorfer, L. Schweikhard, R. Olson, G. V. Brown, S. B. Utter, J. R. Crespo López-Urrutia, and K. Widmann, Phys. Scripta **T80**, 121 (1999).
11. D. Schneider, D. A. Church, G. Weinberg, J. Steiger, B. Beck, J. McDonald, E. Magee, and D. Knapp, Rev. Sci. Instrum. **65**, 3472 (1994).
12. P. Beiersdorfer, B. Beck, St. Becker, and L. Schweikhard, Int. J. Mass Spectrom. Ion Proc. **157/158**, 149 (1996).
13. P. Beiersdorfer, B. Beck, R. E. Marrs, S. R. Elliott, and L. Schweikhard, Rapid Commun. Mass Spectrom. **8**, 141 (1994).
14. P. Beiersdorfer, A. L. Osterheld, V. Decaux, and K. Widmann, Phys. Rev. Lett. **77**, 5353 (1996).
15. P. Beiersdorfer, V. Decaux, and K. Widmann, Nucl. Instrum. Methods **B98**, 566 (1995).
16. P. Beiersdorfer, in *AIP Conference Proceedings 389, X-Ray and Innershell Processes*, AIP, edited by R. L. Johnson, H. Schmidt-Böcking, and B. F. Sonntag (AIP, Woodbury, NY, 1997), p. 121.
17. J. Burgdörfer, R. Morgenstern, and A. Niehaus, Nucl. Instrum. Methods **B 23**, 120 (1987).
18. L. Schweikhard, P. Beiersdorfer, G. V. Brown, J. R. Crespo López-Urrutia, S. B. Utter, and K. Widmann, Nucl. Instrum. Methods **B 142**, 245 (1998).
19. R. E. Olson, J. Pascale, and R. Hoekstra, J. Phys. B **25**, 4241 (1992).

Snapshot of Highly Charged Molecular Ions

H. Shiromaru, T. Nishide, T. Kitamura, F. A. Rajgara*, J. S. Sanderson[†],
Y. Achiba, and N. Kobayashi

Tokyo Metropolitan University, Tokyo, Japan
[*] Tata Institute of Fundamental Research, Mumbai, India
[†] University College London, London, UK

Abstract. Explosive fragmentation of highly charged molecular ions has been studied by
a position sensitive time-of-flight (TOF) technique. The highly charged molecular ions
of CO_2, NO_2 and CD_4 were produced by 90-120 keV collisions of Ar^{n+} (n=6,8). By the
detailed analysis of the 3-dimensional velocity vectors of the fragment ions, molecular
images at the instant of multiple ionization are "reconstructed", which are consistent
with known structure of the neutral molecules. This in turn means that the dissociation of
the highly charged ions is reasonably described by the pure Coulomb explosion scheme.

INTRODUCTION

The dynamics of highly charged molecular ions attracts increasing attention in the
recent years, and has present updated topics in various fields, for examples, matters in
the intense electric field, ionization of an inner electron and subsequent Auger decay,
site selective dissociation, and collisions of highly charged ions with molecules.
Generally speaking, the highly charged small molecular ions are quite unstable and
dissociate into fragment ions very quickly. Besides the interest in these ultra fast
reactions themselves, the application of these reactions to the molecular and cluster
science is also attractive since the information on the target molecules may be obtained.
This technique so called "Coulomb Explosion Imaging" works under the conditions
that the dissociation is practically described by the pure Coulomb explosion scheme.
That is, the point charge is put on each atom, and the motion of it is governed by the
Coulomb repulsion by itself. This is not exactly true unless the electrons in the
molecule are fully stripped. However in many cases, pure Coulomb explosion scheme
is considered to be a good approximation.

In fact, the validity of this assumption has been already certified for highly charged,
but not fully stripped, molecular ions, which are produced by passing through a thin
foil (1). The Coulomb explosion imaging technique combined with the foil-stripping
gives reasonable molecular structure for many kinds of molecular ions, for example,
CH_4^+ (2). Since the molecules to be studied should be accelerated to a high energy, the
application of this technique to neutral molecules is not straightforward. To induce
multiple ionization of the neutral molecules, several methods have been so far

CP500, *The Physics of Electronic and Atomic Collisions*, edited by Y. Itikawa, et al.
© 2000 American Institute of Physics 1-56396-777-4/00/$17.00

employed, one photon ionization using X-ray, field ionization using ultra short pulse lasers, or multiple electron transfer using a projectile beam of highly charged atomic ions. Concerning imaging study of molecular structure, at least triatomic molecule should be studied because these are smallest size for the structural isomers. It has been so far reported that, in some case, dissociation can be explained by the pure Coulomb explosion and in some case can not.

When ultra short pulse lasers are tightly focused, the electric field is high enough to induce efficient field ionization giving rise to explosive dissociation. Theoretically it is predicted that the multiple ionization mainly occurs in the intense electric field when the bond length are much longer than ordinal one (3-5). There are many experimental evidences that the molecules are deformed in the laser field, for example, bent CO_2 molecule is found in the laser field (6,7). Such peculiar behavior of the molecules in the intense electric field itself is quite interesting and may open a new field of the molecular science. However, multiple ionization by the short pulse laser seems to be not suitable for the Coulomb explosion imaging study until much shorter pulse (1 fs) laser will be available (8). In addition to this, the difficulty is arising from the fact that the coincidence technique does not work in the pulsed laser study. The covariance approach is commonly used for channel-resolved studies (9), however, this technique can not be readily combined with the position-sensitive TOF measurement.

The ionization of an inner electron followed by the Auger process is a well-known technique to form multiply ionized molecular ions. However, the details of the Auger electron emission and dissociation are not fully understood. From the viewpoint of the dissociation scheme, it has been controversial whether the dissociation process is pure Coulombic or not (10-12). So far, the Coulomb explosion imaging technique has not been applied to the high charged molecular ions, by the ionization of inner electrons.

The collision of an HCI with a molecular target also provides highly charged molecular ions via multiple electron transfer. Though the HCI collision with molecules has been vigorously studied in these several years, the most of the studies have been dedicated for diatomic molecules, and determination of the cross section. The studies on the dissociation process so far reported have been quite few, especially for the molecules larger than diatomic molecules (8, 13-18).

The velocity vectors of dissociating fragments and the correlation among the vectors strongly depend on the dissociation process. Thus, if the molecular structure is reconstructed from these vectors according to the imaging recipe, the obtained molecular image at the instant of the ionization would be a good criterion whether the pure Coulombic scheme is practically acceptable or not. In 1994, Lutz and coworkers have reported the Coulomb explosion experiments on simple molecules by means of HCI collisions, in which the bond angle of H_2O molecule is reasonably reproduced (14). In the series of their work, however, the deviation from the pure Coulomb explosion scheme is emphasized, especially for the kinetic energy release of the fragment ions (14-17).

Recently we reported that the proper image of CO_2 is obtained by 120 keV collision of Ar^{8+}, especially when the charge states of the molecular ions are high (18). However, there are huge combination of the experimental conditions which may affect the dissociation process, for example, charge state of the projectile, collision energies, dissociation channels, number of electrons and atoms in the target molecules, internal temperature of the target molecules and so on. To determine unknown molecular structure, it is necessary to find the conditions for fulfillment of the pure Coulomb explosion scheme. In the present paper, the results for simple molecules, CO_2, NO_2 and CD_4, are described. Since the structure of these molecules is well known, we can judge the validity of the imaging procedure. For CO_2, projectile with different charge states and kinetic energy was used. A part of the results reported in the present paper has been already reported elsewhere (8, 18).

EXPERIMENTAL

The measurements have been performed in an ECR ion source facility of Tokyo Metropolitan University (TMUECRIS) (19). The apparatus for the Coulomb explosion experiments has been reported already, where the 3-dimensional detection system for the fragment ions is described in detail (20). The HCI beam from the ECR ion source was collimated by an aperture with a diameter of 2 mm and collided with the target gas flow introduced through a multi-capillary plate. In the present study, 90 keV Ar^{6+}, 90 keV Ar^{8+}, or 120 keV Ar^{8+} beam was used. Fragment ions produced by the dissociation were collected by the extraction field (700 V/ 3 cm) perpendicular to the incident HCI beam and to the target flow.

The fragment ions were detected by a 40 mm micro channel plate with the MBWC anode, which divides the charge according to the x-y positions (21,22). A fast digital storage oscilloscope recorded the four outputs of the anode, and the waveforms were stored in the computer. This system enables position sensitive TOF measurements for the fragment ions produced by a single dissociation event, which would hit the detector within short time difference (typically several tens - hundreds ns). The trigger of the TOF measurements was detection of an Auger electron emitted from the projectile, which was extracted to the opposite side from the ions. The triple coincidence signals of the fragment ions (in fact this is a four fold coincidence, electron/ion/ion/ion) were assigned to the corresponding dissociation channels of M^{n+} (M=CO_2, NO_2, n=3-6 and M=CD_4, n$\geq$3). From the TOFs of the ions, z components of the velocity vectors were determined, and from the positional information, x and y components of the velocity vector were determined for each fragment ion formed by a single dissociation event.

RESULTS AND DISCUSSION

In the present study, we analyzed the highly charged molecular ions of n$\geq$m where n is the charge states and m is the number of atoms in the molecules. The pure Coulomb explosion scheme is the approximation neglecting the presence of the remaining electrons and interaction with the projectile before and after the collision. The screening effect by the remaining electrons would make the dissociation "slower", or induce a stepwise dissociation. In the extreme case, this makes the highly charged molecular ions stable. This issue is complicated since the screening depends on the electronic states of the highly charged molecular ions. In fact, the presence of various electronic states has been revealed by the spread of the kinetic energy release of the fragment ions, which is much broader than that expected for pure Coulombic energy. Concerning the latter factor, collision and multiple ionization is not exactly "sudden". From the viewpoint of electric field induced by the projectile, the interacting time is about 1 fs for the collision of 120 keV Ar^{8+}. It is of course longer if collision energy is smaller. Since the projectiles are not neutralized after collision, Coulomb repulsion between target and projectile may also induce deviation from pure Coulomb explosion of the target. It is natural to deduce that the screening effect would be less important for the higher charge states of the molecule, while the interaction before and after collision would be more important. In the subsequent section, dissociation of each molecule will be described, focusing attention to the validity of the pure Coulombic scheme.

Dissociation of highly charged CO_2

CO_2 is a typical linear triatomic molecule readily available commercially. The triple coincidence maps of the CO_2 obtained by 90 keV Ar^{6+} and Ar^{8+} are shown in Figs. 1a and 1b, respectively. The each island in the map corresponds to the dissociation channel $CO_2^{n+} \rightarrow O^{a+} + C^{b+} + O^{c+}$ which will be referred to as "(abc) channel" hereafter. Since these maps are merely 2-dimensional, the information available by the triple coincidence measurements is not fully involved, for example, the dissociation channels of (221) and (222) are superficially not well separated in the C^{2+}/O^{2+} island. To demonstrate how it works for the assignment of coincidence signals to the dissociation channels, the example of (221) channel in the 90 keV Ar^{6+} collision will be shown.

The simple TOF spectrum of the fragment ions is shown in Fig. 1c. This spectrum corresponds to that integrate the points over the vertical line in the map (Fig. 1a). As can be seen in the spectrum, the peak separation is not sufficient. When one goes from left to right at the vertical position of 700 ns in the map, the coincidence signals with C^{2+} are obtained as shown in the Fig. 1d. There are mainly four peaks, O^{2+}_f, O^{2+}_b, O^+_f, O^+_b correlating with C^{2+}, where each peak is _not_ formed by a single channel. The peak of O^{3+}_f is also observed. Here, the subscript f and b indicates the fragment is initially emitted forward and backward to the detector, respectively. The blank between forward and backward peaks means the fragments emitted perpendicular to the TOF axis were not detected because of the acceptance angle of the detector. Among these coincidence signals, those also correlate with O^{2+}_b are shown in fig. 1e. There, the peaks of O^{2+}_f and O^+_f can be seen in the spectrum. The presence of the O^{3+} peak is more convincing. Clearly, the each peak consists of a single channel. In this way we can assign the triple coincidence signals to the (221) channel without ambiguity.

It should be noted that the TOF measurements are triggered by the Auger electron detection, which strongly depends on the number of electrons captured by the projectile. Consequently, the comparison can not be made between the dissociation channels in which the charge states of CO_2 are not the same. However, it can be clearly seen that the branching ratio for lower channels are enhanced when the projectile of lower charge states are used, as naturally expected.

Previously, Mathur et al. reported 100 MeV collision of Si^{8+} with CO_2 molecule, in which the kinetic energy release was measured (13). The results suggest that there may be considerable contribution of sequential dissociation. To examine the validity of the pure Coulombic scheme, first, simultaneity of the fragmentation was checked by means of the angular correlation of velocity vectors of three fragment ions, for various fragmentation channels. If the break-up of two CO bonds occurs sequentially, the trajectory of the fragment ions are not simply determined by the initial nuclear configuration. For this purpose, the angle χ (illustrated in the top of the Fig. 2) is a good criterion, which is defined as following equation.

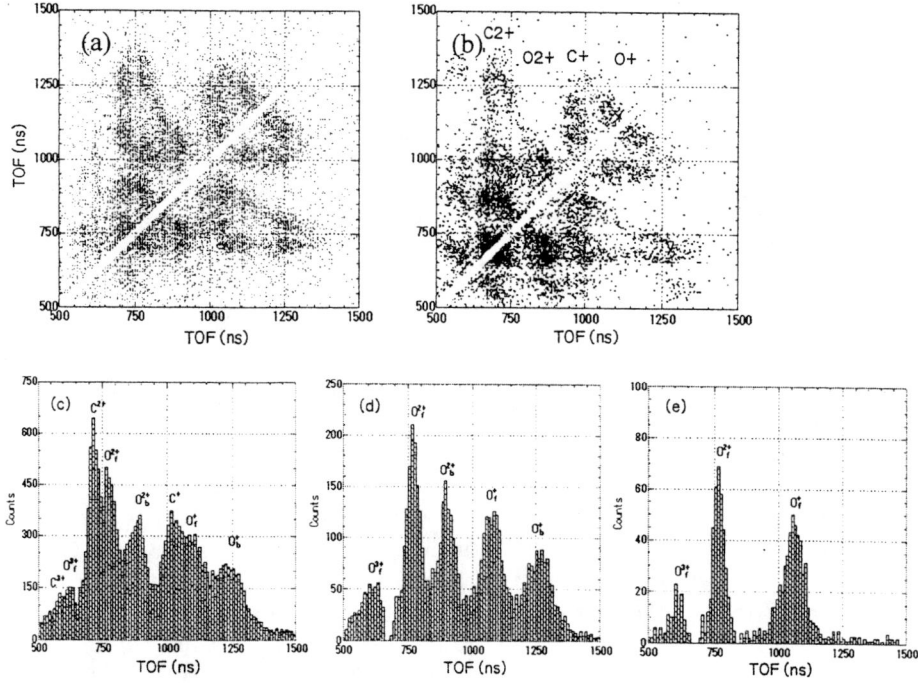

Figure 1. Coincidence map for (a) 90 keV Ar^{6+}/CO_2, and (b) 90 keV Ar^{8+}/CO_2 systems. 1-dimensional TOF spectrum is shown in (c), and that correlated with (d) C^{2+}, (e) C^{2+} and O^{2+}_f are also shown for 90 keV Ar^{6+}/CO_2.

$\vec{v}_c \bullet (\vec{v}_{o1} - \vec{v}_{o2}) = |\vec{v}_c||\vec{v}_{o1} - \vec{v}_{o2}|\cos \chi$, where $\vec{v}_c$, $\vec{v}_{o1}$, $\vec{v}_{o2}$ are the velocity vectors of the fragment ions. Figures 2a-c show the χ angle distribution for (222), (111) and (221) channels obtained for the collision of 90 keV Ar^{6+}. As shown in Figs. 2a and 2b, in which the charge states of two outgoing oxygen atoms are the same (namely the symmetric channel), χ angle distribution centers around 90 degree. This means the dissociation is highly simultaneous. For the (221) channel (asymmetric channel, shown in Fig. 2c), the peak position deviates from the center also being consistent with the fact the carbon ion is kicked more by O^{2+} than O^+.

By the same procedure shown in the χ angle distribution for the collision of 90 keV Ar^{6+}, the simultaneity is also certified for the collision of 120 keV (details are given in

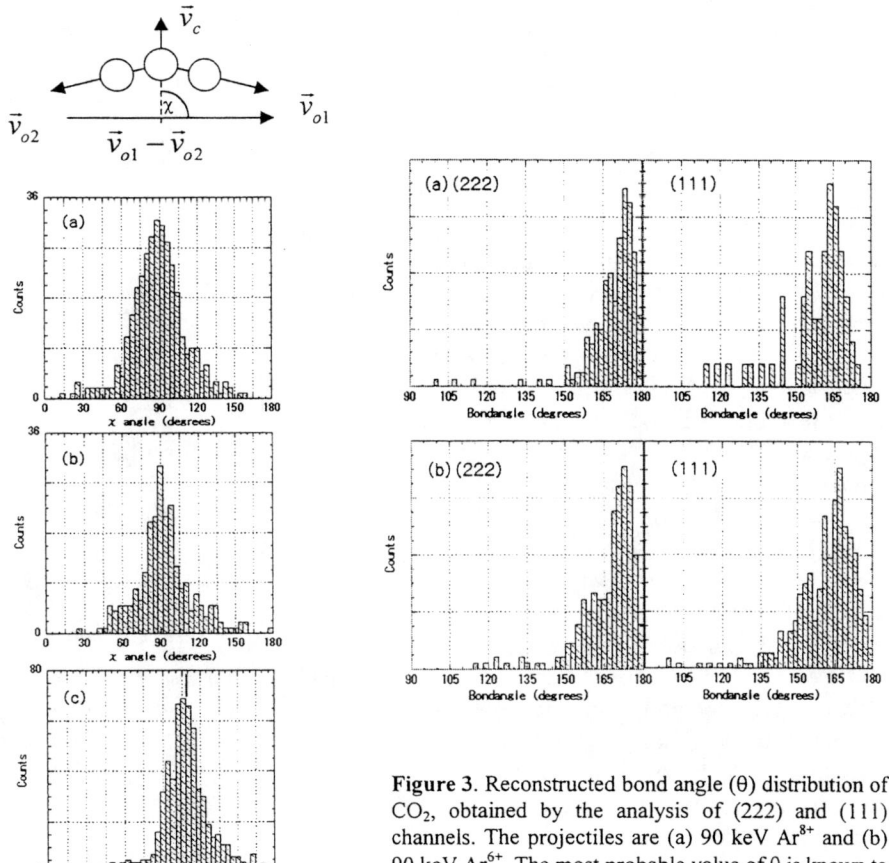

Figure 3. Reconstructed bond angle (θ) distribution of CO_2, obtained by the analysis of (222) and (111) channels. The projectiles are (a) 90 keV Ar^{8+} and (b) 90 keV Ar^{6+}. The most probable value of θ is known to be 174 degree.

Figure 2. χ angle distribution for a: (222), b: (111) and c: (221) channels. The definition of the angle is shown schematically. The vertical line in (c) indicates the χ angle based on the pure Coulomb Explosion model.

Ref. 17) and 90 keV Ar^{8+}. This means bond angle may be reconstructed from the angular correlation of two outgoing oxygen ions. As described in literatures, the most probable value of the bond angle is not 180 degree but 174 degree, due to the zero point vibration of degenerate bending mode (23). Figures 3a and 3b show bond angle (θ) distribution of CO_2 obtained by the collision of 90 keV Ar^{8+} and 90 keV Ar^{6+}. As shown in these figures, the peak of the reconstructed angles reasonably reproduces the known values of neutral CO_2 molecules. Especially in the case of the highly ionized CO_2 (n = 6), the agreement is very good. For the lower charge states, small deviation from 174 degree has been found. This is true irrespective of the charge of the projectile and collision energy (the results of 120 keV collision of Ar^{8+} is shown in Ref. 17). These results suggest that production of higher charge states is favorable for the fulfillment of a pure coulombic dissociation scheme. This indicates the major factor affecting the fragmentation processes of the highly charged CO_2 molecule is the screening of the remaining electrons.

Dissociation of highly charged NO_2

Nitrogen dioxide is a typical bent molecule with C_{2v} symmetry. The bond angle is known to be 134 degree. Although the approach similar to the case of CO_2 molecule was applied to NO_2 molecule, the assignment of the triple coincidence event to the dissociation channel was not easy. The analysis of NO_2 molecule is more complicated because of the difficulty arising from its bent structure and small difference in masses between carbon atom and nitrogen atom. For example, TOFs of N^{n+} dissociated backward from the detector were occasionally larger than heavier O^{n+} dissociated forward to the detector, depending on the molecular orientation respect to the TOF axis. As demonstrated in the Figs. 1a-e, at least one fragment ion should be well assigned to proceed next step (picking up the signals

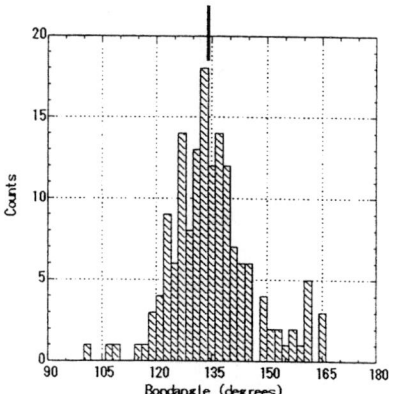

Figure 4. Reconstructed bond angles distribution of NO_2 obtained by the collision of 120 keV Ar^{8+}. The vertical line indicates the known bond angle of NO_2.

correlated with specific ions) of the assignment procedure. Thus, the number of the events clearly assigned is smaller than the CO_2 case. Even so, as shown in Fig. 4, the reconstructed bond angle is well consistent with the known molecular structure. Comparing Fig. 4 with Figs. 3a-c, it can be deduced that we can judge whether a triatomic molecule is linear or bent by means of the Coulomb explosion imaging.

Dissociation of highly charged CD$_4$

Methane is a typical molecule of a high symmetry (T$_d$). To avoid confusion arising from residual gases, deuterited methane, CD$_4$, was employed in the present study. The collision experiment was performed using 120 keV Ar^{8+}. Here, number of events of 5-fold coincidence, that is, coincidence of one C^{n+} and 4 D$^+$s, were not enough for reliable statistics. Thus the triple coincidence events, in which C^{n+} and two D$^+$s are detected, are analyzed based on the following assumption. First, C atom does not possess initial kinetic energy after dissociation event. This would be acceptable since the C atom is much heavier than D atom, and the repulsion of 4 D$^+$ would be effectively cancelled. The other assumption is that all the D atoms are ionized. It would be true when the charge states of C atom are high.

The triple coincidence map of CD$_4$$^{n+}$ is shown in Fig. 5 in which the island of C^{n+} (n=1-3)/D$^+$ and D$^+$/D$^+$ can be seen. The separation of the two D$^+$ islands correlated with C^{n+} becomes larger for higher charge states. Since the number of the events assigned to C^{2+}/D$^+$/D$^+$ dissociation channel, which is actually expected to correspond to the dissociation channel of CD$_4$$^{6+}$ → C^{2+} + 4D$^+$, is relatively large, this channel is analyzed in detail.

The histogram of the reconstructed bond angle is shown in Fig. 6. It should be noted that this simplified procedure is based on the fact that all the DCD angles are equivalent.

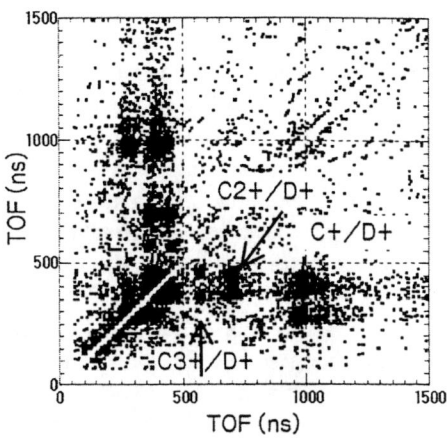

FIGURE 5. Triple coincidence map obtained for CD$_4$. Signals at 300 ns are due to H$^+$ impurity.

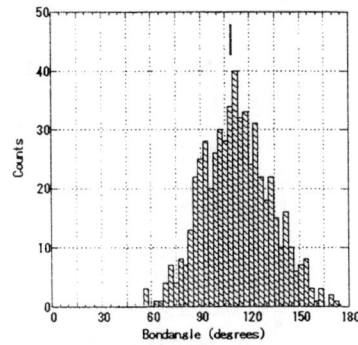

FIGURE 6. Reconstructed bond angles distribution of CD$_4$ obtained by the collision of 120 keV Ar^{8+}. The vertical line indicates the angle for T$_d$ symmetry (109 degree).

As shown in the figure, the peak locates at about 109 degree, again being consistent with the pure Coulomb explosion model, in spite of the fact that two of the fragment ions were not measured. The results suggest that a kind of local structure of the molecule may be obtained by this method.

CONCLUSION

For all target molecules examined in the present study, it is confirmed that the pure Coulomb explosion scheme is a reasonable approximation. The imaging procedure gives proper instantaneous bond angles at the timing of the ionization, namely, a snapshot of the highly charged molecular ions were obtained. This method will be applicable to the structure-unknown molecules in near future. General trend can be seen that the higher charge states of the molecules dissociate more purely Coulombic way. This means the screening of the remaining electrons is the most important factor determining the dissociation scheme. The interaction between the HCI projectile and the target before and after the collision seems to be a minor factor compare to the screening.

Finally, the reconstruction of the bond length should be mentioned briefly. For the systems examined so far, the kinetic energy distribution obtained in the present study is much broader than that expected from zero point vibrations of neutral molecules, probably due to the presence of various electronic states of the highly charged ions. The results are consistent with many previous reports. Thus, the distribution of bond length can not be reconstructed properly from the velocity vectors. However, the mean values of the kinetic energy distributions are close to the values based on the pure Coulomb explosion model. This forms a sharp contrast with the case of the multiple ionization/fragmentation in a strong laser field. Thus, the mean value of the bond length may be obtained within reasonable accuracy, though presently we have been focusing the attention to bond angles, which are critical for determining whether the isomer is really present or not.

ACKNOWLEDGMENTS

We wish to thank professor T. Mizogawa for his help in the use of the MBWC anode. The members in the Atomic Physics Laboratory of Tokyo Metropolitan University are gratefully acknowledged for their help in the operation and maintenance of the ECR ion source. F.A.R. acknowledges support from Japan Society for the Promotion of Science (DST-9815). This work is partly supported by Grant-in-Aid from the Ministry of Education, Science, and Culture (11440178).

REFERENCES

1. Vager, Z., Naaman, R., and Kanter, E, *Science* **244**, 426 (1989).
2. Vager, Z., Kanter, E. P., Both, G., Cooney, P. J., Fabis, A., Koenig, W., Zabranski, B. J., and Zajfman D., *Phys. Rev. Lett.* **57**, 2793 (1986).
3. Posthumus, J. H., Frasinski, L. J., Giles, A. J., and Codling, K., *J. Phys.* **B 28**, L349 (1995).
4. Seideman, T. Ivanov, M. Yu., and Corkum, P. B., *Phys. Rev. Lett.* **75**, 2819 (1995).

5. Zuo, T., and Bandrauk, A. D., Phys. Rev. **A 52**, R2511 (1995).
6. Cornaggia, C., *Phys. Rev.* **A 54**, R2555 (1996).
7. Sanderson, J. S., Thomas, R. V., Bryan, W. A., Newell, W. R., Langley, A. J., and Taday, P. F., *J. Phys.,* **B 31**, L599 (1998).
8. Sanderson, J. S., Nishide, T., Shiromaru. H., Achiba, Y., and Kobayashi, N., *Phys. Rev.* **A 59**, 4817 (1999).
9. Frasinski, L. J., Codling, K., and Hatherly, P. A., *Science* **246**, 1029 (1989).
10. Ankerhold, U., Esser, B., and von Busch, F., *J. Phys.* **B 30** 1207 (1997).
11. Eland, J. H. D. and Sheahan, J. R., *Chem. Phys. Lett.* **223**, 531 (1994).
12. Simon, M., Lavollee, M., Meyer, M., and Morin, P., *J. Electr. Spectrosc. Relat. Phenom.* **79**, 401 (1996).
13. Matur, D., Krishnakumar, E., Rajgara, F. A., Reheja, U. T., Krishnamurthi, V., *J. Phys.* **B 25**, 2997 (1992).
14. Werner, U. Beckord, K. Becker, J. and Lutz, H. O., *Phys. Rev. Lett.* **74**, 1962 (1995).
15. Werner, U. Beckord, K. Becker, J., Folkerts, H. O. and Lutz, H. O., *Nucl. Instr. and Meth.* **B 98**, 385 (1995).
16. Werner, U. Becker, J., Farr, T., and Lutz, H. O., *Nucl. Instr. and Meth.* **B 124**, 298 (1997).
17. Werner, U. and Lutz, H. O., *The Physics of Electronic and Atomic Collisions*, AIP Press, 1995, pp.741-751.
18. Shiromaru H., Nisgide, T., Kitamura, T., Sanderson J. S., Achiba, Y., Kobayashi, N., *Physica Scripta* **T 80** 110 (1999).
19. Tanuma, H., Mizutani, M., Kobayashi, N., and Shiromaru, H., *Proc. 12th Int. Workshop on ECR ion Source* 303 (1995).
20. Shiromaru, H., Kobayashi, K., Mizutani, M., Yoshino, M., Mizogawa, T., Achiba, Y.and Kobayashi, N., *Physica Scripta* **T 73**, 407 (1997).
21. Mizogawa, T., Awaya, Y., Isozumi, Y., Katano, R., Ito S., and Maeda, N., *Nucl. Instr. and Meth.* **A 312** 547 (1992).
22. Mizogawa, T., Sato M., and Awaya, Y., *Nucl. Instr. and Meth.* **A 366**, 129 (1995).
23. Hsieh S. and Eland, J. H. D., *J. Phys.* **B 30**, 4515 (1997).

Electron Emission from Slow Ar^{17+} Ions Interacting with a Si Surface

N. Stolterfoht, V. Hoffmann, D. Niemann, and J.-H. Bremer

Hahn − Meitner Institut Berlin GmbH, Glienickerstr. 100,
D − 14109 Berlin, Germany

Abstract. Experimental and theoretical methods were used to study the formation of hollow atoms during the interaction of Ar^{17+} projectiles with a Si(111) surface. Electron spectra were taken in a wide range of projectile energies from 17 eV to 170 keV. Peak structures due to K- and L-Auger electrons were found to vary significantly with the projectile energy. The angular dependence of the Auger electron intensities were interpreted by means of model calculations yielding information about the decay depth of the hollow atoms within the first layers of the surface.

INTRODUCTION

In the past, considerable interest has been devoted to the interaction of slow and highly charged ions with a surface, see e.g., Burgdörfer [1] and Arnau *et al.* [2]. This interest has been motivated through the search for unique properties of hollow atoms [3,4] created by highly charged ions approaching the surface with a velocity much smaller than the Fermi velocity. A slow and highly charged ion acquires several electrons by resonant charge transfer into Rydberg states from the valence band of the solid. Thus, hollow atoms are produced with many electrons in higher orbitals and empty intermediate shells. Hollow atoms may undergo autoionizing transitions [5] where the electron-electron interaction leads to the relaxation of an electron and the ejection of another electron from a Rydberg orbital. When the projectile further approaches the surface, electrons are captured into lower lying orbitals accompanied by a flow back of Rydberg electrons into the solid or the continuum [6]. Thus, the diameter of the hollow atom gradually shrinks as it approaches the surface.

At the surface the remaining Rydberg electrons are removed (peeled off) or enter into the solid together with the ion to participate in the formation of a strong screening cloud of valence band electrons. This cloud, referred to as the C shell [7], leaves inner shells empty so that a compact hollow atom is formed below the surface. The hollow atom below the surface is significantly smaller than that occurring outside the solid. Although of smaller extension, the C shell is rather dense since it

CP500, *The Physics of Electronic and Atomic Collisions*, edited by Y. Itikawa, et al.
© 2000 American Institute of Physics 1-56396-777-4/00/$17.00

fully neutralizes the multiply-charged ion core [7]. Below the surface, hollow atoms can be verified from screening effects on Auger transition energies revealing both the emptiness of inner shells and the existence of the dense C shell [8]. Finally, when the hollow atoms move within the solid, they transfer their potential energy via Auger transitions and collisional charge transfer [2].

Detailed information about the properties of hollow atoms has been achieved from the spectroscopy of x-rays and Auger electrons ejected in conjunction with the cascading decay of the projectile. In early pioneering work, Briand and collaborators [4,9,10] used the method of x-ray spectroscopy to study highly charged Ar projectiles with an initial K-shell vacancy interacting with a surface. For Ar this method has advantages due to the fact that it has a relatively large fluorescence yield. Alternatively, Auger-electron spectroscopy is particularly suitable for first-row atoms which have been studied in detail by that method [11–15].

A major question that has been raised in various studies is whether hollow atoms decay above or below the surface. For incident Ar^{17+}, Briand et al. [10] attributed the apparent changes of high-resolution x-ray spectra obtained in a wide energy range to different projectile locations during the x-ray emission, i.e., above and below the surface. For lighter ions, such as neon and nitrogen, only certain fractions of the K-Auger spectra have been attributed to above-surface emission [11,13,16], whereas the major part is found to be due to below-surface emission. The latter finding has been supported by detailed cascade models that describe individual configurations of hollow atoms [17,18].

The models for individual configurations involving the stepwise filling of inner-shell vacancies are particularly useful for first-row atoms, as the number of unfilled inner shells is limited. The complexity of the filling cascade strongly increases as the number of empty inner shells increases. Therefore, previous models for the filling of argon ions [19–21] were limited to treatments where mean numbers for the shell occupation have been evaluated. Only recently, configuration models has been set up for the description of hollow argon [22,23].

In the present work, L- and K-Auger electrons from hollow Ar atoms are studied using the method of Auger spectroscopy. The measured Auger spectra are interpreted by means of a cascade model treating individual configurations. The analysis of the measured spectra shows that Auger spectroscopy is well suited to gain information about dynamic properties of hollow argon atoms moving just below the surface.

EXPERIMENTAL METHOD AND RESULTS

The experiments were performed using the 14.5-GHz ECR source at the Ionen-strahl-Labor (ISL) of the Hahn-Meitner-Institut Berlin. The experimental set-up has been described in detail previously by Grether et al. [24] The ion source provides projectiles with energies up to 15q keV where q is the charge state of the extracted ions. The end of the beam line is equipped with a deceleration lens system to extract ions at energies as low as 1q eV. In this case the beam line is set on a high-voltage so that the experimental apparatus can be operated on ground potential. The experiments were conducted in an ultra-high-vacuum chamber including an electron spectrometer and facilities for surface preparation and examination. During the measurements a base pressure of a few 10^{-10} mbar was maintained in the chamber.

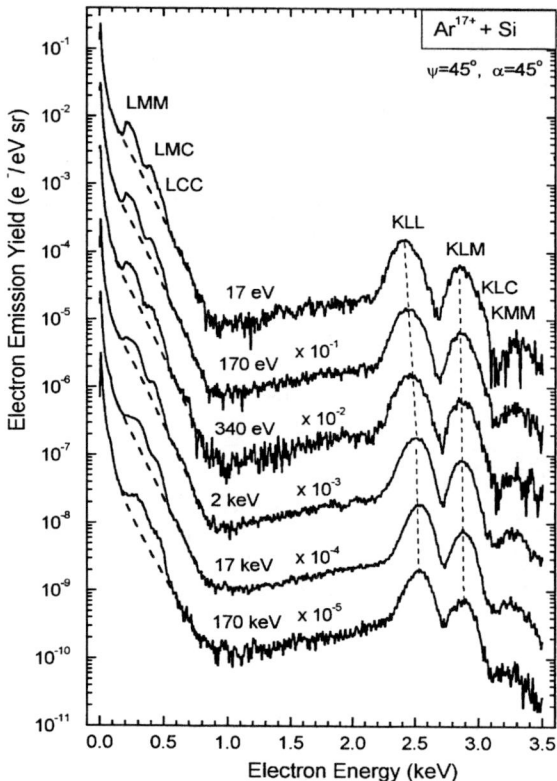

FIGURE 1. Electron spectra produced by Ar^{17+} incident at $\psi = 45^0$ on a Si(111) surface. The electron observation angle is $\alpha = 45^0$. Note that for graphical reasons the spectra are multiplied by the factors indicated in the figure.

In this study we used hydrogen like Ar^{17+} ions for the first time. The ions were magnetically analyzed and collimated to a diameter of about 1 mm at the position of the Si target. Electrons ejected by the interaction of the ions with the surface were measured with an electrostatic parallel-plate spectrometer whose observation angle could be varied. The electron spectra were normalized to an absolute scale taking the acceptance angle, resolution, and transmission of the spectrometer and the efficiency of the channeltron into account.

Figure 1 shows representative examples of electron spectra acquired in a wide range of projectile energies. The Auger spectra exhibit different peaks which can be attributed to the shell structure of the hollow atoms formed near the Si surface. In particular, the spectra indicate peaks due to L- and K-Auger transitions in hollow Ar. These peaks show that both L and M shells exist in hollow Ar atoms when moving inside the solid. In addition, the highly charged Ar ion induces the negative charge cloud C, which coincides with the N shell. The experiment provides clear evidence for the C shell (see for instance the Auger transitions LMC and KLC).

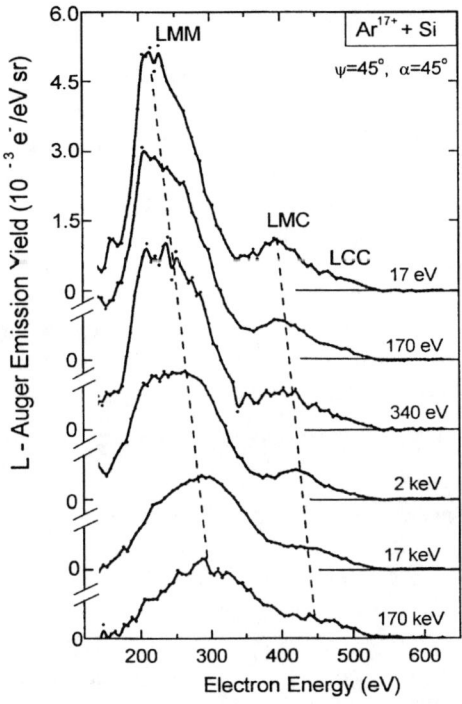

FIGURE 2. Emission yield for L-Auger spectra acquired for Ar^{17+} incident on a Si surface. The data are extracted from Fig. 1. The continuous background was subtracted from the Auger spectra. Note the shift of the peak energies with varying projectile energy indicated at the spectra.

The electron spectra are found to remain essentially the same both in shape and absolute value when the projectile energy is varied. However, closer inspection of the KLL and KLM Auger peaks shows a line shift with varying projectile energy. This line shift can be attributed to the decreasing population of the L shell during the K-Auger transition. The variation of the L-shell population has previously been observed in high-resolution x-ray spectra [10] which are composed of individual lines, each one associated with a specific number of electrons in the L shell. In fact, the number of L-shell electrons increases as the projectile energy increases.

The increased population of the L-shell is produced by an enhancement of LMM Auger transitions following an increasing population of higher orbitals [23]. When the number of L-shell electrons increases with the projectile energy, the centroid energy of the emission spectra is shifted accordingly.

From Fig. 1 it is noted that there is no indication for Si K-Auger transitions in the present electron spectra. Possible peak structures due to Si K-Auger electrons, expected near 2 keV, are covered by the continuous background. The present spectra were used to estimate an upper-limit value of about 0.05 for the ratio of Si to Ar K-Auger electron intensities. This value corresponds to about 0.02 for the Si to Ar K-x ray intensity ratio which agrees with the results recently obtained by Lehnert *et al.* [20].

Figure 2 shows an enlarged portion of the electron spectra in an energy range covering the Ar L-Auger structures. The continuous background of the spectra was subtracted from the L-Auger intensities. It is seen that the centroid energy of the L-Auger peak varies in a similar manner as noted for the K-Auger electrons. This can be attributed to screening effects, since L-Auger transitions at low projectile energies take primarily place after the K-shell filling, whereas at high energies the L-Auger transitions occur before the K-shell filling [10,23]. Furthermore, contrary to the K-Auger structures, the L-Auger peak decreases in intensity as the projectile energy increases. This decrease is attributed to attenuation effects on the Auger electrons ejected in deeper layers of the solid as discussed in more detail below.

ANALYSIS OF THE AUGER SPECTRA

Projectile Energy Dependence

To obtain additional information about the decay properties of hollow atoms, the projectile energy dependence of the spectral structures are studied in more detail. Auger electrons ejected within the solid suffer attenuation effects which increase with increasing projectile energy. Early observation of significant attenuation effects on Ar L-Auger electrons have been made by de Zwart *et al.* [25]. More recently, systematic studies of attenuation effects on Ne K-Auger electrons have been performed by Grether *et al.* [24].

In the following we consider single differential emission yields obtained from the Auger spectra after integration over electron energy. The results for the L- and K- Auger electrons are shown in Fig. 3. To allow for a comparison with the total number of ejected Auger electrons, the emission yields are multiplied by 4π. The total number of vacancies that can be filled per incident ion are 1 and 10 for the K and L shell, respectively. (Note that the K-Auger transition increases the number of L-shell vacancies by 2.)

It is seen that the emission yield for K-Auger electrons is smaller than unity. This finding is due to competing K x-ray transitions. The branching ratio for the x-ray and Auger transitions is governed by the fluorescence yield ω_K or the corresponding Auger yield $a_K = 1 - \omega_K$. The fluorescence yield and, hence, the Auger yield depends on the number of electrons occupying the L shell [26,27]. As mentioned, the number of L-shell electrons increases with the projectile energy [10] so that the increase of the intensity of the K-Auger electrons (Fig. 3) may be understood.

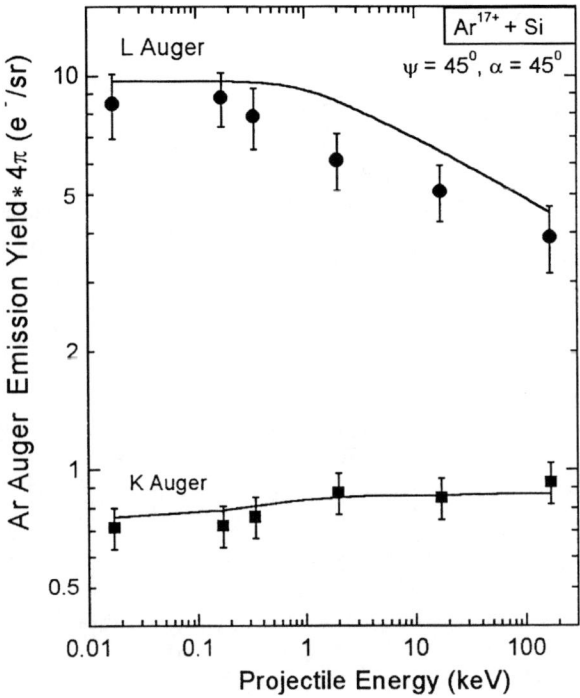

FIGURE 3. Integrated K- and L-Auger emission yields obtained for Ar^{17+} incident on a Si surface at $\psi = 45^0$. Observation angle is $\alpha = 45^0$. The solid curves originate from corresponding model calculations.

As shown further below, attenuation effects for the K-Auger electrons of ~2.5 keV are negligible at an observation angle of 45^0. Since competing processes other than x-ray and Auger transitions are not expected for the filling of the K-shell vacancy, the present results for the K-Auger emission yields y_a can directly be compared with a_K. Similarly, the corresponding x-ray yields $y_x = 1 - y_a$ can be compared with ω_K. The values of y_x are found to vary significantly from 0.28 to 0.08 as the energy increases from 17 eV to 170 keV. These values are consistent with the fluorescence yield determined theoretically as a function of the L-shell occupation number [26,27]. However, some care is required when fluorescence yields are compared with the present data due to the uncertainties of the experimental values (Fig. 3).

For the filling of the L-shell vacancies, competing x-ray transitions can be neglected [27]. Indeed, for low projectile energies the experimental results for the L-Auger emission yields are close to the expected maximum value of 10. However, for increasing projectile energy the L-shell emission yield decreases significantly. This finding provides clear evidence for attenuation effects acting on the L-Auger electrons. With increasing energy the projectile penetrates deeper into the solid so that the emission depth of the L-Auger electrons is increased.

In Fig. 3 the experimental data are compared with calculations modelling the cascading decay of hollow Ar atoms [23]. A detailed description of the cascade model is beyond the scope of the present article so that only a few details can be given here. The model is based on a set of rate equations which describe the occupation probabilities of individual electron configurations involving the K, L, and M shells. The N shell considered above the surface is assumed to merge into the C shell inside the solid. A mean value is used for the occupation number of the N/C shell. The model contains several parameters which have primarily been taken from literature.

For the calculation of the K-Auger emission yields only a few model parameters are relevant. It is plausible from the previous discussion that the K-Auger data are primarily influenced by the branching ratios for K x-ray and Auger transitions. These parameters have been determined using different methods [26,27] and the corresponding results are found to be in good agreement. Accordingly, the theoretical results for the K-Auger electron emission compare well with the corresponding experimental data (Fig. 3).

For the L-Auger electrons the agreement between the experimental and theoretical results is still reasonable although not as good as that for the K-shell electrons. The L-Auger data are primarily influenced by the attenuation length of $\sim$250 eV for the L-Auger electrons [28]. In addition, the L-Auger data depend sensitively on the filling cascade of the hollow atom which, in turn, governs the emission depth of the L-Auger electrons. The reasonable agreement between experiment and theory indicates that the main features of the decay sequence of the hollow atom are fairly well understood from the present model. Additional information about the decay depth of the hollow Ar atoms is given in the next Section.

Electron Angular Distributions

The favorable method to verify the decay depth of hollow atoms is the observation of an anisotropic angular dependence of the Auger electron emission. Electrons observed at an angle normal to the surface travel the smallest distance through the solid, whereas for electrons detected at grazing observation angle, the travel distance through the solid increases so that attenuation effects are significantly enhanced. Thus, anisotropic electron emission provides direct information about the emission depth of the Auger electrons. The attenuation method for analyzing anisotropic electron emission has been applied to multiply charged neon by

Köhrbrück *et al.* [29] and Grether *et al.* [30].

Figure 4 shows Ar Auger spectra measured at different observation angles α relative to the surface plane. The upper and lower figures refer to the projectile energies of 17 eV and 170 keV, respectively, i.e., the lowest and highest values used in this study. The plots at the left hand side show L-Auger spectra (without continuous background) and the plots at the right hand side show the corresponding K-Auger spectra.

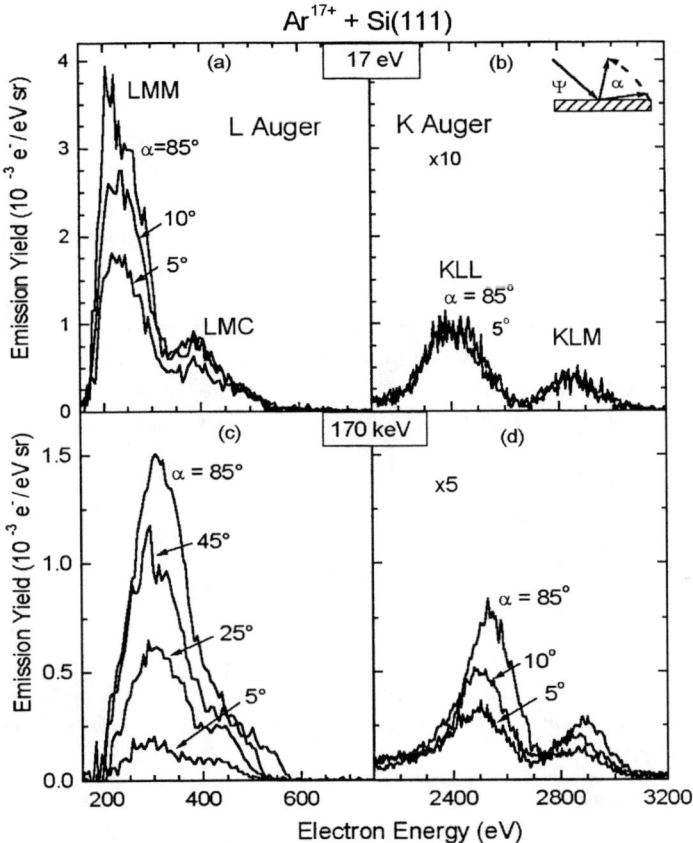

FIGURE 4. Emission yield of K- and L-Auger electrons for Ar^{17+} incident on a Si(111) surface at $\psi = 45^0$. The spectra in the upper and the lower figures are taken with a projectile energy of 17 eV and 170 keV, respectively. The left and right figures show spectra from L- and K-Auger transitions, respectively. The observation angle α relative to the surface plane is varied (see also the schematic plot in the upper right corner).

For the lowest projectile energy one may question whether 17-eV Ar^{17+} ions enter into the solid (at 45^0). From the L-Auger spectra in Fig. 4(a) one can see that the electron intensities detected at grazing observation angles (5^0 and 10^0) are noticeably smaller than those for angles near the surface normal (e.g., 85^0). This indicates that 17-eV Ar^{17+} ions indeed enter into the solid and the L-Auger electrons are ejected in shallow layers just below the surface.

On the other hand, Briand *et al.* [10] stated for 17 eV Ar^{17+} incident on a SiH target that the Ar K-shell vacancy is filled above the surface. For 17 eV projectiles the x-ray spectra [10] show that most of the K-x rays (and K Auger electrons) are ejected before the emission of the L-shell electrons. Hence it follows that the K-Auger electrons originate from outside the solid or from depths smaller than those for L-Auger emission. This conclusion is in agreement with the present result that the K-Auger electrons exhibit an isotropic angular distribution, see Fig. 4(b).

For the highest energy of 170 keV the projectile enters more deeply into the solid and the L and K Auger electron are ejected at a depth of tens of a.u. [23]. Accordingly, the L-Auger electron emission is found to be strongly anisotropic due to attenuation effects. In addition, the K-Auger electrons exhibit an anisotropy in the angular distribution, see Fig. 4(d). The attenuation length of 2.5 keV electrons is equal to $\sim$100 a. u. [28] which results in significant attenuation at the grazing observation angles of 5^0 and 10^0. This shows that even the K-Auger electrons of relatively high energy suffer attenuation when travelling through the solid.

Summarizing, we have shown that energy and angular dependencies of Auger electron yields provide information about dynamic properties of hollow atoms formed near a surface. The relatively fast Auger transitions serve as a unique tool to analyze time-dependent phenomena within the 10^{-15} sec time scale. In the future we plan a more detailed comparison between the experimental results and model calculations revealing life times and decay depth of hollow Ar atoms.

Acknowledgment

We are indebted to Martin Grether for his assistance in the early stage of the present work. We thank Andres Arnau, Ricardo Díez Muiño, Jean-Pierre Briand, John Gillaspy, and John Tanis for helpful discussions concerning various aspects of the present work.

REFERENCES

1. J. Burgdörfer. *Review of Fundamental Processes and Applications of Atoms and Ions*, World Scientific, Singapore, 1993, p. 517.
2. A. Arnau *et al.*, *Surface Sci. Repts.* **27**, 113 (1997).
3. E.D. Donets, *Nucl. Instrum. Methods B* **9**, 522 (1985).
4. J.P. Briand, L. De Billy, P. Charles, S. Essabaa, P. Briand, R. Geller, J. P. Desclaux, S. Bliman, and C. Ristori, *Phys. Rev Lett.* **65**, 159 (1990).
5. U.A. Arifov, E.S. Mukhamadiev, E.S. Parilis, and A.S. Pasyuk, *Zh. Tekh. Fiz.* **43**, 375 (1973). [Sov. Phys.-Tech. Phys. **18**, 240 (1973)].

6. J. Burgdörfer, P. Lerner, and F.W. Meyer, *Phys. Rev. A* **44**, 5674 (1991).

7. A. Arnau, R. Köhrbrück, M. Grether, A. Spieler, and N. Stolterfoht. *Phys. Rev. A* **51**, R3399 (1995).

8. N. Stolterfoht, D. Niemann, M. Grether, A. Spieler, A. Arnau, C. Lemell, F. Aumayr, and HP. Winter, *Nucl. Instrum. Methods B* **124**, 303 (1997).

9. J.P. Briand, L. de Billy, P. Charles, J.P. Desclaux, P. Briand, R. Geller, S. Bliman, and C. Ristori, *Europhys. Lett.* **15**, 233 (1991).

10. J.-P. Briand, G. Giardino, G. Borsoni, M. Froment, M. Eddrief, C. Sébenne, S. Bardin, D. Schneider, J. Jin, Z. Xie, H. Khemliche, and M. Prior, *Phys. Rev. A* **54**, 4136 (1996).

11. H.J. Andrä, A. Simionovici, T. Lamy, A. Brenac, G. Lamboley, J.J. Bonnet, A. Fleury, M. Bonnefoy, M. Chassevent, S. Andriamonje, and A. Pesnelle. *Z. Phys. D* **21**, S301 (1991).

12. L. Folkerts and R. Morgenstern, *Z. Physik D* **21**, S351 (1991).

13. F.W. Meyer, S.H. Overbury, C.C. Havener, P.A. Zeijlmans van Emmichoven, J. Burgdörfer, and D.M. Zehner, *Phys. Rev. A* **44**, 7214 (1991).

14. R. Köhrbrück, K. Sommer, J.P. Biersack, J. Bleck-Neuhaus, S. Schippers, P. Roncin, D. Lecler, F. Fremont, and N. Stolterfoht, *Phys. Rev. A* **45**, 4653 (1992).

15. S. Schippers, S. Hustedt, W. Heiland, R. Köhrbrück, J. Kemmler, D. Lecler, J. Bleck-Neuhaus, and N. Stolterfoht, *Phys. Rev. A* **46**, 4003 (1992).

16. J. Das and R. Morgenstern, *Phys. Rev. A* **47**, R755 (1993).

17. N. Stolterfoht, A. Arnau, M. Grether, R. Köhrbrück, A. Spieler, R. Page, A. Saal, J. Thomaschewski, and J. Bleck-Neuhaus, *Phys. Rev. A* **52**, 445 (1995).

18. H. Limburg, S. Schippers, I. Hughes, R. Hoekstra, R. Morgenstern, S. Hustedt, N. Hatke, and W. Heiland, *Phys. Rev. A* **51**, 3873 (1995).

19. S. Winecki, C.L. Cocke, D. Fry, and M.P. Stöckli, *Phys. Rev. A* **53**, 4228 (1996).

20. U. Lehnert, M.P. Stöckli, and C.L. Cocke, *J. Phys. B: At.Opt. Phys.* **31**, 5117 (1998).

21. W. Huang, H. Lebius, R. Schuch, M. Grether, A. Spieler, and N. Stolterfoht, *Nucl. Instr. Methods in Phys. Res. B* **135**, 1126 (1998).

22. K. Suto and T. Kagawa, *Phys. Rev. A* **58**, 5004 (1998).

23. N. Stolterfoht, J.H. Bremer, and R. Díez Muiño, *Int. J. Mass Spectrom.*, in print (1999).

24. M. Grether, A. Spieler, D. Niemann, and N. Stolterfoht, *Phys. Rev. A* **56**, 3794 (1997).

25. S.T. De Zwart, T. Fried, U. Jellen, A.L. Boers, and A.G. Drentje, *J. Phys. B* **18**, L623 (1985).

26. C.P. Bhalla, *Phys. Rev. A* **8**, 2877 (1973).

27. R. Díez Muiño, A. Salin, N. Stolterfoht, A. Arnau, and P. M. Echenique, *Phys. Rev. A* **57**, 1126 (1998).

28. M. Rösler and W. Brauer. *Particle Induced Electron Emission I*, Springer Tracts in Modern Physics, Springer Verlag, Berlin, 1991, Volume 122.

29. R. Köhrbrück, M. Grether, A. Spieler, N. Stolterfoht, R. Page, A. Saal, and J. Bleck-Neuhaus, *Phys. Rev. A* **50**, 1429 (1994).

30. M. Grether, A. Spieler, R. Köhrbrück, and N. Stolterfoht, *Phys. Rev. A* **52**, 426 (1995).

Coincidence measurements of highly charged ions interacting with surfaces

C. Lemell[1], J. Stöckl[2], J. Burgdörfer[1], G. Betz[2], HP. Winter[2], and F. Aumayr[2]

[1] *Institute for Theoretical Physics, Vienna University of Technology, Wiedner Hauptstraße 8-10, A-1040, Vienna, Austria*
[2] *Institut für Allgemeine Physik, TU Wien, Wiedner Hauptstraße 8-10, A-1040, Vienna, Austria*

Abstract. We have measured electron emission yields for multiply charged ions imping-ing on clean Au(111) and LiF(100) surfaces under varying incidence angles ψ ranging from 3° to 45°. For small angles of incidence we also recorded the impact position of scattered projectiles on a two dimensional detector using the coincidence technique. A clear separation of potential and kinetic electron emission becomes possible and pro-vides insight in electron emitting processes involved in the neutralization dynamics in ion-surface interaction. Model calculations simulating electron emission at and below the surface support our interpretation of the data.

INTRODUCTION

Due to their intrinsic multi-particle nature, ion-surface interactions have received considerable attention by experimentalists as well as theorists. Starting with the landmark papers by Hagstrum [1] and later extended to experiments with higher projectile charge states by the development of suitable ion sources (e.g. ECRIS, EBIT), many signatures of ion-surface interactions have been investigated (X-rays [2], emitted electrons [3–5], image charge acceleration [6,7], and others; for extensive reviews see e.g. [8,9] and refs. therein).

The following picture of ion-surface interactions as it pertains to the interaction with metal surfaces has emerged: At first, the presence of a target surface leads to the image charge acceleration of the incoming highly charged ion (HCI). Cal-culations show that up to the moment of first electron transfer processes between HCI and target the projectile has acquired about 75% of the total energy gain. According to the classical over the barrier (COB) model by Burgdörfer *et al.* [10] the projectile starts to capture electrons from occupied target states at a critical distance (typically some tens of atomic units) which depends on the work function of the target and the charge state of the HCI. This leads to the formation of a so-called "hollow ion". This name originates from the large mean radius of the

CP500, *The Physics of Electronic and Atomic Collisions*, edited by Y. Itikawa, et al.
© 2000 American Institute of Physics 1-56396-777-4/00/$17.00

resonantly populated highly excited Rydberg states which is comparable to the distance between the surface and the HCI and the large number of unfilled inner shells [11]. The following sequence of capture, loss and deexcitation results in the quasi-neutralization of the projectile ("hollow atom") before impact on the surface. Latest experimental efforts aim at the extraction of such "hollow atoms" and "hollow ions" into vacuum [12]. Upon entering the target most of the highly excited projectile levels are promoted above the vacuum level and become unbound due to additional screening of the core charge by conduction band electrons. Within the metal electrons of the conduction band quickly form a quasi-shell around the HCI [13]. Auger-type processes from this shell and fast cascading into the ground state complete the neutralization sequence of the projectile. Modifications to this scenario for insulators are presented in e.g. [14].

One important observable in ion-surface interactions is the emission of electrons. Both kinetic electron emission (KE; the kinetic energy of the projectile is converted into electron emission) and potential electron emission (PE; highly charged ions carry a large amount of potential energy, which also leads to electron emission mainly via Auger processes) measurements suffer from the experimental difficulty to distinguish electrons emitted by PE from those arising from KE. To circumvent this problem and to measure KE and PE separately one has to use either low charge states ($q = 0, 1$) where potential emission is suppressed or reduce the impact velocity of the projectiles ($v \leq v_{th}^{KE}$, with v_{th}^{KE} being the threshold velocity for KE for a specific projectile). Unlike the dependence on the total velocity of the projectile in the case of KE, the number of electrons produced by PE depends on its charge state and on the velocity component normal to the surface.

In order to measure absolute electron yields we use an electron emission statistics (EES) detector [15]. Earlier experiments were carried out under normal incidence conditions with kinetic energies reaching down to the image acceleration limit [16]. The new experimental setup described below further expanded the capabilities of our EES detector by changing to grazing incidence conditions and by adding a channelplate detector with a two dimensional wedge&strip-anode to record scattering distribution and energy loss of reflected projectiles in coincidence with the number of emitted electrons. Using single crystals as target material we now can distinguish between projectile trajectories entering the target and those being surface channeled. This technique not only opens a way to unambiguously separate PE and KE but also to trace back the place of origin of emitted electrons thereby determining the relative importance of above- and below surface contributions to the total electron yield.

The next section contains a detailed description of this new experimental setup. It is followed by results of angular dependent electron yield measurements and coincidence experiments on Au and LiF. The analysis of our data is performed in terms of a theoretical model which is able to explain the large differences in the total number of emitted electrons for different types of trajectories.

EXPERIMENTAL SETUP

Our experiment has been installed at the 5 GHz ECR-ion source at the Vienna University of Technology [17] which provides highly charged ions (for the experiments described here Ar^{q+}, $q \leq 9$; O^{q+}, $q \leq 7$; N^{q+}, $q \leq 6$) with kinetic energies between 1 and $10 \times q$ keV. The extracted ions are mass-to-charge separated, focused, collimated to 1 mm in diameter, and transported through a differentially pumped beam line (base pressure $p < 10^{-8}$ mbar) to an UHV chamber (base pressure $p \simeq 10^{-10}$ mbar).

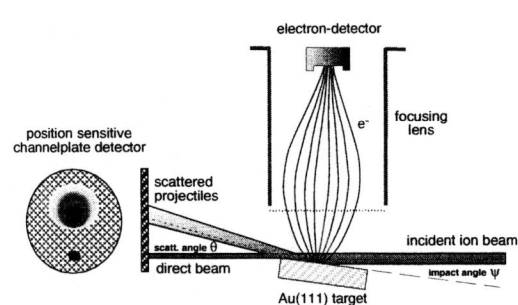

FIGURE 1. Experimental setup (schematically)

The ion beam is directed onto the target surface (Au(111) or LiF(100)) which has been prepared by cycles of sputtering with 1–5 keV Ar^+ ions and annealing at about 500°C (for metal targets only, [18]). The angle of incidence can be varied between 0° and 90°.

For small incidence angles ψ the angular distribution of scattered projectiles as well as the part of the beam passing the target without interaction are recorded on a 2D-position sensitive detector (PSD, see Fig. 1).

Electrons produced in the ion-surface interaction are extracted by a weak electric field and focused on a surface barrier detector biased at +25 kV. The number of electrons emitted in a particular scattering event is determined by the pulse height of the electron detector. From the resulting EES spectrum the total yield γ (being the mean of the EES distribution) can be derived [15].

On the incoming part of their trajectories the projectiles are deflected due to the electric extraction field. This influence has been taken into account in our data evaluation procedure. After the interaction with metal targets most projectiles are neutralized and therefore not subject to deflection [19]. In the case of insulator targets the fraction of negatively charged projectiles becomes comparable to the number of neutrals and cannot be neglected [20].

Data from both detectors can be recorded either in coincidence (synchronized in time, fed into a multi-parameter ADC, and stored as n-tupels) or in non-coincidence mode (for incidence angles $\psi \geq 10°$ when measuring electron emission only). A more detailed description of our experimental setup can be found elsewhere [21].

ANGULAR DEPENDENT ELECTRON YIELD

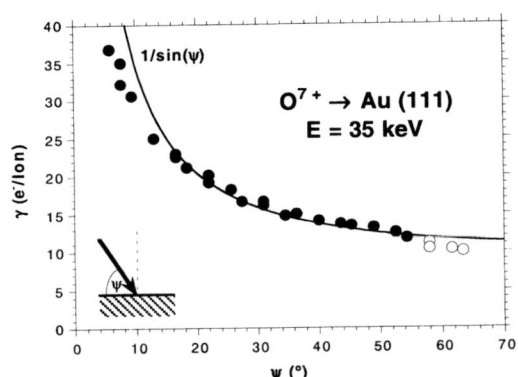

FIGURE 2. Measured electron yields γ for O^{7+} impinging on Au(111) vs. angle of incidence ψ.

It is well known that above the threshold velocity for KE the electron yield γ shows a strong dependence on the projectile's angle of incidence ψ (Fig. 2, [22]).

For smaller angles of incidence electrons emitted along the trajectory within the solid have a larger escape probability. To a good approximation a $1/\sin\psi$ dependence of γ can be expected. In our experiment we find deviations from a $1/\sin\psi$-fit only for very small and large angles of incidence. The latter can be explained by our experimental setup which has been optimized for small ψ and does not have a 100% electron collection efficiency for $\psi > 45°$ while the deviation for $\psi < 10°$ comes from an increasing number of reflected projectiles which produce fewer electrons (see below).

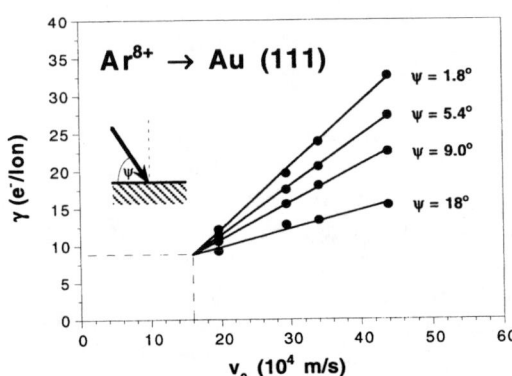

FIGURE 3. Electron yield γ as a function of the ion velocity for specific angles of incidence ψ. The linear fits cross in one point defining the threshold velocity for KE.

Experiments using different charge states of Ar ($q = 6, 8, 9$) show that, despite the additional L-shell hole of Ar^{9+} which might give rise to a changed angular dependence of γ, the electron yield for different charge states displays the same dependence over the hole range of ψ. This is a strong indication that KE is responsible for the dominant part of the ψ-dependence of γ. It was therefore tempting to deduce the well known threshold for KE [23] by comparing total electron yields for a given ion species (in this case Ar^{8+}) with different initial kinetic energies. In Fig. 3 we plotted γ vs. the ion velocity and joined data points for identical impact angles. All lines intersect at about 8.5 e$^-$/ion and a velocity of 1.6×10^5 m/s (~ 135 eV/amu) which is in reasonable agreement with previous data by Eder *et al.* [23].

COINCIDENCE MEASUREMENTS

We have measured the number of emitted electrons in coincidence with the energy loss or the position of the scattered projectile on the position-sensitive detector under grazing incidence conditions. To this end the signals from EES and position sensitive detectors were amplified, delayed in time, and fed into a multiparameter analog-to-digital converter. The gate signal was provided either by the EES- or the PSD-branch of our electronics to allow measurement of both coincident as well as non-coincident events.

Using the information contained in the experimental data it is possible to separate contributions of PE and KE from above and below the target surface.

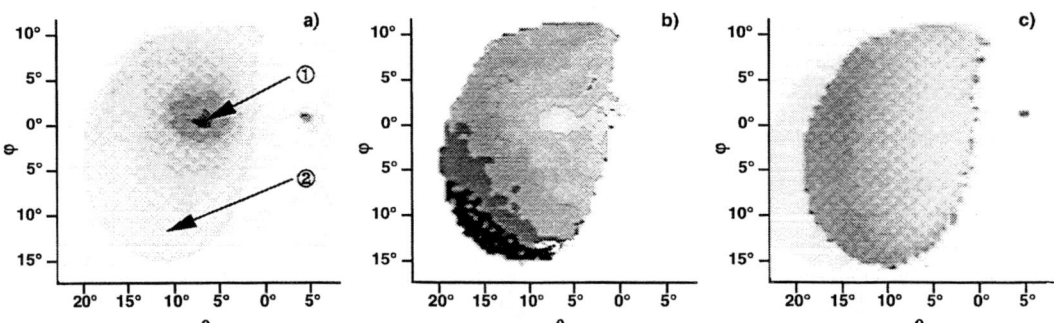

FIGURE 4. Scattering distribution for Ar^{6+} ions impinging on clean Au(111). Figs. a), b), and c) show intensity, number of emitted electrons, and time of flight of scattered projectiles in dependence of their impact position on the PSD, respectively. Dark areas correspond to high count rates in a), high electron numbers in b), and increased time of flight in c).

In Fig. 4a the intensity distribution of scattered projectiles on the PSD is shown for 18 keV Ar^{6+} ions impinging onto a Au(111) single crystal surface under an angle of incidence of $\psi = 3.2°$. The highest count rate is observed at the specular reflection angle. The small feature on the right hand side results from projectiles passing the target without interaction (direct beam). Although these ions did not produce any electrons, the feature appears also in Fig. 4b due to random coincidences with field emission electrons. The width of the scattering distribution points to imperfections of the target surface [24] and can be considerably reduced by increasing the number of cycles of sputtering and annealing or by using crystal targets cleaved immediately before insertion into the vacuum chamber.

In coincidence with projectiles ending up in the central area of the scattering distribution we find a minimum in the number of emitted electrons (Fig. 4b). Projectiles undergoing large angle scattering (recorded in outer areas of the PSD) show an electron yield typically increased by a factor of 2. Also the time of flight (Fig. 4c; corrected for geometrical factors given by our setup) for such projectiles is considerably longer than for those being specularly reflected.

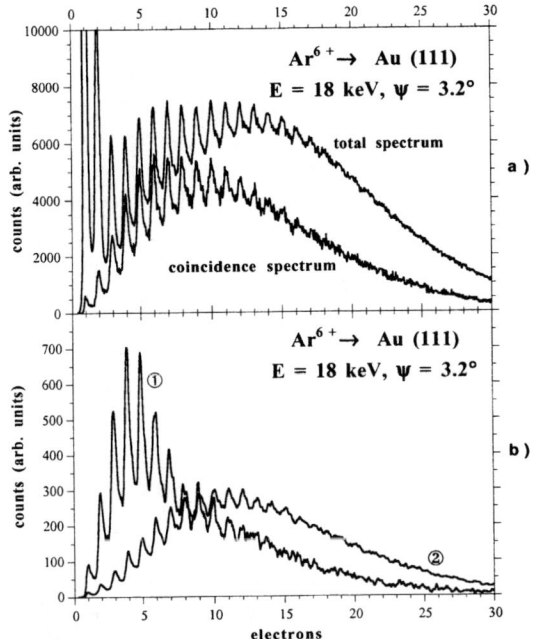

FIGURE 5. a) Electron-emission statistics spectra measured in coincidence and non-coincidence mode for Ar^{6+} ions impinging on clean Au(111). b) EES subspectra for projectiles recorded in the central (labelled 1) and peripheral (labelled 2) areas of the PSD.

In Fig. 5a we compare the total EES spectrum (electron signals measured coincidently as well as non-coincidently) to its pure coincident part. The total spectrum obviously consists of at least three parts. The first one at low electron numbers only present in the non-coincident spectrum comes mainly from field emission and can be easily subtracted. The second one peaks at 6 e^-/ion and appears also in the coincidence spectrum. The third one has a broad maximum at about 14 e^-/ion and is strongly reduced when measured in coincidence with scattered particles. This "missing" part of our total EES spectrum can be attributed to electron production along projectile trajectories which either result in bulk penetration or in scattering into angles too large to be detected by our setup.

Fig. 5b shows EES subspectra recorded in coincidence with projectiles ending up in the central and outer parts of the PSD. While surface channeled (i.e. specularly reflected) projectiles emit only a small number of electrons leading to a narrow distribution peaking at about 4 e^- per ion (labelled 1), projectiles entering the target at surface imperfections and leaving it under slightly larger exit angles produce a much broader multiplicity spectrum with a maximum around 12 e^- per ion. The broad shoulder appearing towards higher electron multiplicities in spectrum 1 can be attributed to projectiles with trajectories of the latter type but ending up "accidentally" in the central area of the scattering distribution. The difference between spectra 1 and 2 corresponds to processes taking place below the topmost atomic layer.

Additional evidence for the consistency of our separation procedure was obtained by recording the projectile time-of-flight spectrum in coincidence with the signals from EES- and PSD detectors. Further increase of the resolution of our time of flight measurements would enable a direct determination of the mean energy necessary to produce one electron in ion-surface interactions. Changes of our setup for this purpose are underway.

Our results which have been analyzed here for 18 keV Ar^{6+} on Au have been confirmed for different projetile-target combinations and/or different projectile energies. In Fig. 6 data for 18 keV Ar^{8+} impinging on LiF(100) are shown. Again

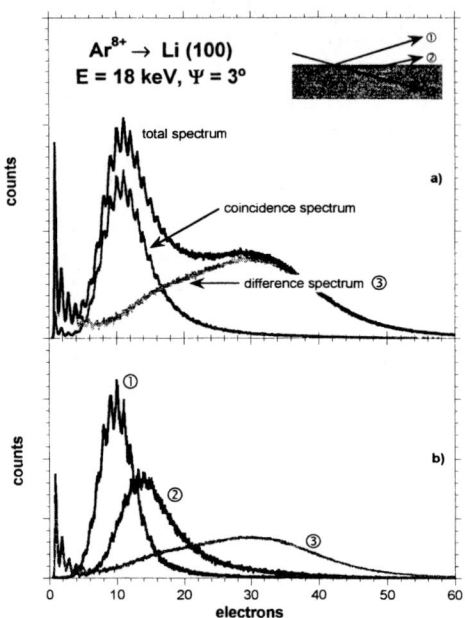

a separation for specularly reflected projectiles from those undergoing subsurface channeling or ending in the target is possible. Due to the good quality of the surface, spectrum 1 does not show a tail towards higher electron multiplicities. It has a mean electron yield for specularly reflected projectiles of about 16 e^-/ion as compared to 8 e^-/ion for interaction with a Au(111) surface under otherwise identical experimental conditions. Distribution 3 can, as before, be divided into two parts. The first one at small electron multiplicities results from field emission. The second one with its maximum at 40 e^-/ion is produced by trajectories ending in the target. Due to the different electronic structure and electron mean free pathes of the two targets the KE distribution has a higher mean for LiF than for Au with experimental conditions being very similar in both cases.

FIGURE 6. a) Electron-emission statistics spectra measured in coincidence and non-coincidence mode for Ar^{8+} ions impinging on clean LiF(100). b) EES spectra for projectiles recorded in the central (labelled 1) and peripheral (labelled 2) areas of the PSD. Spectrum 3 shows the difference between total and coincidence spectra.

THEORETICAL DESCRIPTION

To explain the large difference in γ for specularly reflected projectiles and those entering the target we performed Monte Carlo (MC) calculations describing KE for projectiles at and below the surface. In our simulation KE arises from two different processes: The first one depends on the impact parameter of the projectile-target atom interaction, for the second one we estimated rates for the creation of a hole in a projectile state and its filling.

To simulate projectile trajectories within the metal we used a MD-code [25].

The ZBL-potential was used to calculate projectile-target atom scattering, for interatomic interaction in the crystal we used a tight-binding potential with a cut-off length of 6 Å. The temperature of the target crystal was set to 300 K. 18 keV Ar atoms were directed on a Au(111) target with different surface structures (ideal surface, steps, holes) and under varying angles of incidence. Positions of projectiles were recorded at all times to distinguish between different types of trajectories.

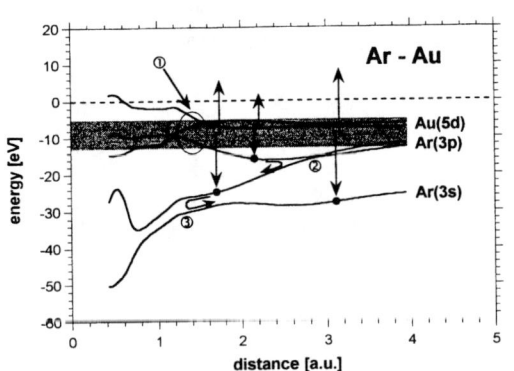

FIGURE 7. Processes entering our simulation (schematically, c.f. text): Impact ionization (1) and electron hole creation and filling (2)

Electron emission was calculated along these trajectories by assuming a probability for impact ionization per scattering event with a target atom to be $Y(b) = Y_0 \cdot \exp(-b/b_c)$. Since the ionization probability for small impact parameters is close to 1, Y_0 describes the probability for an electron set free by this process to escape the solid and contribute to the measured electron yield. The value $Y_0 = 0.4$ was determined based on earlier calculation of escape probabilities for low energy electrons [26]. The characteristic distance b_c was set to 1.3 a.u. which is of the order of the mean radius of the 5d-state of Au ($\langle r \rangle_{5d} \simeq 1.4$ a.u.), most likely the main source for electrons in projectile-target atom collisions, and the position of an avoided crossing appearing in molecular-orbital calculations for gas phase collisions for the system Ar-Au ($r \simeq 1.2$ a.u.).

Additional electrons can be emitted via hole formation and filling. It turns out that for effective electron densities of the conduction band in Au ($r_s \simeq 1.5$ a.u.) the 3p-level of Ar is only very loosely bound. It can therefore be considered to be immersed into the conduction band whenever the local electron density is slightly increased during a close encounter with a target atom. We assigned the 3p-level a probability estimated from similar calculations for the capture rate in the COB-model [10] to contain a hole when becoming atomic again after the collision. This hole can then be transferred to lower lying states of the projectile at avoided crossings of the Landau-Zener-Stückelberg (labelled 2 in Figure 7) and Rosen-Zener types (labelled 3). They can be filled at any time by Auger processes with both active electrons coming from the conduction band of Au. Rates for the filling process have been taken from [13]. The escape probability for these electrons was again set to 0.4. The main source of uncertainty in our simulation is due to the use of the MO-diagram calculated for gas-phase collisions.

Our results show significant differences in electron yield and width between distributions for trajectories of type 1 and 2. Projectiles scattered specularly produce less than 1 electron per ion while those undergoing multiple scattering below the topmost atomic layer produce a broad distribution with a mean of about 9 e⁻/ion.

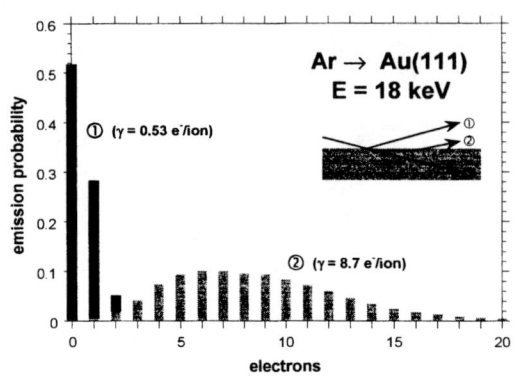

FIGURE 8. Calculated kinetic electron emission for specularly reflected (black bars) and subsurface channeled (shaded bars) projectiles (cf. text).

This makes up for the difference between EES spectra 1 and 2 in Fig. 5. Comparison of the electron yield of specularly reflected projectiles with measured PE yields for Ar^{6+} impinging on clean polycrystalline Au under normal incidence conditions but with identical normal velocity [5] show only a small difference in the mean yield which can be explained by the additional KE yield for trajectories of type 1. The impact ionization process contributes about 75% to the calculated kinetic electron emission. Since the hole creation and -filling mechanism produces only a minor fraction of the emitted electrons (25%) we do not expect more realistic MO-calculations to result in dramatic changes in the outcome of our simulations.

CONCLUSION

We have presented experiments which allow to separate above- and below-surface contributions to the total electron yield of a highly charged ion interacting with a single crystal target. For specularly reflected projectiles we determine the number of electrons originating from potential and kinetic emission. Kinetic electron emission is strongly increased even for relatively slow projectiles undergoing multiple scattering after entering the target at surface imperfections. With a simple model we were able to reproduce the difference between KE for surface channeled projectiles as opposed to those entering the target and undergoing multiple scattering with target atoms. It is shown that the main contribution to the KE yield results from scattering with impact parameters $b \simeq 1.3$ a.u. which is close to the mean radius of the 5d-state of Au. A smaller number of additional electrons is produced in Auger processes after creation of holes in projectile states.

ACKNOWLEDGMENTS

This work has been supported by Austrian Fonds zur Förderung der Wissenschaftlichen Forschung under project numbers P 10164 and P 12470.

REFERENCES

1. H.D. Hagstrum, Phys. Rev. **96**, 325 (1954); Phys. Rev. **96**, 336 (1954).
2. J.P. Briand *et al.*, Phys. Rev. Lett. **65** 159 (1990).
3. D. Niemann *et al.*, Phys. Rev. A **56**, 4774 (1997).
4. I.G. Hughes *et al.*, Phys. Rev. Lett. **71**, 291 (1993).
5. H. Kurz *et al.*, Phys. Rev. Lett. **69**, 1140 (1992).
6. H. Winter, Europhys. Lett. **18**, 207 (1992).
7. F.W. Meyer *et al.*, Nucl. Instrum. Meth. B **98**, 441 (1995).
8. A. Arnau *et al.*, Surf. Sci. Reports **27**, 113 (1997).
9. HP. Winter and F. Aumayr, J. Phys. B: At. Mol. Opt. Phys. **32**, R39 (1999).
10. J. Burgdörfer, P. Lerner, and F.W. Meyer, Phys. Rev. A **44**, 5674 (1991).
11. C. Lemell *et al.*, Phys. Rev. A **53**, 880 (1996).
12. S. Ninomiya *et al.*, Phys. Rev. Lett. **78**, 4557 (1997).
13. N. Stolterfoht *et al.*, Phys. Rev. A **52**, 445 (1995).
14. L. Hägg, C.O. Reinhold, and J. Burgdörfer, Phys. Rev. A **55**, 2097 (1997).
15. G. Lakits, F. Aumayr, and HP. Winter, Rev. Sci. Instrum. **60**, 3151 (1989).
16. F. Aumayr *et al.*, Phys. Rev. Lett. **71**, 1943 (1993)
17. M. Leitner *et al.*, Rev. Sci. Instrum. **65**, 1091 (1994).
18. H. Winter *et al.*, Phys. Rev. Lett. **71**, 1939 (1993).
19. L. Folkerts *et al.*, Phys. Rev. Lett. **74**, 2205 (1995); and erratum **75**, 983 (1995).
20. V.A. Morosov *et al.*, Rev. Sci. Instrum. **67**, 2163 (1996).
21. C. Lemell *et al.*, Rev. Sci. Instrum. **70**, 1653 (1999).
22. R.A. Baragiola, E.V. Alonso, and A. Oliva-Florio, Phys. Rev. B **19**, 121 (1979).
23. H. Eder *et al.*, Rev. Sci. Instrum. **68**, 165 (1997), and Nucl. Instrum. Meth. B **154**, 185 (1999).
24. R. Pfandzelter, Phys. Rev. B **57**, 15496 (1998).
25. G. Betz and W. Husinsky, Nucl. Instrum. Meth. B **102**, 281 (1995).
26. C. Lemell *et al.*, Nucl. Instrum. Meth. B **102**, 33 (1995).

Dynamical and Collisional Approaches to the Transport of Core and Rydberg Projectile States in Solids

D. Vernhet[◊], J-P. Rozet[◊], E. Lamour[◊∇], B. Gervais[∇] and L.J. Dubé[#]

[◊] *Groupe de Physique des Solides, CNRS UMR 75-88, Universités Paris 6 et Paris 7, 75251 Paris Cedex 05, France*
[∇] *Centre Interdisciplinaire de Recherche Ions Lasers, CEA-CNRS-ISMRA, BP 5133, 14070 Caen Cedex 5, France*
[#] *Département de Physique, Université Laval, Québec, Canada G1K 7P4*

Abstract. Experimental studies of the production and transport of projectile excited states in solid carbon targets have been performed for Ar^{17+} and Kr^{35+} at high velocity (respectively v_p=23 and 35.6 a.u.). A range of target thicknesses from single collision condition to equilibrium has been investigated. Charge state distributions, $n\ell$ populations of core and Rydberg projectile states, as well as the population of fine structure substates ($n\ell_j$) are determined. Theoretical predictions have been developed both in a collisional and a dynamical screening picture, using quantum as well as classical descriptions for the transport of projectile excited states. Discussions based on a comparison between experimental results and the different types of calculations are presented.

1. INTRODUCTION

Two extreme representations may be used to describe the interaction between fast highly charged ions and solid targets. They correspond to different approaches for calculating, for instance, the slowing down of a projectile through materials. In the first picture, one assumes that the ion-solid interaction is the result of a *series of binary collisions* with the target electrons, and *ion-atom cross sections* correspond to the data base of the **collisional** calculations. For the ion stopping power, this approach is directly related to the Bethe theory [1] and has recently been extended to the low energy and/or the strong perturbation regime, taking into account explicitly the distortion of target wavefunctions (see for example [2-4]). In the other picture, closer in spirit to the *dielectric* theory first proposed by Bohr [5], the target electrons are considered to *respond collectively* to the passage of the projectile. The polarization of the medium is then described as a wake of electronic density fluctuation trailing the ion. The gradient of the wake potential defines an electric field, and its local value at the projectile can be used to calculate the stopping power [6]. Until recently, most theoretical treatments have performed calculations in the linear response approximation, strictly valid in the high velocity, small perturbation regime, and only

CP500, *The Physics of Electronic and Atomic Collisions*, edited by Y. Itikawa, et al.
© 2000 American Institute of Physics 1-56396-777-4/00/$17.00

free (or quasi-free) electrons were taken into account in the jellium densities [7, 8]. However, in this *dynamical* picture also, there are now new theoretical investigations that go beyond the linear approximation and the uniform electron gas model [9, 10].

Both types of calculations appear to achieve good agreement with stopping power measurements, despite a rather different physical picture for the behavior of the target electrons. If these calculations are used as a starting point to perform *ab initio* predictions of radiation effects in target materials, the final result may very well depend on the choice made between these two types of approaches. There exist in fact very few experimental results that can be used to decide which of these pictures is the most appropriate. Stopping power measurements, for instance, can only give averaged values of the electric field at the projectile: they cannot be used to test the predictions of the wake model concerning the spatial extension of the electronic density fluctuations. Several specific effects have nevertheless been reported to be well accounted for by the wake theory, such as the alignment-ring patterns of fragments of molecular ions [11, 12] or the wake potential induced binding energy shifts of channeled ions [8, 13].

The purpose of the work presented here, is to evaluate the role played by target electrons in these two pictures and to define eventually the validity limits and the applicability of the corresponding models in terms of different experimental observables. The response of the target electrons to the passage of the ion has direct consequences on the projectile ion: not only is it slowed down, but the populations in its excited states are altered by the presence of the environment (the solid). Thus, studies on the production and on the transport of projectile excited states should serve as a probe of the solid response. This is the direction we have chosen to explore.

Several experimental observables have been determined for different systems in the regime of high to intermediate velocity. Our corresponding theoretical expertise has been developed mainly along three complementary and overlapping types of descriptions: the first two are quantum mechanical approaches, a rate-equation model and a density matrix formulation, that allow to follow the population of a given $n\ell$ projectile excited state under the influence of the solid environment. The third one is a classical method, pioneered by Burgdörfer and collaborators [14,15], which describes the motion of the projectile electron on an orbit perturbed by a stochastic force chosen to represent multiple scattering with the surrounding medium.

In this progress report, we restrict the presentation to a selection of our results classified in order of increasing sensitivity to either the individual processes involved (like total or partial ionization and excitation cross sections) and/or to the dynamical screening. Accordingly, we first present charge state distributions, we then discuss the $n\ell$ populations of Rydberg and core states, and finally we examine the evolution of fine structure substate populations ($n\ell_j$).

In the next section, we give the main characteristics of the experimental methods used and describe the techniques applied to determine each observable. Section 3 presents an overview of the results and a comparison with theoretical calculations available so far; special attention is given to classical and quantum descriptions of the transport of projectile excited states in the *collisional* picture. Our conclusions are summarized in the last section with an outline of our present understanding.

2. EXPERIMENTS

All the experiments have been performed with the GANIL facilities in Caen (France), where the ion beam can be extracted either at the exit of the first cyclotron on the SME (Sortie Moyenne Energie) line or at the exit of the second cyclotron using the LISE (Ligne d'Ions Super Epluchés) beam line. These facilities allow to work with very high quality beams from the intermediate to the high energy regime for a wide range of highly charged ions [16].

2.1 Methods

The charge state distributions have been extensively studied by many groups and for a large range of collisional systems. The methods applied are more or less sophisticated but all based on magnetic deflection; for details on the experimental procedure, one can refer to paper [17] and references therein for measurements performed at GANIL.

For the $n\ell$ substate populations of projectile excited states, the significance of the results arises from a simplification of the problem. To perform "test" measurements, the studies have been done on systems where:

• the "primary" production process of the excited states is well identified and the single collision condition is fulfilled even for the thickest target thickness; results, selected here, concern only the capture process where the multiple capture probability is at most 3% for the thickest target.

• the transport of the excited states –mainly due to excitation and ionization processes- can be investigated over a wide range of collision numbers, *from single collision conditions to the equilibrium of populations.*

These conditions imply the use of *fast highly charged ions* and the study of excited states of hydrogen-like or at most helium-like ions: we report on results obtained for Kr^{35+} and Ar^{17+} with carbon targets at a respective velocity of 35.6 and 23 a.u..

2.2 Techniques

We have used x-ray spectroscopy techniques which allow to determine the population of projectile excited states. The experimental set-up is represented Figure 1. Bare ions are directed onto self-supported carbon foils with measured thicknesses and purity. Projectile x-ray emission is recorded by different detection systems depending on the observable one wishes to determine:

• *Rydberg state populations* of Ar^{17+} projectile are deduced, for each target thickness, from long life time Lyman line transitions ($np{\rightarrow}1s$) recorded by a Si(Li) detector placed at 90° with respect to the beam direction. Delay times, behind the target, up to

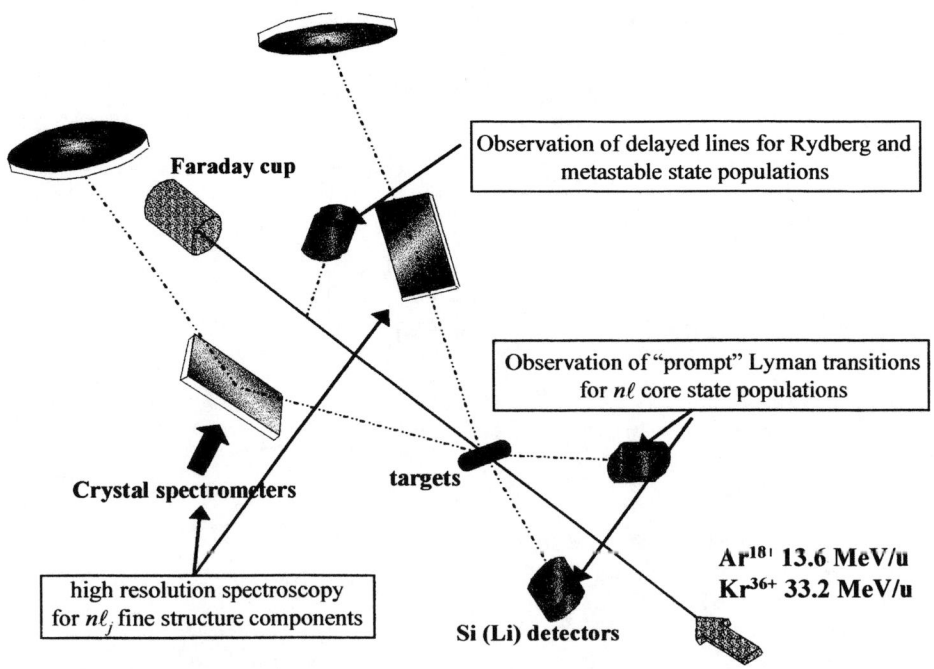

Faraday cup

Observation of delayed lines for Rydberg and metastable state populations

Observation of "prompt" Lyman transitions for $n\ell$ core state populations

Crystal spectrometers

targets

Ar^{18+} 13.6 MeV/u
Kr^{36+} 33.2 MeV/u

high resolution spectroscopy for $n\ell_j$ fine structure components

Si (Li) detectors

FIGURE 1. Schematic representation of the complete experimental set-up.

1 ns are explored with a resolution of 6×10^{-12} s. *Metastable state population*, like $2s_{1/2}$, is also determined using this system by the observation of the two-photon (2E1) transition.

• Prompt Lyman transition intensities are measured with Si(Li) or Ge detectors recording all the emitted x-rays with the target placed at the center of the collision chamber. This allows to obtain the *evolution of np core state populations* as a function of target thickness since a very small fraction ($\leq 3\%$) of the projectile excited states lies above $n = 10$.

For each type of emission lines – delayed or prompt- the energy resolution of the detectors allows us to identify Lyman lines for $n \leq 4$ (as well as the 2E1 decay mode of the 2s metastable state) in the case of Ar ions and for $n \leq 5$ in the case of Kr.

• The effect of transport processes on *fine structure components* is determined using high-resolution high-transmission Bragg-crystal spectrometers equipped with mosaic graphite crystals (HOPG) and position sensitive detectors. Our two crystal spectrometers are used in a vertical geometry, allowing first order correction of the broadening of the emitted lines due to Doppler effect.

Each detection system is placed at a specific angle with respect to the beam direction, either to measure a polarization effect or to be polarization-insensitive. Details on the experimental set-up as well as on the spectra analysis procedure are given in [18-21].

3. RESULTS AND DISCUSSION

3.1 Charge state distributions

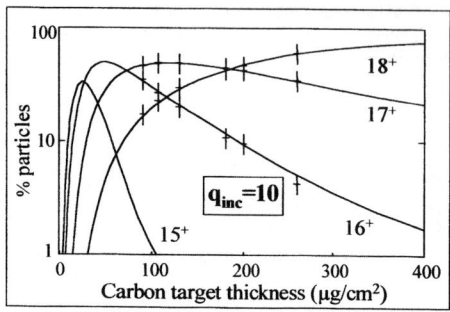

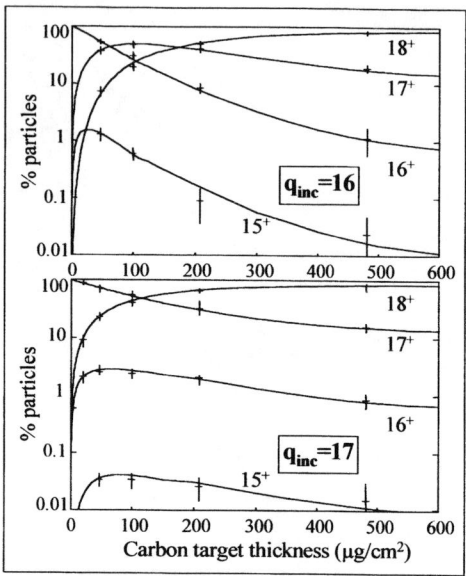

Figure 2. Charge state distributions for Ar $^{q+}$ on C at v_p= 23 a.u.; the full lines correspond to the ETACHA Code

The evolution of charge state distributions as a function of carbon target thickness for different incident charge of Ar ions at a projectile velocity of 23 a.u. (i.e. E_p= 13.6 MeV/u) is reported Figure 2. The full curves correspond to the ETACHA Code [20, 22] which refers to a *collisional* approach. Briefly summarized, this code solves a set of coupled differential rate equations using ion-atom cross sections for capture, ionization and excitation processes. Radiative and Auger de-excitations are included as well. An excellent agreement is obtained with experimental data even for the thinnest target (i.e. in pre-equilibrium conditions) as well as for the lowest fraction of charge state (15^+ here, for example). Indeed, the charge state distributions are not very sensitive to the angular momentum distribution of core states but more or less governed by the total cross sections of a given n depending on the charge state (q) value. Many other measurements can be compared successfully with this code [17, 23, 24] like, for example Pb^{56+} at 28.9 MeV/u [17] on carbon targets where good agreement is found for a wide range of target thicknesses (from 2 to 1200 µg/cm^2) and for very different charge states (55+ to 73+). This ETACHA code is actually limited to ions with a number of electrons ≤ 28 since only n=1, 2 and 3 are considered in the calculations. Furthermore, using perturbative treatment to calculate ion-atom cross sections, this code is well suitable for the high velocity regime. Nevertheless, it has recently been shown in [25], for the case of Ne^{6+-10+} on C at 2 MeV/u, that its validity can easily be extended by simply introducing more accurate cross sections.

3.2 Rydberg state populations

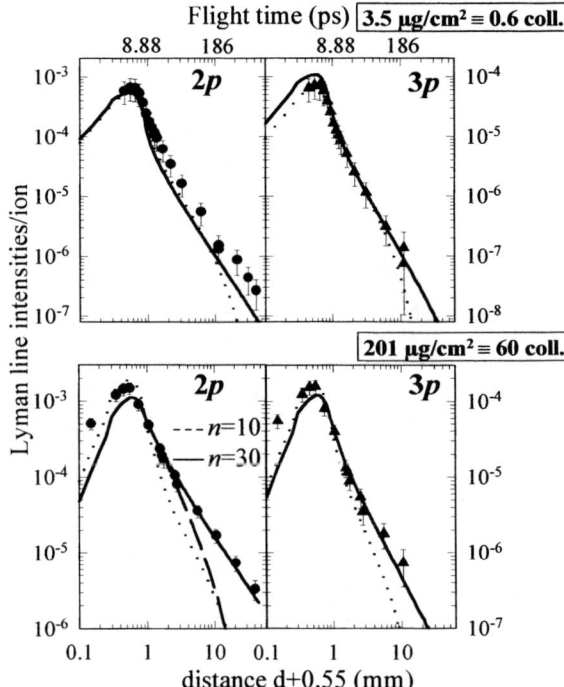

Flight time (ps) $\boxed{3.5\ \mu g/cm^2 \equiv 0.6\ coll.}$

$\boxed{201\ \mu g/cm^2 \equiv 60\ coll.}$

distance d+0.55 (mm)

Figure 3. Evolution of the Ar^{17+} Lyα ($2p\rightarrow1s$) and Lyβ ($3p\rightarrow1s$) intensities as a function of ion time of flight behind the thinnest (3.5 μg/cm^2) and the thickest (201 μg/cm^2) carbon targets studied (*note*: the distance behind the target has been arbitrary shifted by 0.55 mm to clarify the graph). Experiment is given as points with error bars. Predictions of the classical transport simulation, including excited states with $n\leq30$, are plotted as full curves. The dotted lines correspond to the quantum description with the rate-equation model limited, so far, to $n\leq10$. To illustrate the agreement between classical and quantum descriptions of the transport, a dashed line (for the thickest target and the $2p$ state) shows the classical calculations limited to states with $n\leq10$.

The evolution of the Lyα ($2p\rightarrow1s$) and Lyβ ($3p\rightarrow1s$) intensities as a function of ion time of flight behind the thinnest and the thickest carbon foils used in the experiment appears in Figure 3. Keeping in mind that lifetimes of hydrogen-like $n\ell$ states are proportional to $n^3\ell^2$ for a given projectile, the delayed photon emission observed in the decay of $2p$ and $3p$ states comes from the Rydberg state population through cascade contributions. Indeed, the observed $2p$ decay time is much longer than the $3p$ one for all target thicknesses, and this is a direct signature of high angular momentum Rydberg states. Furthermore, by direct observation of the experimental results, one can also see, that the decay slope of a given line is not constant as a function of target thickness; the measurements are therefore sensitive to transport processes in the bulk and not only in the last layers, as was claimed in earlier studies.

The comparison between the experimental data and theoretical predictions shows that classical and quantum collisional approaches give reasonable agreement for the thinnest foil where on average ≈0.6 collision occurs during the transport. For large ion time of flight and thick targets (a number of collisions ≈100 times larger- see Fig. 3) the classical transport model is more suitable and successfully describes the population of high n and high ℓ excited states whereas the rate-equation model is limited by the number of states included (see [20, 26] for details). In fact, with the classical transport model, we can demonstrate that the $2p$ state population, for this system, is sensitive to excited states up to $n = 30$ even if only 3% of the excited states have a principal quantum number $n > 10$. Then, it appears that highly excited state populations are mainly governed by inter- and intra- shell excitation processes.

3.3 Core state populations

We illustrate in Figure 4 the evolution of core state populations as a function of target thickness, or alternatively as a function of ion transit time through solid foils. The absolute populations per incident ion, for np states (with $n \leq 4$) and for the $2s_{1/2}$ state, are presented at different carbon target thicknesses in the case of Ar^{17+} at a velocity of 23 a.u. (details on the extraction of the data can be found in [20, 27]).

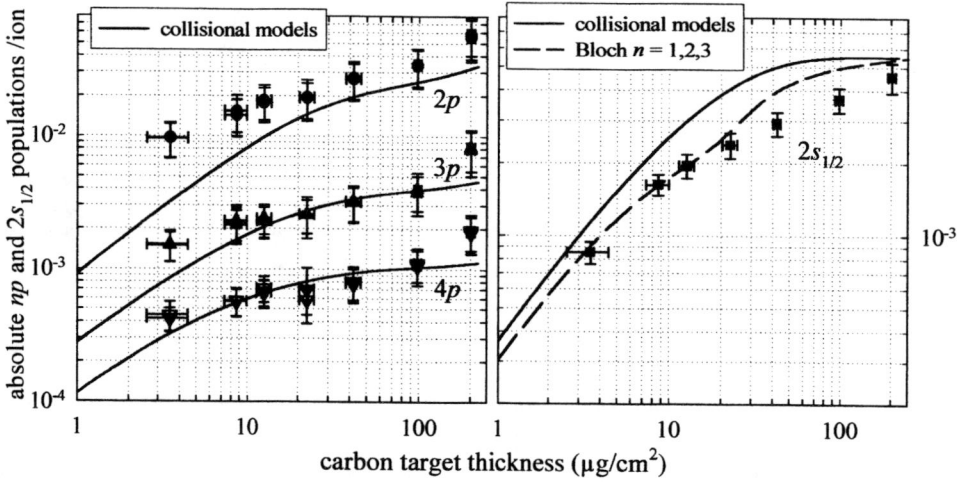

Figure 4. Absolute np and $2s_{1/2}$ populations of Ar^{17+} as a function of carbon target thickness: experimental data (symbols); predictions of quantum collisional model (full lines); the results of the Bloch equations (dashed line).

These inner shell populations are sensitive to direct perturbations during the transport in the solid foils and provide also information on the angular momentum state distribution of higher excited states through their cascades into np and $2s$ states. For example, half of the total $2p$ population comes from cascades involving large angular momentum states with a major (50%) contribution coming from $n=3$. In comparison, the cascade contributions amount to 16 % of the $2s$ population (from $\ell=1$ states) and 16 and 7 % for the $3p$ and $4p$ populations respectively. These contributions are rapidly decreasing as a function of n. For each state studied, the experimental evolution of the population with carbon thickness is compared in Figure 4 with the results of our simulations. We find that for core projectile excited states, as for

Rydberg states, both classical and quantum transport approaches, in a *collisional picture*, give similar predictions [26, 28]. The results of the two calculations essentially overlap and for the sake of clarity only the quantum results are shown in Figure 4. Furthermore, the main effect in the evolution as the target thickness increases, is reasonably well given by collisional models, except in the pre-equilibrium conditions, and this even more clearly for the $2p$ and $2s$ levels. In fact, the observed $2s$-$2p$ mixing is much stronger and faster than predicted. Calculations, including the Stark mixing induced by the wake electric field due to the projectile itself (dynamical mixing [29]), are in progress in a classical as well as in a quantum approach [27, 28]. In particular, preliminary results of quantum calculations, using the Bloch formalism [30], are reported Figure 4. They include so far only the $n = 1$, 2 and 3 levels, and a meaningful comparison can only be assured for the evolution of the $2s_{1/2}$ level where cascade contributions are small. The Bloch equations are in better agreement with the experimental data (Fig. 4) and we conclude that dynamical mixing plays an important role in the evolution of the fine structure populations. Results on the evolution of absolute np populations for different carbon target thicknesses in the case of Kr^{35+} ($v_p=$ 35.6 a.u.) have also been obtained. For such a heavy ion, Stark mixing between $n\ell_j$ components can only occurs for $n \geq 3$ [29]. Therefore, the np populations, which correspond in fact to a sum over the j components of p states, appears to be sensitive to dynamical mixing only for $n \geq 5$.

3.4 Fine structure substate populations of core states

The evolution of fine structure components of the Balmer α line with the carbon target thickness, has been observed for hydrogen-like Kr ions at a velocity of 35.6 a.u..

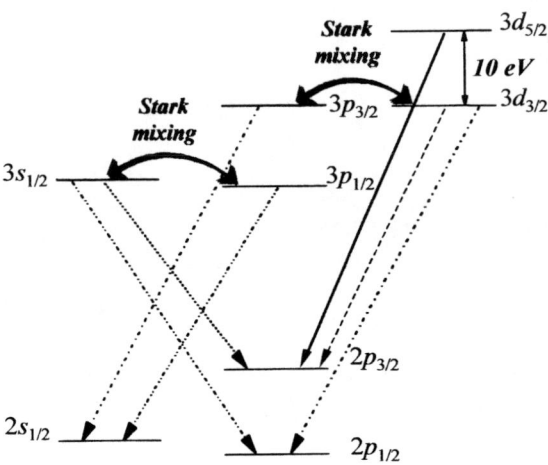

Figure 5. Radiative de-excitation scheme of the Balmer α line for Kr^{35+}

The object of the experiment is to gather quantitative evidence of the effect of the wake field, i.e. a measure of the collective response of the solid [29]. In the case at end, Stark mixing induced by the wake electric field can only occur between the $3s_{1/2}$ and $3p_{1/2}$ components as well as between $3p_{3/2}$ and $3d_{3/2}$ while, in a pure *collisional* approach, the two components of the $3d$ level will be affected in the same way (see Fig. 5). In other words, in a collisional picture, the ratio between $3d_{3/2}$ and $3d_{5/2}$ should remain constant as a function of carbon target thickness. This is clearly not the case as illustrated in Figure 6. For both ratios, we observe a significant deviation due to the *dynamical mixing*, and once again the Bloch formalism gives encouraging results, even in its limited version. Extensive comparisons have been done in [29] with a density matrix approach including *only* the Stark mixing. These studies have shown that the present observables are sensitive not only to the non-diagonal elements describing the production of the initial states (predictions of the CDW theory have been used for the capture process), but also to the value of the effective electric field which acts on a given $n\ell$ level. The values of the effective fields obtained [29] are in good agreement with those of recent calculations [10] that go beyond the uniform electron gas model by including the contribution of target-inner shell electrons.

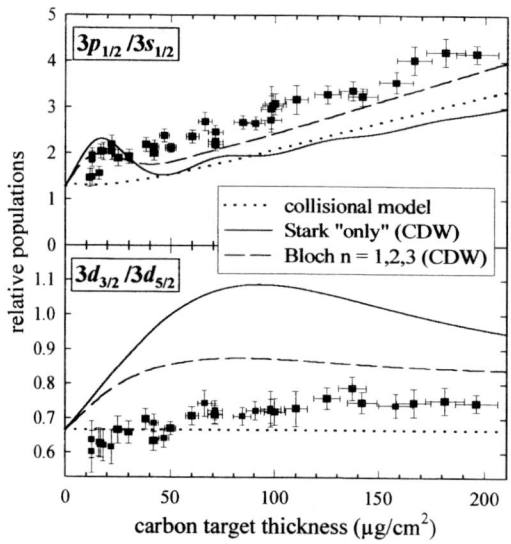

Figure 6. Evolution with carbon target thickness of the relative populations in fine structure substates

4. CONCLUSIONS AND PERSPECTIVES

Using x-ray spectroscopy, core and Rydberg populations of projectile excited states have been determined for fast highly charged ions in interaction with solid targets and used as a tool to probe the solid response. Our results have shown that a "pure" collisional picture can be used to describe many observables, like the charge state distributions, the $n\ell$ Rydberg state populations and the major part of the $n\ell$ core state populations. In this picture, we have found that classical and quantum transport theory gives similar results for heavy ions in agreement with recent studies [31] on hydrogen. Classical and quantum correspondence remains the subject of current investigations, in particular for the transport of heavy ions [27,28]. It is our hope that more detailed (in the sense of less averaging over time, space and contributing collision processes) observations and corresponding calculations will help to identify the parameter regions where deviations are most likely to become apparent. In this respect, studies of the evolution of fine structure substate populations (nl_j) should bring new insight. On the other hand, we have already shown that these very sensitive observables give clear evidence of the limitation of a "pure" collisionnal picture. In a preliminary quantum formulation, including the dynamical mixing due to the wake field, our results [29] show that this effect plays indeed an important role.

New studies have been initiated. On theoretical side, treatments of the effect of the wake field are investigated using both classical and quantum transport approaches in [27, 28, 30, 31]. Our efforts will concentrate first on the inclusion of states n larger than 3 in the Bloch formalism and on calculations with improved *ab initio* cross sections for the primary process. Two new experiments have also been performed. One concerns the studies of the evolution of the fine structure components of the Balmer α of Kr^{35+} (60 MeV/u), where excited states are produced by *excitation* instead of capture (i.e. a change in the primary process). Carbon and aluminum targets have been used to get new information on the wake electric field value. The other experiment, performed very recently (June 99), refers to the studies of Ar^{17+} excited states; great improvements in the resolving power of the crystal spectrometers on one hand, and an optimization of the intensity and quality of the beam, on the other hand, allowed us to record spectra of the entire Lyman series where np states for $n \leq 12$ can be distinguished, as well as the $2p_{1/2}$ substate from the $2p_{3/2}$ one. Also direct comparison is made between solid carbon targets and gaseous CH_4. Data are under analysis, but even the raw spectra already show an enhancement in the population of the $2p_{1/2}$ substate compared to the $2p_{3/2}$ component, a specific solid target effect predicted by our most recent calculations, and consistent with our observations reported in section 3.3 on the $2s$-$2p$ mixing. These new precise experimental results, on solid and gaseous targets, may help the theoretical investigations to quantify the effects of dynamical screening and possibly to reveal quantum properties as well.

ACKNOWLEDGMENTS

We thank all the CIRIL laboratory for their participation in the data acquisition and for their efficient technical support. We are also grateful to the staff at GANIL for providing us with outstanding technical assistance and high-quality beams. This review on our understanding on the ion-solid interaction could not have been done without many enlightening discussions with Joachim Burgdörfer, Camille Cohen, Pedro Echenique, Pablo Fainstein, Robert Gayet, Victor Hugo Ponce, Carlos Reinhold, Joseph Remillieux and Antoine Salin; we thank them for their constant interest in our work. Support for this work has been partly provided by the Laboratoire de Recherhe Correspondant with the CEA/DRECAM under contract n° LRC DSM 97/01.

REFERENCES

1. Bethe, H., *Ann. Phys., Lpz.* **5**, 325 (1930)
2. Olivera, G. H., Martinez, A. E., Rivarola, R. D. and Fainstein, P. D., *Phys. Rev.* **A49**, 603 (1994)
3. Grande, P. L. and Schiwietz, G., *Nucl. Instrum. Methods B* **132**, 264-275 (1997)
4. Schiwietz, G., Wille, U., Díez Muiño, R., Fainstein, P. D. and Grande, P. L., *J. Phys. B : At. Mol. Phys.* **29**, 307 (1996)
5. Bohr, N., *K. Dansk. Vidensk. Selsk. Mat. Fys. Medd.* **18**, n° 8 (1948)
6. Neufeld, J. and Ritchie, R.H., *Phys. Rev.* **99**, 1125 (1955)
7. Echenique, P. M., Ritchie, R. H. and Brandt, W., *Phys. Rev.* **B20**, 2567 (1979)
8. Echenique, P. M., Flores, F. and Ritchie, R. H., *Solid State Physics* **vol 43** (H. Ehrenreich and D. Turnbull, eds.) Academic Press, New York, 1990, p 229 and references therein.
9. Echenique, P. M., García de Abajo, F. J., Ponce, V. H. and Uranga, M. E. *Nucl. Instrum. Methods B* **96**, 583 (1995)
10. Fuhr, J. D., Ponce, V. H., García de Abajo, F. J. and Echenique, P. M., *Phys. Rev.* **B57**, 9329 (1998)
11. Gemmel , D. S., Remillieux, J., Poizat, J. C., Gaillard, M. J., Holland, R. E. and Vager, Z., *Phys. Rev. Lett.* **34**, 1420 (1975)
12. Vager, Z. and Gemmel, D. S., *Phys. Rev. Lett.* **37**, 1352 (1976)
13. Datz, S., Moak, C. D., Crawford, O. H., Krause, H. F., Dittner, P. F., Gomez del Campo, J., Biggerstaff, J. A., Miller, P. D., Hvelplund, P. and Knudsen, H., *The American Physical Society*, **40**, n°13, 843-847 (1978)
14. Burgdörfer, J. and Bottcher, C., *Phys. Rev. Letters* **26**, 2917 (1988)
15. Burgdörfer, J. and Gibbons, J., *Phys. Rev.* **A42**, 1206 (1990)
16. Baron, E., Baelde, J. L., Berthe, C., Bidet, D., Chabert, A., Jamet, C., Gudewicz, P., Loyer, F., Moscatello, M-H., Petit, E., Ricaud, C., Savalle, A. and Senecal, G., *Proc. 15th Int. Conf. on Cyclotrons and their Applications (Caen)* ed. E. Baron and M. Lieuvin, 385-388 (1998); http://www.ganil.fr/
17. Léon, A., Melki, S., Lisfi, D., Grandin, J-P., Jardin, P., Suraud, M. G. and Cassimi, A., *Atomic Data and Nuclear Data Tables* **69**, 217-238 (1998)
18. Despiney, I., *PhD Thesis*, Université Paris 6 (1994)
19. Vernhet, D., *"Populations of projectile excited states act as a probe of ion-matter interaction"*, Nouvelles du GANIL, **60**, 9-16 (July 1997)
20. Lamour, E., *PhD Thesis*, Université de Caen (1997)
21. Vernhet, D., Rozet, J-P., Bailly-Despiney, I., Stéphan, C., Cassimi, A., Grandin, J-P. and Dubé, L. J., *J. Phys. B : At. Mol. Phys.* **31**, 117-129 (1998)
22. Rozet, J-P, Stéphan, C. and Vernhet, D., *Nucl. Instrum. Methods B* **107**, 67-70 (1996)
23. Rothard, H., Grandin, J-P., Jung, M., Clouvas, A., Rozet, J-P. and Wunsch, R., *Nucl. Instrum. Methods B* **132**, 359-363 (1997)
24. Horvat, V., Watson, R.L. and Blackadar, J.M., *Phys. Rev.* **A56**, 1904 (1997)
25. Blaǰeviǆ, A., Bohlen, H.G. and von Oertzen, W., *Phys. Rev. Lett.* (submitted April 1999)
26. Vernhet, D., Rozet, J-P., Lamour, E., Gervais, B., Fourment, C. and Dubé, L. J., *Physica Scripta* **T80A**, 83-86 (1999)
27. Reinhold, C. O., Arbó, D. G., Burgdörfer, J., Gervais, B., Lamour, E., Vernhet, D. and Rozet, J-P., *J. Phys. Lett.* (to be submitted)
28. Arbó, D. G., Reinhold, C. O., Yoshida, S. and Burgdörfer, J., *ICACS Proceedings (Odense)* 1999
29. Rozet, J-P., Vernhet, D., Bailly-Despiney, I., Fourment, C. and Dubé, L.J., *J. Phys. B : At. Mol. Phys.* **32** (scheduled October 1999)
30. Dubé, L. J., Despiney, I., Rozet, J-P. and Vernhet, D., *Abstract of Contributed Papers of the XIXth Int. Conf. on the Physics of Electronic and Atomic Collisions (Whitsler)* vol 2, ed J. B. A. Mitchell, J. W. McConkey and C. E. Brion, 730 (1995)
31. Arbó, D. G., Reinhold, C. O., Kürpick, P., Yoshida, S. and Burgdörfer, J., *Phys. Rev.* **A60**, 1091-1102 (1999)

Molecular Shape and Anisotropy Effects on Collisional Ionization Dynamics

Koichi Ohno

Department of Chemistry, Graduate School of Science,

Tohoku University, Sendai 980-8578 Japan

Abstract. Molecular shape as probed by a test atom from H to Ar has been studied by quantum chemical calculations, and anisotropy effects on collisional ionization dynamics have been disclosed on the basis of two-dimensional Penning ionization electron spectroscopy and trajectory calculations.

INTRODUCTION

Physical substances can be considered to collide on their surfaces where they repel each other in a short distance. In this sense molecular shape can be studied by introducing a test atom on the molecular surfaces; isopotential energy contours constitute certain surfaces which separate the inside from the outside(1, 2). Some interesting characteristics of such isopotential energy surfaces of molecules probed by a test atom are expected to be as follows; (i) probed shapes depend on the kinds and electronic states of the test atom, (ii) probed shapes change their size to become smaller on going from the lower to the higher energies, and (iii) slopes as well as shapes of the repulsive potential wall are anisotropic depending on the anisotropic nature of the molecular structures and electronic states.

Collision dynamics involving molecules are expected to reflect anisotropic nature of the interaction potentials. In the case of collisional ionization called Penning ionization(3), the electron ejection is governed by the local nature of the target orbitals where a metastable rare gas atom, such as He*(2^3S) and He*(2^1S), is in contact with. This propensity was disclosed by the electron exchange model(4). Based on this nature of Penning ionization, branching ratios into ionic states relevant to ionized target orbitals have been connected with the exterior electron densities outside the repulsive boundaries of the molecular surfaces(5, 6).

In this paper, general trends of molecular shapes of diatomic molecules as probed by test atoms by quantum chemical calculations are described, and anisotropic effects on collisional ionization dynamics are studied by two-dimensional Penning ionization electron spectroscopy(7) and classical trajectory calculations(8).

CP500, *The Physics of Electronic and Atomic Collisions*, edited by Y. Itikawa, et al.
© 2000 American Institute of Physics 1-56396-777-4/00/$17.00

MOLECULAR SHAPE

Isopotential energy contours of molecules as probed by a test atom can be obtained by quantum chemical calculations introducing the atom in the vicinities of molecules. In view of collision characteristics of molecules, the energy ranges of our interest are in the region of thermal distributions. In this collision energy region, geometrical relaxation within the molecule is not so fast with respect to the interaction time of the approaching atom. This feature enables us to approximate the molecular structure frozen at its equilibrium geometry in the isolated system. It is of course interesting to compare the frozen structure case with the fully relaxed case in which all atoms are in the optimized structure of the interacting system. It should also be noted that the accuracy of quantum chemical calculations depends on the choice of basis functions and the level of the methods considering the electron correlation effects. Computational details have been described in a previous paper(2).

Figure 1 shows *ab initio* isopotential energy contour of $H_2(^1\Sigma_g^+)$ with (a) the molecular structure frozen at the equilibrium geometry of the isolated molecule in its ground electronic state and (b) the optimized structure of the interacting system including the test atom. Ground-electronic states of atoms from H to Ar are used for the test atom to probe the molecular shape. Figure 2 shows *ab initio* isopotential energy contour of $N_2(^1\Sigma_g^+)$ with (a) frozen and (b) optimized structures. In these figures, potential energies of the contour curves are spaced in a constant interval of 100 meV from the outside; the outermost curves are for the isopotential energy of 100 meV, and the second ones are for 200 meV.

Interesting characteristics of molecular shapes obtained in Figures 1 and 2 are summarized as follows; (i) a round shape(prolate form) like an American foot ball elongated along the diatomic axis are found for many cases in consistent with the common notion of the molecular shape of such diatomic systems, (ii) in some cases the length along the diatomic axis is shortened to become a sphere form or even an oblate form , (iii) some other cases show dented structures along certain directions to produce an apple form, a peanut form, and a lemon form, and these typical shapes are listed in Figure 3.

It should be noted that probed shapes depend on the kinds of the probing atom. This does not however mean that the probed shape is intrinsic to the probing atom. As can be seen in Figures 1(a) and 2(a), the dented structures are different between H_2 and N_2 for the same probing atom of $C(^3P)$; in the case of H_2/C the dented direction is vertical to lead to a peanut form, whereas for N_2/C the axial directions are dented to yield an apple form. Characteristic features of the molecular shapes are due to anisotropic interactions between a molecule and an atom. HOMO-LUMO interactions are found to be important as in the pioneering studies of chemical reactions by Fukui and coworkers(9). In this respect, we must note that the

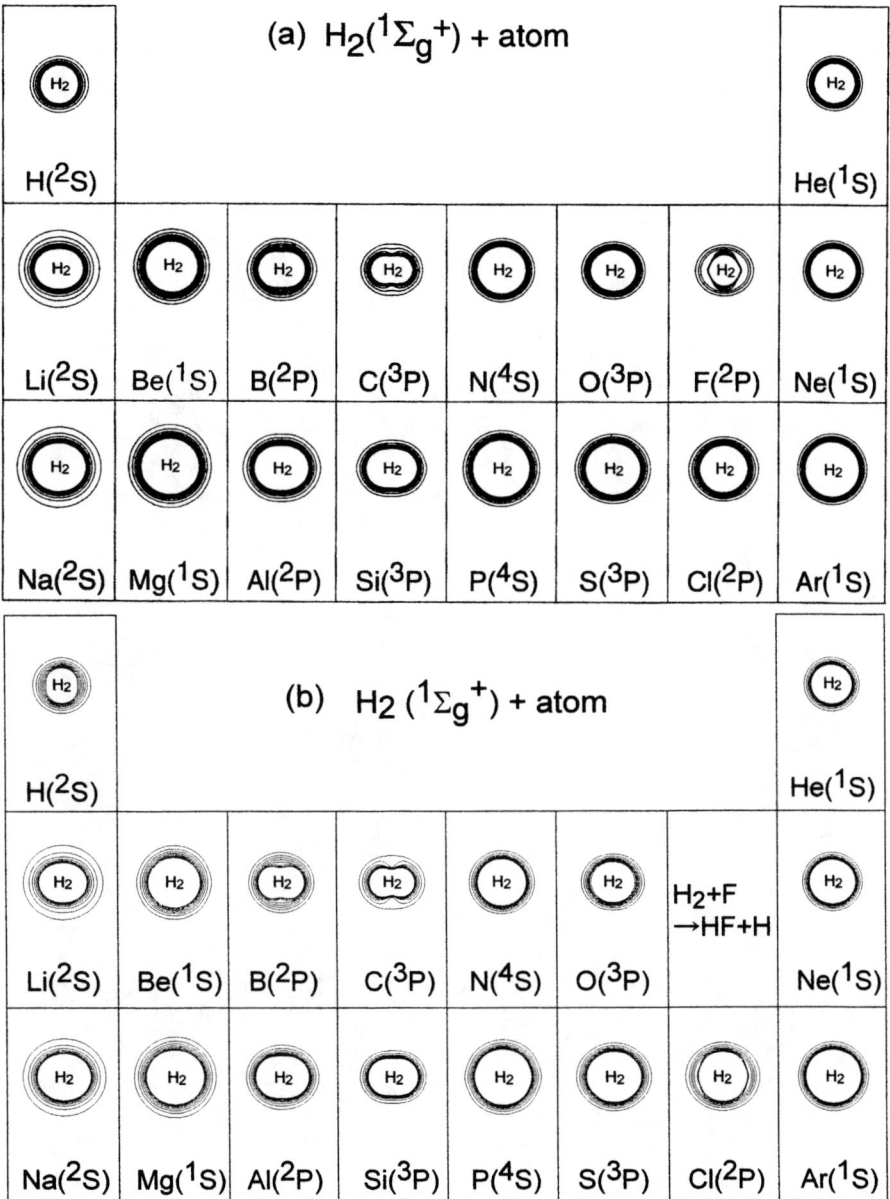

FIGURE 1. *Ab initio* isopotential energy contour of $H_2(^1\Sigma_g^+)$ by a test atom from H to Ar in the ground state for (a) the molecular structure frozen at the equilibrium geometry of the isolated molecule and (b) the optimized structure in the interacting system.

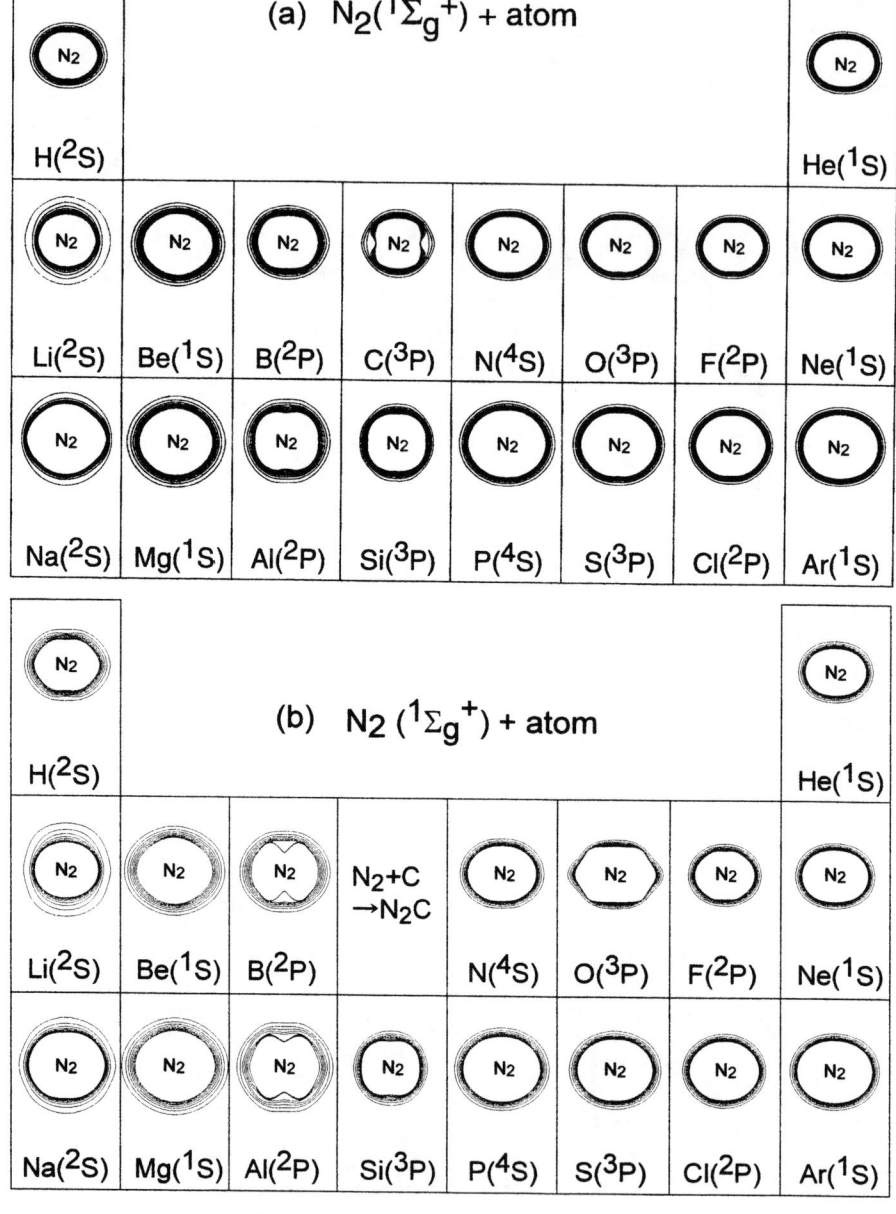

FIGURE 2. *Ab initio* isopotential energy contour of $N_2(^1\Sigma_g^+)$ by a test atom from H to Ar in the ground state for (a) the molecular structure frozen at the equilibrium geometry of the isolated molecule and (b) the optimized structure in the interacting system.

optimized geometry leads to reaction products of HF for H_2/F and N_2C for N_2/C for which contour curves cannot be shown. From a comparison between frozen(a) and optimized(b) shapes, anomalous dents are emphasized in the optimized shapes. Thus, the essential nature in the anisotropic deformation on the molecular surface is related with chemical interactions, although the interactions are not strong enough to lead to chemical reactions in many cases.

The spacing of the contour curves is found to be irrelevant to the directions in most cases. One marked exception can be found for N_2/Li in which the shape of molecules changes dramatically from a large sphere, via an oblate form and a small sphere, to a prolate form. Such a change in the shape of N_2 molecule probed by a Li atom can be related with observed collision energy dependence of partial Penning ionization cross sections.

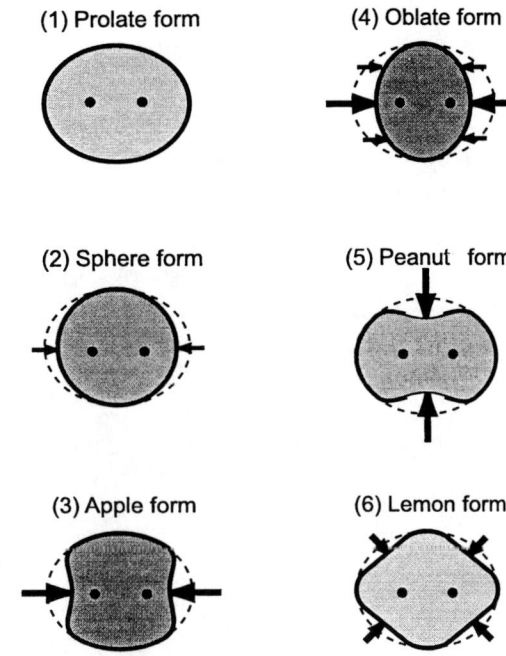

FIGURE 3. Typical Patterns of the outer shape of diatomic molecules probed by an atom. Directions of anisotropic interactions reducing repulsive energies are indicated by arrows.

2D-PENNING IONIZATION ELECTRON SPECTROSCOPY

In Penning ionization processes(3), $A^* + M \rightarrow A + M^+ + e^-$, two important variables are involved, the relative collision energies E_c between the metastable atom A^* and the target molecule M and the ejected electron kinetic energies E_e. Detection of produced ion signals as functions of E_c with using velocity selected A^* gives the collision energy dependence of total ionization cross sections. Electron spectroscopic analyses of E_e provides relative populations (branching ratios) of produced ionic states. Although some pioneering attempts were made to combine these techniques(10,11), experimental difficulties are associated with considerable reduction of signal intensities by a factor of 10^{-5}. After drastic improvements of experimental techniques(12, 13, 14), two-dimensional(2D)-Penning ionization electron spectroscopy(2D-PIES) has been developed(7); electron signals are recorded as functions of both E_c and E_e. Schematic illustration of the experimental apparatus is shown in Figure 4.

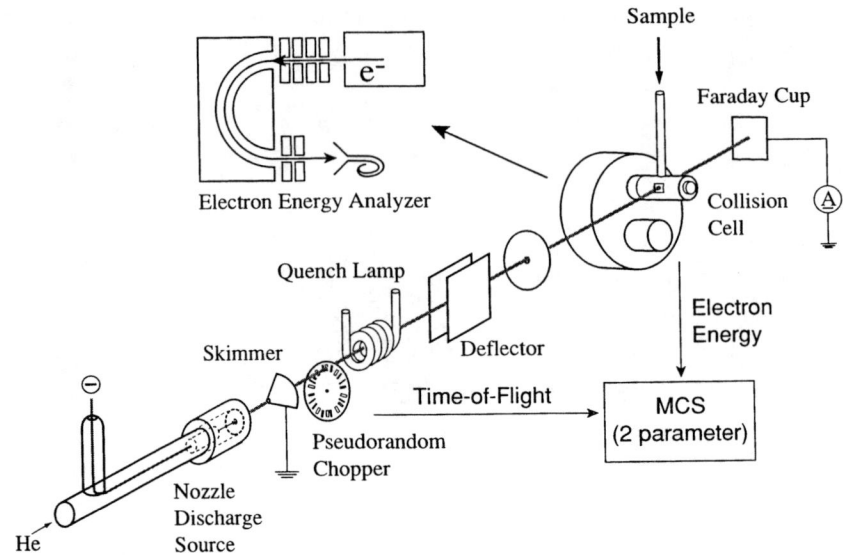

FIGURE 4. Schematic illustration of the experimental apparatus for 2D-PIES.

An example of 2D-PIES is shown in Figure 5 for the system of $H_2O/He^*(2^3S)$. 2D-PIES signals can be converted into collision energy dependence of partial Penning ionization cross sections(CEDPICS) as well as collision energy resolved Penning ionization electron spectra(CERPIES), by simply limiting the energy range of either E_e or E_c, respectively

2D-PIES $H_2O/He^*(2^3S)$

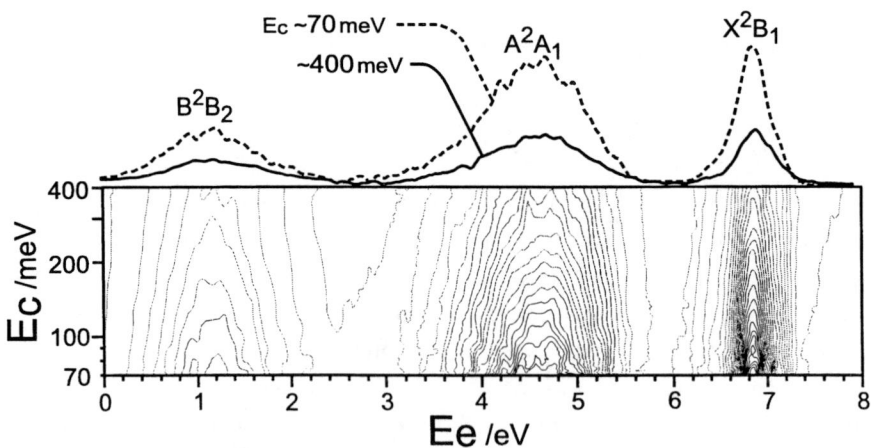

FIGURE 5. 2D-Penning ionization electron spectrum for $H_2O/He^*(2^3S)$.

CLASSICAL TRAJECTORY CALCULATIONS

In order to elucidate collision dynamics involved in Penning ionization processes, we have performed classical trajectory calculations based on quantum chemical calculations of potential surfaces for A*-M and partial ionization widths. Computational details are reported elsewhere(8). One interesting example of trajectory calculations is shown for $N_2/He^*(2^3S)$ in Figure 6; the observed CEDPICS are well-reproduced for three ionic states of N_2^+ when the *ab initio* model potential energy surface of $N_2/Li(2^2S)$ is simply scaled by a factor of 0.50(8). This is not surprising because the similarity between $He^*(2^3S)$ and $Li(2^2S)$ is well known(10). The isopotential energy contour curves shown in Figure 6 can be considered as the molecular shape of N_2 as probed by a $He^*(2^3S)$ atom which is confirmed to be very similar to those for $N_2/Li(2^2S)$ in Figure 2.

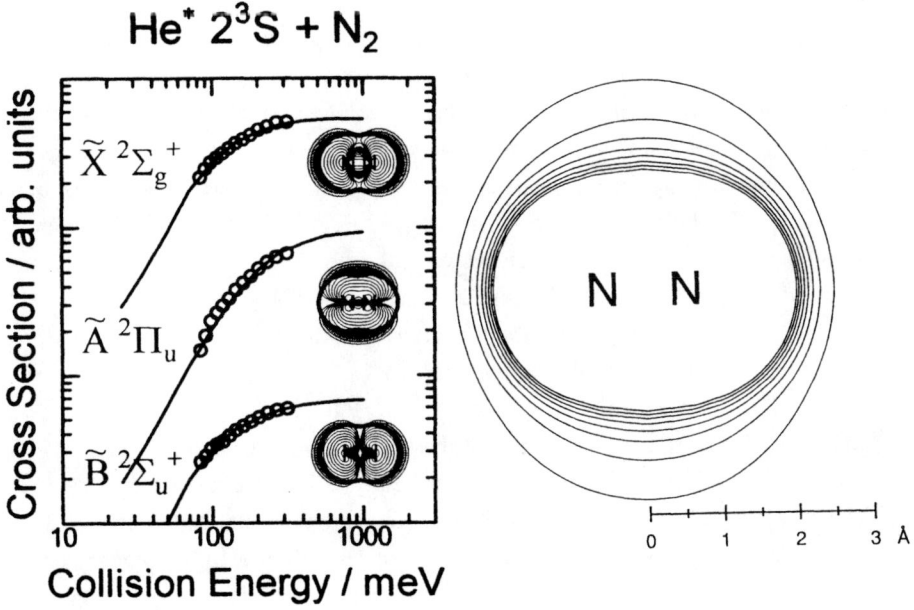

FIGURE 6. Observed collision energy dependence of partial Penning ionization cross sections(open circles) together with calculated curves for $N_2/He^*(2^3S)$ and the isopotential energy contours for the outer shape of a N_2 molecule probed by a $He^*(2^3S)$ atom(the energy spacing is 100 meV).

COLLISIONAL IONIZATION DYNAMICS

In Figure 7 ionization probabilities $P(b, \xi, E_c)$, so called opacity functions, are shown for total and partial ionization for $N_2/He^*(2^3S)$ at collision energies, $E_c = 100$ and 400 meV; b is the impact parameter explicitly shown in the figure and ξ is hidden parameters associated with the initial molecular orientation with respect to the incident He* beam and the initial rotational energy. Many points for the same value of the impact parameter correspond to the different sets of ξ for the initial conditions of the trajectories. The upper and the lower bounds show the maximum and minimum reactivity for a given impact parameter. These boundaries are labeled by the angle from the molecular bond axis where the ionization rate becomes maximum or minimum.

Some interesting features are revealed from the trajectory analyses.

(i) The maximum reactivity is related to the collinear axis for ionization from σ orbitals whose electron distribution extends outside along the collinear directions($0°$). The minimum ionization probability for σ orbitals is related to the vertical directions($90°$) where σ orbitals has the lowest electron densities; in the case of the σ_u orbital, there is a nodal plane vertical to the NN bond which leads to the vanishing reactivity. Because of the similarities of electron distributions of two σ orbitals(σ_g and σ_u) as indicated in the inserted electron density maps, patterns of opacity functions in Figure 7 resemble each other.

(ii) The minimum reactivity for π orbitals is related to the nodal planes including the bond axis($0°$), whereas for the maximum reactivity is not necessarily along with the vertical directions($90°$); at 400 meV the maximum lies vertical depending on the maximum electron distributions of π orbitals, but at 100 meV the maximum is deviated at about $50°$. This interesting behavior can be explained from the drastic change of the outer shape of the molecule as a function of the collision energy(Figure 6).

(iii) The maximum and minimum reactivity for the total ionization cross sections are governed by the σ orbitals. This is an unexpected aspect because the number of orbitals to be ionized are the same for the $\sigma(\sigma_g$ and $\sigma_u)$ and π(doubly degenerate π_u) orbitals. Since the ionization probability for each trajectory is limited to the unity, the total ionization probability shows 100 % ionization for the lower impact parameter values at $0°$ directions. This is consistent with the fact that the sum of two types of σ ionization at $0°$, ca. 0.59 and 0.41, equals to the unity. Correspondingly at $90°$, the sum of π and σ_g explains the lower bound of the total ionization.

(iv) Collision energy dependence is much more drastic for π ionization, reflecting the dramatic change of the boundary shape of the molecule at the vertical directions, as can be seen in Figure 6.

E_c = 100 meV

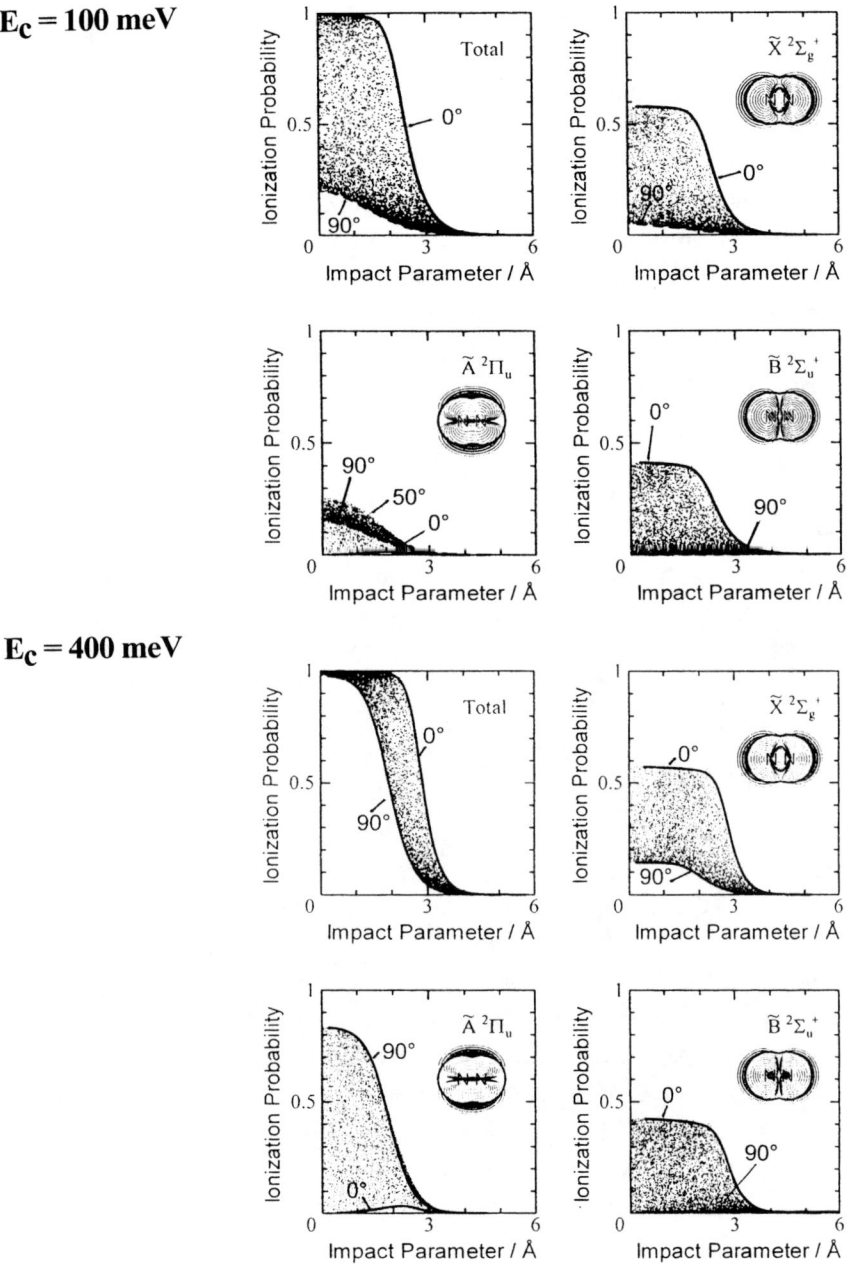

E_c = 400 meV

FIGURE 7. Ionization probabilities as functions of the impact parameter(opacity functions) for statistically sampled trajectories).

CONCLUSION

Outer boundary surface of molecular targets in collision processes is important, since it determines turning points of trajectories where reaction probabilities become the maximum. The outer shape of molecules can be probed by a test atom in quantum chemical calculations of interaction potentials. Varieties of shapes are found even for diatomic molecules, and its origin is related to anisotropic interactions peculiar to the molecule and the probing atom.

Two-dimensional Penning ionization spectroscopy enables us to disclose stereodynamics of collisional ionization. For the system of $N_2/He^*(2^3S)$, anisotropic nature of the molecular shape as functions of collision energies are disclosed by observation in good agreement of classical trajectory calculations based on *ab initio* potential surfaces and ionization widths.

ACKNOWLEDGMENTS

The author has been indebted to H. Tsunoyama, J. Kimura, M. Shiokawa, M. Kurita, and S. Hoshino for calculations of molecular shape and also to T. Ogawa for trajectory calculations. The author would like to thank the contribution of his colleagues for the experimental studies.

REFERENCES

1. Ohno, K. and Sunada, S., *Proc. Indian. Acad.Sci.* **106**, 327-337 (1994).

2. Hoshino, S. and Ohno, K., *J. Am. Chem. Soc.*, **119**, 8276-8284 (1997).

3. Niehaus, A., *Advances in Chemical Physics* **45**, New York: John Wiley & Sons, 1981, ch.5 , pp. 399-486.

4. Hotop, H. and Niehaus, A., *Z. Phys.*, **228**, 68-88 (1969).

5. Ohno, K., Mutoh, H., and Harada, Y., *J. Am. Chem. Soc.*, **105**, 4555-4561 (1983).

6. Ohno, K., Matsumoto, S., and Harada, *J. Chem. Phys.*, **81**, 4447-4454 (1984).

7. Ohno, K., Yamakado, H., Ogawa, T., and Yamata, T., *J. Chem. Phys.*, **105**, 7536-7542 (1996).

8. Ogawa, T. and Ohno, K., *J. Chem. Phys.*, **110**, 3773-3780 (1999).

9. Fukui, K., Yonezawa, T., and Shingu, H., *J. Chem. Phys.*, **20**, 722-725 (1952).

10. Hotop, H., *Radiation Research*, **59**, 379-404 (1974).

11. Hotop, H., Kolb, K., and Lorenzen, J., *J. Electron Spectrosc. Relat. Phenom.*, **16**, 213-214 (1979).

12. Mitsuke, K., Takami, T., and Ohno, K., *J. Chem. Phys.*, **91**, 1618-1625 (1989).

13. Ohno, K. Takami, T. Mitsuke, K., and Ishida, T., *J. Chem. Phys.* **94**, 2675-2687 (1991).

14. Kishimoto, N., Aizawa, J., Yamakado, H., and Ohno, K., *J. Phys. Chem.*, **101**, 5038-5045 (1997).

The quantitative characterization of liquid and solid surfaces with metastable helium atoms

H. Morgner[†]

Institute of Experimental Physics, University of Witten/Herdecke, Stockumer Str. 10, D-58448 Witten, Germany

Abstract. The electron spectroscopy based on the excitation energy of metastable rare gas atoms is a valuable tool for surface analysis because its surface sensitivity does not depend on the limited mean free path of the emitted electrons in condensed matter, but is perfect due to the physics of the emission process. Up to recently, the application of the method has suffered from the fact that a quantitative evaluation of the measured spectra has not been possible. The present contribution describes some progress in this respect which has been made during the last few years.

INTRODUCTION

Metastable helium atoms in their triplet state $He(1s2s;2^3S)$ carry an excitation energy of 19.819 eV. Even though their lifetime is about 4 hours when isolated, they readily transfer this energy as soon as they are in contact with other matter. This energy transfer leads with high probability to the emission of electrons, the spectrum of which is characteristic for the targets. The electron spectroscopy which is based on this process is called PIES (=Penning Ionization Electron Spectroscopy) for gaseous target. The application of the method to surfaces is often called MIES (= Metastable Induced Electron Spectroscopy) or MAES (= Metastable Atom Electron Spectroscopy).

The present article is devoted to the description of this spectroscopy to the characterization of surfaces. Therefore we discuss here the two major mechanisms for electron emission which apply to surfaces and which must be understood before measured electron energy spectra can be converted into information about the surface studied.

RESONANCE IONIZATION AND AUGER NEUTRALIZATION

The 2s-electron of the metastable helium atom may tunnel into unoccupied states of the surface material (RI=Resonance Ionization). If energetically allowed this process is very probable. The remaining $He^+(1s)$ ion continues its trajectory towards the surface. An Auger process driven by the recombination energy $E_{rec}[He^+]$ of the ion leads then to neutralization of the helium ion and the emission of an electron (AN=Auger

---

[†] Present address: Wilhelm-Ostwald-Institut für Physikalische und Theoretische Chemie, Universität Leipzig, Linnèstr. 2, D-04103 Leipzig, Germany, hmorgner@rz.uni-leipzig.de

CP500, *The Physics of Electronic and Atomic Collisions*, edited by Y. Itikawa, et al.
© 2000 American Institute of Physics 1-56396-777-4/00/$17.00

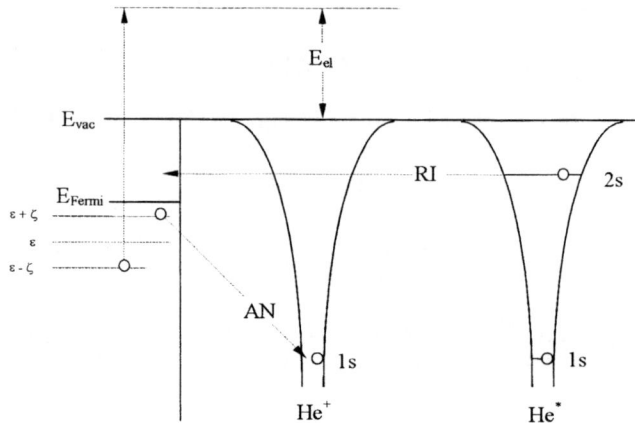

FIGURE 1.a   Resonance Ionization followed by Auger Neutralization as electron emitting step

Neutralization). Both processes are depicted in Fig. 1a. The energy of the emitted electron is given by

$$E_{el} = E_{rec}\left[He^+\right] - 2(\phi + \varepsilon)$$

where ϕ is the work function of the surface and ε the average binding energy of the two electrons involved in the Auger process.
The probability for emission of an electron with energy E_{el} is given by

$$P(E_{el}) \propto \int d\zeta \, N(\varepsilon + \zeta) \, N(\varepsilon - \zeta) \, H(\varepsilon + \zeta, \varepsilon - \zeta)$$

where $N(\varepsilon)$ denotes the density of states and H stands for the operator appropriate to describe the Auger neutralization process.

AUGER DEEXCITATION

If the process of RI (Resonance Ionization) does not take place, e.g. because the surface consists of a wide gap insulator, then the neutral, metastable atom reaches the surface. Again an electron from the surface tunnels into the 1s-hole of the metastable helium atom, but the emitted electron is the former 2s-electron of the atoms, cf. Fig. 1b. The electron energy spectrum which arises in this case (AD=Auger Deexcitation) is easily related to surface properties. The electron energy is given as
$$E_{el} = E\left[He^*\right] - E_{bind},$$
the difference between the excitation energy of He^* and the binding energy. The probability for electron emission as function of energy can be understood as measure for the density of states in front of the surface at the position of the metastable atom.

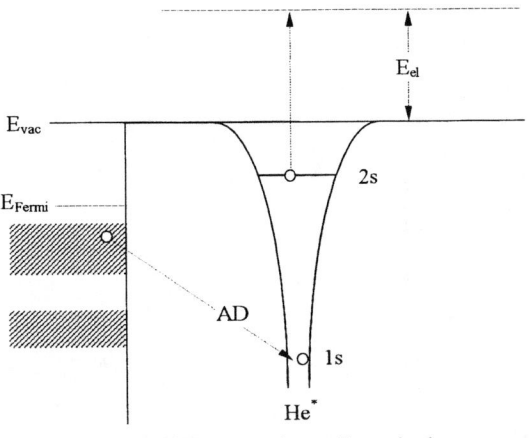

FIGURE 1.b   Auger Deexcitation

$$P(E_{el}) \propto N(E_{bind})$$

It is obvious that the understanding of a MIE-spectrum in terms of the surface properties depends on the mechanism which leads to electron emission. On the other hand, the method MIES is distinguished among other electron spectroscopies in that its surface sensitivity does not depend on the mean free path of electrons in condensed matter, but is always guaranteed due to the tunneling process which accompanies electron emission. This ambiguity may have motivated Harada et al [1] in their review article on MIES to state that 'the information obtained by MAES is unique but rather qualitative compared to other electron spectroscopic techniques'.

The present contribution describes progress with respect to quantitative characterization of surfaces by means of MIES which has been made possible in the last few years [2].

MIES OF GRAPHITE

As first example we discuss the emission of electrons from highly oriented pyrolytic graphite (HOPG) due to metastable helium atoms He(1s2s, 2^3S). Several years ago, Masuda et al. have discussed which mechanism is active during electron emission at this surface [3]. They came to the conclusion that the shape of the MIE-spectrum of graphite is governed by Auger Deexcitation (AD). A few years later we have confirmed this finding by showing that the spectrum can be quantitatively related to the structure of the occupied π-band of graphite [4], cf. Fig. 2. After revealing the nature of the electron emitting process, Masuda et al. found themselves obliged to find the answer to another puzzle [3]: if AD governs the ionization process then the process of resonance ionization is prohibited even though graphite is a semi metal and as such has unoccupied states in resonance with the orbital energy of the 2s-electron of metastable

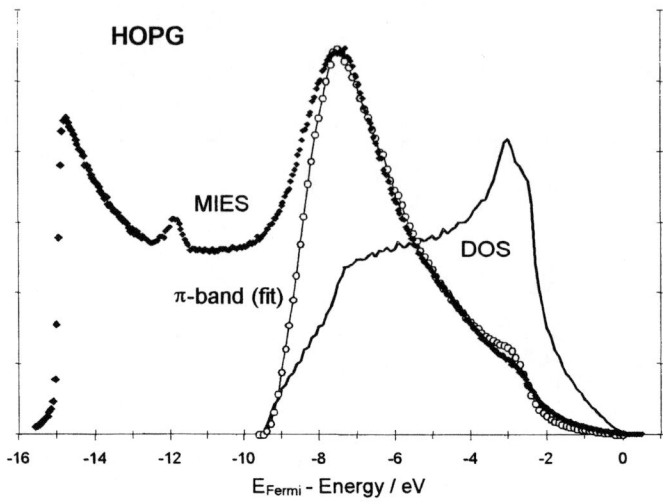

FIGURE 2   He(2^3S)-MIE spectrum of graphite

helium. The shape of the unoccupied π-band of graphite is such that the helium 2s-electron is in resonance with unoccupied π-state near the border of the Brioullin zone, between the M-point and the K-point, fig. 3.   Masuda et al. concluded that the tunneling of the 2s-electron into the states with high k-vector is symmetry forbidden. So far, graphite is the only material which prohibits resonance ionization due to symmetry. Therefore, it appeared interesting to analyze the situation in more detail on the basis of experimental data.

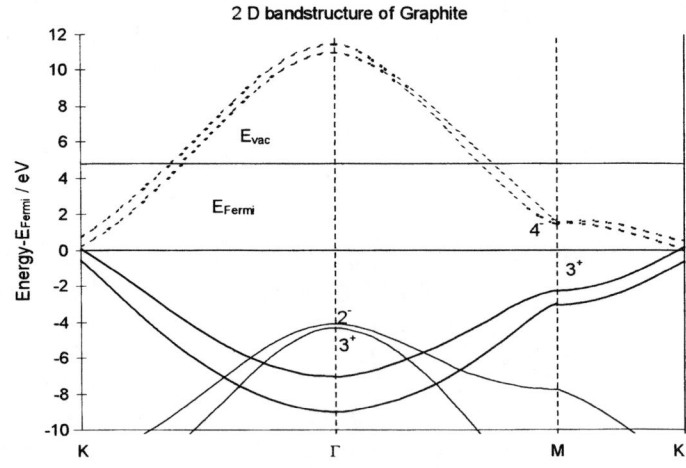

FIGURE 3   Band structure of occupied and unoccupied bands of graphite

The development of appropriate mathematical tools [2] made this attempt possible. A. Hoffknecht [5] has repeated the measurement of the MIE-spectrum of graphite at room temperature and, in addition, he has noticed that the shape of the spectrum varies distinctly with the temperature of the substrate, cf. Fig. 4. This has not been observed before for any metallic or insulating surface, except for trivial reasons like desorption of adsorbed impurities. By means of XPS and UPS, A. Hoffknecht could ascertain that

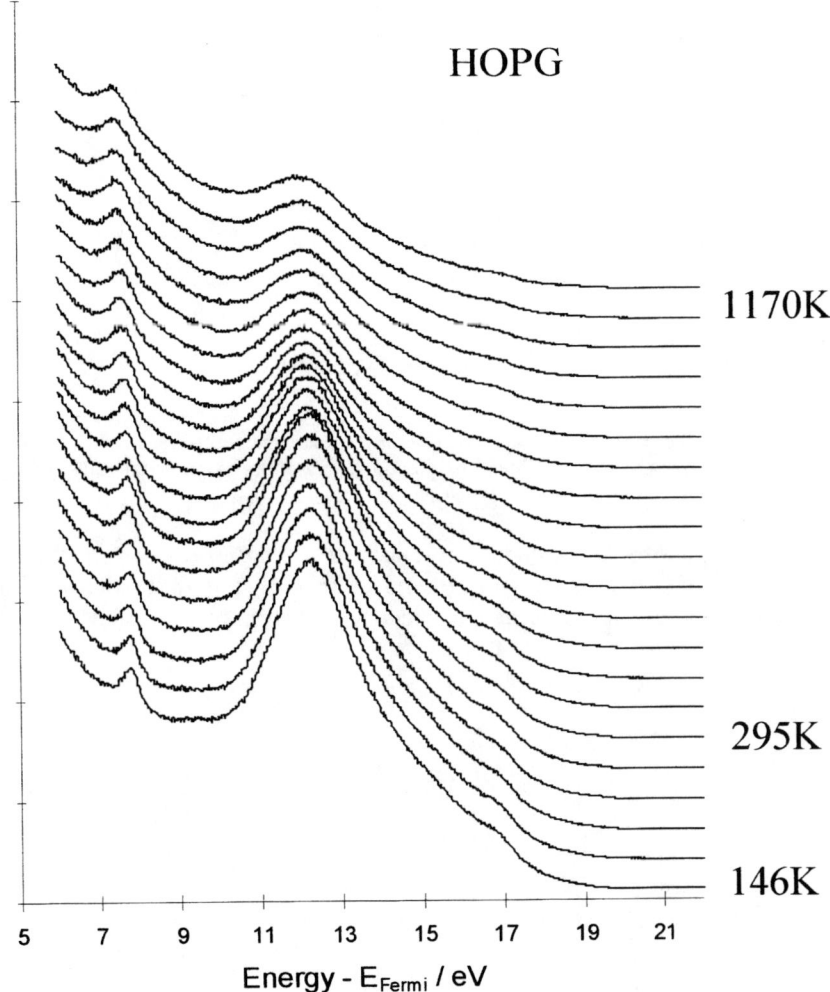

FIGURE 4   Series of MIE-spectra of graphite with substrate temperature varying between 146K (bottom) and 1200K (top)

his graphite sample was free of adsorbates throughout the whole series of spectra for temperatures between 146K and 1170K.

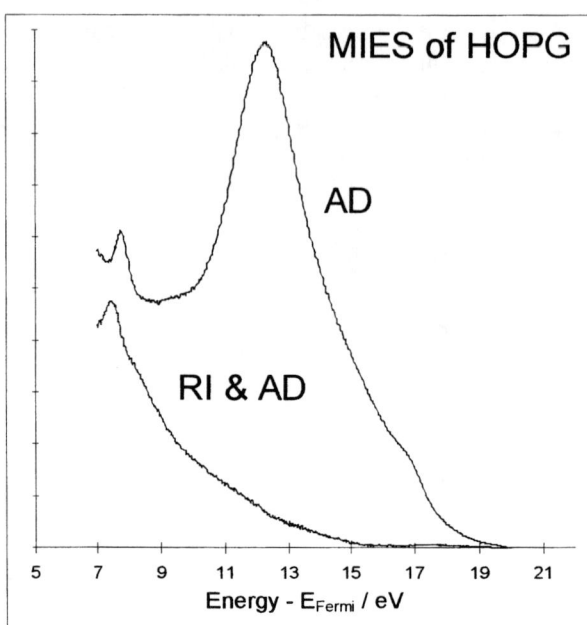

FIGURE 5   The two reference spectra. One of the reference spectra has been identified as due to the AN process. Thus, the second reference spectrum is taken as reference for the shape of the pure AD spectrum.

The mathematical tools for quantitative data analysis described in ref.2 are suited to analyze series of spectra as given in fig.4. Hoffknecht [5] found that the whole series can be represented with good accuracy as linear combination of two reference spectra

$$S(T) = \alpha_A(T)S_A + \alpha_B(T)S_B$$

The shapes of the two reference spectra are displayed in fig. 5.

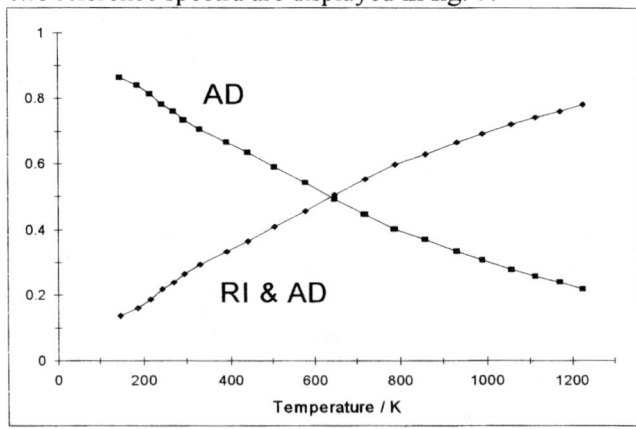

FIGURE 6   Development of the weighting factors $\alpha_{AN}(T)$, $\alpha_{AD}(T)$ with temperature

One of the spectra is labeled RI&AN because its shape is nearly identical to a spectrum obtained by He$^+$ ions of ~50eV kinetic energy impinging at grazing angle onto the graphite surface. Hence, the second reference spectrum is taken as resulting exclusively from AD. The weighting factors of both reference spectra $\alpha_{AN}(T)$ and $\alpha_{AD}(T)$ are shown in fig. 6 as function of temperature. We are now in the position to quantify the previous statement on the relative importance of both processes, AD and AN. At room temperature the probability for the occurrence of AN is only ~20% which makes AD the dominating mechanism. However, we observe that the temperature is an experimental parameter which allows to reverse this trend.

As mentioned above, this temperature dependence of the reaction mechanism is not a general one. So far, it has been found only for graphite.

The effect can be understood if one writes down the probability W_{RI} for resonance ionization as coherent sum over the tunneling amplitudes of the 2s electron into unoccupied 2_{pz}-orbitals of the surface carbon atoms. If the square of this amplitude is given as

$$F\left(\vec{k}, E_{2s}, x, y, z\right) = \left| \sum_1 \left\langle u_1 \middle| V(\vec{r_1}) \middle| 2s \right\rangle \frac{z - z_1}{r_1} \cdot e^{i\left(k_x \cdot x_1 + k_y \cdot y_1 + \delta_1\right)} \right|^2 \tag{1}$$

then the influence of the temperature induced mean square displacement $\left\langle t_m^2 \right\rangle$ yields the expression

$$W_{RI}\left(E_{2s}, x, y, z,\right) = \tag{2}$$

$$= \frac{2\pi}{\hbar} \int\limits_{\{\vec{k}, E(\vec{k}) = E_{2s}\}} d\vec{k} \frac{1}{A_{BZ}} \cdot \frac{dk_g}{dE} \cdot \left\{ F\left(\vec{k}, E_{2s}, x, y, z, t_m\right)\Big|_{(t_m = 0)} + \frac{1}{2} \frac{\partial^2 F\left(\vec{k}, E_{2s}, x, y, z, t_m\right)}{\partial t_m^2}\Bigg|_{(t_m = 0)} \cdot \left\langle t_m^2 \right\rangle \right\}$$

The first term in the parentheses stands for the value in case of ideal order of surface atoms. If this first term is small due to a swiftly varying phase factor between surface atoms then the second term which is proportional to the mean square displacement of the surface carbon atoms takes over with increasing temperature. The expression is evaluated under the assumption of a random phase between phonons which is justified for higher temperature. Hence, we are not surprised to see the weight of the coefficient α_{AN} to rise almost linearly with T at higher temperatures.

SURFACE OF BINARY LIQUID MIXTURE

The two polar solvents formamide (FA) and hydroxipropionitrile (HPN) are miscible in all proportions. Their different molecular structure and, hence, their different electronic structure leads to easily discernible spectra in MIES. The two respective spectra of the surface of the pure liquids are shown in fig. 7. Within experimental accuracy it is possible to represent the spectra of the binary mixture as linear combination

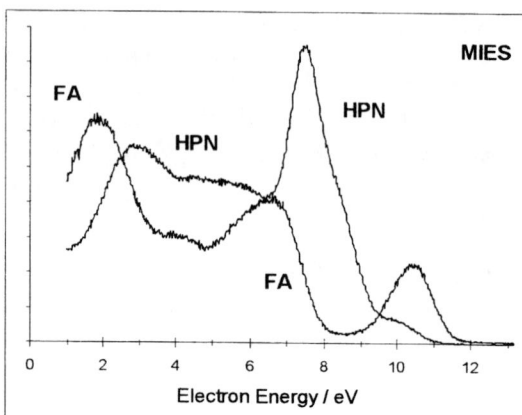

FIGURE 7   MIE-spectra of the surface of the pure liquids formamide (FA) and Hydroxypropionitrile (HPN)

$$S_{mixture} = \alpha_{FA} \, S_{FA} + \alpha_{HPN} \cdot S_{HPN} \quad \text{with} \quad \alpha_{FA} + \alpha_{HPN} = 1 \tag{3}$$

where S_{FA}, S_{HPN} denote the spectra of the pure liquids and α_{FA}, α_{HPN} give the probability that the $He^*(2^3S)$ atoms approach a surface molecule of either species. Thus the coefficients α_{FA}, α_{HPN} can be identified with the fraction of the surface occupied by the respective species.

At the first glance, one could take the validity of eq.(3) as granted since no third species is present. However, we have to take into account that MIE spectra depend sensitively on the orientation (in case of liquid surfaces it is more appropriate to talk about the mean orientation of the molecules).

In ref. 2 this is explicitly demonstrated for formamide, and the following section of the present contribution covers the orientation dependence of alkanes. Therefore, the mere fact that eq.(3) is successful in evaluating the MIE-spectra of the liquid mixture indicates that the mean orientation of both molecular species is not noticeably affected in the surface of the mixture compared to the pure liquids. As one would expect that the presence of a foreign molecule should affect the molecular orientation one is compelled to believe that any molecule is more likely to be surrounded by its own kind rather than to be in contact with the other species. This consideration would lead to the conclusion that both molecules have the tendency to form clusters.

In fig. 8 the coefficient α_{HPN} which is considered to describe the areal fraction of HPN is plotted against the bulk molar fraction. Making use of the estimated areas per molecule it is possible to compute the surface molar fraction x_{HPN}. One recognizes an enhanced abundance of hydroxipropionitrile in the surface compared to the bulk composition. This is to be expected, since the surface tension of HPN is smaller than that of FA by about 10mN/m. Fig. 9 shows that the surface tension of the mixture scales linearly with the composition of the top surface. This result may lead in the long run to a better understanding and, thus, to a better prediction of surface tension of

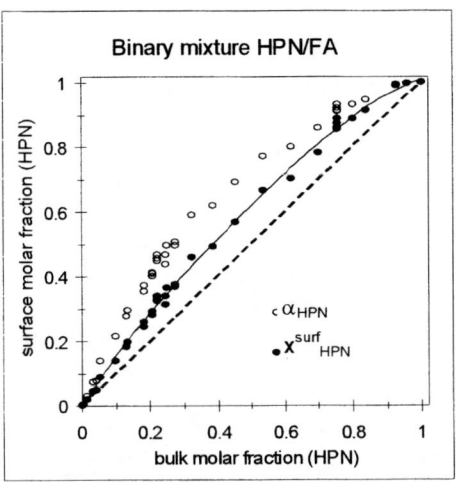

FIGURE 8   Binary mixture of FA and HPN. Areal fraction and surface molar fraction of HPN as function the HPN bulk molar fraction. Evaluated from MIES data.

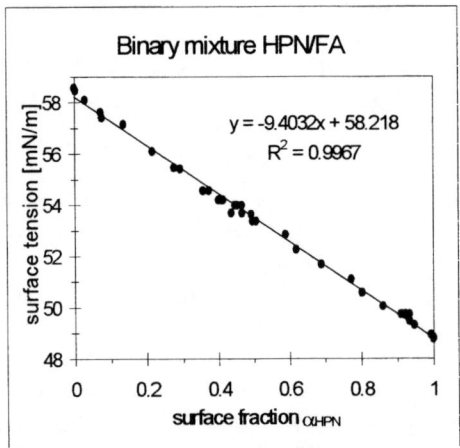

FIGURE 9   Surface tension of the binary mixture HPN/FA as function of the areal fraction of HPN in the surface.

mixtures and solutions. Of course, for liquids composed of only two species this would be merely of scientific interest because the surface tension can be measured with good accuracy. However, in case of ternary or quaternary mixtures the number of necessary measurements increases so much that the possibility to predict interfacial tensions appears to be of relevance for technical applications.

ALKANES

The MIE-spectrum of a surface composed of alkanes depends strongly on the orientation of the molecules. Alkanes represent the best documented example for this orientation dependence. From a large number of MIE-spectra taken from alkane samples it has been possible to derive 4 reference spectra whose linear combination is able to explain all spectra of surfaces which are covered by a closed layer of saturated alkanes [2]. In fig. 10 the reference spectra are shown. Three of them refer to alkanes in all-trans conformation. They are distinguished by the different orientation of the molecules. Spectrum A refers to alkanes lying on the substrate with their molecular plane being parallel to the surface. If the molecules are standing upright, exposing their $-CH_3$ groups, then spectrum B yields the corresponding reference. Alkanes lying on the substrate but with their molecular plane being perpendicular to the surface again yields a different spectrum which is labeled C. Finally, we have to deal with alkanes carrying gauche conformations. Even though the manifold of gauche conformations is very large it was found [2] that one reference spectrum suffices to describe the presence of gauche conformations. The most likely explanation seems to be that if deviations from the all trans conformation occur at all the full spectrum of conformations is found on the area investigated by MIES which typically is several mm^2. All MIE spectra of adsorbed alkanes measured so far (several hundred) can be represented as linear combination involving the reference spectra A, B, C, D

$$S_{adsorbed\ alkanes} = \alpha_A S_A + \alpha_B S_B + \alpha_C S_C + \alpha_D S_D$$

It is interesting to note that neither of the reference spectra has been obtained by direct measurement. Apparently, it has not been possible to experimentally prepare a surface with ideally ordered alkane molecules over a macroscopic area of several square millimeters. The reference spectra are the result of a mathematical procedure described in ref. 2, but based on a large number of experimental spectra, i.e. without any attempt to model the reference spectra by means of theoretical input.

The comparison of the four reference spectra with theory provides an independent test on the identification of the spectra as referring to specific physical situations at the surface. In ref. 2 the comparison with spectra derived from the band structure of polyethylene is shown which is very convincing. A second comparison with spectra simulated on the basis of Hartree-Fock calculations of hexadecane is not quantitative, but shows the correct tendencies. A new calculation performed by C. Engler[6] using density functional theory to describe the correlation energy yields a satisfying agreement with the reference spectra, cf. fig. 10.

Finally, we will discuss how the reference spectra can be applied to characterize a sample. We have prepared a self assembled monolayer (SAM) of hexadecanethiolates on Ag(111) according to recipes found in the literature and claimed to lead to films of perfect order. Nominally, the surface of this film should consist of the terminal $-CH_3$

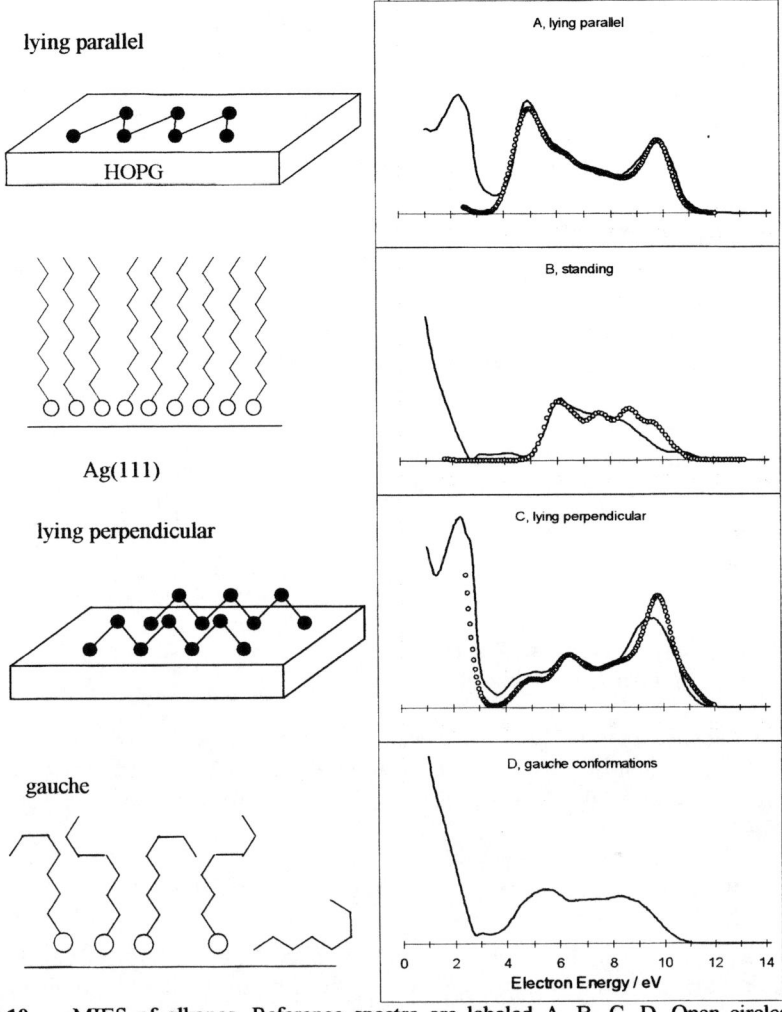

FIGURE 10   MIES of alkanes. Reference spectra are labeled A, B, C, D. Open circles: spectra derived from ab initio (DFT) calculations by C. Engler, ref.6.

groups of the alkyl chains. Fig. 11 shows the result of the analysis: the film –prepared and kept at room temperature– consists only to 76% of methyl groups. Ca. 8% of the surface is covered by molecules having gauche conformations. The remainder of the signal originates from molecules which expose their broad side to the metastable helium atoms, i.e. which contribute to reference spectra A,C.

Now we observe the effect of a temperature program. Cooling down to 110K does not increase the order at the top surface as investigated by MIES. Only after an intermediate heating to 370K does the lowering of the temperature lead to an enhanced fraction of molecules exposing their methyl group in all-trans conformation. Even after

697

this annealing process we find that ca. 15% of the surface do not correspond to the expected order.

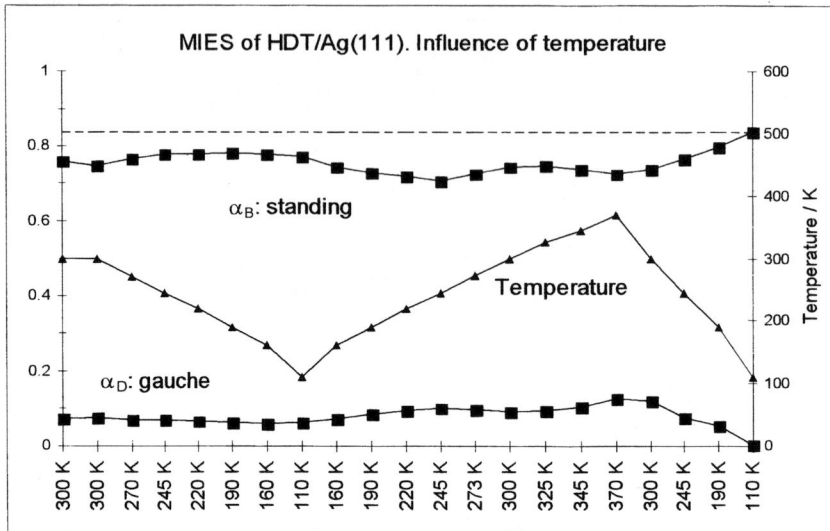

FIGURE 11   Hexadecanethiolate film on Ag(111). Development of surface properties during temperature program as characterized by MIES.

Scanning microscopies (ATM or STM) often show perfect order over microscopic areas, say 100x100nm² or less. Compared to these data, one must keep in mind that an electron spectroscopy like MIES averages over a macroscopic area of several square millimeters.

ACKNOWLEDGEMENT

The work described in this contribution was financially supported by the German Ministry of Research and Technology and by the German Research Foundation (DFG).

REFERENCES

[1] Y. Harada, S. Masuda and H. Ozaki, Chem. Rev. 97 (1997) 1897-1952
[2] H. Morgner, Adv. At. Mol. Physics 42B (1999) 387-488
[3] S. Masuda, H. Hayashi and Y. Harada, Phys. Rev. 42 (1990) 3582-3585
[4] B. Heinz and H. Morgner, Surface Science 372 (1997) 100-116
[5] A. Hoffknecht , work performed as graduate student at the University
             Witten/Herdecke
[6] C. Engler, University Leipzig, DFT calculation of the electronic structure of
             hexadecane (1999) unpublished results

Swarm Study of Kr^{2+} Ions in Helium Gas at Very Low Temperature

Hajime Tanuma, Hiroshi Hidaka, and Nobuo Kobayashi

Department of Physics, Tokyo Metropolitan University
Hachioji, Tokyo 192-0397, Japan

Abstract. Doubly charged helium cluster ions Kr^{2+}-He_n ($n \leq 40$) have been produced when Kr^{2+} ions are injected into helium gas cooled by liquid helium. Magic numbers $n = 12$ and 32 have been observed in the size distribution of these cluster ions. These magic numbers suggest the shell structure in which an icosahedron consisting of twelve helium atoms performs as a core in the cluster ion. Not only the clustering, charge transfer between Kr^{2+} and He also takes place as a competitive reaction of the cluster formation. Behavior of each electronic states of the incident ions in collisions with helium have been discussed on a crude potential crossing model.

INTRODUCTION

In previous works, we had reported formation of helium cluster ions by swarm method, which can be regarded as one of powerful experimental techniques for low energy ion-molecule collisions [1], when several sorts of singly charged ions are injected into a drift tube filled with helium gas at very low temperature of 4.4 K [2]. We had found twelve to be common magic number, which is the number of helium atoms in the cluster ions having particular stability, for all sorts of clusters which have atomic and diatomic ions as their cores. This magic number implies icosahedral structure, in which twelve equivalent helium atoms are surrounding the ionic core, however evidence to prove this suggestion directly has not been found in our experiment yet.

We believe that the cluster ions are formed through the following three-body ion-atom association reactions :

$$X^+ + He + He \rightleftharpoons X^+\text{-He} + He, \tag{1}$$

$$X^+\text{-He}_{n-1} + He + He \rightleftharpoons X^+\text{-He}_n + He, \tag{2}$$

where X^+ denotes the injected ion. In these cluster ions, two different kinds of attractive interaction must be considered: the van der Waals force between neutral helium atoms and the polarization force between the charged ion and the helium

CP500, *The Physics of Electronic and Atomic Collisions*, edited by Y. Itikawa, et al.
© 2000 American Institute of Physics 1-56396-777-4/00/$17.00

atom. We suppose that the latter plays the dominant role in the clusters. The polarization potential V_{pol} as a function of the distance r between the ion and the neutral is given by the following equation :

$$V_{pol}(r) = -\frac{\alpha q^2 e^2}{2r^4},$$ (3)

where α is the polarizability of the neutral atom, which is 0.205 Å^3 for helium, and q is the charge number of the ion. This equation tells us that a doubly charged ion with $q = 2$ has four times as strong attraction as a singly charged one with $q = 1$ at the same distance. Therefore formation of larger clusters can be expected in the case of the doubly charged helium cluster ions. In this paper, we report the result of Kr^{2+} injection into helium gas at very low temperature. We are interested in also the ion mobility [3] and have some preliminary results on it, which should be omitted for want of space here.

EXPERIMENTAL

The incident ions Kr^{2+} are produced with a conventional electron impact ion source. Ions mass-selected by a Wien filter are injected into a drift tube with energy of less than 10 eV. This drift tube with 99 mm long is incorporated into a liquid helium cooled cryogenic system, and is filled with pure helium gas. Inside the tube, the ions quickly thermalize in collisions with helium gas, and begin to drift toward the end of the tube under the influence of weak uniform electric field produced by guard rings. During this drift, the clustering reaction takes place to produce helium cluster ions. The product ions leave the drift space through an aperture at the end plate of the tube, and are detected by a secondary electron multiplier after mass-analysis by a quadrupole mass filter. Brief description about our experimental set-up has also been given elsewhere [4].

RESULTS

A typical mass spectrum showing the size distribution of Kr^{2+}-He_n cluster ions produced in the drift tube filled with helium gas at 4.3 K is presented as a function of the mass-to-charge ratio m/q in Figure 1 (a). Isotopes of krypton have been observed with modified natural abundance because of poor mass resolution of the Wien filter. Since prominent peaks are corresponding to ^{84}Kr which has the largest abundance, viz 57 %, their mass number can be given by $42 + 2n$. Magic number has been observed clearly in the mass spectrum at $n = 12$, which is the same number as that has been observed as the common magic number in several kinds of singly charged helium cluster ions. Not only doubly charged ions but also singly charged helium cluster ions are observed with very weak intensities in the mass spectrum. These Kr^+-He_m ions should be formed by the clustering reaction of Kr^+ ions produced by the charge transfer between Kr^{2+} and He in the drift tube.

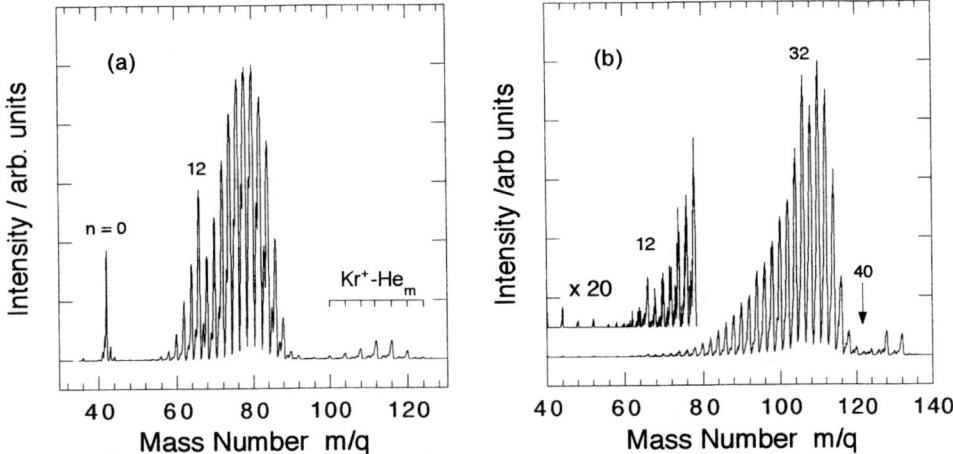

FIGURE 1. Mass spectra of product ions with Kr^{2+} injection into the drift tube filled with helium gas at very low temperature. The numbers beside peaks indicate the number of helium atoms in the doubly charged cluster ion Kr^{2+}-He_n. (a) Gas temperature T, helium gas pressure P, and electric field strength E were 4.3 K, 3.8 Pa, 2.34 V/cm, respectively. (b) $T = 2.1$ K, $P = 3.0$ Pa, and $E = 0.58$ V/cm.

Figure 1 (b) shows a typical mass spectrum of product ions observed with the injection of Kr^{2+} ions into cooled helium gas at 2.1 K. We can see very small peak at $n = 40$ as the largest size of cluster ions in this study, which is much larger than ones in our previous works for singly charged ions. Magic number of $n = 12$ have been observed also in this spectrum, and the $n = 32$ which shows particular irregularity in the size distribution has been considered as another magic. The charge transfer reaction to produce singly charged ions takes place also at this lower temperature, however, the fraction of singly charged ions is always smaller than 7 % when the large clusters with $n \geq 12$ have been observed in the mass spectra.

DISCUSSION

Shell Structure of Helium Cluster Ions

Both magic numbers $n = 12$ and 32 that are observed in mass spectra of Kr^{2+}-He_n imply a shell structure with an icosahedral core, which has twelve apexes and twenty faces. If twelve helium atoms surround the Kr^{2+} ion and make an icosahedral cluster ion with exceptional stability, each face of this icosahedron is a triangle consisting of three helium atoms. The center of each triangle, at which the charge of the core ion is not directly shielded by the electron clouds of helium atoms

701

in the first shell, must be the most stable site on the surface of the icosahedron for neutral helium atoms in the second shell. We therefore can say that the first twelve helium atoms make up the first shell (or layer) and the next twenty atoms compose the second shell.

This simple model seems to be quite reasonable, and another possibility cannot be imagined for the structure of this helium cluster ion with $n = 32$. For the $n = 12$, not only the icosahedral structure, the face-centered-cubic (fcc) and hexagonal-close-packed (hcp) geometries, both of which have eight triangles and six squares on their surfaces, can be candidates because twelve apexes are completely and closely equivalent in these structures, respectively. However, neither the fcc nor hcp structures can explain the extra stability for $n = 32$ since the number of faces is fourteen in both geometries.

Competition of Charge Transfer with Clustering

Electronic configuration np^4 yields five different states, which are $^3P_2$, $^3P_1$, $^3P_0$, $^1D_2$, and $^1S_0$. Because the incident $Kr^{2+}(4p^4)$ ions are produced by the Nier-type ion source with electron impact energy of more than 100 eV, the population of these five electronic states must be determined with the statistical weights. Diabatic potential energy curves for $(Kr-He)^{2+}$ are shown in Figure 2, in which only the Coulomb repulsion between positive ions and the polarization interaction are taken into account. As seen in this figure, four states of Kr^{2+} except $^3P_2$ can take exothermic charge transfer reaction in collisions with helium. This reaction should have the transition probability which is strongly localized around the crossing point of the initial channel with the final one. Hence we can make discussion about the behavior of each initial Kr^{2+} state in the drift tube filled with helium gas according to the approximate potential curves in Figure 2.

The $^1S_0$ state should have a large cross section for the charge transfer reaction because the crossing point locates within the called reaction window. On the other hand, the contribution of the $^1D_2$ state of Kr^{2+} to the charge transfer reaction can be neglected since Koizumi has reported that this state quickly turns to the $^3P_1$ state by collisional de-excitation with helium [5].

The crossing radii for the $^3P_1$ and $^3P_0$ states are more than 20 Å, which should be much larger than the distance between the core ion and the helium atom in the icosahedron of Kr^{2+}-He_{12}. If the clustering reaction takes place very fast and the icosahedral cage forms without electronic transition of these Kr^{2+} states, there is no chance that the charge transfer reaction happen because of the fixed small internuclear distance in the shell structure. This finding means that the clustering can be regarded as a competitive reaction of the charge transfer even if the charge transfer reaction is an exothermic process with no activation energy.

Hence, in the drift tube conditions when large clusters with $n \geq 12$ have produced dominantly, only the $^1S_0$ state with the population of 7 % in the incident beam takes the charge transfer reaction with helium. This should be the reason why the singly

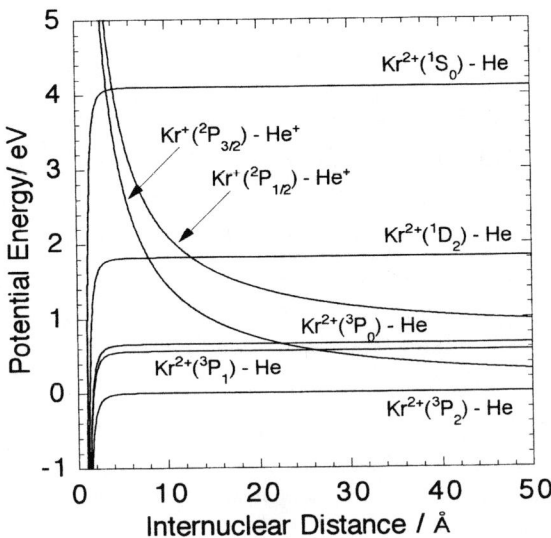

FIGURE 2. Diabatic potential energy curves for $(Kr-He)^{2+}$ systems.

charged ions produced by the charge transfer reaction have very small fraction in the mass spectra. The size distribution of cluster ions strongly depends on the electric field strength in the drift tube. In the high field region, in which helium cluster ions disappears and only the charge transfer reaction can be discussed, the measurement of the charge transfer reaction rate is expected to show each contribution of five states, not only the reactive states but also the non-reactive $^3P_2$ state.

REFERENCES

1. Kaneko, Y., Megill, L. R., and Hasted, J. B., *J. Chem. Phys.* **45**, 3741-3751 (1966).
2. Tanuma, H., Sanderson, J., and Kobayashi, N., *J. Phys. Soc. Jpn.* **68**, 2570-2575 (1999) and its references [3]-[6].
3. Sanderson, J., Tanuma, H., Kobayashi, N., and Kaneko, Y., *J. Chem. Phys.* **103**, 7098-7103 (1995) and its references [11]-[13].
4. Tanuma, H., Fujimatsu, H., Sakamoto, M., Okuno, K., and Kobayashi, N., *Photonic, Electronic and Atomic Collisions*, edited by Aumayr, F. and Winter, HP., World Scientific, 1998, pp. 603-606.
5. Koizumi, T., *J. Phys. B: At. Mol. Opt. Phys.* **25**, L335-L339 (1992).

Charge Transfer between a Laser-cooled Ion and a Thermal Atom in a Radio-frequency Trap

Taro Hasegawa and Tadao Shimizu

Department of Electronics and Computer Science, Science University of Tokyo in Yamaguchi Onoda, Yamaguchi 756-0884, Japan

Abstract. The charge transfer process between a laser-cooled magnesium ion and a thermal barium atom in a radio-frequency ion trap in extremely low energy regime is observed. The merit of using the trap is that the generated ions are also confined in the trap. The occurrence of the charge transfer process is confirmed by the detection of the laser-induced fluorescence from the generated ions. The cross section of the process is estimated to be $10^{-3} \text{Å}^2 \sim 10^{-1} \text{Å}^2$.

INTRODUCTION

Charge transfer is one of the most interesting subjects not only of physics but also of chemistry because of the relation to the chemical reactions. However, the energy regime of the chemical reaction is less than eV, whereas the kinetic energy of usual experiments of the charge transfer process is more than keV. Therefore, the experiments in much lower energy regime are expected to be carried out.

In usual experiments of the charge transfer, the high collisional energy is mainly due to the speed of the ions. In our experiment, the charge transfer between the laser-cooled magnesium ion ($< 0.001\text{eV}$) and the thermal barium atom ($\sim 0.1\text{eV}$) is observed for the first time in a radio-frequency (rf) ion trap [1] and the cross section of the charge transfer process is estimated. In the rf trap, the ions are confined and can be laser-cooled [2]. In such a case the collisional energy of the system is determined by the temperature of the atoms instead of the speed of the ions. Moreover, the generated ions as a result of the charge transfer (barium ions in this case) are also confined in the rf trap with the residual magnesium ions. Consequently, the yield of the generated ions are expected to be large and also the highly sensitive optical detection is applicable. These are the reasons why an experiment in such a low energy regime is possible in the rf trap even if the cross section of the charge transfer process is small.

CP500, *The Physics of Electronic and Atomic Collisions*, edited by Y. Itikawa, et al.
© 2000 American Institute of Physics 1-56396-777-4/00/$17.00

EXPERIMENT

The cross section of the charge transfer process depends on the discrepancy in the internal energy of the colliding system before and after the process, especially in the low energy regime. In this experiment, the charge transfer process is represented as,

$$\text{Mg}^{+*}(3p\,^2P_{3/2}) + \text{Ba}^+(6s^2\,^1S_0) \rightarrow \text{Mg}^*(3s3p\,^1P_1) + \text{Ba}^{+*}(6p\,^2P_{1/2}) + 0.011\text{eV}. \tag{1}$$

Here, the ion and atom with asterisks are in the excited states. As shown in Fig.1, the energy discrepancy is only 0.011eV and this charge transfer process is in a good resonance condition. It should be noted that the excited state of the magnesium ion is highly populated because of the cooling laser.

Experimental outline is as follows. At first, the magnesium ions are prepared in the rf trap and laser-cooled by the laser beam whose frequency is detuned 200MHz below the transition frequency of the magnesium ion. Next, the barium atoms from the thermal oven are introduced into the trap. The charge transfer process may take place at the center of the trap, and then the magnesium atoms and the barium ions are generated. The magnesium atoms are out of the trap and the barium ions

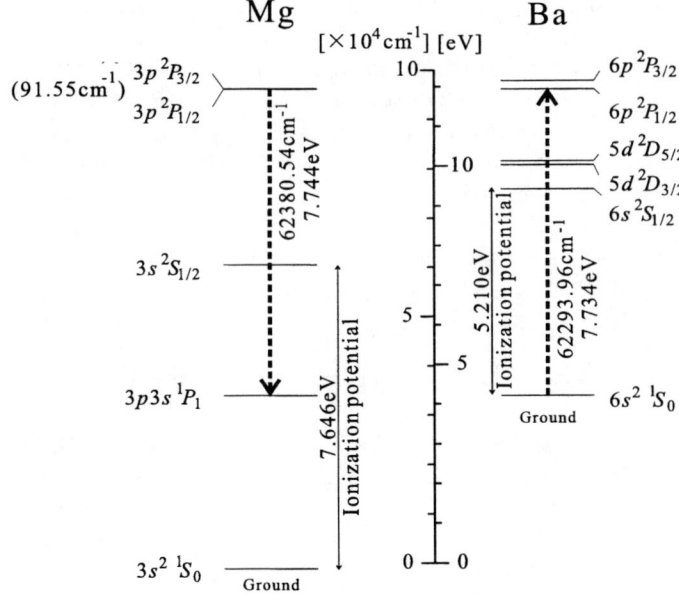

FIGURE 1. Energy diagrams of magnesium and barium in scale. The energy discrepancy between $3p\,^2P_{1/2}$ and $3p\,^2P_{3/2}$ states of Mg^+ is so small that two states are not distinguished in this figure.

are trapped in the rf trap with the residual magnesium ions. Here, the barium ions are indirectly cooled through the collision with the laser-cooled magnesium ions, and this cooling mechanism is called "sympathetic cooling" [3]. Lastly, the laser beam for the detection of the barium ions is irradiated and the fluorescence from the barium ions is detected. It should be noted that the barium atoms are not introduced during the ionization of magnesium atoms and no ionization mechanisms of barium exist except the charge transfer between the magnesium ions and the barium atoms. Therefore, the detection of the fluorescence implies the occurrence of the charge transfer process.

In this experiment, three lasers are necessary. One is for the cooling of the magnesium ions and its frequency corresponds to the transition $3s\ ^2S_{1/2}$–$3p\ ^2P_{3/2}$ (wavelength $\lambda = 280$nm). This laser is supplied by the second harmonic generation of the ring dye laser. Another laser is for the excitation of the barium ions and its frequency corresponds to the transition $6s\ ^2S_{1/2}$–$6p\ ^2P_{1/2}$ ($\lambda = 493.5$nm). This laser is supplied by the second harmonic generation of the laser diode. The other laser is for the repumping of the barium ions from the metastable state back to the fluorescence transition and its frequency corresponds to the transition $5d\ ^2D_{3/2}$–$6p\ ^2P_{1/2}$ ($\lambda = 650$nm). This laser is supplied by the ring dye laser. The fluorescence of the $6p\ ^2P_{1/2} \rightarrow 6s\ ^2S_{1/2}$ transition is detected. When the spectral line of the fluorescence of the barium ions is recorded, the laser frequencies at 280nm and 493.5nm are fixed and that at 650nm is swept.

The experimental setup of rf trap is an ordinary type [1]. The trap electrodes of the hyperboloid of revolution are arranged in the conventional quadruple configuration. The radius of the ring electrode r_0 is 7.5mm, and the spacing between the cap electrodes $2z_0$ is 10.8mm. The trapping electrodes are set in a vacuum chamber of the order of 10^{-7}Pa. Three laser beams are aligned and directed to the center of the trap through the spacing between the ring and the lower cap electrodes. The fluorescence from the barium ions are detected by the photomultiplying tube (PMT) . The optical filter is placed in front of the PMT for the removal of the fluorescence of the magnesium ions at 280nm. The spectral line of the fluorescence of the barium ions is recorded by the x-y recorder.

RESULTS AND DISCUSSIONS

Figure 2 shows the observed spectral line of the barium ions with the spectral line of the iodine molecule for the calibration of the absolute laser frequency. As it has been stated, the observation of the spectral line of the barium ions itself implies the occurrence of the charge transfer process in extremely low collisional energy regime. The number and the temperature of the barium ions are estimated from the intensity of the fluorescence and the Doppler broadened Gaussian lineshape, respectively. As the results, it is found that the number and the temperature of barium ions are about 10 and 500K, respectively. Similarly, the number and the temperature of magnesium ions trapped with the barium ions are estimated to be 100 and 200K, respectively, from the fluorescence line at the 280nm. The difference in the temperatures of simultaneously trapped barium and magnesium ions is explained as a characteristic result of the sympathetic cooling [4].

From this result, the cross section of the charge transfer process is estimated. In our experiment, the relative velocity between the magnesium ions and the barium atoms is determined by the temperature of barium atoms. Then the formula of the vacuum evaporation can be used as,

$$N = \eta \frac{Pt}{\pi r^2 \sqrt{2\pi m k_B T}} dS\, dS'. \tag{2}$$

Here, N is the number of the generated barium ions, η the ratio of trapped barium ions to the generated ions, P the pressure of the barium atom at the thermal oven, t the open-gate time of the introduction of the barium atom, dS the area of the target (in this case, this corresponds to the number of the magnesium ions N_{Mg} times the cross section σ), dS' the area of the barium oven, r the distance from the barium oven to the trap center, m the mass of a barium atom, k_B the Boltzmann constant, T the temperature of barium atoms. By substituting the values to Eq.(2), it is found that the cross section σ is between 10^{-3}Å^2 and 10^{-1}Å^2 in the case of $\eta = 1$, which is an adequate assumption because the kinetic energy of the barium ions is much less than the barrier of the trapping potential (3eV in our experimental setup). The theoretical prediction of the cross section, in which the discrepancy in the internal energy of the system before and after the process is supposed 0.011eV as in Fig.1, is shown in Fig.3 [5] with the observed value. It is found that the

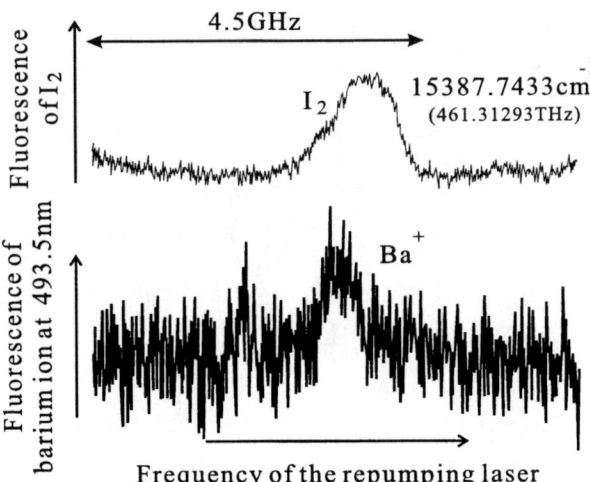

FIGURE 2. Observed spectral line of barium ions generated by the charge transfer in the rf trap (lower trace) and the spectral line of iodine molecule for the calibration of the laser frequency (upper trace).

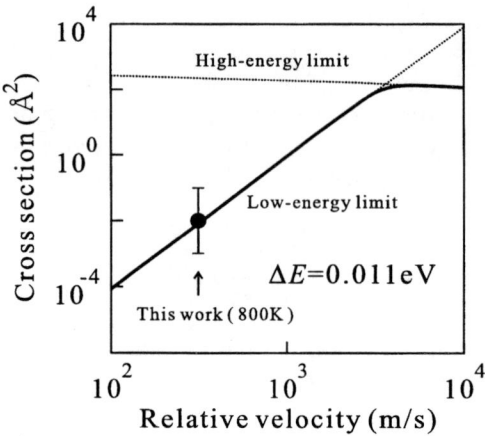

FIGURE 3. Dependence of the cross section of the charge transfer process on the collisional velocity. Lines are the theoretical prediction by the formula of Rapp and Francis, *J. Chem. Phys.* **37**, 2631 (1962).

observation agrees with the theory.

The ambiguity of the cross section in this experiment is mainly due to that of the pressure of the barium atom P. In our case, P is the vapor pressure at the temperature of the barium oven. Therefore the ambiguity of P is large because of the exponential dependence of the vapor pressure on the temperature.

CONCLUSION

The charge transfer process between the laser-cooled magnesium ion and the thermal barium atom in the rf trap is observed in extremely low collisional energy regime for the first time. The cross section of the process in this energy regime is estimated as $10^{-3}\text{Å}^2 < \sigma < 10^{-1}\text{Å}^2$. As the future subject, the atom beam of the barium should be controlled for the purpose of the reduction of the ambiguity of the result.

REFERENCES

1. Wuerker, R. F., Shelton, H., Langmuir, R. V., *J. Appl. Phys.* **30**, 441 (1959).
2. Wineland, D. J., and Itano, W. M., *Phys. Rev. A* **20**, 1521 (1979).
3. Larson, D. J., Bergquist, J. C., Bollinger, J. J., Itano, W. M., and Wineland, D. J., *Phys. Rev. Lett.* **57**, 70 (1986).
4. Oshima, Y., Moriwaki, Y., and Shimizu, T., *Prog. Crystal Growth and Charact.* **33**, 405 (1996).
5. Rapp, D., and Francis, W. E., *J. Chem. Phys.* **37**, 2631 (1962).

Electrons Emitted from 33-TeV Pb Ions During Penetration of Solids

C. R. Vane,[*] U. Mikkelsen,[a] H. F. Krause,[*] S. Datz,[*] P. Grafström,[b]
H. Knudsen,[a] S. Møller,[a] E. Uggerhøj,[a]
C. Scheidenberger,[c] R. H. Schuch,[d] and Z. Vilikazi[e]

[*]Physics Div., Oak Ridge National Laboratory, P.O. Box 2008, Oak Ridge, TN 37831-6377 USA
[a]Institute of Physics, Aarhus University, DK-8000, Aarhus C, Denmark
[b]CERN SPS/SL Division, CH-1211, Geneva 23, Switzerland
[c]Gesellschaft für Schwerionenforschung mbh, Planckstrasse 1, D-64291, Darmstadt, Germany
[d]Atomic Physics Department, Stockholm University, Frescativägen 24,
S-104 05, Stockholm 50, Sweden
[e]Schonland Research Centre for Nuclear Sciences, Johannesburg, 2050 Gauteng, South Africa

Abstract. Electrons ejected in the forward direction from fully-stripped and one-electron 33-TeV Pb ions passing through thin foils of Al and Au have been studied. Spectral peaks centered at a momentum equivalent to the velocity of the projectile have been measured, which are attributed to electron loss to the projectile continuum (ELC). From previous theoretical studies, the ELC 'cusp' shape is expected to reflect the initial state of the electron released from the projectile. The ELC peaks are very narrow even for loss from ground-state incident ions. The observed widths arise primarily from instrumental broadening effects - especially multiple Coulomb scattering (MCS). Correction of the measured electron momentum distributions for instrumental broadening leads to very narrow ELC cusps, which have essentially the same widths for loss from incident one-electron ground-state ions and loss from hydrogenic ions populated by electron capture from pair production (ECPP), where there should be significant excited-state formation.

INTRODUCTION

At ultrarelativistic energies, ionization cross sections exceed electron capture cross sections by several orders of magnitude (1,2). Effectively, all electrons transferred to a highly relativistic heavy ion moving in a solid or gaseous target medium are stripped in a relatively short distance. Above ~20 GeV/nucleon, the principal mechanism for electron capture is from pair production (ECPP) (2). The total cross sections for ECPP are technically important for making reliable

CP500, The Physics of Electronic and Atomic Collisions, edited by Y. Itikawa, et al.
© 2000 American Institute of Physics 1-56396-777-4/00/$17.00

predictions of operating limitations for relativistic heavy-ion colliders, e.g., RHIC and LHC (3). In ECPP, it is expected that ~30% of capture proceeds to excited states of the capturing ion. Some of these relatively weakly bound electrons are radiatively long-lived and easily lost in secondary collisions in solid targets, making measurements of their contributions to total capture experimentally difficult. Electrons lost from high-energy ions in collisions with target atoms form a cusp-shaped spectral peak in the forward direction in the laboratory frame centered at the velocity of the moving ion (4-5). The shape of this electron loss to projectile continuum (ELC) peak has been shown (5,6) to depend on the initial atomic bound state from which the electron is ionized. We have measured and compared ELC electrons from direct ionization of hydrogenlike 33-TeV Pb^{81+}(1s) ions (Lorentz factor $\gamma = 168$) in Al with similar data for electrons created by ECPP for bare Pb^{82+} ions in Au – followed by ionization. Both measured ELC peaks are narrow in momentum and angle and very similar in shape.

For fully-stripped 33-TeV Pb^{82+} ions, production of ELC electrons requires a two-step, double-collision process. In the first step, the Pb^{82+} ion interacts electromagnetically with a target atom forming an electron-positron pair. The electron formed in atomic bound states of the resulting Pb^{81+} ion through ECPP is carried along at the projectile velocity. In a second step, the Pb^{81+} ion is ionized in a long-range collision with another target atom. The low-energy ionized projectile-frame electrons, appear in the lab frame at $\gamma = 168$, or ~86 MeV/c. Free-pair electrons form an underlying continuum covering several hundreds of MeV/c. Corresponding positrons are emitted as free particles in either case.

A schematic of the apparatus used is shown in Fig. 1. Either hydrogenic or fully-stripped 33-TeV Pb ions from the CERN SPS accelerator were passed through thin foil targets in vacuum. Electrons collimated to 0.55° half-angle to eliminate contamination from direct knock-on electrons were deflected in a uniform magnetic analyzing field of ~0.1 T, and passed through two 2D position-sensitive drift chamber detectors located outside the analysis field. The two detector coordinate positions for coincident hits were used to calculate the trajectory for each electron and determine both the corresponding ejection angle and, from the horizontal deflection in the magnetic field, the momentum. Ions were counted in a fast scintillator detector for timing and coincidence. Free positrons were also detected in a large scintillator counter.

The ELC electrons were momentum dispersed by the uniform magnetic analyzing field only in the horizontal plane. The vertical angle distributions, as shown in Fig. 2, obtained from trajectory analysis of the two 2D drift chamber hit positions, agree with calculations for a combination of detector position resolution and accumulated angular broadening due to multiple Coulomb scattering in the targets, the aluminum exit window from the vacuum chamber, and from air between the

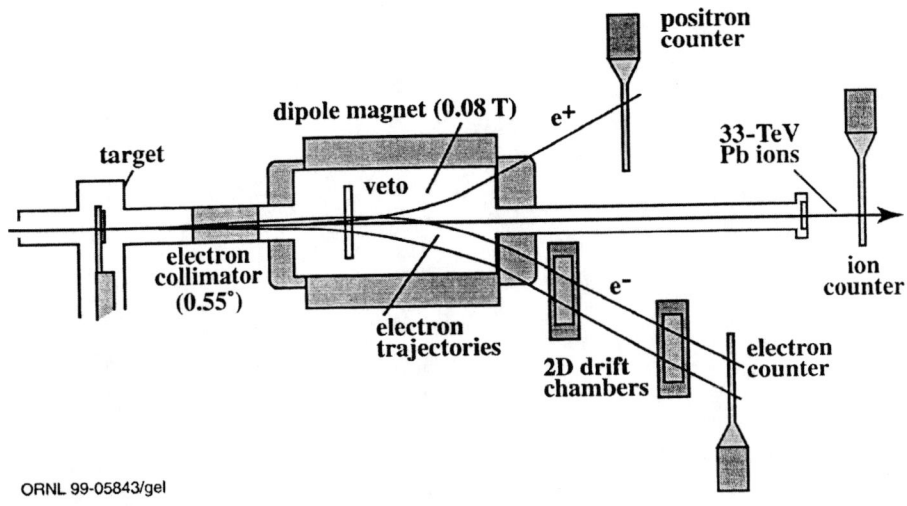

ORNL 99-05843/gel

Figure 1. Experimental apparatus. Magnetic spectrometer for zero-degree ELC electrons.

window and the detectors. This vertical distribution for each target was used to correct the horizontal position distribution data for instrumental broadening to obtain the ELC momentum distributions.

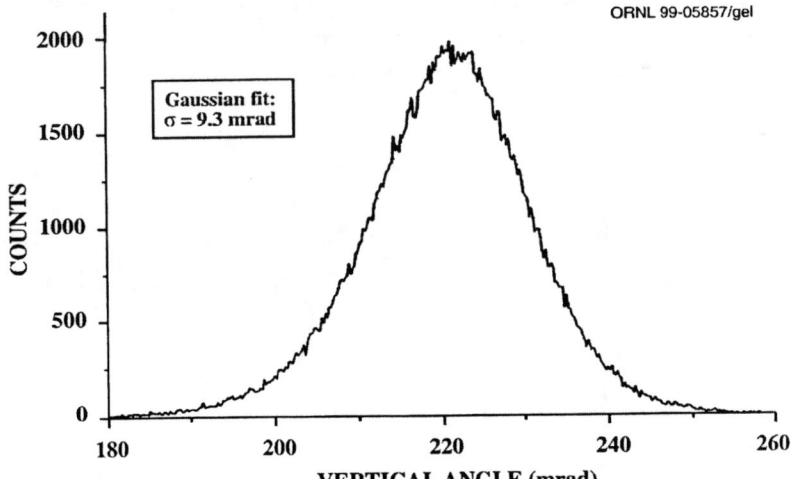

ORNL 99-05857/gel

Gaussian fit:
$\sigma = 9.3$ mrad

COUNTS

VERTICAL ANGLE (mrad)

Figure 2. Vertical angular distribution for ELC electrons from Pb^{81+} + Al.

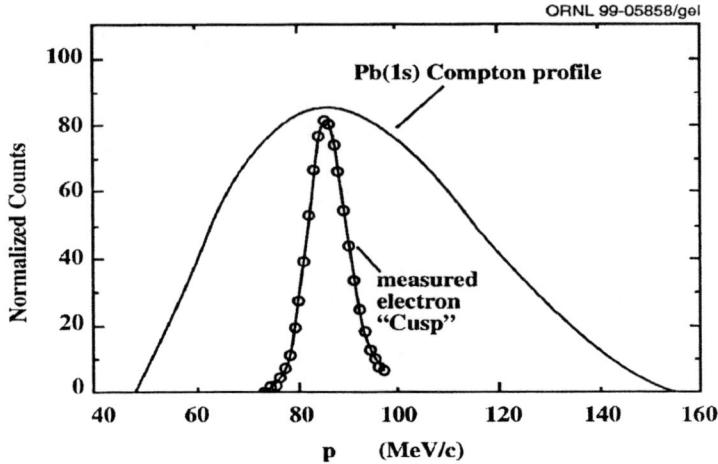

Figure 3. Measured Pb^{81+}(1s) + Al ELC momentum distribution compared with Compton Profile.

Figure 3 shows a comparison of the measured ELC 'cusp' electron momentum distribution for 33-TeV Pb^{81+}(1s) ionized in an Al target with a simple Pb(1s) hydrogenic Compton profile mapped into the laboratory frame. The assumed angular emission distribution in the projectile frame is dipole, i.e., $\sim \sin^2 (\theta_{proj})$. The comparison is made simply to indicate the relatively narrow shape of the ELC peak. In Fig. 4, we show an expanded view of the measured ELC electron momentum distribution from 33-TeV Pb^{81+}(1s) ions passing through a 77-mg/cm^2 Al foil. The figure also shows the maximum width 'cusp'-shaped peak consistent with the measured momentum profile, convoluted with the overall instrument function which was taken directly from the measured vertical electron distribution. The fitting 'cusp'-shape was chosen somewhat arbitrarily to be the 'Dettmann'

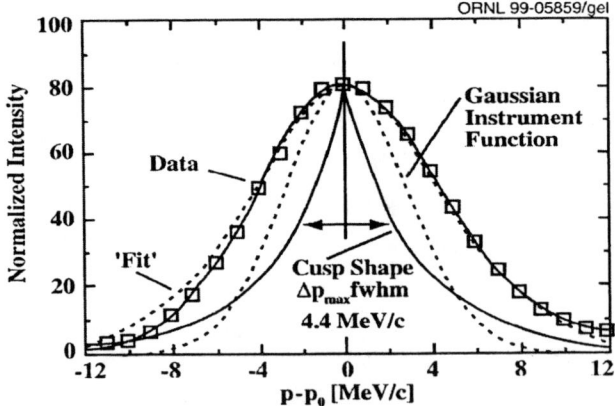

Figure 4. Measured Pb^{81+}(1s) + Al ELC momentum distribution and maximum-width symmetric cusp-shaped spectrum consistent with the data.

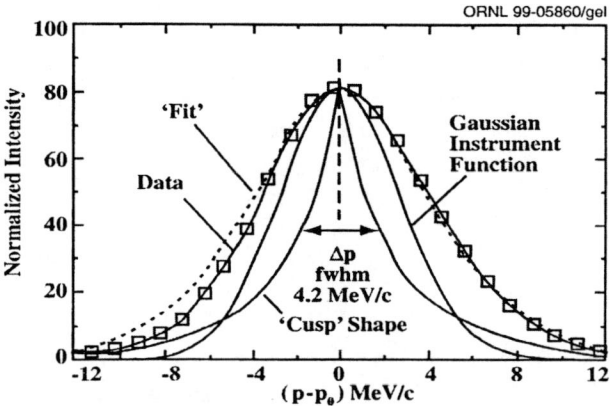

Figure 5. Pb^{82+}+ Au momentum distribution and maximum-width cusp-shaped spectrum.

cusp shape, which is symmetric and proportional to

$$\sqrt{(p-p_0)^2 + p_0^2\theta^2} - |p-p_0|,$$

where here θ_0 is simply a fitting parameter; p is the electron momentum, and p_0 is the electron momentum at the projectile velocity (86 MeV/c).

Figure 5 shows the measured ELC electron momentum distribution from 33-TeV Pb^{82+} ions passing through a 17-mg/cm^2 Au foil. The figure also shows the maximum width 'cusp'-shaped peak. For fully-stripped ions, the ELC electrons must come from ECPP (>95%), followed by ionization. The 'cusp'-shape observed should then somewhat represent ELC from the states populated through ECPP. We see essentially no evidence for a significant population of excited states leading to a narrower peak component than observed for ionized Pb^{81+}(1s) ions.

CONCLUSIONS

We have measured electrons ejected in the forward direction from bare- and one-electron 33-TeV Pb ions passing through thin foils of Al and Au. We observed spectral peaks centered at a momentum equivalent to the velocity of the projectile, and attribute these electrons to ELC. From previous theoretical studies, the ELC 'cusp' shape should reflect somewhat the origin of the electron released from the projectile – 1s electrons leading to broader cusp shapes than more highly excited n-state electrons. The observed ELC peaks are very narrow even for loss from ground-state ions. The peaks are so narrow that the measured widths arise mainly from instrumental broadening effects – especially multiple Coulomb scattering.

These broadening effects are directly and independently measurable in the experiment. Correction of the electron momentum distributions for measured broadening leads us to conclude that the ELC cusps are intrinsically very narrow, and essentially the same for loss from ground-state ions and from ions populated by electron capture from pair production (ECPP). Analysis is continuing to attempt to place limits on the fraction of any excited-state population consistent with these observations.

ACKNOWLEDGMENTS

Authors (HFK,CRV,SD) gratefully acknowledge support by the USDOE, Office of Basic Energy Sciences, Division of Chemical Sciences, under Contract No. DE-AC05-96OR22464 with Lockheed Martin Energy Research Corporation. HK acknowledges the support of the Danish Natural Science Research Council.

REFERENCES

1. Eichler, J. and Meyerhof, W. E., *Relativistic Atomic Collisions,* San Diego: Academic Press, 1995.
2. Krause, H. F. et al., *Phys. Rev. Lett.* **80**, 1190 (1998).
3. Grafström, P. et al., CERN Report SL-99-033EA (1999).
4. Breinig, M. et al., *Phys. Rev. A* **25**, 3015 (1982).
5. Burgdörfer, J. et al., *Phys. Rev. A* **28**, 3277 (1983).
6. Takabayashi, Y. et al., contributed paper TU120, ICPEAC XXI, Sendai, Japan, 1999.

Associative ionization in collisions of C^+ with D^-

A. Naji, K. Olamba, J. P. Chenu, S. Szücs, F. Brouillard and P. Defrance

Université Catholique de Louvain, Département de Physique, Unité de Physique Atomique et Moléculaire, Chemin du cyclotron, 2, B-1348 Louvain-la-Neuve, Belgium

Abstract. The absolute cross section of associative ionization in the collision of C^+ with D^- has been measured in a merged beam experiment for collision energies ranging from 0.01 to 2 eV. The measured cross section is found to be compatible with an E^{-1} law at low energies and presents a relative maximum between 0.8 and 1 eV.

Introduction

Attempts by astrophysicists to produce models taking into account the formation and the destruction of molecules in interstellar clouds have been frustrated by the lack of reliable experimental collisional data, as pointed out by Mitchell and McGowan[1]. The case of CH^+ is particularly interesting, as discrepancy exists between the observed and the calculated abundances of this molecular ion in interstellar clouds. The solution of this problem implies the knowledge of the respective role of all of the individual processes involved both in the formation and in the destruction of CH^+. Its formation by associative ionization between atomic species has however remained unexplored, both theoretically and experimentally. The associative ionization processes are :

$$C^+ + H^- \rightarrow CH^+ + e \qquad (1)$$
$$C^- + H^+ \rightarrow CH^+ + e \qquad (2)$$
$$C^+ + H \rightarrow CH^+ \qquad (3)$$
$$C + H^+ \rightarrow CH^+ \qquad (4)$$
$$C^+ + H_2 \rightarrow CH^+ + H \qquad (5)$$

The destruction of CH^+ is mainly due to dissociative and dielectronic recombination :

$$CH^+ + e \rightarrow C + H \qquad (6)$$
$$CH^+ + e \rightarrow CH^* + h\nu \qquad (7)$$

Dissociative excitation or recombination has been the subject of many theoretical studies that were traced by Amitay et al[2]. It has also been investigated experimentally first by Mul et al[3] and more recently by Amitay et al[4]. The work reported here is dealing with the formation of CD^+ in the collision of C^+ with D^-.

CP500, *The Physics of Electronic and Atomic Collisions*, edited by Y. Itikawa, et al.
© 2000 American Institute of Physics 1-56396-777-4/00/$17.00

$$C^+ + D^- \rightarrow CD^+ + e^- \tag{8}$$

This study, although performed with Deuterium instead of Hydrogen, provides useful information for the understanding of reaction (2), as far as potential energy curves are concerned. Measurements with Hydrogen will be carried out later on, after the improving the resolution of the beam analyzer.

Experimental set-up and procedure

The set-up used in this experiment has been described previously [5]. The apparatus is shown schematically in figure 1.

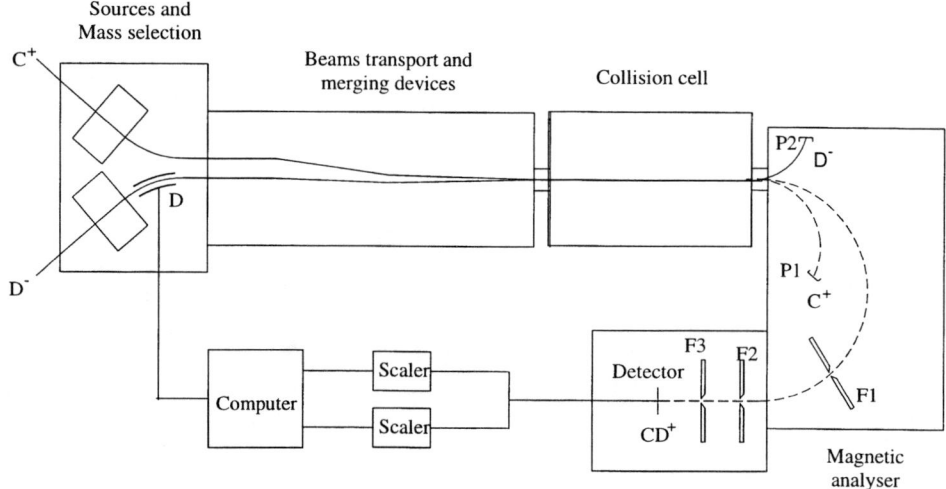

FIGURE 1. A schematic block diagram of the experimental set-up.

Four sections are distinguished. The first one includes the ion sources and the beam forming optical devices. C^+ and D^- ions are produced by an Electron Cyclotron Resonance (ECR) ion source and by a duoplasmatron source respectively. In the second section, the mass selected beams are merged by a set of electrostatic plane deflectors and the interaction takes place in the third section. The interaction region consists of an electrically biased cell called the observation cell and contains some probes and beam profile and position measuring devices. The last section contains the magnetic analyzer where the reaction products are separated from the primary beams. CD^+ ions are detected by a channel electron multiplier. The diaphragms are defined so that total transmission of the products to the detector is achieved.

In merged beam experiments the absolute cross section (σ) is derived [5] from the number of reactions N(T) occurring during a time T by the relation :

$$N(t) = \sigma \frac{v_r}{q_1 q_2 v_1 v_2} \int_0^T dt \int_V I_1(t) I_2(t) F(t) dt \qquad (9)$$

where v_1, v_2, q_1, q_2, $I_1(t)$, $I_2(t)$ are the velocities, charges and intensities of beam 1 and 2 respectively. v is their relative velocity. $F(t)$ is the so-called form factor, which expresses the beam overlap and is defined as :

$$F(t) = \int_L F(z, t) dz \qquad (10)$$

where

$$F(z,t) = \frac{\int_{-\infty}^{+\infty}\int_{-\infty}^{+\infty} j_1(x,y,z,t) j_2(x,y,z,t) dx dy}{I_1(t) I_2(t)} \qquad (11)$$

j_1 and j_2 are the current densities and L is the useful length of the interaction region. The knowledge of the form factor requires the measurement of the density profiles of the two beams. Presently, this is conveniently done thanks to the computer-assisted measurement of the beam profiles [5]. The interaction length L is defined by applying a voltage V on the observation cell. As a result, the kinetic energy of molecular ions formed in the cell is raised allowing their separation, in the magnetic analyzer, from molecular ions formed elsewhere. The observation voltage also determines the barycentric energy E_{cm}.

The molecular ions can only be produced by the mutual interaction of the beams and not by their interaction with the residual gas. However, an important background is created by the elastic scattering of the primary C^+ ions (especially in the magnetic analyzer) and also by photon emission at the impact on surfaces. This background was reduced to an acceptable level (typically three times the signal) by the use of shielding elements as the slits F2 and F3 shown in figure 1.

The background contribution was measured by chopping the negative ion beam by means of a crenel voltage applied on the deflector D and recording the count rate with the negative beam both on and off. It has been checked that the negative beam by itself does not create any background and that the background is identical whether the negative beam is on or off. In these circumstances, the number of molecular ions is obtained as the difference between the number of counts recorded during the same time with the negative ion beam on and off.

Results and comments

The cross section of reaction (8) has been measured at collision energies ranging from 0.01 to 2 eV. Typical experimental conditions and the main factors limiting the accuracy of the present measurements are given in table 1 together with the corresponding errors. Details of errors are given in [5].

TABLE 1. Typical experimental conditions and errors.

Beam acceleration voltages	18 kV (C$^+$)
	4.5 kV (D$^-$)
Observation voltage	-1.3 kV
Beam diameters (common)	0.15 cm
C$^+$ beam intensity	40 nA
D$^-$ beam intensity	40 nA
Background	10 Hz
Signal count rate from 0.01 to 2 eV	from 5 to 0.4 Hz
Interaction length L	6.8 cm
Beam divergence	2 mrad
Error on the detection efficiency	2%
Error on the interaction length	3%
Error on the beam intensities	2%

In figure 2 we show the cross section as a function of the barycentric energy E. It appears that the energy dependence is compatible, at low energy, with an E^{-1} law. The deviation from this law, observed at very low energy, can be explained by the limited energy resolution of our measurements. This energy resolution is limited (i) by

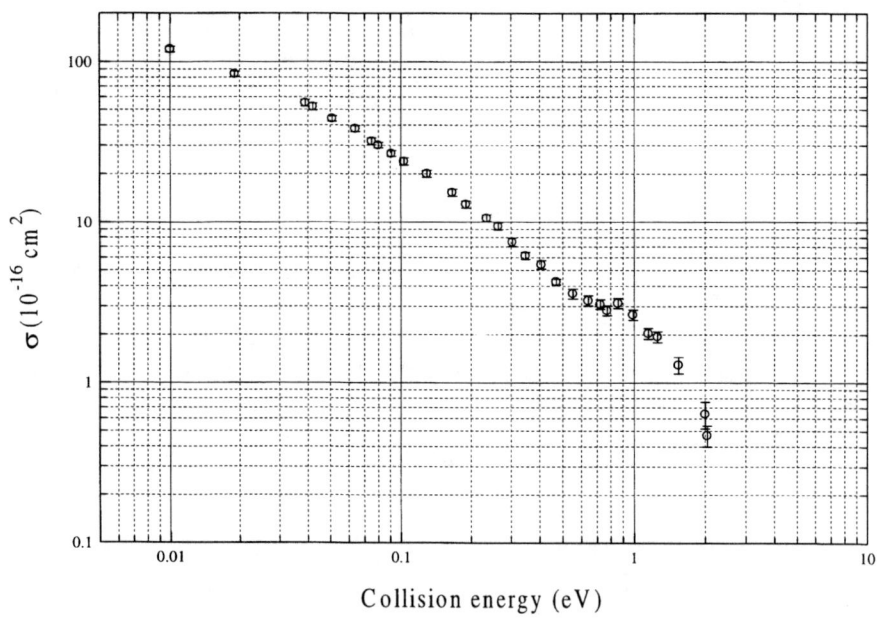

FIGURE 2. Cross section as a function of the barycentric energy.

the energy spread in each of the interacting beams and (ii) by their divergence. The first term is found to be very small (< 0.5 meV) [6]. On the other hand, due to the finite divergence (α = 2 mrad) of the beams, the relative velocity is distributed over the range $0-2\alpha v$. Assuming a uniform angular ion distribution, the average barycentric energy of the collisions taking place between two beams with equal velocities ($\mu\alpha^2 v^2/2$) amounts to 0.013 eV. This value is also the uncertainty affecting the barycentric energy. The values of the cross section at very low energies are thus average values that are not expected to reflect the true energy dependence but will tend to a ceiling value, that of the average value over the range 0–0.013 eV.

The E^{-1} dependence extends up to 0.5 eV. Above this value the cross-section first shows a slight maximum (which could be attributed to the formation CH^{+*} ($^3\Pi$)) and then decreases quickly due to the opening of new dissociative channels. The E^{-1} dependence can be explained in the following way. According to the asymptotic position of the entrance channel potential energy curves, the ionic attractive curve will cross the neutral ($C^*(n) + D$) curves at distances R equal to 87 a_0, 16 a_0 and 3.5 a_0 corresponding to n = 4, 3 and 2 respectively. A classical description of this process predicts the following energy dependence:

$$\sigma = \pi \, a_0^{\,2} \left(\frac{3.5}{E} + 12.2 \right) P_0 \, P_a \tag{12}$$

where P_0 is the probability of forming the neutral molecule at the avoided crossing corresponding to n = 3 which is supposed to be the dominant one. P_a is the probability that the molecule autoionises along the potential curve for low internuclear distances. This classical picture qualitatively describes the CH^+ formation. The present cross-section is found to be almost an order of magnitude larger than cross-sections obtained for similar reactions [7]. The presence of C^+ ions eventually formed in long-lived excited states could affect the result. Detailed calculations should give the explanation of the observed features.

REFERENCES

[1] Mitchell, J. B. A., and McGowan, J. Wm., Ap. J. **222** L77-79 (1978).
[2] Amitay, Z., Zajfman, D., Forck, P., Hechtfischer, U., Seidel, B., Grieser, M., Habs, D., Repnow, R., Schwalm, D., and Wolf, A., Phys. Rev. A **54**, 4032 (1996).
[3] Mul, P. M., Mitchell, J. B. A., Defrance, P., D'Angelo, V. S., McGowan, J. Wm., Froelich, H.R., *J. Phys. B **14**, 1353 (1981).*
[4] Amitay, Z., Zajfman, D., Forck, P., Hechtfischer, U., Grieser, M., Habs, D., Schwalm, D., and Wolf, A., (private communication 1998).
[5] Olamba,K., Szücs, S., Chenu, J. P., Naji, A., and Brouillard, F., *J. Phys. B: At. Mol. Opt. Phys.* **29**, 1996, pp. 2837–2846.
[6] Naji, A., Olamba,K., Chenu, J. P., Szücs, S., Chibisov, M., and Brouillard, F., *J. Phys. B: At. Mol. Opt. Phys.* **31**, 1998, pp. 2961–2970.
[7] Naji, A., Olamba,K., Chenu, J. P., Szücs, S., and Brouillard, F., *J. Phys. B: At. Mol. Opt. Phys.* **31**, 1998, pp. 4887–4894.

Non–adiabatic quantum molecular dynamics: Reaction mechanisms in atom–cluster collisions

Ulf Saalmann[1] and Rüdiger Schmidt[2]

[1] *Atomfysik, Stockholms Universitet, Frescativägen 24, 10405 Stockholm, Sweden*
[2] *Institut für Theoretische Physik, Technische Universität, 01062 Dresden, Germany*

Abstract. Atom–cluster collisions are studied by means of a simultaneous and self–consistent treatment of electronic and atomic degrees of freedom involved. We examine the excitation mechanisms (electronic versus vibrational) and the related relaxation phenomena (phase transitions, fragmentation) for impact energies from eV → MeV. "Cluster transparency" is found in a small window of energies. In addition, the laser–enhanced charge transfer competing with fragmentation in collisions with sodium clusters is investigated. A strong temperature dependence is observed.

Introduction

Collisions with atomic clusters represent a relatively new branch of collision physics as compared to the traditional fields of ion–atom collisions and ion–solid interaction. Recently they have received great interest as they offer the possibility to tackle bridge–building questions (like the continuous transition from individual excitations in the elementary ion–atom collision to the macroscopic stopping power in solids) as well as fundamental problems (like phase transitions in finite systems). In general, the fundamental events accompanying these cluster collisions include the simultaneous occurrence and mutual coupling of electronic transitions and atomic motion in systems with a large but finite number of degrees of freedom. To cope with such situations we have recently developed a microscopic approach which treats electronic (quantum) *and* atomic (classical) degrees of freedom self–consistently by combining time–dependent density functional theory with molecular dynamics: Non–Adiabatic Quantum Molecular Dynamics (NA-QMD) [1].

The coupled equations of motion comprising time–dependent Kohn–Sham equations (treated within ALDA) and Newton equations have to be solved simultaneously. For the studies presented here we use an expansion of the electronic single–particle functions within atomic orbitals (LCAO) which is best suited for collisions. Thereby one gets a set of ordinary differential equations which is, in contrast to

CP500, *The Physics of Electronic and Atomic Collisions*, edited by Y. Itikawa, et al.

the well–known close–coupling equations of ion–atom scattering, highly non–linear. The corresponding classical equations of motion are complicated in the case of an LCAO representation owing to the fact that the basis is finite and local [1]. To allow for systematic studies one has to introduce approximations for the numerical solution: We restrict ourselves on the valence density which is divided into an adiabatic and an explicitly time–dependent part. Thus one can introduce stepwise approximations for the Hartree and the exchange–correlation terms; cf. details in [1]. This approach has been tested against experiments for collision–induced dissociation [2] as well as for absolute cross sections for integral charge transfer [3].

Based on the NA–QMD approach, we present here two studies of collisions involving sodium clusters with emphasis on the competition of the various reaction mechanisms: firstly, excitation mechanisms as well as the related relaxation processes for a wide range of impact energies [4] and, secondly, charge transfer with ground–state as well as laser–excited targets competing with collision–induced dissociation [5].

Excitation and relaxation

Firstly, we study the gradual change of the excitation mechanisms [4] as a function of the impact energy E_{cm} for the collision system $Na_9^+ + Na$. This can be deduced from the total kinetic energy loss $\Delta E := E_{cm} - E_{cm}(t \to \infty)$ giving the amount of energy transferred into internal degrees of freedom, i.e. cluster excitation. In Fig. 1 we compare the NA–QMD calculation with an adiabatic QMD calculation where

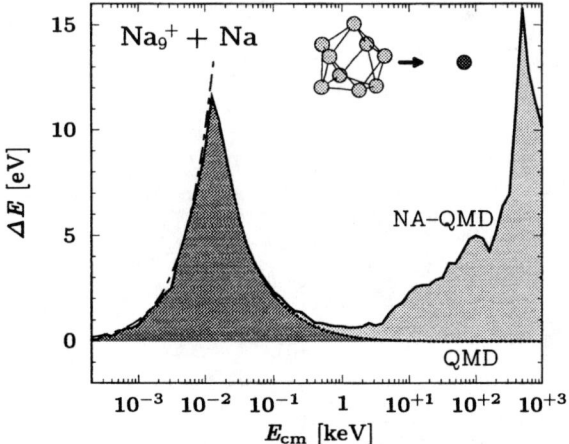

FIGURE 1. Total kinetic energy loss of relative motion ΔE in central $Na_9^+ + Na$ collisions for a fixed collision geometry (see insert) as a function of the center–of–mass impact energy E_{cm} calculated with NA–QMD (solid line) and QMD (dotted line). Vibrational and electronic contributions to ΔE are distinguished by grey and light–grey shaded areas, respectively. The dot–dashed line corresponds to $\Delta E = E_{cm}$.

the electrons are restricted to the instantaneous ground state. Whereas at impact energies $E_{cm} \lesssim 200\,eV$ only vibrational excitations occur (adiabatic regime), for $E_{cm} \gtrsim 5\,keV$ electronic transitions dominate (non–adiabatic regime). The strong excitations at low and high energies, respectively, lead to immediate fragmentation of the cluster. Recently, these two basically different types of collision–induced fragmentation — named "impulsive" and "electronic" — have been experimentally observed and theoretically analyzed [2]. In the transition range around $E_{cm} \sim 1\,keV$ (cf. Fig. 1) both mechanisms compete. Following the relaxation dynamics of the cluster on a picosecond time scale, phase transitions from solid to liquid–like behaviour (induced by electron–vibration coupling) have been observed in this energy range [4].

Surprisingly, we have found that for the system $Na_9^+ + He$ (with a "magic" initial electronic configuration and dominating repulsive forces) the occurrence of "cluster transparency". This mechanism is characterized by a practically undisturbed atom–cluster penetration within the transition regime. Figure 2 shows the total kinetic energy loss ΔE as a function of E_{cm} as calculated for random orientations

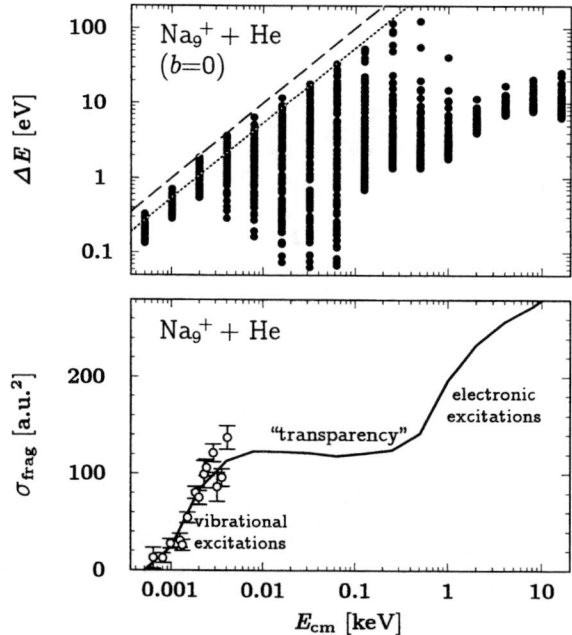

FIGURE 2. Upper part: Total kinetic energy loss ΔE in central $Na_9^+ + He$ collisions as a function of the center–of–mass impact energy E_{cm} calculated with NA–QMD using randomly chosen orientations (dots). The lines correspond to a complete energy loss $\Delta E = E_{cm}$ (dashed) and the maximal energy loss for a binary collision of the He atom with *one* cluster atom (dotted). Lower part: Total fragmentation cross sections as a function of E_{cm} calculated with NA–QMD for the same system (solid line). Available experimental data (circles) are from S. Nonose *et al.* J. Chem. Phys. **105**, 9167 (1996).

with impact parameter $b=0$. Obviously, for specific collision geometries in the range of $E_{cm} \sim 10 \ldots 100 \, \text{eV}$ the relative velocity between projectile and target becomes, on one hand, too large to excite vibrational degrees of freedom and is, on the other hand, still too small to induce electronic transitions. As a consequence, the cluster appears to be transparent even in central collisions.

One possibility to study the change in the excitation mechanisms and in particular the occurrence of transparency in connection with experiments is to consider the total fragmentation cross section σ_{frag} as a function of E_{cm} (shown in the lower part of Fig. 2). An excellent agreement between theory and experiment is found in the adiabatic regime where experimental data are available at present. It is just the transparency effect which leads to a pronounced plateau of $\sigma_{frag}(E_{cm})$ between 10 and 100 eV before electronic excitations induce a further increase.

Charge transfer versus fragmentation

In connection with the first experimental study of charge transfer in atom–cluster collisions with targets in a laser-excited state we have examined the interplay between electron transfer and cluster fragmentation. The experiment measures the neutralization of clusters in

$$Na_n^+ + Na(3s) \rightarrow Na_n + Na^+$$
$$Na_n^+ + Na(3p) \rightarrow Na_n + Na^+$$
$$\text{for } n = 1 \ldots 4$$

discriminating against collision–induced fragmentation of the neutralized clusters [5]. In Fig. 3 the integral cross section ratio for charge transfer σ_{3p}/σ_{3s} in these

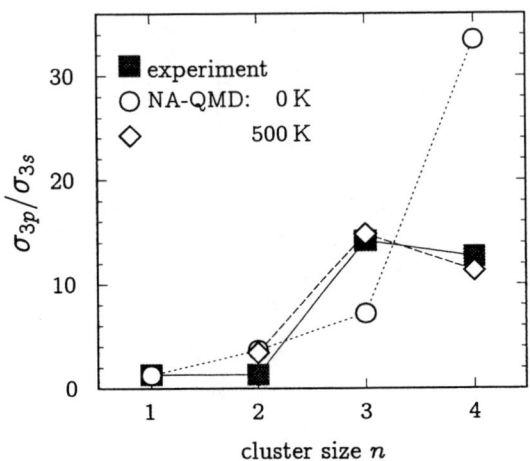

FIGURE 3. Integral cross section ratio for charge transfer in collisions $Na_n^+ + Na(3s,3p) \rightarrow Na_n + Na^+$ at an energy of $E_{lab} = 5 \, \text{keV}$. The experimental data (full squares) are compared with theoretical results obtained from NA-QMD calculations (open symbols) for two initial temperatures of the clusters. The thin lines are only to guide the eyes.

collisions is shown for an energy of $E_{lab} = 5\,\text{keV}$.

To understand the strong dependence of the laser–enhanced charge transfer on the cluster size n, the collisions have been analyzed theoretically within the NA-QMD framework. This analysis is complicated for the case of clusters as compared to the ion–atom case for the following reasons: One has to (i) treat the charge transfer in a real *many–electron system*, (ii) take into account simultaneously the *fragmentation* of the clusters because the experiment discriminates against fragmentation, (iii) consider for a given impact parameter many different *orientations* of the cluster with respect to the beam axis, in order to obtain orientation–averaged quantities, (iv) consider that the clusters in the beam do have a *temperature*, which in the present experiment is unknown. For details of this investigation we refer to [5].

In the calculations we have varied the the initial rovibrational energy content of the clusters (temperature T) before the collision and found that both processes — the charge transfer as well as the ensuing fragmentation — are very sensitive to it. This can also be seen in Fig. 3 where the the measured laser enhancement is compared with the final theoretical ratios σ_{3p}/σ_{3s} for zero–temperature clusters as well as for hot clusters with $T = 500\,\text{K}$, the temperature of the best overall agreement with the experiment [5]. Even though this temperature dependence is the result of a complex interplay of electron transfer and cluster fragmentation such a comparison of experimental data and theoretical results may serve as a kind of a *"cluster thermometer"*.

Acknowledgments

One of us (U. S.) acknowledges a grant by the German Academic Exchange Service and financial support by Stockholm University.

References

1. U. Saalmann and R. Schmidt, Z. Phys. D **38**, 153 (1996).
2. J. A. Fayeton, M. Barat, J. C. Brenot, H. Dunet, Y. J. Picard, U. Saalmann, and R. Schmidt. Phys. Rev. A **57**, 1058 (1998).
3. O. Knospe, J. Jellinek, U. Saalmann, and R. Schmidt, Eur. Phys. J. D **5**, 1 (1999) and to be published.
4. U. Saalmann and R. Schmidt. Phys. Rev. Lett. **80**, 3213 (1998).
5. Z. Roller-Lutz, Y. Wang, H. O. Lutz, U. Saalmann, and R. Schmidt. Phys. Rev. A **59**, R 2555 (1999).

Gwinner, G., 400

V

Vane, C. R., 709
Vernhet, D., 666
Vilikazi, Z., 709

W

Walters, H. R. J., 454
Wang, H., 156
West, J. B., 218
Whelan, C. T., 241
Widmann, K., 626
Wightman, J. P., 146

Williams, E. M., 392
Winkler, K. D., 436
Winter, HP., 656
Wolf, A., 400

Y

Yamanouchi, K., 182

Z

Zhang, Z., 192

$$\frac{4}{9L}$$